Annotated Teacher's Edition

HOLT
LIFE SCIENCE

Annotated Teacher's Edition

Patricia A. Watkins
Science Curriculum Specialist
Dean of College Preparatory Programs
Incarnate Word College
Principal, Incarnate Word High School
San Antonio, Texas

Glenn K. Leto
Biology Teacher
Barrington High School
Barrington, Illinois

SENIOR EDITORIAL ADVISOR
Curriculum and Multicultural Education

John E. Evans, Jr.
Science Education Specialist
Philadelphia, Pennsylvania

HOLT, RINEHART AND WINSTON

Austin • *New York* • *Orlando* • *Chicago* • *Atlanta*
San Francisco • *Boston* • *Dallas* • *Toronto* • *London*

Patricia A. Watkins

Dr. Watkins holds an Ed.D. in Instructional Management from Texas A & M University and is Dean of College Preparatory Programs at Incarnate Word College and Principal of Incarnate Word High School in San Antonio, Texas. Dr Watkins has also served as Associate principal of the Health Careers High School in San Antonio and as Special Education Coordinator for the Northside Independent School District, also in San Antonio. She is an author of General Science from Holt, Rinehart and Winston, as well as a high school biology textbook. In addition, she has written other life science materials and has taught general science, biology, physical science, and special education for over 20 years.

Glenn K. Leto

Mr. Leto received an M.S. in Biology from Northern Illinois University, Dekalb, Illinois. He currently serves as a biology teacher and curriculum developer for Barrington High School, Barrington, Illinois. He is Science Museum cofounder and coordinator for the Science Place, Barrington High School. He has also been a Kellogg Foundation Fellow—Exploratorium Resident. Mr. Leto is a contributor to Biology Today and Modern Biology from Holt, Rinehart and Winston. He is also coauthor of a physics text and a contributor for an electrical systems instructor's resource.

John E. Evans, Jr.

Dr. Evans holds an Ed.D. in Education and a Secondary Principal Certification from Temple University, Philadelphia, Pennsylvania. He is currently a Science Program Advisor to the National Science Foundation, Washington, D.C., and was site coordinator for Project 2061 in Philadelphia. He is a member of the Division of Multicultural Science Education Committee of the National Science Teachers Association. He has previously received the Award of Achievement in Science Education, School District of Philadelphia. He has been involved in education since 1965.

Cover Design: Didona Design Associates

Cover: Bluestriped Snappers. Photo by Fred McConnaughey/Photo Researchers, Inc.

Copyright © 1994 by Holt Rinehart and Winston, Inc.

All rights reserved. No part of this publication may be reproduced or transmitted in any form or by any means, electronic or mechanical, including photocopy, recording, or any information storage and retrieval system, without permission in writing from the publisher.

Requests for permission to make copies of any part of the work should be mailed to: Permissions Department, Holt Rinehart and Winston Inc., 8th Floor, Orlando, Florida 32887.

Some material in this work was previously published in HBJ LIFE SCIENCE, ANNOTATED TEACHER'S EDITION, copyright © 1989 by Harcourt Brace & Company. All rights reserved.

Acknowledgments: See page 611, which is an extention of the copyright page.

Printed in the United States of America   ISBN 0-03-097527-1
1 2 3 4 5 6 7 036 97 96 95 94 93

Acknowledgments

Content Advisors

Jerry Brand, Ph.D.
Professor, Department of Botany
University of Texas
Austin, Texas

John R. Bristol, Ph.D.
Professor of Biology
Department of Biological
  Sciences
University of Texas at El Paso
El Paso, Texas

Robert Fronk, Ph.D.
Head, Science Education Department
Florida Institute of Technology
Melbourne, Florida

Georgia E. Lesh-Laurie, Ph.D.
Professor of Biology
Vice Chancellor for Academic
  Affairs
University of Colorado
Denver, Colorado

Jan M. Ozias, Ph.D., R.N.
Coordinator of Health Service
Austin Independent School
  District
Austin, Texas

Barbara Rothstein, Ph.D.
Environmental Education
  Consultant
Hollywood, Florida

Leon J. Zalewski, Ph.D.
Professor of Science Education
Dean, College of Education
Governors State University
University Park, Illinois

Curriculum Advisors

Charles Kish
Science Teacher
Saratoga Springs
  Junior High School
Saratoga Springs, New York

Lynn Mederos
Life Science Teacher
Glenridge Middle School
Winter Park, Florida

Jim Pulley
Science Teacher
Oak Park High School
North Kansas City, Missouri

Dana Ste. Claire
Curator of Science and History
The Museum of Arts and Sciences
Daytona Beach, Florida

Thomasena H. Woods, Ed.D.
Science Supervisor
Newport News City Schools
Newport News, Virginia

Reading/Language Advisors

Edward C. Turner, Ph.D.
Associate Professor
College of Education
University of Florida
Gainesville, Florida

Philip E. Bishop, Ph.D.
Professor, Department of
  Humanities
Valencia Community College
Orlando, Florida

Annotated Teacher's Edition

Contents

Using the Program
- Introducing *Holt Life Science* — T20
- Pacing Chart — T26

Classroom Techniques
- Thematic Science — T30
- Multicultural Instruction — T31
- Meeting Special Needs — T32
- Cooperative Learning — T34
- Integrating Other Disciplines — T36
- Science/Technology/Society — T37
- Promoting Positive Attitudes — T38
- Assessment — T39

Laboratory Science
- The Role of Inquiry — T40
- Field Trip Guidelines — T41
- Safety Guidelines — T42
- Materials List — T44
- Resources — T46

Unit and Chapter Interleaves

Unit 1	Our Fragile Environment	XIXa
Chapter 1	Humans and the Environment	1a
Chapter 2	Populations, Communities, and Ecosystems	31a
Chapter 3	Life in the Biosphere	57a
Unit 1	Science Parade	84

Unit 2	Studying Living Things	93a
Chapter 4	Cells	95a
Chapter 5	Cell Function	117a
Chapter 6	Natural Selection and Heredity	141a
Chapter 7	Classification	171a
Unit 2	Science Parade	194

Unit 3	Simple Living Things	205a
Chapter 8	Viruses and Monerans	207a
Chapter 9	Protists and Fungi	227a
Unit 3	Science Parade	248

Unit 4	Plants	255a
Chapter 10	An Introduction to Plants	257a
Chapter 11	Plant Growth and Adaptations	285a
Unit 4	Science Parade	306

Unit 5	Animals	315a
Chapter 12	Types of Animals	317a
Chapter 13	Invertebrates	339a
Chapter 14	Vertebrates	369a
Chapter 15	Animal Behavior	397a
Unit 5	Science Parade	420

Unit 6	The Human Body	429a
Chapter 16	Support and Movement	431a
Chapter 17	Digestion and Circulation	455a
Chapter 18	Respiration and Excretion	487a
Chapter 19	Coordination and Control	509a
Chapter 20	Reproduction and Development	541a
Unit 6	Science Parade	564

Contents

Unit 1 Our Fragile Environment — 1

Chapter 1
Humans and the Environment — 2

Solving an Environmental Problem — 4
Skill Designing and Conducting Experiments — 9
Pollution of the Environment — 10
Protecting the Environment — 18
Investigation Modeling Soil Erosion — 28
Highlights — 29
Review — 30

Chapter 2
Populations, Communities, and Ecosystems — 32

Interactions — 34
Energy Flow — 42
Skill Diagramming Food Chains and Food Webs — 47
Changing Ecosystems — 48
Investigation Observing an Ecosystem — 54
Highlights — 55
Review — 56

Chapter 3
Life in the Biosphere — 58

The Biosphere and Water Ecosystems — 60
Investigation Observing Changes in Ecosystems — 70
Land Biomes — 71
Skill Modeling Climate — 80
Highlights — 81
Review — 82

Unit 1
Science Parade
Science Applications Studying the Rain Forest 84
Read About It! An Eagle to the Wind 87
Then and Now Rachel Carson, Akira Okubo 91
Science at Work David Powless, Conservationist 92
Science/Technology/Society A Hole in the Sky 93

Unit 2 Studying Living Things 94

Chapter 4
Cells 96
Tools for Life Science 98
Investigation Using the Microscope 104
Discovery of Cells 105
Structure of Cells 108
Skill Comparing Plant and Animal Cells 114
Highlights 115
Review 116

Chapter 5 118
Cell Function
Transport in Cells 120
Investigation Observing Osmosis 125
Cell Reproduction 126
DNA and Cell Energy 130
Skill Modeling DNA 138
Highlights 139
Review 140

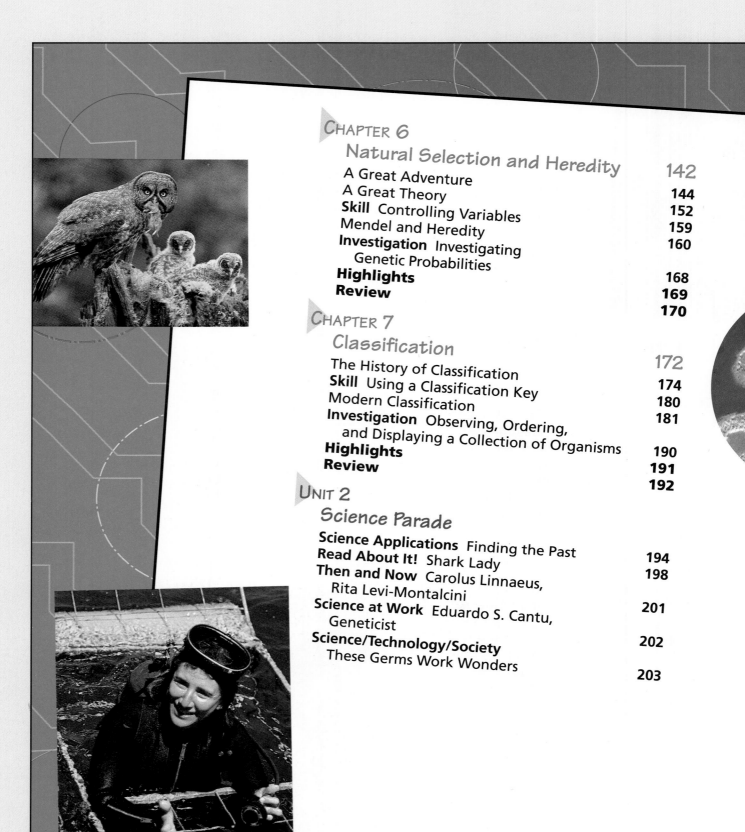

Chapter 6
Natural Selection and Heredity 142
- A Great Adventure **144**
- A Great Theory **152**
- **Skill** Controlling Variables **159**
- Mendel and Heredity **160**
- **Investigation** Investigating Genetic Probabilities **168**
- **Highlights** **169**
- **Review** **170**

Chapter 7
Classification 172
- The History of Classification **174**
- **Skill** Using a Classification Key **180**
- Modern Classification **181**
- **Investigation** Observing, Ordering, and Displaying a Collection of Organisms **190**
- **Highlights** **191**
- **Review** **192**

Unit 2
Science Parade
- **Science Applications** Finding the Past **194**
- **Read About It!** Shark Lady **198**
- **Then and Now** Carolus Linnaeus, Rita Levi-Montalcini **201**
- **Science at Work** Eduardo S. Cantu, Geneticist **202**
- **Science/Technology/Society** These Germs Work Wonders **203**

Unit 3 Simple Living Things — 206

CHAPTER 8
Viruses and Monerans — 208

Viruses	210
Skill Identifying Viral Variables	217
Monerans	218
Investigation Testing Disinfectants	224
Highlights	225
Review	226

CHAPTER 9
Protists and Fungi — 228

Protozoans	230
Investigation Observing Protozoans	235
Algae	236
Skill Organizing Data	239
Fungi	240
Highlights	245
Review	246

UNIT 3
Science Parade

Science Applications AIDS Research	248
Read About It! Through the Microscope	250
Then and Now Robert Koch, Lynn Margulis	252
Science at Work Helene Gayle, Public Health Physician	253
Science/Technology/Society Algal Blooms	254

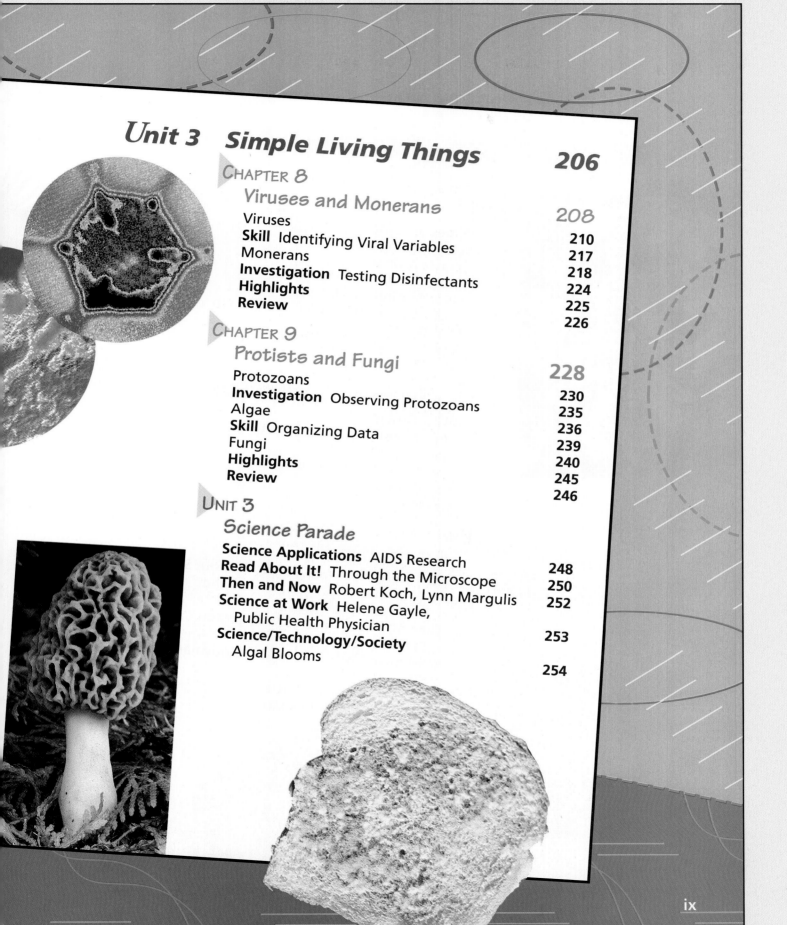

Unit 4 Plants 256

CHAPTER 10
An Introduction to Plants 258

Types of Plants	260
Skill Making a Bar Graph	269
Photosynthesis	270
Reproduction in Seed Plants	275
Investigation Predicting the Effect of Colors of Light on Plant Growth	282
Highlights	283
Review	284

CHAPTER 11
Plant Growth and Adaptations 286

Plant Growth	288
Skill Measuring Plant Growth	295
Plant Adaptations	296
Investigation Examining Seeds	302
Highlights	303
Review	304

UNIT 4
Science Parade

Science Applications Salads Without Soil	306
Read About It! Andy Lipkis and the Tree People	308
Then and Now George Washington Carver, Elma Gonzalez	312
Science at Work Shirley Mah Kooyman, Plant Scientist	313
Science/Technology/Society Growing Plants in Space	315

Unit 5 Animals 316

▸ CHAPTER 12   318
Types of Animals

Introduction to Animals	320
Characteristics of Animals	325
Skill Identifying Symmetry	331
Animal Classification	332
Investigation Comparing Vertebrates and Invertebrates	336
Highlights	337
Review	338

▸ CHAPTER 13   340
Invertebrates

Sponges, Coelenterates, and Worms	342
Investigation Tracing the Life Cycle of a Liver Fluke	353
Mollusks and Echinoderms	354
Insects—The Most Common Arthropods	358
Other Arthropods	362
Skill Classifying Insects	366
Highlights	367
Review	368

▸ CHAPTER 14   370
Vertebrates

Fishes, Amphibians, and Reptiles	372
Investigation Identifying Foods That Tadpoles Need	383
Birds and Mammals	384
Skill Making and Interpreting a Graph	394
Highlights	395
Review	396

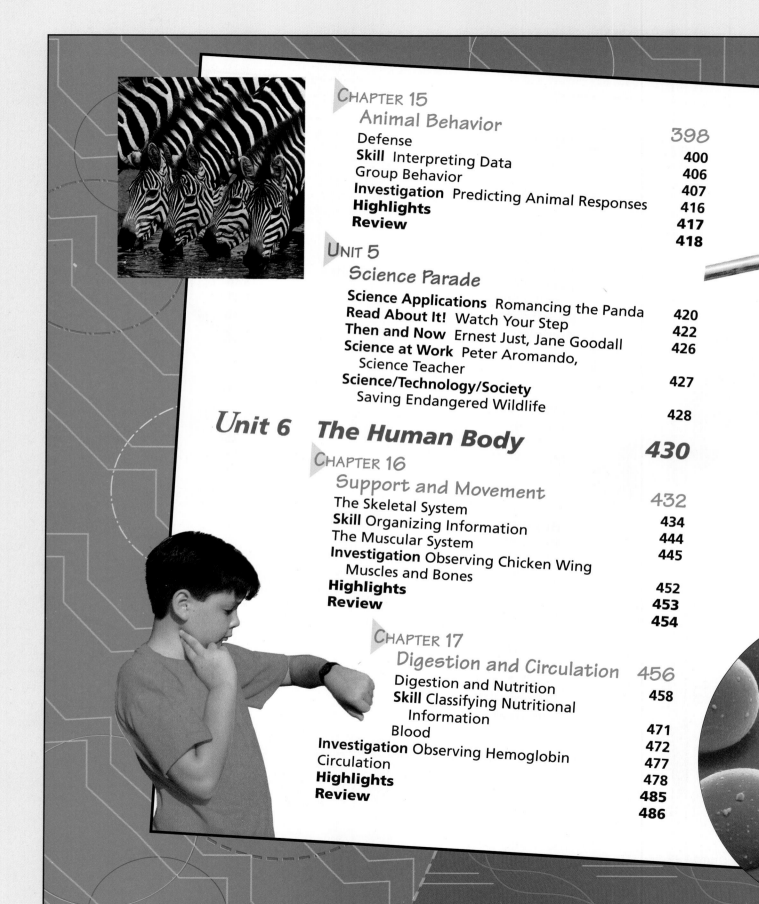

Chapter 15
Animal Behavior
Defense	398
Skill Interpreting Data	400
Group Behavior	406
Investigation Predicting Animal Responses	407
Highlights	416
Review	417
	418

Unit 5
Science Parade
Science Applications Romancing the Panda	420
Read About It! Watch Your Step	422
Then and Now Ernest Just, Jane Goodall	426
Science at Work Peter Aromando, Science Teacher	427
Science/Technology/Society Saving Endangered Wildlife	428

Unit 6 The Human Body 430

Chapter 16
Support and Movement
The Skeletal System	432
Skill Organizing Information	434
The Muscular System	444
Investigation Observing Chicken Wing Muscles and Bones	445
Highlights	452
Review	453
	454

Chapter 17
Digestion and Circulation 456
Digestion and Nutrition	458
Skill Classifying Nutritional Information	471
Blood	472
Investigation Observing Hemoglobin	477
Circulation	478
Highlights	485
Review	486

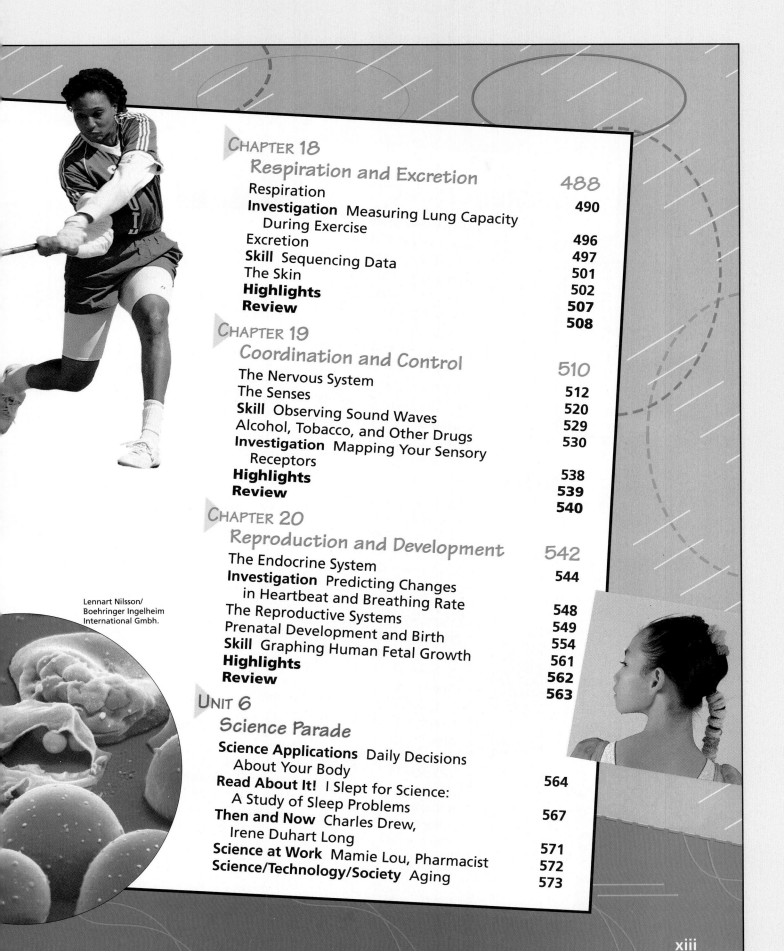

Chapter 18
Respiration and Excretion — 488
- Respiration — 490
- **Investigation** Measuring Lung Capacity During Exercise — 496
- Excretion — 497
- **Skill** Sequencing Data — 501
- The Skin — 502
- **Highlights** — 507
- **Review** — 508

Chapter 19
Coordination and Control — 510
- The Nervous System — 512
- The Senses — 520
- **Skill** Observing Sound Waves — 529
- Alcohol, Tobacco, and Other Drugs — 530
- **Investigation** Mapping Your Sensory Receptors — 538
- **Highlights** — 539
- **Review** — 540

Chapter 20
Reproduction and Development — 542
- The Endocrine System — 544
- **Investigation** Predicting Changes in Heartbeat and Breathing Rate — 548
- The Reproductive Systems — 549
- Prenatal Development and Birth — 554
- **Skill** Graphing Human Fetal Growth — 561
- **Highlights** — 562
- **Review** — 563

Unit 6
Science Parade
- **Science Applications** Daily Decisions About Your Body — 564
- **Read About It!** I Slept for Science: A Study of Sleep Problems — 567
- **Then and Now** Charles Drew, Irene Duhart Long — 571
- **Science at Work** Mamie Lou, Pharmacist — 572
- **Science/Technology/Society** Aging — 573

Lennart Nilsson/
Boehringer Ingelheim
International Gmbh.

Reference Section

Safety Guidelines	576
Laboratory Procedures	578
Reading a Metric Ruler	579
Converting SI Units	579
SI Conversion Table	580
Reading a Graduate	581
Using a Laboratory Balance	581
Using Dissecting Tools	582
Making a Wet Mount	582
Using a Compound Light Microscope	582
Five-Kingdom Classification of Organisms	584
Glossary	586
Index	596

Skill

Chapter 1	Designing and Conducting Experiments	9
Chapter 2	Diagramming Food Chains and Food Webs	47
Chapter 3	Modeling Climate	80
Chapter 4	Comparing Plant and Animal Cells	114
Chapter 5	Modeling DNA	138
Chapter 6	Controlling Variables	159
Chapter 7	Using a Classification Key	180
Chapter 8	Identifying Viral Variables	217
Chapter 9	Organizing Data	239
Chapter 10	Making a Bar Graph	269
Chapter 11	Measuring Plant Growth	295
Chapter 12	Identifying Symmetry	331
Chapter 13	Classifying Insects	366
Chapter 14	Making and Interpreting a Graph	394
Chapter 15	Interpreting Data	406
Chapter 16	Organizing Information	444
Chapter 17	Classifying Nutritional Information	471
Chapter 18	Sequencing Data	501
Chapter 19	Observing Sound Waves	529
Chapter 20	Graphing Human Fetal Growth	560

INVESTIGATION

Chapter	Title	Page
Chapter 1	Modeling Soil Erosion	28
Chapter 2	Observing an Ecosystem	54
Chapter 3	Observing Changes in Ecosystems	70
Chapter 4	Using the Microscope	104
Chapter 5	Observing Osmosis	125
Chapter 6	Investigating Genetic Probabilities	168
Chapter 7	Observing, Ordering, and Displaying a Collection of Organisms	190
Chapter 8	Testing Disinfectants	224
Chapter 9	Observing Protozoans	235
Chapter 10	Predicting the Effect of Colors of Light on Plant Growth	282
Chapter 11	Examining Seeds	302
Chapter 12	Comparing Vertebrates and Invertebrates	336
Chapter 13	Tracing the Life Cycle of a Liver Fluke	353
Chapter 14	Identifying Foods That Tadpoles Need	383
Chapter 15	Predicting Animal Responses	416
Chapter 16	Observing Chicken Wing Muscles and Bones	452
Chapter 17	Observing Hemoglobin	477
Chapter 18	Measuring Lung Capacity During Exercise	496
Chapter 19	Mapping Your Sensory Receptors	538
Chapter 20	Predicting Changes in Heartbeat and Breathing Rates	548

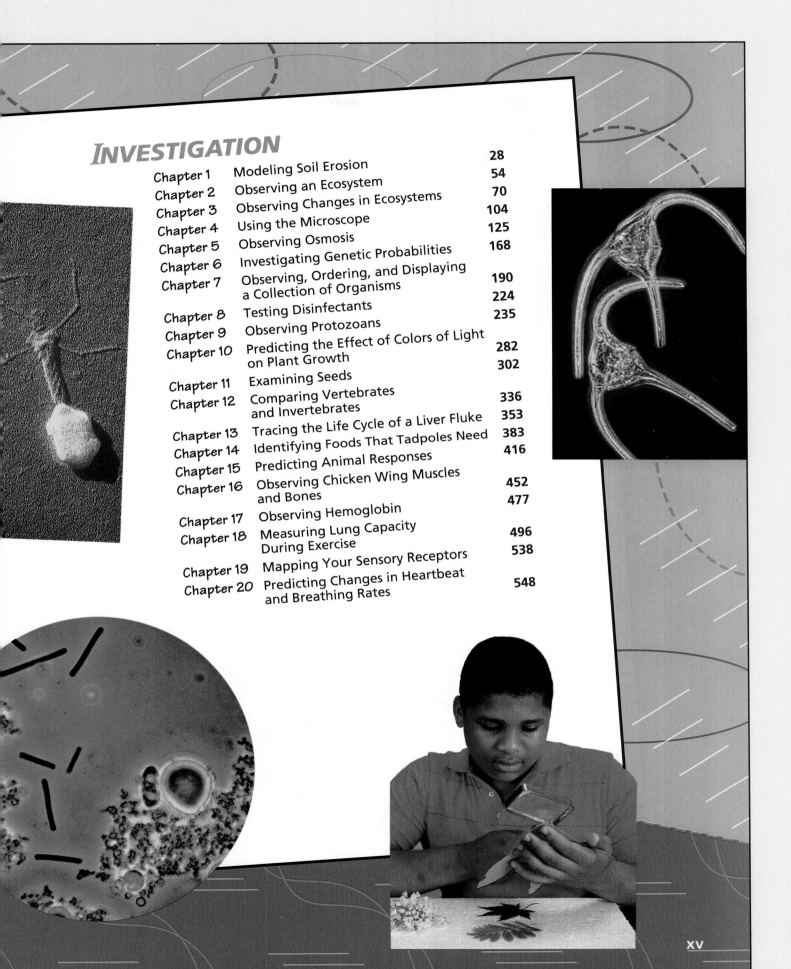

ACTIVITY

Chapter 1	How has the amount of carbon dioxide in the atmosphere changed?	16
Chapter 2	How can you count a population?	38
Chapter 3	What happens when fresh water and salt water meet?	68
	What is the root structure of grass plants like?	78
Chapter 4	How can you measure volume?	100
Chapter 5	How do food substances diffuse through a cell?	123
Chapter 6	What is the role of variation in evolution?	151
Chapter 7	How many different ways can you classify seeds?	186
Chapter 8	What microorganisms cause various diseases?	220
Chapter 9	What are the characteristics of fungi?	243
Chapter 10	What does a leaf look like when observed under a microscope?	272
Chapter 11	What can you learn from a collection of traveling seeds?	298
Chapter 12	Can you match the animals with the facts?	321
Chapter 13	How can you directly observe the characteristics of a typical coelenterate?	347
Chapter 14	How would you redesign a fish for life on land?	377
Chapter 15	How effective is protective coloration?	404
	What behavior can you observe in an ant colony?	411
Chapter 16	How do the muscles, bones, tendons, and ligaments in your hand work?	450
Chapter 17	How is starch digested?	465
Chapter 18	What are some characteristics of skin?	504
Chapter 19	Where are the greatest number of skin receptors located?	519
Chapter 20	What do sperm and ova look like?	552

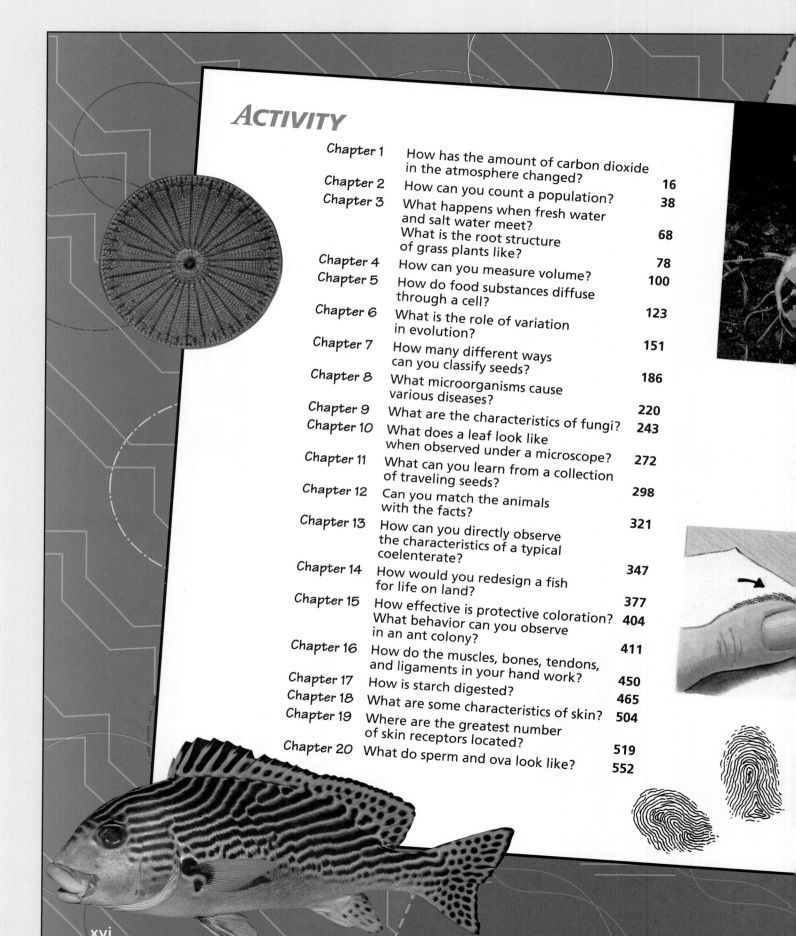

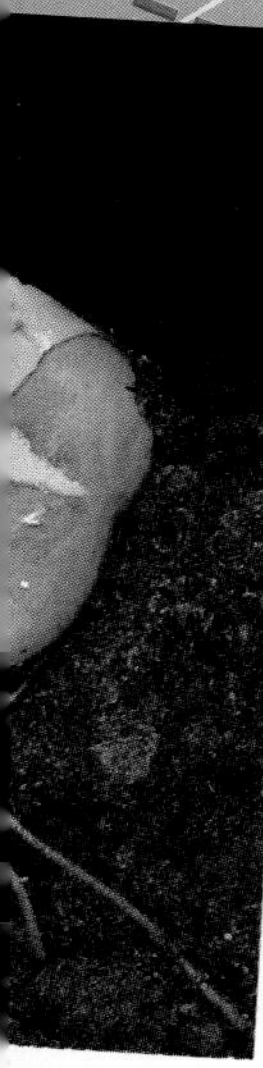

Discover By...

Chapter 1
- Writing 4
- Doing 7
- Observing 12
- Doing 13
- Doing 14
- Doing 20
- Researching 25

Chapter 2
- Doing 34
- Doing 39
- Observing 42
- Doing 45
- Researching 53

Chapter 3
- Doing 61
- Researching 73

Chapter 4
- Doing 98
- Calculating 99
- Doing 105
- Researching 108

Chapter 5
- Observing 121
- Writing 122
- Researching 127
- Problem Solving 129
- Researching 131
- Doing 135

Chapter 6
- Researching 147
- Doing 156
- Doing 167

Chapter 7
- Observing 175
- Researching 178
- Writing 185
- Researching 188

Chapter 8
- Researching 214
- Writing 215
- Calculating 219
- Doing 222

Chapter 9
- Doing 233
- Researching 236
- Observing 237
- Doing 241
- Writing 244

Chapter 10
- Observing 261
- Doing 263
- Doing 267
- Doing 271
- Doing 273
- Doing 278
- Researching 280

Chapter 11
- Doing 288
- Doing 289
- Doing 291
- Doing 293
- Doing 297
- Observing 299

Chapter 12
- Observing 320
- Researching 323
- Writing 326
- Doing 328
- Researching 335

Chapter 13
- Problem Solving 343
- Doing 345
- Observing 346
- Writing 346
- Observing 351
- Observing 356
- Calculating 357
- Researching 361
- Doing 363

Chapter 14
- Observing 373
- Writing 379
- Researching 385
- Observing 388
- Writing 390
- Doing 390

Chapter 15
- Researching 401
- Researching 409
- Researching 412

Chapter 16
- Doing 434
- Observing 437
- Calculating 439
- Observing 441
- Observing 442
- Doing 447
- Calculating 449
- Doing 449
- Observing 451

Chapter 17
- Observing 459
- Doing 460
- Researching 462
- Doing 462
- Doing 466
- Calculating 467
- Researching 469
- Problem Solving 470
- Calculating 474
- Doing 476
- Calculating 478
- Researching 479
- Calculating 484

Chapter 18
- Doing 492
- Doing 493
- Researching 499
- Researching 500
- Observing 505

Chapter 19
- Calculating 513
- Doing 514
- Doing 517
- Problem Solving 518
- Observing 521
- Observing 523
- Doing 527
- Researching 528
- Researching 532
- Researching 533
- Calculating 534

Chapter 20
- Researching 547
- Calculating 553
- Researching 555

Using The PROGRAM

INTRODUCING HOLT LIFE SCIENCE

Science is a field of study that is constantly changing with advancements in technology and medicine. Discoveries are announced with regular and increasing frequency. These advances in science require today's students to prepare themselves to make informed decisions on such questions as preferred medical treatment, genetic engineering, government regulations, and environmental pollution. Success in everyday living depends on the students developing an interest in and understanding of their world.

The general organization of HOLT LIFE SCIENCE proceeds from a survey of our environment to an exploration of living things from the simplest cells and organisms to the most complex. Unit 1 concentrates on the interrelations among all forms of life and investigates the different components of the environment. Unit 2 provides a structure for the examination of all organisms from the simplest monerans to the most complex members of the animal kingdom—human beings.

Unit 3 takes up the study of monerans and the hard-to-classify viruses. The unit also discusses the characteristics of protozoans, algae, and fungi. The two chapters of Unit 4 are devoted to the study of the plant kingdom. In Unit 5, the discussion of the animal kingdom begins with the invertebrates, continues with the vertebrates, and ends with a presentation of animal behavior. Unit 6 discusses the systems of the human body and the interactions of systems necessary for maintaining total health.

The text has been carefully written to provide the students with a readable presentation of science topics. The colorful pages have been thoughtfully designed to integrate explanatory graphics that complement the text.

HOLT LIFE SCIENCE offers intriguing reinforcement of scientific concepts through the use of literature. For example, Chapter 5 relates the scientific efforts to identify basic cellular structure to the puzzle-solving strategies of the master detective Sherlock Holmes. Chapter 13 uses Mary Ann Hoberman's whimsical poem "Spiders" to help the students remember features that distinguish spiders and other arachnids from insects. In Chapter 15, the students encounter unusual animals as they take a fascinating journey through a tropical rain forest with scientist Adrian Forsyth. Interwoven throughout HOLT LIFE SCIENCE, literature focuses and extends the presentation of scientific concepts.

To achieve a balanced presentation, HOLT LIFE SCIENCE integrates all areas of the curriculum including language arts, mathematics, social studies, and earth and physical science. As it investigates the major biomes, the text presents a physical map and geographical concepts of climate. It discusses proper nutrition as it describes the digestive system. It explores earth science concepts of the changing earth as it presents the theory of natural selection. These are just a few examples of how HOLT LIFE SCIENCE integrates the curriculum areas. Through this integration, the text enables the students to understand how science and other curricular areas are related. In addition, the use of real world examples continually ties the study of science to the world of the student, exploring and relating the concepts to day-to-day personal models. These practical applications of the science concepts allow the students to incorporate what they learn into their daily lives.

From the presentation, text, and graphics in the Pupil's Edition to the carefully structured teaching materials, HOLT LIFE SCIENCE is a readable, teachable program that explores the fascinating topics of life science.

BECOMING FAMILIAR WITH THE TEXTBOOK

HOLT LIFE SCIENCE is a user-friendly program designed to aid students in learning and teachers in teaching. For the student, each part of the textbook is an important resource designed to facilitate learning. It is important that you familiarize the students with the learning resources incorporated in the textbook.

Unit/Chapter Organization

Introduce the students to the program by asking them to thumb through a unit of the textbook. The consistency of unit and chapter organization makes it easy for the students to find the features in the textbook. Discuss the unit opener photograph and explore the information in the text box. Then have the students skim the chapter summaries, pointing out the range of material that will be covered in the unit. Explain that the *Science Parade* feature covers a variety of subjects related to the unit content from science applications, reading selections, and information about well-known scientists and careers to science/technology/society issues.

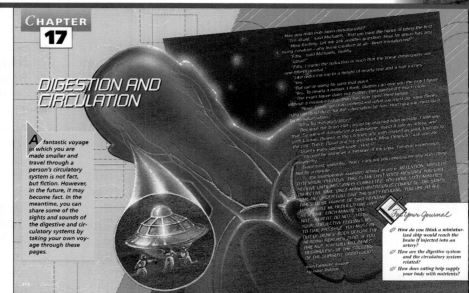

Next, have the students look at the chapter opener and read the introduction. The graphic or literary introductions are designed to provide motivation for the students. The discussion and preview can also provide an informal assessment of the background knowledge that the students bring to each unit and chapter. This assessment will help you fine-tune your planning for the chapters. The *For Your Journal* is a writing connection that introduces the students to a systematic method for recording information and assessing their familiarity with a topic.

Chapter Features

Look through the first chapter with the class, locating such features as the introduction, section objectives, and headings and subheadings. Discuss various methods of organizing information that can be used as the students read and study the chapter. If you encourage them to outline material, the students may use the chapter headings and subheadings as a basis for the development of a detailed outline of chapter material.

While reviewing how the textbook is organized, alert the students to the regular features of the text. Point out the *Discover By...* feature, which involves a brief experiment, research, or hands-on activity. Call attention to the *Ask Yourself* questions that

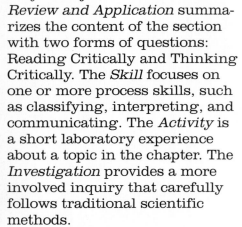

provide opportunities for the students to recall what they have just read. The *Section Review and Application* summarizes the content of the section with two forms of questions: Reading Critically and Thinking Critically. The *Skill* focuses on one or more process skills, such as classifying, interpreting, and communicating. The *Activity* is a short laboratory experience about a topic in the chapter. The *Investigation* provides a more involved inquiry that carefully follows traditional scientific methods.

Chapter Summary and Review

The *Highlights* page begins with *The Big Idea,* which reviews the theme of the chapter and connects the chapter ideas to help the students summarize the important concepts of the chapter. *Connecting Ideas* is a graphic presentation of the concepts/ideas of the chapter. These ideas are presented as a concept map, a table, a chart, an illustration, or a flow chart. *For Your Journal* reviews the students' journal entry from the chapter opener and gives the students the opportunity to revise their original ideas and add new information to their original answers.

The *Chapter Review* summarizes the learnings and provides opportunities for assessing the students' progress. The review also includes *Discovery Through Reading*—a short summary of an interesting book related to the chapter theme.

Unit Closure

The *Science Parade* at the end of each unit is a feature of up-to-date information on unit topics and ties the study of science to people, places, events, and issues. The section includes:

- Science Applications—topical features related to the unit theme and concepts
- Read About It!—articles, book excerpts, or stories related to unit topics
- Then and Now—two biographies, one historical and one modern, of people who have done work in areas related to unit topics
- Science at Work—information on a career in a field related to life science, highlighting a real person engaged in that career
- Science/Technology/Society—information on developments that have caused or may cause tremendous advances in the life sciences and that affect everyday life.

The Reference Section

While helping the students familiarize themselves with the textbook, do not overlook the *Reference Section*, which begins on page 574. The *Reference Section* includes Safety Guidelines for the classroom and the laboratory and Laboratory Procedures with information on SI measurement and conversion tables and laboratory methods. The Five-Kingdom Classification of Organisms is presented in chart form. The *Reference Section* also contains an illustrated Glossary and a cross-referenced Index.

USING THE ANNOTATED TEACHER'S EDITION

The *Annotated Teacher's Edition* is organized to provide all the information you need to teach HOLT LIFE SCIENCE. It features comprehensive strategies for teaching units, chapters, and chapter sections. The textual material presents teaching strategies, teacher demonstrations, background information, skills activities, answers to all questions, and other useful information.

The basic design of the *Annotated Teacher's Edition* includes reduced Pupil's Edition pages bordered by margin columns, which provide a wealth of information for teachers of all experience levels. All pertinent teacher information is located at point of use for ease and convenience. Most notably, the notes pertaining to a Pupil's Edition page will be found in the margins adjacent to the page. The margin notes include a complete plan for teaching each section and subsection and for custom tailoring your teaching style to the ability levels of your students.

Unit Presentation

The *Annotated Teacher's Edition* includes two interleaved pages before each unit. These pages include a Unit Overview and Unit Resources: Print Media for Teachers, Print Media for Students, and Electronic Media. Each unit begins with a discussion activity, which offers suggestions for using the unit opener to set the stage for the chapters that follow and a journal idea to prompt writing about the topic and investigating prior knowledge.

Chapter Organizational Information

Immediately preceding each chapter are two interleaved pages with charts for Planning the Chapter and Chapter Materials. Planning the Chapter includes all chapter sections, chapter features, and program resources that supplement each section, all with appropriate page references and level designations. Chapter Materials lists all materials necessary to perform Investigations, Activities, Discover By and Skill activities, and teacher demonstrations. Below the chart are suggestions for Advance Preparation, Field Trips, and Outside Speakers.

Chapter Teaching Information

The Chapter Opener Each chapter opener begins with the *Chapter Theme*, which describes the major theme of the chapter, lists other thematically related chapters, and includes any supporting themes. The *Chapter Motivating Activity* is provided to introduce the students to a major concept in the chapter. *For Your Journal* helps explore the students' background and prompts them to record information about the topic. The side columns include a *Multicultural Connection* and *Meeting Special Needs* to help tailor the lesson to the students' abilities. The *About the Literature/Photograph/Art/Author* gives interesting background information, which you may wish to share with the students.

Section Lessons Every section has a predictable lesson format that provides section information and a lesson cycle. Section information includes:

- ◆ Overview Chart that identifies process skills, attitudes, terms, media, and resources for the section
- ◆ Focus that provides a short overview of section content

The lesson cycle for each section motivates, reinforces, extends, and reviews to provide a complete instructional plan.

1. *Motivating Activity* focuses the students' attention on the lesson and sets the stage for learning.
2. *Teaching Strategies* develop lesson content through reinforcement and extension. Each strategy uses one or more process skills to foster greater understanding and to develop the students' thinking and communicating abilities.
3. *Guided Practice* provides strategies to ensure that the students have understood the section content.
4. *Independent Practice* allows the students to work on their own as they review the section content and concepts.
5. *Evaluation* provides a questioning strategy to encourage the students to demonstrate their understanding of the concepts through application.
6. *Reteaching* offers an additional strategy for students who need extra help to acquire basic concepts presented in the section.
7. *Extension* provides ideas for exploring beyond the content of the section through application, research, and other instructional methodologies.
8. *Closure* suggests individual or group activities that help the students summarize section concepts.

The lesson cycle is accompanied by a wealth of additional notes in the margins of the lessons, including *Science Background, Did You Know?, Multicultural Connection, Integration* (with cross-curricular concepts), *The Nature of Science, Demonstration, Laser Disc* information, *Reinforcing Themes, Science/Technology/Society,* and *Meeting Special Needs.* The margin areas also contain the answers to any text or activity questions. These answers are printed in red to distinguish them from other notes. In addition, the margin columns have any instructional notes that are needed for the special features in the Pupil's Edition, such as *Discover By, Skill, Investigation,* and *Activity.*

Interspersed throughout the margins are opportunities for assessment: *Ongoing Assessment, Performance Assessment,* and *Portfolio Assessment.* These three types of evaluation are informal tools to help gauge student progress.

Chapter Highlights and Review The *Chapter Highlights* page provides a suggestion for reinforcing the unit theme, notes reflecting the types of revisions that the students may make in their journal entries, and answers for the *Connecting Ideas* graphic. The *Chapter Review* provides answers to the in-text review questions.

USING ANCILLARY MATERIALS

A complete program of supplementary materials accompanies HOLT LIFE SCIENCE. Many of the materials can be found in the Teaching Resources, which are organized into six unit booklets and contain the following Blackline Masters.

◆ Study and Review Guide—pages to reinforce vocabulary and concepts for each section and a chapter review
◆ Record Sheets for Textbook Investigations—pages for recording data and conclusions for the textbook *Investigations*
◆ Transparency Worksheets—blackline illustrations of each transparency with accompanying teaching strategies
◆ Laboratory Investigations—two investigations keyed to each chapter of the textbook and designed to help the students further apply lesson concepts
◆ Reading Skills—pages designed to improve the students' reading strategies
◆ Connecting Other Disciplines—pages that integrate life science with other content areas of the curriculum
◆ Extending Science Concepts—pages that explore related science topics
◆ Thinking Critically—pages that focus on science topics while requiring the use of higher-order thinking skills
◆ Chapter and Unit Tests—materials to evaluate each chapter and unit

Additional laboratory notes and materials lists and other laboratory information and answer keys are also provided.

Other supplementary materials include Instructional Transparencies, Classroom Reference Posters, and Classroom Instructional Posters. There are fifty Teaching Transparencies with accompanying worksheets. The Classroom Reference and Classroom Instructional Posters are designed to supplement unit instruction. A separate Test Generator available for MacIntosh® and IBM® computers provides a complete testing program that enables teachers to test precisely the content they want by choosing questions from a comprehensive bank of test items.

HOLT LIFE SCIENCE is complemented and supplemented by the *Science Discovery* videodisc program, which consists of still images and motion footage. Specific frames are referenced in the chart preceding each chapter and at point of use in the margin notes, and are indicated by .

SCHEDULING

The teacher can arrive at an appropriate course schedule by considering the ability level of the students; their previous science experience; student interest; teaching style; local science-related resources; and local, county, and state requirements.

While HOLT LIFE SCIENCE is designed with the intention that all the main topics be covered during the typical one-year life-science course, the individual teacher can adapt the way in which he or she uses the textbook to the needs of the local teaching situation. No teacher should feel compelled to teach every chapter or to give equal emphasis to all chapters of the textbook. The teacher and the school system should dictate what constitutes appropriate course content.

The pacing chart beginning on Page T26 is designed to assist teachers in making decisions regarding scheduling, course content, and emphasis. The recommended coverage of a particular chapter or section for a basic, average, or honors course is indicated by a ■. The number of recommended class sessions to be devoted to each chapter and section is also indicated. The days indicated in the pacing chart allow for laboratory work related to the pertinent section. Time for chapter and unit tests is also considered. Use of the Pacing Chart may be supplemented by the Planning the Chapter Chart found on the interleaved pages immediately preceding each chapter.

PACING

Unit	Chapter	Section Number and Title	Basic Course	(days)	Average Course	(days)	Honors Course	(days)
1 OUR FRAGILE ENVIRONMENT				25		27		27
	1	**Humans and the Environment**		10		10		9
		1 Solving an Environmental Problem	■	2	■	2	■	1
		2 Pollution of the Environment	■	3	■	3	■	3
		3 Protecting the Environment	■	4	■	4	■	4
		Chapter Review and Test	■	1	■	1	■	1
	2	**Populations, Communities, and Ecosystems**		5		8		9
		1 Interactions	■	3	■	3	■	3
		2 Energy Flow	■	1	■	2	■	2
		3 Changing Ecosystems		0	■	2	■	3
		Chapter Review and Test	■	1	■	1	■	1
	3	**Life in the Biosphere**		9		8		8
		1 The Biosphere and Water Ecosystems	■	3	■	4	■	4
		2 Land Biomes	■	5	■	3	■	3
		Chapter Review and Test	■	1	■	1	■	1
		Science Parade and Unit Test	■	1	■	1	■	1
2 STUDYING LIVING THINGS				30		33		36
	4	**Cells**		8		7		7
		1 Tools for Life Science	■	3	■	2	■	2
		2 Discovery of Cells	■	2	■	2	■	2
		3 Structure of Cells	■	2	■	2	■	2
		Chapter Review and Test	■	1	■	1	■	1
	5	**Cell Function**		7		9		10
		1 Transport in Cells	■	3	■	3	■	3
		2 Cell Reproduction	■	3	■	2	■	2
		3 DNA and Cell Energy		0	■	3	■	4
		Chapter Review and Test	■	1	■	1	■	1

CHART

Unit	Chapter	Section Number and Title	Basic Course	(Days)	Average Course	(Days)	Honors Course	(Days)
	6	**Natural Selection and Heredity**		8		10		11
		1 A Great Adventure	■	3	■	3	■	3
		2 A Great Theory	■	3	■	3	■	3
		3 Mendel and Heredity	■	1	■	3	■	4
		Chapter Review and Test	■	1	■	1	■	1
	7	**Classification**		6		6		7
		1 The History of Classification	■	4	■	3	■	3
		2 Modern Classification	■	1	■	2	■	3
		Chapter Review and Test	■	1	■	1	■	1
		Science Parade and Unit Test	■	1	■	1	■	1
3	**SIMPLE LIVING THINGS**			18		15		16
	8	**Viruses and Monerans**		8		7		7
		1 Viruses	■	4	■	4	■	4
		2 Monerans	■	3	■	2	■	2
		Chapter Review and Test	■	1	■	1	■	1
	9	**Protists and Fungi**		9		7		8
		1 Protozoans	■	3	■	2	■	2
		2 Algae	■	3	■	2	■	3
		3 Fungi	■	2	■	2	■	2
		Chapter Review and Test	■	1	■	1	■	1
		Science Parade and Unit Test	■	1	■	1	■	1
4	**PLANTS**			17		18		16
	10	**An Introduction to Plants**		10		11		10
		1 Types of Plants	■	4	■	3	■	2
		2 Photosynthesis	■	1	■	3	■	3
		3 Reproduction in Seed Plants	■	4	■	4	■	4
		Chapter Review and Test	■	1	■	1	■	1

PACING

Unit	Chapter	Section Number and Title	Basic Course (days)		Average Course (days)		Honors Course (days)	
	11	**Plant Growth and Adaptations**		6		6		5
		1 Plant Growth	■	3	■	3	■	2
		2 Plant Adaptations	■	2	■	2	■	2
		Chapter Review and Test	■	1	■	1	■	1
		Science Parade and Unit Test	■	1	■	1	■	1
5		**ANIMALS**		**41**		**37**		**37**
	12	**Types of Animals**		10		9		8
		1 Introduction to Animals	■	5	■	2	■	2
		2 Characteristics of Animals	■	1	■	3	■	2
		3 Animal Classification	■	3	■	3	■	3
		Chapter Review and Test	■	1	■	1	■	1
	13	**Invertebrates**		14		11		10
		1 Sponges, Coelenterates, and Worms	■	6	■	5	■	5
		2 Mollusks and Echinoderms	■	2	■	2	■	1
		3 Insects—The Most Common Arthropods	■	2	■	1	■	1
		4 Other Arthropods	■	3	■	2	■	2
		Chapter Review and Test	■	1	■	1	■	1
	14	**Vertebrates**		12		10		10
		1 Fishes, Amphibians, and Reptiles	■	6	■	5	■	5
		2 Birds and Mammals	■	5	■	4	■	4
		Chapter Review and Test	■	1	■	1	■	1
	15	**Animal Behavior**		4		6		8
		1 Defense	■	3	■	3	■	3
		2 Group Behavior		0	■	2	■	4
		Chapter Review and Test	■	1	■	1	■	1
		Science Parade and Unit Test	■	1	■	1	■	1

CHART

Unit	Chapter	Section Number and Title	Basic Course	(Days)	Average Course	(Days)	Honors Course	(Days)
6 THE HUMAN BODY				49		50		48
	16	**Support and Movement**		8		9		8
		1 The Skeletal System	■	5	■	5	■	5
		2 The Muscular System	■	2	■	3	■	2
		Chapter Review and Test	■	1	■	1	■	1
	17	**Digestion and Circulation**		11		10		10
		1 Digestion and Nutrition	■	5	■	5	■	5
		2 Blood	■	2	■	2	■	2
		3 Circulation	■	3	■	2	■	2
		Chapter Review and Test	■	1	■	1	■	1
	18	**Respiration and Excretion**		9		9		8
		1 Respiration	■	3	■	3	■	3
		2 Excretion	■	3	■	3	■	2
		3 The Skin	■	2	■	2	■	2
		Chapter Review and Test	■	1	■	1	■	1
	19	**Coordination and Control**		13		11		11
		1 The Nervous System	■	3	■	3	■	4
		2 The Senses	■	5	■	4	■	3
		3 Alcohol, Tobacco, and Other Drugs	■	4	■	3	■	3
		Chapter Review and Test	■	1	■	1	■	1
	20	**Reproduction and Development**		7		10		10
		1 The Endocrine System		0	■	4	■	4
		2 The Reproductive Systems	■	3	■	2	■	2
		3 Prenatal Development and Birth	■	3	■	3	■	3
		Chapter Review and Test	■	1	■	1	■	1
		Science Parade and Unit Test	■	1	■	1	■	1

THEMATIC *Science*

The word *science* has its origin in the Latin word *scientia*, which means knowledge. Therefore, simply stated, science is knowing. Typically, science is considered a process that leads to a collection of facts. Through years of observation, scientists have amassed a body of knowledge that is quite vast.

Due to the sheer magnitude of knowledge, scientists and educators have had to devise schemes for categorizing this large volume of scientific information. Some state and local curriculum frameworks commonly call for the organization and teaching of science into the separate subject matter categories, or disciplines of life, earth, and physical science. These categories are often subdivided into narrower disciplines. While this categorization is a convenient and seemingly logical method, it fails to connect science knowledge in a comprehensible and interrelated fashion.

In an attempt to integrate and link the collection of facts that comprise the disciplines of science, educators and learning theorists have proposed a system of themes for organizing science curricula. These themes serve to integrate ideas and facts into meaningful schema for the students.

Themes span all the disciplines of science. They help make sense out of thousands of disparate science facts, leading to improved student understanding and achievement.

Teaching Themes

The content within HOLT LIFE SCIENCE is organized into major and supporting themes. Each chapter has one of three major themes as a focus. Many chapters have a supporting theme. It is important to note, however, that other themes are possible. You should feel free to introduce other themes that you feel are appropriate.

As they learn the chapter content, help the students understand the relationship of the chapter concepts to the major and supporting themes. The chapter review feature entitled "Reviewing Themes" provides the students with an opportunity to demonstrate their understandings.

The HOLT LIFE SCIENCE teacher's edition provides strategies for reinforcing themes in the margin notes. Reinforcing the chapter themes will contribute to the students' understanding and help the students assimilate the big ideas of science.

Definition of Themes

Environmental Interactions: The living and nonliving components of the environment interact with and depend on each other.

Systems and Structures: The natural world consists of systems and structures at micro and macro levels. Structures interact within systems, and interactions occur between and among systems and structures.

Changes Over Time: The natural world is characterized by patterns of change. Change can occur in cycles, in a steady linear progression, or in irregular patterns.

Energy: Interaction within and among systems requires energy. In living things, energy is needed for maintenance, growth, and reproduction.

Diversity: The natural world is filled with a diversity of systems and structures, which preserves the balance of nature.

Conservation: There is constancy within the living world. Systems and structures have a remarkable ability of conservation despite changes occurring within and around them.

Nature of Science: Science is knowing and understanding through observation and experimentation. It is using knowledge to understand and influence the forces of nature.

Thematic Table of Contents

Chapter	Major Theme	Supporting Theme
1	Environmental Interactions	Conservation
2	Environmental Interactions	Energy
3	Environmental Interactions	Systems and Structures
4	Systems and Structures	Nonspecific
5	Systems and Structures	Energy
6	Changes Over Time	Nature of Science
7	Systems and Structures	Changes Over Time
8	Systems and Structures	Environmental Interactions
9	Systems and Structures	Energy
10	Systems and Structures	Diversity
11	Environmental Interactions	Nonspecific
12	Systems and Structures	Energy
13	Systems and Structures	Diversity
14	Changes Over Time	Systems and Structures
15	Environmental Interactions	Nature of Science
16	Systems and Structures	Nonspecific
17	Systems and Structures	Energy
18	Systems and Structures	Energy
19	Systems and Structures	Nonspecific
20	Systems and Structures	Nonspecific

MULTI MULTI MULTI MULTI Cultural
INSTRUCTION

The nation's classrooms are a reflection of the cultural diversity of our people. Increased recognition of cultural diversity and the opportunities it provides has led educators to incorporate multicultural approaches into the curriculum. As a dynamic, ongoing process, multicultural education helps the students gain a greater understanding of themselves, their heritage, and their values as well as a respect for others. By identifying the contributions and strategies of people from diverse cultures, educators can raise the students' consciousness about the contribution of each individual to a group, to the school community, and to society as a whole. They can promote self-esteem, impart a feeling of pride in who we are, and achieve a sense of belonging for each member of a group. A multicultural approach to education gives each student equal educational opportunity by recognizing differences in cultural heritage and in learning styles.

The students who enter the classroom bring with them widely diverse cultural backgrounds. Such diversity leads to rich and varied experiences that affect the way each individual approaches learning experiences. The teacher can use this diversity to enrich the learning experiences of all students.

As content is taught and strategies are selected, the teacher should keep in mind that the students bring values into the classroom that are a reflection of their heritage and community. The students need an opportunity to share their knowledge and experiences. By using discussion techniques for eliciting prior knowledge and experience, the teacher can detect misconceptions and mold the learning to each student's experiential base. Achievement is greater when the methods and topics connect with who the students are and what they already know. Learning theory suggests that everyone learns new things by finding ways to connect them to what he or she knows and has experienced.

Positive role models from diverse cultural backgrounds are part of the life science curriculum. The curriculum recognizes that approaches to concept development vary from culture to culture. It provides for the use of different strategies to foster understanding, and it also provides opportunities for interactive learning activities including cooperative grouping strategies.

Teachers are encouraged to cultivate a classroom environment that promotes alternative opinions and different interpretations of conceptual information when appropriate. Using current news events to advance the contributions of people from various cultures is also encouraged.

In short, multicultural education supports enhanced self-concept, values the uniqueness of each individual, and promotes respect for others. Teachers should take appropriate steps to advance multicultural education whenever possible. It is the goal of HOLT LIFE SCIENCE to develop multicultural educational opportunities through content presentation, graphics, and activities.

Our nation's future depends on educating all Americans. Ethnic and gender stereotypes have long served as obstacles to the scientific literacy of all students. A central goal of education must be equal educational opportunities for all. Multicultural education contributes to the fulfillment of this goal. Only when you acknowledge and respect the differences in the students can you begin to achieve educational equity.

MEETING Special Needs

The science class offers a variety of students important information that will help them function and succeed in an increasingly scientific and technological society. Guidance counselors, special-education teachers, teachers of the gifted, and the school nurse may be consulted to help work out the best learning environment as well as realistic goals for each student. The following recommendations apply to mainstreamed students with specific types of learning problems.

MAINSTREAMED STUDENTS

Learning Disabled

Allow the student access to the special-education instructor as needed. Where feasible, make use of teaching helps available from such companies as National Teaching Aids, NASCO, and others. These helps include kits, models, and rubber stamps to reproduce diagrams.

- Give simple, clear directions.
- Encourage repeated efforts.
- Allow for group work during oral assignments.
- Provide for simplified rephrasing of concepts, tests, and reviews.
- Give oral examinations.
- Make certain that easy-to-read science reference material is available.
- Provide a daily, unvarying routine so that your expectations are clear.
- Make certain that instructions are understood before the student starts to work.
- Seat the student where distractions are minimized.

- Allow the student to express ideas with drawings or models if the disability permits. Dyslexic students will be able to develop drawings of their own; the dysgraphic students will be more successful at labeling figures that have been prepared for them.
- Within reasonable expectations, evaluate the student's grasp of the concept, not of spelling, punctuation, or intricate sentence structure.
- Provide spelling lists to which the student may refer when writing. Accept the use of printing when cursive writing presents too great a challenge.

Hearing Impaired

Allow the student access to your lecture notes if he or she has serious difficulty in learning auditorily.

- Avoid speaking while facing the chalkboard.
- Provide seating where the student can hear best or lip-read most easily.
- Allow for group work during oral assignments.
- Obtain close-captioned films for the hearing impaired.
- Rephrase instructions. Some sounds may be heard better than others.

Visually Impaired

Check with the special-education instructor or with local or state organizations regarding the availability of large-print textbooks. Seat the student close to the chalkboard.

- Stand facing the windows to avoid putting a demonstration into shadow.
- Encourage the handling of materials before or after a demonstration.

- Assign a student to make copies of notes.
- Use verbal cues rather than facial expressions or nods.
- Provide a sighted student guide.
- Provide high-contrast copies of worksheets or chalkboard diagrams.

Health Impaired The school nurse and special-education teacher can acquaint you with each student's limits.

- Become acquainted with the various symptoms of any health emergencies that might occur.
- Obtain training in first-aid procedures and CPR to be used in the event of a health emergency.

Orthopedically Impaired Arrange for seating that is comfortable for the student.

- Provide rest breaks.
- Cover the student's desk with felt so that materials do not slip.

SECOND LANGUAGE SUPPORT

Many school systems and districts have teachers whose main responsibility is to instruct the limited-English-proficient (LEP) student. Work with this teacher as a resource person in designing lesson plans for the LEP student. If your school system or district has no such resource person, the following suggestions may be of help in dealing with students whose first language is not English. HOLT LIFE SCIENCE offers Spanish language materials for several of its components that can be used by those students whose first language is Spanish.

- Use pictures, diagrams, models, and other props as often as possible.
- Read the captions of illustrations aloud to the student, simplifying the language if necessary.
- Use videotapes or films with accompanying audio to help clarify concepts.
- Prepare read-along tapes for each lesson.
- Provide a word list of key terms (in addition to the new science terms) that may come up often in the discussion of a particular topic.
- Allow the LEP student to work with a partner or in a group whenever feasible.
- Simplify language in descriptions and discussions whenever possible.
- Speak slowly and enunciate clearly. Restate sentences and phrases if necessary.
- Provide for oral testing with oral or pictorial responses from the student. Exercise patience while waiting for a response.

AT-RISK STUDENTS

At-risk students are those who, for any number of reasons, are liable to perform poorly and who have a high probability of dropping out. HOLT LIFE SCIENCE is engaging and interesting throughout, appealing to all students. The clear, easy-to-read prose and straightforward, attractive graphics reduce the potential for the students to grow bored. The style of HOLT LIFE SCIENCE is intentionally friendly and unintimidating.

GIFTED STUDENTS

There are many definitions for gifted students. In the past, definitions focused solely on I.Q. Current definitions identify gifted students as those who are exceptional in an ability area compared to others of the same age. The giftedness includes creative aspects and superior performance in some recognized area.

- Present lessons that accommodate the students' ability to learn faster and remember more.
- Encourage participation in a variety of learning experiences to broaden the students' areas of interest.
- Design curriculum to motivate and enhance the learning process.
- Design independent projects and comprehensive problem-solving tasks.
- Allow student choices to provide options for class activities.
- Avoid comparisons among and between the students in order to develop positive attitudes within the students and toward others.
- Encourage the students to develop leadership skills.

COOPERATIVE LEARNING

Allowing the students to share their learning with one another is a strategy that has been used since the day of the one-room schoolhouse. Throughout every lesson in HOLT LIFE SCIENCE, activities have been developed that are designed to provide worthwhile cooperative learning experiences. These activities can easily be identified by the colorful Cooperative Learning symbol.

COOPERATIVE LEARNING GROUPS

Grouping the students for cooperative learning gives them the opportunity to work toward both group and individual goals. The basic elements of a cooperative learning group include positive interdependence, face-to-face interaction, individual accountability, and interpersonal and small-group cooperative skills.

Positive Interdependence

The students must work with and depend on each other to achieve the group goal. Positive interdependence may be achieved through a division of labor, resources, and roles in the group. Rewards for the group, such as the same grade for every member of the group, may be included.

Face-to-Face Interaction

In cooperative learning, it is important that all group members provide support, encouragement, and help to each other. The members should be encouraged to share and discuss their ideas. For cooperative learning to be successful, the students must interact with each other.

Individual Accountability

Individual accountability means that each group member

is responsible for knowing the assigned material. In addition, each group member is responsible for the learning of other group members.

Interpersonal and Small-Group Cooperative Skills

A basic ingredient of cooperative learning is teaching the students the skills that are necessary for effective collaboration. The teacher needs to specify interpersonal and small-group cooperative skills such as staying with the group, taking turns, looking at the person who is talking, and checking to make sure that other members understand and agree with the answers. It is important for the students to learn what a cooperative relationship should look and sound like. It is also important to allow the students the time to analyze (process) how well their groups cooperate.

It is necessary to provide the students with opportunities to work as members of both small and large groups to encourage sharing, acceptance of responsibility, and decision making. Successfully guiding cooperative learning activities in the classroom requires a complete understanding of cooperative learning strategies. For further information, you may wish to refer to Circles of Learning: Cooperation in the Classroom, Third Edition, (Johnson, Johnson, and Holubec, 1991).

THE TEACHER'S ROLE IN COOPERATIVE LEARNING

In cooperative learning situations, the teacher should identify the group goal; decide on the group size, group makeup, room arrangement, and materials needed for the activity; and sometimes assign specific roles to group members. The teacher will also need to explain the task, structure the activity, monitor the interpersonal and small-group cooperative skills, and evaluate the product as well as the cooperative skills.

Identify the Group Goal

The students need to understand what is expected of them. Identify the group goal, whether it be to master specific objectives or to create a product such as a chart, report, or booklet. In addition, you should identify and explain the specific cooperative skills for each activity.

Decide Group Structure and Arrangement

For each activity, you will need to make some basic decisions. These decisions include group size, group makeup, room arrangement, materials needed for the activity, and specific role assignments for each group.

Group Size Cooperative learning groups may consist of two to six members. If you are using cooperative learning activities for the first time, you may find it easier to keep groups to two or three members. Smaller groups require fewer cooperative skills of the students than do larger groups.

Group Makeup Use heterogeneous grouping of members when possible. Heterogeneous groups are those that include students with high, average, and low ability levels. This type of grouping encourages greater diversity of thinking. Use homogeneous grouping (grouping of students with similar levels of ability) when mastery of specific skills is the goal.

Room Arrangement Have each cooperative learning group sit in a circle so that every member can see every other member.

Materials Encourage interdependence by providing only one set of the materials to the group. By doing this, you can help the members learn to work together to be successful. You may also give different materials to each group member. The completion of the task then depends on how well the members work together.

Role Assignment For some cooperative learning activities, you may want to assign a

recorder, who writes down the group's decisions and edits reports; a *summarizer-checker*, who makes sure all group members understand the material; a *research-runner*, who communicates with other groups and gets materials; a *reader*, who reads the directions and questions; and an *observer*, who keeps a record of how well the students perform the required cooperative skills.

Explain the Task For each activity, make sure that all group members understand the task. To ensure understanding, you may want to follow these suggestions:

- Give clear and specific instructions.
- Explain lesson objectives.
- Help the students see the relationship of the concepts to their past experiences and prior learning.
- Define concepts and provide models of what the students must do to finish a task.
- Ask questions to be sure that the students understand the task.

Structure Positive Goal Interdependence In cooperative learning, it is essential that the students work together. You should stress the "sink or swim together" relationship that is necessary to achieve the group goal. To facilitate positive goal interdependence, you can structure activities in the following ways:

- Have the group produce one product (answer or solution), which each group member signs. Each member must know the reason for the agreed-upon product or answer and be able to explain it.
- Give a group grade. You may evaluate members individually, but reward the group based on the total achievement of the group.
- Evaluate often by giving practice tests, by randomly choosing group members to explain answers or to read papers, and by having members edit each other's work.
- Encourage groups to help other groups by rewarding the entire class if a certain criterion is met.

Monitor Cooperative Skills

As the teacher, you have an important role in cooperative learning. While the students are working together, you can pick up vital information on the task because, in a cooperative setting, learning takes place "out loud." You can actually hear the learning while it is occurring. This is also an opportune time to observe systematically how well the students work together, especially on the skills that you specified before the activity started. You can count and record the number of times appropriate skills occur and provide the group with feedback. Student observers may also be used to record some cooperative skills. It is important that you evaluate the product that each group creates. You may also want to evaluate the group on how well the members worked together.

INTEGRATING OTHER DISCIPLINES

HOLT LIFE SCIENCE presents a complete life science curriculum that integrates science with other disciplines. Science is a part of the students' daily lives, and as such, it is an integral part of the learning process. The Pupil's Edition pages explore connections between areas of science, integrating earth science and physical science topics. The Pupil's Edition also contains connections across the curriculum with integration of literature and related language skills (reading, writing, speaking, and listening), mathematics, history, geography, social studies, health, art, music, and physical education.

Crosscurricular integration is a regular feature of the *Annotated Teacher's Edition* in which margin copy contains teaching tips that weave science concepts into the fabric of the students' knowledge of other curriculum areas. The integrated learning approach is also found in the components of the program. *Connecting Other Disciplines* are specially developed activities that focus on a chapter topic and a related curriculum area. The *Reading Skills, Extending Science Concepts,* and *Thinking*

Critically activities also involve crosscurricular content.

Literature and Language Skills

The organization and presentation of topics tie into real-world events as the students read selections from literature. The science content covering people, places, and events recounts famous milestones in scientific study. Integration of language happens as the students discuss, debate, explore, and record information in their journals. The journal activities help the students gather information, record questions, make predictions, draw conclusions, and summarize data for daily lesson activities. The journal provides a tool for learning; keeping a journal is an effective technique for all areas of study. Throughout the program, emphasis is placed on discovery and the inquiry approach, which provides a natural integration of listening and speaking skills.

Mathematics

Science is closely allied to the field of mathematics. Mathematical measurements and amounts are commonly used for gathering, analyzing, and presenting data. Scientists use mathematics to design experiments, to report their findings accurately, and to make predictions on scientific investigations.

Social Studies

History, geography, and other aspects of the social studies curriculum are a natural crosscurricular tie. As the students study the events of science, they recognize the developments over time and how such developments have affected history. The science aspects of many historical events provide data for understanding the ramifications for the future. Social scientists also use scientific methods to explore current topics, gather data, report findings, and make predictions.

Making Connections

When the students study concepts, it is important for them to understand a topic and how it relates to their world. Learning isolated data without understanding the intricacies and scope of the content will hamper the students from acquiring true knowledge. Every lesson should be presented without curriculum boundaries, and the students should be encouraged and motivated to explore across all curricular areas.

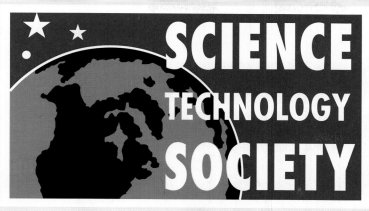

SCIENCE TECHNOLOGY SOCIETY

The problems of dwindling energy resources and the search for alternate energy supplies are addressed almost daily in newspapers and on television. The impact of computers and related technology on our daily lives has been significant. Social and ethical issues are constantly raised as technology progresses at a rapid pace in the fields of medicine, energy production, and genetic engineering.

To prepare for the future, to understand the role of technology, and to make informed decisions about the social implications of new scientific technologies, the students first need a strong foundation in the basic principles and processes of science. They must develop an accurate image of the nature of science and the usefulness of science in solving problems. Finally, it is also important for the students to gain confidence in their ability to identify science-related social issues and to use their own scientific knowledge to resolve these problems.

Teachers can play a vital role in helping the students gain the knowledge and the skills necessary to make responsible decisions about social issues related to science and technology. The sections that follow provide suggestions to help teachers in this role, using features from HOLT LIFE SCIENCE in conjunction with a variety of classroom strategies and outside resources.

STRATEGIES USING HOLT LIFE SCIENCE

HOLT LIFE SCIENCE provides the students with opportunities to analyze science-related and technology-related issues. A discussion of these issues can provide excellent additional opportunities for the development of thinking skills.

"Why should I be concerned about the environment?" and "What causes some plants and animals to become endangered or extinct?" are two questions often asked by the students. Both of these questions are answered in chapters found in Unit 1: Our Fragile Environment

Chapter 6 presents basic information about heredity and the genetic principles associated with it. The chapter includes discussions of the process of evolution and the theory of natural selection.

The chapters in Unit 6: The Human Body provide an overview of the various systems of the human body. The structures of these systems and their functions are discussed in detail appropriate to the life-science student. The chapters also stress the interrelationships of the systems.

The study of HOLT LIFE SCIENCE introduces the students to such topics as environmental pollution and recombinant DNA about which there may be several points of view. A discussion of these topics can provide excellent learning opportunities, especially for the development of higher-level thinking skills.

The students may also be encouraged to gather information from references. Once information has been gathered, a positive learning environment can be created if the students are taught to respect the right of others to express their views.

These are just a few examples of how life science is related to the lives of the students. You will find many opportunities to encourage the students to think about how science affects their lives.

OTHER CLASSROOM STRATEGIES

One popular way to introduce social issues into the science classroom is to have students bring in articles on science and technology from newspapers and magazines. Class time should be set aside on a regular basis to discuss the articles and the students' views. The teacher's role is vital in these discussions. However, the teacher must remember not to make judgments or to reject or praise student responses, but rather to accept all possible answers, ideas, and positions.

Debates are another way to approach the issues of science, technology, and society. For this strategy to be useful, however, the students must do more than repeat the opinions of others. They must be given adequate time and direction to research the issues, collect data, formulate their own opinions, and support their positions.

PROMOTING POSITIVE ATTITUDES

A student's attitude about science will have a strong effect on his or her achievement and future use of scientific knowledge and skills. Your teaching approach and expectations will affect a student's attitude toward science.

For far too many years, the roles of women and people of many cultures in science and math have not been emphasized. Too often, many students have not been able to identify role models in these fields. If this nation is to maximize its human resources and compete effectively in a highly scientific and technical global economy, schools must promote scientific literacy for *all* students.

Teachers can play an influential role in building positive attitudes toward science by viewing instruction from a three-dimensional perspective—concept development, behavior development, and attitude development. Failure to develop any one perspective is likely to lead to difficulty in developing the other perspectives. It is, therefore, important to encourage positive attitudes in your program.

Fostering Positive Attitudes

Beyond helping the students take a positive, can-do approach to science, you must base attitude development on critical thinking, decision making, and problem solving. Attitude development also involves fostering scientific values.

Helping the students believe that they can succeed in science includes offering words of encouragement and praise. You should look for positive behavior and success and acknowledge it. You should create opportunities for success for all students. These opportunities can often be offered effectively in cooperative group settings. You should also be aware that during discussions those with the lowest self-esteem will be slower to respond and to volunteer an opinion. Allow extra time in your questioning sessions and ask open-ended questions that carry less risk of being wrong.

A positive attitude toward science and critical thinking are closely aligned. Critical thinking is often defined as thinking that considers other points of view. Critical thinkers are better decision makers and problem solvers because they consider all the alternatives. A positive attitude helps the students entertain counter-arguments, avoid bias, and consider how others perceive and understand a situation. You can foster such an attitude by creating a classroom environment that openly accepts differing points of view and encourages independent thinking. In addition, asking questions that lead the students to analyze, evaluate, and apply information will contribute to their growth as critical thinkers.

HOLT LIFE SCIENCE encourages the development of positive attitudes by focusing on the following attitudes throughout the program:

- caring for the environment
- cooperativeness
- creativity
- curiosity
- enthusiasm for science and scientific endeavor
- honesty
- initiative and persistence
- openness to new ideas
- precision
- skepticism

Benefits of Positive Attitudes

Appreciation and respect for science are also part of building the proper attitude for science. Scientific knowledge and science skills can be both creative and problem-solving tools. When used in a positive manner, they can benefit the planet and all living things that inhabit it.

Experience supports the premise that a positive attitude leads to positive thoughts that then lead to positive actions. In science, this is an important consideration. Many decisions and actions can have far-reaching consequences. All questions are best considered when approached with a positive attitude and the use of critical thinking skills.

ASSESSMENT

With greater emphasis being placed on higher-order thinking and active learning, traditional assessment with its emphasis on fact recall is no longer compatible with curriculum goals. This is particularly true in science in which an enormous explosion of information has rendered sheer knowledge of facts to be of questionable value. Instead, emphasis must be placed on understanding broad concepts and big ideas and using information to solve problems. The students need to know how to retrieve, interpret, and use information effectively. Assessment, then, must determine if the students are acquiring these skills.

Role of Assessment

Assessment serves several purposes. It determines the value that the students place on information and tasks for which there is accountability. Therefore, you should design instruction with attention to the assessment that accompanies it. Assessment should do more than diagnose a student's ability to demonstrate knowledge or perform tasks. It also should provide an opportunity to measure student attitudes, prior knowledge, social interaction skills, learning styles, and communication skills. Used properly, assessment itself can and should be a learning tool. HOLT LIFE SCIENCE offers an effective assessment program by providing a variety of assessment strategies.

Types of Assessment

Current assessment strategies place emphasis on higher-order thinking, demonstration of understanding, ability to communicate, and task performance.

Performance Assessment

This method of testing requires the students to perform a task or series of tasks rather than choose an answer from a ready-made list of responses. These tests can be in the form of open-ended or response items in which the students are expected to respond to some issue or observation. They can be short or extended tasks that the students have to complete. Or they can consist of a portfolio of representative student work.

Portfolio Assessment

Suggestions in the *Annotated Teacher's Edition* provide many opportunities for accumulating activities and assignments for the students' portfolios. *Portfolio Assessment* designations provide helpful hints for saving documents that demonstrate the students' growth in understanding science concepts and problem-solving techniques.

Ongoing Assessment

The *Ask Yourself* questions in the Pupil's Edition provide opportunities for the students to pause briefly to summarize and review their understanding of the content. These brief assessments, in coordination with the *Section Review and Application*, *Highlights Page*, and *Chapter Review*, promote an ongoing evaluation system to help the students track their progress in understanding the content and science processes.

Checklists

In addition to performance-based assessment, student attitude and student demonstration of scientific process should also be assessed. An observational checklist is a good vehicle for these types of assessment. Samples of such checklists are given below. You may use these as models and expand them to meet your needs.

ASSESSING SCIENTIFIC ATTITUDES

ATTITUDE	POOR	GOOD	EXCELLENT
Shows respect for living things	___	___	___
Displays enthusiasm for science	___	___	___
Cooperates and actively participates in small-group projects	___	___	___

ASSESSING SCIENTIFIC PROCESSES

PROCESS	POOR	GOOD	EXCELLENT
Analyzes data	___	___	___
Observes accurately	___	___	___
Records observations carefully	___	___	___

THE ROLE OF INQUIRY

HOLT LIFE SCIENCE incorporates an inquiry approach to learning within the context of specific content presentations. Each feature of the program is designed to allow the students the time to explore, analyze, and assess information. From the format of the Pupil's Edition to the **Annotated Teacher's Edition,** the materials are organized to provide inquiry-based activities.

The **Investigation** found in each chapter of HOLT LIFE SCIENCE applies chapter content within the context of a scientific method. You are encouraged to augment the textbook Investigations with those found in the *Laboratory Investigations,* which accompany the Pupil's Edition. Pre-Lab strategies encourage students to form hypotheses and make predictions. The students are offered Post-Lab strategies, which prompt them to reflect on their original hypotheses and predictions and relate new information and conclusions gathered in the *Investigation.*

There are one or more **Activities** in each chapter of HOLT LIFE SCIENCE. These activities apply new concepts by having the students use basic laboratory skills. Each *Activity* allows the student to explore a specific question related to the section concept.

Discover features are found throughout each chapter of HOLT LIFE SCIENCE. The Discovers are short activities that apply new concepts in the chapter to everyday life. The Discovers are labeled *Discover By Doing*, *Researching,*

Observing, Calculating, Writing, or *Problem Solving* and may require laboratory work, library research, or manipulative skills. They may be performed in the classroom or at home. Most Discovers require simple, easy-to-obtain materials.

Skill activities enable the student to explore and manipulate the environment, to hypothesize and reason from data, to formulate explanations, and to communicate these explanations to others. In fact, process skills are prerequisites to success not only in science but also in other subjects and in the community and workplace as well.

The acquisition of process skills is a key goal in HOLT LIFE SCIENCE. To accomplish this goal, the program provides numerous opportunities for skills development through engaging narrative style and thought-provoking questions.

The common process skills in the Pupil's Edition and in the *Annotated Teacher's Edition* in HOLT LIFE SCIENCE include:

- Analyzing
- Applying
- Classifying/Ordering
- Communicating
- Comparing
- Construction/Interpreting Models
- Evaluating
- Experimenting
- Expressing Ideas Effectively
- Formulating Hypotheses
- Generating Ideas
- Generating Questions
- Identifying/Controlling Variables
- Inferring
- Interpreting Data
- Measuring
- Observing
- Predicting
- Recognizing Time/Space Relations
- Solving Problems/Making Decisions
- Synthesizing

As they develop and use these skills, the students must also estimate, sequence, describe, record, and draw conclusions. As a result, the students not only acquire basic scientific information but also develop the process skills needed to understand and interpret scientific information.

Field Trip GUIDELINES

Field trips are a desirable part of any science program. Field trips serve to remind the students that the real world exists outside the classroom. It is only in the field, for example, that organisms and physical phenomena can be observed and studied under natural conditions. Whether you go to a museum, a local industry, or an outdoor site, a truly successful field trip should add a new dimension to the students' grasp of science. It should make classroom work more meaningful and more enjoyable. Three things are required of the teacher in order to make any field trip an enriching experience: taking proper safety precautions, setting and evaluating goals, and careful trip planning.

ESTABLISHING SAFETY GUIDELINES

Some precautions in advance of the field trip may eliminate potential safety problems. The following suggestions will help ensure a safe and enriching trip.

- When planning a trip to an outdoor site, visit the site in advance and note any potential safety hazards. Discuss necessary safety measures with the students in advance. Include warnings of any hazards discovered during your advance visit.
- When planning a trip to a museum, local laboratory, or government agency, visit the location and meet with institution staff prior to the trip. Ask what precautions the students need to be given.
- Arrange for additional adult supervision, whenever necessary.
- If any part of the field trip is to take place on private property, get written permission in advance from the landowner.
- Make certain that any consent forms required from parents or guardians or school officials are drawn up, signed, and filed in advance.
- Caution the students to dress in a manner that will keep them comfortable, warm, and dry.
- Pack a basic first-aid kit. Depending on the type of trip, area, and season, consider including an insect repellent. On any trip to a large body of water, be sure to include an adult skilled in water safety and CPR.
- Caution the students to report any injury immediately to an adult supervisor.

SETTING AND EVALUATING GOALS

Assessing the value of a trip means first setting goals so that you have some standard against which to measure student accomplishments. Well-defined goals must then be transmitted to the students to encourage them to regard the field trip as an important learning experience. Use the following guidelines to set and evaluate realistic goals.

- Make sure that the students understand how the field trip relates to the appropriate textbook topic.
- Provide the students in advance with an outline of what they will experience on the field trip. If the trip involves an on-site guide, such as a museum curator or an animal breeder, try to get an advance summary of the material he or she plans to cover.
- If an on-site guide is involved, encourage the students to ask questions.
- Give the students a short list of questions to be answered by them after completing the field trip. This will give the students a concrete goal to work toward and will provide you with a means of evaluating the field trip.
- On trips to natural habitats, encourage the students to make sketches of specimens, such as insects and leaves, rather than to collect specimens.
- Follow up the field trip with a class discussion.
- Send personal or student-written thank-you letters to on-site guides and institution or company representatives for their help in planning and conducting the field trip.

It would be beneficial for you to develop a checklist for implementing the planning and organization of field trips. Your checklist can be developed using the guidelines and suggestions provided.

SAFETY GUIDELINES

Many states have school laboratory regulations covering such topics as eye and body protection, storage of combustible materials, fire protection, and the availability of first-aid supplies. Check with your state department of education. Even in the absence of local regulations, the following safety precautions should be routine procedure in any laboratory. You

can shape and reinforce proper student attitudes toward laboratory safety by setting a good example.

EQUIPMENT

- Microscopes, hot plates, and other electric equipment should be kept in good working order. Three-prong plugs must always be plugged into compatible outlets, or an adapter must be used. Never remove the grounding prong.
- Before each laboratory session, examine glassware for cracks and ask the students to do the same. Never use cracked or chipped glassware. Glassware used for heating should be made of heat-resistant material.
- Broken glassware should be swept up immediately, never picked up with the fingers. Broken glassware should be disposed of in a container specifically denoted for this purpose. If such a container is not available, broken glass should be adequately wrapped in paper and the paper secured. Alert the maintenance staff that the package contains broken glass.
- Never permit the students to operate an autoclave or a pressure cooker for sterilization. Conduct the sterilization yourself.
- Mechanical equipment should be set up according to instructions. The students should read all instructions for use before handling mechanical equipment.

MATERIALS

- Volatile liquids such as alcohol should be used only in ventilated areas and never near an open flame. To heat such substances, warm them on a hot plate or in a water bath.
- For the correct procedures to be used by the students when handling or heating chemicals, see pages 576-577 in the Reference Section.
- Used chemicals should be disposed of in a manner that does not pollute the environment. Check on school and state guidelines for proper chemical disposal.
- When teaching large classes, you may wish to set up materials stations in several parts of the room. The students can obtain chemicals and other supplies there, thus minimizing the distance they must carry supplies to their work areas.
- Chemicals should be stored in a well-ventilated area that is kept locked at all times. Flammable chemicals should be stored in a fire-resistant cabinet. Never store such chemicals in a refrigerator, unless the refrigerator is specifically marked explosion proof.
- Chemical storage areas should be kept clean, orderly, and well lighted.
- Pathogenic bacteria should NEVER be used.
- Dissection of preserved specimens should be conducted in well-ventilated areas. If fresh specimens are to be used for longer than one day, they should be preserved or kept under refrigeration. Specimens preserved in formaldehyde should not be used.
- Petri dishes containing bacterial cultures should be sealed with tape.
- Before washing Petri dishes, the cultures should be killed by heating the dishes in an

autoclave or a pressure cooker or by applying alcohol or a strong disinfectant.

◆ Wire loops used to transfer microorganisms should be flamed before and after the organisms are transferred.

LIVE ANIMALS

When live animals are introduced into the laboratory for observation and experimentation, a double safety standard must be maintained. The safety of the students is one objective; the humane treatment of the animals is the other. In addition to the following guidelines, consult the publication "Guidelines for the Use of Live Animals at the Preuniversity Level," which is available from the National Association of Biology Teachers, 11250 Roger Bacon Drive, No. 19, Reston, VA 22090.

◆ Before introducing animals into the laboratory, complete all plans for the care and feeding, including their maintenance over weekends and school holidays.

◆ Be sure that all mammals used in a school laboratory have been inoculated for rabies unless they have been purchased from a reliable biological supply house.

◆ Wild animals should never be brought into the classroom.

◆ Animals should not be teased or subjected to unnecessary handling.

◆ Experiments with animals should not involve the use of drugs, toxic products, anesthetics, surgery, carcinogens, or radiation.

◆ Make certain that any student who is scratched or bitten by an animal receives immediate attention from the school nurse or a physician.

◆ Remind the students to wash their hands thoroughly after handling animals.

PLANTS

Using plants does not require elaborate caution. However, many plants, or parts of plants, including some common house plants, are poisonous.

◆ Caution the students never to put any part of a plant in their mouths or near their eyes.

◆ The students should always wash their hands thoroughly after handling plants.

GENERAL INFORMATION

◆ Always perform an experiment yourself before asking the students to try it. Cautionary statements, which are printed in boldface type, are included in the procedures given for all laboratory investigations. Refer to those statements when preparing for the experiments. In addition, safety symbols alert the students to procedures that require special care.

◆ Do not permit the students to work in the laboratory without supervision.

◆ Do not allow the students to conduct unauthorized experiments.

◆ When an investigation has been completed, insist that the students clean their work areas; wash and store all materials and equipment; and turn off all water, gas, and electric appliances.

SAFETY EQUIPMENT

Safety equipment commonly found in school laboratories should include fire extinguishers, fire blankets, sand buckets, eyewash fountains, emergency showers, safety goggles, laboratory aprons, surgical gloves, thermal mitts, tongs, respirators, and a first-aid kit.

◆ Note the location of each piece of safety equipment in your laboratory, learn to use it correctly, and teach the students to do the same.

◆ Learn how to use the first-aid equipment. Call the school nurse or a physician immediately in case of injury.

◆ Post the telephone number of a poison-control center in your area.

SAFETY SYMBOLS

The following safety symbols will appear whenever a procedure requires extra caution.

SAFETY SYMBOLS USED IN HOLT LIFE SCIENCE

- wear lab apron
- wear goggles
- flammable
- poisonous chemical
- sharp/ pointed object
- bio hazard
- electric hazard
- rubber golves

MATERIALS

The following materials list has been compiled to help the teacher order supplies for the textbook *Investigations, Activities,* and *Skill* activities. The items are keyed to the *Investigation* (I), *Activity* (A), and *Skill* (S) in each chapter in which the item is needed. The numeral following the letter indicates the chapter number. Additional information regarding amounts and preparation of materials may be found on the interleaved pages immediately preceding each chapter.

APPARATUS AND EQUIPMENT	INVESTIGATIONS, ACTIVITIES, AND SKILLS
air pump (optional)	I14
aquaria	I14, A14
dissecting pan	I16
forceps	I7, I8, I16, A9
hand lens	I11, A3, A7, A13, A18
laboratory apron	I8
lung volume bag	I18
medicine dropper	I3, I5, I9, I17, A5, A9, A13, S4
meter stick	I2
metric ruler	I1, A1, A3, A4, A11, A13, A18, A19, S11

	INVESTIGATIONS, ACTIVITIES, AND SKILLS
microscope, compound light	I3, I4, I5, I7, I9, A9, A10, A13, A20, S4
mouthpiece (disposable)	I18
mouthpiece holder	I18
paper filter	A5
filter disks	I8
probes	I11
safety goggles	I8
scalpel	I11
scissors	I15, I16, A6
scoop	A5

GLASSWARE	INVESTIGATIONS, ACTIVITIES, AND SKILLS
baking dish	A3
beakers, 250 mL	I17
coverslips	I3, I5, I7, I9, A9, S4
dropping bottle	A17
graduates 10 mL	I3
100 mL	A4
Petri dish	I8, A5, A13
slides	I3, I5, I7, I9, A9, A13, S4

LOCAL SUPPLY	INVESTIGATIONS, ACTIVITIES, AND SKILLS
bags, plastic	A15
blindfold	A19
bread	I7
cake pans, aluminum	I1
can opener	S19
carrot, slice	A17
cellophane colored (yellow, red, blue)	I10
chicken wing, raw	I16
clay, modeling	S5
coins	I6, A16
container with lid	I3
cork	A19
cornmeal	A15
cotton swab	I8
crackers, soda	A17
cups, paper	I19, A6
die	I13
dishpan	I1
disinfectant solution	I8
egg white, cooked	A17
envelopes	A11
flashlight	S19
flower pots	S1
food coloring (blue and yellow)	A3

food products	A5, S17
food wrap, plastic	I17
glue	A7, S19
grass dried	I3
samples	A3
seeds	A15
ground beef	I14
hammer	I3
hand trowel	I2, A15
index cards	S2
ink pad	A18
jar with cap small	I2, S11
large	A15
jelly beans	S5
lettuce	I14
markers	A11

permanent, fine-tipped	S11
marshmallows, miniature	S5
mealworms	I14
mirror	A6, S19
nail	I3
newspaper	A15
paper	
black	A15
blue	A6
cardboard	A3, A4, S11
crumpled	A16
drawing	I7
graph paper	A1, S3, S10, S14, S15, S20
notebook paper	I2
poster board	A11
red	A6
paper clips	I19
paper towels	I5, I12, I16, I18, A17, S11
paper tube	A15
pen, felt-tip (water-soluble)	I19
pencils	
colored	A8, A16, S14, S20
dull	I19
yellow	A15
pictures of common objects	S12
pipe cleaners (4 different colors)	S5
plant, bean	S11

plants	I7
plants, aquarium	I14
plants, healthy	I10
potato, white	A17
rocks	I14
rubber band	I18, S19
rubber membrane, thin	S19
salt	A3
screening material	I2
shoe box	I15
shovel	A15
sod	I1
soil, potting	I1, S1
soup can, empty	S19
sponge	I15, A15
stakes	I2
straight pin	I19, S11
straw, dried	I3
straw, plastic drinking	A5

string	I2
sugar	A15
tape	A11, A15, A16
toothpicks	I11
watch with second hand	I20
water	
dechlorinated	I14, I17
distilled	I3
pond	I3, I7, I9, I14
salt	I5
wax pencil	I3, I8
wire, fine	A19, S5

BIOLOGICAL SUPPLIES

INVESTIGATIONS, ACTIVITIES, AND SKILLS

agar	I8, A5
ant colony	A15
chiton	A4
Daphnia	I17, A13
earthworm, preserved	I12
Elodea	I5, S4

fishes	A14
frog, preserved	I12
fungi samples	A9
hydra	A13
insects	I7
seeds	
bean	I11, S1, S11
corn	I11
grass	A15
variety (eight types)	A7
slides, prepared	
human cheek cells	S4
human ova	A20
human sperm	A20
insect leg	I4
insect wing	I4
leaf cross section	A10
sow bugs	I15
stock cultures	I9
tadpoles	I14
worms	I7
yeast suspension	I17

Resources

ELECTRONIC MEDIA SUPPLIERS

Agency for Instructional Television
111 West 17th Street
Box A
Bloomington, IN 47402

AIMS Media
9710 DeSoto Avenue
Chatsworth, CA 91311

Altschul Group Corporation
1560 Sherman Avenue
Suite 100
Evanston, IL 60201

American School Publishers
P.O.Box 408
Hightstown, NJ 08520

Barr Films
12801 Schabarum Avenue
Irwindale, CA 91706

Bergwall Productions, Incorporated
P.O. Box 2400
Chadds Ford, PA 19317

BFA Educational Media
468 Park Avenue South
New York, NY 10016

Biolearning Systems
Route 106
Jericho, NY 11753

Brorderbund Software Direct
P.O. Box 6125
Novato, CA 94948-6125

Carolina Biological Supply Company
2700 York Road
Burlington, NC 27215

Center For Humanities, Incorporated
Communications Park
90 South Bedford Road
P.O. Box 1000
Mt. Kisco, NY 10549

Central Scientific Company
3300 Cenco Parkway
Franklin Park, IL 60131

Charles Clark Company
170 Keyland Court
Bohemia, NY 11716

Churchill Films
12210 Nebraska Avenue
Los Angeles, CA 90025

Clearvue, Incorporated
6465 North Avondale
Chicago, IL 60631

Conduit
The University of Iowa
Oakdale Campus
Iowa City, IA 52242

Connecticut Valley Biological Supply Company
82 Valley Road
P.O. Box 326
Southampton, MA 01073

Coronet/MTI Film and Video
108 Wilmot Road
Deerfield, IL 60015

CRM/McGraw-Hill Films
2233 Faraday Avenue
Suite F
Carlsbad, CA 92008

Cross Educational Software
504 East Kentucky Avenue
P.O. Box 1536
Ruston, LA 71270

Direct Cinema, Ltd.
Box 10003
Santa Monica, CA 90410

Diversified Educational Enterprises, Incorporated
725 Main Street
Lafayette, IN 47901

Educational Activities, Incorporated
1937 Grand Avenue
Baldwin, NY 11510

Educational Images Ltd.
Box 3456 West Side Station
Elmira, NY 14905

Educational Services, Incorporated
1725 K Street NW
Suite 408
Washington, DC 20006

Edutech
1927 Culver Road
Rochester, NY 14609

Encyclopaedia Britannica Educational Corporation
310 South Michigan Avenue
Chicago, IL 60604

Films for the Humanities & Sciences
P.O. Box 2053
Princeton, NJ 08543

Fisher Scientific Company
Educational Materials Division
4901 West LeMoyne Avenue
Chicago, IL 60651

Focus Educational Media, Incorporated
485 South Broadway
Suite 12
Hicksville, NY 11801

Frey Scientific Company
905 Hickory Lane
Mansfield, OH 44905

Human Relations Media
175 Tompkins Avenue
Pleasantville, NY 10570

Instructional Video
P.O. Box 21
Maumee, OH 43537

International Film Bureau
332 South Michigan Avenue
Chicago, IL 60604

Kemtec Educational Corporation
9889 Crescent Park Drive
West Chester, OH 45069

Kons Scientific Company, Incorporated
P.O. Box 3
Germantown, WI 53022

Marty Stouffer Productions
300 South Spring Street
Aspen, CO 81611

Merlan Scientific
247 Armstrong Avenue
Georgetown, Ontario
Canada L7G 4X6

Micro Learningware
Rural Route #1
P.O. Box 162
Amboy, MN 56010

Modern Talking Picture Service, Incorporated
5000 Park Street North
St. Petersburg, FL 33709

Nasco International, Incorporated
901 Janesville Avenue
Fort Atkinson, WI 53538

Nasco West
P.O. Box 3837
Modesto, CA 95352

National Geographic Society
Educational Services
Department 91
1145 17th Street NW
Washington, DC 20036

National Teaching Aids, Incorporated
1845 Highland Avenue
New Hyde Park, NY 11040

Phillips Petroleum Company
Advertising Department
4th and Keeler Avenue
6-A1 Phillips Building
Bartlesville, OK 74004

Sargent-Welch Scientific Company
911 Commerce Court
Buffalo Grove, IL 60089

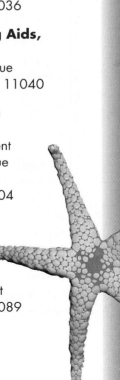

Resources

Scholastic Software
P.O. Box 7502
Jefferson City, MO 65102

Science Kit and Boreal Laboratories
777 East Park Drive
Tonawanda, NY 14150

Sunburst Communications
P.O. Box 660002
Scotts Valley, CA 95067

Time-Life
1450 East Parham Road
Richmond, VA 23280

Videodiscovery, Incorporated
1515 Dexter Avenue North
Suite 600
Seattle, WA 98109

Walt Disney Educational Media Company
108 Wilmot Road
Deerfield, IL 60015

Ward's Natural Science Establishment, Incorporated
P.O. Box 92912
Rochester, NY 14692

West Wind Productions
P.O. Box 3532
Boulder, CO 80307

DIRECTORY OF ORGANIZATIONS

American Association for the Advancement of Science
1333 H Street NW
Washington, DC 20005

American Association of Physics Teachers
5112 Berwyn Road
College Park, MD 20740

American Cancer Society
1599 Clifton Road NE
Atlanta, GA 30329

American Chemical Society
1155 16th Street NW
Washington, DC 20036

American Heart Association
The National Center
7320 Greenville Avenue
Dallas, TX 75231

American Institute of Biological Sciences
730 11th Street NW
Washington, DC 20001

American Institute of Chemists
7315 Wisconsin Avenue
Bethesda, MD 20814

American Institute of Professional Geologists
7828 Vance Drive
Suite 103
Arvada, CO 80003

American Lung Association
1740 Broadway
New York, NY 10019

American Medical Association
Education and Research Foundation
515 North State Street
Chicago, IL 60610

Centers for Disease Control
1600 Clifton Road NE
Atlanta, GA 30333

Institute for Chemical Education
University of Wisconsin
Department of Chemistry
1101 University Avenue
Madison, WI 53706

National Academy of Sciences
Office of News and Public Information
2101 Constitution Avenue NW
Washington, DC 20418

National Association of Biology Teachers
11250 Roger Bacon Drive
Number 19
Reston, VA 22090

National Health Council
1730 M Street NW
Suite 500
Washington, DC 20036

National Science Foundation
1800 G Street NW
Room 520
Washington, DC 20550

National Science Teacher's Association
1742 Connecticut Avenue NW
Washington, DC 20009

Society of Independent Professional Earth Scientists
4925 Greenville Avenue
Suite 170
Dallas, TX 75206

Superintendent of Documents
U.S. Government Printing Office
710 North Capitol Street NW
Washington, DC 20401

U.S. Environmental Protection Agency
Office of Communications and Public Affairs
401 M Street SW
Washington, DC 20460

U.S. Public Health Service
Department of Health and Human Services
200 Independence Avenue SW
Washington, DC 20201

LABORATORY SUPPLIERS

Carolina Biological Supply Company
2700 York Road
Burlington, NC 27215

Central Scientific Company
3300 Cenco Parkway
Franklin Park, IL 60131

Connecticut Valley Biological Supply Company
82 Valley Road
P.O. Box 326
Southampton, MA 01073

Damon Industries
DCM Instructional Systems
82 Wilson Way
Westwood, MA 02090

Difco Laboratories, Incorporated
P.O. Box 331085
Detroit, MI 48232

Eastman Kodak Company
BLDG 701 LRP
343 State Street
Rochester, NY 14652

Edmund Scientific Company
101 East Gloucester Pike
Barrington, NJ 08007

Fisher Scientific Company
Educational Materials Division
4901 West LeMoyne Avenue
Chicago, IL 60651

Forestry Suppliers, Incorporated
205 West Rankin Street
P.O. Box 8397
Jackson, MS 39284

Frey Scientific Company
905 Hickory Lane
Mansfield, OH 44905

Hach Company
P.O. Box 389
Loveland, CO 80539

Harvard Apparatus Company
22 Pleasant Street
Natick, MA 01760

Hubbard Scientific Company
3101 Iris Avenue
Suite 215
Boulder, CO 80301

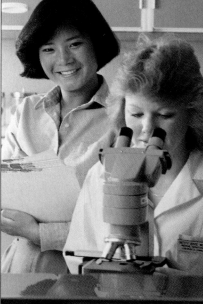
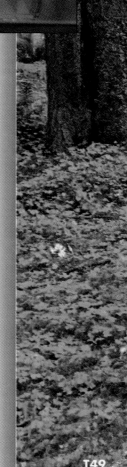

Resources

La Pine Scientific Company
13636 Western Avenue
Blue Island, IL 60406

Leica
Instrument Division
P.O. Box 123
Buffalo, NY 14240

Nasco International, Incorporated
901 Janesville Avenue
Fort Atkinson, WI 53538

Nasco West
P.O. Box 3837
Modesto, CA 95352

Sargent-Welch Scientific Company
911 Commerce Court
Buffalo Grove, IL 60089

Science Kit and Boreal Laboratories
777 East Park Drive
Tonawanda, NY 14150

Swift Instruments, Incorporated
P.O. Box 562
San Jose, CA 95106

Triarch, Incorporated
P.O. Box 98
Ripon, WI 54971

U.S. Geological Survey
12201 Sunrise Valley Drive
Reston, VA 22092

Van Waters & Rogers
5353 Jillson Street
Los Angeles, CA 90040

Ward's Natural Science Establishment, Incorporated
P.O. Box 92912
Rochester, NY 14692

Wilkens-Anderson Company
4525 West Division Street
Chicago, IL 60651

HOLT
LIFE SCIENCE

A MESSAGE TO STUDENTS

About Holt Life Science

Share the Wonder

Have you ever wondered why tropical rain forests are important? How all living things are alike? How you get energy for work and play? These are just a few of the questions that have been asked by life scientists over the years. The answers to these and many of your own questions about life and living things can be found in HOLT LIFE SCIENCE.

Explore the Nature of Science

Have you ever seen a bird catching a bug, a horse eating grass, or an animal taking care of its young? Have you ever grown a vegetable garden? Have you ever eaten yogurt or cheese? If you've ever done any of these things, you have experienced the study of life science. The study of the interactions among living things is what life science is all about. Science is knowledge gained by observing, experimenting, and thinking. Life science is knowledge of living things—their makeup and how they work. It is also knowledge of the relationships between living things and the world around them. You probably have a lot of this knowledge already. HOLT LIFE SCIENCE will help you add to your knowledge through reading, discussion, and activities.

Work Like a Scientist

Much of science is based on observations of events in nature. From observations, you form questions about the things you see around you. By doing experiments, you will gather more information. During your study of life science, you will be working and thinking like a scientist. You will develop and refine certain skills—such as your observation skills. In some cases, you will be asked to make predictions, and through activities you will have a chance to test the accuracy of your predictions. With new information, you will try to answer questions, to think about what you have observed, and to form conclusions. You will also learn how to communicate information to others, as scientists do.

Keep a Journal

One way to organize your ideas is to keep a journal and make frequent entries in it. Like a scientist, you will have a chance to write down your ideas and then, after doing activities and reading about scientific discoveries, go back and revise your journal entries. Remember, it's OK to change your mind after you have additional information. Scientists do this all the time!

Make Discoveries

Life science is an ongoing process of discovery—asking and answering questions about the structure and function of living things. If you choose a career in life science, you may find answers to the questions of students and scientists who lived before you. Through curiosity and imagination, life scientists are able to create a never-ending list of questions about the world in which we live. You can also gain respect and appreciation for all living things.

Prepare yourself for discovery. Through your studies, you will learn many interesting and exciting things. Allow yourself to explore and to discover the ideas, the information, the challenges, and the beauty within the pages of this book. HOLT LIFE SCIENCE can help you understand the world around you and how each living thing is important to the way in which the world works. You will begin to ask new questions and to find new answers. You will discover that learning about life science is fun and helpful.

UNIT 1

OUR FRAGILE ENVIRONMENT

UNIT OVERVIEW

This unit addresses the interactions, within an ecosystem, of plants and animals with their physical environments. The concept of a system in which community populations sustain themselves and how those environments provide for their survival are discussed. Also highlighted is the use of a scientific method to identify and solve environmental problems.

Chapter 1: Humans and the Environment, page 2

This chapter introduces the environment and interactions within the environment. The use of a scientific method to identify and solve environmental problems is explored. The many systems that interact within Earth's environment, producing both positive and negative results, are presented.

Chapter 2: Populations, Communities, and Ecosystems, page 32

This chapter introduces individual populations within an ecosystem. The various populations within a community and how they interact with one another are discussed. The role of the sun as the source of energy for all living things is presented.

Chapter 3: Life in the Biosphere, page 58

This chapter introduces the structures of the biosphere. How plants and animals interact with their physical environments and how those environments provide for the survival of living things are discussed.

Science Parade, pages 84-93

The articles in this unit's magazine deal with the conservation of Earth's environment; specifically, tropical rain forests, endangered species, and the ozone layer. The work and responsibilities of a conservationist are described in the career feature, and the biographical sketches focus on a biologist and an oceanographer.

UNIT RESOURCES

PRINT MEDIA FOR TEACHERS

Chapter 1
Ehrlich, Anne, and Paul Ehrlich. *Earth*. Franklin Watts, 1987. Summarizes Earth's environmental health problems and offers possible solutions to the problems.

Kramer, Stephen P. *How to Think Like a Scientist: Answering Questions by the Scientific Method*. Crowell, 1987. A simple introduction to using the scientific method.

Timberlake, Lloyd. *Only One Earth: Living for the Future*. Sterling, 1987. Presents examples from around the world of people who are using Earth's resources wisely and with regard for future generations' needs.

Chapter 2
Adams, Richard. *Watership Down*. Macmillan, 1972. A group of rabbits sets out in search of a new home after their old one is threatened.

Brooks, Bruce. *Predator!* Farrar Straus Giroux, 1991. Investigates the interaction between predators and prey that forms the basis of a food chain.

Lampton, Christopher. *Forest Fire*. Millbrook, 1991. Examines the causes of forest fires, the damage by and sometimes benefits of fires, and the techniques used to put them out.

Chapter 3
Gentry, Linnea, and Karen Liptak. *The Glass Ark: The Story of Biosphere 2*. Viking, 1991. A miniature replica of Earth is recreated in a sealed greenhouse in Arizona in an attempt to study environmental problems.

Reynolds, Jan. *Far North: Vanishing Cultures*. Harcourt Brace Jovanovich, 1992. Text and photographs reveal the life and culture of the Samis, who herd reindeer in northern Finland.

PRINT MEDIA FOR STUDENTS

Chapter 1
Javna, John. *Fifty Simple Things Kids Can Do To Save the Earth*. Andrews and McMeel, 1990. Explains how specific things in your environment are connected to the rest of the world, how using them affects the planet, and how an individual can develop habits and projects that are environmentally sound.

Chapter 2
Lauber, Patricia. *Summer of Fire: Yellowstone 1988*. Orchard Books, 1991. Describes the season of fire that struck Yellowstone in 1988, and examines the complex ecology that returns plant and animal life to a seemingly barren, ash-covered expanse.

Chapter 3
George, Jean Craighead. *One Day in the Tropical Rain Forest*. Crowell Jr. Books, 1990. From a colony of army ants to an overview of the importance of rain forests in the biosphere, Ms. George takes readers on a memorable journey through a jungle.

ELECTRONIC MEDIA

Chapter 1

Pollution: World at Risk. Videocassette. National Geographic. Shows many causes and effects of pollution of Earth's resources.

Protecting Endangered Animals. Videocassette. National Geographic. Explores reasons why animals are endangered and what humans can do to save those animals.

Scientific Methods: An Introduction and Problems. 2 Filmstrips. American School Publishers. Illustrates steps in a scientific method and the role of new devices and techniques.

Chapter 2

Animal Populations: Nature's Checks and Balances. Videocassette. Britannica. Studies the effect of environmental factors on animal communities.

Photosynthesis: Life Energy. Videocassette. National Geographic. Explains the mechanics of photosynthesis and its importance to life on Earth.

Succession: From Sand Dune to Forest. Videocassette. Britannica. Traces the stages of ecological succession.

Chapter 3

An Ecosystem: A Struggle for Survival. Videocassette. National Geographic. Explains what an ecosystem is by investigating the people, animals, and plants in India's Gir Forest.

The Living Earth. Videocassette. National Geographic. Explores the interdependence of land, water, air, and life on Earth.

UNIT 1

Discussion Point out to the students that the environment of our planet is fragile in the sense that it can be affected or changed for better or worse by the actions of humans. Explain that tropical rain forests are a particularly fragile part of the environment of Earth. Ask the students to name the areas of Earth that contain rain forests. (Rain forests can be found in the Amazon basin of South America, and in parts of Central America, Africa, Southeast Asia, Malaysia, and New Guinea.)

UNIT 1 OUR FRAGILE ENVIRONMENT

CHAPTERS

1 Humans and the Environment 2

People must be environmentally aware. What can you do to protect Earth's environment?

2 Populations, Communities, and Ecosystems 32

People and other living things interact with each other and with the non-living environment.

3 Life in the Biosphere 58

The only thing constant in the biosphere is change. How does change affect the biosphere?

Journal Activity You can extend the discussion by asking the students why tropical rain forests are being destroyed at such an alarming rate. Also ask the students to discuss the seriousness of rain forest destruction—is it a concern only for local ecosystems, or is it truly a global concern, one that affects all the people on the planet? Then have the students describe possible implications of such deforestation with respect to the Unit topics of the environment, ecosystems, and the biosphere. Lead the students to understand that a change in a fragile environment will affect all of the interactions of all of the living things in that environment. In their journals, ask the students to write a list of things they can do to help protect the environment of Earth.

Imagine a dense forest five kilometers long and five kilometers wide, containing hundreds of different trees, many of them over 75 meters tall. Living among the trees are thousands of different mammals, reptiles, amphibians, and insects. Now imagine that in the time it has taken you to read this, all the trees have been cut, or burned to the ground and all the animals have been killed. You can stop imagining now, because all this is true. Is this to be the future for all of Our Fragile Environment?

Science PARADE

SCIENCE APPLICATIONS
Studying the
Rain Forest 84

READ ABOUT IT!
"An Eagle to
the Wind" 87

THEN AND NOW
Rachel Carson
and Akira Okubo . . . 91

SCIENCE AT WORK
David Powless,
Conservationist 92

SCIENCE/TECHNOLOGY/ SOCIETY
A Hole in the Sky . . . 93

CHAPTER 1

HUMANS AND THE ENVIRONMENT

PLANNING THE CHAPTER

Chapter Sections	Page	Chapter Features	Page	Program Resources	Source
Chapter Opener	2	*For Your Journal*	3		
Section 1: SOLVING AN ENVIRONMENTAL PROBLEM	4	Discover By Writing (B)	4	*Science Discovery**	SD
• Identifying Problems (B)	4	Discover By Doing (A)	7	Reading Skills:	
• Conducting Experiments and Drawing Conclusions (B)	6	Section 1 Review and Application	8	Locating Information (B)	TR
		Skill: Designing and Conducting Experiments (A)	9	Connecting Other Disciplines: Science and Art, Making Observations About Paintings (B)	TR
				Study and Review Guide, Section 1 (B)	TR, SRG
Section 2: POLLUTION OF THE ENVIRONMENT	10	Discover By Observing (A)	12	*Science Discovery**	SD
• Pollution and Its Causes (B)	10	Discover By Doing (A)	13	Investigation 1.1:	
• Solid Wastes (A)	11	Discover By Doing (A)	14	Water Pollution (A)	TR, LI
• Acid Rain (H)	13	Activity: How has the amount of carbon dioxide in the atmosphere changed? (A)	16	Thinking Critically (A)	TR
• A Hole in the Sky (A)	14			Study and Review Guide, Section 2 (B)	TR, SRG
• The Greenhouse Effect (A)	15	Section 2 Review and Application	17		
Section 3: PROTECTING THE ENVIRONMENT	18	Discover By Doing (B)	20	*Science Discovery**	SD
• Natural Resources (B)	18	Discover By Researching (A)	25	Investigation 1.2: Recycling Newspaper (A)	TR, LI
• Soil Conservation (A)	19	Section 3 Review and Application	27	Extending Science Concepts: Acidity of Lake Water (A)	TR
• Water Conservation (A)	22	Investigation: Modeling Soil Erosion (A)	28	The Water Cycle (B)	IT
• Wildlife Conservation (A)	23			Record Sheets for Textbook Investigations (A)	TR
• A Success Story (A)	26			Study and Review Guide, Section 3 (B)	TR, SRG
Chapter 1 HIGHLIGHTS	29	The Big Idea	29	Study and Review Guide, Chapter 1 Review (B)	TR, SRG
Chapter 1 Review	30	For Your Journal	29	Chapter 1 Test	TR
		Connecting Ideas	29	Test Generator	

B = Basic **A** = Average **H** = Honors
The coding Basic, Average, and Honors indicates subsections, features, and resources that might be appropriate for different levels of learners. For additional suggestions regarding choice of topic and depth of coverage, see the Pacing Chart on pages T26–T29.

*Frame numbers at point of use
(TR) Teaching Resources, Unit 1
(IT) Instructional Transparencies
(LI) Laboratory Investigations
(SD) *Science Discovery* Videodisc Correlations and Barcodes
(SRG) Study and Review Guide

▶ 1A

CHAPTER MATERIALS

Title	Page	Materials
Discover By Writing	4	(per individual) journal
Discover By Doing	7	(per individual) journal
Skill: Designing and Conducting Experiments	9	(per individual or pair) bean seeds, flower pots of soil (3), water
Discover By Observing	12	(per individual) packaged items
Discover By Doing	13	(per individual) vinegar, chalk, leaves, bowl
Discover By Doing	14	(per individual) clear cup, teaspoon, rainwater, purple grape juice
Activity: How has the amount of carbon dioxide in the atmosphere changed?	16	(per individual or pair) graph paper, ruler
Discover By Doing	20	(per individual) 4 cups of soil, 2 cups water, pebbles, paper, pencil
Teacher Demonstration	22	plant seeds (40), pots (4), sand, clay, topsoil, water
Discover By Researching	25	(per individual) paper, pencil
Investigation: Modeling Soil Erosion	28	(per group of 2 or 3) aluminum cake pans (2), potting soil, ruler, sod, dishpan, liter bottle of water

ADVANCE PREPARATION

For the *Discover By Doing* on page 7, all students might not have access to an animal to observe. You may wish to ask those students with pets to arrange for others to observe their animals. For the *Discover By Doing* on page 14, rainwater should be collected in clean containers and saved to use in the activity. For the *Demonstration* on page 22, use plant seeds that are fast growing such as bean or grass seeds.

TEACHING SUGGESTIONS

Field Trip
Take a trip to a local sewage treatment plant to observe how waste water is purified before it is returned to the environment.

Trips can also be organized to observe environmental pollution at local sites such as streams or lakes and open dumps or landfills.

Outside Speaker
Have a local life scientist come in and discuss his or her work. Local universities, health and environmental protection agencies, agricultural services, research hospitals, and pharmaceutical companies are possible sources for such people. Ask the speaker to describe the activities and problems associated with his or her work.

CHAPTER 1
HUMANS AND THE ENVIRONMENT

CHAPTER THEME—ENVIRONMENTAL INTERACTIONS

This chapter introduces the students to the environment in which they live and the interactions within the environment. While studying the chapter, the students will explore how a scientific method is used to identify and solve environmental problems. The students will also discover that many systems interact within Earth's environment, producing both positive and negative results. The theme of **Environmental Interactions** is also developed in Chapters 2, 3, 11, and 15. A supporting theme developed in this chapter is **Conservation**.

▶ MULTICULTURAL CONNECTION

The students should realize that environmental problems are not simply a local problem—all parts of the earth are affected by such problems. Focus on a specific, relatively simple issue, such as using paper or plastic bags rather than cloth bags to carry store-bought goods. First, have the students discuss why the issue is important. Then ask students who have visited or lived in other nations to explain whether the issue is a problem in those places and, if so, how it is addressed in the other nations.

▶ MEETING SPECIAL NEEDS

Second Language Support
Fluent English-speaking students should be paired with students who have limited English proficiency to work together to create a list that describes many of the ways people can cause harm to the environment.

CHAPTER 1
Humans and the Environment

▶ Visible air pollution and water pollution highlight the ability of humans to cause changes in their environment. Solving environmental problems is a challenge people are now facing. Only then will Earth remain a hospitable home to humans and all other forms of life.

CHAPTER MOTIVATING ACTIVITY

Prior to class, mix a solution of water and food coloring in a shallow pan. Fold a sheet of paper "accordion fashion," dip the folded edges of the paper into the solution, unfold the paper, and let it dry—the paper should be striped. Smooth out the paper. In class, ask the students to look closely at the pattern and hypothesize how the pattern was formed. Help the students test their hypotheses. Explain that making observations, formulating hypotheses, and testing those hypotheses are part of the method that scientists use to answer questions.

For Your Journal

The journal questions provide an opportunity for the students to tell what they already know about environmental problems and to speculate about possible solutions to those problems. The answers given by the students will provide you with an opportunity to note any misconceptions they might have about environmental issues. Before they begin writing, remind the students that their environment influences all parts of their lives and ask them to consider why environmental problems must be solved as a result.

From space, Earth looks like a glimmering blue jewel as white puffs of clouds swirl over its blue oceans. Up close, we can see its majestic mountains, rolling plains, and deep canyons. We can view its bubbling streams, reflecting pools, and massive oceans. Earth is a remarkably beautiful, yet fragile, place. On it, we can also see eroded land, polluted waters, and smog. We can watch forests being destroyed and animal and plant species disappearing. Centuries of human activities have helped create these problems. Now, humans have the responsibility to help solve them.

ABOUT THE PHOTOGRAPH

The NASA photograph of the earth was taken from space by astronauts. From this picture of Earth, a viewer would not suspect that problems such as polluted air, land, and water exist. The vivid blue, green, and white sphere set against the darkness of space looks untouched. Only by focusing on specific areas can a viewer identify environmental problems. Most of these problems have resulted from human activities. The problems must be solved because they threaten the fate of all organisms on Earth; they threaten the fate of Earth itself.

For Your Journal

- List any environmental problems you have observed.
- How do you think the problems can be solved?
- What can individuals do to help solve the problems?

CHAPTER 1

Section 1: SOLVING AN ENVIRONMENTAL PROBLEM

FOCUS

This section focuses on the ways in which scientists answer questions. Each of the processes of scientific investigation, including observing, collecting data, forming a hypothesis, conducting experiments, and drawing conclusions, is examined in an actual case study. Collectively, these processes are called a scientific method.

MOTIVATING ACTIVITY

To demonstrate the importance of making careful observations, have another teacher walk into the classroom, take an object from your desk, and leave. Encourage the students to discuss what they saw in detail. Then have the teacher recreate the scene, and discuss the accuracy and completeness of the students' initial observations.

PROCESS SKILLS
- Solving Problems/Making Decisions • Inferring

POSITIVE ATTITUDES
- Caring for the environment
- Precision • Skepticism

TERMS
- hypothesis • theory

PRINT MEDIA
How to Think Like a Scientist by Stephen P. Kramer (see p. xixb)

ELECTRONIC MEDIA
Scientific Methods: An Introduction and Problems, American School Publishers (see p. xixb)

Science Discovery Scientific method

BLACKLINE MASTERS
Study and Review Guide
Reading Skills
Connecting Other Disciplines

DISCOVER BY Writing

If the students have difficulty describing the steps they might follow when solving a problem, have them first identify a real or hypothetical problem. Then ask them to describe, in their journals, the steps they might follow to solve that problem. Encourage the students to imitate the problem-solving steps used by scientists, tailoring those steps to their particular problem.

SECTION 1

Solving an Environmental Problem

Objectives

List the processes used to solve a problem scientifically.

Analyze problem solving through controlled experimentation.

Compare and contrast a hypothesis and a theory.

You probably have many questions about the world in which you live. Why are plants green? Why is the sky blue? Why do people still pollute the environment? If you ask questions such as these, you are thinking like a scientist. Scientists ask questions about the world in which they live. They also work to find the answers through scientific investigation and observation.

DISCOVER BY Writing

You solve problems every day. Some problems are simple, such as deciding what movie to see. Other problems are more important, such as deciding what courses might prepare you for a certain career. In your journal, describe the steps you might follow when you solve a problem. Are your problem-solving steps like those of a scientist?

Figure 1–1. Asking questions is one way to think like a scientist. What questions might these pictures pose to scientists?

Identifying Problems

There is no list or set of rules that all scientists follow to solve problems. Each problem is different and must be viewed with an open mind. Yet solving a problem scientifically usually involves certain processes, or steps. These processes, called a *scientific method,* are part of the orderly way in which problems are solved.

TEACHING STRATEGIES

● **Process Skills:** *Observing, Applying*

Remind the students that the first step in a scientific method is making observations. Point out that all the senses can be used to make observations. Have the students close their eyes and observe based on senses other than sight.

● **Process Skills:** *Classifying/Ordering, Analyzing*

Suggest that the students begin a chart outlining the steps of a scientific investigation and identifying the steps that were taken during the investigation of the coyote problem. The students can add to the chart throughout their study of this section and later place the chart in their science portfolios.

The first step in using a scientific method is usually to make observations. Once you make observations, you can decide which problem interests you. Suppose a scientist chooses to study coyotes, animals that prey on rabbits and other small wild animals. Coyotes also prey on farm animals. Coyotes cost farmers and ranchers millions of dollars each year. Destroying coyotes reduces the loss of farm animals, but it also disturbs the natural balance of the environment.

A scientist might be concerned about how to protect farm animals without harming coyotes. If it is possible to solve this problem, the solution will help the ranchers and it would help the coyotes.

Once a scientist has identified the problem to be investigated, the next step is to collect more information. Often other scientists are researching similar problems. Scientists share their results so that more can be learned about the problem. Books, magazines, scientific meetings, and computer-databanks can be used to learn about information already available.

Figure 1–2. Although coyotes' natural food source is small wild animals, they also kill and eat farm animals.

Using this information together with new observations, a scientist can then form a question about the problem. A possible question might be stated as follows: "Can coyotes be prevented from killing farm animals without upsetting the natural balance in the wild?"

A scientist's next step is to suggest a possible answer to the question based on known information. This possible answer is called a **hypothesis** (hy PAHTH uh sihs). A scientist studying coyotes, for example, might make the following hypothesis: "Coyotes made slightly ill by eating farm-animal meat treated with chemicals will not kill and eat farm animals in the future."

SCIENCE BACKGROUND

A great deal of science is not experimental. Much of biological science involves accurate description of living things. Hypothesis formulation and experimentation are only part of scientific investigation.

✦ Did You Know?
About half of all the scientific knowledge that is known today has been discovered during the past forty years.

THE NATURE OF SCIENCE

The Italian astronomer and physicist Galileo (1564–1642) was the first scientist to be credited with using a scientific method in problem solving. He recognized the importance of carefully controlled experiments in developing scientific theories. Galileo took a scientific problem, reduced it to basic everyday situations, and used observation and mathematic principles to analyze the results. The steps that he followed became the foundation of the modern scientific method used by scientists today.

SECTION 1 5

TEACHING STRATEGIES, continued

● **Process Skills:** *Applying, Solving Problems/Making Decisions*

Have the students discuss why coyotes might eat farm animals instead of, or in addition to, wild animals. (Farm animals might be more available or easier to catch than wild animals.) Ask the students to think of ways to stop coyotes from eating farm animals without damaging the environment. (Erecting tall or electric-wire fences and using guard dogs might help deter coyotes.)

● **Process Skills:** *Applying, Inferring*

Sheep carcasses treated with lithium chloride were put out as coyote bait. The treated sheep meat made the coyotes ill for a short time but caused them no long-lasting harm. Point out that a basic life characteristic of organisms is their response to stimuli. Ask the students to describe the stimulus and the response of this experiment. (The stimulus is the coyote illness caused from eating the chemically treated meat; the expected response is the future avoidance of sheep by coyotes.)

ONGOING ASSESSMENT
ASK YOURSELF

A scientific method is an orderly, systematic approach to solving problems.

INTEGRATION—Language Arts

On the chalkboard, write a traditional saying such as "An apple a day keeps the doctor away" or "Feed a cold and starve a fever." Encourage the students to record the saying in their journals. Then ask them to go on to conjecture whether the message of the saying could be tested using a scientific method, and if it could, what results do they think such a test might yield?

MEETING SPECIAL NEEDS

Mainstreamed

Remind the students that a hypothesis can take the form of a possible explanation of an event. To practice generating hypotheses, have a student volunteer create a noise while all other students' eyes are closed. Then have the students open their eyes and hypothesize, or guess, how the noise was created. Repeat the activity several times.

▶ **6** CHAPTER 1

This hypothesis states an action to be taken and a possible outcome. Coyotes are to be given chemically treated meat that will make them slightly ill. The scientist expects the coyotes to associate becoming ill with eating the treated meat and, therefore, to avoid preying on farm animals.

ASK YOURSELF

What is a scientific method of solving problems?

Conducting Experiments and Drawing Conclusions

The most important characteristic of a hypothesis is that it must be able to be tested with experiments. An experiment is an organized way of collecting facts, or *data,* through observations.

An experiment often includes the study of two groups. These two groups should be exactly alike, except for one difference, or *variable.* In this case, the hypothesis describes one group of coyotes. This group feeds on a small amount of chemically treated meat that makes them ill. Meat with no chemicals is given to the second group of coyotes. The variable between the two groups in this experiment is the chemical in the meat.

The group given chemically treated meat is the experimental group. The group given untreated meat is the control group.

Figure 1–3. The collection of experimental data involves careful observations during an experiment.

- **Process Skills:**
Classifying/Ordering, Analyzing

After the students have examined the data in Table 1-1, explain that the experiment was conducted during a two-year period, and the treated meat was put out only during the second year. Also point out that as a result, the table reflects only the number of sheep killed during one year. Ask the students in which year of the experiment they think the data in Table 1-1 was collected, and why the experiment took two years to complete, even if the data was collected only during one year. (The first year, the scientists had to make observations, collect data, and formulate a hypothesis; the second year, the scientists had to test their hypothesis in a controlled experiment. The data in the chart was collected during the second year when treated meat was put out for the experimental group of coyotes. This data could then be compared to the data that was gathered from the control group of coyotes.)

A scientist compares the data collected from the two groups. In the case of the coyotes, the scientist wants to see the effect that eating chemically treated meat has on the experimental group. Remember, the two groups are exactly alike except for the single variable. Therefore, any differences seen between the groups must be caused by the variable. An experiment that uses two groups such as these is called a *controlled experiment*.

Now the scientist must set up the experiment, make observations, and record data. In the following activity, you can practice making observations and collecting data.

DISCOVER BY Doing

Observe an animal, such as a dog, a cat, a bird, a lizard, or a fish, for a period of 15 minutes. Write your data—all the things you observed—in your journal, and report your observations to your classmates.

Just as you observed and recorded data on the animal you studied, scientists make and record their observations. For the coyote problem, scientists studied two groups of wild coyotes in Saskatchewan, Canada. The experimental group was given a small amount of chemically treated sheep meat that would make the coyotes slightly ill for a short time. Untreated meat was put out for the other group. The number of lambs and sheep killed by coyotes in each group was recorded for a period of one year. Data from the experiment is shown in Table 1-1.

Table 1-1 Sheep Killed by Coyotes

Subjects of Study	Control Group	Experimental Group
Total Number of Sheep	100 000	100 000
Number of Deaths	10 000	10 000
Deaths Due to Coyotes	400	150

The data shows that the coyotes in the experimental group killed and ate 63 percent fewer sheep than the coyotes in the control group. The data supports the hypothesis. Scientists concluded that feeding coyotes sheep meat that makes them ill reduces their killing and eating of sheep.

DISCOVER BY Doing

Any small animal, including an insect, can be observed for this activity. The observations made by each student might include size, color, markings, type of hair or other body covering, movements, reactions to sounds, feeding behavior, and so on.

PERFORMANCE ASSESSMENT

Ask the students to organize their animal observations in a table. Have each student explain how he or she constructed the table. Check the student's table to evaluate his or her ability to observe, organize data, and communicate using a graphic aid.

INTEGRATION—Social Studies

Cooperative Learning Preventing coyotes from eating sheep is a serious problem for some ranchers. In some states, lethal poison baits have been used to kill coyotes. Have the students form small debate groups. Assign each group a position supporting or opposing state laws that permit the use of poison baits. Ask each group to develop arguments to support their positions. Ask opposing teams to debate the issue.

GUIDED PRACTICE

Ask the students to recall a minor problem that they solved on their own. Have the students recall the steps they took to solve the problem and the order in which those steps were taken. Then ask them to compare their method of problem solving to a method used by scientists.

INDEPENDENT PRACTICE

 Have the students provide written answers to the Section Review and Application questions. Then in their journals, have them identify a problem from everyday life and describe how a scientific method could be used to find a solution to that problem.

EVALUATION

Ask the students to determine whether the data collected in an experiment will always support the hypothesis of the experiment and to explain their reasoning. (No, many experiments fail to support a hypothesis. However, even though an experiment fails to support a hypothesis, the data gathered is still valuable because with it, the researcher can revise the original hypothesis or experiment or generate a new hypothesis.)

LASER DISC
473, 474, 475, 476
Scientific method

ONGOING ASSESSMENT
ASK YOURSELF

Controlled experiments provide comparisons that can be used to support a hypothesis.

SECTION 1 REVIEW AND APPLICATION

Reading Critically

1. A scientific method typically includes observation, identification of a problem, formulation of a hypothesis, an experiment, collection of data, and a statement of results.

2. Observations are usually the first step in a scientific method of problem solving. A scientist then formulates hypotheses and conducts experiments based on the observations.

3. A control group acts as a comparison for the experimental group in an experiment.

Thinking Critically

4. A control group of plants grown in white light and an experimental group grown in blue light might be used to test the hypothesis.

5. The claim that a particular dog food is "better" than another brand is difficult, if not impossible, to prove because the way in which it is better is not stated; that is, no expected results are stated.

▶ 8 CHAPTER 1

Figure 1–4. Controlled experiments lead to meaningful results.

If this experiment had been done without a control group, no comparisons could have been made. The data from the experimental group is meaningful only when compared to the data from the control group. Knowing that 150 sheep were killed by the coyotes given chemically treated meat is of little value by itself. However, when this information is compared to the number of sheep killed by coyotes in the control group, the effect of the variable can be seen.

The results of this experiment were reported in the *Journal of Range Management*. By reading the journal, other scientists can share the information and decide if more investigation should be done. Other studies may be made to test this hypothesis further.

When a hypothesis is supported by the work of many scientists, it may be called a **theory**. A theory is not a fact. A theory is, however, more than a hypothesis. Theories explain why things happen the way they do. Every theory is changed as new data is gathered. A theory can be judged by how well it supports observations and by how much good evidence supports it. Theories are useful because they help us predict what might happen in new situations.

 ASK YOURSELF

Why are controlled experiments needed to support a hypothesis?

SECTION 1 REVIEW AND APPLICATION

Reading Critically
1. What are the processes a scientist uses to solve a problem by a scientific method?
2. In what ways are observations important?
3. Why is it important to use a control group in an experiment?

Thinking Critically
4. How would you design a controlled experiment testing the hypothesis "blue light causes plants to grow taller"?
5. Would it be difficult to investigate scientifically the claim of a dog food company that its product was "better" than a competitor's? Explain.

RETEACHING

List the steps of a simple scientific investigation in random order on index cards. Have the students put the cards in a certain sequence and explain why they ordered the cards that way and why it is essential to follow the steps in sequence.

EXTENSION

Have the students describe the advantages and disadvantages of introducing natural coyote predators into an ecosystem to control a coyote population.

CLOSURE

Cooperative Learning Have the students work in small groups to identify a problem that currently exists in their community, formulate a hypothesis based on the problem, and then design an experiment intended to address the problem.

SKILL

Designing and Conducting Experiments

Process Skills: Observing, Interpreting Data, Experimenting

Grouping: Individuals or pairs

Objectives
- **Formulate** a hypothesis.
- **Design** and **conduct** a scientific experiment.

Discussion

Remind the students that a hypothesis is a possible answer to a problem. Experiments are conducted to test the validity of a hypothesis, not to prove it as fact. A hypothesis that is unsupported often leads a scientist to other discoveries. To foster critical thinking skills, ask the students what could be done if their hypothesis is not supported by the outcome of their experiments. Encourage them to respond that new hypotheses could be formulated and new experiments conducted to test the validity of the new hypotheses.

Application

Responses should reflect accurate experimental procedures, logical thinking, and careful recording of data. The students should state whether their hypotheses were supported by the data.

✸ Using What You Have Learned

Check the students' new hypothesis, data, and conclusions to make sure that they understand how to design and conduct an experiment.

SKILL Designing and Conducting Experiments

▶ **MATERIALS**
- bean seeds • three flower pots of soil • water

▼ **PROCEDURE**

1. State a hypothesis about the effect of light on the length of bean sprouts. Decide on a variable to include in your hypothesis—for example, the effects of different amounts of light or different kinds of light.

2. Using a scientific method, design an experiment that could test your hypothesis. Your design should include your hypothesis, a list of materials you will need, an identification of your experimental and control groups, and your procedure. Be sure that the steps required for the experiment are outlined in your procedure and that all instructions are clear. Your procedure should accurately describe the way in which you will make your observations and the method you will use to record what you find.

3. Have your teacher check your experimental design. Make any necessary changes or revisions in your plan before you begin.

▶ **APPLICATION**

1. After your teacher has approved your design, set up and conduct your experiment. Make daily observations and record what you see carefully.

2. Organize your observations so that other students can understand what you have seen. You may wish to use a table, chart, or graph for this purpose.

3. Analyze your observations. Be sure to state any relationships you observe.

4. State your conclusion. Be sure to explain how your data supports your conclusion. Also indicate if your hypothesis was proved or disproved by your experiment.

✸ Using What You Have Learned

State a new hypothesis that deals with an aspect of sprout growth other than length (number of leaves, for example). Using the sprouts that you have already grown, collect additional data to test your new hypothesis; then state and explain your conclusion.

Section 2:
POLLUTION OF THE ENVIRONMENT

FOCUS

This section describes pollution as making the environment unclean with waste products. Some of the ways that the inhabitants of the earth cause land, air, and water pollution are identified. Some of the effects of pollution, such as ozone depletion, acid rain, and global warming, are discussed.

MOTIVATING ACTIVITY

Cooperative Learning To help the students realize that the amount of garbage created by people is tremendous, have the students work in small groups to list all of the things that they use and throw away during a typical week. Before the activity is completed, point out that it is often said that we live in a "throw-away" society. Have the groups evaluate that statement based on the information in their lists.

PROCESS SKILLS
- Observing • Inferring
- Interpreting Data
- Predicting

POSITIVE ATTITUDES
- Caring for the environment
- Creativity

TERMS
- pollution

PRINT MEDIA
Earth by Anne and Paul Ehrlich (see p. xixb)

ELECTRONIC MEDIA
Pollution: World at Risk, National Geographic (see p. xixb)

Science Discovery Pollution, air Recycling

BLACKLINE MASTERS
Study and Review Guide
Laboratory Investigation 1.1
Thinking Critically

① Waste products include discarded paper, cloth, plastic, glass, metal, and untreated human waste. They fill up land space and contaminate water supplies. Human waste, in particular, can pollute water and spread diseases.

LASER DISC
1458
Pollution, air

SECTION 2

Pollution of the Environment

Objectives

Report about ways humans pollute the environment.

Describe several environmental problems.

Relate the causes and effects of specific environmental problems.

More than five billion people live on Earth today! Every second of every day, three people are added to the world's human population. As the population grows, it requires more and more fuel, housing, food, and clothing. Raw materials used to produce products must be taken from the environment. At the same time, the use of these materials produces more waste that is dumped into the environment.

Pollution and Its Causes

Whether drying your hair with a blow dryer or riding on a bus, you are using energy. Every day, people use electricity, gasoline, batteries, and other forms of energy to meet their needs and to make life more comfortable. Look around you. How many devices require the use of energy to operate? How many of these things do you use daily? How would your life be different without them? Would you be willing to give them up? How would your life change if you did? Answering these questions points out our tremendous need for energy to maintain our lifestyles.

The fuels used to produce energy also produce waste. Making the environment unclean with waste products is **pollution**. Any kind of waste can cause pollution. Other than waste produced by energy use, what other waste products can you identify? How can these waste products pollute the land, water, or air? ①

Figure 1–5. Environmental damage results from the pollution of the environment or the destruction of natural resources.

▶ 10   CHAPTER 1

Manufacturing, agriculture, mining, and transportation help improve the quality of life for humans. However, these activities are also major sources of pollution. The waste products from these activities contribute to the pollution of the earth's water, air, and land. These waste products may be in the form of chemicals, heat, radiation, or even noise. In what ways have manufacturing, agriculture, mining, and transportation improved your life? ②

Waste products can pollute the air you breathe and the water you drink. Pollutants can be found in the soil in which food is grown and on which homes are built. Some pollutants can be tasted, smelled, seen, or heard. Other pollutants cannot be detected by human senses. They must be identified through the use of laboratory instruments.

Some waste materials can be broken down by living organisms. Waste materials that are broken down by microorganisms in the soil and water are called *biodegradable*. Other waste materials, such as glass, metal, plastic, and certain chemicals, are nonbiodegradable—they remain in the environment forever.

Figure 1–6. Dumping wastes into waterways causes serious damage to the environment.

ASK YOURSELF

Give some examples of land, water, and air pollution.

Solid Wastes

One of the most difficult problems facing humans today is the disposal of solid wastes. *Garbage, refuse,* and *litter* are some of the terms used to describe solid waste. In the United States, each person produces about 3 kg of refuse a day. That's about 750 000 000 kg of garbage a day—every day! In addition, industry and agriculture produce many times this much solid waste daily. In the next activity, you can see where some of this waste comes from and what could be done to reduce the amount produced.

TEACHING STRATEGIES, continued

● **Process Skills:** *Inferring, Analyzing*

Point out that many people seem to associate pollution with large industrial centers and major cities. Remind the students that pollution also occurs in rural areas of the United States. Ask the students to think of pollution problems that occur in such rural areas. (The students may suggest agricultural use of fertilizers and pesticides that may leach out of the soil and into nearby ponds, streams, and the underground water supply; topsoil that erodes and may clog streams, rivers, and lakes.)

● **Process Skills:** *Comparing, Solving Problems/Making Decisions*

Since the 1960s, many companies that once packaged their products in paper and glass containers have switched to plastic containers. Have the students discuss reasons why this change may have occurred and then discuss why it would be better for the environment to return to the use of paper and glass containers. (The switch to plastic containers probably occurred for economic reasons: Plastic is less expensive and more durable than glass, and returnable or reusable containers are more trouble for a manufacturer than

---

DISCOVER BY *Observing*

You may wish to have the students bring samples of over-packaged products to class. Ask the students to describe ways in which the packaging of products could be reduced yet remain functional. You may also wish to have interested students choose a product and design a package that is more friendly to the environment.

ONGOING ASSESSMENT
▼ **ASK YOURSELF**

Excess packaging of products is one of the largest sources of solid waste.

BACKGROUND INFORMATION

In March 1987, a barge called the *Mobro* left New York. Loaded with garbage, the *Mobro* headed for North Carolina. Five months later, the *Mobro* was back in New York—still loaded with garbage. What happened? Turned away by officials in North Carolina, Mississippi, Alabama, Florida, Texas, Louisiana, Mexico, the Bahamas, and Belize, the *Mobro* could not find a place to dump its garbage. Back in New York, after much debate, the garbage was finally unloaded, burned, and buried in a landfill. The *Mobro* became a celebrity because of its strange journey. But even though its story seems funny, it also points out an important problem—how to get rid of our ever-increasing garbage when no one wants it.

▶ **12** CHAPTER 1

Figure 1–7.
Barges are used in some places to haul garbage out to sea. In other cases, garbage is dumped in the open or buried in a landfill.

DISCOVER BY *Observing*

Excess packaging is one of the major sources of solid waste. Look at some packaged items you have recently purchased. How much excess packaging is there on the product? Check some other brands of the same product in the store. How much waste could be eliminated by reducing the excess packaging?

Excess packaging contributes to solid waste, and disposing of packaging and other solid waste is a serious problem. Some communities dispose of refuse in an open dump, where the refuse is left on the ground. These open dumps are a health hazard and a threat to the environment. Rain carries materials from the dump into water supplies. Since rats, insects, bacteria, and fungi are found in open dumps, such dumps may be sources of disease.

Other communities solve their disposal problems by dumping solid waste into the ocean. This removes the refuse from sight, but it damages the ocean's environment. The refuse causes changes in the ecosystem that kill or injure living things.

Still other communities dispose of their solid waste in sanitary landfills. Landfills are locations where solid waste is buried with a layer of soil around the waste on all sides. When sanitary landfills are properly made, they are less harmful to the environment than are open dumps or ocean dumping. However, many communities are running out of sites for landfills.

▼ **ASK YOURSELF**

What is one of the biggest sources of solid waste?

throw-away containers. Paper is more environmentally friendly than plastic because paper breaks down in a shorter period of time than plastic, which takes a very long time to break down, if it ever does.)

● **Process Skills:** *Predicting, Analyzing*

Ask the students to estimate what percent of the cost of a supermarket item is spent for packaging—in other words, if a supermarket item costs $1.00, how much of that dollar is spent to package the item? Point out that about 50 percent of the cost of most products bought in supermarkets goes into packaging the item, and this packaging is usually thrown away by the consumer. Ask the students why companies might spend so much to package an item and why the companies manufacture packaging that is larger or more elaborate than is necessary to ship the product. (Boxed items are easy to ship and display. Large packaging takes up more shelf space in a store and makes the product more visible to the consumer. Visibility can help increase sales.)

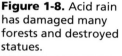

Figure 1-8. Acid rain has damaged many forests and destroyed statues.

Acid Rain

Water is one of the most important needs for life. Water collects on the earth's surface, giving living things the moisture they need to survive. Water that evaporates into the air eventually returns to the earth's surface in the form of rain or other precipitation. Organisms need precipitation for their supply of fresh water. Unfortunately, because of air pollution, rain that falls in some parts of the world is harmful to plants and animals.

Carbon dioxide in the air dissolves in rainwater. Normal amounts of carbon dioxide in the air make rain slightly acidic. Sulfur oxides and nitrogen oxides are other gases that dissolve in water and make rain more acidic. All three gases are produced when high-sulfur coal or oil is burned. Coal and oil are commonly used as fuel to power factories and automobiles and to heat and cool homes. The polluted rain that results from large amounts of these gases combining with rainwater is called *acid rain*.

These photographs show the damage acid rain can do to trees and statues. In addition to destroying forests and eating away statues, acid rain kills fish in lakes and streams. In parts of Scandinavia, Canada, and the United States, hundreds of lakes are now nearly lifeless due to acid rain. In the next activity, you can see how acid affects rocks and plants.

Discover by Doing

Put a few drops of vinegar, which is an acid, on a piece of chalk. Chalk is similar to limestone, a rock used in buildings and statues. Observe and describe what happens. Now put a few small leaves into a bowl of vinegar. Observe the leaves after five or ten minutes. What does the acid do to these plant parts?

INTEGRATION— *Mathematics*

The United States is the largest producer of solid waste in the world. Australia, with a population of about 10 000 000 people, produces only 0.8 kg of solid waste per person per day. Have the students determine the amount of garbage produced each day by homes in Australia (8 000 000 kg) and then compare that amount to the amount of garbage produced each day by homes in the United States. (The United States has about 25 times more people than Australia and produces about 94 times more garbage.)

 About 50 billion kg of the wastes produced in the United States each year are considered to be hazardous to human health. Hazardous wastes are produced by factories, mining operations, hospitals, and nuclear power plants. These wastes may be flammable, poisonous, radioactive, corrosive, explosive, unstable, or infectious. Currently, hazardous wastes that cannot be recycled may be buried in landfills that have clay or plastic liners, pumped into deep wells, burned at high temperatures, or broken down by bacteria. Interested students can investigate the effectiveness and safety of these disposal methods.

TEACHING STRATEGIES, continued

● **Process Skills:** *Inferring, Applying*

Acid rain is the term used to describe any precipitation that is more acidic than normal. Point out to the students that acid rain may also occur in the form of snow, sleet, or fog. Ask the students to predict how acid rain might affect buildings that are made of stone and metal. (The acidic nature of the precipitation will increase the rate at which metal and stone buildings will deteriorate.) Tell the students that famous buildings such as the Washington Monument, the Coliseum (in Italy), the Parthenon (in Greece), and the Taj Mahal (in India) are showing the effects of acid rain.

● **Process Skills:** *Interpreting Data, Applying*

Point out that an automobile is usually the second-most expensive purchase that people make. (A home represents the most expensive purchase.) Then ask the students to describe some of the things that many automobile owners do to protect their vehicles from environmental damage. (Automobile owners will wash and wax their vehicles to remove and protect against corrosive elements such as salts.) Also point out that it is recommended that automobile owners wax their cars once or twice each year, but that it is recommended that automobiles in large cities should be waxed

---

SCIENCE BACKGROUND

Peat bogs are natural ecosystems that have extremely acidic conditions. Since decomposers cannot survive in a peat bog, decay does not take place. The bog gradually fills in with a solid mass of spongy moss and material that may be as much as 10 m thick.

✳ DISCOVER BY *Doing*

The degree of color change exhibited by the grape juice will depend on the concentration of acids in the rainwater sample. The sources of acidic pollutants in rainwater are varied and include the burning of fossil fuels.

★ PERFORMANCE ASSESSMENT

Cooperative Learning Have small groups of students write the *Discover by Doing* activity as an experiment. Review the major elements, including the hypothesis, list of materials, variable and control groups, procedure, data, and conclusions. Check the groups' experiments to see whether they understand how to construct an experimental design.

ONGOING ASSESSMENT
▼ **ASK YOURSELF**

Carbon dioxide, sulfur oxides, and nitrogen oxides are dissolved gases that contribute to acid rain.

▶ **14** CHAPTER 1

In the activity you saw the effect of one type of acid on leaves and chalk. Acid rain can have a similar effect on plants and buildings. Tests have shown that rainfall in some areas is 10 to 30 times more acidic than ordinary rain. In the next activity, you can test rainwater for acid.

Doing

You will need a clear cup, a teaspoon, rainwater, and purple grape juice. Grape juice is an indicator—it turns red in an acid and green in a base, the opposite of an acid. Collect one-half cup of rainwater. Add a teaspoon of grape juice. Observe any color change. Does the rainwater you collected contain acid? If it does, what might the source of the acid be?

Acid rain damages the leaves of plants, reducing a plant's ability to make its own food. Acid rain changes the soil so that some plants can no longer survive. When plants die, the animals that need those plants for food also die. Acid rain also damages wood, metal, and stone structures. Automobiles, bridges, and buildings show the effects of acid rain.

▼ **ASK YOURSELF**

Name three dissolved gases that cause acid rain.

A Hole in the Sky

Ultraviolet waves are invisible light waves emitted by the sun. Exposure to the sun's ultraviolet rays is very harmful to living organisms. Fortunately, high in the earth's atmosphere is a thin layer of gas that absorbs many of the sun's ultraviolet rays. This thin, protective layer of gas is the ozone layer. *Ozone* is a type of oxygen. It is produced when ultraviolet rays strike oxygen.

If the ozone layer did not absorb most of the sun's ultraviolet rays, the rays would reach the earth. Many plants would die. Animals and people would suffer severe sunburn, skin cancer, and blindness. The oceans would warm up, and most sea life would die. Over time, Earth would become a dead planet.

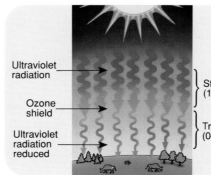

This photograph was taken by a satellite over Antarctica. The dark area in the picture is caused by a hole in the ozone layer. Scientists discovered the hole in 1985, and they have observed it growing larger since then.

Scientists have evidence that the hole may have been caused by chemicals called *chlorofluorocarbons* (klawr uh FLOOR uh kahr buhnz), or CFCs. These chemicals are used for coolants in refrigerators and air conditioners and in aerosol sprays. Some industries, such as the microchip industry and makers of plastic foam, also use CFCs. CFCs have been drifting around in the atmosphere since they were developed in the 1920s. They slowly rise into the stratosphere, where they break down the ozone layer.

The only major hole found in the ozone layer so far is the one over Antarctica. However, a smaller hole has been found over the Arctic Ocean. Recent evidence suggests that the ozone is getting very thin or a hole is developing over populated areas. Scientists are very concerned. If people keep using CFCs, more ozone will be destroyed. The United States and more than twenty other countries have agreed to cut use of CFCs by 50 percent before the year 2000.

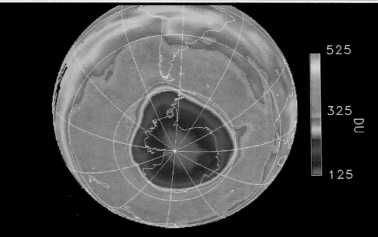

Figure 1–9. The ozone layer over the Arctic may soon have a hole as large as this one over Antarctica.

 ASK YOURSELF

Why do you think it is important that all countries agree to cut down on the use of CFCs?

The Greenhouse Effect

Holes in the ozone layer are not the only problem with Earth's atmosphere. Pollution could cause a global warming of Earth's climate due to the greenhouse effect.

Just what is the greenhouse effect? As the sun shines on the earth, some of its heat is absorbed by the land and by water. Much of the sun's energy is reflected back into the atmosphere. The reflected energy is held in the atmosphere by carbon dioxide and water vapor. This warms the atmosphere and helps the earth keep a fairly constant temperature as it rotates.

ACTIVITY

How has the amount of carbon dioxide in the atmosphere changed?

Process Skills: Interpreting Data, Predicting

Grouping: Individuals or pairs

Hints
As the students create their line graphs, remind them that the units indicated on the vertical and horizontal axes of their graphs should be linear.

▶ **Application**
The amount of carbon dioxide in the atmosphere has increased during the years shown on the graph. Most students will predict that the amount of carbon dioxide in the atmosphere will increase in the future. An estimate of 345 to 355 ppm in the year 2000 is reasonable.

PERFORMANCE ASSESSMENT
As a means of evaluating the students' understanding of the greenhouse effect, ask them to draw a diagram that shows the relationship between increasing levels of carbon dioxide and increasing temperatures in the atmosphere.

 Because of the pollutants given off by gas-powered automobiles, the electric car is receiving more attention as a future method of mass transportation. Have interested students research the status of the development of electric cars and report their findings to the class.

▶ 16 CHAPTER 1

GUIDED PRACTICE
Ask the students to describe how the problems of solid waste disposal, acid rain, ozone depletion, and the greenhouse effect are related.

INDEPENDENT PRACTICE
 Have the students provide written answers to the Section Review and Application questions. In their journals, ask the students to write a paragraph describing how pollution impacts their lives now and how they think it will impact their lives in the future.

EVALUATION
Have the students identify the type of pollution that they believe represents the most serious threat to their environment and explain why that form of pollution is a more serious threat than other forms of pollution. (Answers should give reasons that support the student's selected type of pollution as the most serious threat.)

Figure 1–10. Earth's temperature is balanced by the amount of carbon dioxide and water vapor in the atmosphere.

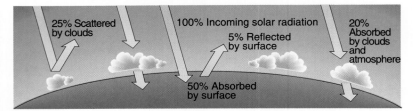

However, too much carbon dioxide can make the earth's atmosphere too warm. The atmosphere is like the inside of a closed car on a sunny day. Sunlight goes through the windows of the car and warms up the interior, but the heat cannot escape. The temperature inside the car can get dangerously high.

Carbon dioxide traps the earth's heat just as a closed car traps heat. A greenhouse has the same problem with overheating as a closed car, especially in the summer. For that reason, the heating effect of carbon dioxide in the atmosphere is sometimes called the *greenhouse effect*. The next activity can show you what has happened to the amount of carbon dioxide in the atmosphere.

ACTIVITY
How has the amount of carbon dioxide in the atmosphere changed?

MATERIALS
graph paper, ruler

PROCEDURE
1. Look at the table. It shows the amount of carbon dioxide (CO_2) in the atmosphere in the years between 1960 and 1990. The amounts are given in parts per million (ppm).
2. Make a line graph using the data from the table.

APPLICATION
What happened to the amount of CO_2 in the atmosphere between 1960 and 1990? Predict what might happen to the amount of CO_2 in the future? Estimate the amount of carbon dioxide in the air in the year 2000. Find out if the average temperature of the atmosphere changed between 1960 and 1990.

Table 1: Amount of Carbon Dioxide in the Atmosphere

Year	ppm of CO_2	Year	ppm of CO_2
1960	317	1980	337
1965	320	1985	340
1970	325	1990	342
1975	330		

RETEACHING

Point out that, in varying degrees, everyone contributes to the pollution problems of the earth. Ask the students to draw pictures that show several activities they do that contribute to the earth's pollution. Then display the pictures and have volunteers describe several things that could be done to help minimize the amount of pollution created by people. (Pictures might suggest, for example, that a willingness to carpool would reduce the number of cars on the road and thus decrease the pollutants released into the atmosphere by cars.)

EXTENSION

Our planet is sometimes called Spaceship Earth because it is a closed ecosystem. Ask the students how the concept of Earth as a spaceship might affect the way humans interact with the environment. (Earth and its resources must be managed for long-term use.)

CLOSURE

Cooperative Learning Have the students work in small groups to think of ways that the pollutants produced by their school could be significantly reduced.

Figure 1–11. What other problems, in addition to rising carbon dioxide levels, are related to the clearing of the rain forests? ①

In the activity, you learned how the amount of carbon dioxide has increased. Today, there is 15 percent more carbon dioxide in the atmosphere than there was in 1980. Most of that results from the burning of fossil fuels, such as oil. This picture shows another source of carbon dioxide. As the tropical rain forests are burned, carbon dioxide is released into the atmosphere. Scientists predict that the amount of carbon dioxide will rise an additional 10 percent by the year 2000.

ASK YOURSELF

What might happen to the earth's climate if the amount of carbon dioxide in the atmosphere keeps increasing?

SECTION 2 REVIEW AND APPLICATION

Reading Critically
1. What is the greenhouse effect?
2. Why is the disposal of solid wastes such a problem?
3. What causes acid rain?

Thinking Critically
4. Why is acid rain found downwind of industrial centers?
5. Why should people be concerned about acid rain?
6. In what ways might you be able to reduce the problem of solid waste pollution?

---

① Destruction of rain forests decreases the amount of oxygen produced globally by photosynthesis and creates a loss of habitats for plants and animals.

ONGOING ASSESSMENT ASK YOURSELF

The earth's climate could become much warmer as increased carbon dioxide traps more and more heat in the earth's atmosphere.

SECTION 2 REVIEW AND APPLICATION

Reading Critically

1. The greenhouse effect is the process by which carbon dioxide traps heat in the earth's atmosphere, much like glass traps heat inside a greenhouse.

2. The disposal of solid wastes takes up valuable land, contributes short-term and long-term pollutants to the environment, can be a source of disease, and increases as the population increases.

3. Acid rain is caused by carbon dioxide, sulfur oxides, and nitrogen oxides in the atmosphere mixing with rainwater.

Thinking Critically

4. The pollutants produced and released by industrial centers are carried by wind currents.

5. Acid rain poisons groundwater, destroys trees and crops, and damages buildings and other structures.

6. Solid waste pollution can be reduced by recycling materials and packaging products in more environmentally friendly ways.

Section 3:
PROTECTING THE ENVIRONMENT

FOCUS
This section identifies the important natural resources of the earth and describes those resources as being either renewable or nonrenewable. The characteristics of resources and methods of conservation are discussed. Endangerment, extinction, and conservation of wildlife are presented.

MOTIVATING ACTIVITY
Ask each student to name an animal and then find out more about that animal. Have the students note interesting facts about their chosen animal and present their findings to their classmates orally. Then encourage the class to discuss what the earth would be like if all the animals they chose became extinct.

PROCESS SKILLS
- Observing • Inferring

POSITIVE ATTITUDES
- Caring for the environment
- Openness to new ideas

TERMS
- conservation • erosion
- habitat

PRINT MEDIA
Only One Earth by Lloyd Timberlake (see p. xixb)

ELECTRONIC MEDIA
Protecting Endangered Animals, National Geographic (see p. xixb)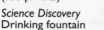

Science Discovery Drinking fountain

BLACKLINE MASTERS
Study and Review Guide
Laboratory Investigation 1.2
Extending Science Concepts

✦ Did You Know?
Our first National Park was established in 1872 and was named Yellowstone. Today, there are 49 national parks in the National Park System of the United States.

SECTION 3

Protecting the Environment

Objectives

Distinguish between renewable and nonrenewable resources.

Explain the need for soil and water conservation.

Describe some wildlife conservation practices.

What do a wooden desk and the pages of a book have in common? Both are products made from a natural resource—timber. Natural resources are materials from the environment that humans use. Timber used to supply wood for a desk and the paper in a book is one natural resource. Soil, water, coal, and oil are also natural resources. Soil and water are needed to grow food. Coal, oil, and gas provide energy.

Natural Resources

Some natural resources are renewable. *Renewable resources* are natural resources that can be replaced by the environment.

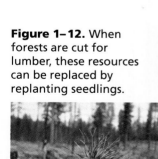

Figure 1–12. When forests are cut for lumber, these resources can be replaced by replanting seedlings.

Forests and animals are examples of renewable resources.

Trees are cut down for timber, but in time, new trees will grow. Seedlings can be planted to replace the trees cut down. Over a period of years, the seedlings become fully grown trees. Livestock, fish, and wildlife are also renewable resources. Through reproduction, these resources are replaced.

▶ **18** CHAPTER 1

TEACHING STRATEGIES

● **Process Skills:** *Observing, Applying*

Emphasize the idea that humans are dependent in many different ways on the earth's natural resources. Ask the students to name the resources they have used today and to classify those resources as renewable or nonrenewable. (Common resources used on a daily basis include fossil fuels and paper. Paper is a renewable resource; fossil fuels are nonrenewable resources.) Then ask the students why it is better to use renewable resources than nonrenewable resources. (With good management practices, renewable resources will eventually be replaced. Once nonrenewable resources are used, they are gone forever.)

● **Process Skills:** *Inferring, Predicting*

Point out that the sun is also a natural resource. Humans use the sun directly and indirectly. Ask the students to explain why the sun is a reliable resource. (The sun can be expected to burn for several billion more years.)

Figure 1-13. Although fish are a renewable resource, fish farms help to prevent certain species from becoming endangered.

ONGOING ASSESSMENT
ASK YOURSELF

Conservation of natural resources is the careful use of renewable and nonrenewable resources.

REINFORCING THEMES—
Environmental Interactions

While discussing the natural resources of the earth, stress the role of conservation. Help the students realize that life on Earth is dependent on the natural resources of the earth, and that it is essential that these resources are managed with intelligent conservation practices. Such practices will benefit human beings in both the short term and the long term.

Renewable resources can be replaced only if they are used slowly. Enough fish and wildlife must survive to produce a new generation. Seedlings may take twenty years or more to grow into a new forest. If resources are used more quickly than they can be renewed, they will disappear.

Other resources are nonrenewable. *Nonrenewable resources* are not replaced by the environment fast enough to build new supplies. Coal, natural gas, oil, metals, and minerals are nonrenewable resources. These resources were formed under very special conditions. The formation of many nonrenewable resources took millions of years. The supply of each nonrenewable resource is limited. It continues to decrease as the resource is used or destroyed.

To make resources last as long as possible requires conservation. **Conservation** is the careful use of resources. Both renewable and nonrenewable resources should be conserved. Conservation allows resources to be used wisely. With proper conservation, the earth's resources can supply the needs of humans for many years to come. You can read about the work of a conservationist in the feature on page 92.

✧ **Did You Know?**
The United States creates and uses about 1/4 of the world's supply of energy. Much of this energy is generated through the burning of nonrenewable fossil fuels.

 ASK YOURSELF
What is conservation of natural resources?

Soil Conservation

Most people think of soil as "dirt," but it is much more than that. Soil is a mixture of both living and nonliving parts. The main part of soil is tiny bits of broken rock. Weather conditions such as wind, rain, heat, and cold produce tiny cracks in the surface of rocks. Water and air get inside these cracks and react with the chemicals in the rock, causing the rock to break down. Water expands as it freezes and breaks the rock into smaller pieces. However, the most important part of soil is composed of material from decaying organisms. Soil also contains water and living organisms.

TEACHING STRATEGIES, continued

● **Process Skills:** *Inferring, Applying*

Ask the students to explain why the subsoil has more rock and less humus than topsoil. (There is more rock in the subsoil because the elements that break down rock into soil do not influence the rock in the subsoil in any way—the subsoil is protected from wind, rain, and temperature changes by the topsoil. There is less humus in the subsoil because the decomposers that produce humus live near the surface of the soil. They are unable to live in the subsoil because of their need for water. In addition, dead and decaying organic material collects on the surface of the soil and in the topsoil rather than underground.)

● **Process Skills:** *Comparing, Evaluating*

Point out that the presence of plants also slows erosion by wind; the root systems of plants help hold soil in place. Use the chalkboard to sketch the two basic root systems of plants—the taproot system and the fibrous root system. Point out

---

DISCOVER BY *Doing*

The size of the soil mound and the amount of water used in each trial should be the same. The students should observe that when the mound is covered with pebbles, the water tends to flow less quickly down the mound, and less soil erodes.

PERFORMANCE ASSESSMENT

Ask the students what would happen to a soil mound (1) if it was covered with leaves or planted with small plants, or (2) if the soil was very dry and a strong wind was blowing. Make sure that the students understand both what causes soil erosion and what can prevent it.

INTEGRATION— *Mathematics*

Have the students use the information about the rate of topsoil formation to answer the following question: If topsoil forms at the rate of 2.5 cm per 500 years, how long might it take the 15 m of topsoil that currently covers the western plains of the United States to form? (300 000 years)

▶ **20** CHAPTER 1

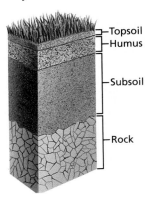

Figure 1–14. This diagram shows the layers of soil.

Soil can be divided into several layers as shown in the illustration. *Topsoil* is the uppermost layer of soil. In addition to bits of rock, topsoil contains humus, or dead and decaying organic material, such as leaves. Usually, the darker the soil, the more humus it contains. Humus holds moisture and supplies the nutrients needed for plants to grow. The layer located below the topsoil is the *subsoil*. Subsoil contains much less humus and more rock than topsoil. Because subsoil has fewer nutrients than topsoil, plants do not grow well in it.

Soil forms very slowly. It can take as long as 500 years to make a 2.5-cm layer of soil. The amount of soil varies greatly from one area to another. The soil of a desert may be only a few centimeters thick. In some parts of the western plains of the United States, the topsoil alone is 15 meters thick. Unfortunately, poor land practices cause soil to disappear faster than it forms. The carrying away of topsoil by water, wind, or glaciers is called **erosion.** Runoff from heavy rain causes a great deal of soil erosion. Water runs quickly downhill, carrying soil with it. The soil is carried into streams and rivers and eventually into the sea. In the next activity you can observe soil erosion.

Figure 1–15. Soil erodes when it is not held in place by plants.

DISCOVER BY *Doing*

You will need four cups of soil, two cups of water, and one cup of pebbles. Make a mound with about two cups of soil. Slowly pour about one cup of water over the mound. Observe and record what happens to the soil.
Now make another mound of soil. Place several pebbles on the mound. Again, pour water over the mound and record your observations. How do the pebbles affect what happens to the soil?

to the students that dandelions are examples of plants with taproots, and grasses in a lawn are an example of plants with fibrous roots. Then ask the students to identify the type of root system that would work best to prevent soil erosion and explain their choice. (The fibrous root system helps prevent soil erosion better because its roots penetrate a wider area of the soil than a taproot system does. A fibrous root system acts like an adhesive or a net that holds soil together.)

● **Process Skills:** *Applying, Predicting*

Point out that two other methods of protecting the soil from erosion are strip farming and terrace farming. In strip farming, rows of crops sold for cash are alternated with strips of grass or other cover crops. Terrace farming is used on hilly land. The land is cut into flat steps or terraces. Ask the students to predict which of these two practices might be less common in the United States and explain why. (Since much of the land in the United States used for agriculture is relatively flat, terrace farming would be used less frequently than strip farming.)

Erosion is a natural process, but it can be slowed down by soil conservation practices. Plants that cover the soil help to slow the runoff of rain. This allows more water to sink into the soil, and less erosion occurs. Some farmers protect their fields by planting a cover crop during the winter months. This crop holds the soil in place and slows erosion. Before spring planting, the cover crop is plowed under to add more humus to the soil.

Figure 1–16. Plowing around a hill slows water flowing downhill and helps prevent erosion.

Each spring as the rainy season begins, farmers plant their crops. Before the plants begin to grow, the soil is exposed to erosion. Contour plowing is a method of protecting the soil until the crops begin to grow. In contour plowing, the land is plowed across a slope rather than up and down it. This form of plowing builds little dams of soil that slow the flow of rainwater.

Soil is also damaged when the same crop is planted in a field every year. This practice results in the same nutrients being removed year after year and never being replaced. To avoid this problem, some farmers change, or rotate, their crops from year to year. For example, crops such as beans and peas can replace nutrients removed from the soil by other crops, such as corn.

In a natural environment, nutrients are returned to the soil as dead plants and animals decay. Farming and harvesting remove plants from the soil. This breaks the natural cycle of nutrients. A common solution to this problem is the use of fertilizers. A fertilizer is a material added to the soil to replace the nutrients used by crops. Some fertilizers are made from animal or plant wastes. Other fertilizers are manufactured from chemicals.

SCIENCE BACKGROUND

In crop rotation, one of the crops that is rotated should be a legume. Beans and peas are legumes, as are alfalfa and clover. A legume is a plant whose roots contain nitrogen-fixing bacteria. These bacteria take nitrogen from the atmosphere and form nitrates, which are then added to the soil. Nitrates are nutrients that are removed from the soil by crops such as corn and wheat.

✧ **Did You Know?**

Which state in the United States produces the most corn? The most wheat? The most cotton? The most potatoes? The most cattle?

Corn—Iowa
Wheat—Kansas
Cotton—Texas
Potatoes—Idaho
Cattle—Texas

TEACHING STRATEGIES, continued

● **Process Skills:** *Observing, Applying*

Direct students' attention to Figure 1–17. Ask volunteers to describe each step in the water cycle and explain how the water cycle purifies water. (In the water cycle, only water evaporates. Salt, pollutants, and other debris are left behind.)

● **Process Skills:** *Ordering, Analyzing*

Ask the students to identify the step in the water cycle during which the water combines with the gases that cause acid rain. (The gases combine with water when the water vapor is in the atmosphere.)

● **Process Skills:** *Comparing, Expressing Ideas Effectively*

Point out that dams are used to control water runoff and to increase water conservation. Water is stored behind a dam until it is needed. Dams also serve as sources of hydroelectric power. Ask the students to suggest reasons why people might not favor the building of dams. (The students might suggest that the land behind a dam becomes flooded, thus destroying habitats; dams interfere with the natural course of a river or stream; dams interfere with the reproductive cycles of some fishes.)

Demonstration

To demonstrate the quality of different soils, plant ten fast-growing seeds in each of four different pots. Fill one pot with sand, another with clay, another with topsoil, and another with a mixture of clay, sand, and topsoil. After watering the soil in each pot, put the pots away for several days and allow the seeds to germinate. Ask the students to predict which soil type will allow the most seeds to germinate. Then after the seeds have sprouted, provide the students with an opportunity to check the results of the experiment and see whether their predictions were correct.

ONGOING ASSESSMENT
ASK YOURSELF

Soil conservation practices include keeping soil covered by plants and using contour plowing. These practices help reduce the amount of soil lost to wind and water erosion.

① Responses may include watering lawns when it is unnecessary, using toilets that require a great deal of water to flush, washing small rather than large loads of clothes or dishes, and letting a faucet run while brushing teeth.

▶ **22** CHAPTER 1

Good soil conservation practices reduce wind and water erosion by keeping the soil covered. Soil conservation also allows more water to be absorbed, reducing runoff. Nutrients removed from the soil by farming are returned by practicing soil conservation.

▼ **ASK YOURSELF**

How do soil conservation practices reduce soil erosion?

Water Conservation

About 97 percent of the earth's water is in the oceans. This sea water is too salty for land animals and plants to use. They need fresh water. Much of the earth's fresh water is locked in glaciers and the polar ice caps. Part of the supply of fresh water is vapor in the air. Liquid fresh water makes up less than one percent of the earth's water supply. This relatively tiny amount of water must be shared by all life on land.

Almost half of all the water used by humans is used to irrigate plants. Water is also needed to manufacture products and to generate electricity. In addition, each person in the United States uses about 500 L of water each day. What does one person do with so much water each day? Each person drinks only about 2 L of water a day, but he or she also uses water for bathing, food preparation, and watering lawns. People also use water to flush away wastes and to fill swimming pools.

The goal of water conservation is to preserve our supply of clean fresh water. The most obvious method of water conservation is to reduce the amount of wasted water. Allowing faucets to leak and taking long showers are common ways of wasting water. What are some other ways in which water is wasted?

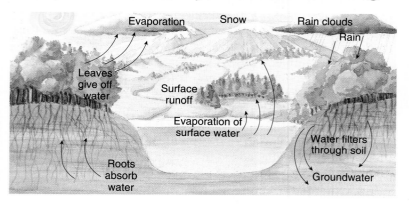

Figure 1–17. Water is continuously removed from and returned to the surface of the earth through the water cycle.

● **Process Skills:** *Comparing, Analyzing*

Remind the students that each person in the United States typically uses about 500 L of water each day. Point out that in less developed countries, each person may use only 40 or 50 L of water per day. Ask the students to explain why this disparity in water use exists between developed and less developed countries. (Responses might suggest that supplies of water are not as readily available in less developed countries as they are in the United States, and consequently, the inhabitants of those countries use the available water only for essential activities, such as drinking. Other responses might suggest that water may be regularly available in less developed countries, but with fewer factories and plants, less water is needed for purposes other than drinking, cooking, and bathing.)

Figure 1-18. Although watersheds seem to have unlimited amounts of water, waste threatens that supply.

INTEGRATION— *Mathematics*

In their journals, have the students list some of the ways in which water is wasted. Then have them describe some water conservation methods they might use to reduce their own use of water. Encourage the students to monitor their water use for a week, note their efforts to conserve water, and calculate how much water they saved.

Did You Know?
Water makes up about 3/5 to 3/4 of your body weight.

Much of the water people use can be recycled for use again and again. Supplies of fresh water must also be restored. Water that is pumped from the ground through wells must be replaced. Swamps and other wetlands allow rainwater to sink back into the ground to replace some of the water being removed. Forests slow runoff and allow water to be absorbed into the ground. A *watershed* is a section of land from which water drains into a river or lake. Conservationists try to protect watersheds whenever possible because watersheds help resupply underground water reserves.

 ASK YOURSELF

How does each person use water every day?

Wildlife Conservation

You probably know that many animals that once lived on Earth are now *extinct,* or completely gone. Dinosaurs and other prehistoric animals became extinct because of natural events. But in the last two hundred years, hundreds of animals have become extinct because of the actions of humans. One example is the passenger pigeon.

In the 1800s, there were billions of passenger pigeons in the United States. Now there are none. What happened to them?

Figure 1–19. A passenger pigeon from a nineteenth-century engraving

ONGOING ASSESSMENT
ASK YOURSELF

Obvious ways in which water is used include eating, drinking, bathing, and excreting. Encourage the students to think of less obvious ways in which people use water each day, such as watering house plants and cooling cars.

 **LASER DISC**
2617

Drinking fountain

SECTION 3

TEACHING STRATEGIES, continued

● **Process Skills:** *Evaluating, Generating Ideas*

Point out that the dodo bird (a clumsy, flightless bird approximately the size of a turkey) has become extinct, the Galápagos tortoise is an endangered species, and the chimpanzee is a threatened species. Ask the students to explain why humans should be concerned that some species are threatened, endangered, or extinct. (Responses might suggest that some wildlife is used for food or may be a food source in the future; many people enjoy observing wildlife in their leisure time; many useful medicines are derived from various plant and animal species; many new uses of plants and animals may be discovered in the future; what happens to other organisms on Earth could happen to humans.)

● **Process Skills:** *Inferring, Predicting*

Write the following animal species on the chalkboard: gorilla, bald eagle, humpback whale, crocodile, pelican. Point out that these animals are currently endangered species. Ask the students to infer why each species might be endangered.

MEETING SPECIAL NEEDS

Second Language Support

 Have students with limited English proficiency work with student tutors to write their own explanations of what it means when a species is threatened, endangered, or extinct. Encourage the students to illustrate their explanations with pictures or drawings of appropriate animals for each designation and place their work in their science portfolios.

BACKGROUND INFORMATION

It is sometimes assumed that human activities are always responsible for the extinction of animal species. Although in the past 200 years, humans have been responsible, either directly or indirectly, for the extinction of many species, it is possible for a species to die out for other reasons. A harsh winter, a disease epidemic, or natural disasters such as fire, flood, or drought can cause the extinction of a species.

Figure 1–20. The last passenger pigeon died in 1914.

Passenger pigeons were hunted commercially for many years. Millions of them were killed every year. They were shipped by train to markets in the eastern United States, where they were sold as food.

Hunters not only shot the passenger pigeons but also attacked the birds in their nests. Trees were cut down, and woods were set on fire to drive the birds out. Without the protective cover of the woods, the birds were easier to find.

After many years, people realized that they were no longer seeing large flocks of passenger pigeons. Every year, fewer and fewer of the birds were found. Finally, on September 1, 1914, the last passenger pigeon, named Martha, died at the Cincinnati Zoo. There will never be another passenger pigeon.

Today, laws in most states protect animals from being hunted to extinction. For example, some animals can be hunted only during certain seasons. Laws often place limits on the number of animals that can be hunted. And *endangered animals,* those identified as being nearly extinct, cannot be hunted at all.

The survival of the American alligator shows that such laws can protect animals. By the 1960s, hunting had endangered the alligator's chances of survival. Then laws were passed making it illegal to hunt alligators. By 1987, the number of alligators had increased, and limited hunting was allowed again in some states.

Figure 1–21. American alligators were once endangered but now are thriving.

(Responses might include: gorilla—loss of forest habitat; bald eagle—excessive hunting; humpback whale—excessive hunting; crocodile—loss of wetland habitat; pelican—diminished food supply due to acid rain poisoning lakes and rivers.)

● **Process Skills:** *Inferring, Applying*

When asked to name extinct, endangered, or threatened species, people usually name animals. Point out that there are also many endangered species of plants. Ask the students to find out species of plants that are endangered. (Some species of cacti, redwoods, and pines are currently endangered.)

Although the American alligator is no longer endangered, it is still threatened in some areas. *Threatened animals* are in danger of becoming extinct if large numbers of them should suddenly die. The next activity can help you find out how threatened and endangered species are protected by laws in your state.

Discover by *Researching*

Visit the library in your school or community. Find out what animals in your state are threatened or endangered. Make a list of them. Find out if these animals are also threatened or endangered in other states. Make a table that shows this information. Then call an agency in your area that sells hunting licenses. Find out what laws there are to protect the animals on your list. Report on your findings to your classmates.

Figure 1–22. This endangered elephant was killed for its valuable tusks.

In some parts of the world, the hunting of threatened and endangered species continues, even though it is against the law. Endangered cheetahs, Siberian tigers, Asian lions and elephants, mountain zebras, snow leopards, and lowland gorillas are still hunted. What are some other ways in which humans contribute to the extinction or near-extinction of wildlife species? ①

The place where a population of organisms lives is called its **habitat.** The destruction of habitat is one of the most common causes of extinction. As the human population grows, it expands into new areas. Forest, prairie, and freshwater habitats are replaced by cities of concrete and steel. People build dams to store water. The dam creates a new habitat for some organisms, but it destroys the habitats of others. The prairie habitat disappears as more land is plowed for food crops. When humans change a habitat, some species' food sources are destroyed and their natural shelters are lost. Many animals and plants adapted to one habitat cannot adapt to a new one.

DISCOVER BY *Researching*

PORTFOLIO ASSESSMENT

Threatened and endangered animals are determined by the federal government, although the numbers of these animals can vary from state to state. For example, the grizzly bear is an endangered species, and although these bears have disappeared entirely from some areas of the western United States, they still exist in relatively large numbers in parts of Alaska. Laws that are established to protect endangered species typically allow an endangered species to be killed only in situations in which a human life is threatened. Have the students place their completed tables in their science portfolios.

① Responses may suggest that pollutants humans add to the environment adversely affect wildlife species. Also, as people take up more and more land, the available habitats for wildlife decrease.

INTEGRATION— *Geography*

Cooperative Learning Assign each group a continent and ask the groups to use reference materials to find out about animals that are currently endangered on their assigned continent. On a large outline map of the world, have each group write the names of those animals on the appropriate continent. Display the map on the classroom bulletin board.

SECTION 3 **25** ◀

GUIDED PRACTICE

Write the following terms on the chalkboard: conservation, erosion, habitat. Ask the students to define each term and explain how the terms relate to the topic of the environment.

INDEPENDENT PRACTICE

 Have the students provide written answers to the Section Review and Application questions. Then have them write a paragraph in their journals in which they express their opinion on the importance of global conservation.

EVALUATION

Ask the students to describe two good wildlife conservation practices. (Responses might include saving natural habitats and controlling the size of wildlife populations.)

MULTICULTURAL CONNECTION

Respect for the land and natural resources is an inherent part of most Native American cultures. For example, the Native Americans who relied on the buffalo for almost all their needs recognized the importance of maintaining the buffalo herds. They killed only as many buffalo as they needed for food and other supplies.

ONGOING ASSESSMENT
ASK YOURSELF

Greed and lack of concern for the long-term survival of wildlife species are reasons why people continue to hunt threatened or endangered animals.

① The remaining condors were captured before they died out completely and there was no hope of trying to breed them in captivity.

The goal of wildlife conservation is to protect the wild animals and plants that live on the earth. Saving natural habitats is one way to protect wildlife. Wildlife refuges, national parks, national forests, and wilderness areas preserve natural habitats. Conservationists also try to control the size of wildlife populations. Overpopulation can destroy a habitat or result in animals starving. Laws that permit hunting and fishing at certain times of the year can help protect wildlife populations. Good conservation practices reduce the effect of humans on wildlife.

ASK YOURSELF

Why do you think some people continue to hunt threatened or endangered animals?

A Success Story

The California condor is America's largest flying bird. An adult condor may weigh more than 45 kilograms and have a 3-meter wingspan. Because of its large size, the California condor needs a great amount of open space. But because more and more land is used by humans, the condor's range is getting smaller. Today there are fewer than 30 condors. The condor is nearly extinct.

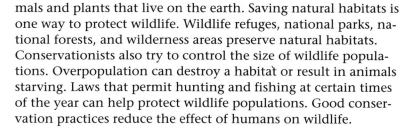

Figure 1-23. The California condor, America's largest bird, is nearly extinct.

Scientists decided that the only hope for saving condors was to try breeding them in captivity. Wild condors were captured and kept in zoos and wildlife parks. In 1987, the last wild condor was captured. Why do you think scientists caught the few remaining wild condors? ①

Breeding condors in captivity proved to be difficult. But in May 1988, the first condor chick was hatched at the San Diego Wild Animal Park in California.

Figure 1-24. Molloko was born at the San Diego Wild Animal Park.

RETEACHING

Point out that the students can touch animals directly by scratching a dog's ears, for example, or indirectly by feeling the softness of a wool sweater. Ask the students to describe other ways that they can directly and indirectly touch animals.

EXTENSION

Have interested students investigate watersheds in their area. The students can visit and write about the watersheds and share their impressions with their classmates.

CLOSURE

Cooperative Learning Have the students work in small groups to design slogans for posters or bumper stickers that could be used to make people aware of the importance of conserving natural resources.

The chick was named Molloko, a Native American word for *condor*. Molloko was taken from her nest and raised by a group of scientists. They wanted to be sure that nothing happened to her. To help her feel secure, the scientists played tape-recorded condor sounds to Molloko. They also handled her with a puppet made to look like a condor mother.

Until Molloko hatched, no one was sure the plan for breeding condors would work. Now scientists have begun releasing condors bred in captivity back into the wild. There they may breed and start a new population of wild condors.

The story of the California condor may have a happy ending. But for many other species, extinction is likely. Only strictly enforced laws and caring people can save them. In 1854, a great Native American chief named Seattle gave his reason for caring about wildlife:

> *What is man without the beasts? If all the beasts were gone, man would die from a great loneliness of the spirit. For whatever happens to the beasts soon happens to man.*

Figure 1–25. A hand puppet was used as a "mother" to help feed Molloko.

ASK YOURSELF

What is the goal of wildlife conservation?

SECTION 3 REVIEW AND APPLICATION

Reading Critically
1. What is a renewable resource? List four examples.
2. How does contour plowing help conserve soil?
3. What are some of the factors that can lead to the extinction of a species?

Thinking Critically
4. Why might the use of fertilizers be a good soil conservation practice but a poor water conservation practice?
5. How can people conserve water in their homes?
6. What are some ways you can help protect threatened or endangered species?

ONGOING ASSESSMENT
ASK YOURSELF

The goal of wildlife conservation is to protect the wild animals and plants that live on the earth.

SECTION 3 REVIEW AND APPLICATION

Reading Critically

1. A renewable resource is a resource that can be replaced by the environment. Examples of renewable resources include trees, wildlife, livestock, and fish.

2. Contour plowing builds little dams that slow the flow of water, so that more water can be absorbed and less soil will be carried away through erosion.

3. Habitat destruction and overhunting are some of the factors that can lead to the extinction of a species.

Thinking Critically

4. The use of fertilizers returns vital nutrients to the soil so that crops can be grown. But the chemicals in some fertilizers can leach out of the soil and contaminate nearby ponds, streams, and underground water supplies.

5. Water conservation practices in the home might include using water more efficiently for cleaning, food preparation, and lawn watering.

6. Supporting comprehensive laws protecting plants and animals, reporting unauthorized hunting or poaching of plants and animals, and not buying products made from threatened or endangered species are three ways to help protect those species.

INVESTIGATION

Modeling Soil Erosion

Process Skills: Formulating Hypotheses, Inferring

Grouping: Groups of 2 or 3

Objectives
- **Formulate** a hypothesis about the relationship between the presence of grass and the rate of soil erosion.
- **Design** an experiment to test a hypothesis.
- **Identify** and **describe** the control and the variable of an experiment.

Pre-Lab
Remind the students that water can cause soil to erode. Then ask the students: If soil is covered with grass, will the grass keep water from eroding the soil?

Analyses and Conclusions
Because the hypotheses formed by the students may vary, the collected data may or may not support their hypotheses. The data collected in the investigation will support the hypothesis that the presence of grass will decrease the rate of soil erosion by water.

Application
Because grass covers the surface of soil, and its root system helps hold soil in place, the presence of grass should decrease the rate of soil erosion by wind.

Discover More
Each experiment should be designed to determine that the presence of grass decreases the rate at which wind erodes soil.

Post-Lab
Remind the students of the question they answered before the investigation. Ask them to revise their answer based on the results of their experiment.

INVESTIGATION

Modeling Soil Erosion

▼ MATERIALS
- aluminum cake pans (2) • potting soil • ruler • sod • dishpan • liter bottle of water

▼ PROCEDURE

1. Form a hypothesis about how grass might affect the rate of water erosion of soil.
2. Using the materials provided, set up an experiment to test your hypothesis.
3. Identify and describe the variable and control in your experiment.
4. List the procedure you will follow. Remember to establish some method for recording your data.
5. Perform your experiment several times. Record the data for each trial.

▶ ANALYSES AND CONCLUSIONS
Examine your data and draw a conclusion based on the data. Was your hypothesis supported by your data? Does grass affect the rate of water erosion of soil in some way?

▶ APPLICATION
Do you think grass would have a similar effect on wind erosion of soil? Explain your answer.

✳ Discover More
Devise an experiment to test your hypothesis about the effect of grass on the rate of wind erosion of soil.

CHAPTER 1 HIGHLIGHTS

The Big Idea—ENVIRONMENTAL INTERACTIONS

Lead the students to understand that the environment of Earth provides all of the materials, or resources, necessary for the survival of Earth's inhabitants. Some of these resources are renewable and some are nonrenewable. The long-term survival of the inhabitants of Earth, both animals and plants, hinges on the conservation practices of people—the intelligent use and management of the resources that belong to all of Earth's inhabitants.

CHAPTER 1 HIGHLIGHTS

The Big Idea

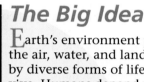

Earth's environment provides the air, water, and land needed by diverse forms of life to survive. Humans depend on the resources of the environment to meet their needs. The use of these resources has led to changes in the environment, which can disrupt the natural balance within environmental systems. Human activities have had a great effect on other organisms and on the air, water, land, and climate of the earth. While some human activities cause changes that are destructive to the environment, others help to protect the environment. The preservation of the earth's environment is dependent on the human ability to recognize and solve problems.

Look back at the ideas you wrote in your journal at the beginning of the chapter. How have your ideas changed? Revise your journal entry to show what you have learned. Be sure to include information about the problems people have created in the environment and the ways we can solve some of these problems.

The ideas expressed by the students should reflect the understanding that even though environmental problems have occurred in the past and will occur in the future, these problems can be reduced and even solved through the intelligent efforts of all people, including themselves. These efforts do not all require great numbers of people; individuals can contribute in positive ways to the protection of their environment.

Connecting Ideas

Copy the following cause and effect chart in your journal.
Complete the chart by identifying the missing cause or effect.

Cause	Effect
Scientists feed a group of coyotes chemically treated sheep meat.	
	The ozone layer is thinning.
	The passenger pigeon became extinct.
Carbon dioxide, sulfur oxides, and nitrogen oxides combine with rain water.	
A California condor is bred in captivity.	

CONNECTING IDEAS

Row 1: The coyotes become ill and thereafter avoid eating sheep meat.

Row 2: Large amounts of CFCs are being released into the atmosphere.

Row 3: Hunting was uncontrolled, and its habitat was destroyed.

Row 4: Acid rain forms, falls, and damages the environment.

Row 5: The actions of humans help an almost extinct species begin to recover.

CHAPTER 1 REVIEW

ANSWERS

Understanding Vocabulary

1. A hypothesis suggests a possible answer to a scientific question, and if enough data can be collected to support the hypothesis, it may become a theory.

2. A hypothesis is tested by an experiment, which may study two groups, an experimental group and a control group.

3. Results of pollution in the air are acid rain and an expanding hole in the earth's ozone layer.

4. Plants help decrease the rate of topsoil erosion caused by water.

5. As available habitats decrease, wildlife must increasingly rely on refuges for survival.

Understanding Concepts

Multiple Choice

6. a

7. d

8. c

9. Row 3: soil; Row 6: intelligent use of fertilizers; protecting watersheds; controlling overpopulation

Interpreting Graphics

10. Approximately 25 L is used each day.

11. Approximately 2625 L would be saved each week.

Reviewing Themes

12. Responses might suggest that although initially it is air pollution that mixes with rainwater to form acid rain, once acid rain falls to the ground and into lakes and rivers, it becomes land and water pollution.

13. Responses might suggest that if CFCs continue to be used, the ozone layer could be destroyed. The amount of harmful radiation reaching the surface of the earth would increase, killing plants and animals, and in time, the planet.

Thinking Critically

14. Designs should be checked to ensure that they adhere to the steps of a scientific method and would reliably measure the effects of softened water on the growth of plants.

15. Fossil fuels such as coal, oil, and gas should be conserved because they are nonrenewable resources that will be needed by future generations.

16. A hypothesis suggests a possible answer to a question that needs to be tested. A theory is a hypothesis that has been tested repeatedly and has data to support it. The students might suggest that a hypothesis describing life on other planets might one

---

CHAPTER 1 REVIEW

Understanding Vocabulary

Explain how the terms in each set are related.

1. scientific question, hypothesis (5), theory (8)
2. control group (6), hypothesis (5), experiment (6)
3. acid rain (13), hole in ozone (15), pollution (10)
4. water, plants, erosion (20), topsoil (20)
5. wildlife, refuges (26), habitat (25)

Understanding Concepts

MULTIPLE CHOICE

6. In an experiment designed to determine how fertilizer might affect the growth of plants, the amount of fertilizer given the control group would be
   a) zero.
   b) less than the experimental group.
   c) more than the experimental group.
   d) the same as the experimental group.

7. Which of the following is *not* a water conservation method?
   a) watering house plants with rainwater
   b) taking a short shower
   c) fixing a leaky faucet
   d) throwing cooking water down the drain

8. Because raccoons seem to be able to adapt easily to new habitats, they are likely to
   a) become extinct.
   b) be placed on the endangered species list.
   c) thrive.
   d) be found only in one habitat.

9. Copy and complete the following concept map.

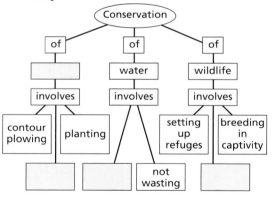

Interpreting Graphics

10. Look at the graph. About how much water is used for cooking every day?

11. If all outdoor use of water were banned during a drought, estimate the amount of water that would be saved in a week.

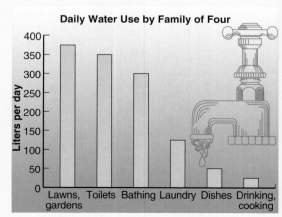

Daily Water Use by Family of Four

day develop into a theory. The conditions necessary for the establishment of a theory include an abundance of supporting evidence from the work of many scientists.

17. Recycling can be a means of conserving both renewable and nonrenewable resources. For example, paper is a renewable resource that can be recycled, and aluminum is a nonrenewable resource that can be recycled.

18. Individual programs should identify specific, logical, and effective ways in which water and energy will be conserved.

19. Responses should describe activities that allow wind and water erosion, such as lack of plant cover and poor farming practices, and activities that slow these processes, such as planting cover crops and using proper plowing methods.

20. The rabbit population would increase, possibly beyond the habitat's ability to supply enough food. The population explosion would negatively affect the other animals and plants in the habitat by upsetting the ecological balance of the habitat. Solutions could include decreasing the population of rabbits by hunting or reintroducing the rabbit's natural predators.

Reviewing Themes

12. *Environmental Interactions*
Acid rain begins as air pollution. Using acid rain as an example, explain how damage to one part of the environment can affect other parts.

13. *Conservation*
The evidence indicates that CFCs are destroying the ozone layer. Explain the changes that are likely to result if the use of CFCs is not reduced or eliminated.

Thinking Critically

14. How would you design a controlled experiment to test the effects of softened water on the growth of plants?

15. Why should humans conserve the supply of coal, oil, and natural gas?

16. How does a theory differ from a hypothesis? Give a specific example of a hypothesis that might later be developed into a theory. What conditions are necessary for the establishment of a theory?

17. Is recycling a means of conserving renewable or nonrenewable resources? Justify your answer.

18. Describe a program that you, as an individual, could follow to conserve both water and energy. What specific steps could you and your family take to implement this program?

19. The formation of soil is a long process, whereas the erosion of soil may be quite rapid. Describe how human activities can contribute to both erosion and conservation of soil.

20. If all of a rabbit's natural predators were eliminated in a wildlife preserve, what might happen to the rabbit population? What effect might this have on the preserve habitat? What might be done about the situation?

Discovery Through Reading

Javna, John. *Fifty Simple Things Kids Can Do To Save the Earth.* Andrews and McMeel, 1990. Explains how specific things in your environment are connected to the rest of the world, how using them affects the planet, and how an individual can develop habits and projects that are environmentally sound.

CHAPTER 2
POPULATIONS, COMMUNITIES, AND ECOSYSTEMS

PLANNING THE CHAPTER

Chapter Sections	Page	Chapter Features	Page	Program Resources	Source
Chapter Opener	32	*For Your Journal*	33		
Section 1: INTERACTIONS	34	Discover By Doing (B)	34	*Science Discovery**	SD
• Living Things Interact (B)	34	Activity: How can you count a population? (B)	38	Investigation 2.1: Using the Microscope (A)	TR, LI
• Characteristics of Populations (A)	35	Discover By Doing (A)	39	Connecting Other Disciplines: Science and Mathematics, Interpreting a Graph (A)	TR
• Biotic Potential (H)	38	Section 1 Review and Application	41		
• Relationships Between Populations (B)	40			Study and Review Guide, Section 1 (B)	TR, SRG
Section 2: ENERGY FLOW	42	Discover By Observing (B)	42	*Science Discovery**	SD
• Habitats and Niches (B)	43	Discover By Doing (A)	45	Investigation 2.2: Producers and Consumers in a Microscopic Community (H)	TR, LI
• Food Chains and Webs (A)	44	Section 2 Review and Application	46	Extending Science Concepts: Diagramming a Food Web (A)	TR
		Skill: Diagramming Food Chains and Food Webs (A)	47	Food Chains and Food Webs (B)	IT
				Study and Review Guide, Section 2 (B)	TR, SRG
Section 3: CHANGING ECOSYSTEMS	48	Discover By Researching (H)	53	*Science Discovery**	SD
• Succession (A)	48	Section 3 Review and Application	53	Reading Skills: Finding the Main Idea (B)	TR
• Climax Communities (A)	51	Investigation: Observing an Ecosystem (A)	54	Thinking Critically (H)	TR
				Record Sheets for Textbook Investigations (A)	TR
				Study and Review Guide, Section 3 (B)	TR, SRG
Chapter 2 HIGHLIGHTS	55	The Big Idea	55	Study and Review Guide, Chapter 2 Review (B)	TR, SRG
Chapter 2 Review	56	For Your Journal	55	Chapter 2 Test	TR
		Connecting Ideas	55	Test Generator	

B = Basic   **A** = Average   **H** = Honors
The coding Basic, Average, and Honors indicates subsections, features, and resources that might be appropriate for different levels of learners. For additional suggestions regarding choice of topic and depth of coverage, see the Pacing Chart on pages T26–T29.

*Frame numbers at point of use
(TR) Teaching Resources, Unit 1
(IT) Instructional Transparencies
(LI) Laboratory Investigations
(SD) *Science Discovery* Videodisc Correlations and Barcodes
(SRG) Study and Review Guide

▶ 31A

CHAPTER MATERIALS

Title	Page	Materials
Discover By Doing	34	(per individual) paper, pencil
Activity: How can you count a population?	38	(per group of 2 or 3) journal
Discover By Doing	39	(per individual) graph paper, pencil
Discover By Observing	42	(per individual) journal
Discover By Doing	45	(per individual) paper, pencil
Teacher Demonstration	45	potted plants (2), water
Skill: Diagramming Food Chains and Food Webs	47	(per group of 3 or 4) index cards, journal
Discover By Researching	53	(per individual) journal
Investigation: Observing an Ecosystem	54	(per group of 3 or 4) meter stick, stakes, string (4.5 m), notebook paper, pencil, hand trowel, small jar with cap, screening material

ADVANCE PREPARATION

For the *Demonstration* on page 45, obtain two green potted plants of similar size. For the *Activity* on page 38 and the *Investigation* on page 54, permission for the field trip and work areas should be arranged in advance.

TEACHING SUGGESTIONS

Field Trip
A trip to a sand dune, rock quarry, or other area that illustrates succession can be very worthwhile. Lead the students through areas that illustrate advancing successional stages, asking them to describe the changes they see in the flora, fauna, and nonliving elements of the ecosystem.

Outside Speaker
Have a local conservationist, forest ranger, or representative from a public conservation agency visit your class. Your visitor may be able to speak about local examples of environmental disturbance or the succession process that follows. Also ask your visitor to speak about the importance of understanding the limitations of the climate in land management and planning.

CHAPTER 2
POPULATIONS, COMMUNITIES, AND ECOSYSTEMS

CHAPTER THEME—ENVIRONMENTAL INTERACTIONS

Chapter 2 introduces the students to individual populations within an ecosystem. The students explore the various populations within a community and identify how they interact with one another. The students discover how energy from the sun becomes the source of energy for all living things. Through interaction, plants and animals within an ecosystem form a system in which the populations of a community can sustain themselves. The theme of **Environmental Interactions** is also developed in Chapters 1, 3, 11, and 15. A supporting theme of this chapter is **Energy**.

MULTICULTURAL CONNECTION

Point out to the students that human populations make use of the resources of their environment. For example, the Inuit of northern Canada hunted caribou and seals, eating the meat and using the skins for clothing and boats. Ask the students to investigate ways people of their own ethnic heritage or another culture used the resources of their environment to make a living. Suggest that the students create a poster or write a report outlining their findings.

MEETING SPECIAL NEEDS

Gifted

As you begin the chapter, suggest that the students choose a particular animal population and research information about it. The students should identify the ecosystem in which the population lives, the interactions it has with other organisms in the ecosystem, the flow of energy to and from their chosen population, and the impact of changes in the ecosystem on the population. Suggest that the students provide written reports, videos, or other media to document their research. The completed project may be placed in students' science portfolios.

▶ 32   CHAPTER 2

CHAPTER 2
Populations, Communities, and Ecosystems

A warm, tropical sea is one kind of environment. Clown fish and sea anemones are two of the organisms that live in that environment. They share it with other living things, such as moray eels, seahorses, and coral, and with nonliving things, such as sand, rocks, and water. All living and nonliving things in an environment are interconnected in many ways.

The sunlight filters through the water in a warm, tropical sea. Many colorful fish dart about a coral reef. Parrot fish nibble at the coral. A moray eel lurks.

A sea anemone can move, but generally it attaches itself to a surface and stays there. Its cylinder-shaped body is topped

CHAPTER MOTIVATING ACTIVITY

In a small, shallow box, place a number of dried beans, marbles, and grains of rice. Ask the students to determine the number of each type of object in the box. Chart the number of beans, marbles, and rice grains on the chalkboard. Explain that the box and its contents represent an ecosystem, the number of each type of object represents a scientific population, and the total number of objects represents a community. Then ask the students to offer definitions of *ecosystem, population,* and *community* based on their observations.

For Your Journal

Before the students make entries in their journals, ask them to think of the relationships of two different kinds of animals that live in the same area—for example, squirrels and birds or lions and antelopes. Through their journal entries, the students can demonstrate their understanding of interactions among populations within a community. Have the students identify ways in which interactions between populations may benefit or harm a particular population.

ABOUT THE PHOTOGRAPH

The colorful clown fish has an orange body with three blue-and-white bands. It grows to a length of about 3 cm and is one of about 26 species of fishes that live among the tentacles of sea anemones. The clown fish can move unharmed among the stinging cells on the sea anemone's tentacles because of a mucous coating on its body. Without the mucus, the clown fish would be paralyzed by the stings and eaten by the sea anemone.

The clown fish is protected from predators by the stinging cells that the sea anemone uses to capture food. It also feeds on the remnants of food captured by the sea anemone. In addition, the clown fish deposits its eggs in the basal disc of the sea anemone to protect them from predators.

With tentacles that wave gently with the water's movement. But be careful! A sea anemone's tentacles have stinging cells that can paralyze small fish and other sea animals. Then the tentacles pull the prey into the sea anemone's mouth to be digested.

Look at the brightly colored clown fish! Will it be the sea anemone's next meal? No, the clown fish is immune to the sea anemone's sting. The clown fish lives in the sea anemone's tentacles as a way of protecting itself from its predators. The sea anemone and the clown fish have a relationship in which one animal benefits and the other is seemingly unaffected.

The clown fish and the sea anemone's relationship is just one of many that exist between the organisms that live in a tropical sea environment. It is just one of countless relationships that exist between living things and between living things and their environments everywhere on the earth.

For Your Journal

- What other interesting relationships between two animals can you name?
- Do those relationships help or harm the animals involved?
- What do the words community and population mean to you?

Section 1: INTERACTIONS

FOCUS

In this section, the scientific definitions of *population*, *community*, and *ecosystem* are provided, and the characteristics of a population are identified. In addition, the types of interactions among various populations within an ecosystem are explored, and factors that limit populations are discussed.

MOTIVATING ACTIVITY

Cooperative Learning Have the students work in small groups to list the living and nonliving objects in the classroom. Remind the students to include themselves and you in the list. After groups have completed their lists, ask them to identify the number of students, teachers, plants, and animals, if any, in the classroom. Then have the students in each group list ways that the various living things in the classroom are alike and different. Also have them tell how the living things depend on one another.

PROCESS SKILLS
• Comparing • Interpreting Data • Predicting

POSITIVE ATTITUDES
• Caring for the environment
• Cooperativeness

TERMS
• population • community
• ecosystem • ecology

PRINT MEDIA
Predator! by Bruce Brooks (see p. xixb)

ELECTRONIC MEDIA
Animal Populations: Nature's Checks and Balances, Britannica (see p. xixb)

Science Discovery Fish; with cleaner shrimp

BLACKLINE MASTERS
Study and Review Guide
Laboratory Investigation 2.1
Connecting Other Disciplines

✦ Did You Know?

The grizzly bears of western North America belong to a species of big brown bears distinguishable by the hump on their backs. An adult grizzly can be as long as 2.4 m and can weigh as much as 230 kg. During the summer months, a grizzly may eat as much as 40 kg of food daily. This omnivore's diet includes animals, berries, roots, and grasses.

SECTION 1

Interactions

Objectives

Distinguish among populations, communities, and ecosystems.

Summarize the characteristics of a population.

Compare the relationships between populations.

A single organism has many levels of organization. Cells group together to form tissues; tissues make up an organ; and groups of organs make up a system. Together, cells, tissues, organs, and systems form an organism. Levels of organization are also found in places where humans, other animals, and plants live.

Discover by Doing

Choose one area near your school or your home. Choose one type of organism that lives in the area. The organism may be a tree, an insect, a bird, or even a human. Through careful observations over several days, estimate the number of individuals of that type of organism living in the area. What factors enable the organisms to live there? How do these organisms affect other kinds of organisms in the area? What name would you give to the group of organisms you observed?

Living Things Interact

Figure 2–1. These grizzly bears are part of a population.

Organisms of the same species living together in a particular place make up a **population.** Each population is described by kind of organism, time, and location. For example, the grizzly bears currently living in Montana make up a population. The grizzly bears that were living in Montana in 1983 made up a different population. Different populations of animals and plants living together in an area make up a **community.** The populations living in a park near your neighborhood or city form a community. In such a community, you might find populations of mice, oak trees, mosquitoes, mushrooms, and many other kinds of organisms living together.

▶ 34 CHAPTER 2

TEACHING STRATEGIES

● **Process Skills:** *Analyzing, Comparing*

Explain to the students that the scientific meanings of *population* and *community* are different from their popular definitions. Ask the students to compare the popular and scientific definitions of *population*. (The popular definition is "the number of people living in a place"; the scientific definition is "the organisms of the same species in a particular place at a particular time.") Supply some general terms, such as *students*, *flowers*, and *animals*. Then ask the students to identify a specific population for each example. (Answers may include seventh-grade students, daisies in the garden, and cows on the farm.) Have the students identify other populations that might be found in the same communities as the examples.

● **Process Skills:** *Observing, Classifying*

Have the students examine the picture on page 34 carefully. Ask them to classify objects in the picture as living and nonliving parts of the ecosystem shown. (Living: bears, ducks, trees, grasses, and bushes; nonliving: water and soil.)

Each community depends on its environment. Nonliving things in the environment, such as minerals in the soil, are necessary for a community to survive. An **ecosystem** is the combination of all living things in a community and its nonliving, or physical, environment.

Figure 2–2. A forest ecosystem includes not only the trees and animals but also the physical environment in which the organisms live.

Consider this example. The Mississippi River runs from Minnesota to the Gulf of Mexico. The river itself forms one type of ecosystem where fishes and other aquatic organisms live. The banks of the river form another ecosystem. When you move away from the river, you can find many other ecosystems. Each is made up of a community of many populations of animals, plants, and other organisms interacting with its physical environment. The study of the relationships between organisms and their environments is called **ecology**.

 ASK YOURSELF

How are the concepts of a population, a community, and an ecosystem related to each other?

Characteristics of Populations

Organisms are described scientifically according to their characteristics, such as body structures, body coverings, and chemical makeup. Populations are also described according to certain characteristics.

One characteristic of a population is the number of organisms it contains. The number of organisms per unit of living space is the *population density*. Notice that population density is not just the number of organisms in the population.

THE NATURE OF SCIENCE

Ecology is considered a branch of biology. Yet, ecologists take a multidisciplinary approach to their work. Earth science, chemistry, computer science, and physics are used by ecologists to help them understand interactions among organisms in an environment and between the organisms and the physical environment. Many ecologists do field work in isolated areas where the interactions among the plants and animals are easier to observe.

INTEGRATION—Art

Cooperative Learning On the chalkboard, list a number of ecosystems, such as saltwater ecosystem, freshwater ecosystem, beach, riverbank, forest, desert, and grassland. Have the students work with partners to research common plants and animals of one of the ecosystems. Partners should create and display a diorama of the ecosystem they research.

ONGOING ASSESSMENT
ASK YOURSELF

A population is a group of organisms of the same species living in a particular place. All the different populations of plants, animals, and other organisms in an area make up a community. The living community and the nonliving environment make up an ecosystem.

SECTION 1

TEACHING STRATEGIES, continued

- **Process Skills:**
Communicating, Analyzing

Have each student make a drawing of the area around his or her home. (If the student lives in a rural area, the drawing should include the area within a one-kilometer radius from the home. If the student lives in an urban area, the drawing should include a four-block square area.) Have the students represent the populations, communities, and ecosystems in the area in their drawings. (The term *ecosystem* varies with the context in which it is used. For example, an entire ecosystem can be defined under a rotting leaf or on a forest floor. A student could define as many ecosystems in a city lot as another student could define in a 50-hectare ranch.) Have the students place their drawings in their science portfolios.

---

① The 10-km² area contains 6 trees—an average of 0.6 trees per km². The 1-km² area also contains 6 trees—an average of 6 trees per km². The 1-km² area has a greater density of trees than the 10-km² area.

② By evenly spreading themselves out, the birds lessen the competition for the same resources within the ecosystem.

SCIENCE BACKGROUND

Terrestrial animals move about and, therefore, have a range. When determining the density of a population, a scientist must consider the entire range of the organism. For example, 100 bison may be found herded together in a 1-km² area, but they have a range of 100 km². Considering the range of the bison, the scientist determines that the density of the herd is only 1 bison per km².

INTEGRATION—
Mathematics

Population densities can vary greatly from place to place. Have the students determine the area in m² or ft² of various rooms in the school, such as their classroom, gymnasium or auditorium, lunchroom, and so on. Then ask them to determine the population density for each place if the entire class were to locate there. (The students' answers should reflect the density when the size of the class population is divided by the area of the room's floor space.)

Describing a population in terms of the number of organisms only can be misleading. The following illustration shows two populations. Each is made up of the same number of oak trees. The trees in Population I are spread over 10 km², while those in Population II occupy only 1 km². Which of these populations has greater population density? Explain why. ①

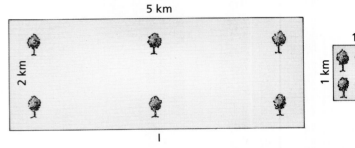

Figure 2–3. Population density is determined by calculating the number of organisms that live in a defined area.

In some cases, a population is spread over a large area, as wildflowers might be spread out over a field. In other cases, a population may be clumped together within a smaller area, with the members of the population spaced at regular intervals. The red-winged blackbirds shown in the photograph are evenly spread throughout the marsh. Why do you suppose they spread out evenly? ②

Another important characteristic of a population is its size. The size of a population is constantly changing. One way populations increase in size is through reproduction of the organisms. A population's birthrate is the number of offspring produced in a certain amount of time, such as the number of people born in the United States in a ten-year period. The size of populations also changes as organisms die. The death rate is the number of organisms that die in a given amount of time.

Figure 2–4. The population density of these red-winged blackbirds is high.

▶ 36 CHAPTER 2

- **Process Skills:**
  Comparing, Inferring

  Refer the students to Figure 2–3 and discuss population density. Ask the students in which area new oak trees would have the greatest chance of survival if the environmental conditions were the same in each area. (New trees would have the greatest chance of surviving in Area 1 because there is less competition for water, light, and nutrients in that area.)

- **Process Skills:**
  Comparing, Interpreting Data

  On the chalkboard, list the following one-year data for two animal populations in the same 1-km² area.

	Mice	Cats
Original Population	45	6
Births	105	4
Deaths	95	1
Immigration	6	2
Emigration	16	0

Which population has the greater density at the beginning of the year? At the end of the year? Which population was larger at the end of the year? (Mice are more densely populated at the beginning and at the end of the year. The cat population grew, whereas the mouse population remained the same size.)

Birthrate and death rate within a population can be analyzed separately. However, only when they are considered together can their real effect on population size be measured. For example, suppose that a population of rabbits lives in a field measuring 1 km². At the beginning of one year, the population included 225 rabbits. At the end of the year, the population had grown to 310 rabbits. Does the increase mean that the birthrate was 85 rabbits a year? Although the population increased by 85 individuals, the birthrate might actually have been 150 rabbits. Perhaps 65 rabbits died.

Consider the same rabbit population the next year. The total population increases from 310 to 400 rabbits. The birthrate for this year is still 150, but the death rate increases to 75. The difference between 150 births and 75 deaths produced a net gain of 75 rabbits. However, the population grew by a total of 90 rabbits. Where did the other 15 rabbits come from? ③

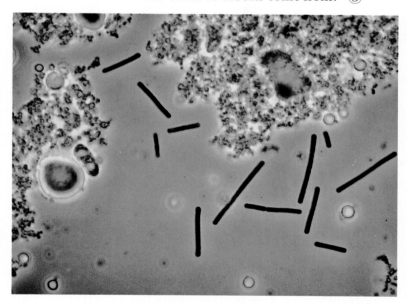

A population also changes size when members enter or leave the area. When new members join a population, the population is increased by *immigration*. If members of a population leave, the population is reduced by *emigration*.

From the birthrate, death rate, immigration, and emigration, the total change in the size of a population can be determined. Every ten years, the United States Census Bureau compiles these figures in order to determine the actual size of the human population in this country. In the next activity, you can practice counting populations.

Figure 2–5. The population of the bees can gain or lose members by immigration and emigration, while the population of the bacteria (left) cannot.

③ The rabbits immigrated to the ecosystem.

BACKGROUND INFORMATION

The United States Constitution, which became the supreme law of the land in 1788, provides that a census of the United States population must be taken every ten years in accordance with laws established by the Congress. The first census began in 1790, and the information gathered was used to determine the number of representatives from each state and the amount of taxation. Today, the census is conducted by the Bureau of the Census, an agency of the Department of Commerce. Information gathered by the bureau includes the age, employment, sex, race, citizenship, and income of residents of the United States as well as housing statistics.

ACTIVITY

How can you count a population?

Process Skills: Observing, Analyzing

Grouping: Groups of 2 or 3

Hints

Encourage the students to avoid disturbing the populations within an ecosystem when making their observations. Suggest that the students concentrate on a smaller area within the locale that they choose for their observations.

You may wish to organize short class outings to a nearby park or nature center over several days for this activity.

▶ **Application**
The students should find that over a period of days, the plant population remains relatively stable. The animal population is likely to vary to a greater degree because the animals have mobility, which enables them to travel in and out of the area. Birthrates and death rates of both plants and animals are likely to be low during the short observational period.

ONGOING ASSESSMENT
▼ **ASK YOURSELF**

Population density describes the concentration of individuals within an area.

TEACHING STRATEGIES, continued

● **Process Skills:** *Inferring, Generating Ideas*

Explain to the students that conditions must be ideal for the biotic potential of a population to be reached. Ask the students to identify factors they think would make conditions ideal. (Plentiful food, space, water, and air, the lack of disease, natural enemies, pollution, or disasters are factors that would help make an environment ideal.) Ask the students whether there would be any environment in which an organism could reach its biotic potential. (An organism could reach its biotic potential only in the controlled environment of a laboratory in which food, space, water, and air availability could be ensured and danger from disease, enemies, or disasters could be avoided.)

ACTIVITY
How can you count a population?

MATERIALS
journal, pencil

PROCEDURE
1. Select a large area that has several different populations, such as a park, field, roof garden, and so on.
2. Write a description of the area you are going to observe.
3. Pick out a plant population and an animal population.
4. Describe each of the organisms you have selected and provide a sketch of each organism.
5. Over a period of two or three days, count the members of the populations several times and take an average

APPLICATION
Did you notice any changes in the populations you chose? What might account for these changes?

 ASK YOURSELF

Why is population density more descriptive than total population?

Biotic Potential

A beehive may contain hundreds of honeybees. Thousands of salmon live in the Great Lakes. Millions of bacteria live in a spoonful of yogurt. Each population reaches a size that can survive in a particular environment. But any population can grow too large for its environment. When that happens, there is not enough space and food for all the members of the population. *Biotic potential* is the number of individuals that could be produced under the best possible conditions.

For a population to reach its biotic potential, each individual must survive and reproduce at the maximum rate. Many animals have the possibility of reproducing in large numbers. A female green turtle, for example, lays about 1800 eggs in her lifetime.

Figure 2–6. The biotic potential of the green turtle is high.

● **Process Skills:**
*Classifying/Ordering,
Expressing Ideas Effectively*

Explain that there are two types of limiting factors. Density-independent limiting factors affect a population in the same way regardless of the population's size. Density-dependent limiting factors have a greater impact on denser populations than on less dense populations. Have the students create a two-column chart with one column labeled "Density-Independent Limiting Factors" and the other labeled "Density-Dependent Limiting Factors." Have the students list each of these limiting factors in the appropriate column: flood, food supply, natural enemies, earthquake, weather, disease, stress, landslide, water supply. (Density-dependent limiting factors are food supply, natural enemies, disease, stress, water supply.) Discuss the charts, and ask the students to explain why they charted the limiting factors as they did. Have the students place their charts in their science portfolios.

Fortunately, few populations ever reach their full biotic potential. If they did, the earth's resources would be gone by now. A population's biotic potential is controlled by such things as food supply and space. The factors that prevent a population from reaching its full potential are called *limiting factors*.

Examples of limiting factors can be seen in the green turtle. As mentioned, a female green turtle lays about 1800 eggs in her lifetime. However, about 1400 of the 1800 eggs never hatch due to genetic defects, injury to the eggs, or their being eaten by predators. Young turtles from the eggs that do hatch find their first hours and days very dangerous. After digging their way to the surface of the beach, the hatchlings must travel as far as 50 m to reach the sea. Along the way, they may be eaten by birds and other animals. If the hatchlings reach the water, they are in danger of being eaten by fishes. More than three-fourths of the hatchlings do not survive their first few months. Of the 1800 possible offspring, only about three turtles survive to reproduce.

While the biotic potential of the green turtle population is high, there are a large number of limiting factors on this population. The combination of the biotic potential and the limiting factors determines the population size that can be supported by the environment. The largest number of individuals that a specific environment can support is called the environment's *carrying capacity*.

Figure 2–7. Due to limiting factors, few young turtles survive.

MEETING SPECIAL NEEDS

Mainstreamed

To help students with learning disabilities visualize the concept of *carrying capacity*, fill a small box halfway with dried beans. Have the students indicate whether the carrying capacity of the box has been reached. (No, the box can hold more beans.) Completely fill the box with beans. Ask whether the carrying capacity has been reached. (Yes, no more beans can be added.) Ask what would happen if you tried to add more beans. (They would spill over.) Relate the carrying capacity of the box to the carrying capacity of an environment.

DISCOVER BY *Doing*

The students should predict that the population will not have doubled by the year 2000. Factors that could affect their predictions include new scientific discoveries or major disasters.

ONGOING ASSESSMENT
ASK YOURSELF

Biotic potential is the number of individuals that can be produced under the best of conditions. Populations are unable to achieve biotic potential because of limiting factors that reduce a population to a size that can be supported by an environment. The largest number of individuals that can be supported by the environment is the environment's carrying capacity.

DISCOVER BY *Doing*

Graph the data about the human population found in Table 2–1. Study your graph. What patterns do you observe? Predict what the population might be in the year 2000. What factors might affect the accuracy of your prediction?

Table 2-1 **Human Population**

Year	Population
1650	500 million
1850	1 billion
1930	2 billion
1980	4.5 billion

ASK YOURSELF

How does carrying capacity relate to biotic potential and limiting factors?

SECTION 1

TEACHING STRATEGIES, continued

● **Process Skills:** *Classifying/Ordering, Comparing*

Have the students identify the type of relationship each of these sets of populations has: bears and fish (predator-prey), sea lamprey and fish (parasitism), woody plants and orchids (commensalism), bees and flowers (mutualism).

GUIDED PRACTICE

Write the following pairs of words on the chalkboard and ask the students to tell how the terms in each pair are related to one another: organism, population; population, community; community, ecosystem; ecosystem, ecology.

INDEPENDENT PRACTICE

Have the students write the answers to the Section Review and Application questions in their journals. Then have them explain biotic potential, limiting factors, and carrying capacity in terms of a specific population.

REINFORCING THEMES—
Environmental Interactions

When discussing the types of interactions among the populations within an ecosystem, reinforce the concept of these environmental relationships as factors that affect not only the individual populations but also the entire ecosystem within which the populations interact. For example, without the predator-prey relationship between lions and zebras, the zebra population might grow, and the increased population might overgraze the grassland, which could lead to soil erosion.

SCIENCE BACKGROUND

Parasitism, commensalism, and mutualism are forms of relationships known as *symbiosis*. A symbiotic relationship is one in which two species of organisms live together with at least one of the organisms benefiting from the relationship.

LASER DISC
3535

Fish; with cleaner shrimp

Relationships Between Populations

Some populations have close, permanent relationships with one another. Sometimes both populations benefit from the relationship, or one population may benefit while the other is unaffected. The relationship may help one population while harming the other. The predator–prey relationship is one of the most common between populations. A *predator* is an animal that hunts and eats other animals. The animal eaten by a predator is its *prey*. Lions are predators that eat many different prey. The zebra and the antelope are common prey of the lion.

Figure 2–8. Predator-prey relationships, such as those shown here, are common between populations.

A relationship between two types of organisms in which one organism benefits while the other one is harmed is called *parasitism* (PAR uh syt ihz uhm). One example of this type of relationship is athlete's foot, an infection caused by a fungus that lives on human skin. The warm, moist conditions created by shoes and socks provide the right habitat for the fungus.

In the case of athlete's foot, the fungus is the parasite. A *parasite* is an organism that lives in or on another organism and is harmful to that organism. The infected organism is the *host*, or the organism that is harmed by the parasite. Other examples of parasites are fleas, ticks, lice, and mosquitoes. These animals live off the blood of the host.

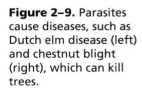

Figure 2–9. Parasites cause diseases, such as Dutch elm disease (left) and chestnut blight (right), which can kill trees.

EVALUATION

Have the students relate population growth to birthrate, death rate, immigration, and emigration of a population and then have the students identify what the impact of a sudden population growth may be on an ecosystem. (The students should note that rapid population growth may tax the resources of the ecosystem and create an imbalance in nature.)

RETEACHING

In their journals, the students should draw ecosystems in which they show these interactions among populations: predator-prey, parasitism, commensalism, mutualism.

EXTENSION

Ask the students to choose a plant or animal population and to research information about the types of ecosystems it lives in and its relationships with other populations.

CLOSURE

Have the students outline this section. Remind them to include the important terms included in the section. Then ask the students to compare their outlines with those of classmates and make revisions if necessary.

A parasite is different from a predator. The parasite does not need to kill the host organism to benefit. In fact, the death of the host usually results in the death of the parasite also.

A relationship between two types of organisms in which one organism benefits while the other is not affected is called *commensalism* (kuh MEHN suh lihz uhm). In many cases, the host organism makes a home for another organism. Remember the example of the clown fish that lives among the venomous tentacles of the sea anemone? The tentacles do not harm the clown fish; they protect it from predators.

A relationship between two types of organisms in which both organisms benefit is known as *mutualism* (MYOO choo wuhl ihz uhm). For example, mutualism exists between a termite and the protozoan that lives in the termite's gut. Termites eat wood, but they cannot digest it. The protozoan living inside the termite digests the wood for the termite and, at the same time, gets food for itself and a home. Both the termite and the protozoan survive because of their cooperation.

Figure 2–10. A remora cleans the mouth of a moray eel and feeds on the eel's leftovers. What type of relationship is this? ①

ASK YOURSELF

How does a parasite differ from a predator?

SECTION 1 REVIEW AND APPLICATION

Reading Critically
1. Describe an ecosystem near your school.
2. How is biotic potential different from carrying capacity?

Thinking Critically
3. What kinds of limiting factors might act on a population of tropical fish living in an aquarium?
4. In what ways, do you suppose, could the human population growth rate be slowed?

---

① Since both organisms benefit (the eel gets a clean mouth; the remora gets food), the relationship is a mutualistic one.

ONGOING ASSESSMENT ASK YOURSELF

A parasite feeds from its host, but it does not need to kill the host to benefit. A predator feeds by killing and eating another organism.

SECTION 1 REVIEW AND APPLICATION

Reading Critically

1. The students' descriptions should include organisms and nonliving factors of the chosen ecosystem.

2. Biotic potential is the maximum number of individuals that can be produced under ideal conditions. Because of limiting factors, biotic potential is not reached under normal circumstances. Carrying capacity is the maximum number of individuals that an environment can support. If the carrying capacity of an environment is reached, individuals will die or must move to a new environment.

Thinking Critically

3. Limiting factors might include population density, availability of food and oxygen, cleanliness of the environment, presence of salt, and water temperature.

4. Human population growth rate could be slowed by increased incidence of disease, war, famine, or other major disasters.

Section 2: ENERGY FLOW

FOCUS

This section examines the places different types of organisms occupy in an ecosystem. It also details how energy is transferred from one organism to another through food chains and webs. The section emphasizes the importance of the interactions within an ecosystem to the survival of all organisms.

MOTIVATING ACTIVITY

Display pictures of people building houses, working in factories, cooking, and so on. Also display pictures of people buying goods in a store, eating meals, reading newspapers, and so on. Label a bulletin board with the words *producers* and *consumers*. Have the students arrange the pictures under the appropriate headings on the bulletin board. Ask the students to explain why they organized the pictures as they did and then to provide definitions for the terms *producers* and *consumers*.

PROCESS SKILLS
- Observing • Comparing
- Constructing/Interpreting Models

POSITIVE ATTITUDES
- Caring for the environment
- Creativity

TERMS
- niche • producers
- consumers

PRINT MEDIA
Watership Down by Richard Adams (see p. xixb)

ELECTRONIC MEDIA
Photosynthesis: Life Energy, National Geographic (see p. xixb)

Science Discovery Food chain

BLACKLINE MASTERS
Study and Review Guide
Laboratory Investigation 2.2
Extending Science Concepts

DISCOVER BY *Observing*

Before the students begin the *Discover* activity, remind them that plants and fungi as well as animals are organisms. Suggest that the students use reference resources, if necessary, to help them identify the types of food each organism uses as a source of energy.

SECTION 2

Energy Flow

Objectives

Define habitat *and* niche.

Distinguish among producers, consumers, and decomposers.

Explain how energy flows through a community.

Every population has a place where it lives in the ecosystem—its *habitat*. Within every ecosystem there are many habitats. For example, a woodland ecosystem has many different populations living in it. Each population lives in a particular habitat within the ecosystem. Bears, squirrels, and foxes hunt for food in the forest. Earthworms live in the soil. Some birds search the forest floor for food. Other birds, such as woodpeckers, spend their time searching for insects higher up, on the trunks of trees. Still higher, hawks might look for food while circling in the sky.

DISCOVER BY Observing

Identify the organisms in the illustration shown here. In your journal, design a table with two columns. In column 1, write the name of the organism. In column 2, write the type of food the organism might find within its community. Include all possibilities. Give your table a title. Below the table, write a statement that describes the relationships among the organisms of this community.

42 CHAPTER 2

TEACHING STRATEGIES

- **Process Skills:** *Comparing, Evaluating*

Emphasize that two populations can share a habitat but cannot occupy the same niche. Tell the students that tree squirrels eat nuts and seeds and live in nests or dens in trees. Chipmunks eat nuts and seeds and live in underground burrows. Ask the students to explain why tree squirrels and chipmunks can live in the same habitat. (Although they eat the same types of food, tree squirrels and chipmunks occupy different niches because they live in different types of shelter.)

- **Process Skills:** *Analyzing, Comparing*

Cooperative Learning Divide the class into small groups. Tell the students that the school is a habitat in which many populations can be found. Have the students identify the niches the habitat's populations belong in. Then ask the students to define *habitat* and *niche* based on their discussion. (The habitat is the school environment; the niches are the roles the populations have within the habitat. The population in each niche has a different function within the school habitat.)

Habitats and Niches

Many populations can occupy the same habitat. The earthworm is not alone as it inches its way through the moist soil. Ants, sow bugs, slugs, mushrooms, trees, and other species also live in the soil. The habitat of one species may even be on or in another species. Mosses and mushrooms live on the trunks of trees. Bacteria live in the digestive systems of many soil organisms.

Each population has a particular function in its habitat. The function of a population includes its lifestyle, the food it eats, and the place where it builds its home. The function of an earthworm, for example, is to dig through the soil by eating the material in its path. The earthworm digests this material, which is then released from its body. This material adds nutrients to the soil. An organism's function in the habitat is somewhat like a human's job or profession. The function an organism plays in the habitat is its **niche** (NIHCH).

Although many populations can share a habitat, only one population can occupy a particular niche. An interesting example of this fact is found in Australia. Two very different animals are both called *kangaroo mice*. The two populations live close together in the dry, sandy regions of the continent.

One population is made up of true mice. They get the name *kangaroo mice* because of their long back legs. When they are frightened, they quickly hop away from danger. These hopping mice dig into the dry, sandy soil to make their homes. They eat seeds, berries, and other plant foods.

Figure 2–11. Many different types of organisms can be found in the same habitat.

Figure 2–12. The hopping mouse (left) and the pouched mouse (right) share the same habitat. Why don't they share the same niche? ①

INTEGRATION—*Social Studies*

Many unusual plants and animals, such as eucalyptus and koalas, are native to Australia. Have the students choose one of the plants or animals of Australia and research information about the chosen organism's habitat and niche within that habitat. Encourage the students to summarize their research findings in their journals.

MULTICULTURAL CONNECTION

Australia is a country with many unusual plants and animals. Organisms vary from one place to another, and some are unique to particular places. Have the students who know their cultural backgrounds research information about the plants and animals that are native to their ancestral homelands. Assign other students a particular country to investigate. Invite the students to share their findings with the class.

① The two different kinds of kangaroo mice do not share the same niche because they feed on different kinds of organisms.

SECTION 2 **43**

TEACHING STRATEGIES, continued

● **Process Skills:**
Classifying/Ordering, Applying

Have the students create a poster with the labels "Producers" and "Consumers." Under each label, the students should paste pictures of appropriate organisms. Have the students further label the consumers as "Herbivores," "Carnivores," "Omnivores," or "Scavengers." After the students present their posters to the class, have them display their posters in the classroom. Later have the students keep their posters in their science portfolios.

● **Process Skills:** *Observing, Classifying/Ordering*

Display pictures of a freshwater or saltwater ecosystem. Ask the students to look at the pictures and to classify the organisms they see as either producers or consumers. (Water plants are producers, and water animals are consumers.)

ONGOING ASSESSMENT
ASK YOURSELF

A habitat is the place where a population of organisms lives. A niche is the function or role organisms serve in their habitat.

REINFORCING THEMES—
Environmental Interactions

The discussion of food chains and food webs demonstrates the environmental interactions among organisms and between organisms and their physical environments. Emphasize the importance of the physical elements of sunlight, water, and soil in the environmental relationships involved in the food chains and webs.

MEETING SPECIAL NEEDS

Second Language Support

Latin is closely related to other languages, such as Celtic, Germanic, Slavic, and Sanskrit languages, and is the origin of many words in other languages, such as Spanish, Italian, and French. As a result, many second-language students may find it helpful to identify the Latin roots of the terms *herbivore* (*herba*—grass, herb; *vorare*—to devour), *carnivore* (*caro*—flesh; *vorare*), and *omnivore* (*omnis*—all; *vorare*).

▶ **44** CHAPTER 2

The second population of kangaroo mice are not really mice at all. They have a pouch like a kangaroo's. They are commonly called *mice* because of their size. These pouched animals eat beetles, cockroaches, termites, spiders, and other small animals. Since these two types of kangaroo mice do not compete for the same niche, they are often found in the same habitat. They have even been found sharing the same burrow.

 ASK YOURSELF

What is the difference between a habitat and a niche?

Food Chains and Webs

Every population within a community needs energy. The source of all this energy is sunlight. Green plants trap light energy and change it into chemical energy during photosynthesis. Because they make their own food, green plants are called **producers.** All communities need food from producers for energy. In addition to green plants, cyanobacteria and a few other types of bacteria are also producers.

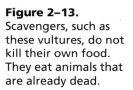
Figure 2–13. Scavengers, such as these vultures, do not kill their own food. They eat animals that are already dead.

Organisms that eat other organisms for food are called **consumers.** There are different types of consumers. Animals that eat only plants are *herbivores*. Cattle, deer, and rabbits are examples of herbivores because they eat only grasses. Animals that eat only other animals are *carnivores*. Wolves, lions, and owls are carnivores. Like the lion that hunts and kills a zebra, the wolf and the owl also hunt and kill their food. Animals that eat both animals and plants are *omnivores*. Bears, raccoons, and pigs are omnivores. Humans are also omnivores. *Scavengers* are animals that eat animals they find already dead. Crows, vultures, and hyenas are scavengers.

- **Process Skills:** *Evaluating, Applying*

Cooperative Learning Have the students work in pairs to identify the foods they ate at their last meal. Once the pairs have identified all of the foods eaten, they should trace the food energy from each type of food back to the sun. (For example, cereal can be traced to plants and then to the sun, and milk can be traced from animals to plants to the sun.)

- **Process Skills:** *Inferring, Generating Ideas*

Explain to the students that in nature simple food chains seldom exist. Food webs are a much more accurate depiction of the relationships between producers and consumers. Ask the students to suggest reasons why a food web is a better depiction of the relationship between producers and consumers than the food chain is. (Most organisms eat more than one type of food.)

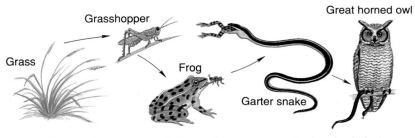

Figure 2–14. A typical food chain is shown here.

Energy is transferred through a community by *food chains*. In a meadow community, for example, grass changes light energy to food energy. The grass uses some of this energy for itself. A grasshopper eats the grass, and, in turn, a meadow frog eats the grasshopper. The frog is eaten by a garter snake that is later eaten by an owl. Energy from sunlight passes through each link in the food chain. Each organism uses some energy to stay alive and stores some energy in its tissues. How might energy be lost from the food chain? ①

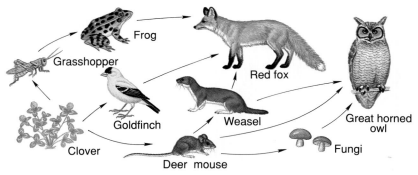

Figure 2–15. Many food chains connect to form a food web.

Many food chains exist within a community. A second meadow food chain might begin with the clover that is eaten by a mouse. The mouse provides food for the weasel that is later eaten by an owl. The owl is part of several food chains. These food chains connect to form a *food web*. The organisms in a food web can eat or be eaten by many other organisms. In the next activity, you can diagram a food web from a community where you live or one that your teacher shows you.

Discover By Doing

Using producers and consumers from a community near where you live, diagram several interconnecting food chains that form a simple food web.

Demonstration

You will need two green potted plants. Place one plant where it has access to sunlight and place the other in a darkened area, such as a cabinet or closet, where it receives no light. Give both plants the same amount of water. Have the students observe what happens to the plants. (The plant that receives no light cannot produce food through photosynthesis and so it becomes weakened and could die. The other plant remains healthy because it receives light and is, therefore, able to produce food.)

① Energy is lost from the food chain as the organism uses energy and as body heat is lost.

LASER DISC
4320

Food chain

SCIENCE BACKGROUND

A few organisms are both producers and consumers. One example is the Venus' flytrap. This plant produces food through photosynthesis. It also consumes insects to obtain the nitrogen that is lacking in its soil.

SECTION 2   **45**

① In an energy pyramid, energy decreases as height increases.

ONGOING ASSESSMENT
ASK YOURSELF

The essential elements that decomposers free from waste and dead organisms are made available to be used again by living members of the community.

SECTION 2 REVIEW AND APPLICATION

Reading Critically

1. A habitat is the place where an organism lives, whereas a niche is the role or function an organism plays in its ecosystem. No two organisms ever play exactly the same role in a habitat, so several niches *must* exist within a single habitat.

2. Parasitism, commensalism, mutualism, and predator-prey relationships can exist between populations in an ecosystem.

3. A producer is an organism, such as a green plant, that makes its own food. A consumer is an organism, such as a human being, that eats other organisms for food.

Thinking Critically

4. Both scavengers and decomposers obtain their nutrition from waste materials or dead organisms. Decomposers return carbon, oxygen, nitrogen, and other substances to the environment—scavengers do not.

5. The students should strive to name one or more organisms at each level of their chain.

6. Used paper is broken down or shredded and made into pulp. New paper is then made from the pulp. This process of paper recycling is similar to the breakdown of materials by decomposers, the production of compounds from those materials, and the reuse of those compounds by other living organisms.

▶ 46 CHAPTER 2

GUIDED PRACTICE

Write the following terms on the chalkboard: producers, consumers, and decomposers. Initiate a discussion in which the students define the terms; provide examples of organisms that are producers, consumers, and decomposers; and explain the roles of producers, consumers, and decomposers in a food web.

INDEPENDENT PRACTICE

Have the students write their answers to the Section Review and Application questions in their journals. Then have them write a paragraph that identifies the concept they think is the most important one developed in this section.

EVALUATION

Have the students create a food web in which humans receive nutrition from the major food groups. (Webs should reflect the need for grains, meats, dairy products, fruits, vegetables, and small amounts of fat in the diet.)

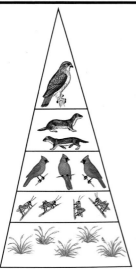

Figure 2–16. This pyramid shows a loss of energy from one consumer level to the next.

Figure 2–17. Decomposers, such as these mushrooms, break down dead organisms and return nutrients to the soil.

A community must have many more producers than consumers because producers provide the energy necessary for both themselves and the consumers. The energy found in the levels of a food web can be shown by a pyramid, such as that shown in the illustration. The base of the pyramid represents the producers. All the producers taken together have the most energy. What happens to the energy as you go higher in the pyramid? ①

The organisms at each level use much of their energy just to stay alive. Because only about 10 percent of the energy can be passed on to the next level, the populations get smaller at each higher level of the pyramid. Only a few animals are found at the top of the energy pyramid. These animals, called *top carnivores*, such as lions, tigers, and wolves, exist in only very small populations. They depend on very large populations of other animals and plants to survive.

The materials that make up the organisms in a community—carbon, oxygen, nitrogen, and others—are constantly reused. *Decomposers* return these chemicals to the ecosystem. Bacteria and fungi are common decomposers. Decomposers use some of the materials for food and return the rest to the soil. Green plants can then reuse these nutrients to make new tissue. This tissue is eaten by animals, and the cycle continues.

 **ASK YOURSELF**
How do decomposers complete the energy cycle?

SECTION 2 REVIEW AND APPLICATION

Reading Critically
1. How does an organism's habitat differ from its niche? How can several niches exist within a single habitat?
2. What are the relationships that can exist between populations in a community?
3. What is the difference between a producer and a consumer?

Thinking Critically
4. How are scavengers and decomposers similar? How are they different?
5. Describe a food chain in an ecosystem near your school. What organism is at the top of the chain? What organism is at the bottom of the chain?
6. Compare recycling a product, such as paper or aluminum, with the natural recycling of chemicals in the environment. How does recycling benefit the environment?

RETEACHING

To reinforce the lesson concepts, tape record the section content and encourage the students to listen to the tape in the media center. After listening, ask the students to respond orally to the Section Review and Application questions.

EXTENSION

Have the students research information about a saltwater ecosystem and create a food web for that ecosystem.

CLOSURE

 Cooperative Learning Divide the class into small groups and have the group members work together to summarize the section content in writing. Then select several groups to present their summaries to the class.

SKILL
Diagramming Food Chains and Food Webs

 Process Skills: Classifying/Ordering, Constructing/Interpreting Models

Grouping: Groups of 3 or 4

Objectives
- **Organize** information about a food chain.
- **Diagram** a food web.

Discussion
If visiting a nearby ecosystem is not possible, display pictures of ecosystems that the students may use as a resource, or have the students research information about such ecosystems as forests, prairies, and tundra. Encourage the students to identify at least 15 different organisms that live in the chosen ecosystem.

▶ **Application**

1. Organisms are identified as producers if they make their own food, consumers if they feed on other organisms, and decomposers if they break down dead organisms and return nutrients to the soil.

2. Herbivores eat plants, and carnivores eat other animals.

3. A complex community is most stable because the loss of any single species is unlikely to have as great an impact as the loss of a species in a simple community would have on the total ecosystem.

✳ **Using What You Have Learned**

1. The carnivores would die of starvation.

2. If the same herbivore were eliminated from the food web, the other animals would feed on different herbivores.

SKILL: Diagramming Food Chains and Food Webs

▼ PROCEDURE

1. As a class, select an ecosystem near your school with which all the class members are familiar. This ecosystem might be a prairie, a woodland, a pond, a city park, or a roof garden.

2. On index cards, write the names of the organisms that you think may be found in this ecosystem. Use one card for each organism. Make cards for as many organisms as you can think of.

3. On additional index cards, write the names of all the nonliving parts of the ecosystem you have chosen. For instance, you could include the sun, wind, water, and so on.

4. Using your cards, arrange the organisms into separate food chains. You should have more than one food chain. Arrange the cards for the nonliving parts of the ecosystem near the organism they affect. Copy your food chains into your journal.

5. Now combine your chains into a food web. Copy your food web into your journal.

6. As a class, make a giant food web by attaching your cards to a bulletin board.

▶ **APPLICATION**

1. Identify each organism in the giant food web as a producer, consumer, or decomposer.
2. Identify the herbivores and carnivores in the giant food web.
3. Some communities are very simple. They have only a few kinds of organisms arranged in a simple food web. Other communities are complex; they have many different kinds of organisms arranged in a complex food web. Which type of community is more stable or more resistant to change? Why?

✳ **Using What You Have Learned**

1. Look at one of the food chains that you developed. What would happen to the other animals in the food chain if the herbivore were eliminated from the community?
2. Find this same herbivore in the giant food web. What would happen to the other animals in the food web if this herbivore were eliminated from the community?

SECTION 2

Section 3: CHANGING ECOSYSTEMS

FOCUS

In this section, the conditions leading to ecological succession are described. The characteristics of successive biotic communities are also identified. In addition, the role of environmental disturbances that lead to succession is detailed.

MOTIVATING ACTIVITY

Point out that in 1992, Hurricane Andrew devastated much of southern Dade County in Florida. Ask the students to consider the stages of succession of plant growth in the area after the hurricane. (The students might suggest that grasses and wild flowers would be the first plants to regrow in the area, followed by bushes, small trees, and larger trees.)

PROCESS SKILLS
- Comparing • Analyzing

POSITIVE ATTITUDES
- Caring for the environment
- Creativity

TERMS
- succession

PRINT MEDIA
Forest Fire
by Christopher Lampton
(see p. xixb)

ELECTRONIC MEDIA
Succession: From Sand Dune to Forest,
Britannica
(see p. xixb)

Science Discovery
Weathering, biological; with lichen
Mt. St. Helens; plant succession

BLACKLINE MASTERS
Study and Review Guide
Reading Skills
Thinking Critically

SECTION 3

Changing Ecosystems

Objectives

Describe the differences between stable and unstable populations.

Distinguish between pioneer and climax communities.

Compare and contrast primary and secondary succession.

A fire rages through a once majestic forest. But what appears to be the destruction of a woodland ecosystem is not, because the damage is not permanent. Over time, the forces of nature will slowly begin the process of regrowth.

Succession

One illustration shows a forest that has been destroyed by fire. Since the fire, however, many changes have taken place. Grasses, shrubs, and seedling trees have begun to grow. The slow changes that occur when a community recovers from an event such as a fire is called **succession.**

Succession also occurs in areas such as a sand dune or bare rock that has not previously had any life. This succession, called *primary succession,* also occurs on land recently exposed by a melting glacier or on lava fields created by a volcano.

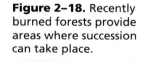

Figure 2–18. Recently burned forests provide areas where succession can take place.

48 CHAPTER 2

TEACHING STRATEGIES

● **Process Skills:** *Inferring, Classifying/Ordering*

Cooperative Learning In random order, write the following stages of succession on the chalkboard: lichens, wildflowers, spruce trees, grasses, aspens, bushes. Have small groups of students work together to determine the order in which the plant species might succeed one another. Have the groups provide reasons for their sequencing of plants. (Plants can be sequenced in this order: lichens, grasses, wildflowers, bushes, aspens, spruce trees.)

● **Process Skills:** *Comparing, Applying*

Have the students define the term *pioneer* in terms of settlement in the United States. (People who were among the first to settle in a territory of the United States were called "pioneers.") Then have the students relate this definition to pioneer plants in ecological succession. (Pioneer plants are those that are among the first to grow in a previously barren area.)

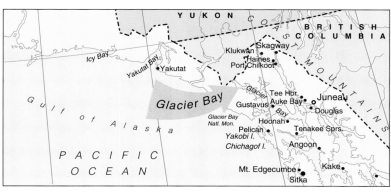

Figure 2–19. Glacier Bay, Alaska, provides an area in which to study primary succession.

SCIENCE BACKGROUND

Succession occurs because a community is unstable. Light, water, temperature, and soil nutrients are some of the natural resources that contribute to stability. As the availability of these and other natural resources changes, the habitat becomes less suitable for some populations and more suitable for others. Thus, the composition of the community changes and succession takes place.

INTEGRATION—*Social Studies*

Glacier Bay National Park became a national park in 1980. It had been designated a national monument in 1925. Just 200 years ago, the park was completely covered by glaciers. As the glaciers receded, primary succession began. The park is located about 160 km northwest of Juneau, Alaska. Ask the students to locate the park on a map of Alaska or the United States. You may wish to point out that Alaska has more national parks than any other state in the Union.

 **LASER DISC**
612

Weathering, biological; with lichen

One interesting example of primary succession can be seen at Glacier Bay National Park in Alaska. This area contains glaciers that formed during the last Ice Age. Since that time, a slow but steady melting has taken place, exposing land that has been covered by ice for thousands of years. This area gives biologists an unusual and excellent opportunity to study primary succession.

The land closest to the glacier has been exposed for only a matter of days. The slow process of primary succession is just beginning to take place on this barren land. Near the mouth of the bay, the land has been exposed for hundreds of years, and the final stages of succession are taking place. Traveling down the bay is like traveling through time.

Pioneer Stage The area closest to the melting glacier is, at first, lifeless. The first living things to develop in a previously lifeless area are *pioneer plants*. The habitat of pioneer plants is very harsh. They must be able to survive high winds, bright sun, and heavy rains. There is usually little or no soil during the first stage of primary succession.

Pioneer plants grow in crushed rock and gravel and are able to survive without soil. Lichens are perhaps the hardiest pioneer plants. A lichen is a combination of a fungus and a green alga, living together in a unique and mutually beneficial relationship. Lichens attach themselves to bare rock and obtain nutrients directly from the rock. The action of the lichens combined with the action of the weather breaks up the rock. The cracks and depressions in the rock hold dead lichens and bits of rock. These materials are the beginnings of soil.

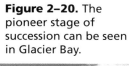

Figure 2–20. The pioneer stage of succession can be seen in Glacier Bay.

TEACHING STRATEGIES, continued

● **Process Skills:** *Classifying/Ordering, Comparing*

Suggest that the students develop a chart identifying the stages of changes in succession and list typical examples of a community in each of the stages. Have the students place the charts in their science portfolios.

● **Process Skills:** *Inferring, Applying*

Discuss how soil forms. Tell the students that the process of soil formation occurs as a result of weathering by water and wind, the action of plant root systems, and the accumulation of decaying material as plants and other organisms die. Ask the students how mosses contribute to formation of soil. (Mosses capture bits of decaying matter from the air and enrich the soil with them.)

The weeds that appear in lawns and in parks are evidence that succession is trying to return the habitat to its climax community. Using weed killers keeps the natural process of plant succession from restoring the climax community. The use of weed killers can also contaminate the soil and possibly the groundwater supply. Some weed killers can harm animals as well.

ONGOING ASSESSMENT
ASK YOURSELF

The pioneer stage of primary succession contains the first living things to develop in a previously lifeless area. During the mossy stage, mosses slowly replace the pioneering lichens. Grasses, flowering plants, and alder trees characterize the grassy stage at Glacier Bay. The final stage, or climax community, has spruce trees, which shade and eventually eliminate the alder trees from the grassy stage.

Figure 2–21. Many stages of succession can be seen in Glacier Bay. The mossy stage of succession (left), the grassy stage of succession (center), and the final stage of succession (right) are shown here.

Mossy Stage A short distance down Glacier Bay, the next stage of primary succession can be observed. After many years of growth, lichens are slowly replaced by mosses. Mosses grow in the soil formed by the lichens and the rock. Mosses capture bits of dead matter brought in by wind or waves. The decomposing remains of sea creatures and bird droppings enrich the thin soil held in place by the mosses.

Grassy Stage Slowly the habitat improves, and the mosses are crowded out by new plants. Farther down Glacier Bay, grasses follow the mossy stage of succession. Grasses have spreading roots that hold soil in place better than mosses do. After the grasses are established, flowering plants begin to grow from seeds carried in by the wind and by birds. These flowering plants attract more animals. The spreading layer of roots traps more of the decaying matter, and the soil continues to thicken and improve.

The grassy stage of succession creates a habitat where trees can grow. At Glacier Bay, the seeds of alder trees sprout in the soil that has been enriched by the grasses. The grasses shelter the seedlings from sun and wind. As years pass, a forest of alder trees slowly forms.

Final Stage The final stage of succession at Glacier Bay comes as spruce trees slowly crowd out the alders. As in the earlier stages, the alders improved the habitat, making it suitable for new populations. Spruce seedlings grow well in the shade of the alders. The alder seedlings, on the other hand, need bright sunlight. Thus, most of the young trees in the forest are now spruce. As time passes, the taller-growing spruce trees eventually shade the older alders, which slowly die out.

ASK YOURSELF

What are the characteristics of each stage of primary succession?

- **Process Skills:** *Comparing, Analyzing*

A climax community is more stable than a pioneer community or any of the other communities preceding the climax community stage. Emphasize the concept that each successive stage is replaced until a climax community is established. Ask the students to compare the stability of pioneer and climax communities. Have them discuss why the plant population of a climax community is not replaced by another plant population. (Pioneer communities cause changes in their habitat that make it less suitable for themselves and more suitable to support subsequent plant and animal populations. Climax communities contain a diverse array of plants and animals that interact with each other to maintain a balanced, stable system.)

Climax Communities

Spruce trees are part of the *climax community,* or final stage of succession, at Glacier Bay. Unlike earlier stages, a climax community is a stable habitat. Spruce seedlings grow well in the shade of older trees. As the older spruce trees die, they are replaced by seedlings of the same species. The population maintains itself without increasing or decreasing in size.

The earlier stages of succession consist of populations that are less stable. Less stable populations change the habitat as they increase in size. Changes make the habitat less suitable for old populations and more suitable for new populations. As the populations of new organisms grow, older populations die off. One community is replaced by another. Eventually, a climax community forms. The populations of the climax community live in the habitat without changing it. Thus, the climax community remains stable unless it is disturbed by some outside force.

Figure 2–22. The climax community at Glacier Bay is a spruce forest.

Succession is often described in terms of its plant communities. Yet, animals also play an important part in the process. Each community has both animal and plant populations. As more plants grow, changes in the habitat occur. Such changes provide new niches for some animal populations, while the niches of others may die out. The growth and decline of these animal populations help to change the habitat.

MEETING SPECIAL NEEDS

Gifted

The specific nature of a climax community is determined by physical factors, such as light, temperature, and moisture, in the region. Have the students choose one of the following regions or a region in their home state and determine what the plant populations of the climax community are: tundra in Alaska, prairie in Illinois, marshland in Florida, forest in Washington State, forest in Vermont, rain forest in Hawaii.

 **LASER DISC**
705

Mt. St. Helens; plant succession

 In the early 1990s, forest fires raged throughout areas in the Pacific West because of drought. Modern fire-fighting techniques helped contain these fires in most cases. Some ecologists suggest that forest fires should be allowed to burn unless humans and their homes are threatened. They think that fires are part of the natural process of succession and renewal. Have the students debate whether modern technology should be used to stop forest fires or whether forest fires should be allowed to burn themselves out if human communities are not threatened.

TEACHING STRATEGIES, continued

● **Process Skills:** *Inferring, Evaluating*

Ask the students to suggest evidence they might look for to determine whether a community experiencing secondary succession is in a pioneer or a climax stage. (Evidence of a pioneer stage might include the presence of saplings among grasses and shrubs. Evidence of a climax community might include the presence of only one or two dominant species.)

GUIDED PRACTICE

Have the students identify the plant populations for each of these stages of development in Glacier Bay National Park: pioneer stage, mossy stage, grassy stage, climax community.

INDEPENDENT PRACTICE

 Suggest that the students write their answers to the Section Review and Application questions in their journals. Then have the students write a description of a pioneer community in primary succession and a pioneer community in secondary succession.

---

① Both secondary and primary succession result in climax communities after several stages of development.

✧ **Did You Know?**
A forest region destroyed by fire can return to a climax community in 100 to 200 years.

★ **PERFORMANCE ASSESSMENT**

Ask the students to compare the stages of primary succession with the stages of secondary succession. Encourage them to make a sequence diagram for each kind of succession. Use the students' explanation of their diagrams to evaluate their understanding of the processes involved in both kinds of succession.

✦ **DISCOVER BY** *Researching* **(page 53)**

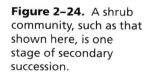 If your local library has its card catalogs on computer files, you may wish to visit the library with the class and have a librarian explain the use of the computer for finding research resources. Suggest that the students write first drafts summarizing their findings before they write the final draft in their journals.

Figure 2–23. How is secondary succession, shown here, similar to primary succession? ①

Natural disasters, such as fires or floods, can disturb a climax community. Disease and insect pests can destroy an entire community of climax species. Humans can also cause major changes in habitats, affecting the climax community. Agriculture and forestry are common practices that destroy climax communities.

Secondary succession is the series of changes in communities that occurs in a disturbed area. This type of succession occurs on land that has not lost its soil. Since soil is already present and does not have to be built up from bare rock, secondary succession takes place more quickly than primary succession.

An example of secondary succession might be a beech-maple forest that has been burned out or been cut down for timber. The pioneer plants in this situation are grasses and weeds, whose seeds are carried to the area by wind and animals. These plants have shallow roots that spread through the soil, protecting it from wind and water erosion. Within a few months, the ground is covered in green. Soon taller plants with deeper roots begin to appear. These plants make it difficult for the pioneer plants to survive. After many years, the grassy field contains tall weeds and shrubs.

The shrub community is a suitable habitat for the seedlings of pioneer trees, such as dogwoods, sumacs, and aspens. These seedlings grow well in bright sunlight. As the trees grow, they capture more and more of the sunlight and available water and nutrients. Gradually, the shrub community is replaced by the pioneer trees.

Figure 2–24. A shrub community, such as that shown here, is one stage of secondary succession.

EVALUATION

Have the students describe the stages of succession that might occur if an empty lot were left undisturbed. Ask the students if the succession is primary or secondary.

RETEACHING

Give the students a series of pictures depicting the stages of primary succession and have them sequence the pictures. Encourage the students to state reasons for the sequence they choose.

EXTENSION

Have the students investigate the fires that destroyed forest and meadow communities in Yellowstone National Park in 1988. Suggest that the students locate magazine articles or books that describe the succession taking place in Yellowstone.

CLOSURE

Cooperative Learning Have the students work in small groups to outline the information in this section. Each group should write a summary based on the outline and then read the summary to the class.

Beech and maple seedlings grow well in the shade of pioneer trees. The taller beech and maple trees also have deeper root systems, which make them better able to capture water than the pioneer trees. Slowly, the pioneer trees die and are replaced by more beech and maple seedlings. After many years, succession will reestablish a climax community of beech and maple trees. In the next activity, you can find out how rapidly succession occurs in an area devastated by a volcano.

Figure 2–25. In some areas, a beech and maple forest is the climax community.

DISCOVER BY Researching

The eruption of Mount Saint Helens in 1980 was a major disturbance of the ecosystem. Using your library, find books and articles on this or another environmental disaster and on the succession that followed the destruction. Try to find "before" and "after" pictures that you can use. Record your findings in your journal.

ASK YOURSELF

What are the stages of secondary succession?

SECTION 3 REVIEW AND APPLICATION

Reading Critically
1. How is a stable population different from an unstable population?
2. Describe the stages of primary succession.
3. How is secondary succession different from primary succession?

Thinking Critically
4. In which stages of succession are the populations most stable? Why?
5. Suppose a wooden shed was built in a grassy field fifteen years ago. The shed gets knocked down. What would the ground look like where the shed had been standing? When plants start to grow, would this be primary or secondary succession? Why?

ONGOING ASSESSMENT
ASK YOURSELF

Stages of secondary succession include pioneer plants such as grasses, shrub communities such as tall weeds and shrubs, and pioneer trees such as maples.

SECTION 3 REVIEW AND APPLICATION

Reading Critically

1. A stable population maintains its size. An unstable population experiences large changes in size.

2. Pioneer plants such as lichens are the first living things to occupy an area. Mosses and then grasses replace the pioneer plants, and animals move into the area. Finally, various animal and plant populations replace each other until a stable climax community is achieved.

3. Secondary succession occurs in areas where communities have been disturbed. Primary succession occurs in areas where living things have not recently been present. Primary succession includes the building up of soil. In secondary succession, soil is already present.

Thinking Critically

4. Populations are the most stable in the climax stage of succession because many different plant and animal populations interact to create a balanced community.

5. The ground under the shed would be barren of plant life and most animal life. Secondary succession would occur in this area because soil is already present.

INVESTIGATION

Observing an Ecosystem

Process Skills: Observing, Comparing

Grouping: Groups of 3 or 4

Objectives
- **Observe** the relationships between plants and animals.
- **Interpret** observed plant and animal data.
- **Infer** and **describe** the impact that human beings have had on an ecosystem.

Pre-Lab
After the students have identified their study area, ask them to write a list of the plants and animals they predict they will find there.

Hints
As an alternative activity, arrange for a field trip to a nearby botanical garden, forest preserve, or park. Have the students record their observations of plants and animals as they walk through the location. Caution the students not to disturb any part of the environment.

Analyses and Conclusions
1. The students should suggest that the types of plants in the area help determine the kinds of animals found.
2. Similar plants and animals may be found in similar study areas.
3. The students' answers should reflect an understanding of the interactions among the organisms.
4. Answers should reflect an understanding of organisms and their niches within a habitat.
5. Humans can affect the environment both negatively and positively.
6. Food chains should identify producers and consumers as well as decomposers, if any.

Application
The students might suggest that humans can positively affect the area by avoiding direct pollution of the area. They can negatively affect the area through dumping.

✷ Discover More
The students might describe how some homeowners trim and thin trees, creating healthier grass by allowing more light to reach the ground, but destroying potential habitats for birds and squirrels.

Post-Lab
Have the students compare their final list with the list they made before the investigation to check the accuracy of their predictions.

 Have the students place their notes and written responses to the *Investigation* in their science portfolios.

INVESTIGATION

Observing an Ecosystem

▶ MATERIALS
- meter stick • stakes • string, 4.5 m • notebook paper • pencil • hand trowel
- small jar with cap • screening material for sifting soil

▼ PROCEDURE

1. **CAUTION: Do not disturb your study area any more than is necessary. Do not grab any small animal with your hands. Identify poisonous plants before you touch anything.**
2. Using the meter stick, mark off an area of study 1 m by 1 m. Use the stakes and string to mark the boundaries of the area as shown.
3. While standing beside the study area, carefully observe your area for animals or animal signs, such as leaves eaten by insects. Write a description of the study area that includes information about the plants and animals or evidence of animals.
4. Dig for insects, small worms, and other living things in your study area. Record names or draw pictures of the types of plants and animals found in your study area.
5. Gather a soil sample by gently digging or scraping a small soil sample from the surface of your study area. Sift the soil through a piece of screening. Collect any animals or animal parts and place them in the jar. Keep the jar cap loose.
6. List all of the animals found on the surface or immediately below the surface of the soil.

▶ ANALYSES AND CONCLUSIONS
1. How does your study area differ from other study areas? What might be the reasons for the differences?
2. How are the study areas the same? Explain.
3. What characteristics do the animals and plants have that help them live in the study area?
4. What are the roles of the plants and animals found in your area? For example, which animals dig into the soil? Which feed on the grasses or weeds?
5. How have humans affected your study area? Has the impact improved or harmed the area? Explain.
6. Draw a possible food chain for the plants and animals found in your area.

▶ APPLICATION
Describe how humans could positively or negatively affect the area you have studied.

✷ Discover More
Investigate the ways in which humans have both helped and harmed the plants and animals in an environment that is different from your study area.

CHAPTER 2 HIGHLIGHTS

The Big Idea—ENVIRONMENTAL INTERACTIONS

Lead the students to understand that populations within a community interact with one another and with the nonliving parts of the ecosystem. Emphasize the importance of the sun as the source of all energy flow in food webs. Help the students identify primary and secondary succession as a natural process in which change occurs in an ecosystem.

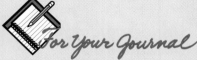

The students' ideas should reflect the understanding that differing relationships exist among the populations within a community. These interactions and relationships help keep the ecosystem balanced. The destruction of one relationship could lead to an imbalance in the system. The students may also express an understanding of the impact of the physical environment on populations and their interactions.

CONNECTING IDEAS

This concept map should be completed with words similar to those shown here.

Row 3: Living things; Nonliving things

Row 5: Community; Population

CHAPTER 2 HIGHLIGHTS

The Big Idea

Imagine a huge spider web made of many connecting threads. All living and nonliving things in an ecosystem are connected together as if they were all parts of the same spider web. Just as a movement in one part of the web affects the entire web, a change in one part of the ecosystem affects everything in the ecosystem.

There are many different kinds of interactions and relationships that can exist among living things and among living and nonliving things in an ecosystem. These interactions and relationships are like the threads that make up the spider web. Energy and food are two of the threads in the web. The ways in which they are transferred from one organism to another are vital to the survival of the ecosystem.

Look back at the relationships and definitions you wrote in your journal at the beginning of the chapter. How have your ideas about these concepts changed? Revise your journal entry to show what you have learned. Be sure to include information on the different kinds of relationships and interactions among living things and their environment.

Connecting Ideas

Copy this unfinished concept map into your journal. Complete the concept map by writing the correct term in each blank.

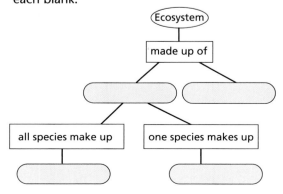

CHAPTER 2 **55**

CHAPTER 2 REVIEW ANSWERS

Understanding Vocabulary

1. Ecology is the study of the relationships between organisms and their environments.

2. Ecosystems consist of populations that live in various habitats and occupy different niches.

3. Succession is the evolution of a community through primary and secondary stages of development.

Understanding Concepts

Multiple Choice

4. b
5. a
6. c
7. d
8. a
9. d

Short Answer

10. The food chain should begin with a producer and end with a human being.

11. Limiting factors include food supply, water supply, natural disasters, and weather.

12. The students should describe a mutualistic relationship in which the family and pet both benefit. For example, a family is protected by a dog, and the dog receives food and shelter from the family.

Interpreting Graphics

13. Biotic potential is the number of individuals that could be produced given ideal conditions. The factors that prevent a population from reaching its biotic potential are known as limiting factors. The largest number of individuals that can be supported by an environment is called the carrying capacity of that environment. When a population exceeds the carrying capacity of an area, individuals will die or must migrate. The impact of limiting factors on biotic potential determines the carrying capacity of an area.

Reviewing Themes

14. All three groups are important in maintaining the balance of the sealed ecosystem. Producers trap light energy as chemical energy. Consumers feed on the plants and provide the carbon dioxide that the plants need to make food for themselves. Decomposers release materials for recycling.

15. Producers, such as grasses and trees, obtain energy from sunlight. Consumers, such as bears and raccoons, obtain energy by eating plant producers or other animals. Decomposers, such as bacteria and fungi, obtain energy by breaking down dead organisms.

CHAPTER 2 REVIEW

Understanding Vocabulary

Explain how the words in each set are related to each other.

1. ecology (35), organisms, environment
2. niche (43), habitat (42), ecosystem (35), population (34)
3. succession (48), community (34), primary, secondary

Understanding Concepts

MULTIPLE CHOICE

4. Which term best describes the relationship between a lion and an antelope?
   a) commensalism
   b) predator–prey
   c) mutualism
   d) parasitism

5. Which of the following does *not* contribute to changes in population size?
   a) organization
   b) birthrate
   c) death rate
   d) immigration

6. Because an earthworm and a mushroom both live in the soil, they share the same
   a) niche.
   b) population.
   c) habitat.
   d) species.

7. Which of the following is an omnivore?
   a) wolf
   b) rabbit
   c) vulture
   d) pig

8. During primary succession, lichens appear in the area during the
   a) pioneer stage.
   b) mossy stage.
   c) grassy stage.
   d) final stage.

9. Which term best describes the relationship between a flea and a dog?
   a) commensalism
   b) predator–prey
   c) mutualism
   d) parasitism

SHORT ANSWER

10. Give an example of a food chain involving human beings.

11. What are two types of limiting factors?

12. What type of relationship exists between a family and its pets? Explain your answer.

Interpreting Graphics

13. Look at the following equation. Define the terms and explain how they are related to each other and to the concept of population.

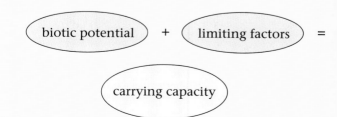

56 CHAPTER 2

Thinking Critically

16. The grassy stage of succession is shown. Before the grassy stage, the area contained mosses. After the grassy stage, the area will contain young and mature trees—characteristics of the final stage of succession.

17. Coal, oil, and natural gas supplies were created from the remains of plants and animals. When these plants and animals were alive, they obtained energy directly or indirectly from the sun. Plants used the sun's energy to grow, reproduce, and provide a source of food for animals, and animals obtained the sun's energy by eating plants or other animals that ate plants. The sun, therefore, is the source of the energy found in organic fuels.

18. Both predators and parasites obtain food from other organisms. A predator kills and eats the organisms; a parasite feeds on the live organisms and does not need to kill them to benefit.

19. Descriptions of the food web should identify the wild onions, grass, blackberry plants, and oak trees as producers; mushrooms and bacteria as decomposers; the wolf, rabbits, mice, grasshoppers, lizards, snakes, the owl, the hawk, and spiders as consumers.

20. If a herbicide were to enter the pond, the plants would be killed. Without these producers, the first level of consumers would die. Then the second level of consumers would die, and so on.

Reviewing Themes

14. *Environmental Interactions*
Imagine that you have a small ecosystem sealed off under a plastic bubble. Your ecosystem includes producers, consumers, and decomposers. Which of these groups of organisms is the least important to the continued survival of the ecosystem? Why?

15. *Energy*
Describe the three ways in which organisms obtain energy in an ecosystem. Name an organism that plays each energy role.

Thinking Critically

16. Which stage of primary succession is shown in the photograph? How did the area look before this stage? What will it look like in the next stage?

17. Explain how the energy that heats your home came to you from the sun, through a food chain. (Hint: Coal and oil are used to heat most homes. These fuels formed from the remains of plants and animals that lived millions of years ago.)

18. How are predators and parasites alike? How are they different?

19. Describe a food web that includes the following organisms: a wolf, wild onions, rabbits, grass, mice, grasshoppers, lizards, snakes, an owl, blackberry plants, a hawk, mushrooms, bacteria, spiders, and oak trees. Identify each type of organism in your food web as a producer, consumer, or decomposer.

20. Explain what would happen to the pond community in the photograph if a herbicide accidentally entered the pond.

Discovery Through Reading

Lauber, Patricia. *Summer of Fire: Yellowstone 1988*. Orchard Books, 1991. Describes the season of fire that struck Yellowstone in 1988, and examines the complex ecology that returns plant and animal life to a seemingly barren, ash-covered expanse.

CHAPTER 3
LIFE IN THE BIOSPHERE

PLANNING THE CHAPTER

Chapter Sections	Page	Chapter Features	Page	Program Resources	Source
Chapter Opener	58	*For Your Journal*	59		
Section 1: THE BIOSPHERE AND WATER ECOSYSTEMS	60	Discover By Doing **(A)**	61	*Science Discovery**	SD
• What is the Biosphere? **(B)**	60	Activity: What happens when fresh water and salt water meet? **(A)**	68	Investigation 3.1: Making Pond Cultures **(A)**	TR, LI
• Marine Ecosystems **(A)**	63	Section 1 Review and Application	69	Investigation 3.2: The Oxygen-Carbon Dioxide Cycle in a Freshwater Ecosystem **(A)**	TR, LI
• Freshwater Ecosystems **(A)**	66	Investigation: Observing Changes in Ecosystems **(A)**	70	Thinking Critically **(H)**	TR
				Extending Science Concepts: Ecosystems **(A)**	TR
				Earth's Major Biomes **(B)**	IT
				Zones of Ocean Life **(B)**	IT
				A Freshwater Ecosystem **(B)**	IT
				Record Sheets for Textbook Investigations **(A)**	TR
				Study and Review Guide, Section 1 **(B)**	TR, SRG
Section 2: LAND BIOMES	71	Discover By Researching **(A)**	73	*Science Discovery**	SD
• Tropical Rain Forests **(B)**	71	Activity: What is the root structure of grass plants like? **(B)**	78	Reading Skills: Mapping Concepts **(A)**	TR
• Deciduous Forests **(B)**	74	Section 2 Review and Application	79	Connecting Other Disciplines: Science and Geography, Reading a World Map **(A)**	TR
• Boreal Forests **(B)**	75	Skill: Modeling Climate **(A)**	80	Study and Review Guide, Section 2 **(B)**	TR, SRG
• Tundra **(B)**	76				
• Grasslands **(B)**	77				
• Deserts **(B)**	78				
Chapter 3 HIGHLIGHTS	81	The Big Idea	81	Study and Review Guide, Chapter 3 Review **(B)**	TR, SRG
Chapter 3 Review	82	For Your Journal	81	Chapter 3 Test	TR
		Connecting Ideas	81	Test Generator	
				Unit 1 Test	TR

B = Basic   **A** = Average   **H** = Honors
The coding Basic, Average, and Honors indicates subsections, features, and resources that might be appropriate for different levels of learners. For additional suggestions regarding choice of topic and depth of coverage, see the Pacing Chart on pages T26–T29.

*Frame numbers at point of use
(TR) Teaching Resources, Unit 1
(IT) Instructional Transparencies
(LI) Laboratory Investigations
(SD) *Science Discovery* Videodisc Correlations and Barcodes
(SRG) Study and Review Guide

▶ 57A

CHAPTER MATERIALS

Title	Page	Materials
Discover By Doing	61	(per individual) notebook paper, tape, marker or colored pencil
Teacher Demonstration	67	dishpan, plastic wrap, short glass, salt water, stones
Activity: What happens when fresh water and salt water meet?	68	(per group of 2 or 3) glass baking dish, cardboard, salt, blue and yellow food coloring
Investigation: Observing Changes in Ecosystems	70	(per group of 3 or 4) container (5L) with lid, dried straw or grass, distilled water, wax pencil, graduate (10 mL), medicine dropper, pond water, hammer, nail, microscope, slides, coverslips
Discover By Researching	73	(per individual) magazines, newspaper, journal
Activity: What is the root structure of grass plants like?	78	(per group of 3 or 4) hand lens, grass, ruler
Skill: Modeling Climate	80	(per group of 2 or 3) graph paper, pencil

ADVANCE PREPARATION

For the *Demonstration* on page 67, prepare a solution of salt water. For the *Investigation* on page 70, obtain appropriate dried straw or grass and pond water samples. For the *Activity* on page 78, you will need grass samples with the roots intact.

TEACHING SUGGESTIONS

Field Trip
Plan a trip to freshwater ecosystems. Select one location at a lake or pond and another at a river or stream. Help the students identify plants, animals, and other characteristics. Remind the students to record amount of sunlight, soil composition, and temperature, as well as other environmental aspects. The students can then compare the two locations.

Outside Speaker
Invite a local traveler who has visited different biomes such as tropical rain forests (for example, in the Amazon basin) or tundra (for example, in Alaska) to speak to the class. Local libraries often know of people who would be willing to talk about their travels. Suggest that they use slides and/or show artifacts.

Chapter 3
Life in the Biosphere

CHAPTER THEME—ENVIRONMENTAL INTERACTIONS

This chapter introduces the students to the structures of the biosphere. While studying the chapter, the students will discover how plants and animals interact with their physical environments as well as how those environments provide for the survival of living things. The students will also learn how living things can survive only in certain areas on Earth. This theme is also developed through concepts in Chapters 1, 2, 11, and 15. A supporting theme of this chapter is **Systems and Structures.**

MULTICULTURAL CONNECTION

Many people realize that if they lived on another planet, their lives would be different than they are on Earth. What many do not realize is that in order to meet the climatic challenges on Earth, cultures have developed different types of housing, clothing, and food. Have the students choose a category (housing, clothing, or food) and a climatic region. Encourage the students to choose topics that offer differences in culture and climate. Suggestions include Native Americans of the Southwest United States, Africans such as the Zulu, or nomadic peoples of the Middle East. You may wish to ask the students who know their ethnic backgrounds to explore their own heritage. Ask the students to present their research in the form of a report, poster, or model.

MEETING SPECIAL NEEDS

Second Language Support
Fluent English-speaking students should be paired with students who have limited English proficiency to work together to design a poster. The poster should illustrate the differences between Mars and Earth as described on page 59.

▶ 58   CHAPTER 3

Chapter 3
LIFE IN THE BIOSPHERE

Have you ever thought about living on another planet—maybe Mars? It's not out of the question. Humans first landed on the moon nearly 30 years ago. Thirty years from now there will likely be colonies on the moon, and perhaps on Mars as well. What will it be like, living on Mars? How is Mars different from Earth?

58   Chapter 3

CHAPTER MOTIVATING ACTIVITY

Obtain photos or examples of plants from different climates to show the students. Ask them to describe the factors of climate, such as temperature and precipitation, that they believe affect and determine the growth of each plant where it is found. You may wish to have the students write their responses in their journals.

For Your Journal

By answering the journal questions, the students can demonstrate what they know about the conditions necessary for life to exist on Earth or on another planet. Also you can use the students' answers to note any misconceptions they may have about the importance of environmental conditions in the survival of living things.

ABOUT THE PHOTOGRAPH

The large photograph shows the surface of Mars. The photograph was taken by a *Viking* spacecraft, which visited Mars in the mid-1970s. The characteristic red color of the Martian soil is caused by the amount of iron oxide, or rust, in the soil. This fact was revealed when the *Viking* spacecraft scooped up samples of the soil and tested them. The spacecraft also took many photographs of Mars, both while on the surface and while orbiting in space. The small illustration shows an artist's view of the *Viking's* arrival at Mars.

In 1992, another United States spacecraft, the *Mars Observer*, was launched on an eleven-month journey to Mars. The *Mars Observer* is programmed to map the planet's surface and study its composition, topography, and atmosphere in the most thorough investigation yet. This information is vital to future expeditions to Mars by both robots and humans.

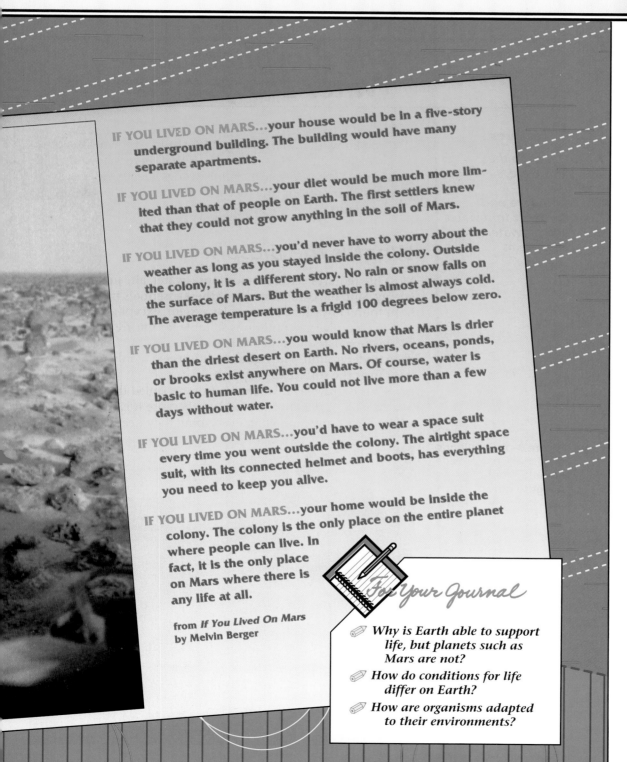

IF YOU LIVED ON MARS...your house would be in a five-story underground building. The building would have many separate apartments.

IF YOU LIVED ON MARS...your diet would be much more limited than that of people on Earth. The first settlers knew that they could not grow anything in the soil of Mars.

IF YOU LIVED ON MARS...you'd never have to worry about the weather as long as you stayed inside the colony. Outside the colony, it is a different story. No rain or snow falls on the surface of Mars. But the weather is almost always cold. The average temperature is a frigid 100 degrees below zero.

IF YOU LIVED ON MARS...you would know that Mars is drier than the driest desert on Earth. No rivers, oceans, ponds, or brooks exist anywhere on Mars. Of course, water is basic to human life. You could not live more than a few days without water.

IF YOU LIVED ON MARS...you'd have to wear a space suit every time you went outside the colony. The airtight space suit, with its connected helmet and boots, has everything you need to keep you alive.

IF YOU LIVED ON MARS...your home would be inside the colony. The colony is the only place on the entire planet where people can live. In fact, it is the only place on Mars where there is any life at all.

from *If You Lived On Mars* by Melvin Berger

For Your Journal

- Why is Earth able to support life, but planets such as Mars are not?
- How do conditions for life differ on Earth?
- How are organisms adapted to their environments?

CHAPTER 3 59

Section 1:

THE BIOSPHERE AND WATER ECOSYSTEMS

FOCUS

In this section, the dimensions of the biosphere and what it provides for living things to survive are presented. Then the two major water ecosystems are introduced. The ability of both marine and freshwater biomes to support life and their interaction in estuaries are discussed.

MOTIVATING ACTIVITY

Cooperative Learning Have the students work in small groups to make a poster that illustrates the environmental conditions necessary for life. The students can cut pictures from magazines and arrange them with labels that describe how each pictured item is important to living organisms. The groups can display their environmental posters in the classroom. Ask the students to compare the pictures that the groups chose to represent necessary environmental conditions.

PROCESS SKILLS
- Comparing • Observing

POSITIVE ATTITUDES
- Caring for the environment
- Curiosity

TERMS
- biosphere • biomes
- climate

PRINT MEDIA
The Glass Ark: The Story of Biosphere 2 by Linnea Gentry and Karen Liptak (see p. xixb)

ELECTRONIC MEDIA
The Living Earth, National Geographic (see p. xixb)

Science Discovery Intertidal; California Estuary; mangroves

BLACKLINE MASTERS
Study and Review Guide
Laboratory Investigation 3.1, 3.2
Thinking Critically
Extending Science Concepts

INTEGRATION—Language Arts

After the students have discussed what the biosphere is, challenge them to look up the word *biosphere* in a dictionary and find out what the two parts of the word mean (*bio:* life; *sphere:* zone, globe, area). Then call on a volunteer to use the information to create a definition of his or her own.

SECTION 1

The Biosphere and Water Ecosystems

Objectives

Describe the biosphere.

Compare and contrast freshwater and saltwater ecosystems.

You know that Mars is very different from Earth. One of the biggest differences, of course, is that there is no known life on Mars. On Earth, living things are found nearly everywhere—from the deepest oceans to the highest mountains.

What Is the Biosphere?

Most life requires certain conditions—water, air, and a source of energy. Suitable combinations of these essentials have not yet been found anywhere in our solar system, except on Earth.

Even on Earth, conditions suitable for life are not found everywhere. Life cannot exist high in the upper atmosphere or deep underground. The right conditions exist only in a narrow layer near the surface of the earth.

This narrow layer where life exists is called the **biosphere.** Life exists only within the biosphere because it is the only place where organisms can obtain everything they need to survive.

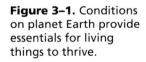

Figure 3–1. Conditions on planet Earth provide essentials for living things to thrive.

▶ 60 CHAPTER 3

TEACHING STRATEGIES

- **Process Skills:** *Comparing, Expressing Ideas Effectively*

Point out to the students that to us the limits of the biosphere may seem broad, but in comparison to the total size of the earth, this area is really quite small. Ask the students for examples of places on Earth where they think that life cannot exist and to explain their reasons for choosing those places. (Answers may include areas of intense heat and cold or areas that lack oxygen, water, and food.)

- **Process Skills:** *Communicating, Generating Ideas*

After the students complete the *Discover By Doing* activity, ask them to think of another visual method they could use to show the relative size of the biosphere. Have them describe and sketch their idea in their journals.

An explorer to Mars leaves the biosphere behind. But an explorer's needs are the same on Mars as they are on Earth. On Earth, the biosphere provides all the things a person needs to survive. On Mars, these things must be provided for the explorer. Mars explorers must take a substitute for the biosphere with them.

Even on Earth, life is not found everywhere. A few organisms live as high as 8 km above sea level, while a few others live 10 km below sea level. However, most life is found from about 110 m above to about 110 m below sea level. The activity that follows can help you visualize the relative size of the biosphere.

Figure 3–2. How does a spacesuit provide an astronaut with an Earthlike environment? ①

SCIENCE BACKGROUND

Some students may assume that the term *below sea level* refers only to the oceans. You may wish to explain that some land areas may also be below sea level. For example, Death Valley in the United States is 86 m below sea level. About two-fifths of the Netherlands in Europe is below sea level.

① A spacesuit environment duplicates the conditions of Earth by regulating temperature, pressure, and air composition.

⭐ DISCOVER BY *Doing*

You may wish to do this activity as a group or whole class activity to save paper and thus further demonstrate wise interaction with the environment. Be sure the students understand that the strips of paper should be taped together end to end. When the students finish, they should have one long, thin piece of paper that is only one strip wide. The students should note that the portion of their diagram from 110 m above sea level to 110 m below sea level, or about 1 percent of the colored area, is the portion of the biosphere where most life is found.

⭐ Doing

You will need notebook paper, tape, and a marker or colored pencil. Cut several sheets of notebook paper lengthwise into strips about 5 cm wide. The distance between lines on the strips of paper will represent 0.1 km. Tape enough strips of paper together end to end to represent 10 km. Mark the top line *sea level*. This represents the depth of the oceans. Add other sheets of paper to represent 10 km above sea level. This additional paper represents part of the atmosphere.

Using the marker or colored pencil, color the area of your paper that is between 10 km below sea level and 8 km above sea level. The colored area represents the biosphere. In what percent of the "biosphere" is most life found? ✏️

SECTION 1

TEACHING STRATEGIES, continued

● **Process Skills:** *Classifying/Ordering, Communicating, Applying, Evaluating*

Show the students a set of Russian nesting dolls or another set of items that fit inside of each other. Ask the students to apply the idea of the nesting dolls to the biosphere. (Accept all reasonable responses that refer to a progression from large to small. A possible response is the earth, the United States, a state, a county, a city or town, and a yard.)

● **Process Skills:** *Inferring, Analyzing*

Emphasize that weather is the present condition of the atmosphere, and climate is the general annual weather pattern. Ask the students what environmental conditions are associated with climate. (amount of annual precipitation, average annual temperature, available sunlight, and winds)

---

① The dominant plants are grasses.

INTEGRATION—Art

Cooperative Learning Divide the class into groups of four or five students. Members of each group should work cooperatively to make murals that show how organisms, populations, communities, and ecosystems go together to form the biosphere. Suggest that the groups brainstorm their ideas and agree on the idea they like best. Then individual members could take on specific responsibilities. For example, one member could make a sketch of the mural. One or two others could decide how best to make it. The remaining members could gather the materials. All members might come together to actually complete the mural.

▶ 62 CHAPTER 3

The biosphere is so vast and complex that it would be difficult to study it all at one time. However, the biosphere can be divided into smaller segments to make studying it simpler.

Remember, organisms of the same species living together make up a population, and different populations living together make up a community. Remember too that a community and its physical environment describe an *ecosystem*.

Major land ecosystems are called **biomes.** Each is identified by the dominant plants found there. For example, trees are the dominant plants in a forest biome. What do you think the dominant plants are in a grassland biome? ①

The factor that most affects the plants in a biome is climate. **Climate** is the long-term weather patterns that occur in an area. Climate depends mostly on temperature and precipitation. Generally, climates become colder as you move farther away from the equator. Coastal climates are usually warmer and wetter than inland climates. And one side of a mountain usually receives more precipitation than the other side.

Figure 3–3. Earth has many different kinds of climates.

Temperature and precipitation also determine the kinds of living things found in a biome. Plants and animals are adapted for certain climates. Cacti, for example, grow well in dry ecosystems, while ferns grow best where it is wet.

The biosphere is made up of six major land biomes plus all the water ecosystems. Look at the map of major land biomes in Figure 3–4. In which biome do you live? What is the climate like where you live? ②

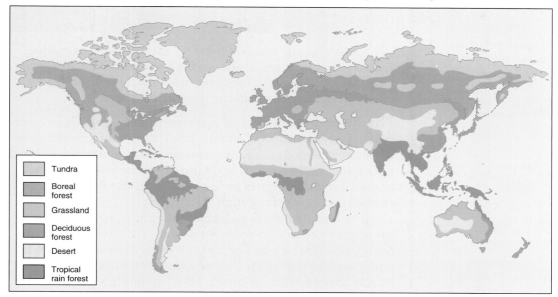

Figure 3–4. Earth's major biomes

 ASK YOURSELF

What factors help to determine what biome will be found in a particular area?

Marine Ecosystems

Biomes are defined as major land ecosystems. But nearly three-fourths of the earth is covered by water. While not true biomes, bodies of water are important ecosystems that contain a wide variety of living things. Oceans, seas, lakes, and rivers provide food and other resources.

Although water ecosystems, like biomes, have dominant plants, they are most often identified as freshwater ecosystems or saltwater ecosystems. The water in lakes, ponds, rivers, and streams is fresh, while the water in oceans and seas is salty.

TEACHING STRATEGIES, continued

● **Process Skills:** *Comparing, Inferring, Analyzing*

The students' interest in marine and freshwater ecosystems may be tied to recreational pursuits. Capitalize on this interest by asking the students to share relevant experiences. They may be able to describe animal and plant life they have observed. Ask the students how limiting factors affect the ability of organisms to thrive in a freshwater or marine ecosystem. (The students should indicate awareness of the adaptive ability of organisms to a narrow range of conditions.)

● **Process Skills:** *Classifying/Ordering, Inferring*

Point out to the students that the intertidal zone supplies many land mammals and birds with food. Have the students explain why the intertidal zone can be thought of as part of the base of a food pyramid. (Small marine organisms are often found in pools of water that remain during low tide. These organisms are fed upon by larger organisms, such as crabs and sea birds.)

 Involve the students in a discussion of how the marine ecosystem is being polluted by the dumping of sewage, garbage, industrial waste, and agricultural chemicals. Encourage the students to think about how this pollution affects the organisms that live in the ocean, the food that we eat, and the beaches that we visit. Suggest that the students look for magazine and newspaper articles about the subject, bring them to class, and share them with classmates. Encourage the students to focus on ways that marine pollution is being fought and, in particular, on ways that the students themselves can help with the problem.

 **LASER DISC**
4313
Intertidal; California

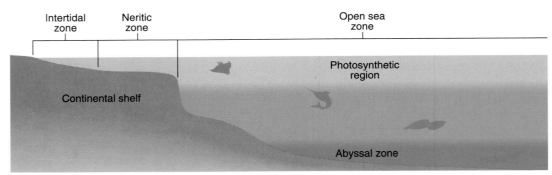

Figure 3–5. Zones of ocean life

Figure 3–6. Organisms of the intertidal zone are adapted to harsh conditions.

Earth's oceans and seas make up the saltwater, or marine, ecosystems. In these ecosystems, life is found in four zones: the *intertidal zone,* the *neritic* and *open sea zones,* and the *abyss*. Locate each of these zones in the illustration above.

The Intertidal Zone If you've ever walked along an ocean beach or seen pictures of surfers, you know that the water near the shore is seldom calm. As waves rush in, the beach is covered with water, only to be left high and dry as the water falls back again. Here communities are exposed to air during low tide and covered by water during high tide.

The intertidal zone is closest to the shore, running down the edge of the land to the low-tide line. This zone can be a difficult place in which to live. Organisms must be adapted for life underwater as well as for exposure to air. These organisms must be able to withstand the drying effects of bright sunlight and the pounding of ocean waves. This zone is rich in nutrients that are washed down from shoreline communities. Barnacles, clams, crabs, and seaweed are some of the organisms that live in the intertidal zone.

● **Process Skills:** *Identifying/Controlling Variables, Applying*

Discuss the abundance of life forms in and the importance of the neritic zone to the marine ecosystem. Have the students enumerate the tremendous variety of organisms, including plants, fishes, reptiles, and mammals, that inhabit this ecosystem. Have the students describe the conditions in the neritic zone that account for its abundance of organisms. (The medium depth and mildness of the waves help maintain constant temperatures; nutrients are brought in from the intertidal zone.)

● **Process Skills:** *Inferring, Analyzing*

Phytoplankton, also known as green or plant plankton, form the basis of the marine food web. Ask the students to explain why phytoplankton are not normally found below 100 m in the ocean. (Below 100 m there is not enough sunlight for photosynthesis to take place in phytoplankton, so very few of them are present at that depth.)

The Neritic and Open Sea Zones The neritic zone starts where the intertidal zone ends. Plants and animals of the neritic zone have a more constant environment than those of the intertidal zone. Water always covers the neritic zone. There are no breaking waves, and the temperature is fairly constant. The photographs below show some of the organisms found in the neritic zone.

Figure 3–7. Plankton like those shown here float in the neritic and open sea zones.

Beyond the neritic zone is the open sea zone. The open sea zone begins where the continental shelf drops off. The neritic and open sea zones share many of the same organisms. The most important organisms of these zones are the microscopic plankton, which drift with the ocean currents.

In the upper, or photosynthetic, region of the open sea zone, green plankton make their food as land plants do—by photosynthesis. Plankton are the basis for all the ocean food chains. Green plankton are also responsible for much of the oxygen in Earth's atmosphere.

Animal-like plankton eat green plankton. They, in turn, are eaten by larger, free-swimming organisms. Through these actions, energy from the sun, converted by green plankton to food energy, is made available for a large and complex web of marine life. Many kinds of free-swimming organisms—mammals, turtles, and fish—live in the neritic and open sea zones.

INTEGRATION—*Art*

Ask the students to write their own descriptions of each zone in the marine ecosystem in their journals. Suggest that they also draw pictures of organisms that live in each zone. Encourage them to use reference sources to find organisms in addition to those mentioned in the text. Point out that the students should try to organize the information in a way that will help them remember the characteristics of each zone. In addition, the students can create a diagram that illustrates the importance of green plankton in ocean food chains. Ask the students to share their diagrams with a partner and comment on the clarity and effectiveness of each other's diagram.

REINFORCING THEMES— *Environmental Interactions*

When discussing the students' journal entries made in the *Integration* activity, stress the role of green plankton in the environmental relationships of the marine ecosystem's food chain. Also encourage the students to recognize the plankton's impact on land biomes. Guide the students to an understanding of green plankton's environmental relationships within the entire biosphere.

TEACHING STRATEGIES, continued

● **Process Skills:** *Formulating Hypotheses, Generating Ideas*

Point out to the students that warmer surface waters and colder deep waters do not mix. The denser cold waters of the abyss support a lesser variety of life forms than other marine zones. Ask the students to suggest adaptations of abyssal organisms to their cold, dark environment. Encourage the students to use reference sources. (Answers might include dark coloration, some sort of bioluminescence, or, in some cases, a lack of eyes.)

● **Process Skills:** *Inferring, Analyzing*

Ask the students why a fast-moving stream supports less life than a quieter body of fresh water. (Fast-moving water carries away the nutrients needed to support life.) Ask the students why little photosynthesis occurs in a fast-moving water ecosystem. (Most nutrients needed for plants to produce food and grow are washed away by fast-moving waters.)

THE NATURE OF SCIENCE

The development of remote-operated vehicles has broadened exploration of the ocean depths. Aided by spin-off space technology, oceanographers, such as Robert Ballard, finder of the shipwrecks *Titanic* and *Bismarck*, are now going where no one has gone before. ROVs equipped with cameras, manipulator arms, and remote sensors now give both scientists and students a view into a world hidden in great depths and darkness. Oceanographers using this technology have already found strange deep-sea animals, undersea volcanoes, thermal vents, and other previously unseen features of the abyss.

Ongoing Assessment
 ASK YOURSELF

The neritic and open sea zones receive more light, are warmer, and have a greater diversity of living organisms than the cold and dark abyss region of the ocean.

SCIENCE BACKGROUND

Although fast-moving streams have fewer nutrients than quiet waters, they have a higher oxygen content. Waters are continually oxygenated as they churn and bubble. When they deliver water to quiet bodies, such as slow rivers, ponds, or lakes, they provide minerals and oxygenated water.

▶ **66** CHAPTER 3

Figure 3–8. This strange-looking fish is found in the abyss.

The Abyss Below 2000 m, the temperature of the ocean remains just above 0°C. No sunlight penetrates there, so photosynthesis cannot take place. This is the abyss. The abyss, which extends to the ocean bottom, is the largest zone in the marine biome. However, there is little life here. The few animals of the abyss are mostly scavengers, feeding on dead organisms that rain down from above. Jacques Cousteau describes the abyss:

> *Half of the earth's surface is covered by more than 13,000 feet of water. It is the abyss—vast plains interrupted by volcanoes, rugged mountain ranges, and great scarring fractures. No humans have walked here as they have on the moon. Only a few humans in bathyscaphs have visited it. Our knowledge and understanding of life in this hostile world is derived almost entirely from a limited number of automatic photographs or samples taken almost at random in trawl nets.*

▼ **ASK YOURSELF**

How is the abyss different from the neritic and open sea zones?

Freshwater Ecosystems

If all the water in the oceans is salty, how could rivers and lakes be fresh water? In his book *Secrets of Rivers and Streams,* naturalist Peter Swensen describes the water cycle.

> *Fresh pure water is constantly evaporating from the oceans and from the land masses of our earth. When the concentration of this water vapor becomes large enough in the atmosphere, it condenses around wind-blown grit and dust particles. These droplets of water are carried by winds and eventually return to the earth as rain or snow. Some of this precipitation immediately soaks into the soil to become ground water or falls at an elevation cold enough to keep it frozen as glacial ice or snow. The rest of the precipitation fills ponds and lakes or runs into streams.*

Taken together, the freshwater ecosystems make up only a tiny fraction of the earth's water ecosystems, but they provide many different environments. Fast-moving bodies of water, such as streams, contain little life. Slow-moving bodies of water, such as rivers, contain more living things.

● **Process Skills:** *Comparing, Predicting*

Ask the students whether they would expect fishing to be better in a lake or a fast-moving stream. Why? (Fishing would probably be better in a lake. A fast-moving stream carries away more nutrients than a nonmoving lake. Less food means there will be less life, including fish, in the fast-moving stream.)

● **Process Skills:** *Comparing, Predicting*

Ask the students whether they would expect to find more water plants growing in the shallow or the deep areas of a lake. Why? (More water plants would grow in the shallow areas of a lake because that is where photosynthesis can occur, and plants need photosynthesis in order to live and grow. Also, since most plants need to be anchored in soil, those with shorter stems would be growing in shallower waters.)

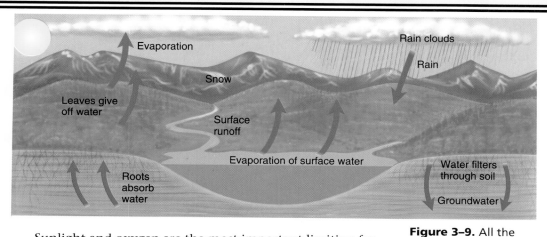

Sunlight and oxygen are the most important limiting factors in quiet waters. Lakes and ponds are rich in nutrients because dead organisms fall to the bottom and decompose. Also soils and sediments washed from their banks remain in the water. Plants near the surface grow well in the nutrient-rich water, but these plants keep the light from reaching the deeper parts of the water. Therefore, photosynthesis occurs only at shallow depths.

Decomposers remove oxygen from the deeper waters and add carbon dioxide. The low oxygen and high carbon dioxide levels make the deeper water less suitable for animal life. Changes in temperature during each spring and fall cause water currents to bring the nutrients toward the surface. With a fresh supply of nutrients, the producers continue to grow, keeping the ecosystem in balance.

Figure 3–9. All the water on Earth is recycled over and over.

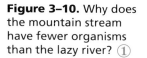

Figure 3–10. Why does the mountain stream have fewer organisms than the lazy river? ①

Demonstration

You will need a dishpan, a large piece of plastic wrap that will fit over the top of the dishpan, tape, a short glass, some salt water, and two small, clean stones. Put the salt water in the dishpan, the glass in the middle of the dishpan, and a stone in the glass to keep it from floating. Stretch the plastic wrap over the dishpan and tape it in place. Put the other stone on the plastic wrap over the glass. Put the setup in the sunshine, and leave it there for two days. Have the students observe what forms on the plastic wrap. (fresh water)

① In a mountain stream, fast-moving water carries nutrients downstream. Organisms will be found where nutrients are plentiful, as in a lazy, slow-moving river.

MEETING SPECIAL NEEDS

Second Language Support

Limited-English-Proficient students can make a diagram of a freshwater ecosystem that is common in their country of origin. Encourage them to use a foreign language dictionary to look up words and phrases they want to use in their diagrams. Have the students save their diagrams in their science portfolios.

ACTIVITY

What happens when fresh water and salt water meet?

Process Skills: Observing, Measuring

Grouping: Groups of 2 or 3

Hints

You may wish to cut the cardboard ahead of time. Be sure it fits snugly between the walls and bottom of the baking dish. If the baking dish has rounded edges between the bottom and sides, you may wish to place a strip of clay along the bottom into which to fit the cardboard.

After the cardboard has been removed, the students should observe that the color of the water changes to green as the blue and yellow water mix.

Application

Where the freshwater river enters the ocean, the ocean water immediately surrounding the river becomes less salty.

LASER DISC

4317

Estuary; mangroves

INTEGRATION—Social Studies

Have the students study a map of the United States and identify places where there might be estuaries. Suggest that they use reference sources to confirm their ideas. Have them use their information to make a table, fact sheet, or bulletin board display that shows where estuaries in the United States are located.

GUIDED PRACTICE

Write the following terms on the chalkboard: biosphere, biomes, and climate. Ask the students to define each term and explain how the terms relate to one another.

INDEPENDENT PRACTICE

 Have the students provide written answers to the Section Review and Application questions. In their journals, the students can write a paragraph about a marine or freshwater organism in which they describe its environment.

EVALUATION

Have the students describe the characteristics of the saltwater and freshwater ecosystems that make them a part of the biosphere. (Answers should include the idea that both have conditions suitable for life.)

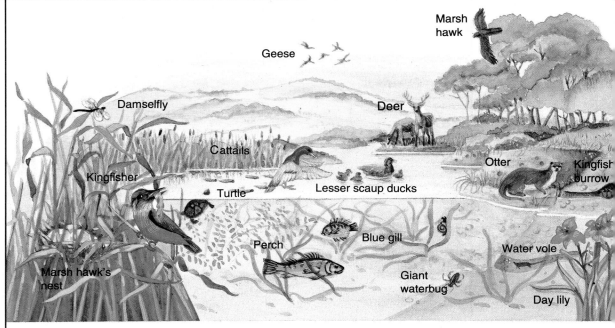

Figure 3–11. A pond is a freshwater ecosystem that contains many different populations.

Fish, turtles, and insects live in and around ponds, feeding on the numerous plants growing there. Birds are also a part of freshwater ecosystems. Even though they don't live in the water, they feed in it and are considered part of the community.

What do you think happens where fresh water empties into the ocean? The next activity can give you an idea.

ACTIVITY

What happens when fresh water and salt water meet?

MATERIALS

glass baking dish, a strip of cardboard that fits across the dish, salt, blue and yellow food coloring

PROCEDURE

1. Fill the baking dish with water. Then place the cardboard across the middle of the dish.
2. Add a teaspoon of salt and a few drops of blue food coloring to the water on one side of the cardboard. Gently stir the blue, salty water.
3. Add a few drops of yellow food coloring to the water on the other side of the cardboard. Gently stir this too.
4. Carefully remove the strip of cardboard. Describe what happens where the waters meet.

APPLICATION

How is the mixing of fresh water and salt water in the dish similar to the mixing of fresh water and salt water where a river enters an ocean?

RETEACHING

Have the students draw or construct an imaginary biosphere that includes the essential environmental conditions necessary for life. (water, source of energy, air, nutrients)

EXTENSION

Ask the students to choose one estuary and research to find out its importance to marine life.

CLOSURE

Cooperative Learning Have the students work in small groups to write a summary of the section. Then ask the groups to present their summaries to the class in any form they wish.

Where a river and an ocean meet, an *estuary* forms. Estuaries can be mud flats, salt marshes, or swamps, but in all estuaries fresh water and salt water mix, producing water that is called *brackish*. Brackish water is more salty than fresh water but not as salty as the salt water of an ocean.

The unique conditions of an estuary allow many kinds of organisms to live there. The brackish water and changes in tides create many habitats. Both freshwater and saltwater organisms inhabit estuaries.

The waters of estuaries contain many nutrients. Marsh grasses and algae are the main producers for the community. Marine animals—such as clams, crabs, shrimp, and oysters—feed on the producers. Many small fish live in estuaries, and birds, which live near estuaries, feed in the shallow waters.

Figure 3–12. Estuaries often form where slow-moving rivers flow into the ocean.

 ASK YOURSELF

How are stream and pond ecosystems different?

SECTION 1 REVIEW AND APPLICATION

Reading Critically
1. What is the biosphere?
2. How are freshwater ecosystems and saltwater ecosystems alike? How are they different?
3. Why do fast-moving waters, such as streams, have less life than still waters, such as ponds?

Thinking Critically
4. Why is the water of an estuary less salty than the water of an ocean?
5. Why do estuaries support such a wide variety of living things?
6. What do you think would happen if you placed a saltwater organism, such as a starfish, in fresh water?

Ongoing Assessment
ASK YOURSELF

Stream ecosystems tend to support life only in those places where the water is relatively slow-moving. Pond ecosystems with their still water tend to support life throughout the ecosystem.

SECTION 1 REVIEW AND APPLICATION

Reading Critically

1. It is a narrow layer near the surface of the earth where organisms can obtain everything they need to survive.

2. Both have water and organisms, but the presence of salt in one predetermines which organisms can live there.

3. The fast-moving waters of streams do not allow many green plants to grow, which results in less food for animals.

Thinking Critically

4. The water of an estuary is less salty than the water of an ocean because it is a mixture of ocean water and fresh water.

5. The freshwater and saltwater mix and changing water level in estuaries create a variety of habitats that allow many different organisms to live there.

6. An organism adapted to live in salt water could not live in fresh water. The starfish would die.

SECTION 1   **69**

INVESTIGATION

Observing Changes in Ecosystems

Process Skills: Measuring, Observing, Analyzing

Grouping: Groups of 3 or 4

Objectives
- **Observe** changes over time in pond water.
- **Relate** pond water observations to ecological changes.

Pre-Lab
Have the students read the procedure and then predict what they think they will see in the pond water at the beginning and at the end of the week.

Hints
When collecting pond water, you may wish to take samples from different layers of the pond in order to compare the organisms found at the bottom of the pond with those at the surface. If possible, take the students on a field trip and have them collect their own samples.
CAUTION: If leeches are present in water samples, instruct the students not to handle them.

Analyses and Conclusions
1. Distilled water and boiled, dried grass or straw were used to avoid contaminating the pond water samples with foreign organisms.

2. The students should notice that the number of organisms in the culture will increase.

3. Responses should include that the water becomes darker over the seven-day period.

Application
Prey populations were consumed by predators. The numbers of prey were reduced by predation; predator populations decreased and were then replaced by other organisms.

Discover More
Responses should reflect a prediction that the number of organisms would be reduced. Guide the students as necessary on setting up the additional experiment.

Post-Lab
Ask the students to compare their predictions to their observations and conclusions. Were their predictions correct?

After the students have completed the *Investigation*, ask them to place their recorded observations and answers in their science portfolios.

INVESTIGATION

Observing Changes in Ecosystems

MATERIALS
- container with lid, 5L • dried straw or grass that has been boiled • distilled water at room temperature • wax pencil • graduate, 10 mL • medicine dropper • pond water • hammer • nail • microscope • slides and coverslips

▼ PROCEDURE

1. Place the previously boiled grass or straw in a clean 5-L container.
2. Fill the container two-thirds full with distilled water. Mark the water level with a wax pencil. On the side of the container, write the names of your lab group members.
3. Add 5 mL of pond water to the container. Using the hammer and nail, punch holes in the lid and loosely place it on your container.
4. Make a slide using the water from your culture. Follow the directions for making a wet mount in the Reference Section on page 582. Observe under both low power and high power of your microscope. Record your observations.
5. Each day for seven days, observe the pond water in the container and record your observations. If the water level drops, add more distilled water to keep the level at the mark.
6. After two days, make another slide of the water from the container. Observe under low power and high power, and record your observations.

7. Make at least three more slides of the water from the container on different days. Be sure to record your observations, especially any increase in the number of organisms observed or changes in the types of organisms you see.

▶ ANALYSES AND CONCLUSIONS
1. Explain why the water was distilled and the dried grass or straw was boiled before being put into the container.
2. What happened to the water when the pond culture was added to the container?
3. How did the appearance of the water in the container change in the seven days?

▶ APPLICATION
Explain what happened in the container in terms of ecological changes over the length of the investigation.

✶ Discover More
What do you think would happen if 100 mL of ice water or hot water were suddenly added to the container? Find out.

Section 2: LAND BIOMES

FOCUS

In this section, the relationship between climate and biome characteristics is explored. The plants and animals common to various biomes, including tropical rain forests, deciduous forests, boreal forests, tundra, grasslands, and deserts, are identified.

MOTIVATING ACTIVITY

Before the students begin to read this section, have them look at the photograph of the tropical rain forest. Elicit from the students a description of this biome in terms of its living and nonliving components. Point out that climate plays a determinant role in the characteristics of an ecosystem. Then ask the students to describe other ecosystems in terms of their living and nonliving components and to discuss the role that climate plays in determining the characteristics of each. Use this discussion as a means of introducing biomes.

PROCESS SKILLS
- Comparing • Constructing/Interpreting Models

POSITIVE ATTITUDES
- Caring for the environment
- Curiosity

TERMS
None

PRINT MEDIA
Far North: Vanishing Cultures by Jan Reynolds
(see p. xixb)

ELECTRONIC MEDIA
An Ecosystem: A Struggle for Survival, National Geographic
(see p. xixb)

Science Discovery Rainforest, tropical; Ecuador Tundra; musk ox

BLACKLINE MASTERS
Study and Review Guide
Reading Skills
Connecting Other Disciplines

Land Biomes

SECTION 2

Objectives

Identify each biome's dominant plants.

Relate a biome's dominant plants to its climate.

Ecosystems found in similar climates look alike. For example, the deserts of Australia are similar to those in the southwestern United States. Forests, grasslands, or tundra also look much the same wherever they occur. Scientists disagree about the number of distinct biomes on Earth, but the six most generally agreed upon are described in this section.

Figure 3–13. The desert in Texas (left) is similar in many ways to the desert in Australia (right).

SCIENCE BACKGROUND

Tropical rain forests have many layers, each with its own community of plants and animals. The uppermost layer, or upper canopy, is made up of the widely spaced tops of the tallest trees. Below this layer, the tops of smaller trees make up one or two lower canopies. The canopies provide so much shade that the forest floor receives less than 1 percent as much sunlight as the upper canopy. As a result, only a few bushes and herbs grow there.

Tropical Rain Forests

In a hot land near the equator, where winter never comes, a new day is beginning. The climbing sun looks close enough to touch as it turns the sky pink. Out of the mist a vast ocean of leaves appears, slashed with yellow, orange, and violet blossoms. It is the roof of the jungle.

High in the trees, the birds are about to begin their morning chorus. On the branches, monkeys sit motionless, their long tails hanging down behind them, like dark quarter notes dotting the gray dawn. Soon the sun will chase the mist and another day will begin in the mysterious green world below.

TEACHING STRATEGIES

● **Process Skills:** *Interpreting Data, Comparing*

Have the students use an almanac or another resource to find out how much annual precipitation their area receives. Ask them to compare that amount of precipitation to the amount tropical rain forests receive. (Most students will determine that their area receives far less rainfall than the rain forests do unless they live within a rainforest biome.)

● **Process Skills:** *Analyzing, Synthesizing*

Ask the students why they think there are so many different kinds of plants and animals in a tropical rain forest. (Accept reasonable responses, which may include: There is an abundant food supply; the climate is stable; plants and animals may reproduce throughout the year; the environment is stable.)

> ✦ **Did You Know?**
> Rain forests can be found in biomes outside the tropics. Olympic National Park in the state of Washington is one such rain forest. There good soil and exceptionally heavy rainfall have resulted in a temperate deciduous rain forest.

 Yes, there are tropical rain forests in Central America.

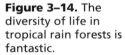 **LASER DISC**
4311, 4312

Rainforest, tropical; Ecuador

SCIENCE BACKGROUND

Some students may assume that a jungle and a tropical rain forest are the same thing. In fact, a jungle is an area of dense undergrowth within a tropical rain forest. Jungles grow in areas of the tropical rain forests where large amounts of sunlight reach the ground. Most jungles are found near rivers or in former clearings.

▶ **72** CHAPTER 3

The South American rain forest contains tall trees with large leaves, brilliant orchids, and long climbing vines. A tropical rain forest in Africa looks very much the same. The species that make up each of these forests may differ, but the effect is the same—a tropical rain forest

The Jungle, by Helen Borton, describes the mist rising from the canopy, or roof, of the forest. The term *jungle* is often used when describing a tropical rain forest. However, the jungle is only one community within this biome.

During the year, a tropical rain forest receives more than 400 cm of rain, falling nearly every day in short, intense cloudbursts. Temperatures in tropical rain forests stay about the same all year. Locate the tropical rain forests on the world biomes map on page 63. Are there any tropical rain forests in North America? ①

Scientists hypothesize that more kinds of organisms live in the tropical rain forest biome than in all the other biomes combined. Tropical rain forests probably contain thousands of plants and animals that have not yet been discovered. Tall, flowering trees shade the forest floor and provide homes for many different kinds of animals. Monkeys, apes, and lemurs are common, as are reptiles and amphibians. All flourish in the warm, wet climate.

Figure 3–14. The diversity of life in tropical rain forests is fantastic.

- **Process Skills:** *Formulating Hypotheses, Inferring*

Point out to the students that plants known as *epiphytes* grow on trees in the tropical rain forests. Explain that the roots of these plants never touch the soil and that the epiphytes depend on the trees for support but not for nutrients or water. Ask the students to hypothesize how these plants absorb water. (The students might hypothesize that the epiphytes absorb rain as it falls or that they absorb water directly from the humid air.)

- **Process Skills:** *Inferring, Predicting*

Ask the students how the loss of the Amazon rain forest might affect the entire planet. (The tropical rain forests contain a significant percentage of Earth's biological diversity and gene pool. If the rain forests vanish, so will many species. These species could have important medical or environmental purposes. The destruction of the rain forests could affect global weather patterns through the greenhouse effect.)

Tropical rain forests are being destroyed at an alarming rate. Farmers need the forest land to grow food crops, but the soil of the rain forest is very poor, so their crops do poorly. After the crops fail, the farmers clear more land, and the cycle continues. Others cut down the trees for the valuable hardwoods they contain—the lumber is often used for making furniture. Once destroyed, it takes a very long time to regrow a tropical rain forest.

Figure 3–15. Hundreds of hectares of tropical rain forest are destroyed every day.

Tropical rain forests are important to the biosphere. During photosynthesis, forest plants take carbon dioxide from the air and release oxygen. If many trees are cut down, the atmospheric balance between oxygen and carbon dioxide changes. What changes might result from an increase in the amount of carbon dioxide in the atmosphere? ②

Saving the tropical rain forests is necessary for preserving the biosphere. By doing the next activity, you can make people aware of the importance of saving this biome.

DISCOVER BY Researching

Use headlines and pictures from magazines and newspapers to make a poster about saving the tropical rain forests. In your journal, describe efforts being made to save the rain forests. Share your information and your poster with your classmates and with the people of your community.

ASK YOURSELF

Do you think it is important to save the tropical rain forests? Explain your answer.

MEETING SPECIAL NEEDS

Second Language Support

You may wish to remind the students that pictures can sometimes offer helpful clues about what a text is talking about. Have a volunteer explain what course of events is depicted in the picture on this page.

② An increase of carbon dioxide would result in an imbalance of carbon dioxide and oxygen in the atmosphere. The increased carbon dioxide might contribute to ozone depletion or global warming. In addition, a reduced amount of oxygen might not be able to support life as we know it.

DISCOVER BY Researching

Remind the students that they can use the *Readers' Guide to Periodical Literature* to identify magazines that include articles about the destruction of rain forests. Magazines such as *Time*, *Newsweek*, and *National Geographic* are possible sources.

ONGOING ASSESSMENT ASK YOURSELF

The students might suggest that the plants and animals that live in the rain forest are important resources, which need to be conserved, and that rain forests help maintain the balance between carbon dioxide and oxygen.

TEACHING STRATEGIES, continued

● **Process Skills:** *Observing, Comparing*

You may wish to ask the students how the tropical rain forest and the deciduous forest are different. (The climate of a deciduous forest is less rainy and milder, or more temperate, than that of a tropical rain forest, which is hot and humid; the deciduous forest has four seasons whereas the tropical rain forest has only one.)

● **Process Skills:** *Analyzing, Synthesizing*

Cooperative Learning Have the students work together to research and report on the ways that one plant and one animal have adapted to the deciduous forest. (Some animals hibernate in winter. Many plants die and grow from seeds in the spring or sprout again from roots. Other plants, such as trees and shrubs, become dormant.)

---

① Deciduous forests are found in the eastern United States and Canada.

SCIENCE BACKGROUND

The shorter days and cooler temperatures of autumn cause the green chlorophyll in leaves to break down. As the chlorophyll breaks down, the other colors once hidden by it are revealed. Without the chlorophyll, the leaves can no longer produce food and begin to die. The cells in the leaves' stems become weakened and the dying leaves hang from the stems by only a few strands.

ONGOING ASSESSMENT
▼ **ASK YOURSELF**

Winters are cold and snowy, whereas summers are warm with occasional rain.

Deciduous Forests

Red oaks, red maples, and ash trees were everywhere. On the forest floor [were] . . . seedlings . . . trees that like the deep shade. Every autumn the trees lost their leaves. They fell to the ground with dead twigs and branches. All of these things decayed and made a rich layer of stuff called humus. The beeches and sugar maples were the kings of the forest. It is home for many wild animals—for foxes, bobcats, wood turtles, chipmunks, bears, deer, squirrels, mice, porcupines, and many other creatures.

Have you ever sat under a big shade tree on a hot summer day? In his book *How the Forest Grew,* William Jaspersohn mentions the shade of a deciduous forest. Deciduous forests are shady and damp. They receive about 75 cm of precipitation each year. The climate of deciduous forests is temperate—winters are cold and snowy, while summers are warm with occasional rain. Look again at the world biomes map on page 63. Where in North America are deciduous forests located? ①

Figure 3–16. Fall colors in a deciduous forest

Oak, hickory, beech, maple, and other useful hardwood trees make up the dominant plants of the deciduous forest biome. Deciduous trees lose their leaves each fall.

A wide variety of shrubs and many types of ferns and fungi live on the forest floor. The animals of the deciduous forest include bears, deer, bobcats, snakes, foxes, squirrels, and many birds.

 **ASK YOURSELF**

How do winter and summer differ in the deciduous forest?

- **Process Skills:** *Analyzing, Comparing*

Discuss with the students the relationship between temperature and precipitation and the types of vegetation that are dominant in each of the three types of forests. Ask the students which of the three forests is most biologically productive and why. (The tropical rain forest receives the most precipitation and has the widest variety of plant and animal life. Decomposition and growth are more accelerated due to the high temperatures and amounts of moisture.)

- **Process Skills:** *Comparing, Observing*

Ask the students how boreal forests differ from deciduous forests. (Accept reasonable responses, such as: Conifers are the dominant trees in boreal forests; broadleaved trees are the dominant plant in deciduous forests. The boreal forest has less rainfall and a colder climate than the deciduous forest. Boreal forests are located farther north than deciduous forests.)

Boreal Forests

They opened the bars at the entrance to an old wood-road, and stepped into the cool shadow of the pines. It was quiet in the woods, and dark, except where filtered patches of sun came through to sprinkle the brown needle carpet with flecks of gold. A soft sound, like the hush-sh-sh. . . hush-sh-sh *of waves on a beach, descended from the far green roof of the forest.*

In his book *Lumberjack,* Stephen W. Meader describes the hush of the pine forest. Try to find a stand of pines near your home and listen to the gentle sounds of wind blowing softly through the trees. *Conifers*—trees that produce cones—are the dominant plants of the boreal forest biome, or taiga (TY guh). The leaves of most conifers are modified into needles. Unlike the leaves of deciduous trees, needles are replaced throughout the year, so most conifers are evergreens.

Figure 3–17. The moose is so common in boreal forests that they are sometimes referred to as "spruce-moose" forests.

The climate of the boreal forest is cold and snowy in the winter and cool in the summer. Annual precipitation is only about 60 cm.

Animals of the boreal forest include snowshoe hares, moose, bears, beavers, deer, and many kinds of birds. Look again at the world biomes map on page 63. At what latitudes are most boreal forests found? ②

▼ ASK YOURSELF

Do boreal forests receive more or less precipitation than deciduous forests?

SCIENCE BACKGROUND

The boreal forest, also called the taiga or northern coniferous forest, is the largest land biome in the world. The productivity of coniferous trees provides the bulk of the world's lumber and pulp wood.

② They are located in the high northern latitudes in the northern hemisphere.

SCIENCE BACKGROUND

Moose, the largest members of the deer family, live in boreal forests in North America, Europe, and Asia. In Europe, moose are called "elk." Like the solitary moose in the picture, most moose live alone. Moose do not travel in herds as many other members of the deer family do.

ONGOING ASSESSMENT
▼ ASK YOURSELF

Boreal forests receive less precipitation (60 cm) than deciduous forests (75 cm).

TEACHING STRATEGIES, continued

● **Process Skills:** *Inferring, Analyzing*

The students generally associate the arctic regions of the earth with deep snow. The tundra may be the best example of how temperature and precipitation interact to determine the amount of available water. Ask the students why so little water is available in the tundra if cold temperatures slow the evaporation of water. (There is little precipitation, and much of it remains frozen most of the year.)

● **Process Skills:** *Comparing, Analyzing*

Ask the students to describe the differences between the tundra and the boreal forest that account for the ability of trees to grow in one biome and not the other. (Summers in the boreal forest are long enough for the ground to thaw, the temperatures are warmer year round, and there is more precipitation than in the tundra. These factors make it possible for trees to thrive.)

MULTICULTURAL CONNECTION

 Point out to the students that there are native peoples who live in the tundra. Suggest that they do some research to find out how the natives of the tundra have adapted to their harsh environment. How do they protect themselves from cold? How do they get their food? How do they travel from one place to another? Have the students write their answers in their journals.

LASER DISC

4268

Tundra; musk ox

BACKGROUND INFORMATION

The Call of the Wild, Jack London's most famous novel, is the story of a dog named Buck. Buck is taken to the Yukon Territory in Canada where he strives to survive in the harsh environment. The dog eventually is accepted by and leads a wolf pack in the wild. London wrote another novel, *White Fang*, and a collection of stories, *The Son of the Wolf*, that also have a tundra setting.

① There are fewer landmasses near the South Pole than near the North Pole.

ONGOING ASSESSMENT
ASK YOURSELF

Tundra plants grow only a few centimeters high because of the cold dryness and the short growing season.

▶ **76** CHAPTER 3

Tundra

It was beautiful spring weather, but neither dogs nor humans were aware of it. Each day the sun rose earlier and set later. It was dawn by three in the morning, and twilight lingered till nine at night. The whole long day was a blaze of sunshine. The ghostly winter silence had given way to the great spring murmur of awakening life. This murmur arose from all the land, fraught with the joy of living. It came from the things that lived and moved again, things which had been as dead and which had not moved during the long months of frost.

Remember how you felt on the first warm day last spring? In *The Call of the Wild*, Jack London describes the coming of spring to the Arctic tundra. The tundra biome is found near the earth's poles. Find the tundra on the world biomes map on page 63. Why is there little tundra in the Southern Hemisphere? ①

Winters on the tundra are long, dark, dry, and very cold—temperatures may drop below –40°C. Precipitation, mostly snow, totals only about 25 cm a year. The ground, which is frozen solid during most of the year, thaws a little on top during the summer. Below this thin layer of soil is the *permafrost,* so called because it never thaws.

Due to the extreme cold, strong winds, and scant precipitation, tundra plants grow only a few centimeters above the ground. Grasses and shrublike trees, mostly willows and birches, are the dominant plants of the community. The grasses and tundra wildflowers must sprout, grow, and reproduce quickly during a summer season that may last only a month or two.

Tundra animals include caribou, musk oxen, and Arctic hares, which feed on the grasses and shrubs. Birds such as puffins and ptarmigans (TAHR muh guhns) live on the tundra, while others, such as Canada geese and Arctic terns, nest there only during the summer. Many insects also inhabit the tundra during the short summer.

Figure 3–18. For all its harshness, the tundra has a large variety of plant and animal life.

 ASK YOURSELF

Why are tundra plants usually only a few centimeters high?

Grasslands

The prairie was glorious, ablaze with wildflowers and overrun with game. The buffalo grass was the best, alive as an ocean, waving in the wind for as far as his eyes could see. It was a sight he knew he would never grow tired of. Everything was immense. The great, cloudless sky. The rolling ocean of grass. Nothing else, no matter where he put his eyes. No road. No trace of ruts for the big wagon to follow. Just sheer, empty space.

Looking at a lawn or the neatly trimmed grass of a park, you may find it hard to imagine a place where a person has to mount a horse just to see over the grasses. But if you have read *Dances With Wolves* by Michael Blake, you know that the prairie, or grassland biome, is dominated by two-meter-tall grasses and similar plants. Corn, wheat, sorghum, and other grains also grow well in the rich prairie soil. Locate the grasslands on the map on page 63. In addition to the prairies of North America, what other continents have large grasslands? ②

Figure 3–19. Most of the world's major food crops come from the grassland biome.

Winters on the prairies are cold and snowy, while summers are hot and dry. Grasslands receive about 50 cm of precipitation each year, but much of it falls as winter snows. There is generally not enough water for large forests to grow on the grasslands, except near rivers and streams.

Bison and antelope were once common on the American prairies. Today, these populations have been replaced by herds of domestic cattle and sheep. Many small animals, such as ground squirrels, prairie dogs, birds, and insects, also inhabit grassland biomes. In the next activity, you can find out about the root structure of grasses.

ACTIVITY

What is the root structure of grass plants like?

Process Skills: Observing, Comparing

Grouping: Groups of 3 or 4

Hints

If the students are collecting the grass samples, suggest that they dig up the grass to keep the root system intact instead of pulling it out of the soil. If grasses are collected ahead of time, keep their roots moist until the students are ready to use them. Rinse the dirt from the roots to make it easier for the students to observe them.

▶ **Application**

1. The fine, densely packed roots of grasses spread under the soil, holding it in place.

2. The roots of grasses are made up of many densely packed shoots that work together to keep the soil in place. *Grassroots* refers to common people working together to achieve a common goal.

ONGOING ASSESSMENT
▼ **ASK YOURSELF**

Grasslands do not receive enough rainfall for trees to grow.

❖ Did You Know?

About one seventh of the earth's land area is covered by deserts. The largest desert is the Sahara in northern Africa. It covers about 9 million km². Other large deserts include the Australian Desert, Arabian Desert, and the Gobi in China and Mongolia. In the United States, deserts cover about 1.3 million km².

GUIDED PRACTICE

Have the students look at a physical map of the United States and compare the biomes on the east and west sides of the Sierra Nevada range.

INDEPENDENT PRACTICE

Have the students provide written answers to the Section Review and Application questions. Have the students summarize the characteristics of each of the six major land biomes by writing descriptions in their journals.

RETEACHING

Have the students construct a chart in their journals that identifies plants, animals, and precipitation for the six land biomes.

ACTIVITY

What is the root structure of grass plants like?

MATERIALS
hand lens, several samples of grass, ruler

PROCEDURE

1. Use a hand lens to examine each sample of grass. Observe the roots. How are the roots shaped?
2. Use a ruler to measure several roots.
3. Observe and measure several blades of grass. How do the measurements of the roots and blades compare?
4. Draw each grass sample and label the roots and blades.

APPLICATION

1. What makes grasses especially useful for keeping soil in place?
2. Look up the term *grassroots* in the dictionary. Describe how the structure of the roots of grasses might have led to the use of this term.

▼ **ASK YOURSELF**

Why are trees unable to grow in the grassland biome?

Deserts

The first explorers found the desert a vast and varied place. They crossed flat white valleys shimmering with heat. Spidery gray creosote bushes dotted the bare ground, each off by itself as if guarding its space. Strange poisonous snakes and insects flourished there. On the horizon, the dark, wrinkled mountains looked scorched. The early explorers rode under giant arches of red rock. They climbed vast blocks of stone so flat they called them mesas—"tables." In other places, the land sank into canyons with rivers twisting at their bottoms.

Sheila Cowing describes the vastness of the American Southwest in *Searches in the American Desert,* but not the climate. Most people think of deserts as dry and hot! But there are two kinds of deserts—hot and cold. Hot deserts have an average temperature of 21°C, while cold deserts have an average temperature of 10°C. All deserts receive less than 25 cm of precipitation during the year.

EVALUATION

Have the students use the world biomes map on page 63 and a world map or globe with longitude and latitude lines to help them answer these questions. Which continent does not have any deserts? Which continents do not have a tundra biome? Which continents have all six biomes? (Europe does not have deserts; Africa, South America, and Australia do not have tundra biomes; Asia and North America have all six biomes.)

EXTENSION

Have the students research how human activities have affected the earth's biomes. (The students should show an awareness of the disruption of ecosystems as a result of urbanization, farming, ranching, industrial pollution, and harvesting flora and fauna.)

CLOSURE

Ask the students to draw a concept map that illustrates the characteristics of two land biomes. Have them save their concept maps in their science portfolios.

Deserts are found in many parts of the world. They are often found in the shadows of tall mountains that cut off moisture-laden winds. Find the deserts on the world biomes map on page 63. Which ones are near tall mountains? ①

While some hot deserts support no life, most deserts support a wide variety of living organisms. The dominant plants of a desert community include cacti, small shrubs, and a few trees. Many desert plants have spinelike leaves, and all have extensive root systems to get any available water. Some desert plants store water to be used during prolonged droughts.

Figure 3–20. What adaptations do desert plants and animals have to keep them from drying out? ②

Desert animals, like desert plants, are adapted to the dry environment. Some animals rest underground or in the shade during the day, becoming active during the cool desert evening. Lizards, scorpions, snakes, and birds, such as the comical road runner, are common in the deserts of North America.

ASK YOURSELF

Describe the characteristics of a desert.

SECTION 2 REVIEW AND APPLICATION

Reading Critically
1. Name and describe the six major land biomes.
2. How are deciduous trees different from coniferous trees?
3. In what ways are the desert and tundra biomes similar?

Thinking Critically
4. In Greek mythology, Boreas was the god of the north wind. How might this be related to the name *boreal forest*?
5. In the grassland biome, you may occasionally see clumps of trees growing near rivers or ponds. Why do grassland trees grow only near bodies of water?
6. Cactus plants are called *succulents*. The first definition of *succulent* in the dictionary is "full of juice; juicy." How does this word apply to cacti?

---

① Deserts in North America, South America, and Asia are near tall mountains.

② Desert plants and animals have adaptations to help them catch, store, or conserve vital and limited water.

ONGOING ASSESSMENT
ASK YOURSELF

Deserts receive less than 25 cm of precipitation each year and may be hot or cold.

SECTION 2 REVIEW AND APPLICATION

Reading Critically

1. Answers should include descriptions of the tundra, boreal forest, deciduous forest, grassland, desert, and tropical rain forest biomes. The biomes should be described in terms of temperature, precipitation, and vegetation.

2. Deciduous trees lose their leaves each autumn; coniferous trees do not.

3. Both have small amounts of precipitation, extremes in temperature, and plants and animals that have adaptations for survival.

Thinking Critically

4. Boreal forests are located in the high northern latitudes in the northern hemisphere.

5. Grassland biomes do not receive enough precipitation to support the growth of trees. Trees are usually found only near constant sources of water such as rivers, streams, and lakes.

6. The fleshy stems of cacti are full of water (juice).

SKILL
Modeling Climate

Process Skills: Constructing/Interpreting Models, Comparing

Grouping: Groups of 2 or 3

Objectives
- **Construct** climatograms.
- **Compare and contrast** climate data.
- **Interpret** climate data.

Discussion
The emphasis in this activity is on modeling the climate of an area using a graph of climatic data. A review of graphing techniques may be appropriate before beginning this activity. Point out that precipitation is represented by the scale on the left side of the graph, while temperature is represented by a different scale on the right side of the graph.

Application
1. Both areas show similarly low precipitation, but region B has a much higher temperature than region A.

2. The climate in region C indicates an area of four seasons, but the summer months occur in December through February. The region has more precipitation than either region A or B. Region C is warmer than region A, but cooler than region B. This region is located south of the equator, which explains the reversal of the seasons.

✷ Using What You Have Learned
The students may find temperature and precipitation data in an almanac. After they have made the climatogram for the area in which they live, have the students conclude from their data how the climate influences their biome.

★ PERFORMANCE ASSESSMENT
To evaluate the students' ability to construct line and bar graphs, check the climatograms the students made for the *Skill* activity. To evaluate their ability to interpret data on graphs, ask the students questions they can answer using their climatograms.

SKILL Modeling Climate

▶ **MATERIALS**
- graph paper • pencil

▼ **PROCEDURE**

1. A model of a region's climate can be made into a graph called a *climatogram*. A climatogram is a graph of both temperature and precipitation. Temperature is shown on a line graph, while precipitation is shown on a bar graph. You can use climatograms to relate climate to biomes.
2. The scale across the bottom shows the months of the year.
3. Precipitation is read from the scale along the left side of the graph.
4. Temperature is read from the scale along the right side of the graph.
5. Make a climatogram for each of the three regions described in the table.

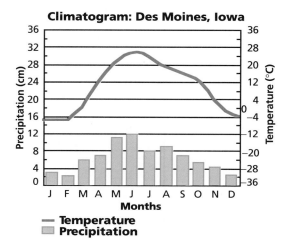

TABLE 1: REGIONAL PRECIPITATION AND TEMPERATURE													
Region	Month												
	J	F	M	A	M	J	J	A	S	O	N	D
A Precipitation (cm)	1	1	0	0	2	2	2	1	1	1	1	1
Temperature (°C)	−22	−24	−20	−16	−4	2	4	4	1	−4	−16	−20
B Precipitation (cm)	0	0	1	1	1	0	0	0	1	1	1	0
Temperature (°C)	24	26	27	28	28	27	26	28	28	28	27	24
C Precipitation (cm)	12	14	8	3	2	2	1	1	1	4	10	12
Temperature (°C)	20	18	16	14	12	10	10	12	14	16	18	20

▶ **APPLICATION**

1. Compare the climatograms for regions A and B. How are they similar? How are they different?
2. Describe the climate of region C over the course of a year. What is unusual about this climate compared to the other two? How can you explain this difference?

✷ **Using What You Have Learned**
Obtain data on temperature and precipitation for the past year in the area where you live. Make a climatogram for your area. How do you think the climate influences the biome in which you live?

CHAPTER 3 HIGHLIGHTS

The Big Idea—ENVIRONMENTAL INTERACTIONS

Point out that climate—temperature and precipitation—determines the types of plants and animals that can survive in each biome. Also help the students identify the characteristics of plants and animals that enable them to adapt to a biome.

For Your Journal

The students' ideas should reflect the understanding that all living things require water. Plants and animals have adapted to the availability of water and the temperatures in their biomes.

CONNECTING IDEAS

The students' answers should include accurate information about the climate, seasons, precipitation, plants, and animals in each of the six biomes—(1) tropical rain forest, (2) deciduous forest, (3) boreal forest, (4) tundra, (5) grassland, (6) desert.

CHAPTER 3 HIGHLIGHTS

The Big Idea

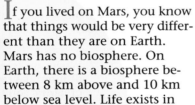

If you lived on Mars, you know that things would be very different than they are on Earth. Mars has no biosphere. On Earth, there is a biosphere between 8 km above and 10 km below sea level. Life exists in the biosphere because it has everything organisms need to survive.

Within the biosphere there are six major land biomes and two water ecosystems. Water ecosystems may be either fresh water or salt water. Within a biome, the climate and the dominant plants are related. The animals as well as the plants are adapted to the environmental conditions of the biome. Tropical rain forests, deciduous forests, boreal forests, tundra, grasslands, and deserts are the major land biomes.

Look again at the answers you wrote about why planets such as Mars cannot support life and whether all places on Earth can support the same kind of life. How have your ideas changed? How are living things adapted to their environment?

Connecting Ideas

The pictures show six land biomes found on Earth. In your journal, identify the biomes and list three physical characteristics of each. Name some of the living things that are adapted to the environmental conditions in each biome.

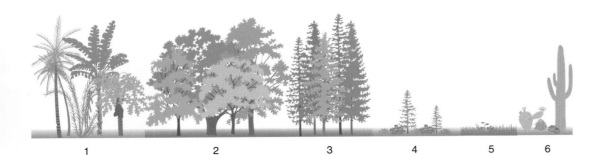

1   2   3   4   5   6

CHAPTER 3 REVIEW

ANSWERS

Understanding Vocabulary

1. The biosphere is the area of the earth that supports life. Within it are biomes, which are the major land ecosystems.

2. Biomes are the major land ecosystems. Within a biome are water ecosystems and smaller land ecosystems.

3. Biome characteristics including plants and animals are determined by climatic factors such as precipitation and temperature.

Understanding Concepts

4. (a) tall trees (b) hardwood deciduous trees (c) 60 cm (d) grasses, small shrubs, and lichens (e) 50 cm (f) cacti, small shrubs, and a few trees

Multiple Choice

5. a
6. d
7. c

Short Answer

8. Tropical rain forests are found near the equatorial region of Earth. These forests receive about 400 cm of precipitation each year. Vegetation is very dense and animal life is diverse because of the wet, warm conditions.

9. Answers should reflect an understanding of the climate and plant and animal life in each biome.

10. Each plant species within a biome is adapted to the specific temperature and precipitation provided in the biome.

11. Both are water ecosystems that have constant temperatures. Saltwater ecosystems are salty and that determines the kinds of organisms found there.

12. Desert and tundra biomes each receive small amounts of precipitation annually. Tundra biomes have lower annual temperatures than deserts. The types of animals and plants common to the biomes are different.

CHAPTER 3 REVIEW

Understanding Vocabulary

Explain the relationships between the following sets of terms.

1. biome (62), biosphere (60)
2. biome (62), ecosystem (62)
3. biome (62), climate (62)

Understanding Concepts

4. Write the letters *a* to *f* on a sheet of paper. Next to each letter, fill in the missing information from the table.

BIOME SUMMARY TABLE

Biome	Rainfall	Plants	Animals
Tropical Rain Forest	400 cm	(a)	monkeys
Deciduous Forest	80 cm	(b)	deer
Boreal Forest	(c)	conifers	moose
Tundra	25 cm	(d)	caribou
Grassland	(e)	grasses	prairie dogs
Desert	20 cm	(f)	lizards

MULTIPLE CHOICE

5. Which two biomes are similar in the amount of available water?
   a) tundra and desert
   b) boreal forest and desert
   c) desert and grassland
   d) grassland and deciduous forest

6. Which biomes are similar in temperatures?
   a) tundra and desert
   b) boreal forest and tundra
   c) desert and grassland
   d) grassland and deciduous forest

7. Where are conditions most stable for life?
   a) intertidal zone b) abyss
   c) neritic zone d) 2000 m below sea level

SHORT ANSWER

8. Describe the characteristics of a tropical rain forest.

9. Describe the differences among the various forest biomes.

10. Why are temperature and precipitation important factors in determining the dominant plants of a biome?

11. Compare and contrast saltwater and freshwater ecosystems.

12. In what ways are deserts and tundra similar? In what ways are they different?

Interpreting Graphics

13. Look at the plants in the drawing. In which biomes might you find these plants?

14. Look at the trees in the photograph. In what biome are they likely to be found? What factors of that biome influence their growth?

82 CHAPTER 3

Interpreting Graphics

13. The plants might be found in desert biomes.

14. The trees would be found in a tundra biome. Low annual precipitation, low temperatures, and a short growing season affect the growth of the trees.

Reviewing Themes

15. Deserts receive low amounts of precipitation and can experience tremendous temperature changes. The plants and animals that survive in this biome have adaptations that enable them to exist with little water and varying temperatures.

16. Organisms of the intertidal zone have adaptations, such as gelatinlike coatings, shells, and other resistant body coverings, that enable them to survive in the changing conditions of the zone.

Thinking Critically

17. The changes in temperature and precipitation as one ascends a mountain simulates conditions of various biomes. As altitude increases, temperature decreases and generally precipitation also increases. At the mountaintop where the tundra is reached, precipitation is too low to support woody plants.

18. The abyss has more in common with a quiet freshwater ecosystem because oxygen and sunlight are the primary limiting factors in both ecosystems. Both also lack movement, tides, and wind.

19. The populations of organisms living in a normally wet area that has suffered a drought would decline as would the populations of organisms living in a normally dry area that has had too much rain. If the changes lasted only a short time, the populations probably would not be affected significantly. If the changes were permanent, the plants and animals would have to adapt or move or they would die.

Reviewing Themes

15. Environmental Interactions
Explain the relationship between climate and the kinds of plants and animals that can live in the desert.

16. Systems and Structures
How are the organisms of the intertidal zone adapted to the conditions there?

Thinking Critically

17. Suppose you were hiking on a mountain and found the following pattern of biomes: a grassland at the base of the mountain, deciduous trees higher up, conifers still higher, and grasses and dwarf trees near the mountaintop. Explain why this pattern of vegetation exists.

18. Does the abyss have more in common with quiet freshwater ecosystems or with intertidal saltwater ecosystems? Explain your answer.

19. One of the areas shown below is a normally wet area that is in a drought condition, while the other area shown is a normally dry area that has been flooded by heavy rains. Explain how changes in weather patterns, such as those shown, would affect the plant and animal life in these areas. What would happen if the changes were not temporary conditions but permanent climatic changes?

Discovery Through Reading

George, Jean Craighead. *One Day in the Tropical Rain Forest.* Crowell Jr. Books, 1990. From a colony of army ants to an overview of the importance of rain forests in the biosphere, Ms. George takes readers on a memorable journey through a jungle. This is one of a series of books on nature by the same author that includes *One Day in the Alpine Tundra, One Day in the Prairie, One Day in the Woods,* and many others.

Science Parade

SCIENCE APPLICATIONS

Discussion

Ask the students to imagine and describe the sights and sounds they might encounter while walking through a tropical rain forest. Help the students conclude that rain forests contain a rich diversity of plant and animal life. Explain that one of the many reasons why scientists are concerned with the destruction of rain forests is because approximately one-half of the more than two million different plant and animal species on Earth live in the rain forests. Point out that continued destruction of rain forest habitat could lead to the extinction of many of these plant and animal species. However, extinction is not scientists' only concern.

Explain that because many rain forest organisms have never been collected, studied, or identified, scientists fear they may never have the opportunity to

---

Science Parade

SCIENCE APPLICATIONS

Studying the Rain Forest

Remember the great old jungle movies—the ones with explorers hacking their way through thick vines? You heard cries of ferocious animals and saw giant insects and millions of flowers. You saw all sorts of exotic animals and plants. Jungles, or tropical rain forests, do exhibit a great variety of life. Perhaps half of all plant and animal species on Earth live in the rain forests. For many years, scientists from all over the world have come to study rain-forest life. Now scientists are studying ways to save the tropical forests. Just fifty years ago, one-tenth of Earth's land area was covered by rain forests. Today less than half of these forests remain. Fifty years from now, there may be none.

Exciting Work

Rain forests are so large that scientists have been able to study only small portions of them at a time. Unfortunately, scientists may not have the chance to study the rain forests much longer. Large sections of rain forest are destroyed every day. Some areas are being cleared by lumber companies. In other areas, farmers are clearing rain-forest land for cattle grazing and planting crops.

Poor Soil

When farmers first began clearing the rain forests, they thought the soil would be rich in nutrients, since it supports such thick plant growth. However, farmers soon found out that the soil is too poor to grow crops for more than a year or two.

discover thousands of rain forest organisms if the destruction of the forests is allowed to continue. Encourage the students to think about and describe how the loss of rain forest organisms might affect the remaining plant and animal species on Earth.

● **Process Skills:** *Inferring, Predicting*

Remind the students that after a rain forest is destroyed and the forest soil has lost its nutrients, the land is abandoned because using artificial fertilizers to restore nutrients is too expensive. Ask the students to describe what other problems might occur if fertilizers were less expensive and could be used to supply nutrients for crops. (The students might suggest that fertilizers could contaminate the underground supply of drinking water.)

The trees of the rain forest are able to survive because their roots have certain kinds of fungi growing on them. The fungi use the nutrients in dead leaves, twigs, and fruits that have fallen to the ground. This process helps to decay materials on the forest floor. The nutrients from the decaying plants are absorbed quickly by the roots of the trees. But when the trees are harvested, most of the nutrients are taken away with them. After crops are raised for a few years, all the nutrients in the soil are used up. There is no way to replace lost nutrients, except with expensive fertilizers.

More Problems

To make the soil richer, farmers now use the slash-and-burn method of clearing the land. Trees are cut and burned. Then their ashes are used to supply nutrients for crops. Even with this method, most of the nutrients are used up after a few years, and the farmers abandon the land.

A variety of bromeliads, or air plants, grow on the rooted plants of a rain forest.

UNIT 1 85

Discussion

Point out to the students that even though the destruction of rain forests is a serious problem, it is not too late to repair much of the damage that has been done through deforestation, and that there is still hope for the future of the rain forests.

Journal Activity In a journal entry, ask the students to describe parts of Earth's environment other than the rain forests that are increasingly affected by the actions of humans. (The students might describe pollution of the oceans and contamination of the atmosphere.) Have the students review the list they wrote at the beginning of the Unit and evaluate how their suggestions would help the parts of Earth's environment that they described in their entry.

EXTENSION

Remind the students that one-tenth of our planet was once covered by rain forests, but less than one-half of these forests remain. Explain that the land area of Earth comprises approximately 200 000 000 square miles. Ask the students to determine how many square miles of rain forests still remain on Earth.

There are many kinds of animals in the rain forest, including poison dart frogs, anoles, and bark beetles.

Then the heavy rains that occur every day wash away any remaining soil.

The burning of trees also releases carbon dioxide into the atmosphere. This gas traps heat, adding to other sources of global warming.

Hope for the Future

Scientists who study rain forests know how important they are, both for the organisms living there and for the forests' influence on Earth's climate. If scientists can convince world leaders that rain forests are more valuable for themselves than for lumber or farmland, perhaps the forests and all the species that live there can be preserved for the future. ◆

This farm was once a tropical rain forest.

Read About It!

Discussion

Have the students discuss whether they think Rob and Matt are right or wrong to help the eaglet. Ask the students to describe the positive things that might happen if Rob and Matt help the eaglet. (The eaglet might survive; the experience might prove rewarding for Rob and Matt.) Then ask the students to describe the negative things that might happen if Rob and Matt help the eaglet. (The eaglet might get hurt or die; Rob and Matt might fall from the tree.)

Remind the students that the best course of action in such a scenario might be to alert authorities such as game wardens, wildlife biologists, or a local department of natural resources. These officials are experienced in dealing with wild animals, and their expertise will increase the likelihood that an animal will survive if intervention is required.

Read About It!

An Eagle to the Wind

by Nancy Ferrell from *Young World*

Rob Simon first heard the squeal as an urgent call. Though weak, the cry forced attention through the other normal sounds along the ocean shoreline.

"Listen," he said sharply, motioning for his friend Matt Mercer to stop. Squinting his eyes, Rob's head swiveled slowly, trying to pinpoint the direction. "You hear that?" he asked.

"Yes," came the reply, "but what is it?"

"Sounds like something's hurt," answered Rob. Again he listened intently until his eyes focused on a tall, scraggly Sitka spruce. Following the trunk upward with his eyes, Rob put a hand on his friend's shoulder, and pointed. In a crotch two-thirds up the spruce was something that resembled a large haystack, and, Rob felt sure, was the source of the cry.

"Why, it's an eagle's nest," he said. With a slight turn, his eyes gazed around the area from tree to tree and finally out beyond the beach. "Must be some young birds up there," he continued, "but I don't see any eagles around. Do you suppose something's happened to them?" "Don't ask me," Matt answered. "They're probably flying around getting food."

Rob shook his head doubtfully. "Maybe," he replied, "but usually one of them stays close if there are eaglets in the nest."

"Ah, come on," Matt said, hitching his pack higher on his back. "They'll be all right. Besides, we have to keep going if we want to make Yankee Cove tonight."

"There's plenty of time," Rob answered. "We've got all afternoon, and we can't just leave. What if the birds don't get any food? They'd die for sure." He let his pack drop to the ground and sat down on a boulder. "Let's stay here for a while and just watch." Matt shrugged and sat down with his friend.

For over an hour the two kept watch on the spruce and the surrounding waters, but during that time no mature eagle approached the nest.

Suddenly a raven, attracted by the young squeals, began circling the spruce high above.

"That does it!" Rob cried. He sprang up and ran shouting toward the tree

Discussion

● **Process Skills:** *Inferring, Applying*

Explain to the students that the word *aerie* refers to the lofty nest of any large bird or bird of prey. Ask the students why some birds such as eagles build their nests high in trees or on cliffs. (The students might suggest that a high nest offers greater protection for the birds' offspring by lessening the likelihood of predation by natural predators or hunters.)

● **Process Skills:** *Inferring, Applying*

Encourage the students to explain why Rob pauses to put on a pair of gloves before handling the eaglet. (The gloves protect Rob from being injured by the beak and sharp claws, or talons, of the eaglet.)

● **Process Skills:** *Analyzing, Generating Ideas*

Remind the students that Matt and Rob are worried that the eaglet may not learn how to fly without the help of its parents. Have the students consider whether they would have learned to walk without the help of their parents.

---

as Matt followed. The startled raven did not fly away, but settled in a nearby tree.

"Something must have happened to the parents," Rob stated. "They'd never let a raven this close to the nest. I'm going up to see."

Matt's mouth dropped open. "But look at that tree!" he exclaimed. "I bet you'd have to have special equipment to get up there."

"I'll have to try," Rob answered. "I have to find out."

His friend stopped him. "Wait a second," Matt said. "I think I read someplace that an eagle won't fly unless his parents teach him. If you take them out of the nest, they may never fly. You ever think of that?"

Staring at the aerie, Rob answered, "No—I didn't." Undecided, he thought for a minute. Finally, he made up his mind. "You could be right," he said, "but we can't just leave them here. Better live eagles who can't fly than no eagles at all. I'm going up."

With that, he began climbing. There were plenty of footholds on the old spruce, but several cracked when pressure was applied, and Rob had to use caution.

At last he reached the base of the aerie and was surprised at the size. The top, or platform, spread nearly six feet across, and Rob realized he had several more feet to go before he could even look into the nest.

Eventually, he stretched over a final bough and peered at the platform. There, staring with fierce dark eyes, stood a month-old eaglet, its brown flight feathers mixed with shedding, light-gray down. The fledgling half-stretched its wings nervously, and Rob caught sight of the talons gripping the nesting material. Scattered nearby and mixed with fluffy down were bits of animal bones or fish bones.

There was a bird in the nest!

"There's one here," he called down to Matt. But as he finished, his eyes caught the shape of another, smaller eaglet pushed to the rim of the aerie. Rob sensed immediately that the smaller one was dead.

Easing up and over, Rob advanced cautiously upon the young eaglet, who danced backward at his approach—never taking his eyes from the intruder. Rob stopped, fumbled in his pocket, and removed a pair of gloves. Abruptly, he lunged forward and gripped the young legs firmly. The bird fought, pecking Rob's hands, but he was too weak and too young to do any serious damage.

"I've got him," Rob called again. With some effort, he covered the talons with the gloves as best he could. Enclosing the wings with one arm, he drew the bird to his chest and zipped his jacket over the eaglet. In the darkness of this pouch, the bird calmed. Then, using great care, Rob made his way over the edge of the aerie and down the tree. Finally reaching bottom, he gently lifted out the bird, and placed him on the ground.

"The poor guy must be starving to death," he said.

"What kind of eagle is it, do you know?" Matt asked, examining the bird from a distance.

"Probably a bald eagle."

"But his head's not white."

"They don't get white until they're about four or five years old."

"How come you know so much about eagles?"

Rob smiled. "You don't have a biologist for a father without picking up something along the way," he replied.

● **Process Skills:** *Inferring, Applying*

Ask the students why the eaglet does not eat the herring until Rob breaks it into small pieces. (The eaglet is used to its parents tearing up its food; it is not old enough to tear up a whole fish for itself.)

● **Process Skills:** *Evaluating, Expressing Ideas Effectively*

Explain that when a bird *preens*, it uses its beak or bill to smooth and clean its feathers and to bring oil from its skin to its feathers. The oil helps waterproof the feathers. Preening is vital to the maintenance of a bird's feathers. Ask the students why birds spend a great deal of time preening. (Birds care for their feathers because they need the feathers for protection, warmth, and flight.)

Journal Activity

Ask the students to imagine that they are Rob and write a journal entry in which they describe what it is like to watch and care for the eaglet for one day on the camping trip. Encourage the students to include their feelings as well as their observations.

With Matt's help, Rob again tucked the bird in his jacket, "Come on," he said, heading for the water. "Tide's out, and we have to find something for the bird."

Searching the tide flats, the two found sea urchins, the meat of which the bird ate ravenously. With some nourishment, the eagle perked up, and began exploring, pecking among the exposed rocks. At one spot Rob found a dead herring, and, holding his nose, flicked it toward the bird with a stick. But it was only after Rob himself broke the meat and offered it in tiny pieces that the bird would eat. At last the fledgling seemed temporarily satisfied. It was then Matt asked, "What are we going to do with him?" "Well," Rob considered, "we can't leave it here. An animal will get it for sure. I just know something happened to the parents, or they would have returned by now." He shrugged his shoulders. "I guess we'll have to take him with us."

"You mean baby-sit an eagle?" Matt asked, shaking his head but smiling at the same time. "That ought to be a first."

Throughout the following days of camping and hiking, the eaglet rode in Rob's jacket. When free, the bird spent hours preening itself, shedding more down, while the flight feathers strengthened. Occasionally it would instinctively jump skyward in short hops, flexing its wings in preparation for flying. It was at these times that a question darkened Rob's mood—would the eagle be able to fly now that it was gone from its parents?

On the day they were to return home, Matt put the question into words. "You think the little guy will ever fly without someone showing him how? It's for sure I'm not going to run around flapping my arms to show him."

Rob laughed. "That sure would be a picture," he said. Still, his face grew serious as the worry persisted. It was later that afternoon when the two finished their trip and reached Rob's home—a comfortable house away from town. Mrs. Simon, happy at their safe return, fed them while the eaglet perched on the back porch.

"Nice-looking bird," she commented. "You'll have to fix something outside for him. Oh," she added, "better call Fish and Game, too. They'll want to know about the eagle, I'm sure."

Later, after discussing the eaglet with the Fish and Game Department, Rob received permission to take care of the bird. He was assured that the eagle should be able to fly when it was old enough—probably near the end of July or early August.

However, Rob was not entirely convinced.

During the rest of June and July, Rob set aside time each day to exercise the bird on the gravel driveway. The down of babyhood disappeared, and the bird feathered out in full, brown flight dress. The eagle allowed itself to be handled by Rob, who had learned that a heavy glove on his hand made a safe perching place, but the bird

Rob set aside time each day to exercise the eagle.

Discussion

● **Process Skills:** *Evaluating, Expressing Ideas Effectively*

Remind the students that after arriving home from the camping trip, Rob calls the Fish and Game Department to report his rescue of the eaglet. Ask the students why this was the correct thing to do. (Bald eagles are an endangered species and are protected by law; Rob may have lacked the expertise necessary to properly care for the eaglet.)

● **Process Skills:** *Inferring, Analyzing*

Ask the students why it is safer for an eaglet to be in its nest than in someone's yard. (A nest provides protection from predators; without the ability to fly, an eaglet on the ground is vulnerable to cats, dogs, and other predators.)

EXTENSION

Many plant and animal species are endangered in the United States. The bald eagle is an endangered bird species, as are the California condor, American peregrine falcon, Hawaiian hawk, and ivory-billed woodpecker. Ask the students to investigate the status of an endangered bird species.

---

kept a reserved, often unpredictable attitude.

Each day the eagle would hop about the yard, jumping first one foot, then two feet—three feet into the air, beating its wings trying to gain altitude but never really succeeding. On these occasions, which grew more frequent, Rob wondered if he had done the right thing by taking the bird, and this dark worry haunted him more and more.

"You've got to fly," he would repeat under his breath. "You've got to fly."

On the last day of July, Rob brought the bird outside, perching it on a low shed roof. Again the eagle leaped up and down, stretching its wings desperately, jumping five feet—six feet—eight feet into the air. It was at the zenith of one jump that a gust of wind caught the bird. At first clumsily, the eagle lifted into the air, rocking uncertainly as it hovered, and then with a tremendous beat of its wings, he was flying! Up, up he climbed, testing wind currents, judging wing movements, gaining confidence. Though unpracticed, the eagle demonstrated excellent muscle coordination, giving indication of the strong, proud bird he would become.

Rob laughed from pure joy as he watched. It was as if he, too, were up there. He could almost feel the rushing wind, the smooth, singing freedom of nothing to touch. He didn't want the moment to end.

Eventually, the eagle, wings spread, talons down, dropped to earth a few yards from Rob. They looked at each other for a few moments while the bird rested. But when Rob took a step forward, the eagle turned and was soon airborne again. He circled once, gliding gracefully past Rob, and then headed toward the sea.

Rob watched until the speck disappeared in the sky. Though he sensed a loss inside, he was happy, too. The eagle was free, and somehow, he felt the same way. ◆

The eagle was free, and somehow, he felt the same way.

Then and Now

Discussion

Have the students read the two biographies and discuss how the contribution of each person helped further our understanding of the environment. Ask the students to compare the background, training, and work of these two scientists.

Point out to the students that the views expressed by Rachel Carson were sometimes challenged or disputed by other scientists. Ask the students to discuss how such scientific disagreement can be healthy for the scientific community.

When discussing the work of Akira Okubo, ask the students to discuss how the study of land animals might lead to a better understanding of marine animals.

Journal Activity
Ask the students to outline the information in one of the biographies, adding any information they research on their own. Then have them plot the career and scientific contributions of either Carson or Okuba on a time line in their journals.

Then and Now

Rachel Carson
(1907—1964)

Rachel Carson published a best-selling book, *The Sea Around Us*, in 1951. In 1962 she published her most important book, *Silent Spring*, which introduced the public to the dangers of toxic chemicals in the environment. Carson stressed the dangers of the pesticide DDT, which was responsible for the deaths of birds and beneficial insects.

Carson was born in Springdale, Pennsylvania, in 1907. She attended the Pennsylvania College for Women (now Carlow College), where she was able to combine her two favorite subjects—biology and writing. She also studied genetics and zoology at Johns Hopkins University, where she received an M.A. in 1932, and spent her summers doing research work at the Marine Biological Laboratory in Woods Hole, Massachusetts.

Carson worked as an aquatic biologist for the United States Bureau of Fisheries. In addition to her books, she wrote conservation bulletins for the government. Despite the attempts of the agricultural chemical industry to discredit her and her books, many people believe that Carson was responsible for beginning the environmental protection movement. ◆

Akira Okubo
(1925—)

Akira Okubo has spent many years studying the ecology of the oceans. Through his studies, he has added much to the understanding of topics such as why and how fish live in schools. He has also studied how some large sea animals are able to survive by eating only plankton, and why plankton group together in large patches.

Okubo was born in Tokyo, Japan, in 1925. He studied at the Tokyo Institute of Technology and at Johns Hopkins University, where he received his doctorate in oceanography in 1963. He has served as a research scientist at the Johns Hopkins Chesapeake Bay Institute and has taught physical ecology at the Marine Science Research Center of the State University of New York at Stony Brook.

Although he is an oceanographer, Okubo has also studied land animals to learn more about marine life. He hopes to form mathematical models that would help him develop laws of the behavior of organisms in their environments. These laws would be similar to the laws of physics that predict the behavior of particles. ◆

SCIENCE AT WORK

Discussion

Ask the students to suggest reasons why conservationists such as David Powless perform a valuable service. Also ask the students to describe how global environmental conditions and local ecosystem conditions might deteriorate without the actions of conservationists.

Lead the students in a discussion of local, state, and federal pollution laws. Ask the students whether they think such laws are or are not a good idea, and remind them to offer reasons for their opinions.

Journal Activity In their journals, have the students write additional questions that they would like to ask David Powless. Ask volunteers to read some of their questions aloud. Encourage the students to choose several questions that they think are good interview questions.

EXTENSION

Have interested students research additional careers related to the Unit topics. Possible careers include botanist, game warden, ornithologist, oceanographer, science writer, and laboratory technician.

SCIENCE AT WORK

DAVID POWLESS, CONSERVATIONIST

David Powless knows the importance of natural resources. He is a conservationist in Green Bay, Wisconsin.

What kinds of things do you do?

I'm involved in environmental testing. I do complete chemical analyses of soil, air, and water. In addition, I test waste water, ground water, and drinking water for pollutants. I also test hazardous waste and toxic substances and do some air-quality monitoring.

In your job, do you work with other people?

Yes, I work with 23 technicians. These technicians have a variety of degrees in chemistry, biology, geology, and biochemistry. Twelve of these people are Native Americans. My company is owned by the Oneida Tribe of Wisconsin, of which I am a member. This company is a tribal venture based on science and technology. We believe that people are responsible for taking care of the earth and for doing something about the environmental problems.

Are computers and other instruments important in your technical work?

Yes. Computers are used to record and store test information. Computers are also used to prepare reports. Other instruments measure hazardous and toxic substances. With these instruments, I'm able to determine all the pollutants in materials.

What kinds of career opportunities are there in this field?

There are many career opportunities in conservation, especially in chemistry. Environmental groups and agencies are setting very strict guidelines for the levels of pollution to be allowed in manufacturing. Trained technicians are needed to operate the new machines to measure the ultralow levels of pollutants allowed under the guidelines. In addition, more technicians are needed in the laboratory to help prepare the samples of test materials.

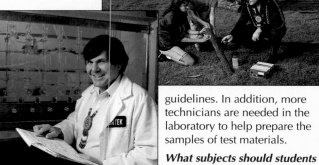

What subjects should students study for this type of work?

I would suggest that students study biology, chemistry, geology, and zoology. At the higher levels of science, mathematics is an absolute requirement. Computer skills will also be important.

What is your educational background?

I studied petroleum engineering at the University of Oklahoma and then got a degree in marketing and economics at the University of Illinois. I first worked in the steel industry; then I received a National Science Foundation grant, the first given to a Native American. I used this grant to develop a process to recycle hazardous waste in steel mills.

Do you think people will be able to stop pollution and clean up the earth?

Yes, I do. We are coming to a point of choice. The earth will always be here, but will we always be able to live on it? The choice is up to us. I believe that we can do it. ◆

Discover More

For more information about careers involving conservation and the environment, write to the

National Wildlife Federation
1400 16th St.,
N.W. Washington, DC 20036-2266

▶ 92 UNIT 1

SCIENCE/TECHNOLOGY/SOCIETY

Discussion

Have the students read the article and discuss the ramifications of the continued erosion of the ozone layer.

When discussing how ultraviolet radiation can cause skin cancer and cataracts, stress the importance of using sunscreens whenever the students are exposed to sunlight for an extended period of time. Also have the students note that they should purchase sunglasses that indicate the percentage of ultraviolet light that is reflected by the lenses.

Ask the students to discuss why CFCs are still used even though they are known to be dangerous and suitable substitutes exist.

Journal Activity

Have the students recall the topics that were presented in the Unit 1 Science Parade. In a journal entry, ask the students to explain why Earth can be described as having a "fragile environment."

SCIENCE/TECHNOLOGY/SOCIETY

A Hole in the Sky

"Make sure you use lots of sunscreen!" Has anyone ever said that to you? We're often told to use sunscreen to filter out the sun's ultraviolet rays. These rays give people a suntan, but they also cause sunburn and some kinds of skin cancer. The sunscreen we use may become even more important in the future because the natural "sunscreen" in our atmosphere is rapidly disappearing.

A Serious Problem

One of the most serious problems the world is facing today is the formation of a "hole" in the ozone layer over Antarctica. Scientists have been aware of this hole for several years, but in recent years it has been growing rapidly.

What is ozone, and why are scientists concerned that the hole is getting larger? Ozone is a relative of oxygen. It forms a layer above the earth and shields the earth from the harmful ultraviolet rays of the sun. A hole in the ozone layer allows more of these rays to reach the earth. If ultraviolet radiation reaches the earth in greater amounts, several things could happen.

The size of the hole in the ozone layer has been increasing recently.

Ultraviolet Hazards

First, ultraviolet rays are damaging to human skin. More ultraviolet rays would mean more cases of fatal skin cancer—up to 200,000 more cases each year in the United States alone. There could also be 60,000 additional cases each year of cataracts, a clouding of the eye.

Second, there could be increased global warming. A temperature increase of as little as one degree would cause some polar ice to melt. This melting would raise sea levels and put many coastal areas underwater.

Third, the increased ultraviolet radiation could increase the amount of smog in the atmosphere. Smog is a hazard to human health and to food crops.

What Caused the Hole?

Many scientists believe that chlorofluorocarbons (CFCs), which chemically destroy ozone, are the major cause of the hole. The largest sources of CFCs are Freon leaks from car air conditioners and the burning of foam cups and foam insulation.

The simplest way to stop the loss of ozone is to limit the use of CFCs. Many scientists would like to have the use of CFCs stopped entirely. Acceptable substitutes to CFCs are already available. It is hoped that these substitutes will soon be put in use, since scientists have recently observed a thinning of the ozone layer over heavily populated northern Europe. ◆

UNIT 1

UNIT 2

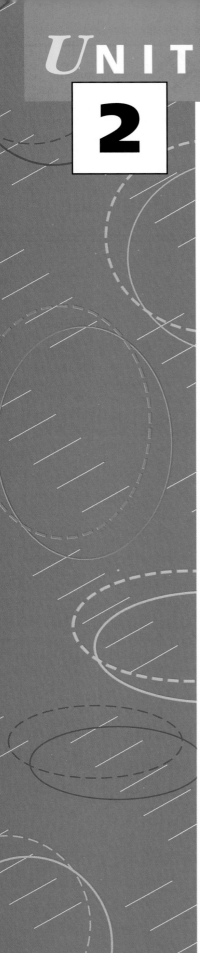

STUDYING LIVING THINGS

UNIT OVERVIEW

This unit covers a wide range of topics including scientific tools, measurement, cell theory, the structures of cells and their functions, cellular reproduction, and classification. The theory of evolution and Darwin's theory of natural selection are discussed as well as the history of genetics, the work of Mendel, and human genetics.

Chapter 4: Cells, page 96

This chapter introduces cells, cell structures, and the scientific tool that made the discovery of cells and their structures possible—the microscope. Microscopic studies led to the cell theory, or the idea that organisms are made of the same basic units, called *cells*.

Chapter 5: Cell Function, page 118

This chapter presents the interactions among cells and their environment. The process by which cells can take in materials from their environment, reproduce, and produce energy is examined. The cooperation of cells, the role of DNA in cell reproduction, and the production of energy through respiration and fermentation are discussed.

Chapter 6: Natural Selection and Heredity, page 142

This chapter presents the roles of natural selection and heredity in evolution. The development of the theory of natural selection by Charles Darwin is illustrated, and the influence of this theory on modern scientific explanations of how things change is discussed. The role of genes in determining the characteristics that are passed from parents to offspring is also examined.

Chapter 7: Classification, page 172

In this chapter, the concept of classification is introduced. The evolution of classification systems over time is presented, as is the importance of the ancestral, or evolutionary, background of organisms in the modern classification system.

Science Parade, pages 194-205

The articles in this unit's magazine focus on archaeology's contribution to the study of human evolution, on studies of sharks, and on the role of microbes in biotechnology. The work of a geneticist is described in the career feature, and the biographical sketches present the creator of binomial nomenclature and a cell biologist.

UNIT RESOURCES

PRINT MEDIA FOR TEACHERS

Chapter 4

Bleifeld, Maurice. *Experimenting with a Microscope*. Franklin Watts, 1988. Discusses the history, mechanics, and use of the microscope.

Thomas, Lewis. *The Lives of a Cell: Notes of a Biology Watcher*. Viking, 1974. Essays reflecting on the connections among science, life, human beings, nature, and language.

Young, John K. *Cells: Amazing Forms and Functions*. Franklin Watts, 1990. Describes different cells of the human body and explains how each is suited to its particular function.

Chapter 5

Crick, Francis. *What Mad Pursuit: A Personal View of Scientific Discovery*. Basic Books, 1988. Crick gives his account of the discovery of the structure of DNA.

Fichter, George S. *Cells*. Franklin Watts, 1986. Presents the discovery of cells, their structures and functions, and research involving cells.

Leone, Francis. *Genetics: The Mystery and the Promise*. TAB Books, 1992. Traces past developments and future hopes in the field of genetics.

Chapter 6

Byczynski, Lynn. *Genetics: Nature's Blueprints*. Lucent Books, 1991. Topics include the discovery of the gene, genes' role in disease, genetic engineering, and past and future genetic research.

Darwin, Charles. *The Voyage of the Beagle*. Bantam Books, 1972. The original 1839 title said it all: *Journal of researches into the geology and natural history of the various countries visited by the H.M.S. Beagle*.

Skelton, Renee. *Charles Darwin and the Theory of Natural Selection*. Barron's, 1987. The life of Charles Darwin, including his development of his theory of natural selection.

Chapter 7

Gutfreund, Geraldine Marshall. *Animals Have Cousins, Too*. Franklin Watts, 1990. Introduces the principles of evolution and taxonomy by focusing on changes in animals over time.

Margulis, Lynn. *Diversity of Life: The Five Kingdoms*. Enslow, 1992. Describes the five kingdoms of the classification system and gives characteristics of each.

PRINT MEDIA FOR STUDENTS

Chapter 4

Bender, Lionel. *Through the Microscope: Atoms and Cells*. Gloucester Press, 1990. Text and photographs introduce microscopic plant and animal life, viruses, microspores, and other life forms that can be viewed through a microscope.

Chapter 5

Bender, Lionel. *Through the Microscope: Frontiers of Medicine*. Gloucester Press, 1991. Examines the advances in medical science made possible by the use of microscopes.

Chapter 6

Hyndley, Kate. *The Voyage of the Beagle*. Wayland Limited, 1989. An account of the five years that English naturalist Charles Darwin traveled around the world.

Chapter 7

Stidworthy, John. *Mammals: The Large Plant-Eaters*. Facts on File, 1989. Articles focusing on individual species of mammals, their structure and size, and other information.

ELECTRONIC MEDIA

Chapter 4

Cell Structure. Videocassette. Britannica. Traces the role of modern technology in unlocking the secrets of the cell.

Imaging a Hidden World: The Light Microscope. Videocassette. Coronet. Defines the parts of a microscope and tells how to use it.

The Living Cell. Videocassette. Coronet. Looks at the structures and functions of cells.

Chapter 5

DNA: Laboratory of Life. Videocassette. National Geographic. Explains how DNA works in the cell.

Membranes and Transport. Videocassette. Britannica. Looks at the cell's ability to transport substances.

Mitosis. Videocassette. Britannica. Demonstrates how the basic process of mitosis takes place in cells.

Chapter 6

Darwin's Finches. Film. Time-Life. Presents the observations Darwin made about finches and their beaks.

Evolutionary Biology. Videocassette. Coronet. Explains how biologists have used studies to develop modern evolutionary theory.

Mendel's Laws. Videocassette. Coronet. Presents Mendel's original work and later work by other scientists.

Chapter 7

Grouping Living Things. Videocassette. Britannica. Carolus Linnaeus discusses the hierarchy of classification.

How We Classify Animals. Videocassette. AIMS. Illustrates the system that scientists use to classify animals.

UNIT 2

Discussion Explain to the students that archaeologists are scientists and researchers who study the ways in which humans and their cultures have developed over time. The work of many archaeologists is carried out through excavation, discovery, and examination of artifacts or other material objects that civilizations leave behind as they evolve. Ask the students to describe the contributions they think archaeologists might make to the body of scientific knowledge. (The students might suggest that archaeological discoveries help create an orderly time line of history.)

UNIT 2: STUDYING LIVING THINGS

CHAPTERS

4 *Cells* 96
What are living things made of? What we know today is the result of scientific study and discovery.

5 *Cell Function* 118
Cells use energy, multiply, and perform other functions in a continuous process.

6 *Natural Selection and Heredity* 142
Life follows a pattern of change as one living organism passes on part of itself to the next generation.

Journal Activity Extend the discussion by explaining to the students that the first archaeologists were interested in finding ancient artifacts solely for their beauty and financial value. Point out that, in contrast, modern archaeologists are interested in the history of artifacts—who made them, when were they made, and what were the people like who made them. Remind the students that the goal of an archaeological excavator is to discover material objects that were left behind by different civilizations. Ask the students to create a journal entry in which they describe the types of things that might be left behind by today's civilization to become tomorrow's artifacts.

In an underground tunnel, workers discover parts of an ancient city. There are many questions to be answered. Archaeologists and other scientists are fascinated by the discoveries they make. Through careful excavation, the ancient Roman city of Pompeii emerges from a slumber of almost 17 centuries. Buried beneath volcanic ash, Pompeii provides a glimpse of what life was like in ancient times. Tools, wooden artifacts, and works of art are discovered. By thorough examination, the scientists are able to find out what the people wore, what they ate, how they lived, and how healthy they were.

Science PARADE

7 Classification 172
People have organized, grouped, and ordered the multitudes of living things they've observed. How has this helped society today?

SCIENCE APPLICATIONS
Finding the Past... 194

READ ABOUT IT!
"Shark Lady" 198

THEN AND NOW
Carolus Linnaeus and Rita Levi-Montalcini........ 201

SCIENCE AT WORK
Eduardo Cantu, Geneticist......... 202

SCIENCE/TECHNOLOGY/ SOCIETY
"These Germs Work Wonders" 203

UNIT 2 95

CHAPTER 4

CELLS

PLANNING THE CHAPTER

Chapter Sections	Page	Chapter Features	Page	Program Resources	Source
Chapter Opener	96	*For Your Journal*	97		
Section 1: TOOLS FOR LIFE SCIENCE	98	Discover By Doing (A)	98	*Science Discovery* *	SD
• A System for Measurement (B)	98	Discover By Calculating (B)	99	Reading Skills: Making an Outline (A)	TR
• Tools of a Life Scientist (B)	102	Activity: How can you measure volume? (A)	100	Extending Science Concepts: Measuring with a Microscope (H)	TR
		Section 1 Review and Application	103	Some Common SI Units (B)	IT
		Investigation: Using the Microscope (B)	104	Compound Light Microscope (B)	IT
				Record Sheets for Textbook Investigations (B)	TR
				Study and Review Guide, Section 1 (B)	TR, SRG
Section 2: DISCOVERY OF CELLS	105	Discover By Doing (B)	105	Investigation 4.1: Comparing Cells (H)	TR, LI
• Early Observations (B)	105	Section 2 Review and Application	107	Plant and Animal Cells (A)	IT
• Cell Theory (B)	106			Study and Review Guide, Section 2 (B)	TR, SRG
Section 3: STRUCTURE OF CELLS	108	Discover By Researching (A)	108	*Science Discovery* *	SD
• Cell Membrane and Nucleus (B)	108	Section 3 Review and Application	113	Investigation 4.2: Modeling a Cell (B)	TR, LI
• Other Cell Parts (B)	110	Skill: Comparing Plant and Animal Cells (A)	114	Connecting Other Disciplines: Science and Social Studies, Cell Communities (A)	TR
• Levels of Organization (A)	111			Thinking Critically (H)	TR
				Plant and Animal Cells (A)	IT
				Study and Review Guide, Section 3 (B)	TR, SRG
Chapter 4 HIGHLIGHTS	115	The Big Idea	115	Study and Review Guide, Chapter 4 Review (B)	TR, SRG
Chapter 4 Review	116	For Your Journal	115	Chapter 4 Test	TR
		Connecting Ideas	115	Test Generator	

B = Basic   A = Average   H = Honors
The coding Basic, Average, and Honors indicates subsections, features, and resources that might be appropriate for different levels of learners. For additional suggestions regarding choice of topic and depth of coverage, see the Pacing Chart on pages T26–T29.

*Frame numbers at point of use
(TR) Teaching Resources, Unit 2
(IT) Instructional Transparencies
(LI) Laboratory Investigations
(SD) *Science Discovery* Videodisc Correlations and Barcodes
(SRG) Study and Review Guide

▶ 95A

CHAPTER MATERIALS

Title	Page	Materials
Discover By Doing	98	(per individual) paper clip, pencil, index card, journal
Discover By Calculating	99	(per individual) paper, pencil
Activity: How can you measure volume?	100	(per group of 3 or 4) metric ruler, chiton, cardboard (2 pieces), 100-mL graduate
Investigation: Using the Microscope	104	(per group of 3 or 4) compound light microscope, prepared slides of an insect leg and wing
Discover By Doing	105	(per individual) leaves, magnifying glass, paper, pencil
Discover By Researching	108	(per class) reference materials (per individual) paper, pencil
Teacher Demonstration	112	model of human skeletal system
Skill: Comparing Plant and Animal Cells	114	(per group of 3 or 4) slide, coverslip, medicine dropper, *Elodea* leaf, compound microscope, prepared slide of human cheek cells

ADVANCE PREPARATION

For the *Activity* on page 100, you will need to order chitons from a biological supply house. For the *Investigation* on page 104, obtain prepared slides of an insect leg and a wing from a biological supply house. For the *Discover By Doing* on page 105, the students will need to obtain several different plant leaves. For the *Skill* on page 114, you will need to obtain *Elodea* and prepared slides of human cheek cells from a biological supply house.

TEACHING SUGGESTIONS

Field Trip
Visit a food processing plant or sewage treatment plant that employs fermentation, diffusion, or osmosis as part of the process. Fermentation is used in the baking, dairy, and brewing industries. Diffusion and osmosis may be employed in preserving certain foods.

Outside Speaker
Have a cell biologist from a local university, or a medical researcher from a hospital, discuss his or her work. Ask the speaker to discuss the importance of cell study to the diagnosis and treatment of diseases.

CHAPTER 4
CELLS

CHAPTER THEME—SYSTEMS AND STRUCTURES

This chapter introduces the students to cells and cell structure and to the scientific tool that made the discovery of cells and their structures possible—the microscope. The students will learn that organisms can be studied on both the microscopic and macroscopic scales.

Microscopic studies led to the cell theory, or the idea that organisms are made of the same basic units, called cells. The theme of **Systems and Structures** is also developed through concepts in Chapters 5, 7, 8, 9, 10, 12, 13, 16, 17, 18, 19, and 20.

MULTICULTURAL CONNECTION

Cooperative Learning Over the centuries, many different people from around the world have contributed to the development of the microscope or have used the microscope to make interesting and important discoveries. Have small groups of students research either the contributions of different people to the development of the microscope, or some of the discoveries made possible by the microscope. Then ask the groups to share their findings with their classmates.

MEETING SPECIAL NEEDS

Second Language Support
Fluent English-speaking students should be paired with students who have limited English proficiency to create a poster using text and pictures to explain how to use a microscope properly.

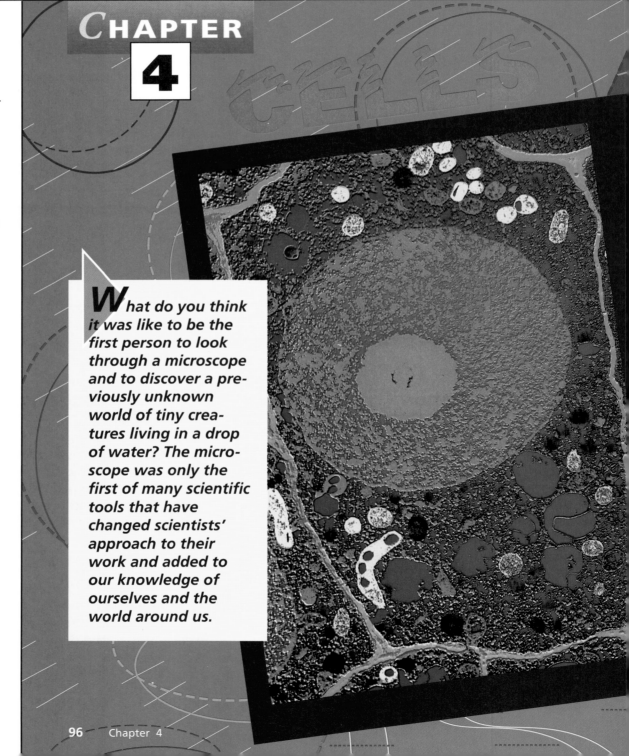

CHAPTER 4

What do you think it was like to be the first person to look through a microscope and to discover a previously unknown world of tiny creatures living in a drop of water? The microscope was only the first of many scientific tools that have changed scientists' approach to their work and added to our knowledge of ourselves and the world around us.

CHAPTER MOTIVATING ACTIVITY

Ask the students to name the characteristics of all living things. List their suggestions on the chalkboard. Then discuss the suggested characteristics with the students, and discard those characteristics that do not accurately define all living things. (For example, if the students name *movement*, point out that *movement* does not accurately define a living thing because some living things such as plants do not move.) Narrow the list to only those characteristics that are applicable to all living things: they experience growth and development, they reproduce, they are composed of cells, they use energy, they respond to environmental stimuli.

The students can demonstrate their knowledge of the microscope, microscopic research, and the microscope's value to science by answering the journal questions. The answers given by the students will also provide you with an opportunity to correct any misconceptions the students might have about the microscope, its use in studying cells, and its relationship to other scientific endeavors.

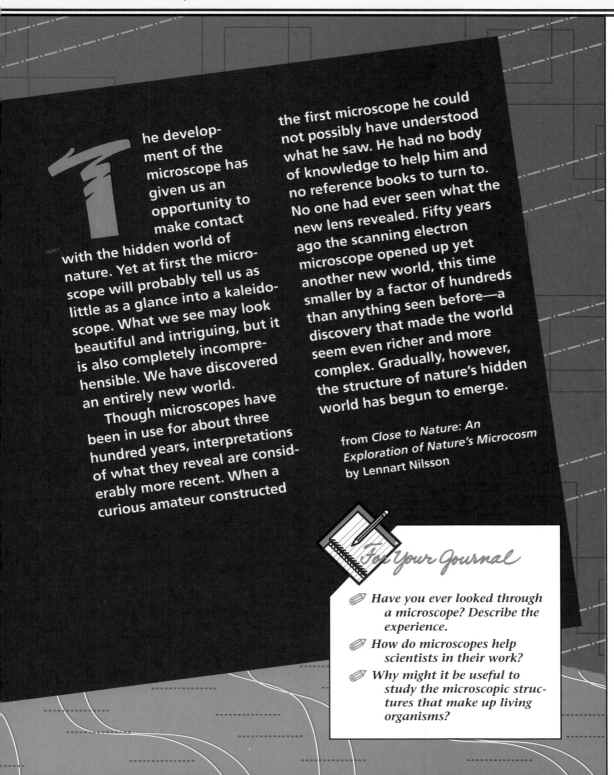

The development of the microscope has given us an opportunity to make contact with the hidden world of nature. Yet at first the microscope will probably tell us as little as a glance into a kaleidoscope. What we see may look beautiful and intriguing, but it is also completely incomprehensible. We have discovered an entirely new world.

Though microscopes have been in use for about three hundred years, interpretations of what they reveal are considerably more recent. When a curious amateur constructed the first microscope he could not possibly have understood what he saw. He had no body of knowledge to help him and no reference books to turn to. No one had ever seen what the new lens revealed. Fifty years ago the scanning electron microscope opened up yet another new world, this time smaller by a factor of hundreds than anything seen before—a discovery that made the world seem even richer and more complex. Gradually, however, the structure of nature's hidden world has begun to emerge.

from *Close to Nature: An Exploration of Nature's Microcosm* by Lennart Nilsson

For Your Journal

- Have you ever looked through a microscope? Describe the experience.
- How do microscopes help scientists in their work?
- Why might it be useful to study the microscopic structures that make up living organisms?

ABOUT THE AUTHOR

Lennart Nilsson is a pioneer in the field of scientific and medical photography. Born in Sweden in 1922, he began his career as a photojournalist. But his interest in science and close-up photography led him to develop special lens systems and photographic techniques, which he used to take startling photographs.

Nilsson's pictures of a human fetus in the womb were published in *Life* magazine and then incorporated into his classic book, *A Child Is Born* (1965), which has been translated into eighteen languages. A film, *The Miracle of Life*, based on Nilsson's photographs of the human reproductive systems and the development of the fetus, has been shown on public television in the United States. In 1983 Nilsson received an Emmy award for the film.

Nilsson expanded his interest from human reproduction to the workings of the human body in *Behold Man* (1973) and *The Body Victorious* (1987). He used light microscopes, scanning electron microscopes, and endoscopic equipment to capture the human body's systems in action. Stockholm University in Sweden made Nilsson an Honorary Doctor of Medicine for his contributions to medical research. Nilsson has been honored throughout the world for increasing public awareness of human biology.

Section 1:
TOOLS FOR LIFE SCIENCE

FOCUS

This section introduces the International System of Units, or SI. Using common SI units such as the meter, liter, and gram allows scientists who speak many different languages to communicate their scientific findings to one another. Some of the tools of scientists, including the microscope, are also presented in the section.

MOTIVATING ACTIVITY

Cooperative Learning Make several slides of objects typically found in the classroom. These objects might include a speck of dirt, a strand of carpeting, or a pencil shaving from a sharpener. Then set up several microscopes, one for each slide.

Cover each microscope so that the students cannot see its slide. Then tell them that when they look through each microscope, they will see something found in the classroom. Allow small groups to rotate from microscope to microscope and try to name the object shown on each slide.

PROCESS SKILLS
• Measuring • Interpreting Data • Comparing
• Inferring

POSITIVE ATTITUDES
• Precision • Enthusiasm for science and scientific endeavor • Curiosity

TERMS
• meter • liter • gram
• compound light microscope

PRINT MEDIA
Experimenting with a Microscope
by Maurice Bleifeld
(see p. 93b)

ELECTRONIC MEDIA
Imaging a Hidden World: The Light Microscope, Coronet
(see p. 93b)

Science Discovery
Metric measurements
How to use a microscope

BLACKLINE MASTERS
Study and Review Guide
Reading Skills
Extending Science Concepts

DISCOVER BY Doing

The students' journal entries should reflect the importance of standard units of measurement to ensure the ability to replicate results.

SECTION 1

Tools for Life Science

Objectives

Identify the SI units of distance, volume, mass, and temperature.

Explain how to use a compound light microscope.

Compare and contrast the SI and customary systems of measurement.

Imagine that you are a member of a team of scientists investigating a new life form. Everything that you discover about this life form will be of interest to your scientific colleagues around the world. As a result, it is important that you begin by making accurate measurements. Studying a living organism requires many different measurements and special tools. Common scientific measuring tools include rulers, balances, and scales. Each type of measuring tool uses one or more special units of measurement.

You are probably already familiar with many of the units of measurement scientists use in their work. The work of a scientist depends on the *quality* of the observations made. For observations to be correct, measurements must be accurate. Scientists all around the world must be able to share the results of their experiments. So it is also important that all scientists agree to use the same units of measure. What do you think would happen if scientists did not use the same measurement system? See for yourself in this activity.

DISCOVER BY Doing

Measure the length and width of your textbook, but do not use any standard unit of measurement. Choose your own unit, such as the length of a paper clip, a pencil, or an index card. Your teacher will record all the measurements on the board. In your journal, explain why standard units of measurement are important.

A System of Measurement

In 1960, scientists from around the world met and decided that a common system of measurement was needed to allow them to communicate more easily with one another. The scientists agreed to use a system of measurement called the *International System of Units (SI)*.

TEACHING STRATEGIES

● **Process Skills:** *Inferring, Analyzing*

On the chalkboard, write the sentence "Mathematics is the language of science," and have the students discuss the meaning of the sentence. (Help the students understand that numbers explain scientific theories.) Also point out that in addition to scientists, many international manufacturers and other entities that operate on a worldwide basis have adopted SI as their measurement system. Ask the students why manufacturers would want to adopt a common measurement system. (The students should recognize the same need for standardization that scientists have.)

● **Process Skills:** *Comparing, Applying*

Have the students examine Table 4–1 and note that the same prefixes are used for length, area, volume, and mass. Point out that, in addition to *kilo-* and *deci-*, scientists also use *centi-* and *milli-* as prefixes to name units of measurement.

Have the students predict the meaning of *centi-* and *milli-*. (*Centi-* means "one-hundredth"; *milli-* means "one-thousandth.") Ask the students what part of a meter is represented by a centimeter and what part of a gram is represented by a milligram. (A centimeter = 1/100th of a meter; a milligram = 1/1000th of a gram)

Measure Up In SI, length or distance is measured in meters. A **meter** (m) is about as long as a baseball bat or as high as a doorknob. The meter can be divided into smaller units. One advantage of working in SI is that all the units are based on the number 10. Prefixes are used to show larger and smaller amounts. For example, the prefix *kilo-* means "one thousand." A *kilometer* (km) is equal to 1000 meters. The prefix *deci-* means "one-tenth." A *decimeter* (dm) is one-tenth of a meter (0.1 m). To convert measurements in SI units, you need only to multiply or divide by 10. When would you multiply? Divide? The next ① activity will give you practice measuring in SI units.

Calculating

Estimate the height in SI units of several objects in your classroom. Record your estimates in a chart. Then measure the height of the same objects in centimeters, decimeters, and meters. What is the relationship among these units?

In your work in this course, you will need to measure quantities such as length, area, volume, mass, and temperature. The following table shows some common units of measure for these quantities.

Figure 4–1. Common measurements are usually used in the construction of most homes. For example, most doorknobs are about 1 meter from the floor.

① In SI, divide to find a larger unit; multiply to find a smaller unit.

DISCOVER BY *Calculating*

The students should determine that the relationship among meters, decimeters, and centimeters is based on 10: 1 meter = 10 decimeters and 100 centimeters; 1 decimeter = 0.1 meters and 10 centimeters; and 1 centimeter = 0.1 decimeters and 0.01 meters. Help the students realize that it depends on the size of the object being measured whether it is better to use meters, decimeters, centimeters, or millimeters when measuring in SI.

SCIENCE BACKGROUND

Tell the students about these additional SI units:
● for length—micrometer (μm)
● for volume—kiloliter (kL)
● for mass—megagram (Mg)
● for density—gram per cubic meter (g/m^3)

LASER DISC

450

Metric measurements

Table 4-1 Some Common SI Units

Measurement	Common Unit	Symbol
Length	kilometer meter centimeter millimeter	km m cm mm
Area	square kilometer hectare square meter square centimeter	km^2 ha m^2 cm^2
Volume	liter milliliter cubic meter cubic centimeter	L mL m^3 cm^3
Mass	kilogram gram milligram	kg g mg
Temperature	Kelvin degrees Celsius	K °C

ACTIVITY

How can you measure volume?

Process Skills: Measuring, Inferring

Grouping: Groups of 3 or 4

Hints

The chiton is a mollusk that most students are likely to find interesting and unusual. Pieces of heavy cardboard, 8 cm by 8 cm to 10 cm by 10 cm, will make the measurement of the chiton's dimensions easier. If the specimens are too large to fit easily in the graduate, substitute a small beaker. The measurements obtained by each student will vary with the specimen.

▶ Application

1. The dimensions of an irregular object are difficult to measure in centimeters. Because of this, milliliters is a more accurate way to determine the volume of an irregular object.

2. Measuring and multiplying the dimensions of a cube, and determining the volume of water displaced by the cube will both yield reasonable estimates of the volume of the cube.

★ PERFORMANCE ASSESSMENT

To evaluate the students' ability to measure volume, ask them to show you, step by step, how they used two different methods to find the volume of the cube. Then have them compare the two methods in terms of accuracy and convenience.

TEACHING STRATEGIES, continued

● **Process Skills:** *Interpreting Data, Generating Ideas*

Ask the students to describe situations in their lives outside of school in which they are exposed to metric units. (Responses might suggest that the speedometers of automobiles can often be read in miles per hour or kilometers per hour; nutritional labels will often contain information such as the number of grams of fat and the number of milligrams of cholesterol; the nuts and bolts with which items such as bicycles and automobiles are assembled are often measured in millimeters.) Then ask the students to describe the importance of being able to convert metric units to units such as feet, miles, and gallons. (Responses should reflect the importance of being able to convert from one system to another. For example, an American who rents a car in Europe would note that road signs give distances in kilometers. If the driver does not understand the relationship between miles and kilometers,

---

Notice three of the units in Table 4–1 that do not have prefixes—the meter, the liter, and the gram. You might think of the SI system as being "built" from these three basic units of measure—the meter for distance, the **liter** for capacity or volume, and the **gram** for mass.

More Than One There is something different about the row of the table for volume. What is the difference? This row uses *two* of the basic units, the liter and the meter. The two units are used because there are two different ways to find volume. In the next activity you will explore these two ways of finding volume.

ACTIVITY

How can you measure volume?

MATERIALS
metric ruler, chiton, cardboard (two sheets), 100-mL graduate

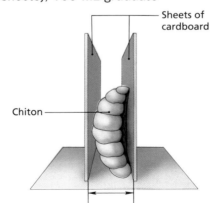

PROCEDURE
Part A
1. Using a metric ruler, measure and record the length of the chiton (KYT uhn). Record your data in cm.
2. Find the maximum and minimum width of the chiton by placing it between two pieces of cardboard as shown.
3. Calculate the volume of the chiton by multiplying the length, the maximum width, and the minimum width. Record your answer in cm^3.

Part B
1. Fill the graduate with 50 mL of water. Record the initial volume of the water as 50 mL.
2. Carefully place the chiton in the graduate and record the new volume.
3. Subtract the initial volume of water from the final volume. The difference is the volume of the chiton in mL. Record this volume.

APPLICATION
1. In SI, 1 cm^3 is equal to 1 mL. Compare your two measurements of the volume. Which do you think was the more accurate? Why?
2. Repeat the activity using a small metal cube. Estimate the volume before you measure and compute. Was your estimate more accurate in Part A or in Part B?

he or she will not be able to estimate the length of time it would take to travel to a city, say, 300 kilometers away. Also a gallon of fuel in the United States is often different from a gallon of fuel in the other countries; a gallon of fuel in the United States is only 83.3% as large as a gallon of fuel in the United Kingdom.)

● **Process Skills: *Comparing, Applying***

Point out that in the United States, the Fahrenheit scale is still commonly used to measure temperature. On the chalkboard, write the conversion formula for degrees Fahrenheit to degrees Celsius: $C = \frac{5}{9}(F - 32)$.

Ask the students to convert a typical Fahrenheit measurement such as room temperature (68 degrees) to degrees Celsius by subtracting 32 degrees from the Fahrenheit temperature, then multiplying by 5 and dividing by 9. (68 degrees Fahrenheit = 20 degrees Celsius)

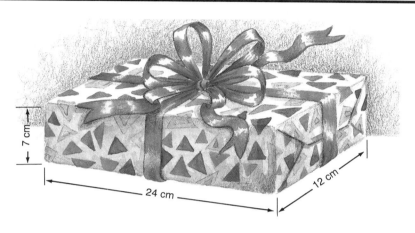

Figure 4–2.
The box has a length of 24 cm, a width of 12 cm, and a height of 7 cm. The volume is found by multiplying the dimensions together. What is the volume of the box in cm^3? ①

① 2016 cm^3

② On the Celsius scale, water boils at 100 degrees and freezes at 0 degrees; on the Fahrenheit scale, water boils at 212 degrees and freezes at 32 degrees.

ONGOING ASSESSMENT
▼ **ASK YOURSELF**

Not all scientists speak the same language. SI allows different scientists from different countries to easily communicate and exchange numerical data in a way they can all clearly understand.

It often happens that you will have a choice of units in recording scientific data. One example is volume. A second example is temperature. The table shows two different units for temperature, *Kelvin* and degrees *Celsius*. The SI unit for measuring temperature is Kelvin (K). Water freezes at 273 K and boils at 373 K. Because the Kelvin scale is difficult to use in the laboratory, the Celsius (°C) scale is used more often. On the Celsius scale, water freezes at 0°C and boils at 100°C. Normal body temperature is about 37°C.

Another factor in choosing the best unit to use is the size of the object. The gram—approximately equal to the mass of one paper clip—is too small for most everyday uses. In measuring mass, the kilogram is most useful for ordinary objects.

SI is both logical and flexible. Units can be found to measure the extremely large distances between objects in space, as well as the very small distances involved in organisms too small to be seen without the aid of a microscope.

▼ **ASK YOURSELF**

Why is it an advantage for scientists in the United States to use SI measurement?

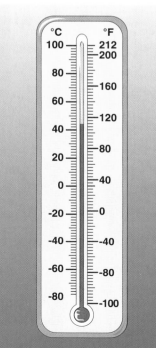

Figure 4–3. The convenience of the Celsius scale can be easily seen when the boiling point and freezing point of water are compared. At what temperature does water boil and freeze on each of these scales? ②

INTEGRATION—*Physical Science*

Temperature is an important measurement in science because many substances are affected by temperature changes. In their journals, have the students suggest several examples of substance changes due to changes in temperature, such as wax melting or cake batter turning into cake. Then ask the students to share their examples with their classmates.

BACKGROUND INFORMATION

Unlike other metric measurements, the Celsius scale is named after a person—Anders Celsius, a Swedish astronomer who developed the scale in 1742. Similarly, the Kelvin scale is named after Lord Kelvin, a nineteenth-century British physicist who suggested the use of the scale.

SECTION 1

TEACHING STRATEGIES, continued

● **Process Skills:** *Inferring, Generating Ideas*

Point out that the microscope is one of the most important tools used by life scientists. Ask the students to describe some of the ways in which the microscope has helped expand knowledge in the life sciences. (Scientists can learn more about nature, anatomy, cell structure, and diseases because the microscope allows observation of minute objects such as unicellular plants and animals, bacteria, viruses, cell structures, and chromosomes.) Also ask the students to describe other situations in which microscopes are used extensively. (Responses might include medical laboratories.)

GUIDED PRACTICE

Have the students describe a situation in which the basic SI units—meter, liter, and gram—are used.

INDEPENDENT PRACTICE

 Have the students provide written answers to the Section Review and Application questions. In their journals, have the students write a paragraph that describes the value of SI to scientists and their work.

---

① Glasses, contact lenses, magnifying glasses, microscopes, binoculars, and telescopes are tools that improve the ability to observe.

MEETING SPECIAL NEEDS

Second Language Support

Write the following terms on the chalkboard: eyepiece, low-power objective, stage, body tube, revolving nosepiece, arm, fine adjustment, high-power objective, and coarse adjustment. Then distribute several compound light microscopes throughout the classroom and ask the students to match the names on the chalkboard with the actual parts on a microscope.

LASER DISC
26906-29346

How to use a microscope

Scanning electron microscopes, or SEMs, transmit a beam of electrons that move across a surface in a back-and-forth motion. Electrons that bounce off the scanned surface are then synthesized by a detector, and a visual image is created. Important discoveries have been made with the help of SEMs. Have interested students research important SEM discoveries and collect pictures taken using SEMs.

▶ 102 CHAPTER 4

Tools of a Life Scientist

Besides accurate measurements, you and your fellow scientists want to make careful and detailed observations of the new life form. What tools could you use to improve your ability to observe? ① One very important tool of the life scientist is the *microscope*. Microscopes allow scientists to see small organisms, cells, and parts of cells that otherwise are too small to be seen. A magnifying glass is the simplest kind of microscope. It can make an object look 2 to 20 times its actual size.

The most widely used kind of microscope is the compound light microscope. A **compound light microscope** uses two magnifying lenses to form an image. The lens closest to the object being studied is the *objective lens*. Most compound light microscopes have a high-power objective lens and a low-power objective lens. The lens

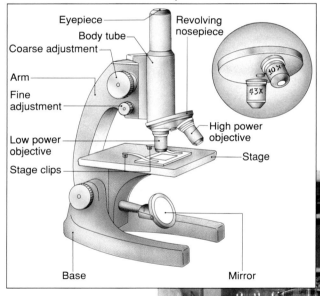

Figure 4–4. A compound light microscope is primarily two lenses mounted inside a tube. The compound light microscope is an important tool used by scientists.

Figure 4–5. A microscope is a delicate instrument and must be handled carefully. When carried, a microscope should be supported as shown.

EVALUATION

Ask the students to describe the relationship of the microscope to scientific research. (The microscope has contributed significantly to the body of scientific knowledge and to our understanding of the world in which we live.)

RETEACHING

Prepare a diagram of a compound light microscope that does not include labels for its parts. Ask the students to label the microscope's parts and to explain the function of each part.

EXTENSION

The Kelvin temperature scale is rarely used in everyday life. Zero on the Kelvin scale is equivalent to –273.2 degrees Celsius and –459.7 degrees Fahrenheit. The branch of physics that deals with very low temperatures is known as *cryogenics*. Have the students investigate the field of cryogenics and present their findings to the class.

CLOSURE

 Cooperative Learning Working in small groups, ask the students to describe as many everyday situations as possible in which people are exposed to SI units.

---

closest to the eye is the *ocular lens,* or eyepiece. The objective and the eyepiece are at opposite ends of the body tube.

The body tube of the microscope is fastened to the arm. The arm is supported by a heavy base. Always carry a microscope upright by the arm, with your other hand under the base.

A microscope works somewhat like a pair of binoculars. You turn an adjustment knob to bring the object into focus. However, most microscopes have two adjustment knobs. The coarse adjustment is used only when the low-power objective is in place. The fine adjustment is used for final focusing under low power, and it is the only focusing adjustment that is used when the high-power objective is in place.

Over several hundred years, microscopes have been improved. The most powerful microscopes, electron microscopes, can magnify objects hundreds of thousands of times. Electron microscopes have made it possible for scientists to study extremely small organisms, viruses, molecules, and even atoms.

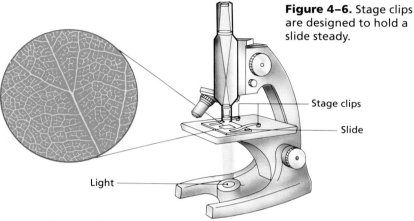

Figure 4–6. Stage clips are designed to hold a slide steady.

Figure 4–7. This dust mite has been magnified by an electron microscope.

▼ **ASK YOURSELF**

Why are microscopes important tools for life scientists?

SECTION 1 REVIEW AND APPLICATION

Reading Critically
1. What are the SI units of measure for distance, volume, mass, and temperature?
2. How does using a prefix change the value of a basic unit in SI?

Thinking Critically
3. What types of information about plants and animals would be unavailable to a scientist who did not have a microscope?
4. When measuring an object, why is it important to use both a number and a unit for measurement?
5. How is a magnifying glass like a microscope? How is it different?

ONGOING ASSESSMENT
▼ **ASK YOURSELF**

Microscopes give scientists an opportunity to observe and collect data about a world that is invisible to the unaided eye. This data helps scientists understand and explain the visible world.

SECTION 1 REVIEW AND APPLICATION

Reading Critically

1. meter; liter; gram; degree Celsius

2. When moving from prefix to prefix, the value changes by multiplying or dividing by ten or exponential multiples of ten such as 100 or 1000.

Thinking Critically

3. A scientist without a microscope would be limited to making observations with the unaided eye. The scientist would not be able to see small organisms, cells, and parts of cells, all of which can be seen using the microscope.

4. The number indicates the quantity, while the unit tells what is being measured—volume, mass, distance, area, or temperature. A measurement that does not have both a number and a unit is meaningless.

5. A microscope and a magnifying glass both use a lens to magnify images. A microscope, however, typically uses more than one magnifying lens and can magnify to much greater detail than a magnifying glass.

SECTION 1 **103** ◀

INVESTIGATION

Using the Microscope

Process Skills: Observing, Comparing

Grouping: Groups of 3 or 4

Objectives
- **Compare** the movement of two images.
- **Experiment** using different magnifications of a microscope.
- **Observe** and **sketch** various images displayed by a microscope.

Pre-Lab
Ask the students to share their knowledge of a microscope, its parts, and its operation with each other.

Hints
Remind the students to not position their microscope or its mirror in direct sunlight.

▶ **Analyses and Conclusions**

1. The movement of the image is opposite to the movement of the slide.

2. Using high power, the image of the wing may require additional light, appear larger, and have better resolution or detail than the image of the wing using low power.

3. Generally speaking, examining the parts of the insect should increase scientists' knowledge of the structure and function of the parts and hence increase their knowledge of the insect.

▶ **Application**

The amount of magnification, expressed as a number, is determined by multiplying the eyepiece magnification by the magnification of the high-power objective.

✴ **Discover More**
If possible, have the students collect and examine some of the objects they listed. Then have the students share their findings with the class.

Post-Lab
Ask the students to compare what they knew about a microscope before the investigation with what they know now. Ask them what they discovered about the microscope during the investigation.

PORTFOLIO ASSESSMENT
After the students have completed the *Investigation*, ask them to place their drawings in their science portfolios.

INVESTIGATION

Using the Microscope

▶ **MATERIALS**
- compound light microscope
- prepared slides of an insect leg and an insect wing

▼ **PROCEDURE**

Part A

1. **CAUTION: Do not position the microscope or its mirror in direct sunlight. Direct sunlight may damage your eyes.**
2. Check to see that the low-power objective is in place and ready for use. Place the prepared slide of the insect leg on the stage and secure the slide with the stage clips.
3. Predict what the insect leg will look like under the microscope.
4. Use the coarse adjustment knob and then the fine adjustment knob to focus on the leg.
5. Make a drawing of what you see in the low-power field of view. Compare your observations with your predictions.
6. While looking at the leg, move the slide slowly to the left. Record your observations. Repeat, moving the slide upward.

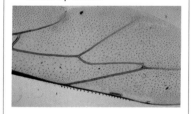

Part B

1. Remove the slide of the insect leg and replace it with the slide of the insect wing.
2. Using the low power, focus on the wing. Make a drawing of the low-power field of view.
3. Move the slide so that the part you want to see under high power is in the center of the field of view.
4. While looking at the slide from the side and at eye level, slowly rotate the revolving nosepiece until the high-power objective is in place.
5. Using only the fine adjustment, focus on the wing. Make a drawing of what you see.

▶ **ANALYSES AND CONCLUSIONS**

1. How does the movement of the image viewed with the microscope compare with the actual movement of the slide?
2. How does the appearance of the insect wing differ under low power and high power?
3. How does examining an insect leg and wing help scientists to better understand the insect?

▶ **APPLICATION**
The amount of magnification on a microscope is shown by numbers marked on the eyepiece and the objective lens. If your eyepiece is marked 10 X and your low-power objective is also marked 10 X, the object you view under low power will be magnified 100 times.

Examine your high-power objective to find out the magnification under high power. How is this number computed?

✴ **Discover More**
Make a list of five objects you would like to view under a microscope. Predict what you might discover about each object. How could you confirm your predictions?

Section 2: DISCOVERY OF CELLS

FOCUS

This section presents the contributions of several scientists to the development of the cell theory and explains that its development was important because the cell theory showed that all living things share similar structures. The section also focuses on the value of contributions by different scientists to the development of a theory.

MOTIVATING ACTIVITY

Cooperative Learning Prepare or obtain microscope slides of thin slices of cork. Instruct the students to work in small groups. Distribute a slide to each group and have the students observe the cork using both high and low magnification of a compound microscope. Ask the students to sketch and write descriptions of their observations. Then use their drawings and descriptions as a basis for a discussion to introduce the concept of cells and the work of Robert Hooke.

PROCESS SKILLS
- Observing • Comparing
- Applying

POSITIVE ATTITUDES
- Enthusiasm for science and scientific endeavor • Initiative and persistence • Skepticism

TERMS
- cells • cell theory

PRINT MEDIA
Cells: Amazing Forms and Functions by John K. Young (see p. 93b)

ELECTRONIC MEDIA
The Living Cell, Coronet (see p. 93b)

BLACKLINE MASTERS
Study and Review Guide
Laboratory Investigation 4.1

Discovery of Cells

SECTION 2

As part of your investigation of the new life form, you decide that you must make two kinds of observations. You first make observations on a macroscopic scale. A macroscopic object is large enough to be observed with the unaided eye. You record data about how and where the life form lives, what it eats, how it moves, and how it interacts with others of its own kind.

For a better understanding of your new discovery, you also make microscopic observations. In this way, you look at the fundamental structure of the life form and decide if it is similar to any on Earth. Your observations will be added to those made by other scientists.

Objectives

Identify the contributions of Schleiden, Schwann, and Virchow to the development of the cell theory.

Summarize the three parts of the cell theory.

Evaluate the importance of the cell theory.

Discover By Doing

Observe a leaf with your unaided eye. Then observe the leaf by using a magnifying glass. How many times does the magnifying glass enlarge the leaf? What can you see with the magnifying glass that you could not see before? Which of your observations surprised you? Discuss with a classmate the best way to record your observations.

PERFORMANCE ASSESSMENT

Have small groups of students discuss the best way to record their leaf observations from the *Discover By Doing* activity. Ask the groups to decide on a method and then explain why they chose that method. Use the discussion to evaluate the students' ability to solve problems and make decisions.

Early Observations

Have you ever wondered how the first microscope was made? Did the inventor just put two magnifying glasses together?

Actually, that is close to what really happened. At the end of the sixteenth century, a Dutch lens maker, Zacharias Janssen, put two magnifying lenses together in a tube and invented the compound microscope. About eighty years later, another Dutch scientist, Anton van Leeuwenhoek, made many simple microscopes, each with a single lens. Leeuwenhoek used lenses that could magnify objects up to 270 times their actual size. He made careful records of the things he observed with the microscope. You can read more about Leeuwenhoek on page 250 at the end of Unit 3.

SECTION 2 **105**

Magnifying an object helps you discover more about its structure. During the 1660s, an English scientist, Robert Hooke, constructed a microscope to study thin slices of cork. Cork is the material under the bark of certain trees. The open spaces Hooke discovered in the cork reminded him of the tiny rooms in which monks lived. So Hooke gave the tiny spaces in the cork tissue the same name, **cells.**

Scientists now know that Hooke was not looking at living cells. He was looking at the walls that had once surrounded the living cells of the cork. However, Hooke's observations opened the door to many new scientific studies.

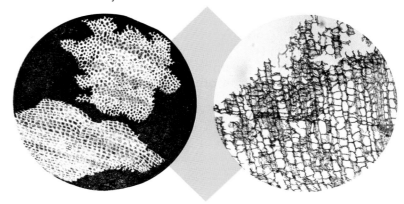

Figure 4–8. Cork examined under a modern microscope (right) looks very similar to what Hooke saw (left).

 ASK YOURSELF

Why do you think the early scientists were interested in making microscopic observations?

Cell Theory

Nearly 200 years after Hooke's discovery, a German biologist, Matthias Schleiden, was using a microscope to study living plants. Over the years, he looked at the parts of many different plants and found that they were alike in many ways. All the plant parts that Schleiden studied contained cells. His observations led him to conclude that all plants are made of cells.

About the same time, Theodor Schwann, another German biologist, studied many different animals. His many years of observations led him to conclude that all animals are made of cells. The work of Schleiden, Schwann, and many other scientists was studied by Rudolf Virchow, another German scientist. Virchow concluded from his research that all new cells come from other living cells.

EVALUATION

Assign six students one of the people from the section (Janssen, Leeuwenhoek, Hooke, Schleiden, Schwann, Virchow). Ask the students to describe their person's contribution to the development of the microscope or the cell theory. Have the rest of the class listen to each description and identify the person being talked about.

RETEACHING

Have the students orally restate the three parts of the cell theory using their own words.

EXTENSION

Remind the students that many scientists contributed to the development of the cell theory. Ask the students to describe other scientific investigations in which the contributions of many scientists have combined to increase science knowledge. (The students might suggest moon missions or medical research of diseases.)

CLOSURE

In their journals, have the students write a paragraph that explains how the cell theory suggests that all living things are related.

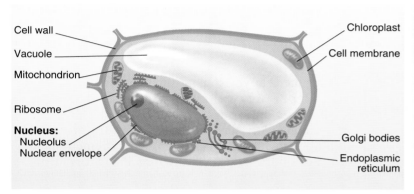

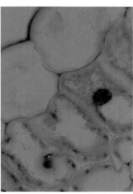

Figure 4–9. The photograph and diagram above show the variety of structures found in plant cells.

The work of Schleiden, Schwann, and Virchow led to one of the most important theories in biology—the cell theory. The **cell theory** is composed of three basic statements.

1. All organisms are made of cells.
2. Cells are the basic units of structure and function for all living things.
3. All cells come from other cells.

The cell theory is important because it shows that all living things share similar structures. This similarity suggests that all organisms are related. In addition to providing answers, however, the cell theory also led to more questions. For example, what is the structure of a cell? How do cells work? How do they reproduce? Although many of these questions have been answered, many others are still being studied by life scientists.

▼ ASK YOURSELF

Why would a scientist studying a newly discovered living organism expect to see cells under a microscope?

SECTION 2 REVIEW AND APPLICATION

Reading Critically

1. State the three parts of the cell theory.
2. Which statements in the cell theory were a result of Schleiden's, Schwann's, or Virchow's work?

Thinking Critically

3. Why was the microscope an important discovery?
4. Why is it important for scientists to study the structure of a cell?
5. A book is made up of chapters, paragraphs, sentences, and letters. What is the basic unit of structure of a book? Use two parts of the cell theory to create a "book theory" similar to the cell theory. Which part of the cell theory were you not able to use? Why?

ONGOING ASSESSMENT
▼ ASK YOURSELF

Cell theory states that all living organisms are made of cells and that cells are the basic units of structure and function for all living organisms.

SECTION 2 REVIEW AND APPLICATION

Reading Critically

1. Cell theory states that all organisms are made of cells, cells are the basic units of structure and function for all living things, and all cells come from other cells.

2. The statement that all organisms are made of cells resulted from the work of Schleiden and Schwann. Schleiden concluded that all plants and parts of plants are composed of cells, while Schwann concluded that the same was true for animals. Virchow concluded that all new cells come from other living cells.

Thinking Critically

3. Responses should include a reference to the ability to observe objects that are too small to be seen with the unaided eye.

4. Understanding the structure of a cell gives clues to the ways in which cells function.

5. Responses might suggest that all books are made of letters and that letters are the basic unit of structure for all books. The component of cell theory that states all cells come from other cells is not usable in this analogy; all letters do not come from other letters.

SECTION 2   **107** ◀

Section 3: STRUCTURE OF CELLS

FOCUS

This section examines cell structures and their functions. The structures and functions of the organelles found in the cytoplasm of cells are also presented. This section describes the levels of cellular organization, emphasizing how the cells of multicellular organisms are organized into different levels of specialization.

MOTIVATING ACTIVITY

Cooperative Learning Ask small groups to design an activity that could be completed by the group as a whole but would be difficult for an individual member to complete. Point out that the design should identify the specific role of every group member. Then have volunteers from each group share their design with other groups. Ask the groups to decide what factors are common to all the designs. (the division of labor and the spirit of cooperation) Ask the students to explain how the concepts of the division of labor and the spirit of cooperation exist inside the human body. (Responses should reflect the understanding

---

PROCESS SKILLS
- Observing • Comparing
- Inferring

POSITIVE ATTITUDES
- Cooperativeness

TERMS
- organelles • cell membrane
- nucleus • tissues • organ
- system

PRINT MEDIA
The Lives of a Cell: Notes of a Biology Watcher by Lewis Thomas
(see p. 93b)

ELECTRONIC MEDIA
Cell Structure,
Britannica
(see p. 93b)

Science Discovery
Cell shapes

BLACKLINE MASTERS
Study and Review Guide
Laboratory Investigation 4.2
Connecting Other Disciplines
Thinking Critically

DISCOVER BY *Researching*

The students should discover that there is a wide variety of cell shapes and cell sizes, depending on the cell's function, and they should determine that all cells share three similar structures: a definite boundary, a control center or nucleus, and cytoplasm.

SECTION 3

Structure of Cells

Objectives

Summarize the function of each organelle.

Recognize different levels of organization within organisms.

The first time that you look at the new life form under the microscope is very exciting. You expect to see cells because all living organisms are made of cells. Each cell functions along with the other cells to keep the organism alive and healthy. But the cell, although extremely small in size, is not the smallest structure in an organism. Within a cell are different structures, called *organelles* (AWR guh nehlz), that help the cell to carry out its functions. **Organelles** are structures within a cell that have certain jobs to do for the cell, much like each organ in your body has a job to do. So organelles can be thought of as the "organs" of the cell.

The parts of the cell make it possible for each cell to grow, to release energy from food, to get rid of wastes, and to divide to create new cells. You might think of a cell as a small factory. Different people in the factory are all busy at different jobs. But working together, they create a smooth-running and efficient team.

DISCOVER BY *Researching*

Using reference materials in a library, look for pictures of cells. Make sketches of cells with interesting shapes to share with the class. Do all cells have the same shape? What types of structures might all cells have in common?

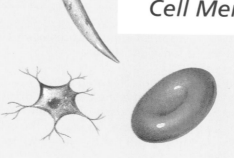

Cell Membrane and Nucleus

Although cells have a wide variety of shapes, sizes, and colors, every cell has an outer covering, much as skin covers the outside of your body. This covering that surrounds the cell is called the **cell membrane.** The structure of the cell membrane allows certain materials to pass through it. Like a security guard at a factory, the cell membrane controls everything that goes into and comes out of the cell.

▶ 108   CHAPTER 4

TEACHING STRATEGIES

that individual cells of the body contribute to the life processes performed by and for the whole body.) Point out that in this section, the students will discover that not only the cells but also the structures of the body formed by cells work together for the well-being of the entire human.

● **Process Skills:** *Comparing, Analyzing*

Develop an analogy between a large manufacturing plant and an animal cell by writing the terms *nucleus, cell membrane, nuclear envelope, cytoplasm, endoplasmic reticulum, Golgi bodies, ribosomes, mitochondria,* and *vacuoles* in a column on the chalkboard. Begin a second column by writing "= plant manager" next to the term *nucleus,* and "= plant security guard" next to the term *cell membrane.* Discuss how the cell nucleus is like a manufacturing plant manager and how the cell membrane is like a security guard at a manufacturing plant. Have the students write both columns in their journals. Then ask them to think of other jobs to complete the list as they study the remaining terms in the chapter.

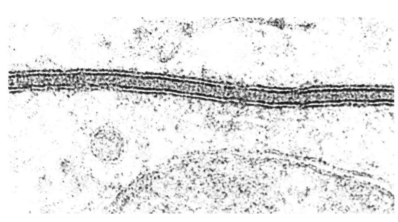

Figure 4–10. An entire red blood cell (right) shows the three-dimensional structure of the cell membrane. The electron micrograph (left) shows the layers of the cell membrane.

The cell membrane has tiny openings that let water enter the cell. Other openings let other small molecules pass through the membrane but usually keep large molecules out.

As you look inside a cell, the most prominent object you see is usually the nucleus. The **nucleus** controls the cell's activities, much as your brain controls the activities of your body or the owner of a factory controls its operations. The instructions for all of the cell's activities are in the material that is found inside the nucleus. Around the nucleus is a special covering called the *nuclear envelope.* Like the cell membrane, this envelope controls the movement of materials into and out of the nucleus.

ASK YOURSELF

How are the cell membrane and the cell nucleus important to the activities of the cell?

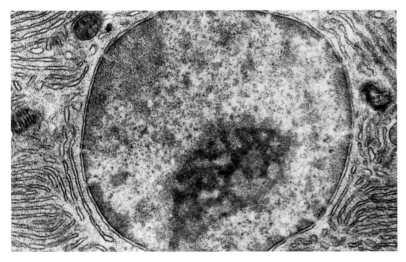

Figure 4–11. The nucleus of a cell is surrounded by the nuclear envelope. How is the nucleus of a cell similar to the human brain? ①

SCIENCE BACKGROUND

The cell membrane separates the living material inside the cell from the nonliving material in the environment. If the cell membrane became damaged or torn, the contents of the cell would spill out into the cell's environment.

ONGOING ASSESSMENT
ASK YOURSELF

The cell membrane retains the contents of a cell, protects the cell, and controls the materials that move into and out of the cell. The cell nucleus controls the cell membrane and the activities of the cell organelles.

① The nucleus of a cell controls the activities of that cell in the same way that the human brain controls the activities of the human body.

✧ Did You Know?
Cobra venom dissolves cell membranes, causing the death of the cells. When too many cells are damaged, the victim dies.

SECTION 3

TEACHING STRATEGIES, continued

● **Process Skills:** *Inferring, Comparing*

Explain to the students that each organ in a body performs a specific function that helps keep the body alive; the same is true for the organelles in cells. Ask the students to name the part of the human body that is similar to the endoplasmic reticulum. (The network of blood vessels in a human body is similar to the tubelike network of the endoplasmic reticulum.) Ask the students to name the part of the human body that is similar to cytoplasm. (Blood in the human circulatory system is similar to the cytoplasm of a cell.)

● **Process Skills:** *Comparing, Interpreting Data*

Ask the students to describe how mitochondria are like batteries. (Mitochondria produce the energy necessary for the life processes of a cell.) Point out that the number of mitochondria in a cell varies from cell to cell, then ask the students to relate the number of mitochondria in a cell to the amount of activity going on in a cell. (Cells performing functions that require a great deal of energy will have more mitochondria than cells performing functions that do not require a great deal of energy.)

INTEGRATION—Health

Ribosomes manufacture proteins from amino acids. Humans need to eat foods that contain amino acids. The best supplies of amino acids can be found in eggs, meat, milk, cheese, and fish. Because the body cannot store proteins for later use, it must receive a steady supply of amino acids in its diet, at least 0.8 g per kg of body weight per day.

MEETING SPECIAL NEEDS

Gifted

Point out that there are two types of endoplasmic reticulum, or ER—rough ER and smooth ER. Have the students research the function of rough and smooth ER, and report their findings to the class. (Rough ER is covered with ribosomes and often carries proteins made by the ribosomes to the outside of the cell. Smooth ER is not covered with ribosomes and helps maintain the cell membrane and break down poisons that enter the cell.)

① The materials produced by ribosomes can be moved directly into the endoplasmic reticulum for transport throughout the cell.

▶ 110 CHAPTER 4

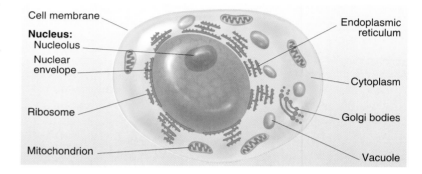

Figure 4-12. This illustration shows the variety of structures found in an animal cell.

Other Cell Parts

The material between the cell membrane and the nucleus is called *cytoplasm* (SY tuh plaz uhm). The largest organelle in the cytoplasm is the *endoplasmic reticulum* (ehn duh PLAZ mihk rih TIHK juh luhm). This organelle is a network of tubelike canals that run through the cytoplasm.

The endoplasmic reticulum is used much like streets or roads. One of its jobs is to transport materials from place to place in the cell. But before the endoplasmic reticulum can move the materials, they must be "packaged." This packaging is the function of the *Golgi bodies*.

Most of the cell and much of the body are made of proteins. The organelles that make proteins are the *ribosomes* (RY buh sohmz). Ribosomes are like small machines that take raw materials from the cytoplasm to make the proteins. The ribosomes are found throughout the cytoplasm. Many are also attached to some parts of the endoplasmic reticulum. Since ribosomes function to produce materials for the cell, what advantage occurs when they are attached to the endoplasmic reticulum? ①

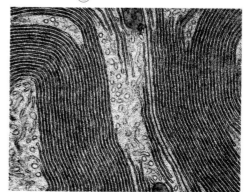
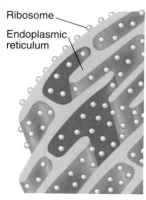

Figure 4-13. The ribbonlike structures in the electron micrograph (left) are the endoplasmic reticulum. In the illustration (right), ribosomes can be seen covering the endoplasmic reticulum.

- **Process Skills:** *Classifying/Ordering, Applying*

Use an overhead projector or slide projector to display pictures of different types of cells. Have the students identify the major organelles in each cell and discuss the differences in the structure of the cells. Explain that a difference in cell structure usually reflects a difference in cell function. (For example, cells of the skin and those that line openings in the body are flat and thin and can cover a large surface area; some muscle cells are tapered and fit into one another; nerve cells have long extensions and can carry impulses for long distances in the body.)

- **Process Skills:** *Inferring, Applying*

Have the students describe the relationship between the location of railroads and the location of older manufacturing facilities in their community. Then ask the students to compare the relationship between manufacturing facilities and railroads to the relationship between ribosomes and the endoplasmic reticulum. (The endoplasmic reticulum, like railroads, provides ribosomes, like manufacturing plants, with a way to directly transport materials from the source to other destinations.)

The organelles that release energy are *mitochondria* (myt uh KAHN dree uh). For this reason, the mitochondria are often called the "powerhouses" of the cell. Food molecules are broken down inside these organelles, and energy is released. This energy is used to keep the cell working properly. Energy from trillions of mitochondria in billions of cells keeps the body working properly.

Another organelle found in cells is the *vacuole* (VAK yu ohl). Vacuoles are sacs filled with liquid. Some vacuoles store water or food. Other vacuoles are used to store waste material until it can be removed from the cell. In a plant cell, there is a large central vacuole that makes up most of the cell. This vacuole is filled with water. By comparison, the vacuoles found in animal cells are much smaller than those found in plant cells.

There are two structures that are found only in plant cells. *Chloroplasts* (KLAWR uh plasts) are large, green, oval-shaped organelles in which photosynthesis takes place. Chloroplasts contain chlorophyll, which, together with light energy, enables the plant to make food. The *cell wall* is the other structure found in plant cells but not in animal cells. The cell wall is found outside the cell membrane, where it helps support the plant and protect the cell. Unlike the cell membrane, the cell wall does not stop materials from entering the cell. The cell wall is made of cellulose, which is nonliving. Why would Hooke see the cell walls of cork cells even after the cells had died? ②

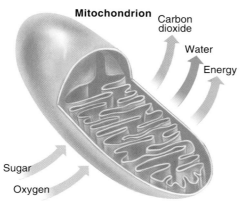

Figure 4–14. Mitochondria, as seen here, are responsible for releasing all the energy the cell needs to function.

Figure 4–15. Chloroplasts convert light energy into chemical energy.

 ASK YOURSELF

Why do cells have so many organelles?

Levels of Organization

Before manufacturing became widespread in the nation, many people lived on farms far away from other farms. Each family raised its own food, made its own clothes, and built its own shelter.

SCIENCE BACKGROUND

The centriole is an organelle found in the cytoplasm of the cell during cell division. Centrioles are found in all animal cells but only in some plant cells. Centrioles are found in pairs. Two pairs of centrioles can be seen outside the nuclear envelope just before cell division begins.

② Since the cell walls are nonliving cellulose, they did not deteriorate after the cell died.

ONGOING ASSESSMENT
ASK YOURSELF

Organelles perform activities for a cell. Because a cell performs many different activities, it must have many organelles to perform those activities.

REINFORCING THEMES—
Systems and Structures

While discussing the various organelles of a cell, make sure the students understand that the presence of specialized cells is an important characteristic of multicellular organisms, and that specialized cells function in a unique way, both for the benefit of the cell and for the benefit of the entire organism.

 **LASER DISC**
2706

Cell shapes

SECTION 3

TEACHING STRATEGIES, continued

● **Process Skills:** *Inferring, Applying*

When discussing levels of organization, point out that *organs* can belong to more than one *system*. For example, bones are classified as organs and belong to the skeletal system, yet their production of red blood cells, lymphocytes, and platelets also classifies them as part of the immune system. Ask the students to think of other organs that belong to more than one system. (The pancreas belongs to the digestive system because of its production of digestive enzymes and to the endocrine system because of its production of insulin.)

GUIDED PRACTICE

Have the students identify the function of the following organelles: endoplasmic reticulum (transports materials throughout the cell); Golgi bodies (package materials for transport); ribosomes (manufacture proteins); mitochondria (release energy); vacuoles (store materials); chloroplasts (make food) cell wall (supports the plant)

INDEPENDENT PRACTICE

Have the students write answers to the Section Review and Application questions. In their journals, ask the students to draw a diagram that shows the relationship among systems, organs, tissues, cells, and the organism.

Demonstration
Use a model of the human skeletal system to reinforce the concept of levels of organization. Point out that cells, which are the first level of organization, make up bone tissue. Tissues are the second level of organization. Bone tissue makes up bones. Organs, such as bones, are the third level of organization. All of the bones of the body make up the skeletal system. Systems are the fourth level of organization. Ask the students what makes up the highest level of organization. (All the cells, tissues, organs, and systems make up the organism.)

BACKGROUND INFORMATION

The concept of division of labor developed during the Industrial Revolution, which began in Great Britain in the 1700s. Before that time, manufacturing was done mostly by hand and at home. One worker would make each product from start to finish. With the development of machines to manufacture goods, the machines and the workers had to come together in one place—the factory. To increase production and lower costs, the factory used division of labor—each worker performed only one step in the process of manufacturing a product.

Figure 4–16. A single-celled organism must perform all life functions. In a multicellular organism, cells specialize to perform certain functions.

Figure 4–17. The shapes of cells vary with their functions. Shown here are muscle cells (left), blood cells (center), and plant transport cells (right).

As time went on, a kind of exchange system developed. The skilled blacksmith would shoe the farmer's horse. The farmer would provide the blacksmith with food. The carpenter would build an office for the doctor, and the doctor would provide medical care for the carpenter's family. As years passed, each person became a specialist. Then factories began to manufacture goods. Each worker in a factory had a specific job. Together the workers produced a finished product.

Cells Organize Single-celled organisms are similar to the independent farming families. Like each family, each cell must perform all the life functions by itself. Each part of the cell has a certain function.

Most organisms are not made of one cell; they are made of many cells. The organisms are *multicellular.* Each cell in a multicellular organism has all of the basic cell parts. However, each of these cells has also become specialized. Each has a function that benefits the other cells. In this way, the cells of a multicellular organism depend on one another for survival just as workers in a factory depend on one another to produce a finished product.

The specialized cells of multicellular organisms are organized into different levels. The cell itself is at the first level of organization. Most cells have a certain size and shape and certain organelles related to the cell's purpose. For example, in humans there are two types of blood cells and three types of muscle cells. There are also covering cells, nerve cells, bone cells, and fat cells. In plants, there are cells specialized for absorbing water from the soil. There are also cells in plants for covering and protecting, transporting, and growing.

Tissues, Organs, and Systems The second level of organization is tissues. **Tissues** are groups of similar cells that perform a specific function. Blood cells form blood tissue. One of the functions of blood tissue is to carry oxygen to the body cells. Muscle cells form muscle tissue. Muscle tissue is responsible for movement.

Plant cells also form specialized tissues. Covering cells on leaves are called *epidermal tissue.* Its function is to protect the cells beneath it.

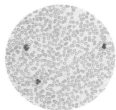
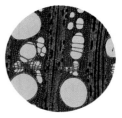

EVALUATION

Ask the students to identify the most important organelle in a cell and explain why other organelles of the cell are not as important. (Responses might suggest that the nucleus is the most important cell organelle because it directs the activities of all of the organelles in a cell.)

RETEACHING

On the chalkboard, draw a small box and then four more boxes, each box enclosing the previous box. Ask the students to identify each of the five boxes as one level of organization by writing the term *cells, tissues, organs, systems,* or *organism* on the appropriate box.

EXTENSION

Point out that organelles can reach a point in their life cycle when they do not function properly. When that occurs, they are broken down and removed from the cell by specialized organelles. Ask the students to find out the name of these specialized organelles that act as cleanup crews in the human body. (lysosomes)

CLOSURE

 Cooperative Learning Assign each group a system of the human body (nervous, digestive, circulatory, muscular, or respiratory). Ask the groups to explain and give examples of the different levels of organization in their assigned system.

Organs represent the third level of organization. An **organ** is a group of different tissues working together to perform a certain task. The human heart is one such organ. The heart is made up of muscle, nerve, covering, and blood tissues. Together these tissues have a unique function: they pump blood through the body. Other examples of human organs are the eye, the stomach, and the lung.

A leaf is an example of a plant organ. A leaf contains several tissues. Each performs a certain task. Together, as an organ, the tissues function to make food. Roots and stems are other plant organs.

A group of organs working together is known as a **system**. Systems are the fourth level of organization. In humans there are many systems. These include the digestive, respiratory, nervous, circulatory, and reproductive systems. In plants, systems include transport and reproductive systems.

The highest level of organization is the *organism*. A single living organism is the combination of all its systems, organs, tissues, and cells.

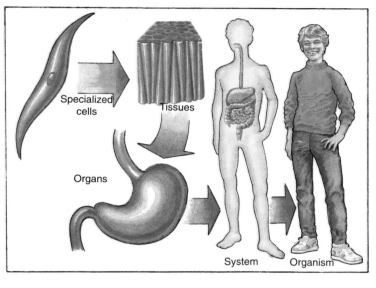

Figure 4–18.
Multicellular organisms exhibit different levels of organization. This results in division of labor, or sharing of functions, within the organism.

ASK YOURSELF
What is the advantage of specialized cells?

SECTION 3 REVIEW AND APPLICATION

Reading Critically
1. How does the structure of the cell membrane help it perform its function?
2. What does the nucleus do for the cell?
3. Make a chart showing each organelle. Next to the organelle, describe its function in a cell.

Thinking Critically
4. Why would a muscle cell have more mitochondria than a skin cell?
5. The chloroplasts in plant cells help the plant make its own food from light energy. How do animals get their food?
6. How might the presence of different organelles help plant cells and animal cells perform different functions? How would this help the organism?

ONGOING ASSESSMENT
ASK YOURSELF
Specialized cells allow an organism to become more complex.

SECTION 3 REVIEW AND APPLICATION

Reading Critically

1. Tiny openings in the cell membrane allow certain molecules to pass through while keeping other molecules out.

2. The nucleus is the control center of a cell, regulating the functions of the cell and its organelles.

3. The students' charts should include descriptions of the functions of the following organelles: cell membrane, cell wall, chloroplast, cytoplasm, endoplasmic reticulum, mitochondria, nuclear envelope, nucleus, ribosomes, and vacuole.

Thinking Critically

4. A muscle cell has more mitochondria because, in order to perform its work, it requires a greater supply of the energy that mitochondria provide.

5. Animals obtain food by eating plants or eating animals that eat plants or other animals.

6. Responses might suggest that different organelles perform different functions, and the greater the diversity of organelles within an organism, the greater the diversity of the organism and the greater the likelihood it will survive and reproduce.

SKILL

Comparing Plant and Animal Cells

Process Skills: Observing, Comparing

Grouping: Groups of 3 or 4

Objectives
- **Observe** the organelles of plant and animal cells.
- **Compare** the organelle structures of plant and animal cells.
- **Communicate** observed data in the form of a table.

Discussion
Stress to the students that the ability to make precise and clear scientific drawings and diagrams is a valuable skill. Also remind them that even though their artistic abilities will vary, all of them can learn to be neat and to record data accurately.

Application
1. Plant cells are represented by the cells of the *Elodea* leaf, and animal cells are represented by the cheek cells. Some organelles (chloroplasts, cell walls) are present in plant cells but not in animal cells.

2. Structures that may be observed include: cell membrane, cell wall, chloroplast, cytoplasm, endoplasmic reticulum, mitochondria, nuclear envelope, nucleus, ribosome, and vacuole.

3. Check each table for accuracy and completeness.

✷ Using What You Have Learned
Additional observations and data will vary. You may wish to have the students compare and discuss their tables.

★ PERFORMANCE ASSESSMENT
Ask the students to refer to their tables as they identify the differences and similarities between plant and animal cells. Have pairs of students exchange and check each other's table.

 Have the students place their tables from the *Skill* activity in their science portfolios.

SKILL Comparing Plant and Animal Cells

▶ MATERIALS
- slide • coverslip • medicine dropper • *Elodea* leaf • compound microscope
- prepared slide of human cheek cells

▼ PROCEDURE

1. Place a drop of water on a clean slide. Place an *Elodea* leaf in the drop of water and cover with a coverslip.
2. Observe the *Elodea* leaf under low power of the microscope. Observe one cell and make a sketch of what you see.
3. Switch to high power. Again observe one cell and sketch what you see.
4. Obtain a prepared slide of human cheek cells from your teacher. Place the slide on the microscope and focus on one cell under low power. Sketch what you see.
5. Switch to high power and again sketch what you see.

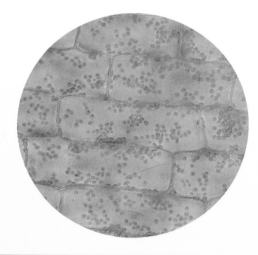 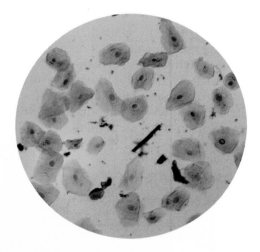

▶ APPLICATION
Which cells that you observed were animal cells? Plant cells? How can you tell the difference through your observations?

What structures were you able to observe in each of the cells you viewed?

Make a table in which to write a comparison between the plant and animal cells you observed. Give your table a title and remember to title the columns.

✷ Using What You Have Learned
Obtain some other prepared slides of plant and animal tissues. Examine these slides under a microscope. Add to your table any new observations you were able to make.

▶ 114 CHAPTER 4

CHAPTER 4 HIGHLIGHTS

The Big Idea—SYSTEMS AND STRUCTURES

Lead the students to understand that even though sophisticated equipment such as the compound light microscope allows scientists to study and learn about the very small structures inside of a cell, observations on a larger scale, performed without the use of scientific equipment, are also valuable. Also remind the students that the life processes of a cell and of an entire organism are dependent on each organelle or structure of each cell performing its allotted task.

CHAPTER 4 HIGHLIGHTS

The Big Idea

Life scientists study plants and animals using a wide variety of tools and measurements. They look at an organism and its external structure, how it interacts with other parts of its environment, and—on a microscopic scale—how its internal structures work together.

By using a common system of measurement, SI, scientists around the world can share their findings. In the same way that the cell theory was the result of the work of several different people, scientists today can read and study each other's findings to gain new knowledge about living organisms.

In fact, scientists working in different countries many thousands of kilometers apart are actually part of a common "team." Just as the structures that make up plant and animal cells all coordinate activities, life scientists work together as they gain new knowledge of the living world.

Review the questions you answered in your journal before you read the chapter. How would you answer the questions now? Add an entry to your journal that shows what you have learned from the chapter. What new ideas do you have about why microscopic studies are important to understanding living organisms?

 *For Your Journal*

The ideas expressed by each student should reflect the understanding that the microscope has revolutionized the scientific view of the world in which we live. You might wish to suggest that volunteers share not only their observations of things they have seen using a microscope, but also their rationale concerning the usefulness of microscopic study.

CONNECTING IDEAS

 The concept map should be completed with words similar to these:

Row 3: all living things; cell membrane; nucleus

Row 5: organelles

Connecting Ideas

Copy this unfinished concept map into your journal. Complete the concept map by writing the correct term in each blank.

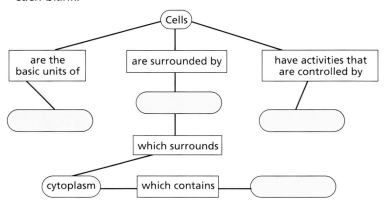

CHAPTER 4 REVIEW

ANSWERS

Understanding Vocabulary

1. *Kilogram* does not belong. Length, volume, and mass are quantities that can be measured; a kilogram is an SI unit of measure for mass.

2. *Millimeter* does not belong. A magnifying glass, compound light microscope, telescope, and binoculars are instruments used to optically enlarge images; a millimeter is an SI unit of measure for length.

3. *Cell theory* does not belong. Mitochondria, cell membrane, cell wall, and nucleus are organelles of a cell; cell theory shows how living things share similar structures.

4. *Microscope* does not belong. Tissues, organ, and system represent levels of multicellular organization; a microscope is an instrument used to magnify very small objects.

Understanding Concepts

Multiple Choice

5. b
6. c
7. a
8. d

Short Answer

9. Responses might suggest that many people do not want to give up a system they are familiar with or learn a new system, while changing to SI would bring the United States in line with many other countries.

10. Molecules that supply energy and raw materials necessary for cell activities must move into a cell. Waste products must move out of a cell.

Interpreting Graphics

11. Responses might suggest that the plant stem and the cork are similar in that both are composed of cells. They are different because the cork is composed of empty cells, while the plant stem cells contain inner structures.

Reviewing Themes

12. Responses might suggest that information about behavioral traits, food preferences, and social interactions cannot be collected during microscopic investigations.

13. Responses should indicate that the other parts of a cell would be adversely affected if water and food were not stored and readily available. Also waste materials would collect throughout the cell if they were not stored in vacuoles before removal.

Thinking Critically

14. Equations should include 12 inches = 1 foot, 3 feet = 1 yard, 36 inches = 1 yard; and 1 kilometer = 1000 meters, 1 meter = 100 centimeters, and 100,000 centimeters = 1 kilometer. SI's advantage is that each SI conversion is accomplished by multiplying or dividing by 10, 100, or 1000.

CHAPTER 4 REVIEW

Understanding Vocabulary

Choose the word or phrase that does not belong in each group. Then explain why that term is different from the others. What do the other terms have in common that makes them alike?

1. length, kilogram, volume, mass
2. millimeter, magnifying glass, compound light microscope (102), telescope, binoculars
3. mitochondria (111), cell membrane (108), cell theory (107), cell wall (111), nucleus (109)
4. tissues (112), organ (113), microscope (102), system (113)

Understanding Concepts

MULTIPLE CHOICE

5. Scientists have agreed to use a common system of measurement
   a) because SI is the most accurate.
   b) so that they can communicate more easily with each other.
   c) so that all scientific experiments will have the same results.
   d) because SI is based on the number 10.

6. The early scientist Matthias Schleiden used a microscope to discover that all living plants
   a) come from other living plants.
   b) can make their own food.
   c) contain cells.
   d) include chloroplasts.

7. Which statement is *not* included in the cell theory?
   a) Cells cannot be studied without using a microscope.
   b) All cells are from other cells.
   c) Cells are the basic units of structure and function for all living things.
   d) All organisms are made of cells.

8. Which of the following activities would occur in a plant cell but *not* in an animal cell?
   a) Materials are transported from one place to another.
   b) Food molecules are broken down and energy is released.
   c) Waste material is stored until it can be removed.
   d) Food is manufactured using light energy.

SHORT ANSWER

9. List reasons for and against using SI in the United States.
10. The cell membrane controls the passage of materials into and out of the cell. Give examples of materials that would be entering and leaving a cell.

Interpreting Graphics

11. The photograph shows a slice of stem from a plant. How is this photograph like the cork that Hooke saw under his microscope? How is it different?

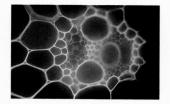

▶ 116 CHAPTER 4

15. Two hundred plant cells is a reasonable estimate.

16. Responses should indicate that those tools have enabled scientists to see structures and organisms previously invisible to humans.

17. If certain materials such as water and nutrients were not allowed to move through the cell membrane, a cell would die because it needs those materials to survive, and it has no other way to obtain the materials.

18. A cell in a unicellular organism must perform all the life functions by itself. A cell in a multicellular organism is specialized; it performs only one particular function and depends on other specialized cells for survival.

19. Models should be checked for accuracy.

20. Examples might describe how each member of a family contributes to preparing and cleaning up after a meal by performing one task that is a part of the whole job. For example, someone prepares the food, someone sets the table, someone clears the table, someone washes the dishes, someone dries the dishes, and someone puts the dishes away.

Reviewing Themes

12. *Systems and Structures*
Animals are studied on both the microscopic and macroscopic scales. What information about an animal *cannot* be collected during microscopic investigations?

13. *Systems and Structures*
How would the other parts of a cell be affected if the vacuoles were no longer able to store water and food?

Thinking Critically

14. Write three equations to show the relationships between inches and feet, feet and yards, and inches and yards. Then write three equations showing the relationships among centimeters, meters, and kilometers. Explain how your equations show an advantage built into SI.

15. A scientist measured a plant cell and found that it was about 50 micrometers (μm) across. Estimate the number of these plant cells that would fit across your thumbnail. Use 1 cm for the width of your thumbnail and the fact that 1 m = 1 000 000 μm.

16. The photograph below is an electron micrograph of a Mediterranean fruit fly magnified 21 times. How have both the compound light microscope and the electron microscope added to our knowledge of living things?

17. Why does a cell's survival depend on the movement of certain materials through its cell membrane?

18. Unicellular living organisms are made of only one cell; multicellular organisms have many cells. Compare the function of a cell in a unicellular organism with the function of a cell in a multicellular organism.

19. Make a three-dimensional model of an animal cell or a plant cell, showing the structures described in this chapter. Use modeling clay or gelatin to form the cytoplasm of the cell. Use modeling clay to form the organelles.

20. Use an example other than a pioneer community to compare division of labor in society with levels of organization in a multicellular organism.

Discovery Through Reading

Bender, Lionel. *Through the Microscope: Atoms and Cells.* Gloucester Press, 1990. Text and photographs introduce microscopic plant and animal life, viruses, microspores, and other forms of life that can be viewed only through a microscope.

CHAPTER 5

CELL FUNCTION

PLANNING THE CHAPTER

Chapter Sections	Page	Chapter Features	Page	Program Resources	Source
Chapter Opener	118	*For Your Journal*	119		
Section 1: TRANSPORT IN CELLS	120	Discover By Observing (B)	121	*Science Discovery**	SD
• Diffusion and Active Transport (B)	120	Discover By Writing (B)	122	Investigation 5.1: Permeable Membranes (H)	TR, LI
• Osmosis (B)	123	Activity: How do food substances diffuse through a cell? (A)	123	Extending Science Concepts: Using Models (H)	TR
		Section 1 Review and Application	124	Record Sheets for Textbook Investigations (A)	TR
		Investigation: Observing Osmosis (A)	125	Study and Review Guide, Section 1 (B)	TR, SRG
Section 2: CELL REPRODUCTION	126	Discover By Researching (A)	127	*Science Discovery**	SD
• Methods of Cell Reproduction (B)	126	Discover By Problem Solving (A)	129	Investigation 5.2: Mitosis (A)	TR, LI
• Mitosis (B)	127	Section 2 Review and Application	129	Reading Skills: Identifying a Scientific Experience (A)	TR
				Mitosis (A)	IT
				Study and Review Guide, Section 2 (B)	TR, SRG
Section 3: DNA AND CELL ENERGY	130	Discover By Researching (B)	131	*Science Discovery**	SD
• DNA and Replication (A)	130	Discover By Doing (B)	135	Connecting Other Disciplines:	
• Respiration and Fermentation (H)	134	Section 3 Review and Application	137	Science and Social Studies, Making a Model of DNA (A)	TR
• Photosynthesis (H)	136	Skill: Modeling DNA (A)	138	Thinking Critically (H)	TR
				DNA Replication (A)	IT
				Respiration and Photosynthesis (A)	IT
				Study and Review Guide, Section 3 (B)	TR, SRG
Chapter 5 HIGHLIGHTS	139	The Big Idea	139	Study and Review Guide, Chapter 5 Review (B)	TR, SRG
Chapter 5 Review	140	For Your Journal	139	Chapter 5 Test	TR
		Connecting Ideas	139	Test Generator	

B = Basic   **A** = Average   **H** = Honors
The coding Basic, Average, and Honors indicates subsections, features, and resources that might be appropriate for different levels of learners. For additional suggestions regarding choice of topic and depth of coverage, see the Pacing Chart on pages T26–T29.

*Frame numbers at point of use
(TR) Teaching Resources, Unit 2
(IT) Instructional Transparencies
(LI) Laboratory Investigations
(SD) *Science Discovery* Videodisc Correlations and Barcodes
(SRG) Study and Review Guide

▶ 117A

CHAPTER MATERIALS

Title	Page	Materials
Discover By Observing	121	(per class) perfume
Discover By Writing	122	(per individual) paper, pencil
Activity: How do food substances diffuse through a cell?	123	(per group of 3 or 4) agar-filled Petri dish, plastic straw, scissors, filter paper, empty Petri dish, medicine dropper, scoop, food substance, water
Investigation: Observing Osmosis	125	(per group of 3 or 4) water, *Elodea* leaf, slide, coverslip, microscope, salt water, medicine dropper (2), paper towel
Teacher Demonstration	126	modeling clay
Discover By Researching	127	(per individual) dictionary, paper, pencil
Discover By Problem Solving	129	(per group of 3 or 4) slides of mitosis, paper, pencil, microscope
Discover By Researching	131	(per individual) poster board, pencil (per class) reference materials
Discover By Doing	135	(per individual) yeast, beaker, warm water, sugar
Skill: Modeling DNA	138	(per group of 3 or 4) fine wire (2), miniature marshmallows (6), jelly beans (8), four colors of pipe cleaners (6 of each color), modeling clay

ADVANCE PREPARATION

For the *Discover By Observing* on page 121, obtain one bottle of inexpensive perfume. For the *Investigation* on page 125, you will need to obtain *Elodea* and prepare a saltwater solution. For the *Demonstration* on page 126, locate modeling clay. For the *Discover By Problem Solving* on page 129, the students will need sets of slides showing the stages of mitosis. These can be obtained from a biological supply house. For the *Discover By Doing* on page 135, the activity needs small packages of yeast. For the *Skill* on page 138, you will need fine wire, miniature marshmallows, jelly beans, four different colors of pipe cleaners, and modeling clay.

TEACHING SUGGESTIONS

Field Trip
Local companies doing pharmaceutical, agricultural, or medical research may be involved in investigations using recombinant DNA. A visit to these laboratories may be possible, or a guest speaker may be available to speak to your class.

Outside Speaker
The process of mitosis is a very important aspect of some areas of cancer research. Contact the local office of the American Cancer Society to arrange for a scientist involved in such research to speak to your class.

CHAPTER 5
CELL FUNCTION

CHAPTER THEME—SYSTEMS AND STRUCTURES

This chapter presents the interactions among cells and their environment. The students examine the processes by which cells can take in materials from their environment, reproduce, and produce energy. The students learn how cells work together to keep an organism functioning. They investigate the role of DNA in cell reproduction and the production of energy through respiration and fermentation. The theme is also developed through concepts in Chapters 4, 7, 8, 9, 10, 12, 13, 16, 17, 18, 19, and 20. A supporting theme is **Energy.**

MULTICULTURAL CONNECTION

Research on cell structure and functions has been an ongoing process conducted by scientists from many different cultural backgrounds. Have the students research information about one of the following scientists: Robert Hooke, Matthias Schleiden, Gregor Mendel, Johann Friedrich Miescher, Walter Sutton, Rosalind Franklin, Maurice Wilkins, Albrecht Kossel, Ernest Just, Christian de Duve, Elma Gonzalez, Barbara McClintock, and Rita Levi-Montalcini. Encourage the students to find out about other scientists who have studied cells. Have the students write a brief summary of the researcher's contribution to cellular biology and place it in their science portfolios.

MEETING SPECIAL NEEDS

Second Language Support
Fluent English-speaking students should be paired with students who have limited English proficiency to work together to create a list of one-word characteristics that describe a scientist. You might choose to provide the students with the words *observant* and *curious* to help them get started.

▶ 118 CHAPTER 5

CHAPTER 5
CELL FUNCTION

*T*he great detective Sherlock Holmes has a puzzle to solve. He thinks that the dancing men are part of a secret code. But how can he decipher the code and find out what the message means? Like Sherlock Holmes, scientists are always trying to solve mysteries. For example, understanding what happens inside cells was once a scientific puzzle. Today, even though much is known about the function of cells, there are puzzles that scientists are still working to solve.

118 Chapter 5

CHAPTER MOTIVATING ACTIVITY

Several days before beginning the chapter, peel a hard-boiled egg and ask a student volunteer to measure its circumference around the middle of the egg. Record the circumference on the chalkboard. Then place the egg in a beaker of water. Every day for five days, have a volunteer measure and record the circumference of the egg. Ask the students to write journal entries describing what happened to the egg's circumference and giving possible explanations. (The students should note that the circumference of the egg has increased. They might explain the increase by saying the egg absorbed water from its environment.)

Answering the journal questions provides the students with an opportunity to display their knowledge about cells and their ideas about the characteristics that detectives and scientists might have in common. These answers will also provide you with an opportunity to note any misconceptions that the students might have about cell processes.

ABOUT THE LITERATURE

"The Adventure of the Dancing Men" is one of a series of stories that Sir Arthur Conan Doyle wrote about Sherlock Holmes and his partner in crime solving, Dr. Watson. After introducing Holmes in 1887, Doyle went on to write over fifty short stories and four novels featuring Holmes, who became one of the best-known fictional detectives of all time. In pursuing all of his cases, Holmes solved mysteries through careful observation, scientific investigation, and logical thinking. In "The Adventure of the Dancing Men," Holmes is asked to interpret, or decipher, a coded message represented by dancing figures. Holmes deduces that the figures represent letters. Since the letter e is the most common letter, he determines that the most frequently appearing figure must represent e. Then through trial-and-error methods and knowledge of the alphabet, Holmes is able to decode the messages and solve the mystery.

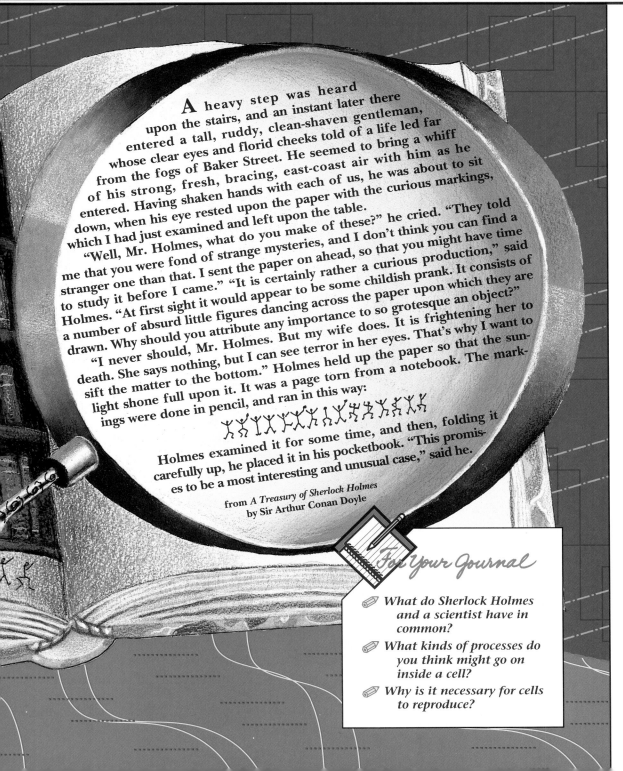

A heavy step was heard upon the stairs, and an instant later there entered a tall, ruddy, clean-shaven gentleman, whose clear eyes and florid cheeks told of a life led far from the fogs of Baker Street. He seemed to bring a whiff of his strong, fresh, bracing, east-coast air with him as he entered. Having shaken hands with each of us, he was about to sit down, when his eye rested upon the paper with the curious markings, which I had just examined and left upon the table.

"Well, Mr. Holmes, what do you make of these?" he cried. "They told me that you were fond of strange mysteries, and I don't think you can find a stranger one than that. I sent the paper on ahead, so that you might have time to study it before I came." "It is certainly rather a curious production," said Holmes. "At first sight it would appear to be some childish prank. It consists of a number of absurd little figures dancing across the paper upon which they are drawn. Why should you attribute any importance to so grotesque an object?"

"I never should, Mr. Holmes. But my wife does. It is frightening her to death. She says nothing, but I can see terror in her eyes. That's why I want to sift the matter to the bottom." Holmes held up the paper so that the sunlight shone full upon it. It was a page torn from a notebook. The markings were done in pencil, and ran in this way:

Holmes examined it for some time, and then, folding it carefully up, he placed it in his pocketbook. "This promises to be a most interesting and unusual case," said he.

from A Treasury of Sherlock Holmes by Sir Arthur Conan Doyle

For Your Journal

- What do Sherlock Holmes and a scientist have in common?
- What kinds of processes do you think might go on inside a cell?
- Why is it necessary for cells to reproduce?

Section 1:
TRANSPORT IN CELLS

FOCUS

In this section, the concepts of diffusion and active transport are presented. The process of osmosis and its relationship to plants are also explored.

MOTIVATING ACTIVITY

Cooperative Learning Have the students work with partners. Provide each partner team with a beaker of water and a medicine dropper containing several drops of concentrated food coloring. Point out that if the food coloring drops were put into the water and the water was stirred with a stirring rod, the entire solution would be likely to change color. Explain that in this activity, the students are to add the food coloring drops to the water but not to stir. Ask the students to predict the outcome of the activity, then allow them to test their predictions through observation.

PROCESS SKILLS
• Observing • Inferring
• Applying • Evaluating

POSITIVE ATTITUDES
• Curiosity • Skepticism

TERMS
• diffusion • active transport
• passive transport
• osmosis

PRINT MEDIA
Cells by George S. Fichter (see p. 93b)

ELECTRONIC MEDIA
Membranes and Transport, Britannica (see p. 93b)
Science Discovery Osmosis experiment

BLACKLINE MASTERS
Study and Review Guide
Laboratory Investigation 5.1
Extending Science Concepts

MEETING SPECIAL NEEDS

Mainstreamed

To help the students understand the concepts of *permeable* and *impermeable*, have them find dictionary definitions of the terms. Then demonstrate the concepts by pouring a spoonful of water on paper and a spoonful of water on plastic wrap. Have the students observe what happens to the water and then identify the permeable and impermeable substances.

▶ 120 CHAPTER 5

SECTION 1

Transport in Cells

Objectives

Describe diffusion through a cell membrane.

Explain the function of a selectively permeable membrane.

Compare osmosis and diffusion.

The first thing Sherlock Holmes noticed about the dancing men message was that the figures could be grouped into words. He thought each figure holding a flag could be the last letter in a word.

In the same way that letters combine to make words, atoms combine to make molecules. But unlike the letters in the message, atoms and molecules have the ability to move from one place to another! In fact, molecules have two different ways in which they can move into and out of cells.

Diffusion and Active Transport

All living things are made of matter. Matter is anything that has mass and takes up space. Just as buildings are made of individual bricks, matter is made from small particles called *atoms*. Atoms are considered the smallest building blocks of matter. When two or more atoms join together, molecules are formed. Atoms and molecules are always moving. When moving molecules hit each other, they bounce and change direction like rubber balls that are moving around in a box.

Think about what happens when a drop of red ink is added to water. The molecules of ink are bunched together as the drop enters the water. An area with many molecules of a substance is called an area of *high concentration*. As molecules of water and ink bump into each other, the ink begins to mix with the water. After a while, the ink and water molecules are mixed equally throughout the beaker, and the water turns red.

Figure 5–1. Diffusion is movement from an area of high concentration to an area of low concentration. The molecules of ink will continue to diffuse throughout the water until the water is an even red color.

TEACHING STRATEGIES

● **Process Skills:** *Observing, Generating Ideas*

Remind the students that atoms and molecules are always in motion. Point out that it is easy to detect such movement when water turns to steam, for example. Note that the atoms and molecules in solid objects are also in motion, although that motion is more difficult to detect. One way to visualize the movement of atoms and molecules is to observe dust particles moving in a beam of light. Ask the students to describe the motion of the dust particles. (The motion is random; the direction of a particle changes as it bumps into other dust particles.)

● **Process Skills:** *Observing, Inferring*

Have the students examine Figure 5–2 and describe some of the factors that might contribute to the rate of diffusion. (the size of the molecules, the size of the container, and the temperature of the solution)

① Examples may include milk diffusing in coffee or chlorine diffusing in a swimming pool.

DISCOVER BY *Observing*

Before performing this activity, make sure that none of the students are allergic to perfumes. If some students are allergic to perfume fumes, substitute a sliced and squeezed orange or lemon. The diffusion demonstration will work best if drafts are eliminated by closing windows and doors. The students closest to the demonstration should detect the odor first because they are closest to the area where more molecules are bumping into each other. Point out that the smell is detected as the molecules diffuse through the air.

PERFORMANCE ASSESSMENT

Cooperative Learning Divide the class into small groups, and ask each group to generate a list of other examples of diffusion in everyday life. Evaluate the lists by asking the students to explain or demonstrate one or two of the examples.

Figure 5–2. The movement of water molecules into and out of a cell is controlled by the concentration of water inside the cell. If the water concentration is higher inside the cell, water will move out. If the concentration of water is higher outside the cell, water will move in.

The ink spreads through the water by the process of diffusion (dih FYOO zhuhn). **Diffusion** is the movement of molecules from an area of high concentration to an area of low concentration. The movement of many molecules within cells is the result of diffusion. The following activity shows you one example of diffusion. What other examples can you think of? ①

Open a bottle of perfume in one corner of the classroom. How does the fragrance diffuse through the classroom? Which students were able to smell the fragrance first? Later? Explain your observations.

Diffusion is a process used by cells to take in materials from their environment. To function properly, animal cells need food molecules for energy. Plant cells make their own food, but they need water and minerals for the food-making process to take place. Waste molecules must be removed to keep the cells healthy. Some materials, like the food made in plant cells, need to be moved to other parts of the organism. Many of these molecules enter and leave cells by diffusion.

The Gate to the Cell The cell membrane controls the movement of molecules into and out of the cell. This membrane lets only certain molecules pass through its openings. This is similar to the way in which a kitchen strainer lets water pass through but not the spaghetti you are draining. A membrane that lets molecules pass through it is called a *permeable* (PUR mee uh buhl) *membrane*. Cell membranes are selectively permeable because only certain molecules can pass through.

SECTION 1

TEACHING STRATEGIES, continued

● **Process Skills:** *Comparing, Applying*

Explain to the students that the energy of molecules moving from an area of higher concentration to an area of lower concentration is like the energy of a ball rolling downhill. It is the energy of motion, or kinetic energy. Also point out that the energy required to get a ball to roll uphill (it must be pushed) is like the energy needed to move a molecule through a cell membrane from an area of lower concentration to an area of higher concentration. During active transport, the cell must provide energy to "push" the material from an area of lower concentration to an area of higher concentration.

DISCOVER BY *Writing*

Remind the students of the importance of presenting information in a clear and organized manner. You might wish to have the students use bar graphs to show the number of times they used "active" transport and the number of times they used "passive" transport.

 Have the students include their lists and graphs from the *Discover* activity in their science portfolios. Also ask the students to integrate their knowledge of fossil fuels by noting each entry that required either direct or indirect consumption of fossil fuels. Then ask them to include a general statement describing their reliance on fossil fuels for transportation purposes.

ONGOING ASSESSMENT
ASK YOURSELF

The movement of molecules during active transport requires cell energy. Because diffusion does not require cell energy, it is not a form of active transport.

SCIENCE BACKGROUND

The pressure that builds in a plant as a result of osmosis is called *turgor pressure*. Excess water is stored in the large central vacuole of the plant cells. The excess water causes the vacuole to swell, forcing the cytoplasm and cell membrane against the cell wall, making the cell stiff.

▶ **122** CHAPTER 5

Figure 5–3. Canoeing can be much like passive and active transport. If the canoe is being carried along by the current (left), no energy is being used, as in passive transport. However, if the canoe is being moved against the current (right), the people paddling would have to use energy, as in active transport.

Passive or Active? The following activity will introduce you to the concept of active and passive transport.

DISCOVER BY *Writing*

Keep a record of the trips you take for one week. For each trip, write down the type of transportation you used. Then classify your trips into two types: those in which you took an "active" role in your transportation, such as walking or bicycling, and those in which you took a "passive" role, such as riding in a car or on a bus.

When you are using a car or a bus for transportation, you are not using your own energy. Your role as a rider is passive. Diffusion in a cell is also a type of **passive transport**. This means that no energy is used by the cell during diffusion. However, diffusion is not the only way molecules move across the cell membrane. For instance, salt molecules move from your bloodstream into your cells until almost all the salt molecules are out of the bloodstream. To do this, the salt molecules must move from an area of low concentration to an area of high concentration. The movement of molecules across the membrane from an area of low concentration to an area of high concentration is called **active transport**. During active transport, the cell uses energy to move the molecules across the cell membrane.

 **ASK YOURSELF**

Why is diffusion not a form of active transport?

● **Process Skills:** *Observing, Applying*

Begin the discussion of osmosis by showing the students two strips of potato, one in tap water, the other in a 10-percent salt solution. (These should be prepared at least 60 minutes prior to class.) Explain that the concentration of water inside most plant cells is usually less than the concentration of water outside the cells. Therefore, water usually diffuses into plant cells, as shown by the potato strip in tap water. However, if the concentration of water is greater inside the cells, water will diffuse out of the cells, and the potato will become limp, as shown by the potato strip in salt solution. Ask the students to describe what will happen if the limp potato strip is placed in tap water. (The concentration of water inside the cells will be less than the concentration of water in its environment, and water will diffuse into the cells.)

ACTIVITY

How do food substances diffuse through a cell?

Process Skills: Observing, Inferring

Grouping: Groups of 3 or 4

Hints
The agar should be prepared according to package directions. Use either glass or disposable plastic Petri dishes.

Food substances diffuse best through freshly prepared agar. To lessen the time needed to complete this activity, make the hole in the agar with a straw or cork borer before class. You can use five straws cut at 45-degree angles as scoops. Remind the students that each scoop should be used for only one type of food crystal. Keep the food crystals dry until they are applied to the agar or filter paper. If possible, provide each group of students with a different substance—liquid food coloring, cherry drink mix crystals, and coffee crystals all diffuse very well through agar.

▶ **Application**
1. Diffusion through the moistened filter paper occurs very quickly. Diffusion occurs more slowly through the agar.

2. The rate of diffusion depends on the food substances used by the various groups.

3. The results of the activity may not be very consistent from group to group. Variations might be attributed to differences in amounts of water on the filter paper or amounts of food substances used.

ACTIVITY

How do food substances diffuse through a cell?

MATERIALS
agar-filled Petri dish, plastic straw, scissors, filter paper, empty Petri dish, medicine dropper, scoop, food substances to be compared, water

PROCEDURE
1. Using a straw, make five holes in the agar in one Petri dish, forming an X-shaped pattern as shown. Put one hole in the middle and the other four holes at the ends of the X.
2. Cut the filter paper so that it fits into the top half of a Petri dish, and put it into a dish. Mark circles on the filter paper that are just like the holes in the agar.
3. Using the medicine dropper, moisten the filter paper with 10 to 15 drops of water.
4. With the scoop, put a small amount of one of the food substances into each hole in the agar. Add three drops of water to each hole.
5. Put the same amount and kind of food substance on the circles on the filter paper. Do not add more water to the filter paper.
6. Watch the agar and the filter paper for two or three minutes. Keep track of the time it takes the food substance to reach the edge of each dish. Record your observations.
7. Cover both Petri dishes and observe them after 24 hours. Record your observations.

APPLICATION
1. How does the rate of diffusion through the agar compare with the rate through the moistened filter paper?
2. Compare your results with those of your classmates. Which food substance diffuses the fastest? Which food substance diffuses the farthest?
3. In your experiment, were the results the same for both the agar and the filter paper? Did all groups have the same results? Explain your results.

Osmosis

The activity demonstrates how molecules are moved during diffusion. Another movement of molecules into and out of a cell is called *osmosis* (ahs MOH sihs). **Osmosis** is a type of diffusion that occurs only when water moves across a selectively permeable membrane. During osmosis, water molecules move from an area of high concentration to an area of low concentration just as in diffusion. Like diffusion, osmosis is a passive process. The movement of water into and out of cells is so important to living things that the diffusion of water through a membrane is given this special term—osmosis.

GUIDED PRACTICE

Ask the students to explain how the process of osmosis can prevent a plant from wilting. (The concentration of water in the soil surrounding a plant is usually greater than the concentration of water in the cells of the plant. The diffusion of water into the plant cells from the soil causes plant cells to swell as water moves into them.)

INDEPENDENT PRACTICE

Have the students provide written answers to the Section Review and Application questions. Then ask the students to write a paragraph in their journals comparing and contrasting active transport and passive transport.

EVALUATION

Ask the students to compare one bicyclist pedaling uphill and another coasting downhill to passive and active transport. (Pedaling a bicycle uphill is like active transport—energy is being used by the bicyclist; a bicycle coasting downhill is like passive transport—no energy is used by the bicyclist.)

RETEACHING

Have the students watch as you pour orange juice with pulp through a funnel. Then pour the orange juice through a sieve. Ask the students to identify the utensil that has selective permeability. (The sieve has selective permeability because the pulp in the orange juice does not pass through it.)

LASER DISC
2523
Osmosis experiment

ONGOING ASSESSMENT
ASK YOURSELF

Diffusion is the movement of molecules from areas of higher concentration to areas of lower concentration. Osmosis is the movement of molecules across a selectively permeable membrane.

SECTION 1 REVIEW AND APPLICATION

Reading Critically

1. A cell membrane is selectively permeable because only some molecules are allowed to pass through the membrane.

2. Active transport is the movement of molecules from regions of lower concentration to regions of higher concentration.

3. Diffusion is the movement of molecules from areas of higher concentration to areas of lower concentration. Examples might include the smell of cooking food wafting through a home.

Thinking Critically

4. Water would move out of the cell. At the same time, other materials may move into the cell through the process of diffusion.

5. As water moves into plant cells, the cells swell and gradually push the cytoplasm against the cell walls. As a result, the stems and leaves of the plant stand up straight.

▶ 124 CHAPTER 5

Figure 5–4. As the concentration of water outside a cell decreases, water diffuses out of the cell. As a result, cell membranes collapse. A lack of water will cause a plant to wilt. However, if water is added to the soil, it diffuses into the plant's cells, and the plant stands up straight.

Have you ever seen a wilted plant straighten up after water has been added to the soil? The movement of water molecules into the plant causes it to straighten.

Living cells get most of their water by the process of osmosis. Osmosis keeps plants from wilting. Usually, there is more water in the soil than there is in the cells of a plant. This concentration of water causes plant cells to swell as water flows across the membrane and into the cells. Eventually, the cytoplasm is pushed tightly against the cell walls, causing the stems and leaves to stand straight.

Osmosis also occurs if there is more water in the roots of a plant than there is in the soil around them. Water will leave the cells of the roots and cause the roots to shrivel. The plant may die because of this loss of water. This can happen if the soil has too many nutrient molecules as a result of too much fertilizer. If there are more nutrients in the soil than in the roots, the water will leave the roots.

 ASK YOURSELF

Compare osmosis with diffusion.

SECTION 1 REVIEW AND APPLICATION

Reading Critically

1. Why is the cell membrane described as being selectively permeable?
2. What is active transport?
3. Define *diffusion* and give an example of it from your everyday life.

Thinking Critically

4. What might happen if the concentration of water inside a cell is greater than the concentration of water outside the cell? Explain why this would occur.
5. What causes the stems and leaves of a plant to stand up straight?

EXTENSION

Have the students devise their own equipment setup to demonstrate osmosis. Then have the students present their demonstration.

CLOSURE

Cooperative Learning Have the students work with partners to define the following terms: diffusion, active transport, osmosis, and selective permeability. Then have the partners relate the concepts to one another.

INVESTIGATION

Observing Osmosis

Process Skills: Observing, Comparing

Grouping: Groups of 3 or 4

Objectives
- **Predict** the effect of osmosis.
- **Observe** changes caused by osmosis.
- **Interpret data** from the investigation.

Pre-Lab
Ask the students to describe the process of osmosis. Have them draw a simple diagram showing the movement of water by osmosis.

Analyses and Conclusions

1. The students should observe that the cell membranes pulled away from the cell walls.

2. The concentration of water inside the cell was greater than the concentration outside the cell. Water molecules moved out of the cell by the process of osmosis. As a result, the cell became smaller. Since the cell wall remained the same, the cell membrane pulled away from the cell wall.

Application
If pure water were injected into the bloodstream, the nearby cells might swell and burst as a result of osmosis. (This event is not likely to occur because the pure water would probably mix with the liquid portion of the blood before the cells would burst, unless massive amounts of water were injected.)

Discover More
The *Elodea* cells appear filled and more rigid because water entered the cells of the leaf by osmosis.

Post-Lab
Have the students add labels representing the *Elodea* leaf, tap water, and salt water to their osmosis diagram.

INVESTIGATION

Observing Osmosis

▶ MATERIALS
- water • *Elodea* leaf • microscope slide • coverslip • microscope • salt water
- medicine droppers (2) • paper towel

▼ PROCEDURE

1. Using tap water, make a wet mount of an *Elodea* (el uh DEE uh) leaf. Observe the *Elodea* leaf under the microscope with low power. Draw what you see.

2. Switch to high power. Pay careful attention to the location of the chloroplasts within the cells. Draw what you see. Label the cell wall, the cytoplasm, the chloroplasts, and the vacuole.

   Show the location of the cell membrane, even though it is not visible.

3. What might happen if you put salt water on the leaf? Discuss your predictions with a classmate before you continue.

4. Without moving the slide, place two or three drops of salt water along the right edge of the coverslip. Hold a torn edge of the paper towel along the left edge of the coverslip. The paper towel will absorb some of the tap water, allowing the salt water to flow beneath the coverslip.

5. Observe the *Elodea* leaf for 3–4 minutes. Describe the changes that occurred. How could you make a record of the results of the experiment?

▶ ANALYSES AND CONCLUSIONS
1. Compare the change that took place in the *Elodea* leaf with what you predicted might happen.
2. Explain how this change occurred.

▶ APPLICATION
Blood cells contain approximately one percent salt. Explain what would happen to the nearby blood cells if pure water were injected into the bloodstream.

✴ Discover More
Continue the investigation by flooding the slide with fresh water. Compare the leaf with the drawings you made at the beginning of the investigation.

SECTION 1 125 ◀

Section 2:
CELL REPRODUCTION

FOCUS

This section of the chapter describes cell reproduction. The four phases of mitosis, including the changes that occur in a cell during each of the four phases, are explained. The time between the phases of mitosis, interphase, is also explained.

MOTIVATING ACTIVITY

Cooperative Learning Divide the class into several small groups. Give each group an original copy of a typewritten page and several photocopies of the same page. Ask the students to identify the original copy. After a time, ask the students if they have had difficulty distinguishing the copies from the originals and why. (Depending on the quality of the copier, the students may have difficulty distinguishing the originals because they are duplicated exactly by the photocopier.)

PROCESS SKILLS
• Comparing • Applying
• Solving Problems/Making Decisions

POSITIVE ATTITUDES
• Curiosity • Precision

TERMS
• mitosis • DNA
• chromosomes

PRINT MEDIA
Genetics: The Mystery and the Promise
by Francis Leone
(see p. 93b)

ELECTRONIC MEDIA
Mitosis, Britannica
(see p. 93b)

Science Discovery
Cell, animal; mitosis

BLACKLINE MASTERS
Study and Review Guide
Laboratory Investigation 5.2
Reading Skills

---

① The cells will wear off and be replaced.

INTEGRATION—
Mathematics

The bacterium *Escherichia coli* can reproduce every 20 minutes. Ask the students to use a scientific calculator to determine how many daughter cells can be produced from one *E. coli* cell in 3 hours. (512)

126 CHAPTER 5

SECTION 2

Objectives

Define the term mitosis.

Explain why interphase is not a part of mitosis.

Summarize the four steps that occur during cell division.

Cell Reproduction

Think back to the dancing men puzzle in the Sherlock Holmes story. Until Holmes collected more messages showing the dancing men, he did not have enough information to solve the puzzle of the code. Holmes wanted to understand how the code in the message worked so that he could write a message of his own.

Many scientists have worked on the puzzles involved in cell function. By studying a great number of cells under the microscope, they have been able to describe the key functions of cells. One very important function of cells is reproduction.

Methods of Cell Reproduction

Look at your hand. Each square centimeter of skin is made of more than 150 000 cells. Most of these cells will be gone by tomorrow. During the next 24 hours, two complete generations of skin cells will live, reproduce, and disappear. What do you suppose will happen to the old skin cells on your hand during the next 24 hours? ①

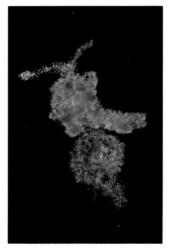

Figure 5–5. All living things reproduce. Single-celled organisms, such as this amoeba, reproduce by dividing into two new cells.

In other parts of your body, other cells are also reproducing rapidly. Red blood cells are made at the rate of about 100 million cells a minute. If you have a broken bone, new bone cells are being made so that the broken bone can heal. Since you are getting taller, your bones are growing in length by adding new cells at the ends of the bones. There is much cellular activity taking place in your body.

Cells reproduce by dividing into two cells. The original cell is called the *parent* cell. The two new cells that are formed are called *daughter* cells.

TEACHING STRATEGIES

● **Process Skills:** *Inferring, Generating Ideas*

Explain to the students that different types of cells divide at different rates. For example, some types of human cells, such as blood cells and skin cells, reproduce very rapidly; connective tissue, such as muscle or bone, takes much longer; and other cells, such as nerve cells, do not reproduce at all. Also point out that cells divide at different rates during different conditions. Ask the students to think of conditions during which skin cells would divide faster than their normal rate of growth. (The students might suggest that tumorous tissue is the rapid growth of cells.)

● **Process Skills:** *Comparing, Applying*

Review the organization of a cell and the structure and function of cell organelles by using an overhead projector to display a picture of a cell. Ask the students why it is necessary for the nucleus to be duplicated exactly during cell division. (The nucleus contains the genetic material, or hereditary information, of the organism.)

Cells reproduce in several different ways. One type of cell division occurs when a single-celled organism splits into two organisms. Another type of cell division occurs in multicellular organisms when cells, such as skin cells, make exact duplicates of themselves. A third type of cell division occurs when sex cells, or reproductive cells, are produced.

 ASK YOURSELF

What is the purpose of cell division?

Mitosis

The type of cell division by which two daughter cells are formed is called *mitosis* (my TOH sihs). **Mitosis** produces daughter cells that are exactly the same as the parent cell. Mitosis is actually the process by which the cell nucleus duplicates. After the nuclear material is duplicated, the rest of the cell simply divides in two.

Mitosis is a continuous process. However, the events of mitosis are easier to understand if the process is broken down into steps, or phases. There are four phases in mitosis. They are *prophase, metaphase, anaphase,* and *telophase.* In the following activity, you will learn more about these four terms.

 *Researching*

Notice that the names of the four stages of mitosis—prophase, metaphase, anaphase, and telophase—all end with *phase.* Use a dictionary to analyze the meanings of the four different prefixes. Then, as you study the process of mitosis, check to see whether the meaning of each word is a good description of what is happening inside the cell at that stage.

Preparing for Mitosis Before mitosis can begin, several events have to occur in the nucleus. First, the hereditary material in the cell must be duplicated. *Hereditary* means "passed on from parents to offspring." The hereditary material, called **DNA,** is found within the nucleus. DNA makes up threadlike structures called **chromosomes** (KROH muh sohmz). Since the DNA duplicates, the chromosomes also duplicate. The duplicated chromosomes are joined together at a point called the *centromere.* Identical hereditary instructions are carried on the two chromosomes.

Figure 5–6. Through mitosis, an embryo steadily increases the number of its cells.

 **LASER DISC**

2678

Cell, animal; mitosis

ONGOING ASSESSMENT
ASK YOURSELF

Organisms have a constant need for new cells. Cell division supplies the new cells.

DISCOVER BY *Researching*

The students should discover that *pro-* means "earlier," and prophase is the first stage of mitosis; *meta-* means "later in time," and metaphase is the second stage of mitosis; *ana-* means "backward," and in anaphase, the third stage of mitosis, chromosomes move away from the center; and *telo-* means "end," and telophase is the last phase of mitosis.

PERFORMANCE ASSESSMENT

After the students have examined each phase of mitosis, have them recall the meaning of the prefixes and ask them why each prefix is a good description of what is happening inside the cell during that phase. Evaluate the students' descriptions based on their application of the prefixes to the events in each phase.

TEACHING STRATEGIES, continued

● **Process Skills:** *Comparing, Expressing Ideas Effectively*

Ask the students to list the similarities and differences between plant and animal mitosis. (Similarities: The movement of the chromosomes is the same in both types of cells, and both animal and plant cells have spindle fibers. Differences: Most plant cells do not have centrioles; after the cytoplasm has divided in a plant cell, a cell plate forms to unite the cell walls around the daughter cells.)

GUIDED PRACTICE

Have the students draw pictures of the stages of mitosis and summarize in their own words what occurs in each stage. Remind the students to include interphase in their drawings. Have them place their pictures and summaries in their science portfolios.

INDEPENDENT PRACTICE

Have the students write answers to the Section Review and Application questions. Then ask the students to write a paragraph in their journals describing the importance of the process of mitosis to living things.

MEETING SPECIAL NEEDS

Gifted

Cancer is characterized by the uncontrolled cell growth. Ask the students to discover more about cancer and share their findings with the class.

SCIENCE BACKGROUND

At the end of telophase, the cytoplasm divides and forms two distinct cells. This division of the cytoplasm is called *cytokinesis*.

Demonstration

To show how a cell's size relates to its ability to take in materials, use modeling clay to make three cubes of increasing size. Help the students determine and record the total surface area and volume of each cube, using the formulas $6s^2$ = surface area and s^3 = volume. (s = length of side) Explain that a cell's surface area is closely related to its ability to take in materials, and a cell's volume is closely related to its need to take in materials. Point out that the volume increases more rapidly than the surface area. Then ask the students why this growth pattern keeps a cell from growing larger than microscopic size. (The cell's volume would become too large for its surface area to support.)

Figure 5-7. Mitosis takes place in four stages—prophase, metaphase, anaphase, and telophase. Each stage blends into the next stage. Interphase is not part of mitosis. However, all life activities of the cell, except mitosis, take place during interphase.

The time before mitosis, when the DNA and chromosomes duplicate, is called *interphase*. *Inter* means "between." Interphase is not a part of mitosis; it is the time between the end of one mitosis and the beginning of the next. Most of a cell's life is spent in interphase. However, interphase is a time of much activity. During interphase, the cell performs all life activities except mitosis. During this time, a cell stores the extra energy that mitosis will require.

Prophase The first phase of mitosis is called *prophase*. Prophase begins as the membrane around the nucleus breaks apart. The chromosomes inside the nucleus begin to twist and thicken. As the nucleus breaks apart, the chromosomes begin to move toward the center of the cell. As mitosis continues, the nucleus disappears, and thin tubes begin to form between organelles called *centrioles*. These tubes are called *spindle fibers*.

Metaphase During metaphase, the spindle fibers seem to push and pull the duplicate chromosomes until they are arranged in a line across the middle of the cell. The centromere of each pair of chromosomes is attached to a spindle fiber.

Anaphase During anaphase, the spindle fibers shorten and pull each chromosome pair apart at the centromere. The spindle fibers continue to shorten, pulling the chromosomes through the cytoplasm.

Telophase During telophase, the last phase of mitosis, the cells complete their division. As the separate chromosomes reach the opposite ends of the cell, telophase begins. The nuclear envelope forms again as the chromosomes untwist and become longer and thinner. At the end of telophase, the new cells separate. In animal cells, the cell membrane pinches together, dividing the cytoplasm in two. The number of

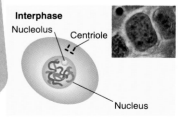
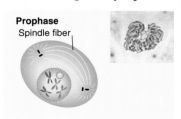

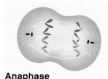

EVALUATION

Point out that interphase is sometimes referred to as the resting stage. Ask the students why that reference is not accurate. (During interphase, the cell is still carrying out all life functions. Also chromosome duplication and energy storage occur during interphase.)

RETEACHING

 Cooperative Learning Ask one partner to describe a cell and its contents during any phase of mitosis. Have the other partner sketch that cell and its contents based on the description. Then have the partners reverse roles.

EXTENSION

Ask the students to discover more about chromatids and their relationship to the process of mitosis. (Each strand of a duplicated chromosome is called a *chromatid*. The molecular structure of a chromosome allows it to reproduce itself exactly. Therefore, the chromatids of a duplicated chromosome are identical.)

CLOSURE

 Cooperative Learning A sequence of events occurs during each phase of mitosis. List these events in random order on the chalkboard, then ask each group to list the events of mitosis in the order that they occur.

---

chromosomes in each daughter cell is the same as it was in the parent cell. The daughter cells now enter interphase.

The stages of cell mitosis can be seen under a microscope. In the next activity, you will have an opportunity to see for yourself what happens during mitosis.

DISCOVER BY Problem Solving

Obtain from your teacher a set of slides showing the stages of mitosis. Before looking at them, ask a classmate to arrange the slides so they are not in order. Use a microscope to examine each slide. Place the slides in the correct sequence. What problem-solving strategies did you use to order the slides? On a separate sheet of paper, sketch and name each stage of mitosis.

The division of cytoplasm in plant cells is different from the division in animal cells. The thick cell wall of a plant cell is too stiff to pinch together. Instead, a structure called the *cell plate* forms between the daughter cells. The cell plate begins forming in the middle of the cell and moves outward until the daughter cells are separated from each other.

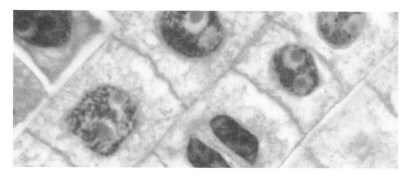

Figure 5–8. In both plant and animal cells, the cell membrane pinches off in the middle to form two new cells. However, in plant cells, a cell plate forms to divide the two new cells.

ASK YOURSELF

Describe what happens in a cell during mitosis.

SECTION 2 REVIEW AND APPLICATION

Reading Critically
1. Why is interphase not a part of mitosis?
2. How is mitosis different in plant cells and animal cells?

Thinking Critically
3. How would a chemical that prevented spindle fibers from forming affect the cell division of an organism?
4. What would happen if DNA did not duplicate?

DISCOVER BY Problem Solving

You may wish to have the students discuss strategies they use to remember the characteristics of each phase of mitosis.

 Have the students place their pictures from the *Discover* activity in their science portfolios.

ONGOING ASSESSMENT ASK YOURSELF

During mitosis, the nucleus of a cell duplicates and the cell divides, forming two daughter cells.

SECTION 2 REVIEW AND APPLICATION

Reading Critically

1. During interphase, a cell is not dividing.

2. In an animal cell, the cytoplasm divides as the cell membrane pinches inward, separating the daughter cells. In a plant cell, after the cytoplasm has divided, a cell plate forms in the center of the cell and moves outward.

Thinking Critically

3. In the absence of spindle fibers, the chromosomes would not separate properly in anaphase. The resulting daughter cells would have an abnormal number of chromosomes.

4. The daughter cells would not receive all the necessary hereditary material from the parent cell.

SECTION 2 **129**

Section 3:
DNA AND CELL ENERGY

FOCUS
This section examines the structure of the DNA molecule. How four nucleotide nitrogen bases pair for replication—the process by which DNA makes exact copies of itself—is explored. Research with recombinant DNA and its uses is also described as are the processes of energy release through respiration and fermentation.

MOTIVATING ACTIVITY

Cooperative Learning Have the students work in small groups to devise a simple coded message. Ask each group to write a brief message and then code the message so that its meaning is not readily apparent. Codes can be easily created by substituting one letter for another or by substituting numbers for letters. Encourage the students to develop codes that are relatively easy to decipher in a short period of time. Have the groups exchange and decode the messages. Then ask the students to describe the techniques and methods that they used to break the codes.

PROCESS SKILLS
• Communicating • Inferring
• Applying

POSITIVE ATTITUDES
• Skepticism • Initiative and persistence • Openness to new ideas

TERMS
• replication • cellular respiration • fermentation

PRINT MEDIA
What Mad Pursuit by Francis Crick (see p. 93b)

ELECTRONIC MEDIA
DNA: Laboratory of Life, National Geographic (see p. 93b)

Science Discovery DNA; model Photosynthesis; compared with aerobic cellular respiration

BLACKLINE MASTERS
Study and Review Guide
Connecting Other Disciplines
Thinking Critically

THE NATURE OF SCIENCE

Johann Friedrich Miescher, a Swiss biochemist, discovered the existence of nucleic acid in 1869. He called the substance "nucleic" because of its location in the nucleus. However, until the 1940s, it was thought that proteins, not DNA, carried the hereditary information. Direct evidence that DNA is the hereditary material did not occur until after Watson and Crick presented their model of DNA structure in 1953.

▶ 130 CHAPTER 5

SECTION 3

DNA and Cell Energy

Objectives

Summarize the process of DNA replication.

Compare respiration and fermentation.

Describe the process and purpose of photosynthesis.

With a moderate amount of effort, Sherlock Holmes unraveled the puzzle of the dancing men. The message read, "AM HERE ABE SLANEY." Holmes then used his knowledge of the dancing men code to trap Abe Slaney.

One of the most exciting scientific puzzles to be solved in recent times has been that of understanding how the genetic material in cells is passed on from one generation to the next. The scientists who unraveled this puzzle faced problems considerably more difficult than the ones Sherlock Holmes faced, but in many ways, they approached the problems in the same way.

DNA and Replication

Have you ever put together a model of an airplane or an automobile? Would you be able to put all the tiny pieces of the model together correctly if you lost the instructions? Scientists often have to work with no instructions. For example, scientists had tried to figure out the structure and function of DNA. They knew that DNA was contained in chromosomes and that both DNA and chromosomes duplicated before mitosis. But how did DNA work? How was it made? These are questions that puzzled scientists for many years.

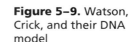

Figure 5–9. Watson, Crick, and their DNA model

The Double Helix The shape and makeup of the DNA molecule were finally worked out by James Watson, an American biologist, and Francis Crick, an English physicist. Watson and Crick made a three-dimensional model of what they thought the molecule was like. Their model was based on information collected by many different scientists over many

TEACHING STRATEGIES

● **Process Skills:** *Inferring, Generating Ideas*

Watson and Crick used a three-dimensional model to communicate their findings about DNA structure. Ask the students why Watson and Crick used a three-dimensional model. (DNA is a three-dimensional structure.) Then ask the students to think of other instances in which a three-dimensional model would be useful. (The students might suggest that three-dimensional models are useful when designing buildings, automobiles, or planes and when trying to communicate the structure of complex molecules, such as proteins.)

● **Process Skills:** *Classifying/Ordering, Applying*

Review the names of the nitrogen bases and the order in which the bases pair. Ask the students to determine the base sequence on the complementary strand if one piece of a DNA molecule has the base sequence A-T-G-C-T-T-G-C. (T-A-C-G-A-A-C-G)

---

years. Watson and Crick used the pieces of information like the pieces of a puzzle to build their model and to prove the structure of DNA. Now scientists know that DNA molecules are very long and thin. Each molecule is made of two chains formed from many small parts. The chains are arranged side by side and are connected by other smaller molecules. Together the chains form a molecule shaped like a ladder that has been twisted into a spiral. This twisted ladder is called a *double helix*. Watson later said of the discovery, "It seemed almost unbelievable that the DNA structure was solved, that the answer was incredibly exciting, and that our names would be associated with the double helix . . . ."

Using poster board, draw and label a time line showing the events of DNA research since DNA was first removed from a cell nucleus in 1869. Use reference materials from the library to include the most recent discoveries.

The Letters in the Code To fully understand how the DNA molecule is formed, scientists looked at its internal structure. The pieces that form the DNA molecule are made of molecules called *nucleotides* (NOO klee uh tydz). Each nucleotide is made of three smaller molecules: a sugar, a phosphate, and a nitrogen base. There are four different nitrogen bases that may be found in DNA: adenine (AD uh neen), guanine (GWAH neen), cytosine (SY tuh seen), and thymine (THY meen). These bases are often abbreviated A, G, C, and T, respectively. The bases form pairs, but each base can pair with only one other. Adenine (A) always pairs with thymine (T), and guanine (G) always pairs with cytosine (C).

The sugars and the phosphates that accompany the bases form the sides of the DNA ladder. The bases of the nucleotides form the rungs. Each kind of DNA has a different sequence of bases. The sequence of the base pairs forms the hereditary code.

If we compare this code to our own language, we see that the hereditary language of DNA is a language built on only a four-letter alphabet—A, T, G, and C. But by using those letters in different combinations and in different lengths, an endless number of DNA "words" and "sentences" can be written. These can be grouped into endless numbers of different DNA "stories." Each of us received our DNA combination from our

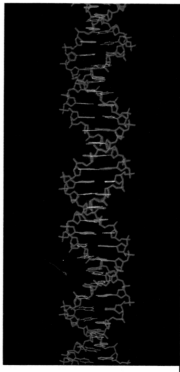

Figure 5–10. A DNA molecule is composed of four different kinds of nucleotides. Each nucleotide contains a sugar called *deoxyribose*, a phosphate molecule, and a nitrogen base.

BACKGROUND INFORMATION

In 1962, James Watson and Francis Crick shared the Nobel Prize for physiology or medicine with British biophysicist Maurice H. Wilkins. Watson and Crick had used experimental data provided by Wilkins and his associate Rosalind Franklin for constructing their DNA model. Franklin died in 1958 and was not awarded the Nobel Prize because it is given only to living scientists.

DISCOVER BY *Researching*

Remind the students that time should be divided into standard units on a time line. For example, if one inch represents 10 years on one part of the time line, one inch should represent 10 years on all parts of the time line.

 Have the students place their posters in their science portfolios.

MEETING SPECIAL NEEDS

Mainstreamed

To help the students remember how the nitrogen bases pair, have them think of the word *at*. Adenine always pairs with thymine and their initial letters spell *at*. That leaves guanine to pair with cytosine.

SECTION 3 **131**

TEACHING STRATEGIES, continued

● **Process Skills:** *Inferring, Predicting*

Remind the students that our present knowledge of DNA is due to the efforts of *many* scientists, and these efforts were characterized by the use of a scientific method. Have the students recall the steps of a scientific method. Then ask the students to tell if they think genetic findings were immediately accepted or skeptically questioned by the scientific community at the time. (Many scientific findings are questioned by skeptical scientists who then duplicate the original research of other scientists. If these duplicated efforts support the original findings, support for the hypothesis increases and the findings evolve to be more universally accepted as a scientific theory.)

✦ **Did You Know?**
The DNA in a single human cell has about 3 billion base pairs. Altogether the DNA in the nucleus of a single human cell is about 2 m long.

 A recent innovation in the field of genetics has been DNA "fingerprinting." DNA "fingerprinting" is a new identification technique used in criminal investigations. Crime laboratories analyze the DNA structure of physical evidence, such as skin cells or blood cells, from a crime scene to determine if the DNA structure of the samples matches the DNA structure of samples taken from an accused individual.

LASER DISC
4233

DNA; model

① The base-order pair must remain the same so that the hereditary code is not changed.

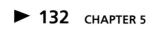 **132** CHAPTER 5

Base pairs in DNA molecule separating

Figure 5-11. Before mitosis can take place, the hereditary code must be duplicated so that both new cells are identical. The code is duplicated when the DNA replicates.

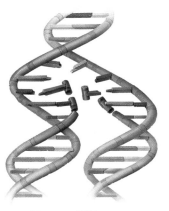

Two new DNA molecules forming

Two new identical DNA molecules

parents. But because we have a unique combination of DNA, our DNA "story"—our own inherited characteristics—are like no other individual's DNA story—past, present, or future.

An Exact Copy   Have you ever seen a copy machine duplicate the information on a piece of paper? DNA can make exact duplicates of itself in a manner that can be compared to the duplication of information by a copy machine. **Replication** is the process by which DNA makes copies of itself. Replication is necessary to provide each daughter cell with a complete and exact copy of the DNA in the parent cell.

This is really the most important idea in cellular reproduction. In order for your cells to be replaced, exact duplicates must be made. Any mistake in the DNA instructions could be, and usually is, fatal for the daughter cells.

Replication begins when a DNA molecule begins to come apart at one end. The paired bases separate, and the separation spreads along the DNA molecule as if the molecule were being "unzipped." As the chains come apart, each side of the original DNA becomes half of a new DNA molecule. New nucleotides move into position to take the place of those that separated. Remember, each base may pair with only one other kind. C will always pair with G, and A will always pair with T. As the new bases come together with the bases in each half of the old molecule, the base-pair pattern of the original DNA forms again. The two new molecules of DNA have exactly the same base-pair order as that found in the old molecule. Why is this important? ①

- **Process Skills:** *Observing, Expressing Ideas Effectively*

Have the students examine the illustration of the DNA molecule on page 132 and explain what is occurring. Have the students give the code for the illustrated parent model based on the knowledge that red represents adenine, yellow represents thymine, green represents guanine, and blue represents cytosine. (The pattern is T, A, C, G, A, T, G, C, C, G, G, C, T, A, A, T, A, T, G, C, C, G, A, T.)

- **Process Skills:** *Predicting, Generating Questions*

Explain to the students that another name for recombinant DNA is combined DNA. Recombinant DNA is created when DNA from two sources is combined. This "combined" DNA is then injected back into living cells. Point out that one of the goals of genetic engineers is to use recombinant DNA to create a variety of plants and animals that differ from those we have now. For example, perhaps someday a genetic engineer could "program" the reproductive cells of a plant to create a food crop that grows well in drought-prone areas. Ask the students to debate the advantages, disadvantages, and possible implications of recombinant DNA.

Although replication usually results in exact copies of DNA, changes, or mutations, in DNA can occur accidentally. In recent years, scientists have learned how to make deliberate changes in DNA. DNA from one cell can be combined with the DNA in another cell. The DNA that is formed from this combination is called *recombinant* (ree KAHM bih nuhnt) *DNA*.

Recombinant DNA can turn bacteria into chemical factories that produce important substances. For example, people whose bodies cannot produce enough of a chemical called insulin suffer from a disease known as diabetes. Many of these people must take daily shots of insulin to stay healthy. Insulin is expensive because it is made from the organs of cows and sheep. Recombinant DNA research has helped develop insulin-manufacturing bacteria. It is easier to produce large quantities of insulin made in this way, and animals do not have to be used. This technology has made insulin less expensive for diabetics to purchase.

In the future, recombinant DNA may also be used to replace incomplete or undesirable DNA instructions in some plants and animals. Scientists are now using recombinant DNA to make food crops that are more nutritious and to make farm animals that are more resistant to disease.

 ASK YOURSELF

How does DNA replicate itself?

SCIENCE BACKGROUND

Diabetes mellitus is a disease in which the body does not process the sugar glucose properly. The glucose builds up in the blood. Type I diabetics lack sufficient amounts of insulin, which is produced by the pancreas. Insulin enables the body to use and store glucose effectively. Type I diabetics must receive daily doses of insulin to prevent the buildup of glucose. Most diabetics, however, are Type II diabetics. They produce insulin in their bodies, but their bodies do not respond properly to the insulin. Most Type II diabetics can control their disease through diet.

ONGOING ASSESSMENT
ASK YOURSELF

After the chains of a DNA molecule move apart, or become "unzipped," new nucleotides take the place of the separated nucleotides, forming two new DNA molecules.

Figure 5-12. Models of the DNA molecule are used in classrooms and laboratories to help students and scientists understand the molecule.

TEACHING STRATEGIES, continued

● **Process Skills:** *Inferring, Applying*

Point out that respiration and breathing are terms that are sometimes used synonymously, although strictly speaking, respiration is the chemical process that releases energy from food, and breathing is the inhalation of oxygen and the exhalation of carbon dioxide. The terms are, however, related in the sense that the oxygen required by respiration and the waste carbon dioxide produced by respiration are supplied or discarded by the process of breathing. Have the students suggest several activities during which their breathing rate would increase. (Responses should suggest any type of movement.) Then ask the students to explain why their breathing rate increases during strenuous activities. (Responses should suggest that strenuous activity requires energy. The rate of respiration must increase to supply this energy. The breathing rate increases because oxygen is required for respiration.)

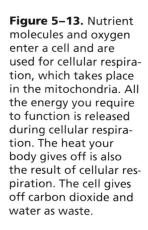

◇ Did You Know?
Respiration releases 18 times more energy from a food molecule than fermentation does.

SCIENCE BACKGROUND

The carbon dioxide released during the process of yeast fermentation in bread causes gas bubbles to form in bread. The bubbles are trapped by a substance in the dough called gluten. The alcohol produced by fermentation in bread dough evaporates during baking, and the yeast is also destroyed by baking.

INTEGRATION—Health

Many people today are involved in some type of exercise program. Point out that the term *aerobic exercise* is used to refer to any type of exercise that requires the heart, lungs, and blood vessels to deliver a large amount of oxygen to the muscles over a long period of time. Regular aerobic exercise tends to increase the efficiency of the circulatory system and strengthen the heart. Ask the students to list ten different forms of exercise in their journals, then classify each as aerobic or not aerobic. (Lists might include: aerobic—bicycling, jogging, walking, jumping rope, swimming; not aerobic—baseball, bowling, golf, shuffleboard.)

▶ **134** CHAPTER 5

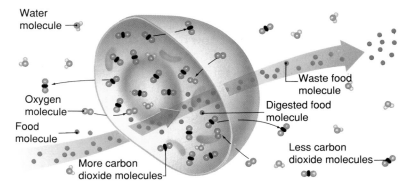

Figure 5–13. Nutrient molecules and oxygen enter a cell and are used for cellular respiration, which takes place in the mitochondria. All the energy you require to function is released during cellular respiration. The heat your body gives off is also the result of cellular respiration. The cell gives off carbon dioxide and water as waste.

Respiration and Fermentation

You need energy for your body to work properly. Running, walking, breathing, and sleeping are just a few of the things you do that require energy. Your body receives the energy it needs from the food you eat. After your body breaks down food into nutrients, the nutrient molecules enter the cells. Sugar is the nutrient that cells use as a source of energy. The sugar molecules are broken down in the cells, and energy is released. Some of this energy is used right away, and some of it is stored.

When Cells Need Oxygen . . .
In the cell, oxygen combines with the sugar molecules, forming carbon dioxide and water. This process is called **cellular respiration.** Respiration releases a great deal of energy. Respiration takes place in a series of small steps inside the mitochondria. Each step gives off a tiny bit of energy. Some of this energy escapes as heat, but most of the energy is used to do work in the cell.

Figure 5–14. Just as each rower in a racing shell provides a little of the energy needed to propel the shell through the water, each of the mitochondria in a cell provides a little of the energy needed by the cell.

● **Process Skills:** *Inferring, Applying*

Explain to the students that *fermentation* is the general term used for describing the process of obtaining energy from glucose (a sugar) in the absence of oxygen. There are, however, different specific types of fermentation that release different waste products including carbon dioxide, alcohol, and lactic acid. In the process of using fermentation to create bread, ask the students to suggest a reason why sugar is added to bread dough. (Responses should suggest that when the yeast in the bread breaks down glucose (sugar) by fermentation, it releases carbon dioxide. Carbon dioxide is a gas that causes the bread to rise.)

. . . And When Cells Don't Need Oxygen

Have you ever seen bread being made? The person making it takes advantage of the fermentation process, which causes the bread to rise. Like cellular respiration, **fermentation** (fur muhn TAY shuhn) is a process that gives off energy as nutrients (mostly sugars) are broken down. Fermentation, however, does not use any oxygen.

Many organisms get their energy from fermentation. For example, yeast is used to ferment sugar. This process releases carbon dioxide and alcohol as waste products. Therefore, yeast cells can be used to produce the alcohol in some beverages and the carbon dioxide that makes bread rise. Some bacteria also use fermentation to get energy. The bacteria give off carbon dioxide and acetic acid as waste products. The vinegar in your salad dressing is mild acetic acid that is made this way. Follow the directions in the next activity to observe the process of fermentation in action.

Figure 5–15. When yeast (above) is used to ferment sugar, carbon dioxide and alcohol are released as waste products. The carbon dioxide causes the dough (left) to rise. The alcohol evaporates as the bread is baked.

Discover BY Doing

Open a package of yeast and put it into a beaker. Add a small amount of warm water to the yeast. Gently swirl the yeast and water mixture inside the beaker. Carefully observe the beaker. What happens to the yeast? Add a small amount of sugar to the beaker. Observe carefully for several minutes. What changes do you see? What causes these changes?

MULTICULTURAL CONNECTION

Historians believe that the ancient Egyptians were the first people to make yeast bread. The Egyptians probably learned to make yeast bread about 2600 B.C. They taught the ancient Greeks how to make yeast bread, and the Greeks taught the Romans how to make the bread. The Romans, in turn, taught other European peoples how to make yeast bread by A.D. 100s. If the students in your classroom have varying ethnic backgrounds, you might have them bring in bread commonly identified with their cultural background. Ask the students to note whether or not the bread is a yeast bread.

DISCOVER BY *Doing*

Dissolve the yeast in 75 mL of warm water (about 30 Celsius). No changes will occur until the sugar is added. The students should observe that the yeast solution begins to bubble several minutes after the sugar is added and should infer that the bubbling occurs because the yeast breaks down the sugar into carbon dioxide, which causes the bubbling, and alcohol.

GUIDED PRACTICE

 Cooperative Learning Ask the students to use their own words to explain one of the following processes to a partner: diffusion, active transport, passive transport, mitosis, replication, respiration, fermentation, and photosynthesis. Have the partners reverse roles until all of the concepts have been explained.

INDEPENDENT PRACTICE

 Have the students provide written answers to the Section Review and Application questions. Then ask the students to create a journal entry describing the importance of DNA to living things.

ONGOING ASSESSMENT
ASK YOURSELF

Cells need energy to maintain all of the cell activities that sustain life. Without energy, cells would die.

REINFORCING THEMES—
Energy

As you discuss photosynthesis, emphasize the interactions that are needed for photosynthesis to take place. Also encourage the students to consider the importance of photosynthesis to the survival of all plant and animal life. Explain to the students that the source of all energy released through respiration and fermentation is the sun. The process of photosynthesis converts the solar energy into chemical energy.

LASER DISC
3057

Photosynthesis; compared with aerobic cellular respiration

Figure 5–16. Fatigue after exercise is the result of lactic acid formation.

When There Isn't Enough Oxygen Fermentation and respiration are similar because both processes can be used by cells to release energy. Have you ever exercised so much that your muscles seemed to burn? Exercise requires large amounts of energy. Normally, your muscles receive all the energy they need from respiration. However, respiration requires oxygen. During heavy exercise, oxygen cannot be brought to your muscle cells fast enough to keep up with the demand.

When oxygen is in short supply, muscle cells have the ability to release energy from sugar through fermentation. Fermentation releases energy from sugar molecules without the use of oxygen. However, less energy is released from a sugar molecule during fermentation than during respiration. In addition, during fermentation muscle cells form lactic acid instead of the carbon dioxide and water formed during respiration. The lactic acid is what causes the "burning" sensation in the muscles.

You may notice a stiffness and soreness in your muscles the day after heavy exercise. This stiffness and soreness will go away as your body gets rid of the lactic acid. Fermentation in muscle cells will stop after exercise ends and enough oxygen becomes available for the cells to again begin respiration. The table shows a comparison of respiration and fermentation.

Table 5-1	A Comparison of Respiration and Fermentation
Respiration	**Fermentation**
Uses oxygen	Does not use oxygen
Releases much energy	Releases little energy
Produces carbon dioxide and water	Produces carbon dioxide and alcohol or acid

ASK YOURSELF
Why do cells need energy?

Photosynthesis

Can your body make its own food? Of course not. Plants, however, can. The process by which plants make food is called *photosynthesis*. Photosynthesis requires light, carbon dioxide, water, and chlorophyll. Chlorophyll traps energy from sunlight and

EVALUATION

Ask the students to compare fermentation and respiration. How are they similar? How are they different? (Both processes release energy and are necessary for survival; respiration occurs in the presence of oxygen, and fermentation occurs in the absence of oxygen.)

RETEACHING

Tape record descriptions or definitions of the concepts and processes presented in this section. Give the students a list with the names of the concepts and processes on it. Have them listen to the tape and match the taped description to the appropriate listed term.

EXTENSION

Encourage the students to locate the chemical equations that describe the processes of photosynthesis and respiration. Then ask the students to compare the equations and explain their findings to the class. (The equation for photosynthesis is opposite the equation for respiration.)

CLOSURE

Cooperative Learning Divide the class into several groups. Have each group work together to create a diagram of a DNA molecule and write an explanation of DNA's structure and function.

---

uses it to break down water molecules into atoms of hydrogen and oxygen. The oxygen is given off as a byproduct that can be used by organisms during respiration. The rest of the energy is used to combine the hydrogen atoms with the carbon dioxide. Eventually, sugar is formed and stored by the plant until it is needed.

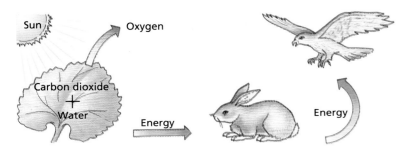

Figure 5-17. Photosynthesis converts solar energy into chemical energy, which is stored by plants. When an animal eats a plant, the animal uses that energy. When one animal eats another animal, energy that was originally solar energy is transferred.

The sugar made during photosynthesis may be used by the plant as food. The sugar can also be converted into starches, fats, and other compounds and stored or used to build new plant structures. An animal that eats a plant takes in the stored molecules and uses them for food. The animal uses the energy from these molecules to do work. Any extra energy is stored in the animal's body. As one animal eats another animal, the energy that came to the plants from the sun is passed on again and again. In this way, all the energy that is used by all the living things on Earth can be traced back to sunlight.

 ASK YOURSELF

What would happen to the animals on Earth if all the plants suddenly disappeared?

SECTION 3 REVIEW AND APPLICATION

Reading Critically

1. How is a DNA molecule like a ladder?
2. Describe the pairing of bases in a DNA molecule.
3. How are respiration and fermentation different from each other?

Thinking Critically

4. What do you suppose would happen if DNA replication stopped before a complete copy of the DNA material had been made?
5. If a cell contained fewer mitochondria than normal, what do you suppose would happen to the cell? Explain.
6. Someday space travel may take people to other solar systems. Why is finding a planet inhabited only by plants much more likely than finding a planet inhabited only by animals?

ONGOING ASSESSMENT
ASK YOURSELF

All of the animals on Earth would eventually become extinct if all of the plant life of Earth suddenly disappeared because plants produce the foods that are the source of animal energy.

SECTION 3 REVIEW AND APPLICATION

Reading Critically

1. The DNA molecule is shaped like a ladder with bases forming the rungs and sugar-phosphate units forming the sides.

2. Adenine (A) always pairs with thymine (T), and guanine (G) always pairs with cytosine (C) in a DNA molecule.

3. Respiration requires oxygen and fermentation does not. Respiration releases more energy than fermentation does.

Thinking Critically

4. If replication stopped before completion, the daughter cells would receive incomplete copies of the hereditary instructions. This would most likely lead to severe abnormalities or death of the cells.

5. A cell that lacked an adequate number of mitochondria would die because all cells require a level of energy that is sufficient to perform their life activities.

6. Plants are capable of both photosynthesis and respiration, while animals perform only respiration. Therefore, plants can make their own food and survive without animals. However, animals could not survive without plants.

SKILL
Modeling DNA

Process Skills: Constructing/Interpreting Models, Classifying/Ordering

Grouping: Groups of 3 or 4

Objectives
- **Assemble** models of a DNA molecule.
- **Arrange** patterns of DNA nitrogen bases.

Discussion
You may wish to display a completed DNA model for the students to examine before this activity begins. **CAUTION:** Advise the students not to eat the marshmallows and jelly beans used in this activity.

▶ Application
1. The DNA molecule would have 5000 molecules of thymine because adenine and thymine are always paired.

2. No, the number of adenine bases is unrelated to the number of cytosine bases.

✴ Using What You Have Learned
1. Models are able to be clearly seen and manipulated, allowing for greater understanding of the structure and function of an object or a substance.

2. The molecule might have too many or too few nitrogen bases, which could result in a deformed cell or in the death of the cell. Also the wrong base might pair with another base, which would change the hereditary code of information.

▶ 138 CHAPTER 5

SKILL
Modeling DNA

▶ MATERIALS
- 25-cm pieces of fine wire (2) • miniature marshmallows (6) • jelly beans (8) • four different-colored pipe cleaners cut into thirds (6) • modeling clay

> wire = sugar-phosphate backbones
> marshmallows = sugar groups
> jelly beans = phosphate groups
> pipe cleaners = bases

▼ PROCEDURE

1. Lay two pieces of wire side by side on your desk.
2. String miniature marshmallows and jelly beans on the wires as shown. Lay the two chains side by side with the larger ends of the jelly beans facing out as shown.
3. Now attach the bases to the sugar groups on each chain. Copy the table and record your color code for each base. Remember, adenine pairs with thymine; cytosine pairs with guanine.
4. Twist the ends of the paired bases together to form the ladderlike structure of DNA.
5. Make a modeling-clay base and carefully stick one end of your DNA molecule into the base. Start at the bottom of the model and gently twist it into the characteristic spiral.

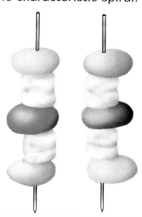

TABLE 1	
Base	Pipe Cleaner Color
Adenine	
Thymine	
Cytosine	
Guanine	

▶ APPLICATION
1. If there are 5000 molecules of adenine in one DNA molecule, how many molecules of thymine will there be?
2. Must DNA molecules have the same number of adenine bases as cytosine bases? Explain your answer.

✴ Using What You Have Learned
1. Why does using a model often make it easier to understand the structure and function of an object or a substance?
2. Because of the base pairings, DNA replicates exactly. However, sometimes a change, or mutation, occurs. Using your model as a DNA molecule, how do you think the molecule might be changed?

CHAPTER 5 HIGHLIGHTS

The Big Idea—SYSTEMS AND STRUCTURES

Lead the students to understand that all of the cells of an organism work together to contribute to the well-being of that organism. Also emphasize that the countless functioning cells of an organism require energy, and the energy of an organism is derived by its cells from the processes of respiration and fermentation.

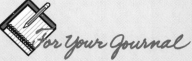

The ideas expressed by the students should reflect the understanding that all of the life processes of a person involve energy. The reproduction of cells is a necessary prerequisite for the survival of a person because the human body has a constant need for new cells. Cell division is the process that supplies the human body with new cells. The students should also compare the methods used by a scientist studying DNA to those used by a detective deciphering a code.

CONNECTING IDEAS

The concept map should be completed with words similar to those shown here.

Row 2: transport materials through

Row 4: diffusion, fermentation, photosynthesis

Row 5: replication

CHAPTER 5 HIGHLIGHTS

The Big Idea

In the same way that the symbols in a coded message work together to convey the message, the cells in a living organism work together to keep the organism healthy and functioning. But cells cannot perform their function in a plant or animal unless each individual cell can function properly on its own. Cells carry out a wide variety of tasks, with the types of activity depending on the role of the specific cell.

Examples of important cell functions include the transportation of materials by diffusion and osmosis, the release of energy through respiration and fermentation, and the manufacture of food in plant cells through photosynthesis. Reproduction of cells and replication of DNA are other important cell functions.

Look back at the ideas you wrote in your journal at the beginning of the chapter. Summarize the cell processes you have studied. Do all of the processes involve energy? Why must cells reproduce? How is a scientist studying DNA like a detective trying to break a secret code?

Connecting Ideas

Copy this unfinished concept map into your journal. Complete the concept map by writing the correct term in each blank.

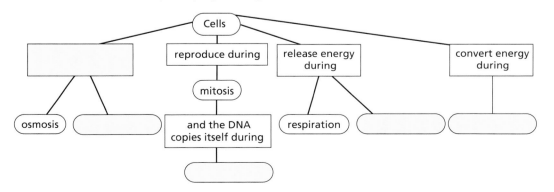

CHAPTER 5   **139**

CHAPTER 5 REVIEW

ANSWERS

Understanding Vocabulary

1. DNA, mitosis, chromosomes, and replication are all components or processes of cell reproduction.

2. Active transport, passive transport, osmosis, and diffusion are all forms of molecule transport in the environment or in cells.

3. Photosynthesis converts energy, active transport requires the use of energy, and respiration releases energy. Energy is used in all of these processes.

4. Respiration and fermentation are processes that release energy for use by an organism.

Multiple Choice

5. c
6. a
7. c
8. d
9. b

Interpreting Graphics

10. The process of photosynthesis performed by green plants uses carbon dioxide from the atmosphere to manufacture glucose.

Reviewing Themes

11. The entire cast works together to perform a school play in a prescribed sequence of acts. In the same way, the phases of mitosis are sequenced to produce new cells.

12. Diffusion and other types of passive transport can occur in a cell in the absence of energy.

▶ 140 CHAPTER 5

Short Answer

13. Molecules that supply energy and raw materials necessary for cell activities must move across the cell membrane from the external environment. Waste molecules must move across the membrane to the external environment, preventing cell pollution and poisoning.

14. A selectively permeable membrane regulates the flow of materials, allowing only some materials to pass. A permeable membrane allows all materials to pass.

15. During interphase, a cell performs basic life activities and stores energy required for mitosis. Without the interphase stage, mitosis could not occur, cells would die, and replacement daughter cells would not be created.

16. Daughter cells are exactly the same as the parent cell because the daughter cells receive an exact copy of the DNA, or hereditary material, carried by the parent cell.

17. The cytoplasm of animal cells divides as the cell membrane pinches inward, separating the daughter cells. In plant cells after the cytoplasm has divided, a cell plate forms in the center of the cell and moves outward.

CHAPTER 5 REVIEW

Understanding Vocabulary

In the problems below, the terms in this chapter have been put into different categories. For each term, give a reason justifying its membership in the given category.

1. **Category: reproduction**
   DNA (127), mitosis (127), chromosomes (127), replication (132)
2. **Category: transportation**
   active (122), passive (122), osmosis (123), diffusion (121)
3. **Category: uses energy**
   photosynthesis (136), active transport (122), respiration (134)
4. **Category: releases energy**
   respiration (134), fermentation (135)

Understanding Concepts

MULTIPLE CHOICE

5. Osmosis is a type of diffusion in which
   a) the cell manufactures its own food.
   b) all types of molecules are allowed to pass through the cell membrane.
   c) water is moved through the cell membrane.
   d) the cell uses energy.

6. Diffusion is *not* a form of active transport because
   a) no energy is required from the cell.
   b) the cell membrane allows only some types of molecules to pass through.
   c) oxygen is not required for diffusion.
   d) food molecules are not created.

7. During mitosis
   a) the cell prepares for the next cell division.
   b) the cell's hereditary instructions are modified.
   c) two new daughter cells are formed.
   d) two cells are combined into one.

8. The hereditary instructions of a cell are formed by
   a) a combination of different chromosomes.
   b) mitosis.
   c) scientific research involving DNA.
   d) a particular sequence of bases in the DNA molecule.

9. Which of the following does *not* describe a difference between respiration and fermentation?
   a) Respiration produces water.
   b) Respiration releases energy, but fermentation does not.
   c) Respiration requires oxygen.
   d) Fermentation gives off less energy.

Interpreting Graphics

The diagram shows a simplified picture of the carbon cycle. What role does photosynthesis play in this cycle?

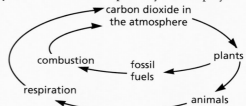

Thinking Critically

18. Responses should indicate that breathing is the process necessary for getting oxygen into the lungs and, therefore, into the cells. Once in the cells, the oxygen is used for cellular respiration.

19. Recombinant DNA might enable scientists to identify disease-resistant genes. The scientists could then insert the disease-resistant genes into farm animals and breed these animals to develop a genetically superior strain.

20. Fermentation releases energy from food molecules as does respiration, but fermentation is much less efficient than respiration. As a result, more food molecules are required for fermentation to produce the same amount of energy that can be produced by respiration.

21. A great variety of characteristics can be coded by changing the length or the sequence of bases of a DNA molecule.

22. A helix is a spiral. Because the DNA molecule consists of two spiraling strands, the DNA molecule is often described as a double helix. The students might note that some bacteria, coiled springs, and some pastas have spiral shapes.

Reviewing Themes

11. *Systems and Structures*
Compare the events of mitosis with the activities of a group of students putting on a school play.

12. *Energy*
What functions can still occur in a cell in the absence of energy?

SHORT ANSWER

13. Why does a cell's survival depend on the movement of certain materials through its cell membrane?

14. How is a selectively permeable membrane different from a permeable membrane?

15. Give reasons why this statement is false: Nothing much happens in a cell during interphase because the most important events all occur during mitosis.

16. Why are daughter cells exactly the same in structure and function as their parent cell?

17. How is division of cytoplasm in plant cells different from division in animal cells?

Thinking Critically

18. Explain how external respiration, or breathing, is related to respiration in the cell.

19. How might recombinant DNA research help scientists find a way to make farm animals more resistant to disease?

20. Why would a cell that uses fermentation require more food molecules than a cell that uses respiration?

21. All organisms are controlled by the DNA in their cells. How does the structure of a DNA molecule help to account for the great variety of life that exists on Earth?

22. The photograph below shows a model of a DNA molecule. Explain why its shape is described as a "double helix." Then give other examples from life science of things that have the shape of a helix or a spiral.

Discovery Through Reading

Bender, Lionel. *Through the Microscope: Frontiers of Medicine.* Gloucester Press, 1991. Examines the advances in medical science made possible by the use of microscopes, from the first studies of plants and insects in the seventeenth century to the discovery of bacteria and viruses two centuries later.

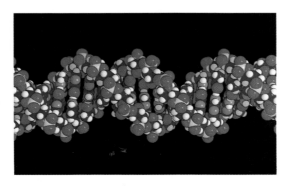

CHAPTER 6

NATURAL SELECTION AND HEREDITY

PLANNING THE CHAPTER

Chapter Sections	Page	Chapter Features	Page	Program Resources	Source
Chapter Opener	142	For Your Journal	143		
Section 1: A GREAT ADVENTURE	144	Discover By Researching (A)	147	Science Discovery*	SD
• Darwin's Voyage (B)	144	Activity: What is the role of variation in evolution? (A)	151	Connecting Other Disciplines: Science and Language, Studying the Dinosaurs (H)	TR
• Darwin's Hypothesis (B)	150	Section 1 Review and Application	151	Study and Review Guide, Section 1 (B)	TR, SRG
Section 2: A GREAT THEORY	152	Discover By Doing (A)	156	Science Discovery*	SD
• The Origin of Life (B)	152	Section 2 Review and Application	158	Investigation 6.1: Fossil Evidence (A)	TR, LI
• Developing the Theory (A)	155	Skill: Controlling Variables (A)	159	Extending Science Concepts: Natural Selection in Peppered Moths (A)	TR
• Support for Darwin's Theory (A)	158			Study and Review Guide, Section 2 (B)	TR, SRG
Section 3: MENDEL AND HEREDITY	160	Activity: Are your traits dominant or recessive? (A)	164	Science Discovery* Investigation 6.2:	SD
• The Story of Gregor Mendel (B)	160	Discover By Doing (A)	167	Genes and Traits (B)	TR, LI
• Dominant and Recessive (A)	162	Section 3 Review and Application	167	Reading Skills: Determining Similarities and Differences (A)	TR
• Reproduction and Heredity (A)	165	Investigation: Investigating Genetic Probabilities (A)	168	Thinking Critically (H)	TR
• The Hereditary Code (H)	166			Seven Traits of Garden Peas (B)	IT
				Mendel's Hybrid Crosses (A)	IT
				Some Human Hereditary Characteristics (A)	IT
				Meiosis (A)	IT
				All Organisms Are Linked by DNA (B)	IT
				Record Sheets for Textbook Investigations (A)	TR
				Study and Review Guide, Section 3 (B)	TR, SRG
Chapter 6 HIGHLIGHTS	169	The Big Idea	169	Study and Review Guide, Chapter 6 Review (B)	TR, SRG
Chapter 6 Review	170	For Your Journal	169	Chapter 6 Test	TR
		Connecting Ideas	169	Test Generator	

B = Basic   **A** = Average   **H** = Honors
The coding Basic, Average, and Honors indicates subsections, features, and resources that might be appropriate for different levels of learners. For additional suggestions regarding choice of topic and depth of coverage, see the Pacing Chart on pages T26–T29.

*Frame numbers at point of use
(TR) Teaching Resources, Unit 2
(IT) Instructional Transparencies
(LI) Laboratory Investigations
(SD) *Science Discovery* Videodisc Correlations and Barcodes
(SRG) Study and Review Guide

CHAPTER MATERIALS

Title	Page	Materials
Discover By Researching	147	(per class) globe or map, encyclopedia or atlas
Activity: What is the role of variation in evolution?	151	(per group of 3 or 4) scissors, red paper, blue paper, paper cups (2)
Teacher Demonstration	153	paper cup (small), plaster of Paris, small object (e.g., seashell)
Discover By Doing	156	(per group of 2) paper punch, construction paper (4 colors), multicolored cloth, watch with a second hand
Skill: Controlling Variables	159	(per group of 2) materials to be identified by students (soil, sand, marigold seeds, flower pots (2), water)
Activity: Are your traits dominant or recessive?	164	(per group of 2) mirror
Discover By Doing	167	(per individual) paper, pencil
Investigation: Investigating Genetic Probabilities	168	(per group of 2) paper, pencil, coins

ADVANCE PREPARATION

For the *Demonstration* on page 153, you will need to obtain plaster of Paris and a small object to use as a fossil (such as a seashell). For the *Skill* on page 159, the students will identify the materials needed. Soil, sand, marigold seeds, and flower pots will be required.

TEACHING SUGGESTIONS

Field Trip
Plan a trip to a nearby zoo and arrange to have someone speak to the class about other interesting facts about the flamingo, giraffe, various tortoises, and other animals studied by the scientists in the chapter. The students can investigate subjects, such as the diet used to keep the flamingo's feathers pink. If there is no local zoo, write to the nearest zoo and ask for written material on the same topics.

Outside Speaker
Invite someone from a blood bank in your area to speak to the class about the different blood types and the way they are inherited.

Invite a science historian from a local college or university to come and speak with the students about the history of discoveries in human evolution. The presentation should concentrate not only on the discoveries and when they were made but also on the theorized relationships between the discovered organisms.

CHAPTER 6
NATURAL SELECTION AND HEREDITY

CHAPTER THEME—CHANGES OVER TIME

This chapter introduces the students to the roles of natural selection and heredity in evolution. The students will discover that the theory of evolution attempts to explain how living things evolve, or change over time. The development of the theory of natural selection by Charles Darwin is illustrated, and the students will find that this theory has greatly influenced modern scientific explanations of how things change. The students will also learn how genes influence the characteristics that are passed from parents to offspring. This theme is also developed through concepts in Chapter 14. A supporting theme of this chapter is **Nature of Science**.

MULTICULTURAL CONNECTION

The theory of natural selection is sometimes paraphrased and described as "only the strong survive." Have students from different geographical areas describe some of the changes in the plant and animal life of those areas that enable organisms to better survive. For example, students from northern latitudes might describe how predators have difficulty seeing animals that change colors as seasons change, or students from tropical latitudes might describe how ground plants have adapted to diffused light because the canopy of a rain forest prevents direct sunlight from reaching the ground. Then encourage these students to describe how these plant and animal adaptations have influenced the people and culture of each area.

MEETING SPECIAL NEEDS

Second Language Support

Fluent English-speaking students should be paired with students who have limited English proficiency to work together using reference materials to locate pictures showing how plants or animals change over the course of time. Examples of pictures include fossil evidence of the extinction of dinosaurs or the similarities and differences of parents and their children. Have the students display their findings to their classmates.

▶ 142 CHAPTER 6

CHAPTER 6
Natural Selection and Heredity

Charles Darwin was born in Shrewsbury, England, on February 12, 1809—the same day as Abraham Lincoln. Everyone thought he would become a doctor, like his father and grandfather, but Darwin preferred collecting stones and minerals and watching birds. He entered Cambridge University to become a minister, and although he passed his courses, the only thing Darwin really enjoyed was observing nature.

In 1831 an invitation was sent to Charles Darwin. He was invited to join HMS Beagle on a surveying trip around the world as the ship's naturalist. The offer came from Captain Robert Fitzroy who was to lead the expedition. They were to make detailed maps of the coast of

Galápagos Islands

CHAPTER MOTIVATING ACTIVITY

Remind the students that they are made up of many characteristics, which have been given to them by their parents, and ask them to record a variety of these characteristics in their journals. Then have the class discuss the characteristics that they listed, and eventually guide the discussion to include characteristics of plants and animals other than humans. Encourage the students to think about characteristics such as flower color, hair color, height of plants, and height or size of animals.

Answering the journal questions gives the students the opportunity to speculate or display what they already know about the role of genetics and natural selection in the process of evolution. The answers provided by the students will give you an opportunity to point out that some of their answers may change as they study the material in this chapter. Remind the students to periodically check their answers to the journal questions.

ABOUT THE LITERATURE

The excerpt in the chapter opener is from the book *The Voyage of the Beagle* by Kate Hyndley. This book is one of a series of books called *Great Journeys*. The series focuses on accounts of men and women who made pioneering journeys. In addition to Darwin's voyage, the books tell about the first humans on the moon, the first transatlantic flight, the race to the South Pole, Marco Polo's travels, and Columbus's voyages. An interesting discussion can begin with the question of what these journeys have in common.

Charles Darwin wrote extensive notes during his five-year journey. Three years after his return to England, he published *Journal of the Voyage of the Beagle*. This fascinating account of the day-to-day events during the voyage was a great success in its day and can still be found in the travel section of most libraries.

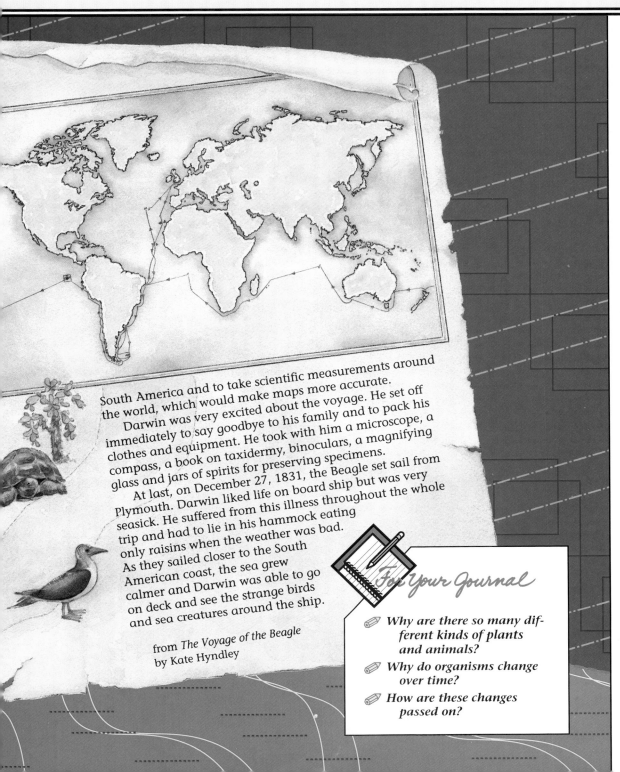

South America and to take scientific measurements around the world, which would make maps more accurate.

Darwin was very excited about the voyage. He set off immediately to say goodbye to his family and to pack his clothes and equipment. He took with him a microscope, a compass, a book on taxidermy, binoculars, a magnifying glass and jars of spirits for preserving specimens.

At last, on December 27, 1831, the Beagle set sail from Plymouth. Darwin liked life on board ship but was very seasick. He suffered from this illness throughout the whole trip and had to lie in his hammock eating only raisins when the weather was bad.

As they sailed closer to the South American coast, the sea grew calmer and Darwin was able to go on deck and see the strange birds and sea creatures around the ship.

from *The Voyage of the Beagle* by Kate Hyndley

For Your Journal

- Why are there so many different kinds of plants and animals?
- Why do organisms change over time?
- How are these changes passed on?

CHAPTER 6 143

Section 1: A GREAT ADVENTURE

FOCUS

In this section, the observations of Charles Darwin are presented. Darwin observed, then hypothesized, that living things change during the course of time. Even though these observations were made during the 1800s, Darwin's work still helps scientists today explain how things evolve, or change over time.

MOTIVATING ACTIVITY

Obtain drawings of a giraffe, a horse, or another animal during its evolutionary stages of development. Without displaying the drawings in the correct order of development, ask the students to organize the drawings in order of the appearance of the animal on Earth. Have the students discuss some of the changes that the animals have experienced during the course of time.

PROCESS SKILLS
- Interpreting Data
- Inferring

POSITIVE ATTITUDES
- Curiosity • Enthusiasm for science and scientific endeavor

TERMS
- evolution

PRINT MEDIA
The Voyage of the Beagle by Charles Darwin (see p. 93b)

ELECTRONIC MEDIA
Darwin's Finches, Time-Life (see p. 93b)
Science Discovery Observation and inference
Darwin's finches

BLACKLINE MASTERS
Study and Review Guide
Connecting Other Disciplines

BACKGROUND INFORMATION

Charles Darwin studied medicine and theology, not biology or natural history. But he was a dedicated amateur naturalist who had collected and studied specimens of animals and plants from the time he was a boy. It was that expertise, as well as his friendship with botanist John Stevens Henslow, that got Darwin the post of naturalist on the *Beagle* expedition. Almost five years later, Darwin returned to England where he spent the rest of his life studying and writing about the specimens he collected on that trip. He never went on another scientific expedition. In fact, he never left England again.

▶ 144 CHAPTER 6

SECTION 1

A Great Adventure

Objectives

Describe the observations Darwin made on his voyage.

Summarize Darwin's inferences about why organisms change.

Relate Darwin's inferences to his observations.

Although Charles Darwin was never a good student, he was an avid reader. One of the books that influenced him the most was a book on geology by Charles Lyell. In this book, *Principles of Geology,* Lyell presented many new ideas about the earth, and he hypothesized that the earth was constantly changing. Although most of the scientists of the day disagreed with Lyell, Darwin thought there was some merit to Lyell's hypothesis.

Darwin's Voyage

Lyell said that not only does the earth change, but there are also two kinds of change—fast change, such as that caused by earthquakes and volcanoes, and slow change, such as that caused by erosion.

The evidence for fast change was all around Darwin. While visiting the city of Concepción, Chile, in South America, he got to see firsthand the effects of an earthquake.

> *The most remarkable effect of this earthquake was the permanent elevation of the land.... There can be no doubt that the land around the Bay of Concepcion was upraised two or three feet.*

Figure 6–1. Lyell's hypotheses seem reasonable; perhaps the earth is older than most people believe.

TEACHING STRATEGIES

● **Process Skills:** *Communicating, Applying*

Darwin observed the result of an earthquake and thought about how the earthquake was changing the earth's surface. Ask the students to describe how a volcanic eruption might change the surface of the earth. (Descriptions may include molten, flowing rock that cools and forms a new surface layer of the earth.)

● **Process Skills:** *Predicting, Expressing Ideas Effectively*

Point out that volcanic disturbances change the earth's surface naturally and that other changes can be unnatural or caused by humans. Ask the students to predict how the actions of people might change the surface of the earth in the future. (Predictions may include the effects of pollution or war on a given geographical area.)

It became obvious to Darwin that the earth does indeed change. Volcanoes created mountains where there had been none before, and earthquakes altered the shape of the land. But the slow changes that Lyell spoke of were not so so obvious. Most people of the 1800s believed that the earth was only a few thousand years old. Slow change, such as Lyell described, could have taken place only if the earth were much older.

One of Darwin's most important discoveries was made while digging in the gravel near Punta Alta in Argentina. He uncovered fossilized bones of large animals that had probably died out a long time before. Darwin wondered why these animals died out, since at one time they must have existed in large numbers.

One day, Darwin observed fossils of seashells at the top of a cliff. He wondered how the shells of sea organisms could have gotten there. Darwin reasoned that the layer of rock containing the shells must have been under the sea at one time. Perhaps some force of nature had lifted the rock to its present position. Maybe this was evidence of the slow change that Lyell had written about.

Wherever the *Beagle* anchored, Darwin went ashore to explore. In Brazil, he was amazed by the beauty of the tropical rain forests. Flowers and brightly colored birds were everywhere. To Darwin, the huge trees covered with vines and the lush vegetation were like a paradise. He spent many hours studying plants and animals no scientists had seen before.

Figure 6–2. The earth does indeed change. Here is an example, just as Lyell said.

SCIENCE BACKGROUND

Charles Lyell was a geologist, and his book *Principles of Geology* caused Darwin to look more carefully for geologic changes during his trip. Darwin had experienced an earthquake and observed volcanic eruptions. He then inferred that if the surface of the earth changes over time, so too might living things change.

THE NATURE OF SCIENCE

Throughout his voyage around the world, Darwin carefully observed and recorded information about plants and animals. He also collected hundreds of specimens for later study. His observations and other data helped him develop his theory of evolution. Darwin followed a scientific method—first he made observations and collected facts, then he developed a theory that unified and explained the data.

Figure 6–3. These rocks must have been under the sea at one time.

SECTION 1   **145** ◀

TEACHING STRATEGIES, continued

● **Process Skills:** *Formulating Hypotheses, Applying*

Remind the students that Darwin observed plants and animals that were well-suited to the environments in which they lived. Ask the students to choose a plant or animal, then hypothesize what evolutionary changes might occur in that organism if the environment in which the organism lived changed significantly over time. (The students' hypotheses may suggest that organisms that could not adapt to the new conditions might become extinct. Those organisms that have the ability to successfully adapt to the new conditions might produce more offspring, and the diversity of populations would change over time.)

▲ MEETING SPECIAL NEEDS

Second Language Support

Ask the students to collect pictures from magazines and other disposable sources and create a collage of the plant and animal life that can be found in their country of origin. Encourage them to use a foreign language dictionary if necessary and label each organism with its English language and native language name. Use the completed collages to discuss reasons why plant and animal life is different from country to country.

Figure 6–4. There are plants and animals here that scientists have never seen before.

As the *Beagle* sailed around the coast of South America, Darwin found many fossils. One skull of particular interest was of an extinct animal known as a *toxodon*. It reminded Darwin of the capybara he had seen in the Brazilian forests.

Darwin continued to wonder why so many animals had become extinct. Darwin was further puzzled after finding the fossils of a horse near Buenos Aires, Argentina. When the Spanish colonists arrived in South America in 1535, there were no living horses on the continent.

Figure 6–5. These animals greatly resemble this ancient skull I have found.

● **Process Skills:** *Inferring, Evaluating*

Darwin collected many things, including the bones of extinct animals. Ask the students to suggest several natural events that might have caused the extinction of dinosaurs and record these events on the chalkboard. (The students might suggest that ice ages, abnormally warm temperatures, or the impact of very large meteorites may have caused the extinction of dinosaurs.) Then have the students discuss and debate the likelihood of each event and rank the events in order of decreasing likelihood.

In September 1835 the Beagle *left Lima to sail west across the Pacific Ocean. The party intended to return to England by sailing around the Cape of Good Hope at the southern tip of Africa. Their first port of call was a group of islands 1000 kilometers west of Peru called the Galápagos Islands.*

At first sight, the islands were not very attractive—black volcanic rocks sticking up from the sea with only a few scorched bushes growing on them. Darwin soon realized that the islands had once been underwater volcanoes, which had now risen above sea level.

DISCOVER BY Researching

Locate the Galápagos Islands on a globe or map. Note the position of the islands in relation to the United States. Near what country in South America are they located? Use an encyclopedia or atlas to find out more about the islands. What are their elevations? What is their climate? Are they inhabited by people? What kinds of plants and animals live there? Write a paragraph summarizing your findings.

DISCOVER BY Researching

The students should discover that the Galápagos Islands are located on the equator at about 90 degrees west longitude, or approximately 100 km west of Ecuador. St. Louis, Missouri, is located about as far west as the Galápagos Islands but is located about 40 degrees farther north. Since the students are likely to discover and summarize different characteristics of the islands, have interested students share and compare their summaries with those of other students. Then ask the students to place their summaries in their science portfolios.

Figure 6–6. The Galápagos Islands support a wide variety of living things such as the iguana (top) and the blue-footed booby (bottom).

TEACHING STRATEGIES, continued

● **Process Skills:** *Inferring, Generating Ideas*

Darwin wondered why the huge marine lizards seemed unafraid of humans. Have the students suggest reasons why the lizards might have been unafraid of Darwin and his companions. (Responses may include the idea that over time, the lizards had no natural predators. Despite the fact that humans could be considered predators, they were not recognized as such by the lizards, and no defensive actions were taken.)

● **Process Skills:** *Observing, Expressing Ideas Effectively*

Have the students study Figure 6–8 and note the caption. Allow them time to discuss and debate reasons why the tortoises differ from island to island. (The students might suggest that although the islands are close to each other, the characteristics of each island environment are somewhat different, requiring different adaptations in the tortoises from island to island.)

BACKGROUND INFORMATION

The terms *turtle* and *tortoise* are sometimes used interchangeably. Turtles live on land or in water, while tortoises live only on land. Tortoises are one group of turtles. The largest tortoises live on the Galápagos Islands and on the Aldabra Islands in the Indian Ocean off the east coast of Africa. Three species of tortoises live in areas of the southern United States.

> ✦ **Did You Know?**
> Turtles live longer than any other vertebrate. Most turtles have a life expectancy of at least 50 years. Some turtles can live to be over 100 years old.

LASER DISC
472

Observation and inference

Figure 6–7. These animals seem completely unafraid of humans.

On the beaches lived black lizards that were about one meter long. They could swim well but never stayed in the sea for long. These sea-living lizards ate seaweed. The land-living lizards were more brightly colored and fed on cactus. They were very gentle animals and did not attack or bite anything, even when provoked.

The word *galápagos* means giant tortoise in Spanish, and these huge animals thrived on the islands. Some were so large that it took eight men to lift one of them. Darwin walked inland to a spring where he watched the tortoises drink greedily. Sometimes they would drink only once a month. He also observed that on each island the tortoises had different shell markings.

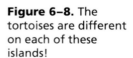

Figure 6–8. The tortoises are different on each of these islands!

▶ 148   CHAPTER 6

● **Process Skills:** *Analyzing, Applying*

Point out that the changes Darwin hypothesized about—that living things evolve, or change over time—are part of a very slow process, and that living things change collectively, not individually, over time. Ask the students to describe ways in which physical changes can occur over several generations of organisms in a species. (There is variation in the structure of all living organisms in a species. For example, giraffes show variation in neck length—some have longer necks than others. If the longer neck has an advantage in feeding on the leaves of trees used for food, giraffes with longer necks will tend to be more successful and have more offspring. The longer neck will appear more and more frequently with each generation. Eventually, all giraffes will display the long neck.)

Darwin noticed that nearly all living things show differences among members of the same species. Some differences seem to have little importance. But was that true of all differences? And why would there be so many differences among tortoises living on islands within sight of each other? Darwin began to notice that the birds, too, differed from one island to another.

Darwin counted thirteen species of finches on the islands. Each species had a different kind of beak. He was puzzled as to why these variations occurred.

Figure 6–9. There seem to be variations among all the animals living on these islands.

Darwin saw the different finches as evidence that, like the earth itself, organisms change over time. He reasoned that several kinds of finches had come to the islands from the mainland of South America. Variations in their beaks allowed some finches to feed more effectively than other finches in this new environment. Therefore, certain finches were likely to survive, while others died out. Those that survived reproduced and slowly established themselves on certain islands.

Figure 6–10. Finches with certain kinds of beaks will survive better on this island.

 **ASK YOURSELF**

What observations did Darwin make about the tortoises and finches of the Galápagos Islands?

 LASER DISC
4253, 4254
Darwin's finches

INTEGRATION—*Social Studies*

 Cooperative Learning As Darwin's visit to the Galápagos Islands ended, he and his companions sailed for home across the Pacific Ocean. Divide the class into groups of three or four, and using maps of the world or atlases, have the groups describe routes the *Beagle* may have sailed and sights that the travelers may have seen along those routes. Have interested groups share their descriptions with other groups.

ONGOING ASSESSMENT
▼ **ASK YOURSELF**

Darwin observed that the tortoises varied in size, shape, and markings. He also observed that the beaks of the finches varied from species to species.

SECTION 1

ACTIVITY

What is the role of variation in evolution?

Process Skills: Recording Data, Inferring

Grouping: Groups of 3 or 4

Hints

This activity will require groups to record a substantial amount of data. Allow sufficient time for each group to study the procedure and devise an organized system for recording the data. After the activity has been completed, have each group compare and contrast its data with the data of the other groups.

If you have students who are interested in mathematics and who understand how to determine the probability of an event, have them determine the probability of obtaining long-necked offspring in each generation. Have the students share their results with their classmates.

▶ Application

Because the four long-necked tortoises did not mate with each other, all offspring would be short-necked. During drought conditions, short-necked tortoises would not be able to reach the only available food—shrubbery leaves—and would eventually die.

★ PERFORMANCE ASSESSMENT

You can use the *Activity* to evaluate the students' ability to work cooperatively and to organize data. Ask the groups to explain how they recorded and interpreted their data.

GUIDED PRACTICE

Ask the students to define the term *evolution* and name an organism that evolves. Then ask the students to describe how their chosen organism has evolved over time.

INDEPENDENT PRACTICE

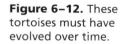 Have the students provide written answers to the Section Review and Application questions. In their journals, ask the students to write a paragraph describing the length of time it takes an organism to show distinct evolutionary changes.

EVALUATION

Darwin hypothesized that living and nonliving things change over the course of time. Ask the students to identify observations that Darwin used to develop and support this hypothesis. (Responses may include volcanic eruptions and the great variety of finch beaks.)

Figure 6–11. I know that species do change. Now I must determine how.

Darwin's Hypothesis

When he returned home, Darwin began the long process of classifying all of the specimens he had collected. He also began to sort out his ideas on how organisms can change over time. Darwin knew that farmers selected and bred plants and animals with certain desirable traits. However, he didn't know what could cause such selection in nature. Again, something Darwin read influenced him.

In *An Essay on the Principle of Population*, Thomas Malthus explained that the human population increases when food and other necessities are sufficient. The population decreases when necessities are in short supply. He suggested that people would overrun the earth if starvation, disease, crime, and war did not keep their numbers down. Malthus called this the struggle for existence.

The struggle for existence was the key to Darwin's own hypothesis. He realized that all organisms must compete for food, water, and other necessities in order to survive. Only the organisms that are better able to compete will survive. Those that are not will die. Now Darwin had the mechanism for **evolution,** or the change in organisms over time. In the next activity, you can see how changes may occur.

Figure 6–12. These tortoises must have evolved over time.

RETEACHING

 Cooperative Learning Remind the students that human beings today are different to some degree from human beings who lived thousands of years ago. Ask small groups of students to discuss ways in which human beings of the past may have been different from human beings of today.

EXTENSION

Ask the students to speculate about what human beings might be like thousands of years in the future.

CLOSURE

 Cooperative Learning Have the students work in small groups to write a summary of the section using their own words. Then ask each group to present its summary orally to the class, and ask other groups to listen and evaluate the accuracy of each summary.

ACTIVITY

What is the role of variation in evolution?

MATERIALS
scissors, red paper, blue paper, 2 paper cups

PROCEDURE
1. Cut 20 small squares of red paper to represent 20 female tortoises. Cut 20 small squares of blue paper to represent 20 male tortoises.
2. Write the words *long neck* on two male and on two female "tortoises." Place all the female tortoises in one cup and all the male tortoises in the other.
3. Suppose your tortoises live on an island. The tortoises are all healthy and full grown, and they begin to pair up for mating.
4. Without looking, select one male tortoise and one female tortoise from the cups. Check to see if either of the tortoises you selected has a long neck.
5. Record the results of the mating. Each mated pair of tortoises has two offspring. All offspring will have short necks unless both of their parents have long necks. Continue selecting male and female tortoises and recording the results until all the tortoises have been mated.
6. Return the tortoises to their containers and repeat the entire process five times. Of the 100 matings, how many were between two long-necked tortoises? How many long-necked offspring resulted?
7. Suppose there is a drought and the only food remaining is shrubs. Half the tortoises die. Remove 10 tortoises from each of the paper cups, but don't remove any with long necks.
8. Repeat steps 4 and 5, recording 100 more tortoise matings. In the second round of matings, how many were between pairs of long-necked tortoises? How many long-necked offspring resulted?

APPLICATION
Suppose the four long-necked tortoises did not mate with one another. If the food problem on the island got worse, what would happen to the tortoise population?

ASK YOURSELF

What did Darwin hypothesize about changes in animals?

SECTION 1 REVIEW AND APPLICATION

Reading Critically
1. Give several examples of variations that Darwin observed.
2. What did Darwin mean when he said that animals change over time?

Thinking Critically
3. How might a variation in the size of a bird's beak help the species survive?
4. Create a concept map relating Darwin's observations and inferences.
5. If long-necked tortoises mated only with other long-necked tortoises, how would the population be affected during long periods of food shortage?

ONGOING ASSESSMENT
ASK YOURSELF

Darwin hypothesized that as time passes, animals slowly change, or evolve.

SECTION 1 REVIEW AND APPLICATION

Reading Critically
1. Examples may include the variety of finch beaks and the variations in the sizes, shapes, and markings of giant tortoises.

2. Darwin thought that the characteristics of animals change slowly over time. In other words, the characteristics of a species today may be different from the characteristics of that species thousands of years ago.

Thinking Critically
3. One possible answer is that if seeds are the only available food source for birds, birds with beaks that can break open seeds would find sufficient food, while birds with beaks that cannot open seeds would probably die.

4. Check concept maps to ensure that they correctly describe both Darwin's observations and inferences.

5. During extended periods of food shortage, long-necked tortoises would be able to reach sources of food both on and above the ground. However, if the period of food shortage is sufficiently long, the tortoises would use up their food supply and die.

SECTION 1   **151**

Section 2: A GREAT THEORY

FOCUS

In this section, Darwin's theory of natural selection is explored. The section also focuses on how the work of other scientists such as Wallace and Mendel supported Darwin's theory that organisms change or evolve through natural selection.

MOTIVATING ACTIVITY

Ask the students to consider a predator-prey relationship such as that between a hungry wolf and a herd of deer. Ask the students to assume that the wolf is successful in catching a deer. Reminding them that Darwin's theory states that only the "most fit" survive, have the students think about what characteristics the caught deer might have had. (The students should suggest that the deer may have been young and inexperienced, sick, diseased, or slow to react or run. The strongest and fastest deer escape to survive and reproduce, while the weaker deer are more likely to be caught by predators.)

PROCESS SKILLS
• Measuring • Interpreting Data

POSITIVE ATTITUDES
• Openness to new ideas
• Initiative and persistence

TERMS
• fossils • natural selection

PRINT MEDIA
Charles Darwin and the Theory of Natural Selection by Renee Skelton (see p. 93b)

ELECTRONIC MEDIA
Evolutionary Biology, Coronet (see p. 93b)

Science Discovery
Fossil; jawbone & teeth
Moth, peppered

BLACKLINE MASTERS
Study and Review Guide
Laboratory Investigation 6.1
Extending Science Concepts

 LASER DISC
1092
Fossil; jawbone & teeth

SECTION 2

A Great Theory

Objectives

Describe what is meant by the "survival of the fittest."

Compare Darwin's and Lamarck's theories of evolution.

List the evidence that supported Darwin's theory.

You know that a library is a building where books are kept. An archive is a place where photographs, public records, and other important historical materials are kept. Anyone curious about the past can go to an archive to get information. Because the earth contains a record of history, it may be compared to an archive.

The Origin of Life

Scientists before and after Charles Darwin have hypothesized about the origin of life on Earth and how living things have changed over time. Many have made observations and done experiments that provide possible solutions to these problems.

Fossil Evidence A great deal of information about the history of the earth can be found in the form of fossils in rocks. **Fossils** are traces of once-living organisms. A leaf, a footprint, a bone, a shell, a skeleton, or even a complete organism could be a fossil. Have you ever made a footprint in wet mud? When the mud dries, the print remains. Some fossils look something like a print you might leave in the mud.

Fossils formed in many ways. Sometimes insects were trapped in plant sap that later hardened. This hardened sap is called *amber.* Some larger animals were trapped and preserved in tar. Other fossils formed when animals were buried in mud. As years passed, the bodies of these animals were covered with layers of soil and dead plants. As these layers became deeper and heavier,

Figure 6–13. The fossils shown here—a trilobite (top), an ammonite (center), and an insect larva in amber (right)—are a record of Earth's past.

TEACHING STRATEGIES

● **Process Skills:** *Inferring, Expressing Ideas Effectively*

Remind the students that fossils present important evidence of events that occurred in Earth's history. Ask the students to explain why this is true. (The students might mention that fossils provide evidence of many plants and animals that are now extinct, as well as those that appeared later in Earth's history. Fossils also illustrate the similarities and differences between primitive and more recent species, the chronology of plants and animals, and give indications of when and where these plants and animals existed. The environmental conditions that occurred at various times in Earth's history are also reflected by the kinds of plants and animals that are part of the fossil record.)

---

pressure built up. The pressure eventually changed the layers into rock. The organisms trapped by this process left their shapes in the rock. Sometimes, however, parts of an organism were replaced by minerals and the organism became *petrified*, literally turned into stone. The petrified forests in Arizona are examples of this type of fossil. The toxodon skull that Darwin found in the Galápagos Islands was a fossil too.

Scientists have learned much about the early history of the earth by knowing the ages of certain fossils. The age of a fossil can be found by using a chemical process to date it. Another way to date fossils is to examine the rock in which the fossils are found. The older the rock, the older the life forms found in it must be. If you took a trip through the Grand Canyon, you could see rock layers that are 2 billion years old. Fossils of reptiles, insects, fishes, and plants have been found in these layers of rock. Fossils such as these are Earth's archives. They tell about the animals and plants that lived when Earth was younger.

SCIENCE BACKGROUND

Sandstone and limestone are types of sedimentary rock that may contain fossils. Most sedimentary rock can be found in an area that was once covered by water. Carbon-14, a radioactive isotope of carbon-12, can be used to date fossils because carbon-14 and carbon-12 are both absorbed into an organism in the same proportions during the life of the organism. When the organism dies, the carbon-14 decays over time, but the carbon-12 remains stable. Scientists can determine the age of a fossil by comparing the amount of carbon-14 to the amount of carbon-12 in that fossil.

Figure 6–14. The Grand Canyon contains a great deal of Earth's history.

Demonstration

Using a small paper cup, plaster of Paris, and a small object such as a seashell, make a "fossil." Pour a small amount of plaster of Paris into the paper cup. Press the seashell into the mixture. Carefully lift the shell out of the plaster and let the plaster dry completely (24–48 hours). Slowly tear the cup away from the plaster. Then provide the students with the opportunity to inspect the "fossil" imprint and compare it to the original seashell.

The fossil record does not provide all the pieces to the puzzle of evolution. Sometimes there are "gaps" in the fossil record. Fossils have not been found for every organism that has ever lived on Earth. Since no humans were present, it is impossible for us to know for certain how life originated. However, by studying the fossil record, it is possible to see how life on Earth has changed.

SECTION 2 **153** ◀

TEACHING STRATEGIES, continued

● **Process Skills:** *Interpreting Data, Analyzing*

Remind the students that one way to determine a fossil's age is by the age of the rock in which the fossil is found and its relative position in the rock layers. Ask the students why this method of fossil dating may be inaccurate. (Older rocks and the fossils they contain are usually found in the bottom layers of a rock formation. However, an earthquake or a buckling of Earth's crust may have caused the rock layers to shift. Older layers may be thrust higher in the rock formation, which would give a misleading indication of the age of the rock and any fossils in that layer.)

● **Process Skills:** *Formulating Hypotheses, Applying*

Point out that plants and animals adapt in response to their surroundings. Ask the students to hypothesize about what evolutionary changes might occur if the climate became several degrees warmer over time. (Hypotheses may suggest that organisms unable to adapt to the new climate might become extinct. Those organisms that display an ability to adapt will survive, supporting Darwin's theory of natural selection.)

SCIENCE BACKGROUND

The experiments of Stanley Miller were based on hypotheses first presented by Aleksandr Oparin. In 1938, Oparin proposed that the primitive atmosphere must have lacked free atmospheric oxygen. Any free oxygen would have readily combined with other molecules. Oparin's argument had first been presented in 1929 by British mathematician J.B.S. Haldane (1892–1964). In the early 1950s, Stanley Miller was a graduate student at the University of Chicago, working under Harold Urey. Miller tested the Haldane-Oparin hypothesis using the apparatus shown in Figure 6–15.

ONGOING ASSESSMENT
ASK YOURSELF

The ages of fossils help scientists estimate the ages of rock layers in which fossils are found. Fossils also provide data about the size and structure of plants and animals that lived in the past.

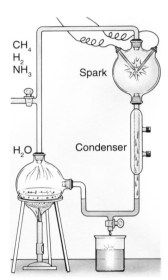

Figure 6–15. Stanley Miller used the apparatus shown here to prove that Oparin's theory was possible.

Chemical Evidence No one can ever know for certain how life first appeared on Earth. However, Aleksandr Oparin (1894–1980), a Russian biologist, proposed a theory to explain how life began. Oparin's theory states that life developed slowly when the molecules in the oceans were heated by the sun and electrified by lightning. The most abundant chemicals at that time were water vapor, methane, hydrogen, and ammonia.

In 1952, an American scientist, Stanley Miller, demonstrated Oparin's theory by making the instrument shown in the picture. Miller combined water vapor with methane, ammonia, and hydrogen and then passed an electric spark through the mixture. He was able to make the basic molecules necessary for life. However, Miller was never able to create "life."

No one knows exactly how these molecules changed to form cells, but the process took many millions of years. The earliest cells were probably one-celled organisms that needed no oxygen. They could make their own food from the chemicals around them. Later, primitive bacteria developed. These organisms were able to produce food through the process of photosynthesis. Since photosynthesis releases oxygen, the earth's atmosphere slowly changed and became able to support life that needed oxygen for survival. Most of these changes occurred very slowly over a long time. Gradually, more and varied organisms developed.

ASK YOURSELF

How are fossils used to study the past life of the earth?

- **Process Skills:** *Comparing, Generating Ideas*

Remind the students that Jean Baptiste Lamarck thought that characteristics acquired during an organism's lifetime could be passed on to its offspring. Ask the students to name characteristics that humans can acquire during their lifetime that are not passed along to their offspring. (The students might mention that the loss of a limb, increased muscle mass because of weight training, and changes in hair color due to dyes are characteristics that are not passed on to the offspring of humans.)

Developing the Theory

In 1809, well before the studies of Charles Darwin, Jean Baptiste Lamarck, a French scientist, presented his theory of evolution. Lamarck's theory stated that a change in a structure followed a change in the job. Lamarck said that an animal might acquire a new trait because of an environmental need. In addition, the animal would then pass on the newly acquired trait to the next generation. For example, Lamarck thought that giraffes grew longer necks because they needed to reach leaves growing higher in the trees.

Figure 6–16. Giraffes were the subject of Lamarck's theory. Lamarck believed that giraffes stretched their necks in order to reach higher leaves. The longer neck would then be passed on to offspring.

Both Lamarck and Darwin theorized that evolution occurs. However, they each proposed a different explanation for how evolution occurs. Lamarck theorized that organisms developed new structures to replace old structures that were not being used. He also believed that acquired traits were passed directly to the organisms' offspring so that new generations could immediately benefit from such changes. This could be compared to saying that if your mother had dyed her hair blue, you would have been born with blue hair!

Darwin agreed with Lamarck that the job comes first. However, Darwin said that some organisms in a group are better able to perform the new job because of some difference in their physical characteristics. These organisms are more likely to survive and reproduce. Since only these organisms reproduce, gradually all organisms in the group have the characteristic.

Darwin's theory of evolution is based on natural selection. **Natural selection** is the process by which those organisms best suited to their environment will survive and reproduce. A change in an organism that makes it better able to survive in its environment is called an *adaptation*.

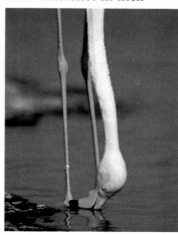

Figure 6–17. The flamingo has adaptations that enable it to feed in a unique manner. It feeds upside down. In the process its upper jaw moves instead of its lower jaw. How does your jaw move? ①

MEETING SPECIAL NEEDS

Second Language Support

Cooperative Learning Have pairs of students work together to find pictures of animals that show a specific adaptation that enables the animals to better survive in their environment. Ask the students to label each picture in English and any other language they know, identifying the animal and explaining the function of the adaptation.

LASER DISC

4252

Moth, peppered

① In humans, the lower jaw moves and the upper jaw is fixed.

SECTION 2 155

TEACHING STRATEGIES, continued

● **Process Skills:** *Predicting, Evaluating*

Write this statement on the chalkboard: "The greater the number of predators an organism has, the greater the number of eggs it tends to lay." Ask the students to debate the validity of the statement and to support their conclusions with real-life examples. (Most students will reason that if an organism with many predators did not lay many eggs, the organism would die out. Examples might include a small fish that has many larger fish—potential predators—in its environment. The offspring of these fish are less likely to survive than, say, the offspring of whales.)

● **Process Skills:** *Inferring, Predicting*

The theories of evolution suggest that structures that have no survival value disappear in a species over time. Ask the students to identify some parts of the human body, in addition to the appendix and wisdom teeth, that humans have now but may not need in the future. (The students should realize that it is difficult to predict such physical changes without considering changing environmental conditions. In humans, the task is

---

BACKGROUND INFORMATION

An English philosopher, Herbert Spencer, first used the phrase "the survival of the fittest" in 1864. Spencer was an early proponent of the idea of evolution, actually predating Darwin and Wallace. Spencer later accepted Darwin's theory of natural selection and attempted to apply that biological theory to fields such as psychology and sociology.

DISCOVER BY *Doing*

The number of dots picked up by each predator-student will vary, but results should indicate that the colored dots with the strongest contrast to the colors in the background were most frequently chosen and picked up. The students should conclude that animals whose color contrasts strongly with the colors of their environment are more likely to be targets of predators than animals whose color blends with the colors of their environment.

 Invite the students to discuss the differences between natural selection and artificial selection. Begin the discussion by explaining that cows with characteristics such as perfectly horizontal spines are selected by breeders to produce offspring for cattle shows. Then have the students generate other examples of artificial selections such as hybrid corn and "seedless" oranges.

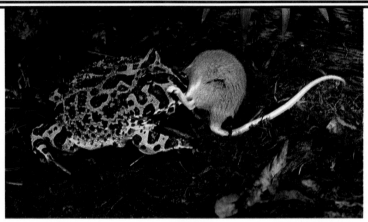

Figure 6–18. This animal was not fast enough to escape its predator.

As Darwin was exploring the forests of Brazil, he observed animals in a struggle for survival. Those that were best suited to their environments lived, while those that were not died. Darwin hypothesized that this struggle to survive might be true for all living organisms. Later, this struggle was termed "the survival of the fittest."

Since there is only so much food, water, and space for living, only the best-adapted individuals from each species survive. These selected few, therefore, are those that will reproduce, passing on their adaptations to their offspring. Those individuals not well adapted to their environment do not survive, so undesirable characteristics are slowly eliminated from the population. In the next activity, you can see how undesirable variations may disappear completely.

DISCOVER BY *Doing*

You will need a paper punch, four different colors of construction paper, a piece of multicolored cloth, and a watch with a second hand. With the paper punch, make 25 dots from each color of paper. Place the piece of multicolored cloth on your desk. Mix up the dots and spread them randomly over the cloth. The colors of the dots represent differences in one animal species.

Ask a classmate to represent an animal predator that eats the dot "animals." He or she should pick up as many dots as possible—one dot at a time—in 20 seconds. Count how many dots of each color your classmate picked up. Which dots were picked up most often? Which dots were picked up least often? Why were some dots not picked up as often as others? How might color variations help an animal survive?

Darwin worked for many years organizing his ideas into a theory. In 1859, he published *On the Origin of Species by Means of Natural Selection*. In this book, Darwin outlined, in great detail, the way organisms evolve through natural selection. The main points of Darwin's theory follow. As you study them, notice how they differ from the theory of Lamarck.

made more difficult because of the social, as well as biological, significance of some characteristics. Some possibilities might include the little finger and toe, hair, and muscles that allow ear movement.)

● **Process Skills:** *Inferring, Applying*

One point in Darwin's theory of natural selection states that individuals of a species struggle to survive. Ask the students to choose different animals and describe ways in which those animals struggle to survive. (Suggested animals may include mice—which not only must forage for food, but also must remain wary of land-based predators such as foxes and air-based predators such as owls.)

The Theory of Natural Selection

1. **Each species produces many more offspring than can survive and reproduce.** For example, female fishes lay enormous numbers of eggs. If all of these eggs hatched and the young survived, the waters would be overrun with fish. Even giraffes sometimes produce more offspring than can survive.

2. **The overproduction of offspring leads to a struggle for survival.** The individuals of a species must compete for the necessities of life. All giraffes must compete for food, water, and space to live in their environment. Some will not be successful and will die before they have young of their own.

3. **All organisms of the same species are somewhat different from one another.** Except for identical twins, individuals are not exactly alike in all of their traits. Some giraffes may be stronger or run faster than others. Some may have longer necks than others.

4. **Individuals with certain traits have a better chance of surviving and reproducing.** This is the "survival of the fittest" idea. Giraffes with long necks are better able to survive because they can eat leaves from tall trees, which are plentiful. Giraffes with short necks cannot.

5. **Those organisms that survive and reproduce pass their traits on to their offspring.** These offspring also have a better chance of survival. Giraffes with long necks survive and reproduce. They pass the trait for long necks on to their offspring, which also survive and reproduce because they too can eat leaves from tall trees.

▼ **ASK YOURSELF**

What is natural selection?

REINFORCING THEMES— *Changes Over Time*

While the students are discussing the main points of Darwin's theory, remind them that evolution is the process by which organisms change and become different from their ancestors. Help the students recognize that such change is not noticeable on a day-to-day basis, but rather it is change that occurs slowly over an extended period of time, such as thousands or millions of years.

THE NATURE OF SCIENCE

Darwin's theory of evolution is generally accepted by scientists because of the vast amount of evidence that supports the theory. Some people, however, have other ideas about evolution. Often these ideas are supported not by scientific evidence but by religious beliefs.

ONGOING ASSESSMENT
▼ **ASK YOURSELF**

Natural selection is the process by which those organisms best suited to their environment will survive and reproduce.

SECTION 2   **157** ◀

GUIDED PRACTICE

Ask the students to define the terms *natural selection*, *adaptation*, and *traits*. Then ask the students to explain the concept of natural selection using their own words, describe a plant or animal adaptation, and name a synonym for the word *traits*.

INDEPENDENT PRACTICE

Have the students write answers to the Section Review and Application questions. Ask the students to describe one way in which a plant has adapted to its environment and one way in which an animal has adapted to its environment.

EVALUATION

Darwin's theory of natural selection contains five main points. Ask the students to provide an example that supports each point.

ONGOING ASSESSMENT
ASK YOURSELF

Alfred Wallace and Gregor Mendel developed ideas that supported Darwin's theory of evolution and natural selection.

SECTION 2 REVIEW AND APPLICATION

Reading Critically

1. Darwin's theory of natural selection states that most species of organisms produce larger numbers of offspring than will survive, individuals of a species must struggle to survive in their environment, individuals of a species are not all alike, and individuals that do survive may pass on differences that give their offspring a better chance of survival.

2. Most organisms must find suitable shelter from weather changes, locate food, and avoid predators. These behaviors may be taken for granted by other organisms such as human beings.

Thinking Critically

3. Any change in conditions over an extended period could cause a population to evolve.

4. Lamarck and Darwin both hypothesized that evolution occurs. But Lamarck theorized that organisms acquired new traits and passed them directly to their offspring. Darwin theorized that some organisms possess certain traits that better enable them to survive and reproduce, thus passing those important traits to their offspring.

Support for Darwin's Theory

It was 20 years after the *Beagle* returned to England before Darwin began to write out his theory. As he was writing his book, he received an essay in the mail from Alfred Wallace. Wallace, also a naturalist, worked in the jungles of Indonesia. He and Darwin had been writing to each other about their research and ideas. In the essay, Wallace stated the main points of the theory of evolution that Darwin had worked on for so long. However, both men had questions about the way traits are passed from one generation to the next.

Figure 6–19. Mendel's work added meaning to Darwin's theory.

The answer involves heredity—the way traits are passed from parents to offspring. Little was known about heredity before 1900. In that year, the work of Gregor Mendel was found. Mendel had published his research on heredity in garden pea plants in 1865, but scientists did not realize the importance of his findings at that time. You will learn more about Mendel and heredity in the next section.

 ASK YOURSELF

Who else had ideas about evolution that were similar to Darwin's?

SECTION 2 REVIEW AND APPLICATION

Reading Critically
1. List the main points of Darwin's theory of natural selection.
2. Why is life a struggle for most organisms?

Thinking Critically
3. What conditions could cause a population to evolve?
4. Compare and contrast Darwin's and Lamarck's theories of evolution.

RETEACHING

Darwin's theory of natural selection states, among other things, that individuals of a species must struggle to survive. Have the students construct a chart describing ways in which humans struggle to survive and then compare those ways to the ways in which plants and other animals struggle to survive.

EXTENSION

Ask the students to describe ways in which humans have made the struggle for survival more difficult for many plant and animal species.

CLOSURE

Cooperative Learning Have the groups choose a plant or an animal other than a human being. Tell the groups to assume that their plant or animal has the ability to think as people do. Then have the groups describe what their plant or animal might "think about" in a day.

SKILL

Controlling Variables

Process Skills: Observing, Comparing

Grouping: Groups of 2

Objectives
- **Design** and **conduct** a scientific investigation.
- **Identify** controlled and manipulated variables.
- **Evaluate** the observed data from the investigation.

Discussion

Discuss the importance of being exact in any investigation—the reliability, validity, and general worth of an experiment hinges on exactness.

▶ **Application**

The students should achieve results that support the hypothesis. The characteristic for marigold tallness is affected by environmental conditions such as soil type.

✳ **Using What You Have Learned**

Ask the students to consider each variable and the ways that it can be controlled.

SKILL Controlling Variables

▶ **MATERIALS**
You will decide what materials you will need.

▼ **PROCEDURE**

1. Here is the hypothesis you will test in this activity: *When seeds of tall marigolds are planted in sand, the plants will not grow as tall as those from seeds of tall marigolds planted in soil.*
2. Design an experiment that will test this hypothesis. Decide what your *manipulated variables* will be. These will be the things that you will vary. For example, you will probably vary the kind of material in which the seeds are planted.
3. Decide what your *controlled variables* will be. These are things you will keep the same. For example, you will need to give each pot the same amount of water.
4. Decide what materials you will need. Then try your experiment.
5. Chart your plant-growth information.

▶ **APPLICATION**
Was the hypothesis correct? What did you find out about seeds of tall marigolds? What inherited trait was affected by the environment?

✳ **Using What You Have Learned**
Compare your experiment with those designed by your classmates. Find out whether you controlled the same variables. Discuss how any differences in the variables that were controlled may have affected the results of each experiment.

PERFORMANCE ASSESSMENT

Have the students write a report about their *Skill* experiment. Ask them to state their hypothesis, identify their controlled and manipulated variables, include their data, and give their results. Use the students' reports to evaluate their knowledge of an experimental setup.

SECTION 2 159

Section 3: MENDEL AND HEREDITY

FOCUS

This section presents the work of Gregor Mendel and explains the genetic principle of dominance and the chromosome theory. The hereditary code, or the way in which different genetic combinations are produced, is also described.

MOTIVATING ACTIVITY

Obtain two pea plants—one with yellow seeds and one with green seeds. Remove the male reproductive structures from each plant. Ask the students to observe the characteristics of each plant, such as seed and pod color and shape, plant height, and so on. Then dust the pollen from the male reproductive structure of the plant with green seeds onto the female reproductive structure of the plant with yellow seeds, and vice versa. Based on this genetic "cross," ask the students to predict what the first generation of pea plant offspring will look like. Have the students record their predictions, germinate the seeds, and compare their predictions to the actual results.

PROCESS SKILLS
• Observing • Inferring
• Interpreting Data
• Classifying/Ordering

POSITIVE ATTITUDES
• Enthusiasm for science and scientific endeavor • Initiative and persistence

TERMS
• dominant • recessive
• genes • chromosome theory • gamete • meiosis

PRINT MEDIA
Genetics: Nature's Blueprints by Lynn Byczynski
(see p. 93b)

ELECTRONIC MEDIA
Mendel's Laws, Coronet
(see p. 93b)

Science Discovery
Pollination, cross
DNA; model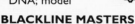

BLACKLINE MASTERS
Study and Review Guide
Laboratory Investigation 6.2
Reading Skills
Thinking Critically

LASER DISC
4234
Pollination, cross

SECTION 3 — Mendel and Heredity

Objectives

Describe the work of Gregor Mendel.

Summarize the chromosome theory.

Propose the outcome of certain genetic combinations.

Gregor Mendel was an Austrian monk who lived from 1822 to 1884. He is best known for his studies in heredity. What made Mendel's work especially valuable were the scientific methods he used in his studies. He chose his subjects carefully, used many samples, and kept accurate records of all his experiments. After he had carefully recorded all his results, he checked them using mathematical principles.

The Story of Gregor Mendel

Figure 6–20. Mendel studied the inheritance of traits in garden peas.

As a young boy, Mendel had been interested in the plants that grew on the family farm. He later became the gardener for the abbey in which he lived, and he studied the pea plants he grew in the abbey garden.

Mendel experimented with the heredity of certain traits found in peas. Look at the chart of these traits. Mendel studied each trait separately and discovered certain patterns in the way traits are inherited in peas.

Figure 6–21. Here are seven traits that Mendel studied.

	Seed Shape	Seed Color	Flower Color	Pod Shape	Pod Color	Flower Position	Stem Length
Dominant	Round	Yellow	Purple	Inflated	Green	Side	Tall
Recessive	Wrinkled	Green	White	Constricted	Yellow	End	Short

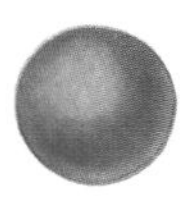

TEACHING STRATEGIES

● **Process Skills:** *Classifying/Ordering, Applying*

This section provides a good opportunity to review a scientific method with the students. Ask them to name the steps that are a part of a scientific method, and make a list of the students' answers on the chalkboard. (These steps should include defining a problem, collecting background information, formulating hypotheses, testing a hypothesis, making and recording observations, and drawing conclusions.) Encourage volunteers to provide an example of each step.

● **Process Skills:** *Observing, Inferring*

Have the students observe the characteristics of pea plants shown in Figure 6–21. Ask the students to think of other characteristics of pea plants that are not observable in the diagram. (Answers may include root structure, germination time, and the amount of sunlight and water required for proper growth.)

Figure 6–22. Mendel crossed tall plants with short plants.

Mendel chose peas because they had obvious traits and usually produced seeds that grew into plants that were similar to the parents. For example, tall plants produced seeds that grew into tall plants. These plants were called *pure* tall.

For each characteristic, Mendel would cross two different traits. In one experiment, he fertilized pure short plants with pollen from pure tall plants. He planted the resulting seeds and counted the types of offspring that grew. He was surprised to find that all the offspring were tall. He got the same results when he fertilized pure tall plants with pollen from pure short plants. Any cross between a pure tall plant and a pure short plant produced only tall plants.

Mendel next crossed the tall offspring—those produced by the cross of the pure tall and pure short plants—with each other. The seeds from this second cross produced mostly tall plants, but one-fourth of the plants were short!

▼ ASK YOURSELF

Why do you think Mendel was surprised by the one out of four short plants?

Figure 6–23. Three-fourths of the plants were tall like the parent plants.

BACKGROUND INFORMATION

Mendel left his monastery in Brunn to study science at the University of Vienna. He then returned to Brunn and in 1856 began his experiments on pea plants. Mendel studied pea plants for eight years, and in 1866 his findings were published in the *Proceedings of the Brunn Society for Natural History*. However, this paper was not widely read and made little impact in the scientific community. It was not until 1900 that Mendel's paper was rediscovered and the implications of his important study were understood by the scientific community.

✧ Did You Know?

To develop his principles of genetics, Mendel had to observe many pea plants as well as individual peas. The number of individuals Mendel counted for each trait in one study is listed here.

Trait	Count
Seed shape	7324
Seed color	8023
Flower color	929
Pod shape	1181
Pod color	580
Flower position	858
Stem length	1054

ONGOING ASSESSMENT
▼ ASK YOURSELF

Mendel was unaware that the characteristics of plants were controlled by factors, or genes.

SECTION 3

TEACHING STRATEGIES, continued

● **Process Skills:** *Observing, Applying*

Bring several packages of familiar seeds, such as daisies, carrots, or pumpkins, to class. Have the students examine the seeds and the information included on the packages. Ask the students to explain how the distributors of the seeds know such information as the color of the flowers, the size each plant will become, and the amount of sunlight and water needed for growth. (Answers should reflect an understanding of the idea that the plants that produced the seeds passed on their hereditary characteristics to those seeds. In other words, by knowing the characteristics of the parent plants, distributors can accurately predict the characteristics of the next generation of plants represented by the seeds.)

THE NATURE OF SCIENCE

Prior to the work of Mendel, people had realized that characteristics were passed from parents to children. After all, it had always been relatively easy to see some of the characteristics of mothers and fathers in their sons and daughters. But it was thought that the characteristics of mothers and fathers blended together in their offspring. As more details about heredity were discovered by Mendel and other biologists, the notion of how characteristics are passed from one generation to another changed radically.

① One-half are pure and one-half are hybrid.

MEETING SPECIAL NEEDS

Mainstreamed

Ask the students to imagine that they want to explain the concept of dominant and recessive genes to someone who does not know these terms. Have the students design a diagram or any other visual aid they wish and write an explanation of dominant and recessive genes using their own words. Encourage the students to try out their diagrams and explanations on a classmate.

Dominant and Recessive

Mendel found similar results when he crossed plants that were pure for round seeds with those that were pure for wrinkled seeds. All offspring of the pure parents showed only the trait of the parent with round seeds. But if these plants, called *hybrids*, were crossed, one-fourth of the offspring were like the grandparent with wrinkled seeds, and three-fourths were like the grandparent with round seeds.

It appeared to Mendel that for each characteristic in peas, one trait was stronger than the other. He called the stronger one the **dominant** trait and the "hidden" one the **recessive** trait. When two pure plants were crossed, all the offspring showed the dominant trait. When two hybrid plants were crossed, three-fourths of the offspring showed the dominant trait and one-fourth showed the recessive trait. Look at the diagram. Even though three-fourths of the offspring of a cross between two hybrid plants show the dominant trait, what fraction are pure and what fraction are hybrid? ①

Figure 6–24. Mendel grew several generations of pea plants.

Figure 6–25. Mendel's hybrid plants were plants with one recessive factor and one dominant factor.

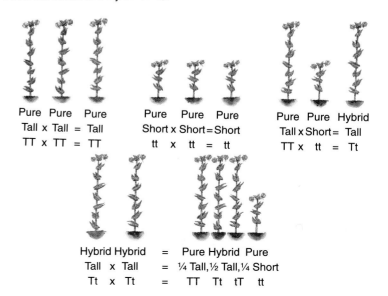

• **Process Skills:** *Classifying/Ordering, Applying*

Reinforce the idea that the factors Mendel studied occur in contrasting pairs—a general characteristic, such as height, is governed by a pair of contrasting factors, tall and short. Ask the students to create a list of general human characteristics and the pairs of contrasting factors that are associated with each characteristic. (Responses might include characteristics such as straight or curly hair, and brown or blue eye color.)

Mendel also did experiments on flower color with his pea plants. Some results are shown in the following table. Study the table, and determine what factors each set of parents had. Mendel's ideas about dominant and recessive factors may help.

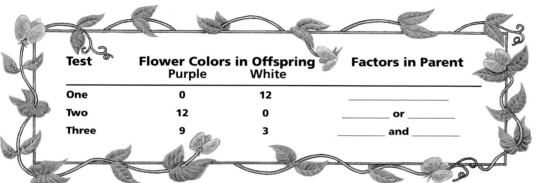

Test	Flower Colors in Offspring		Factors in Parent
	Purple	White	
One	0	12	_____
Two	12	0	_____ or _____
Three	9	3	_____ and _____

Figure 6–26. Offspring inherit recessive traits only if both parents contribute recessive factors.

Has anyone ever told you that you look just like your mother, your sister, or another close relative? The game of "who do you look like" probably began shortly after you were born.

Figure 6–27. It is common for family members to look alike.

Mendel's ideas about factors also apply to human heredity. Your traits are controlled by the factors, now called **genes**, that you inherited from your parents. In the next activity, you can find out which human traits are dominant and which traits are recessive.

MULTICULTURAL CONNECTION

Characteristics inherited by offspring can vary from people to people and from culture to culture. For example, the majority of offspring in certain cultures may have dark-colored eyes. Have the students from different cultures describe some of the inherited characteristics that tend to make the people of one culture physically different from those of other cultures.

INTEGRATION—*Language Arts*

The term *heterozygous* is often used to describe an organism that is hybrid for a specific trait. For example, the organism contains a dominant gene for tallness and a recessive gene for shortness. The term *homozygous* is used to describe an organism that is pure for a specific trait. For example, the organism contains two dominant genes or two recessive genes. Have the students use a dictionary to look up the prefixes *homo-* and *hetero-*. (*Homo-* means "same" and *hetero-* means "different.") Ask the students how knowing the meanings of the prefixes can help them understand and remember the meanings of the words.

ACTIVITY

Are your traits dominant or recessive?

Process Skills: Observing, Inferring

Grouping: Groups of 2

Hints
The students should not have difficulty identifying some of the characteristics used in this activity. For example, it should be clear whether someone can or cannot roll his or her tongue. Other characteristics, such as straight or curly hair, are, however, less clear. In these cases, have groups discuss the variations and reach a consensus decision on the particular characteristic in question.

▶ **Application**
The students might conjecture that the dominance is incomplete.

 Have the students place their charts in their science portfolios.

ONGOING ASSESSMENT
▼ **ASK YOURSELF**

A dominant trait prevents a recessive trait from showing itself. A recessive trait is prevented from showing itself by a dominant trait.

TEACHING STRATEGIES, continued

● **Process Skills:** *Inferring, Analyzing*

Review the structure of chromosomes with the students. Explain that prior to Sutton's work, it was hypothesized that genes were contained somewhere in the nucleus, because the nucleus is divided precisely during cell division. Point out to the students that Sutton studied the nuclei of cells of grasshoppers and observed that the chromosomes occurred in pairs. Ask the students why Sutton made the inference that, since chromosomes occur in pairs, they are likely carriers of genes. (Both genes and chromosomes occur in pairs.)

● **Process Skills:** *Communicating, Inferring, Applying*

Review the process of meiosis by drawing diagrams that include sex chromosomes marked in a different color than other chromosomes. Emphasize that each gamete, or reproductive cell, receives only one sex

---

ACTIVITY

Are your traits dominant or recessive?

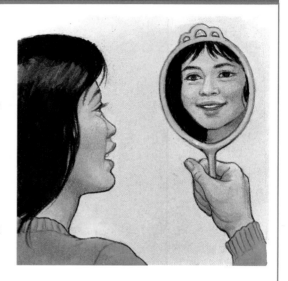

MATERIALS
mirror

PROCEDURE
1. Look carefully at the following chart.
2. Use a mirror, or have a classmate observe these traits for you. Copy the data table to record your findings. For each of these traits, you have inherited one factor from your mother and one from your father. Two factors control each trait.

APPLICATION
Having six toes is a dominant characteristic of cats. Why do you think there aren't more cats with six toes?

TABLE 1: HUMAN CHARACTERISTICS			
Characteristic	**Dominant**	**Recessive**	**Your Trait**
Hair	Curly	Straight	
Hair whorl	Clockwise	Counterclockwise	
Hair color	Brown, black	Blonde, red	
Hair at forehead	"Widow's peak"	None	
Eyelashes	Long	Short	
Dimples	Yes	No	
Nose	Turned up	Not	
Ear lobes	Free	Attached	
Hair on middle section of finger	Yes	No	
Freckles	Yes	No	
Eye color	Dark	Blue, gray, green	
Tongue	Can roll	Can't roll	

▼ **ASK YOURSELF**

What is the difference between dominant and recessive traits?

chromosome. Then ask the students to determine what sex chromosomes must be found in sperm cells and in egg cells if all human males have an X and Y chromosome and all human females have two X chromosomes. (Sperm cells may contain either an X or a Y chromosome, but an egg can contain only an X chromosome.)

● **Process Skills:** *Classifying/ Ordering, Constructing/ Interpreting Models*

Use an assortment of beans, buttons, or other similarly sized objects to illustrate the principle of independent assortment. For example, use a red button and a black button to represent the pair of factors for one characteristic, and a navy bean and a kidney bean to represent the pair of factors for a second characteristic, and a round button and a square button to represent the pair of factors for a third characteristic. Have the students suggest possible combinations of these characteristics. Remind the students that only one factor for each characteristic will be seen in an individual organism. (The students should discover that many different combinations can be created using three of six objects.)

Reproduction and Heredity

Throughout his studies and experiments, Mendel called the characteristics he studied "factors" rather than "genes." He did not know what the factors were or exactly where they were found in living things. When modern scientists discovered those things, Mendel's work became even more impressive.

In 1902, Walter Sutton, a professor at Columbia University, was doing research based on Mendel's work. From this research, Sutton wrote a paper stating a new theory of genetics. The **chromosome theory** says that Mendel's factors are the same as genes and that genes are located on the chromosomes. The chromosome theory also states that traits are passed to offspring by the chromosomes. Moreover, each reproductive cell, or **gamete,** contains these chromosomes, which are located in the nucleus of the cell.

Mendel knew that factors, or genes, separate when gametes are formed. Sutton made it clear that chromosomes, carrying the genes, separated. When fertilization occurs, gametes from each parent join to form a new cell. This new cell divides many times by mitosis to form a complete organism made of billions of cells.

If each gamete had the full number of chromosomes after fertilization, the new cell would have twice that number. In each following generation, the number of chromosomes would double. Since this is not the case, there must be some process to reduce the number of chromosomes in gametes to half. This process is meiosis (my OH sihs). **Meiosis** is the type of cell division by which gametes are formed. Meiosis results in the formation of four daughter cells, each with one-half the number of chromosomes found in the parent cell. Table 6–1 summarizes the differences between meiosis and mitosis.

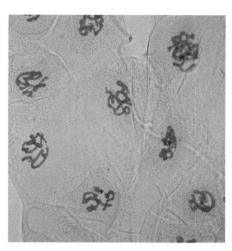

Figure 6–28. The darkly stained objects in this photograph are the chromosomes of a fruit fly.

SCIENCE BACKGROUND

The students may sometimes find it difficult to make a distinction between mitosis and meiosis. The distinction becomes clearer when the result of each process is emphasized rather than the stages of each process. Mitosis results in cells that have a full complement of chromosomes. Meiosis results in cells that have half the number of chromosomes.

✦ Did You Know?

A cell from the human body contains 46 chromosomes—23 from each parent. This list shows the number of chromosomes found in a body cell of other organisms.

Fruit fly	8
Fly	12
Onion	16
Cat	38
Dog	78
Goldfish	94

Figure 6–29. The phases of meiosis are similar to those of mitosis. However, in meiosis there are two cell divisions.

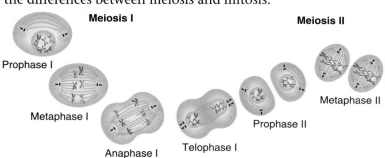
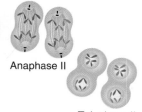

SECTION 3 **165**

GUIDED PRACTICE

Write the terms *dominant* and *recessive* on the chalkboard. Ask the students to define the terms and explain how the terms relate to one another.

INDEPENDENT PRACTICE

Have the students provide written answers to the Section Review and Application questions. In their journals, ask the students to describe the work of Mendel and explain its importance to the world of science.

EVALUATION

Ask the students whether it is possible for two tall parents to produce offspring that are short. Have them explain their reasoning. (Yes, if both parents carry a recessive gene for shortness.)

BACKGROUND INFORMATION

If the students are not familiar with the international Morse code, ask them whether they can remember hearing or reading about the distress message *SOS*. This message is sent using Morse code during times of emergency. Also each day thousands of amateur radio operators, or "hams," communicate with each other using Morse code.

ONGOING ASSESSMENT
ASK YOURSELF

Meiosis is a type of cell division by which gametes are formed.

LASER DISC
4233

DNA; model

DISCOVER BY *Doing* (page 167)

Answers to the activity questions should reflect the idea that combinations of genes, as with combinations of dots and dashes, can be arranged in a number of combinations—all different. Because the number of these possible combinations is so great, no two people are exactly alike (except for identical twins, who are created from identical genetic material).

▶ **166** CHAPTER 6

Table 6-1 A Comparison of Mitosis and Meiosis

Mitosis	Meiosis
One cell division	Two cell divisions
Two daughter cells	Four daughter cells
Daughter cells have the same number of chromosomes as the parent cells	Daughter cells have half the number of chromosomes as the parent (one member of each pair)

ASK YOURSELF

What is meiosis?

The Hereditary Code

By 1953, it was clear to biologists that chromosomes were made up of large molecules of DNA. At that time, James Watson and Francis Crick cracked the code of DNA. Have you ever sent a message to someone in code? A code contains information that only certain people know. DNA contains the code of hereditary information. Instead of using letters, as in a written code, the hereditary code is in the nucleotides of DNA.

Parts of the hereditary code—the DNA molecule—are the same in all organisms, including humans. However, no two people, except identical twins, have exactly the same DNA. The order of nucleotides in a DNA molecule controls what a trait will be. The next activity allows you to practice sending and receiving information in code. It may also help you to understand how DNA can be the hereditary code for so many different traits.

Figure 6–30. DNA determines the hereditary code in all organisms.

RETEACHING

Have the students create a diagram showing the possible offspring for two parents that are each dominant and recessive for the hair color black. (Each of the offspring in the first generation will have black hair.)

EXTENSION

Ask the students to imagine a scenario in which Mendel never existed. Then ask them to debate whether we would still know as much about genetics as we do today.

CLOSURE

Cooperative Learning Have the students work in small groups to create a simple alphabet code and then write a brief message using the code. Encourage the groups to exchange messages and attempt to solve each other's coded message.

DISCOVER BY Doing

Shown here is the international Morse code that represents the 26 letters of the alphabet. Write a short message in Morse code. Trade messages with a classmate. Then try to read your classmate's message. How is sending messages in Morse code similar to a gene sending a message in the hereditary code? How is it different?

INTERNATIONAL MORSE CODE			
A	·-	N	-·
B	-···	O	---
C	-·-·	P	·--·
D	-··	Q	--·-
E	·	R	·-·
F	··-·	S	···
G	--·	T	-
H	····	U	··-
I	··	V	···-
J	·---	W	·--
K	-·-	X	-··-
L	·-··	Y	-·--
M	--	Z	--··

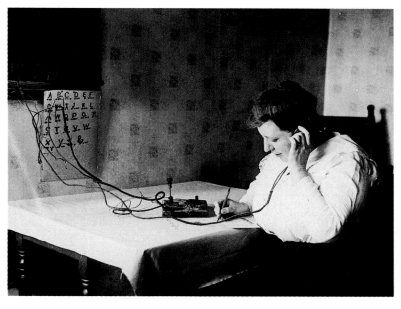

ASK YOURSELF
How does the hereditary code work?

SECTION 3 REVIEW AND APPLICATION

Reading Critically
1. What are genes? What are chromosomes?
2. Why was Mendel's method of studying peas so important to the study of genetics?
3. Why is it necessary for gametes to have only half as many chromosomes as other cells?

Thinking Critically
4. If you crossed a pea plant that was hybrid tall with one that was short, what would the results be?
5. What do you think would happen if chromosomes did not separate during meiosis?
6. If DNA is the same in all organisms, what accounts for the great variety of living things?

ONGOING ASSESSMENT
ASK YOURSELF

The order in which nucleotides are arranged on a DNA molecule controls what a particular trait will be.

SECTION 3 REVIEW AND APPLICATION

Reading Critically

1. Genes are factors that control inherited traits; chromosomes are threadlike structures that are made of DNA and that carry genes.

2. Mendel's method of studying pea plants was important because Mendel counted the number of offspring, thus applying mathematics to biology.

3. In order to maintain the number of chromosomes in succeeding generations, meiosis must produce gametes that have one-half as many chromosomes as other cells.

Thinking Critically

4. It would be expected that one-half of the offspring would be tall and one-half of the offspring would be short.

5. The students may suggest that mutations or cell death would occur.

6. Great variety is possible because of the many different ways the nucleotides in DNA can be arranged.

SECTION 3   **167**

INVESTIGATION

Investigating Genetic Probabilities

Process Skills: Interpreting Data, Classifying/Ordering

Grouping: Groups of 2

Objectives
- **Observe** patterns of dominance.
- **Order** dominant and recessive characteristics.
- **Formulate** predictions using genetic data.

Pre-Lab
Review the meanings of *dominant* and *recessive* with the students. Ask them how many coin combinations would yield a recessive trait.

▶ **Analyses and Conclusions**

1. A dominant trait is inherited over a recessive trait.

2. Each trait is inherited independently of other traits.

▶ **Application**

Check each drawing to ensure that it accurately describes all of the possible offspring that could occur from a cross of the characteristics of each creature.

✳ **Discover More**

Check each possibility for accuracy.

Post-Lab
The students should realize that only one coin combination out of four possible combinations would yield a recessive trait (tails-tails).

After the students have completed the Investigation, ask them to place their test results and drawings in their science portfolios.

▶ **168** CHAPTER 6

INVESTIGATION

Investigating Genetic Probabilities

▶ **MATERIALS**
- pencil
- paper
- coins

▼ **PROCEDURE**

1. Make a table like Table 2.
2. Flip two coins at once to determine the gene pair for each trait. Heads means a dominant gene; tails means a recessive gene.
3. In your table record the gene types and appearance of the trait. For example, suppose from the first coin flips you got one heads and one tails—one dominant gene and one recessive gene. Using the trait of fur color, you would write B (for brown) and b (for green) in the gene type column. The appearance is brown fur.
4. Repeat steps 2 and 3 for each trait.
5. Draw the creature that resulted from this investigation.

TABLE 1: TRAITS

Trait	Dominant	Recessive
Fur	Brown (B)	Green (b)
Tail	Straight (S)	Curly (s)
Horns	Absent (A)	Present (a)
Ears	Pointed (P)	Rounded (p)
Teeth	Double Row (D)	Single Row (d)
Leg Shape	Wide (W)	Narrow (w)
Leg Length	Long (L)	Short (l)
Tongue	Hairy (H)	Smooth (h)
Eye Color	Red (R)	White (r)

TABLE 2: TEST RESULTS

Trait	Gene type	Appearance
Fur		
Tail		
Horns		
Ears		
Teeth		
Leg Shape		
Leg Length		
Tongue		
Eye Color		

▶ **ANALYSES AND CONCLUSIONS**

1. What genetic principle of Mendel's is demonstrated by the dominant and recessive nature of the traits?
2. What genetic principle of Mendel's allows the creature to inherit a dominant gene for one trait and a recessive gene for another?

▶ **APPLICATION**

Assume that you have two creatures with the gene types listed in your table. Make a drawing to show the possible offspring of a cross between the two creatures. Make a different drawing for each trait.

✳ *Discover More*

For each trait, determine the possibility of an offspring inheriting recessive traits if each parent had opposite gene types of those listed in Table 2.

CHAPTER 6 HIGHLIGHTS

The Big Idea—CHANGES OVER TIME

Help the students to understand that we live in a world that is constantly affected by change. A short-term change occurs, for example, when a storm causes a tree to fall. Stress that evolution is also change but change of a long-term nature. Plants and animals evolve, or change, and become different from their ancestors as thousands and millions of years pass by.

CHAPTER 6 HIGHLIGHTS

The Big Idea

After he observed variations among members of the same species, Charles Darwin inferred that these variations allowed some members of a species to have an advantage over other members of the species. These advantageous variations would then be "selected" by nature. Darwin further hypothesized that living things evolve, or change over time. He reasoned that evolution is the result of natural selection and that variations must be passed on to future generations.

Mendel's discoveries gave new meaning to Darwin's theory of natural selection. Mendel discovered that inheritance of variations, or traits, is controlled by factors that we now call genes. The inheritance of genetic traits can be predicted.

Think again about how organisms change over time and about how these changes are passed from generation to generation. How have your ideas changed?

The students' responses should reflect an understanding of the idea that no organism is immune from change over the course of time. Many changes experienced by organisms are passed from parents to offspring, because no offspring is exactly like its parents.

CONNECTING IDEAS

The students' responses should paraphrase the ideas that most organisms will produce more offspring than will survive in order to ensure survival of the species; all living things—to varying degrees—compete to survive in their environments; no two members of the same species, no matter how much they look alike, are exactly alike; and individuals with characteristics that enable them to survive tend to pass those characteristics on to their offspring.

Connecting Ideas

After many years of study and thought, Darwin formed a theory about evolution by means of natural selection. The major points of Darwin's theory are shown in the illustrations. Use your own words to explain these points.

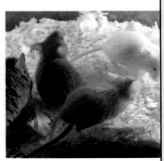

CHAPTER 6 REVIEW ANSWERS

Understanding Vocabulary

1. Within the evolution of organisms, either organisms have the ability to survive and reproduce (natural selection), or the organisms die (extinction).

2. Using pea plants, Gregor Mendel discovered that traits or characteristics can be dominant or recessive.

3. The chromosome theory states that traits or genes are located on chromosomes and that traits are passed to offspring by chromosomes.

4. Gametes are formed during meiosis, and each gamete has half the number of chromosomes found in the parent cell.

Understanding Concepts

Multiple Choice

5. d
6. a
7. c
8. b
9. b

Short Answer

10. Those individuals whose genetic traits fit well with their environment survive and mate with other survivors. The desirable traits for survival are then passed to their offspring, just as Mendel demonstrated how characteristics are inherited by pea plants.

11. Meiosis is a type of cell division that reduces the number of chromosomes of parent cells by one-half. If this reduction did not occur, and gametes had the same number of chromosomes as other cells, the cells that resulted from their union would have twice the number of chromosomes their parent cells had.

Interpreting Graphics

12. No; although the parents could have wrinkled offspring, it is not possible for a hybrid tall parent and a pure tall parent to have a short offspring.

13. The probability of being tall is 4/4; the probability of being wrinkled is 1/4; the probability of being tall and wrinkled is 4/4 x 1/4 or 4/16.

Reviewing Themes

14. The toxodon may not have been able to pass characteristics to its offspring that would give the offspring a better chance of survival—hence the toxodon's extinction. The capybara, though similar to the toxodon, must have had characteristics that enabled it to survive and produce offspring with those vital characteristics.

15. If a pea plant that was hybrid for a particular characteristic was crossed with a pea plant that was hybrid for that same characteristic, it would be expected that 1/4 of the offspring would be pure for that characteristic, 1/2 would be hybrid for that characteristic, and 1/4 would be recessive for that characteristic.

CHAPTER 6 REVIEW

Understanding Vocabulary

For each set of terms, identify how the following terms are related to the first term.

1. evolution (150): natural selection (155), extinction
2. traits: recessive (162), dominant (162), pea plants
3. chromosome theory (165): genes (163), chromosomes
4. gamete (165): meiosis (165), number of chromosomes

Understanding Concepts

MULTIPLE CHOICE

5. Which of the following combinations would result in inheriting the recessive trait of red hair (b) over the dominant trait of brown hair (B)?
   a) BB   b) Bb   c) bB   d) bb

6. Which of these statements does *not* support the theory of natural selection?
   a) All members of a species are alike.
   b) Most species produce more offspring than will survive.
   c) A parent may pass on a trait that will give an offspring a better chance of surviving.
   d) Individuals of a species struggle to survive in their environment.

7. To have an offspring with the recessive trait of blue eyes, what must be true of the parents?
   a) Both parents must have brown eyes.
   b) One parent must have blue eyes.
   c) Both parents must have a recessive gene for blue eyes.
   d) One parent must have a recessive gene for blue eyes.

8. Which of the following statements is *not* part of the chromosome theory?
   a) Genes are located on the chromosomes.
   b) Changes in a species result from natural selection.
   c) Traits are passed on from parent to offspring.
   d) Mendel's factors are the same as genes.

9. If one parent has two recessive genes and the other parent has one dominant and one recessive gene, what is the chance that an offspring will inherit the dominant trait?
   a) one in four   b) two in four
   c) three in four   d) four in four

SHORT ANSWER

10. Explain how natural selection depends on genetic principles.

11. Explain why gametes must be formed through the process of meiosis.

Interpreting Graphics

12. Look at the figure below. Could an offspring of these parents be short with wrinkled seeds? Explain.

13. Look at the figure below. How many chances out of four does an offspring have of being tall and wrinkled. Why?

170 CHAPTER 6

Reviewing Themes

14. *Changes Over Time*
Using Darwin's theories of evolution and natural selection, identify what might be the relationship between the toxodon and the capybara.

15. *Nature of Science*
Using one of the traits of a pea plant studied by Mendel, explain how plants exhibiting dominant and recessive factors can be bred to develop plants that exhibit only recessive factors.

Thinking Critically

16. Some cockroaches have become resistant to pesticides. What might happen if this resistance to known pesticides continues and no new pesticides are developed?

17. The primary source of the giant panda's diet is bamboo. Giant pandas sometimes also eat fish and small rodents. What might happen to the giant panda if the bamboo is destroyed? How might the panda survive the change in its environment?

18. Mendel studied seven characteristics of pea plants. Quite accidentally, each of the seven was carried by a different chromosome. Imagine that Mendel had found that all tall plants had green peas and all short plants had yellow peas. What conclusions might Mendel have drawn about the relationship between plant height and pea color?

19. In some cattle, a cross between a red parent and a white parent will result in a calf that has both white and red hairs. What conclusions can you draw about the traits of white hair and red hair as a result of the offspring inheriting both traits?

20. Straight hair is a recessive trait in humans. If both parents have straight hair, could the offspring have curly hair? Explain your answer.

21. How is the breeding of show animals for particular traits similar to natural selection? How is it different?

Discovery Through Reading

Hyndley, Kate. Wayland Limited, 1989. *The Voyage of the Beagle.* An account of the five years that English naturalist Charles Darwin spent traveling around the world.

CHAPTER 7

CLASSIFICATION

PLANNING THE CHAPTER

Chapter Sections	Page	Chapter Features	Page	Program Resources	Source
Chapter Opener	172	*For Your Journal*	173		
Section 1: THE HISTORY OF CLASSIFICATION	174	Discover By Observing **(B)**	175	*Science Discovery**	SD
		Discover By Researching **(A)**	178	Reading Skills: Solving Problems **(A)**	TR
• Early Attempts to Classify **(B)**	174	Section 1 Review and Application	179	Connecting Other Disciplines: Science and Music, Classifying Instruments **(A)**	TR
• Improving the System **(B)**	176	Skill: Using a Classification Key **(A)**	180	Extending Science Concepts: Classifying Insects **(A)**	TR
• A Classification Key **(B)**	178			Using a Classification Key **(A)**	IT
				Study and Review Guide, Section 1 **(B)**	TR, SRG
Section 2: MODERN CLASSIFICATION	181	Discover By Writing **(B)**	185	*Science Discovery**	SD
• Changes Since Linnaeus **(B)**	181	Activity: How many different ways can you classify seeds? **(B)**	186	Investigation 7.1: Classifying Microscopic Organisms **(H)**	TR, LI
• What Is a Species? **(B)**	182	Discover By Researching **(A)**	188	Investigation 7.2: Studying Variations Within Species **(H)**	TR, LI
• Beyond Species **(A)**	185	Section 2 Review and Application	189	Thinking Critically **(H)**	TR
		Investigation: Observing, Ordering, and Displaying a Collection of Organisms **(A)**	190	Seven Categories of Classification **(B)**	IT
				Five-Kingdom Classification of Organisms **(B)**	IT
				Record Sheets for Textbook Investigations **(A)**	TR
				Study and Review Guide, Section 2 **(B)**	TR, SRG
Chapter 7 HIGHLIGHTS	191	The Big Idea	191	Study and Review Guide Chapter 7 Review **(B)**	TR, SRG
Chapter 7 Review	192	For Your Journal	191	Chapter 7 Test	TR
		Connecting Ideas	191	Test Generator	
				Unit 2 Test	TR

B = Basic **A** = Average **H** = Honors
The coding Basic, Average, and Honors indicates subsections, features, and resources that might be appropriate for different levels of learners. For additional suggestions regarding choice of topic and depth of coverage, see the Pacing Chart on pages T26–T29.

*Frame numbers at point of use
(TR) Teaching Resources, Unit 2
(IT) Instructional Transparencies
(LI) Laboratory Investigations
(SD) *Science Discovery* Videodisc Correlations and Barcodes
(SRG) Study and Review Guide

CHAPTER MATERIALS

Title	Page	Materials
Discover By Observing	175	(per group of 3 or 4) different plants (2), hand lens, paper, pencil
Teacher Demonstration	175	paper, pencil
Discover By Researching	178	(per class) encyclopedia, science reference book
Skill: Using a Classification Key	180	none
Discover By Writing	185	(per individual) journal
Activity: How many different ways can you classify seeds?	186	(per group of 3; in large classes, group of 6) types of seeds, water, hand lens, paper, glue
Discover By Researching	188	(per group of 2) encyclopedias, science reference books, paper, pencil
Investigation: Observing, Ordering, and Displaying a Collection of Organisms	190	(per group of 3 or 4) microscope, slide, coverslip, pond water, drawing paper, pencil, bread mold, forceps, plant parts, worms and insects

ADVANCE PREPARATION

For the *Discover By Doing* on page 175, you will need to obtain two different plants for the activity. For the *Activity* on page 186, obtain eight different types of seeds. Half of the seeds will need to be soaked in water in advance. For the *Investigation* on page 190, locate pond water samples, bread mold, and a houseplant. Also needed are worms and insects.

TEACHING SUGGESTIONS

Field Trip

Arrange a field trip to a botanical garden or a large plant nursery in your area. Have the students take notes during their visit. When they return to the classroom, have them write a journal entry. Ask them to include sketches, as well as the scientific and common names of each plant they sketch.

Outside Speaker

Invite an expert from a university, a county extension office, or a zoo to speak about an animal group that has many species. Ask the speaker to bring actual specimens, if possible, of some of the species. A good example is owls, which have approximately 130 different species.

CHAPTER 7 CLASSIFICATION

CHAPTER THEME—SYSTEMS AND STRUCTURES

In this chapter, the students are introduced to the concept of classification. They note that classification systems have evolved, or changed, over time. The chapter points out that the modern classification system considers the ancestral, or evolutionary, background of organisms to place them in appropriate categories. **Systems and Structures** is also developed as a major theme in Chapters 4, 5, 8, 9, 10, 12, 13, 16, 17, 18, 19, and 20. The supporting theme of this chapter is **Changes Over Time.**

MULTICULTURAL CONNECTION

Cooperative Learning Most grocery stores carry varieties of ethnic foods. These may include breads, sauces, condiments, vegetables, and canned goods. Have the students visit their local grocery store, make a list of ethnic foods, and group, or classify, the foods according to their national origin. Encourage the students to add any ethnic foods that they are familiar with to the appropriate categories on their lists.

MEETING SPECIAL NEEDS

Mainstreamed

Students with learning disabilities can be paired with student tutors for this activity. Give each pair of students this list of food products: milk, ground beef, wheat rolls, cabbage, cheese slices, pork chops, carrot cake, apples, yogurt, chicken wings, corn muffins, squash, margarine, lemons. Ask the partners to decide how the products might be grouped together in a grocery store by making a chart that shows the products listed under labeled categories.

▶ 172 CHAPTER 7

CHAPTER 7 CLASSIFICATION

Imagine a world without organization. It would be a very confusing place. Fortunately, we classify and group just about everything. We group books in a library, shoes in a shoe store, tapes in a music store, and vegetables in a garden. And with over two million living things on Earth, scientists also have found it necessary to classify them into groups.

172 Chapter 7

CHAPTER MOTIVATING ACTIVITY

Cooperative Learning Have groups of four or five students prepare the following set of blue construction-paper figures: a large square, a large square with a yellow dot in its center, a small square, a small square with a yellow dot in its center, a triangle, a triangle with a yellow dot in its center, a circle, and a circle with a yellow dot in its center. Have the groups prepare a second set of figures using red construction paper. Then ask the groups to arrange the figures in at least 10 categories. (Possible categories include all blue items, all red items, all squares, all small squares, all triangles, and so on.)

For Your Journal

By answering the journal questions, the students can demonstrate what they know about the way items are classified and the purposes of classification systems. The students' answers may reveal any misconceptions that they have about today's modern scientific classification system. You may wish to address these misconceptions as you and the students progress through the chapter.

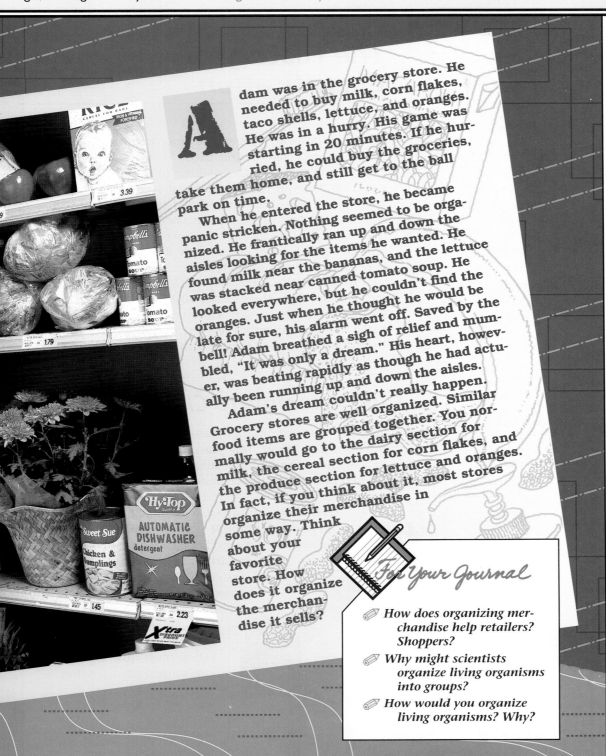

ABOUT THE PHOTOGRAPH

The confusion that Adam found in the grocery store in his dream simply could not happen in real life—in a grocery store, or in any other kind of store. For merchants, organization is not just a matter of neatness; it is a matter of money and survival. In an organized store, customers can see the merchandise and easily find what they want. Satisfied customers will buy products and hopefully come back; frustrated customers will go somewhere else. In an organized store, the merchant can see what products are selling and need to be restocked. Adequate supplies of popular items, as well as supplies of new items, mean more sales for the merchant. Sales bring in money, and money is necessary for the survival of the store.

Adam was in the grocery store. He needed to buy milk, corn flakes, taco shells, lettuce, and oranges. He was in a hurry. His game was starting in 20 minutes. If he hurried, he could buy the groceries, take them home, and still get to the ball park on time.

When he entered the store, he became panic stricken. Nothing seemed to be organized. He frantically ran up and down the aisles looking for the items he wanted. He found milk near the bananas, and the lettuce was stacked near canned tomato soup. He looked everywhere, but he couldn't find the oranges. Just when he thought he would be late for sure, his alarm went off. Saved by the bell! Adam breathed a sigh of relief and mumbled, "It was only a dream." His heart, however, was beating rapidly as though he had actually been running up and down the aisles.

Adam's dream couldn't really happen. Grocery stores are well organized. Similar food items are grouped together. You normally would go to the dairy section for milk, the cereal section for corn flakes, and the produce section for lettuce and oranges. In fact, if you think about it, most stores organize their merchandise in some way. Think about your favorite store. How does it organize the merchandise it sells?

For Your Journal

- How does organizing merchandise help retailers? Shoppers?
- Why might scientists organize living organisms into groups?
- How would you organize living organisms? Why?

CHAPTER 7 173

Section 1: THE HISTORY OF CLASSIFICATION

FOCUS
In this section, a brief history of taxonomy is presented. Carolus Linnaeus's role in developing a system that serves as the basis of the modern classification system is outlined, and Linnaeus's use of binomial nomenclature is emphasized. The use of a classification key for the identification of organisms is also examined.

MOTIVATING ACTIVITY
Have the students bring in a copy of the classified ads from the local newspaper. Ask the students why the ads are called "classified" and how the ads are grouped. (All the ads listed are organized within a classification system. The ads may be grouped according to services offered, jobs available, houses for sale, and so on.) Then ask the students why they think the classification system is useful. (Because the ads are classified, finding certain types of ads, such as the ads for selling used computers, is easier.)

PROCESS SKILLS
- Observing
- Classifying/Ordering

POSITIVE ATTITUDES
- Curiosity • Enthusiasm for science and scientific endeavor

TERMS
- taxonomy • binomial nomenclature

PRINT MEDIA
Animals Have Cousins, Too by Geraldine Marshall Gutfreund (see p. 93b)

ELECTRONIC MEDIA
Grouping Living Things, Britannica (see p. 93b)

Science Discovery Grouping; all living things

BLACKLINE MASTERS
Study and Review Guide
Reading Skills
Connecting Other Disciplines
Extending Science Concepts

① Foods might be classified by food group, packaging, or uses.

MULTICULTURAL CONNECTION
Four thousand years ago, the government of China encouraged the pursuit of science and the application of science to practical problems. In this respect, China was far ahead of Europe. The Chinese's interest in science was based on their belief in an inherent order. Their search for that order among living things led to early classification systems.

SECTION 1

The History of Classification

Objectives

State the need for a common classification system.

Summarize how the classification system for living things has evolved over the years.

Organize objects into appropriate groups.

In Adam's dream, none of the groceries he wanted were where he expected to find them. If Adam had actually been shopping in a supermarket, he would have found the corn flakes in the boxed cereal section. Milk would have been in a section with other dairy products. The lettuce and the oranges would have been found in the produce section. The aisles of the stores would have been labeled so that Adam could easily find the kinds of groceries that he was looking for.

In much the same way that grocers organize the items they sell in their stores, scientists have organized living things into groups. Grouping organisms helps to make studying them easier just as grouping groceries helps to make finding them easier.

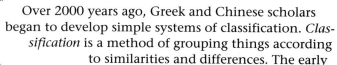

Figure 7–1. How do you think the supermarket would classify these foods? ①

Early Attempts to Classify

Evidence of early systems for organizing plants and animals into groups dates back about 7000 years. These systems were based on people's need to communicate about how plants and animals could be used to provide basic needs such as food, clothing, and shelter.

Over 2000 years ago, Greek and Chinese scholars began to develop simple systems of classification. *Classification* is a method of grouping things according to similarities and differences. The early systems of animal classification were based mainly on body coverings and body parts, such as legs, tails, and teeth. An ancient Chinese scholar noted that a system of classification must truly distinguish one animal from another.

174 CHAPTER 7

TEACHING STRATEGIES

● **Process Skills:** *Inferring, Expressing Ideas Effectively*

Have the students read the quotation again. Then ask them when it might be useful to use a broad classification like "hand" or a narrow classification like "five fingers." (Answers should include the idea that both can be used to classify organisms into groups. A narrow classification may help us understand differences in activities, such as the ability to pick up food, that cannot be identified through broader classifications.)

● **Process Skills:** *Formulating Hypotheses, Inferring*

Remind the students that Aristotle and Theophrastus's classification of plants was based on size and appearance. Ask the students to hypothesize whether Aristotle and Theophrastus would have classified the following pairs in the same categories: redwoods and bonsai trees, roses and dandelions. (The students should infer that Aristotle and Theophrastus would not have classified the pairs in the same category because of differences in size and structure.)

Figure 7–2. Early Chinese scholars used body parts to classify animals such as the horse and the ox. What other characteristics could they have used? ②

The horse and the ox are different, but if someone says that a horse is not an ox because the ox has teeth and the horse a tail, that will not do. In fact, both have teeth and tails—one has to say that horses are not the same as oxen because the ox has horns and the horse does not. . . . Animals of four legs form a broader group than that of horses and oxen. Everything may be classified in broader and narrower groups. It is like counting fingers; each hand has five, but one can take one hand as one thing.

Quoted in Order: In Life *by Edmund Samuel*

Around 350 B.C., Aristotle, a Greek philosopher, was also developing a system for classifying animals. He identified over 500 animals and placed the different animals into 10 groups based on their body parts, life histories, activities, and character. He then organized the groups into two categories. Six of the groups he called "blooded," and four he called "bloodless." Into the blooded group he put humans, birds, and fish. He considered insects and mollusks bloodless. Why do you think he placed these animals in the groups he did? ③

Together with his student Theophrastus, Aristotle also classified plants. They divided plants into groups based on their size and appearance. Their classification systems continued to be used with some changes for about 2000 years.

Discover by Observing

Carefully examine two different plants. If possible, use a hand lens to observe their leaves, stems, and roots. You may want to examine two very different types of plants, for example, a house plant and a small tree. Make a chart identifying the similarities and differences in the structures of the plants. Based on your comparison, tell whether you would place the two plants in the same category.

ASK YOURSELF

What was the basis of most early classification?

Figure 7–3. Aristotle's and Theophrastus's classification systems are simple by today's standards but quite imaginative for the time.

---

② Characteristics for classification might include how the animals reproduce, whether they live on land or water, and whether the animals are useful or harmful.

③ Insects and mollusks do not have red blood; humans, birds, and fish do.

Discover by Observing

The students should be prepared to explain their decisions. Have them place their charts in their science portfolios.

ONGOING ASSESSMENT
ASK YOURSELF

Early classification systems were based on physical characteristics.

Demonstration

Set up this imaginary situation: The students are scientists who have "beamed down" to a planet where they encounter unfamiliar life forms. Ask the students to list characteristics they could use to classify these life forms. (Some possible characteristics include number of legs, color, feeding habits, and so on. The students' answers should reflect an understanding of classification and the need to identify differences in behavior and physical characteristics.)

SECTION 1   175

TEACHING STRATEGIES, continued

● **Process Skills: Formulating Hypotheses, Inferring**

Point out that John Ray tried to classify animals on the basis of behavior as well as structure. Ask the students to suggest reasons why this type of classification might lead to confusion. (The students might suggest that many animals can change their behavior rather quickly in response to changes in their environment, without becoming a different kind of animal.)

● **Process Skills: Generating Ideas, Applying**

John Ray classified mammals into two groups—toed and hoofed animals. How would he classify the following animals: cow, human, gorilla, deer? (hoofed—cow, deer; toed—human, gorilla.) If Ray knew that whales were mammals, how would he have classified them? (He would have had to establish a new classification that did not focus on the organisms' feet.)

INTEGRATION—Language Arts

Have the students look up the prefix *bi-* and the word *nomen* in a dictionary. Ask how an understanding of these word parts will help in remembering the meaning of *binomial nomenclature*. Have the students list other words that have the prefix *bi-* and interpret what the words mean. (The prefix *bi-* means "two"; *nomen* means "name." *Nomenclature* means "naming." Other words with the prefix *bi-* include *bicycle*, which means "two cycles," (two wheels), and *bicentennial*, which refers to a 200-year anniversary.)

BACKGROUND INFORMATION

The fact that Carolus Linnaeus in Sweden learned about the work of John Ray in England can be credited to the influence of the Enlightenment, or Age of Reason, a philosophical movement of the late 1600s and the 1700s. During that time, scientists around the world founded scientific institutes and published many reports about their research as part of the Enlightenment's emphasis on the need to advance knowledge. For the first time, scientists in many countries had access to their colleagues' recent work. It was the beginning of the international scientific community.

① Possible answers include dogs, horses, rabbits, and cattle.

Figure 7–4. In Ray's hierarchy, each of these animals would be placed in separate subcategories of hoofed animals.

Improving the System

As more and more differences in organisms were recognized, scientists began to realize that they needed a better classification system. Eventually, some scientists began devoting full time to classifying living things. Soon classifying became a science. The science of classification has now become known as **taxonomy**.

A Place for Everything During the late 1600s, English biologist John Ray set up a system that attempted to place living things in a hierarchy. In a hierarchy, items are placed in categories and subcategories based on their characteristics.

Think about the supermarket. Grocers identify different categories of food, such as meat, vegetables, bread, and so on. They further identify the different forms and types within each category. For example, bread products include rolls and buns as well as loaves of bread. Different kinds of bread—wheat, white, rye, raisin, and multiple grain—are also identified.

John Ray's hierarchy of organisms worked in the same way. He placed organisms into a hierarchy by dividing and subdividing categories. He examined organisms in detail. He also recognized that differences could exist among organisms of the same kind. For example, all house cats share common characteristics but vary in color, size, and shape. What other examples can you name of differences among animals of the same kind? ①

Ray's hierarchy put mammals into two groups. One group had toes and the other had hoofs. He then divided the hoofed mammals into those with undivided hoofs, two-toed hoofs, and three-toed hoofs. He further divided the animals with two-toed hoofs into three more groups. One group was cud chewers that had permanent horns. Another group consisted of animals that chewed their cud and shed their horns each year. And still another consisted of animals that did not chew their cud. Ray used structure and behavior to describe organisms.

- **Process Skills:** *Analyzing, Inferring*

Remind the students that by the time Linnaeus began his work, Latin was no longer spoken. Instead, it was a written language used mainly for academic works. Then ask the students why using a "dead" language might be an advantage for scientists. (Spoken languages tend to change quickly and vary from place to place as people make up new words to describe new situations. Because Latin was no longer spoken, it would remain the same indefinitely.)

- **Process Skills:** *Comparing, Inferring*

Explain that giving people first and last names is a type of binomial nomenclature. A person's first name identifies the individual, and the person's last name identifies the person's family. Ask the students to offer reasons why the use of first and last names might be advantageous. (It enables someone to identify specific individuals.)

What's in a Name? In the late 1700s, a young Swedish scientist named Carolus Linnaeus (lih NAY us) picked up on the work done by John Ray. While in medical school, Linnaeus had a job taking care of a small botanical garden. He observed and collected plants and insects, writing detailed descriptions of each one. Later he was given a grand total of $50.00 to go to a region north of the Arctic Circle to collect and study plants. Through activities like these, Linnaeus began to develop his classification system for organisms. His system is the basis for today's system of classification. As a result, he is known as the father of taxonomy. You can read more about Linnaeus's life on page 201 at the end of this unit.

Linnaeus's classification system used just twelve Latin words to describe each kind of organism. The last two Latin words in each description were the specific name of an organism. Using two words to name an organism is called **binomial nomenclature** (by NOH mee uhl NOH muhn klay chuhr).

Why did Linnaeus use Latin instead of the Swedish language to classify organisms? At the time, Latin was the international language. By using Latin, he knew that the names he gave each living thing could be understood by scientists throughout the world, not just those who could read Swedish.

Look at the following table. It gives the word for *dog* in several different languages. You can imagine the confusion that would result if all animals were named without a single common language. Yet because of Linnaeus's nomenclature, scientists all over the world can recognize the word for *dog* by its Latin name *Canis familiaris*.

Figure 7–5. Carolus Linnaeus developed the classification system that is the basis of the modern system.

SCIENCE BACKGROUND

Latin is the main language used in taxonomy, but Greek words are also used. An example is the Black rhinoceros, whose name is *Diceros biconis*. *Diceros* means "two horns" in Greek, and *biconis* means "two horns" in Latin. Latinized English words, as well as names of scientists, can also be used for scientific names.

MEETING SPECIAL NEEDS

Second Language Support

Ask the students to make a poster showing some of the plants and animals found in their country of origin. Encourage them to use dictionaries and encyclopedias to label each plant or animal with its name in their first language, its name in English, and its scientific name. Then use the posters to involve the class in a discussion that focuses on how the use of scientific names makes it easier for scientists from many countries to discuss plants and animals.

Table 7-1 The Word *Dog* in Various Languages

English	dog	French	chien	Russian	sabaka
Spanish	perro	Japanese	inu	Polish	pies
Hebrew	kelev	German	hund	Italian	cane

Some languages have several different names for the same animal. For example, in English the mountain lion is also called a panther, a cougar, or a puma. But all scientists know it by its Latin name *Felis concolor*. Even today, some 300 years later, scientists still use Linnaeus's methods to name newly discovered organisms.

Figure 7–6. What is the common name you use for *Felis concolor*? ②

② Mountain lion, panther, cougar, or puma are possible answers.

GUIDED PRACTICE

Ask the students to choose five common animals. Then help them create a classification key for identifying these animals.

INDEPENDENT PRACTICE

 Have the students write answers to the Section Review and Application questions in their journal. Then ask the students to classify the following items from most general to most specific: sweat socks, white sweat socks with colored bands, clothing, socks, white sweat socks with blue bands, white sweat socks. (The students should classify the items in the following order: clothing, socks, sweat socks, white sweat socks, white sweat socks with colored bands, white sweat socks with blue bands.)

EVALUATION

Have the students define the following terms and use them in a paragraph discussing the classification system: taxonomy, classification, hierarchy, binomial nomenclature. (The students' answers should reflect an understanding of the complexity of the classification system.)

DISCOVER BY *Researching*

 Cooperative Learning You may wish to have the students work in small groups to name and describe the animals. To facilitate the class discussion, write the names of some common animals on the chalkboard, and ask the groups who chose those animals to write their descriptions on the board.

ONGOING ASSESSMENT
ASK YOURSELF

Ray and Linnaeus both classified living things according to the similarities and differences in their structures.

THE NATURE OF SCIENCE

Although there have been many changes, Linnaeus's basic classification system is still used by scientists today because of its simplicity and logic. Linnaeus's accomplishment also demonstrates the use of a scientific method, which is based on systematic observations of nature, the collection of data, and the use of logic to draw conclusions from data.

 LASER DISC
2806
Grouping; all living things

DISCOVER BY *Researching*

Make a list naming ten different animals. Find a picture of each animal. Then try to describe each animal using just twelve words as Linnaeus did. Don't use sentences. Instead use brief descriptions for each characteristic. Once you have described an animal, locate its Latin name in an encyclopedia or a science reference book. Share your descriptions with your classmates. Did any of your classmates describe the same animals? How are your descriptions alike? How are they different?

Linnaeus grouped plants and animals according to likenesses and differences in their structures. He did his work about 150 years before the theory of evolution by natural selection was known. Today, scientists classify organisms according to how closely related they are to one another in their evolutionary development. The further two organisms are from a common evolutionary ancestor, the more unlike they will be.

 ASK YOURSELF

How was the work of John Ray similar to that of Carolus Linnaeus?

A Classification Key

By now you are probably saying to yourself, "So what if scientists have described and classified different kinds of organisms? How could I determine what something is if I have never seen it before?" That's a good question. Not even the best scientist can know even a small fraction of all the living things. In fact, any one scientist is lucky if he or she can know and identify a thousand of the over two million organisms. However, scientists have a way to help them identify living things. They have developed keys that can be used for identifying living things.

Unlocking Identity The most common key used to identify organisms is made of paired descriptions. To identify something, you have to decide which description fits the thing you are trying to identify. That description will either give the name of the organism or send you to another number where there is another set of paired descriptions. When you go to those paired descriptions, you will select the one that best describes the thing you are trying to identify. That description will either give you the name or send you to yet another pair of descriptions. You continue in the same way until the

Figure 7–7. Scientists could use a classification key to identify this evening primrose as *Ranunculus septentrionalis*.

RETEACHING

 Cooperative Learning Have the students work with partners to list four important ideas presented in this section. Then have partners read their list to the class. Ask the class to choose the listed ideas that best summarize the section.

EXTENSION

Have the students list several plants and animals that they have seen and research the scientific names of the organisms. Encourage them to use a Latin dictionary to determine the English meaning of the names.

CLOSURE

 Cooperative Learning Have the students work in small groups to list the advantages of using a system of classification to describe plants and animals. Then ask each group to present a summary of its work to the class.

1. Object is round .... go to #2.
   Object is not round ..... go to # 3.
2. Object is red .... a tomato.
   Object is orange .... an orange.
3. Object is long and narrow and is orange in color .... a carrot.
   Object is shaped like a football and is green in color .... a watermelon.

description you choose from a pair is followed by the name of the organism.

Look at the illustration of the four objects. They can be identified using the simple key. To learn how the key works, pick out one of the objects. Then start at the top with the first paired descriptions and go to the next step until you have identified the object. Use the key to identify all the objects.

Scientists have written hundreds of different classification keys. There are keys for birds of the western states, mammals of the Rocky Mountain region, insects of North America, trees of the eastern United States, and so on. Scientists who specialize in certain organisms such as insects would use the key that describes the organisms they are studying.

 ASK YOURSELF

How does a classification key help you to narrow your choices and to identify an organism?

SECTION 1 REVIEW AND APPLICATION

Reading Critically
1. Why is Latin used to name organisms today?
2. What does the story of classification tell you about the system of classification?

Thinking Critically
3. Choose four items in your classroom. Create a system for classifying these items. What other items would you place in the same classifications?
4. Cite two examples of classification systems you use in everyday life, and describe the hierarchy within those systems.

ONGOING ASSESSMENT
ASK YOURSELF

A classification key helps eliminate possible identities of a living thing.

SECTION 1 REVIEW AND APPLICATION

Reading Critically

1. Latin was the language used for scholarly works during Linnaeus's time. Also it provided a common language for all scientists to use, regardless of their country or native language. The same is still true today.

2. The students might suggest that the story of classification demonstrates the need for a system to organize the huge number of living things and give scientists a common frame of reference.

Thinking Critically

3. The students should be able to explain the categories in their classification systems.

4. The students might suggest alphabetized telephone directories and categorized classified advertisements.

SECTION 1

SKILL

Using a Classification Key

Process Skills: Observing, Classifying/Ordering, Applying

Grouping: Groups of 2

Objectives
- **Identify** characteristics of leaves from different trees.
- **Distinguish** among leaves by using a classification key.

Discussion
Classification keys are used in field guides to aid in the identification of living things. The key in this activity provides the user with a step-by-step process that will lead to the positive identification of various leaves. In each step, the user chooses between two descriptions of an organism. By choosing the description that most closely resembles the leaf in question, the user is provided with either an identification or instructions to proceed to another step.

Application
The students should demonstrate a knowledge of how a classification key works and how to use one.

Using What You Have Learned
Most students should find the descriptions contained in the classification key useful.

★ PERFORMANCE ASSESSMENT
Ask each student to explain, step by step, how he or she used the classification key to identify a particular leaf. Use the explanation to evaluate the student's understanding of paired descriptions.

SKILL — Using a Classification Key

▶ **MATERIALS**
none

▼ **PROCEDURE**

1. Examine the illustrations of the leaves from different trees.
2. Use the classification key to identify the tree that produces each of the leaves shown in the illustration.

CLASSIFICATION KEY

1. Leaf is needlelike................. go to #2.
   Leaf is broad and flat........... go to #3.
2. Needlelike leaves are grouped in bundles of two or three...... go to #6.
   Needlelike leaves are not in bundles............................. go to #7.
3. Edge of leaf is smooth................ go to #4.
   Edge of leaf is not smooth....... go to #5.
4. Leaf is made of several leaflets. Fruit is a round nut......................... black walnut, *Juglan nigra*.
   Leaf is heart-shaped. Fruit is a long pod....... northern catalpa, *Catalpa speciosa*.
5. Edge of leaf has rounded indentations and sharp points at the end. An acorn is the fruit............................... pin oak, *Querus palustris*.
   Edge of leaf looks like the edge of a saw blade. Its seed is surrounded by a thin wing...................... American elm, *Ulmus americana*.
6. Needles arranged around the branch. Cone has scales that are open........................ Douglas fir, *Pseudotsuga taxifolia*.
   Needles arranged around the branch. Cone has scales that are tightly closed.............. white fir, *Abies concolor*.
7. Needles grouped in threes............................... loblolly pine, *Pinus taeda*
   Needles grouped in fives....... white pine, *Pinus monticola*.

▶ **APPLICATION**
Obtain a copy of a classification key for trees in your area. Use the key to identify trees around your school and home.

※ **Using What You Have Learned**
Were your descriptions helpful to you? Were you able to identify the leaves by using the key? If not, make changes to improve the key.

Section 2: MODERN CLASSIFICATION

FOCUS

In this section, the major advances in classification systems since Linnaeus's time are discussed. An explanation of the concept of species is presented and compared to the other categories used to classify organisms, with emphasis on the five kingdoms.

MOTIVATING ACTIVITY

Cooperative Learning Have the students work in small groups to make a poster showing examples of organisms in the five kingdoms. The students can cut pictures from magazines, mount them on poster board, and write captions showing the common and scientific names of the organisms. Ask each group to present its poster to the class. Point out to the students that examples of monerans and protists may be easier to find in magazines such as *Current Science* or *Discover*. Articles in these magazines will often include both common and scientific names for organisms.

Modern Classification

SECTION 2

Supermarkets receive many new products each week. Before displaying the new items, the grocers examine the products. They determine what each product is. Is it a new kind of pasta? They examine the packaging. Is the product in a box or a bag? They think about where the new product will fit on the shelves best. They most likely place the new product near products that are similar.

Objectives

Define the word species.

Differentiate among the five kingdoms of living things.

Evaluate the worth of a system of classification.

Changes Since Linnaeus

Grocers sort products based on what the products are and how similar the products are to other grocery items. Scientists classify organisms in much the same way. Classification, as we know it today, is similar to what Linnaeus was doing over 250 years ago. But there have been significant changes since Linnaeus's time.

So Many First of all, there are far more organisms known today than were known during Linnaeus's time. While they don't know how many different organisms there are, scientists estimate that there are over 2 000 000 kinds of organisms alive on Earth today. New discoveries add to this number.

Look at This The second change is that there are better instruments, tools, and methods to use for identifying organisms. This is particularly true when you consider the microscope. Improvements in microscopes have allowed us to see far more detail. This means that characteristics otherwise unseen can be used to describe living things. With the improved tools and instruments, scientists are able to examine organisms in great detail.

Figure 7–8. These bacteria have been magnified by an electron microscope.

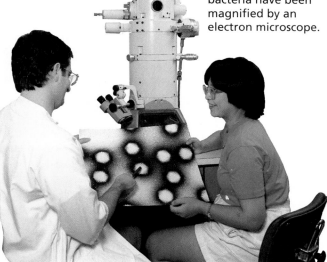

PROCESS SKILLS
- Comparing • Observing
- Classifying/Ordering

POSITIVE ATTITUDES
- Initiative and persistence
- Precision

TERMS
- species • kingdom
- chlorophyll

PRINT MEDIA
Diversity of Life: The Five Kingdoms by Lynn Margulis (see p. 93b)

ELECTRONIC MEDIA
How We Classify Animals, AIMS Media (see p. 93b)

Science Discovery Classification; The five kingdoms

BLACKLINE MASTERS
Study and Review Guide
Laboratory Investigations 7.1, 7.2
Thinking Critically

✧ **Did You Know?**

From Linnaeus's time until the 1960s, scientists acknowledged only two kingdoms—plants and animals. In 1969, R.H. Whittaker, an American biologist, proposed a five-kingdom classification system based partly on findings made with the electron microscope.

TEACHING STRATEGIES

● **Process Skills:** *Generating Ideas, Evaluating*

Have the students identify the three scientific advances that led to improved classification of organisms. (discovery of new organisms, improved scientific tools and methods, and acceptance of concepts of evolution by many scientists) Ask the students which advancement they believe was most important to the development of the classification system and why. (Many students will cite improved methods and tools as the most important development, because discoveries of new organisms and advancement of theories would have been impossible without them.)

● **Process Skills:** *Applying, Analyzing*

Remind the students that in the 1700s organisms were classified as plants or as animals. Bacteria had been observed, but scientists did not realize they were organisms until the 1800s. Were they plants or animals? Scientists could not agree. Ask the students what this meant for the classification system. (The students might suggest that this meant the classification system needed to be reevaluated.)

 The world's tropical rain forests have more kinds of trees and animals than any other area on Earth. Scientists estimate that half of all the species of plants and animals found on Earth live in tropical rain forests. Many species have not yet been discovered and classified. The survival of the largest rain forest, located in the Amazon Basin in Brazil, is threatened by slash-and-burn cultivation in which the trees are cut down and the land is cleared for farming. Environmentalists around the world oppose this practice. They argue that the rain forests should be conserved not only to protect the thousands of species that live there, but also to preserve the rain forests' role in regulating the world's climate. Ask the class to debate and draft a proposal for the future treatment of the Amazon rain forest that will appeal to both the environmentalists and the leaders of Brazil.

ONGOING ASSESSMENT
ASK YOURSELF

The scientific community continues to discover more organisms; it has developed better tools, instruments, and identification methods; and it has taken into consideration the theory of evolution.

It's Relative Scientific knowledge has grown tremendously since Linnaeus's time. Today, scientists have tools, instruments, and methods that enable them to examine the chemical makeup of organisms. They know that the more alike the chemical makeup of two organisms is, the more closely the two organisms are related.

The third significant change is that much of what scientists know now is based on evolution. This means that to identify organisms, scientists look at factors other than obvious physical and structural characteristics.

All of these factors have influenced our modern system of classification. What is more important to realize is that continued developments in science will likely lead to further changes in our modern classification system.

 ASK YOURSELF

What advances have contributed to the improvement of the classification system?

What Is a Species?

Grocers don't simply use the term *soup* to classify all soups. Instead, they label and arrange soup by variety. There are canned and dry soups. Tomato, noodle, vegetable, bean, split pea, and dozens of other soups are arranged on the shelves. Just as grocers don't call all soup *soup*, neither does a scientist classify all feathered creatures as just birds.

Figure 7–9. These animals are all birds, but they must be classified further to note differences among them.

● **Process Skills:** *Comparing, Expressing Ideas Effectively*

Have the students look at Figure 7–9. Ask them to list the characteristics that the animals share as birds. Then ask the students why these animals are *not* classified as the same species. (These birds cannot mate and produce fertile offspring.)

● **Process Skills:** *Classifying/ Ordering, Analyzing*

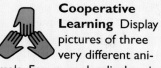
Cooperative Learning Display pictures of three very different animals. For example, display pictures of a dog, an antelope, and a gorilla. Have the students work in small groups to identify similarities and differences among the animals. Then have the students decide, based on their comparisons, whether the animals belong to the same species. Have the groups check their conclusions by consulting reference materials. (Conclusions should note that the animals are not members of the same species because they do not naturally interbreed.)

Figure 7–10. All dogs are members of the same species even though they may look different.

As with varieties of soup, there is a term to use when speaking of different organisms. The term is *species*. But it is not enough to define species as different kinds of organisms. In fact, the concept of species is pretty complicated. In the study of life science, it is important to know what a species is. You might ask what makes one species different from another.

The obvious answer is that they do not look alike. But while that is often the case, it is not always true. As a matter of fact, some members of the same species look less alike than members of a different species look. For example, all dogs are members of the same species. But you know that all dogs do not look alike. Certainly, the poodle in the picture doesn't look like the dachshund. But, nonetheless, they are the same species. There are even major differences in the appearance of the male and female organisms within some species. For example, look at the male and female pheasants in the picture.

Figure 7–11. Males and females of the same species can look quite different.

MEETING SPECIAL NEEDS

Gifted

Explain that when scientists use structures to classify organisms, they distinguish between *homologous* and *analogous* structures. Ask the students to investigate the meaning of those terms, explain the importance of the structural concept to classification, and find examples of both kinds of structures. Encourage the students to share their findings with their classmates.

INTEGRATION—*History*

To reinforce the concept of classification systems, have the students construct a family tree in their journals. Explain that they can begin by writing their name either at the top or the bottom and then adding their parents' and grandparents' names, and so on. Point out that this kind of branching chart, which is also called a concept map, can be used to show the relationships among concepts. Extend the activity by asking the students to draw a family tree for the major branches of science.

TEACHING STRATEGIES, continued

- **Process Skills:** *Observing, Inferring*

On a bulletin board, display a collage of pairs of pictures of animals. The pairs should show animals that belong to the same species but look different from one another. Number the pictures and ask the students to identify the animals that would belong in the same species. Remind the students that members of a species may differ in looks from one another. You might display pictures of a male and a female lion or peacock or pictures of Siamese and Angora cats, for example.

- **Process Skills:** *Applying, Inferring*

Have the students look up the word *hybrid* in the dictionary and provide a definition in their own words. (the offspring of two different species, breeds, or genera of plants or animals) Have the students identify any fruits or vegetables that they think are hybrids and name the "parent" plants of the hybrid. (The students may suggest the tangelo, a hybrid produced by crossing a tangerine plant and a grapefruit plant.) Emphasize that in nature hybrids that produce offspring are very rare. Successful hybridization is a phenomenon created by humans.

SCIENCE BACKGROUND

A hybrid is the organism that results when species crossbreed. Hybrids can occur naturally, but the deliberate breeding of hybrids did not become a science until scientists discovered more about genetics and heredity. Although some animal hybrids are bred, the most successful hybrids have been plants. In fact, nearly every kind of plant used by humans has been crossbred for improvement. Hybrids include grain crops such as corn, sorghum, rice, barley, and wheat; vegetables such as carrots, squash, cabbage, tomatoes, and broccoli; and ornamentals such as roses, petunias, and marigolds. Hybrids are most often bred for characteristics such as higher yields, disease resistance, hardiness in cold or dry conditions, and greater size and strength.

ONGOING ASSESSMENT
ASK YOURSELF

No; human beings, for example, are members of the same species, and except in the case of identical twins, no two human beings look exactly alike.

Figure 7–12. The Asian and African elephants look alike, but they are members of different species.

Look at the two elephants. They look alike, yet they are members of different species. The one on the right is an African elephant, and the one on the left is an Asian elephant.

So then, what makes an organism a certain species? This remains a difficult question to answer. Generally, a **species** is a group of organisms that naturally mate with one another and produce fertile offspring. By "fertile," we mean that the offspring are able to reproduce. Take dogs, for example. While there are many types of dogs, they all can mate and produce fertile young. This is because they are members of the same species. Yet, the African and Asian elephants are not members of the same species and, therefore, do not mate with one another.

Figure 7–13. The lion and tiger do not mate in their natural environment. However, they can mate under artificial conditions to produce infertile offspring such as this liger.

Under captivity, members of different species are sometimes able to crossbreed and produce offspring. For example, a male lion and a female tiger can crossbreed and produce offspring that carry the characteristics of both species. A male donkey and a female horse can also mate under artificial conditions to produce a mule. But usually, the offspring of this crossbreeding are not fertile. Different species of plants have also been artificially crossbred. Unlike animals, however, plants resulting from such crossbreeding can be fertile.

Figure 7–14. Do you think this mule looks more like a donkey or a horse?

ASK YOURSELF

Is it accurate to say that all members of a given species look exactly alike? Explain why or why not.

● **Process Skills:** *Classifying/Ordering, Comparing*

Display a number of different books, such as a telephone directory, address book, textbook, encyclopedia, cookbook, paperback novel, and so on. Ask the students to identify similarities and differences among the displayed items. Then have the students identify a general category that would include each of the books. (The students should note that all of the items are books.) Explain that, in much the same way, scientists have classified organisms into kingdoms based on common characteristics of the organisms.

● **Process Skills:** *Applying, Synthesizing*

Remind the students that mnemonics, or memory devices, can help them remember the seven levels of scientific classification. Read them the sentence, "Katie played chess on Frank's green ship." Help the students note that the initial letter of each word in the sentence is the same as the initial letter of each category in the upside-down pyramid. Have the students repeat the sentence and use the mnemonic device to name the seven levels of classification. Then ask them whether they can make up their own mnemonic device.

Beyond Species

Living things are not classified into species alone. In the same way grocers divide and subdivide products they sell, scientists have built a hierarchy for grouping living things. Take, for example, the produce section. In one part you have fruits, and in another part you have vegetables. The fruits could be divided into different types, such as apples, oranges, bananas, grapes, and so on. Then the apples could be further divided by types, such as Red Delicious, Golden Delicious, McIntosh, and Granny Smith.

Figure 7–15. Even apples are subdivided according to type.

Visit two different types of stores and analyze how they classify their products. Do they arrange products by style or type? Do they put all brand-name products together? Do they place products that would be used together near one another? Use your journal to record your findings. Construct a diagram to represent each system of classification.

The classification system for living things is made up of seven separate categories. These categories range from broad and general categories to narrow and specific categories. The broader the category, the more species you will find in it. Species, of course, is the narrowest category. The largest category is a **kingdom.** You might think of the categories as an upside-down pyramid with species at the bottom. The drawing of the upside-down pyramid lists the seven categories of living things. The list next to it traces the human being from kingdom through species.

Granny Smith, Winesap, Fuji
Rome, Golden Delicious, Red Delicious

Figure 7–16. The upside-down pyramid shows the seven categories of classification. As you move up through the categories, you find more and more types of organisms as members of the categories.

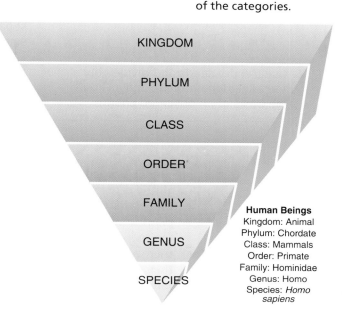

Human Beings
Kingdom: Animal
Phylum: Chordate
Class: Mammals
Order: Primate
Family: Hominidae
Genus: Homo
Species: *Homo sapiens*

DISCOVER BY *Writing*

Before the students visit the stores, point out that making a rough sketch of the store's layout while they are there will help them in constructing their diagram. As an extension, encourage the students to critique the effectiveness of each store's system of classification.

PERFORMANCE ASSESSMENT

Have each student explain the classification system shown in one of his or her diagrams. Ask questions about the divisions and subdivisions to evaluate the student's understanding of hierarchical groups.

SCIENCE BACKGROUND

Even though scientists have added levels of classification, Linnaeus's system of binomial nomenclature is still used in scientific names, which are made of two parts: a genus name and a species name. Both names are Latinized. The first term, the genus name, is always a noun. The second term, the species name, is used to identify members of the same genus. The organism's color or the place it lives may become the species name. Sometimes the discoverer's name is Latinized and used as the species name. For example, *Neutrotrichus gibbsii*, the shrew mole, is named after its discoverer, Gibbs.

SECTION 2 **185** ◀

ACTIVITY

How many different ways can you classify seeds?

Process Skills: Observing, Comparing, Classifying/Ordering

Grouping: Groups of 3; in large classes, groups of 6

Hints

Choose four seeds from each list: **List 1**—lima beans, kidney beans, blackeyed peas, new peanuts, peas; **List 2**—rice, field corn or popcorn, barley, oats, grass seed. The two lists distinguish seeds that split into two halves (dicots) from those that do not.

Place 20 to 30 seeds of each type into the moistened dishes a few hours before the activity. Locate the seeds centrally in the classroom in labeled containers.

Point out that the word *phyla* is plural and the word *phylum* is the singular of *phyla*.

Substitutions can be made for the seeds listed. However, because certain seeds contain toxic chemicals, exercise caution when using seeds with which you are unfamiliar.

▶ **Application**
1. The students should separate the seeds into two "phyla"—those seeds that split into two halves and those that did not. Other characteristics may include color, markings, shape, texture, and size.

2. The variety of characteristics used to group seeds illustrates the many alternative paths of classification. The students should be able to explain their choices.

TEACHING STRATEGIES, continued

● **Process Skills:** *Comparing, Classifying/Ordering*

To reinforce the concept of *kingdom*, write the following numbered sets of items on the chalkboard. Ask the students to identify a general category, or a "kingdom," to which all of the items in a set could belong.

1. buttons, safety pins, snaps, zippers, hooks and eyes (fasteners)
2. house, apartment, tent, mobile home, cabin (shelter)
3. television, radio, newspaper, magazine, book (mass media)
4. bread, rice, broccoli, eggs, meat (foods)
5. automobile, airplane, bicycle, ship, train (transportation)

● **Process Skills:** *Classifying/Ordering, Communicating*

Cooperative Learning Have pairs of students design a chart identifying each kingdom, naming a member of each kingdom, and listing characteristics common to the members of the kingdom.

ACTIVITY

How many different ways can you classify seeds?

MATERIALS
eight types of seeds (half of the seeds should be soaked in water), hand lens, blank paper, glue

PROCEDURE

1. Begin the activity with eight seeds, one from each of the containers in the classroom. Carefully observe each seed. Look for similarities in structure, shape, color, texture, and size. Take the seeds that have been soaked in water and split them in half. Use a hand lens to observe seed coats and internal structure.

2. Use the dry seeds and divide your seed kingdom into two "phyla" on the basis of an observed characteristic. You do not need the same number of seeds in each phylum.
3. Design a chart on which to record your classification. Give your chart and columns or rows a title.
4. Glue the seeds next to their descriptions.
5. Obtain another set of the same eight seeds. Look for more specific characteristics that will help you further divide these two seed "phyla" into smaller categories. Write the description of the subdivision in your chart, and glue the seeds under the appropriate description. Obtain more seeds as needed. Continue writing new descriptions and gluing seeds until each seed is in its own group. Give a name or characteristic to each unique category.

APPLICATION

1. What characteristics did you use to separate the eight seeds into smaller and smaller categories?
2. Look at the completed seed charts of your classmates. Discuss the grouping characteristics that you think are the best.

The broadest category of classification is the kingdom. Of all the categories, it has by far the most organisms. For a long time, scientists placed all living things in just two kingdoms—plants and animals. But with the development of better and better instruments, tools, and methods, they have learned that there are greater differences between certain organisms than they first thought there were. Today, we classify all living things into five kingdoms. The chart shows the five kingdoms and examples of organisms found in each.

- **Process Skills:** *Classifying/Ordering, Expressing Ideas Effectively*

 Cooperative Learning Have the students work in small groups to identify the kingdom to which each of the following organisms belongs: *Rangifer tarandus* (animals), *Daucus carota* (plants), *Clostridium botulinum* (monerans), *Amanita phalloides* (fungi), *Amoeba proteus* (protists). Suggest that the students use science reference books and biology texts as resources for this activity.

- **Process Skills:** *Formulating Hypotheses, Expressing Ideas Effectively*

Tell the students that scientists have just discovered a new organism on the side of a rotting log. They have not observed the organism moving. When they viewed it under the microscope, they saw that its cells have nuclei but do not contain chlorophyll. Ask the students to use what they have learned about the five kingdoms to decide how they would classify the new organism. (The students should classify the organism as a fungus.)

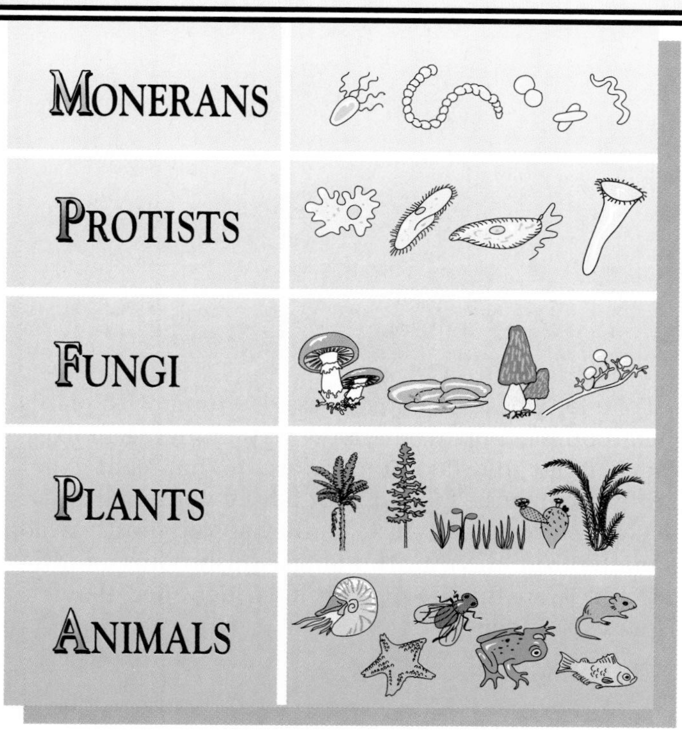

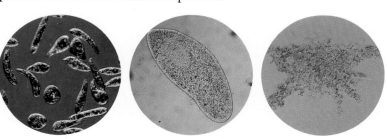

Figure 7–17. Which of the five kingdoms do you think has the most species? ①

① The animal kingdom contains the most species.

② The bacteria are ball-shaped.

 **LASER DISC**

2807

Classification; The five kingdoms

REINFORCING THEMES— *Systems and Structures*

Use the general overview of the five kingdoms in the text to emphasize that the similarities and differences among organisms are the basis for grouping organisms into kingdoms. As the students read the descriptions, encourage them to note the characteristics that are used to classify living things into the five kingdoms.

THE NATURE OF SCIENCE

Point out to the students that scientific classification ultimately depends on interpretation of facts. A biologist will study many specimens of an organism before classifying it, but even so, that decision is still based on the biologist's opinion and judgment. Most biologists agree on the basic structure of the classification system; however, they do not agree on the placement of all organisms within that structure. As with most scientific theories, classification is an ongoing debate.

Monerans The monerans became a separate kingdom when scientists, using powerful microscopes, discovered that some cells do not have a distinct nucleus. Bacteria, which often are incorrectly called "germs," are the most common members of this kingdom. They are one-celled organisms that live as separate cells or as many cells joined to form a colony. Scientists estimate that there are over 10 000 species of bacteria. Even so, the monerans make up the smallest kingdom.

Protists Like the moneran kingdom, the protist kingdom is made up of simple one-celled organisms. But unlike monerans, the protists have distinct nuclei in their cells. Thanks to the microscope and other tools and methods, scientists have been able to do a better job of classifying these tiny organisms. The picture shows three common protists.

Figure 7–18. Bacteria are classified into three shapes—rod-shaped, ball-shaped, and spiral-shaped. What shape are these bacteria? ②

Figure 7–19. The *Euglena*, the paramecium, and the amoeba are common protists.

SECTION 2 **187** ◀

GUIDED PRACTICE

 Have the students create a classification key that will help them determine the kingdom to which an organism belongs. Have the students place their completed keys in their science portfolios.

INDEPENDENT PRACTICE

 Have the students answer the Section Review and Application questions. Then have students write a journal entry explaining how scientific advancements have led to changes in the classification system.

EVALUATION

 Cooperative Learning Using the paper figures from the Chapter Motivating Activity on page 173, have the groups of students develop a classification system based on the seven levels of the modern classification system.

INTEGRATION—*Health*

Some infections, such as athlete's foot, are caused by fungi. However, many fungi are useful rather than harmful to people. Some types of mushrooms and truffles are fungi that are eaten. Other fungi, such as yeast, are used in the preparation of foods.

Because some mushrooms are poisonous, caution the students that they should never eat unfamiliar mushrooms or other unfamiliar fungi unless they are sure the fungi are safe to eat.

DISCOVER BY *Researching*

 Make a master of a blank inverted pyramid and distribute enough copies for both partners to fill out. Ask volunteers to write their answers on the chalkboard. Have the students place their pyramids in their science portfolios.

Figure 7–20. Mushrooms, which are classified as fungi, help decompose dead plants.

Fungi For a long time, scientists classified fungi with plants, calling them nongreen plants. In many ways, fungi seem like plants. For example, they do not move around but seem to be rooted. But unlike plants, fungi do not have chlorophyll and cannot make their own food. Instead, they absorb mostly dead, decaying matter from their surroundings. Mushrooms, molds, yeast, and mildew are fungi. Scientists have identified about 100 000 species of fungi.

DISCOVER BY *Researching*

 With a partner, use encyclopedias and science reference books to identify the seven levels of classification for each of the following organisms: bacillus, moss, bread mold, rose, dog. Write the common name of each organism. On an inverted pyramid, write the kingdom, phylum, class, order, family, genus, and species for each organism.

Plants With over 350 000 known species, the plant kingdom is the second largest kingdom. These species range from tiny green mosses to giant trees. Plants are made of many cells. Their cells contain chlorophyll, which they use to make their own food. Like fungi, they cannot move around from place to place.

Figure 7–21. Plants are both large and small.

RETEACHING

As you show the students a series of flash cards illustrating various organisms, ask them to identify the organism's kingdom and explain their answer. (Answers should indicate that the students can distinguish characteristics of organisms in the five kingdoms.)

EXTENSION

Explain that the five-kingdom system of classification was recommended by R. H. Whittaker in 1969. Have the students research information about Whittaker and his proposal for the five-kingdom system.

CLOSURE

Cooperative Learning Have the students work in small groups to list the advantages (and any disadvantages) of the present classification system. Then ask each group to present a summary of its work to the class.

Animals The animal kingdom is the largest kingdom with well over 1 000 000 known species. Animals have many cells. They are different from plants in that they cannot make their own food and most can move around from place to place. The animal kingdom is divided into two major groups—those with a backbone and those without. The pictures show examples of animals from these groups.

Figure 7–22. The bat and the polar bear are examples of animals with a backbone.

Figure 7–23. The starfish, the mosquito, and the oyster are animals without a backbone.

ASK YOURSELF

List the five kingdoms. For each, tell one way in which it is different from the others.

SECTION 2 REVIEW AND APPLICATION

Reading Critically

1. What three factors contributed to changes in the classification system?
2. Explain the difference between species and kingdoms.
3. What contribution did Carolus Linnaeus make that continues to influence the science of taxonomy today?

Thinking Critically

4. Think about the living things you have seen since waking up this morning. From which kingdom did you see the greatest number of different species? Prepare a list to support your answer.
5. Explain why dogs that look so different can belong to the same species whereas doves and pigeons, which look so much alike, are members of different species.
6. Write your prediction of how the classification system might change in the next 500 years. Give reasons to support your prediction.

ONGOING ASSESSMENT
ASK YOURSELF

Monerans lack nuclei; protists contain nuclei and are typically one-celled; fungi have no chlorophyll and cannot make their own food; plants are multicellular and contain chlorophyll; animals can move from place to place and cannot make their own food.

SECTION 2 REVIEW AND APPLICATION

Reading Critically

1. The development of advanced instruments, the improvement of identification methods, and the theory of evolution have contributed to changes in the classification system.

2. A kingdom is the broadest classification of organisms. A species is the narrowest classification; members of a species can mate with each other and produce fertile offspring.

3. Linnaeus contributed the system of binomial nomenclature.

Thinking Critically

4. Answers are likely to be either the animal or the plant kingdom.

5. The students might suggest that genetic variations within a species can cause two members of the same species to look different and two members of different species to look similar.

6. The students might predict that the classification system will gain new categories as more new organisms are discovered.

INVESTIGATION

Observing, Ordering, and Displaying a Collection of Organisms

Process Skills: Comparing, Classifying/Ordering

Grouping: Groups of 3 or 4

Objectives
- **Identify** characteristic features of organisms.
- **Compare** characteristic features of organisms.
- **Distinguish** among organisms by using a classification system.

Pre-Lab
Have the students list the five kingdoms and name several organisms that belong in each kingdom. Ask the students what the organisms in each kingdom have in common.

Hints
Point out that the pond water may contain microscopic animals.

Analyses and Conclusions

1. The students should have observed organisms from every kingdom except the monerans.

2. Many of the organisms are too small to be seen without the aid of the microscope.

3. Organisms from the plant or animal kingdom are easiest to find because they are plentiful and usually large enough to be seen with the unaided eye.

Application
Some students might classify the organism as a protist, because the protist kingdom contains organisms such as the *Euglena*, which combines plant and animal characteristics.

✳ Discover More
Most organisms that the students observe will belong to either the animal or plant kingdom. Animals and plants form the two largest kingdoms, and many plants and animals are large enough to be easily seen.

Post-Lab
Ask the students to compare the organisms they observed to the ones they listed before the investigation. Ask them to explain their classification of their observed organisms.

 Have the students place their drawings from the *Investigation* in their science portfolios.

INVESTIGATION

Observing, Ordering, and Displaying a Collection of Organisms

▶ MATERIALS
- microscope • slide • coverslip • pond water • drawing paper
- pencil • bread mold • forceps • plant parts • worms and insects

▼ PROCEDURE

1. List characteristics of organisms in each of the five kingdoms. After you examine each of the specimens in this investigation, try to place it in the appropriate kingdom.

2. Use the microscope to view the pond water. Are there any organisms in the water? If so, draw the organisms. Label the drawing with the kingdom you think the organisms belong to.

3. Examine some bread mold under a microscope. What kingdom do you think it belongs to? Why?

4. Examine small parts of a house plant, such as leaves or stems. Sketch the parts of the plant you observe. In what kingdom does this organism belong?

5. Look for worms, ants, or other small organisms in areas near your home or school. Sketch these organisms and label your sketches with the kingdom to which the organisms belong.

6. Make a display of your drawings.

▶ ANALYSES AND CONCLUSIONS
1. Did you observe organisms from each kingdom? Why or why not?
2. Why is it necessary to display drawings rather than the actual organisms?
3. For which kingdom were organisms easiest to find? Why?

▶ APPLICATION
Suppose you have an organism that has chlorophyll and moves. How would you classify it? Why?

✳ Discover More
As you walk or ride to school, look for different types of organisms. Keep a list of your observations. What kingdom do most of the organisms you see belong to? Why do you think you see organisms from this kingdom more than others?

Chapter 7 HIGHLIGHTS

The Big Idea—SYSTEMS AND STRUCTURES

Lead the students to understand how the discovery of new organisms, improved scientific methods and tools, and acceptance of the theory of evolution have led to changes in the classification system over time. Emphasize that today's system of classification is based on the evolutionary ancestry of organisms as well as on the characteristics of the organisms.

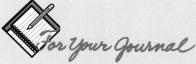

For Your Journal

The students' journal entries should reflect an understanding of today's classification system and its development over time. The entries should also note that classification keys can be used to identify and classify unfamiliar organisms.

CONNECTING IDEAS

The concept map should be completed with words similar to those shown here.

Row 3: seven categories
Row 5: species
Row 7: monerans, fungi, plants, animals

Chapter 7 HIGHLIGHTS

The Big Idea

Humans have been classifying living things for years. Because of the wide range of organisms, scientists classify organisms so that study of them can be simplified. As our knowledge of living things improved, our systems for classifying living things evolved into better systems.

New tools and instruments and new theories about living things not only changed the way we classify things, but also gave us many more organisms to classify. The classification system as we know it today is the result of the work of many scientists and scholars. The continued work of scientists today and in the future will likely bring even greater improvements in the system.

Think about what you have learned about classification in this chapter. Review what you have written in your journal, and revise or add to the entries to reflect your new understandings about scientific classification. Don't forget to include information about how the classification system has changed and what factors have influenced the changes.

Connecting Ideas

Copy the unfinished concept map into your journal. Complete the map by filling in the blanks.

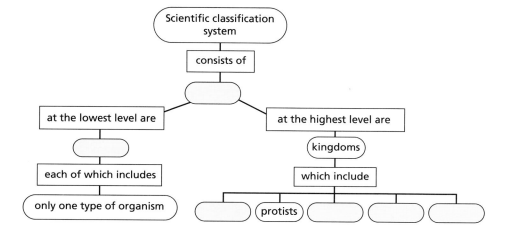

CHAPTER 7 **191**

CHAPTER 7 REVIEW ANSWERS

Understanding Vocabulary

1. The science of classification is known as taxonomy.

2. A species is a group of organisms that can mate with each other and produce fertile offspring.

3. A kingdom is the broadest, most general category in the classification of living things.

4. Monerans do not have nuclei in their cells.

5. Protists are usually one-celled organisms with a distinct nucleus.

6. Fungi are organisms that have no chlorophyll and cannot make their own food.

7. Binomial nomenclature is the use of two Latin words to name an organism.

Understanding Concepts

Multiple Choice

8. d
9. c
10. a
11. a
12. d

Short Answer

13. Poodles and collies can mate and produce offspring because they are members of the same species. Poodles and collies cannot mate with coyotes and produce offspring because they belong to a different species than coyotes.

14. The system for classifying living things will change in the future because more species are being discovered, species will become extinct, and improvements in scientific methods and tools are likely.

Interpreting Graphics

15. The students should describe the divisions and subdivisions of a transportation system.

16. The students might divide cars and trucks according to capacity and number of wheels, trains and subways according to the way in which they are powered, ferries and ships according to their cargo capacity, and airplanes and helicopters according to their landing and takeoff characteristics.

Reviewing Themes

17. Science must be open to the change and evolution of ideas because scientific knowledge increases with improved methods and technology and the discovery of new organisms.

18. The students might suggest that classification must take into account chemical makeup and the impact of evolution.

Thinking Critically

19. The students should be able to record many first names that are shared by students and, as a result, reveal the need for a two-name system.

CHAPTER 7 REVIEW

Understanding Vocabulary

Write a sentence that demonstrates your understanding of each of the terms below.

1. taxonomy (176)
2. species (184)
3. kingdom (185)
4. monerans
5. protists
6. fungi
7. binomial nomenclature (177)

Understanding Concepts

MULTIPLE CHOICE

8. Which of the following statements is *not* true of bacteria?
   a) Bacteria are monerans.
   b) Bacteria are single-celled organisms.
   c) Bacteria may live in colonies.
   d) Bacteria have a distinct nucleus.

9. One reason scientists no longer classify fungi as plants is because
   a) fungi do not have nuclei in their cells.
   b) fungi do not move around.
   c) fungi cannot produce their own food.
   d) fungi cannot grow.

10. Which of the following statements is true?
    a) Some members of different species are similar looking.
    b) Different species can never crossbreed.
    c) All members of the same species look alike.
    d) More than one kind of organism can be in the same species.

11. All of the following developed after Linnaeus classified organisms *except*
    a) binomial nomenclature.
    b) the theory of evolution.
    c) improved tools such as the microscope.
    d) biochemical analysis of organisms.

12. Organisms that are not in the same family *cannot* be in the same
    a) kingdom.   b) class.
    c) order.   d) genus.

SHORT ANSWER

13. Explain why poodles can mate with collies and produce offspring, but collies and poodles cannot mate with coyotes to produce offspring.

14. The system for classifying living things will probably change in the future. Explain why.

Interpreting Graphics

15. What does the diagram below tell you about the things being classified? Give a detailed explanation.

16. Use the diagram below and add another level giving more specific descriptions.

```
                    Transportation
        ┌───────────┬──────────┬──────────┬──────────┐
     Highway     Railway    Waterway    Airway
      /  \       /   \       /   \       /    \
  Cars  Trucks  Train Subway Ferry Ship Airplane Helicopter
```

192 CHAPTER 7

20. Because there are so many different species of ants, scientists would be confused when trying to study and communicate information about ants if a system of classification did not exist.

21. Each key should accurately distinguish the organisms.

22. The students might suggest that different names are used because the human beings of earlier times were very different in some ways from humans today.

23. Although both protists and monerans are one-celled organisms, they have other characteristics such as the lack or presence of nuclei that make them different from each other.

24. If Linnaeus had known and accepted the theory of evolution by natural selection, his classification system might have classified organisms according to how closely they are related to one another in their evolutionary past.

Reviewing Themes

17. *Systems and Structures*
Systems of classification have changed and evolved over the years. Why is it important for science to be open to the change and evolution of ideas?

18. *Changes Over Time*
The observable structure of living things has always played an important role in classification. Cite two reasons why scale and structure are not enough when classifying organisms.

Thinking Critically

19. Select a first name you hear often at school. Prove the need for a two-name system by making a list of all the people you know who have the same first name.

20. There are over 6000 kinds of ants. How does this fact help show the need for a classification system?

21. Devise a paired description key to identify the following organisms: pine tree, daisy, dog, elephant.

22. *Homo sapiens* is used to identify modern humans, and *Homo erectus* is used to identify an earlier form of humans. The English translation of these Latin terms is: *Homo*—"human," *sapiens*—"wise," and *erectus*—"upright." Give reasons why these two different names are used to identify different species of humans.

23. Both monerans and protists are one-celled organisms. Why do you think scientists have placed them in separate kingdoms?

24. You learned about the theory of evolution in Chapter 6. During Linnaeus's time the theory of evolution was not an accepted theory. How might his system for classifying living things have been different had he known about and accepted the theory of evolution?

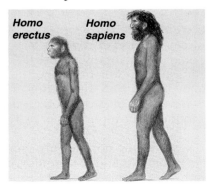
Homo erectus   Homo sapiens

Discovery Through Reading

Stidworthy, John. *Mammals: The Large Plant-Eaters.* Facts on File, 1989. Articles focusing on individual species of mammals, their structure and size, and background information.

Science PARADE

SCIENCE APPLICATIONS

Background Information

To help the students comprehend the violence of the Mt. Vesuvius eruption, point out that Pompeii was located almost 10 km from the volcano, yet it was buried under 7 m of volcanic ash.

Discussion

● **Process Skills:** *Inferring, Applying*

Remind the students that, on a global scale, volcanic eruptions have occurred since the catastrophe at Pompeii, and volcanic eruptions are likely to occur in the future. Ask the students to compare the similarities and differences between the eruption of Mt. Vesuvius in the past and the eruptions of present-day volcanoes. (The scale of eruptions is different; large-scale catastrophes have not occurred. The frequency of eruptions is similar; eruptions have occurred in the past and are likely to occur in the future.)

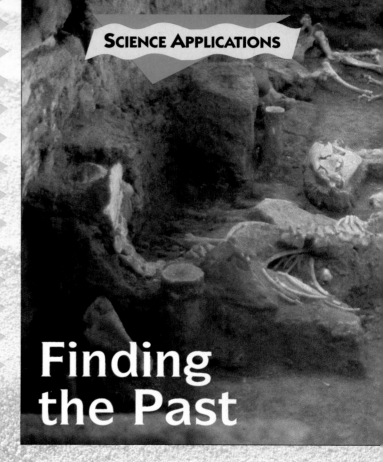

SCIENCE APPLICATIONS

Finding the Past

For nearly 17 centuries, the Roman city of Pompeii lay buried under 7 m of volcanic ash from an eruption of Mt. Vesuvius. The ruins of Pompeii, in southern Italy, were first discovered in the early 1700s by workers building an underground water pipeline.

Accidental Discovery

Pompeii, like many archaeological discoveries, was found by accident. Some important discoveries that have provided a great deal of information about the past have been made by people who were hiking or camping. Some finds have been made by workers building roads or digging foundations for houses or by farmers plowing fields.

Such accidental discoveries are not the only source of information for archaeologists and paleontologists. These scientists actively look for places where people

● **Process Skills:** *Comparing, Applying*

Remind the students that the Roman city of Pompeii was buried under 7 m of volcanic ash. Ask the students to describe modern-day volcanic eruptions that have also produced volcanic ash either in the atmosphere or on the surface of the earth. (The students might suggest the eruption of Mount St. Helens in Washington or the eruption of Mount Pinatubo in the Philippines. In both cases, large amounts of volcanic ash covered surrounding areas and entered the atmosphere.)

Background Information

Ensure that the students understand the differences between an archaeologist and a paleontologist. Explain that an archaeologist relies on nonliving or once-living physical remains to investigate and understand cultures of the past; a paleontologist relies on fossils to study ancient life forms. You might also choose to explain the function of an anthropologist to the students. An anthropologist studies a culture by interacting directly with the culture, such as living with the people and recording their language.

Archaeological sites are excavated, explored for artifacts, and mapped.

may once have lived. They carefully study old maps and aerial photographs for clues to the locations of likely places of habitation, abandoned villages, or burial sites.

Careful Work

When scientists find an area that looks as if it might have been inhabited, they explore it to see if they can find any evidence of human occupancy. They look for such things as building foundations, walls, pottery, and human remains. Once an actual site has been discovered, a report is made before any digging begins. Then a few test pits may be dug to find out what soil layers are present and how large an area the site covers. Wooden stakes and lengths of twine are used to divide the site into sections measuring one or two square meters. The site then looks a bit like a large piece of graph paper.

The actual digging process is slow. A hand trowel is used to remove the soil a little at a time so that any item that may be buried is not disturbed. Loose soil around delicate items is swept away by using a small brush. A screen might be used to sift through soil to find tiny pieces of bone or fragments of objects. Photographs of each

UNIT 2 **195**

Discussion

● **Process Skills:** *Inferring, Applying*

Ask the students to explain the need for precision when archaeologists and paleontologists excavate historical sites. (The students might suggest that precise data provides archaeologists, paleontologists, and other scientists with the greatest likelihood of defining and understanding the historical record. Once a site is disturbed by excavation, a precise and comprehensive record of that site must be created to allow the evidence to be later considered in the context in which it was found.)

Background Information

Explain to the students that the tomb of King Tutankhamen (Tut) in Egypt was thought to be intact for more than 3000 years before its discovery in the 1920s. Since that time, the contents of the tomb have been removed and distributed to museums, and the tomb itself is a tourist stop that people can enter and view. Lead the students in a discussion of the pros and cons of removing artifacts from a historical site in the name of science.

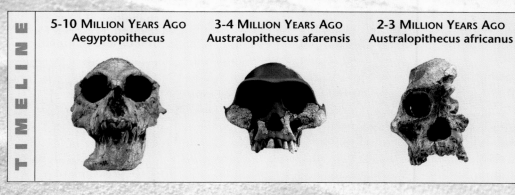

TIMELINE

5–10 MILLION YEARS AGO	3–4 MILLION YEARS AGO	2–3 MILLION YEARS AGO
Aegyptopithecus	Australopithecus afarensis	Australopithecus africanus

item found are taken, and the position of each object in every layer of a square is carefully noted on a grid.

New Insights

These excavation and research skills have provided new insights into the origin of modern humans. Archaeologists and paleontologists have painstakingly applied these procedures at various locations around the world. These scientists have used their findings to trace the evolution of humans. Discoveries of bones, tools, and other artifacts have aided scientists in establishing a timeline of human evolution.

Humanlike fossils have been found that date back millions of years. A 2.5-million-year-old skull called *Proconsul*, found by Mary Leakey in Kenya, Africa, is believed to be the oldest early ancestor of humans. The remains of early humans, called *hominids*, were first discovered by scientist Raymond Dart at a limestone quarry in South Africa in 1924. His findings have been supported by the additional work of Mary and Louis Leakey and their son Richard.

In 1977 in Ethiopia, Africa, Donald Johanson found part of a female skeleton that he named "Lucy" or *Australopithecus afarensis* (aws trah loh PITH uh kuhs AF uh rehn sihs). Lucy walked on two legs and stood slightly over 1 m tall. She is the oldest evidence of a hominid found so far.

Pieces of pottery provide clues to how a people lived.

EXTENSION

Journal Activity Have the students recall the similarities and differences between archaeologists and paleontologists. In a journal entry, ask the students to explain why archaeologists, paleontologists, and other scientists desire a clear, comprehensive picture of our distant past.

Challenge small groups of students to choose an area of the world and then use reference materials to gather information about one or more archaeological discoveries in that part of the world. Then ask each group to compile its findings into a brief report and share that report with the class.

1.9 MILLION YEARS AGO
Homo erectus

35,000–150,000 YEARS AGO
Homo sapiens (Neanderthal)

PRESENT–100,000 YEARS AGO
Homo sapiens (modern humans)

Closest Ancestors

The closest ancestors to modern humans are the hominids named *Homo erectus*. *Homo erectus* means "upright human." Richard Leakey and his co-workers made the largest *Homo erectus* discovery in 1984. This was a skull and nearly complete skeleton of a 12-year-old male. The fossil remains of *Homo erectus* show that they had an ability to stand fully upright and also had a larger brain than other previous hominids had had. Remains of shelters, tools, and cooking fires have been found at the sites where *Homo erectus* was discovered.

The evolutionary history of humans is still under study. In 1987, a group of molecular biologists proposed that mitochondrial DNA could be used to trace the development of humans. Mitochondrial DNA is genetic material that is located outside the nucleus of a cell. This DNA is passed on only by females. It was theorized that by using a computer to trace backward, scientists could determine the original "mother" of all humans. In 1991, computers were used to analyze DNA similarities among people from around the world and to produce one hundred possible ways to find one common human ancestor. Scientists, however, are in disagreement over this research. Some think that DNA inside the cell's nucleus would provide different results. Still others contend that any number of possible outcomes could result depending on the DNA mix.

So the search for our past goes on. There are still gaps in time and many questions that remain to be answered. But thanks to the discoveries of archaeologists and paleontologists, we now have a much better picture of our distant past. ◆

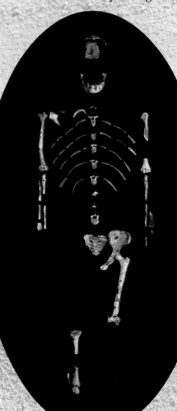

The early hominid skeleton called "Lucy."

READ ABOUT IT!

Discussion

Before reading the selection, have the students pool their knowledge of sharks by sharing and discussing things they know about sharks. As information is contributed to the discussion, ask the students to debate whether a particular piece of information represents a fact or a myth about sharks. Then ask the students to read the selection.

Background Information

After they have had an opportunity to read the selection, the students might be interested to learn other interesting facts about sharks. Explain to the students that more than 350 different species of sharks exist today, ranging in size from about 15 cm to more than 12 m. One species of shark—the whale shark—is much larger than a school bus.

Not all sharks are gray-blue in color; some are derivations of

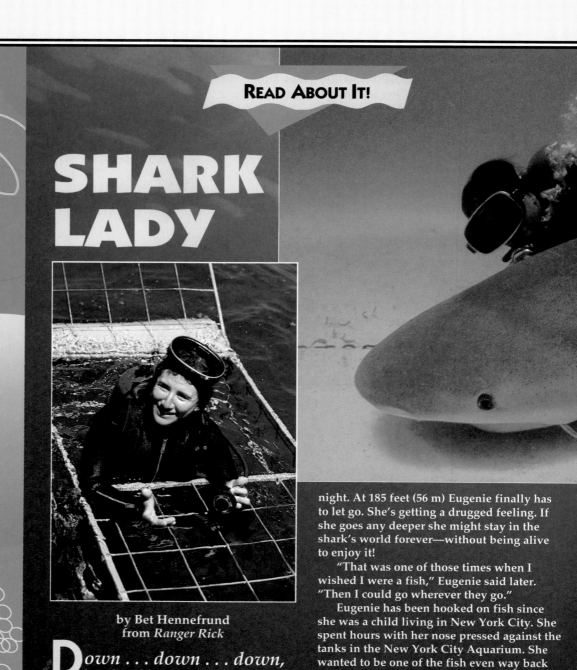

READ ABOUT IT!

SHARK LADY

by Bet Hennefrund
from *Ranger Rick*

Down ... down ... down, into the cold, dark ocean depths, a huge whale shark glides like a giant toboggan. On the shark's back rides a small, dark-haired woman. She's Eugenie Clark, "the shark lady."

Eugenie holds tight to the shark's big back fin. Down the two go together ... 100 ... 130 ... 150 feet. The water becomes dark as night. At 185 feet (56 m) Eugenie finally has to let go. She's getting a drugged feeling. If she goes any deeper she might stay in the shark's world forever—without being alive to enjoy it!

"That was one of those times when I wished I were a fish," Eugenie said later. "Then I could go wherever they go."

Eugenie has been hooked on fish since she was a child living in New York City. She spent hours with her nose pressed against the tanks in the New York City Aquarium. She wanted to be one of the fish even way back then.

When she was in college, Eugenie put on her first diving mask and flippers. Then she dove into the sea, where the fish live free. At the same time, she was studying fish. She learned enough about them to become *Doctor* Eugenie Clark, an expert fish biologist. Today she is a professor at the University of Mary-

black, green, and gray. Others, such as the lantern dogfish shark, have luminous organs that appear to glow in the dark. The typical shark species lives about 25 years, although some species can live for almost 100 years.

The students may also be interested to learn that sharks seem to have a sense of taste. When biting a fish that releases a foul or toxic substance, a shark will invariably stop eating and swim away with "a bad taste in its mouth."

Discussion

● **Process Skills:** *Inferring, Applying*

Explain to the students that even though only a few people die each year from shark attacks, it is still important to know how to act when encountering a shark. Talk about Dr. Eugenie Clark's advice for how to act when meeting a shark. Ask the students to infer reasons why being polite, backing away slowly, and staying away from feeding grounds is probably good advice.

Eugenie will never forget her first time under water with great white sharks. In her metal diving cage she didn't have to worry about being eaten — but the hungry sharks really tried their best to eat the cage!

land. She's written books and articles and has been on television.

How to Meet a Shark

To Eugenie, there's nothing more wonderful than swimming underwater for hours, nose-to-nose with angelfish and eels, octopuses and . . . sharks!

Eugenie says sharks aren't "mean," as many people believe. They attack only to eat or because they feel threatened. And if they sometimes mistake a person for a sea lion or an enemy, that's not the sharks' fault. Eugenie is very careful in the water and she has no fear of anything that lives there. And she doesn't see why *anyone* who is diving carefully should be afraid either.

"You just have to realize that you are a visitor in the fish's backyard," she says.

Her advice to anyone who happens to meet a shark: Be polite. Back slowly away. And in general: Stay clear of sharks' feeding grounds, especially if fish blood is in the water. Don't wear shiny things that might look like flashing fish movements to a shark. And don't thrash around if you're swimming where sharks may be.

Eugenie met her first big shark in the southern Pacific Ocean. She was swimming near a coral reef when a shark came straight toward her. It was so close she could have reached out and touched it. But she didn't. She just stayed perfectly still and stared. She couldn't understand why she wasn't afraid. As the shark gracefully turned and swam off, all Eugenie could think was, *How lucky I am to see that sleek, beautiful creature so close up!*

After that, Eugenie met lots of sharks face to face in many oceans. Once she spent ten days observing great white sharks from a metal diving cage. Her first moments underwater with the giant fish were unforgettable. One shark came straight at the cage with its mouth open so wide that Eugenie could see right down its throat. As the shark's jaws crunched down on the cage, its nose poked through the space between the bars. Eugenie pressed her body against the opposite side of the cage. But at the same time, another shark swam by and brushed against her back!

Even when there were five sharks around the cage at one time, Eugenie was too excited to be afraid. She just aimed her camera right into the sharks' faces or down their throats. She wasn't at all worried about those big, gaping jaws and sharp teeth.

Eugenie *has* been bitten though: One time she reached inside the body of a living, pregnant tiger shark she was examining in the lab. When she pulled out her hand, a baby shark was hanging onto her fingers with its tiny teeth!

Here a Shark, There a Shark

Eugenie got to know sharks best when she was the director of a marine lab in Florida. She and the other scientists caught and studied many different kinds of sharks. There were hammerheads, sometimes a white or sandtiger shark, bull and sand-

Discussion

● **Process Skills:** *Applying, Expressing Ideas Effectively*

Ask the students to recall some of the information presented in the article that is intended to dispel myths that people may believe about sharks. Ask the students to summarize the kinds of activities that Dr. Clark must perform in her work. Then ask the students to describe the kinds of academic courses that Dr. Clark probably completed to become a fish biologist.

Journal Activity In a journal entry, ask the students to describe how Dr. Clark's work might be considered rewarding, dangerous, and necessary. Would they like to do that kind of work themselves? Why or why not?

EXTENSION

The students have learned many interesting facts about sharks during their study of Dr. Clark. Challenge small groups of students to discover other interesting and unique facts about sharks and to present their findings to their classmates.

In an underwater cave, one of Eugenia's students studies a "sleeping" shark. Because they stay very still, the sharks look as if they're asleep. But they're really not sleeping at all.

bar sharks, nurse sharks, tiger sharks, dusky sharks, and lemon sharks.

Eugenie kept the live sharks in big swimming pens so that she could study how they behave. She watched them eat, swim, rest, and communicate with each other. Then she wrote about what she learned. Soon scientists came from all over to learn from her.

Sharks with "Smarts"

Sharks, everybody once thought, were not only mean, but stupid too. That was before Eugenie proved for the first time what sharks can learn. At her marine laboratory, two lemon sharks and a nurse shark were taught to press a target to get food. They even learned to tell white targets from red, and striped ones from plain-colored ones.

Eugenie discovered that sharks can remember things too. After *not* working with the targets for two months, the sharks went right back to pressing targets for their food again—just as if they had been doing it every day.

The Mystery of "Sleeping" Sharks

Several times Eugenie has watched sharks "sleeping." She has found them in underwater caves off the coasts of Mexico and Japan, and in the Red Sea. The sharks aren't really sleeping, but they look as if they are. They keep very still, except for their mouths opening, closing, opening, closing. And their eyes follow every move divers make.

Sharks are supposed to *have* to keep moving so that water flows over their gills. That's how they get the oxygen they need to breathe. But these sharks proved they could breathe even when still.

To find these "sleeping" sharks seemed very strange to Eugenie, though. Why were they there? Did they come to the caves to be cleaned by *remoras*? Remoras are fish that often eat *parasites*, pesky creatures that cling to other fish and take food from them.

Eugenie studied the animals and tested the water. She found that the cave water was less salty than water in the open ocean. And she thinks it's likely that the water may act like a kind of drug on the sharks and their parasites. In less salty water, parasites loosen their grip. And that would make the cleaning easier.

The Shark's Best Friend

Eugenie believes that the more we know about sharks, the less we'll fear them. Her children grew up unafraid of sharks. They have even helped their mother in her work.

By writing her books and articles, teaching, making films, and giving talks, Eugenie spreads the word about sharks. She wants everyone to know what she knows: Sharks are magnificent animals that deserve respect, not fear. ◆

Then and Now

Discussion

To help the students better understand the importance of the classification work of Carolus Linnaeus, have them describe some of the difficulties they might encounter if certain aspects of their daily lives were not organized in an orderly and predictable fashion. Examples might include clothes in a closet, food in a refrigerator, and books on shelves.

When discussing the work of Rita Levi-Montalcini, ask the students to discuss why the stereotype exists that sometimes views the pursuit of science and scientific research as a predominantly male domain. After dispelling the idea that scientists are always male, stress to the students that all scientific fields are open to qualified people of both genders and all races.

Journal Activity Ask the students to write a journal paragraph in which they describe how the work of scientists and researchers affects their everyday lives. Then encourage volunteers to read their paragraphs to the class.

Then and Now

Carolus Linnaeus (1707—1778)

In his book *Systemae Naturae*, Carl von Linné suggested a standardized system for naming plants and animals. Linné introduced the idea of "binomial nomenclature." He assigned to each organism a two-part name in Latin. In keeping with his system, he even changed his own name to Carolus Linnaeus.

Carolus Linnaeus was born in Roshult, Sweden, on May 23, 1707. He studied medicine and the natural sciences at several universities before receiving his medical degree.

Linnaeus based his classification on newly discovered facts about stamens and pistils. He grouped plants into 24 classes according to the number of stamens. He then broke these classes down according to the number of pistils. Counting the number of visible parts let scientists identify plants quickly and classify newly discovered plants. Linnaeus's binomial system quickly became the standard way to classify and name plants and animals. ◆

Rita Levi-Montalcini (1909—)

As the head of the cell research laboratory at Rome's Institute for Cell Biology, Rita Levi-Montalcini studies the growth and function of cell types in the body. Her work has been important for cancer research and may lead to advances in the treatment of Parkinson's disease. It may help explain the process by which one fertilized egg cell develops into an organism made of billions of cells.

Born in 1909 in Turin, Italy, Rita Levi-Montalcini grew up with an interest in neurology. In 1936, she began her career as a research biologist at the University of Turin Medical School. Since she was Jewish, Levi-Montalcini was forced by the Fascist regime to leave the university, but she continued her research in secret during the Nazi occupation.

Washington University in St. Louis, Missouri, offered her a research position in 1947. In St. Louis, Levi-Montalcini discovered the material that causes the growth of nerve cells. This material, found in mammals, birds, reptiles, and fish, is known as *nerve growth factor*. Her research lead to a Nobel prize in physiology and medicine in 1986. She is the fifth woman to receive the prize in this category since 1901. ◆

SCIENCE AT WORK

Discussion

● **Process Skills:** *Inferring, Generating Ideas*

After the students have read the selection detailing the work and life of Dr. Eduardo S. Cantu, ask them to name the tools that Dr. Cantu might use in his work.

(The students should suggest that because the work of a geneticist involves examining and studying chromosomes, which cannot be seen by the unaided eye, compound light microscopes and scanning electron microscopes are important tools.)

● **Process Skills:** *Inferring, Applying*

Remind the students that Dr. Cantu believes that genetic research is a growing field in need of more researchers. Ask the students to infer reasons why Dr. Cantu is optimistic about the future of genetic research.

Journal Activity
Ask the students to discover more about genetic disorders such as sickle-cell disease, phenylketonuria, Huntington's disease, and cystic fibrosis. Ask the students to write information about the disease in their journals.

SCIENCE AT WORK

Eduardo S. Cantu, Geneticist

Dr. Eduardo Cantu studies the role that genes and chromosomes play in determining a person's health. He is a geneticist at the Medical University of South Carolina.

What type of work do you do as a geneticist?

I am the Director of the Medical University of South Carolina Cytogenetic Section. We run a variety of tests that are helpful to physicians in diagnosing diseases, particularly genetic diseases. Most of what we do is chromosome work. Most of our chromosome work involves diagnosis of genetic disorders before birth.

How many genetic disorders are there?

There is a catalog of about 3000 to 3500 genetic disorders. Most of these are extremely rare. I would say 50 to 100 could be called common genetic disorders.

Where do you get the cells you examine?

The sample for a diagnosis before birth is taken by *amniocentes*is (am nee oh sehn TEE sihs). A physician inserts a hollow needle through the abdominal wall into the uterus of a pregnant woman. Some of the amniotic fluid is collected and then is examined under a microscope.

How did you prepare for a career as a geneticist?

I attended Pan American University in the lower Rio Grande Valley in Texas, where I was born and raised. I went to graduate school at the University of Michigan because I was interested in birds, and a famous bird expert was there. His course was wonderful, but I also took a genetics course, which was so fascinating that it replaced my interest in birds. I was happy that I could then go to the University of Hawaii to work with one of the pioneers in genetics. Later, I spent a year at Baylor College of Medicine in Houston before coming to South Carolina.

Is genetics a growing field that students should consider?

Yes, physicians are relying more and more on genetics. They realize that many diseases and disorders have a genetic component. It's a very promising, challenging, and exciting field to get into. You can do that by studying medicine and specializing in genetics, or by doing graduate work in biology and specializing in cell biology. I certainly recommend it. ◆

Discover More

For more information about careers involving genetics, write to the
Genetics Society of America
9650 Rockville Pike
Bethesda, MD 20814

SCIENCE/TECHNOLOGY/SOCIETY

Discussion

● **Process Skills:** *Inferring, Generating Ideas*

After reading the selection, ask the students to describe several reasons why germs are often thought of in a negative way and seldom in a positive way. (The students might suggest that germs often cause sicknesses that not only can influence the quality of a person's life, but also can influence the length of a person's life.)

● **Process Skills:** *Inferring, Expressing Ideas Effectively*

Remind the students that the bacteria, or microbes, that have been made to produce insulin normally have a function different from the production of insulin. Ask the students to debate whether it is fair to consider such microbes as "intelligent" because they have been "taught" to perform a different activity. (No, such microbes simply follow, in a robotic fashion, the instructions that they have been given.)

SCIENCE/TECHNOLOGY/SOCIETY

These Germs Work WONDERS

by Doug Stewart
from *Reader's Digest*

We know germs as hostile little things that spoil meat and spread disease. But to scientists, germs, or microbes, can be obedient servants, ready to cure illness, clean up oil slicks, and even make delicious chocolates. This microscopic labor force has come about thanks to biotechnology, which uses engineering techniques to study living organisms. Ten years ago this field didn't exist. Today it's a billion-dollar-a-year industry. Princeton University professor Freeman Dyson has predicted that biotechnology will change the world more than the Industrial Revolution did.

Dyson's prediction is based on the work of gene splicers like Stephen Lombardi, a microbiologist at the Army's Research, Development, and Engineering Center in Massachusetts. Lombardi can snip and rejoin strings of chemicals inside living cells. These chemical strings, or genes, map out how a cell will behave. They determine, for instance, whether a microbe will turn milk into yogurt or into cheese.

Ten years ago this field didn't exist. Today it's a billion-dollar-a-year industry.

In the past 15 years, gene splicers have learned to remove microscopic pieces of genes from cells lining a human pancreas, for example. The scientists place the pieces of code inside ordinary bacteria cells. The microbes then have the ability of the pancreas cells to produce insulin. In the late 1970s, some 50 million animal pancreases a year were ground up for insulin, which is used to treat diabetes. Rumors spread that a pancreas shortage was on the way. Today a unit the size of a small refrigerator containing altered bacteria cells can pump out more insulin in a day than a hundred animals could provide.

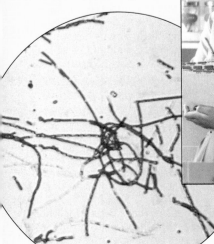

Once a batch of altered microbes starts to multiply, it can serve as a chemical factory, continually producing the desired product without error. Amazingly, the altered microbes mass-produce

UNIT 2   203

Discussion

● **Process Skills:** *Comparing, Generating Ideas*

Ask the students to consider the comment of Earle Harbison, who said that comparing old biology to new biotechnology was like comparing a mule to a tractor. What does he mean? (The students might suggest that Harbison uses the comparison to indicate the kind of swift progress he thinks biotechnology can bring.)

● **Process Skills:** *Inferring, Predicting*

Remind the students that if microbes can be instructed to perform functions that are beneficial to people, then microbes can also be instructed to perform functions that may be dangerous to people. Ask the students to consider how microbes might be used to harm people. (through germ or biological warfare)

● **Process Skills:** *Evaluating, Applying*

You might challenge students interested in mathematics to consider a scenario in which a bacterium takes one minute to duplicate itself. Then ask the students to determine how

---

Microbes can convert waste material into inexpensive fuel.

perfect copies of themselves. A single microbe can become two, then four, then eight, then a billion in less than a day. All that's needed is enough food, which is easily provided.

The first generation of gene-spliced products—nearly all of them medicines—has been a trickle. The next generation will be a flood. The biotechnology industry is hoping eventually to make everyday products like anti-freeze, animal feed, and detergent by the tank load. Earle Harbison, president of a huge chemical company, expects biotechnology to be as important in the 21st century as chemistry and physics have been in the 20th. "To compare the old biology to the new," Harbison says, "is like comparing a mule to a tractor."

Microbes may supply us with clean-burning fuel for our cars. At the University of Florida, microbiologist Lonnie O. Ingram has engineered a new microbe. With a bit of gene-shuffling, Ingram "brainwashed" bacteria into eating the sugars from saw-mill and paper-mill wastes. They eat this waste and produce ethanol, a chemical that can be used as a fuel. With oil prices climbing once again, ethanol could become a cheap alternative for running the family car. Its production could reduce our waste-disposal problems at the same time.

Douglas Dennis, a biologist at James Madison University, in Virginia, has long known about a naturally occurring microbe that produces plastic. Unfortunately, this talented creature thrives only on an expensive sugar diet. So Dennis has spliced the plastic-making part into a microbe that eats whey—a waste product of cheese-making. Plastics made in factories, by contrast, have oil as their raw material. There is even a bonus for Dennis's plastic—every bit of it decomposes.

Microbes' appetites for scraps is one reason manufacturers find them so attractive. Even candy can be made from throwaway ingredients. Since cocoa beans aren't grown in the United States, chocolate makers must pay to import them from the tropics. But scientists have discovered a strange little fungus, related to bread mold, that converts animal fats and vegetable oils into something almost identical to cocoa butter. Unfortunately, the fungus is hard to grow in bulk. So scientists have taken a gene from the fungus and transferred it to a microbe that can mass-produce the "cocoa" all day long.

Some microbes are more prized for what they eat than for what they make. After the

> *Microbes' appetites for scraps is one reason manufacturers find them so attractive.*

many minutes would be required to create one million bacteria. (21 minutes) One billion bacteria? (31 minutes) One trillion bacteria? (41 minutes)

Journal Activity Explain to the students that only some of the future benefits of products derived from microbes were detailed in this article. Encouraging the students to be creative, ask them to predict other kinds of products that might be produced by microbes.

EXTENSION

Because hundreds of large and small biotechnological companies exist in the United States, it is likely that one or more of these companies are located in your state. Ask interested students to research the names of such companies, then write to each company and ask for a description, brochure, or annual report that details and explains the functioning of the company. The collected materials can then be examined and discussed by the entire class.

Exxon Valdez oil spill in Alaska, the Environmental Protection Agency used some naturally occurring bacteria with a craving for oil to help in the cleanup. Early studies suggest that the bacteria may have done as good a job in cleaning some of the lightly soiled beaches as high-pressure hoses and detergents would have.

Often, what nature's hard-working microbes can do well, a custom-made microbe

> *Often, what nature's hard-working microbes can do well, a custom-made microbe can do better.*

can do better. Bacteria with altered genes may someday clean up an oil spill before it has a chance to hit the beach. Ananda Chakrabarty, a microbiologist at the University of Illinois at Chicago, identified bacteria that had a taste for industrial wastes. Chakrabarty implanted genes from them into bacteria suitable for spraying. In powdered form, these new microbes could be sprinkled onto oil slicks that are just starting to spread.

Just 20 years ago the thought of microbes making medicines, plastics, or chocolates would have seemed like science fiction. No one is predicting that microbes will ever be smart enough to do your homework. But in the next century, thanks to the work of biotechnology, the tiny creatures you carefully wash off your hands may be responsible for a new industrial revolution. ◆

Bacteria were used in the Exxon Valdez cleanup.

UNIT 3

SIMPLE LIVING THINGS

UNIT OVERVIEW

This unit discusses the smallest and simplest of all living things—viruses, monerans, protists, and fungi. The characteristics and importance of each organism are presented.

Chapter 8: Viruses and Monerans, page 208

This chapter describes the major characteristics of viruses and monerans, including their structures, shapes, and methods of reproduction. The role of viruses and bacteria as disease-causing agents is explained. Also included is a description of some of the beneficial uses of monerans.

Chapter 9: Protists and Fungi, page 228

This chapter presents the characteristics of protozoans, algae, and fungi, and the criteria by which members of each group are classified. The uses of these organisms and their interaction with the environment and with humans are discussed.

Science Parade, pages 248-255

The articles in this unit's magazine relate the study of viruses, monerans, protists, and fungi to AIDS research, to the invention of the microscope, and to the effects of algal blooms. The career feature describes the work of a public health physician. The biographical sketches focus on a bacteriologist and a biologist.

UNIT RESOURCES

PRINT MEDIA FOR TEACHERS

Chapter 8

Angel, Ann. *Louis Pasteur: Leading the Way to a Healthier World*. Gareth Stevens, 1992. Traces the life of the French scientist who invented pasteurization and founded the field of microbiology.

Nourse, Alan E. *Viruses*. Watts, 1983. Describes the structures of viruses, viral diseases, vaccines, and viral research.

Chapter 9

Greenaway, Theresa. *The First Plants*. Steck-Vaughn, 1991. Presents different kinds and uses of algae.

Lincoff, Gary H. *North American Mushrooms*. Knopf, 1981. A photographic field guide of mushrooms found in North America.

Stidworthy, John. *Simple Animals*. Facts on File, 1990. Introduces invertebrate animals, including protozoa.

PRINT MEDIA FOR STUDENTS

Chapter 8

Nourse, Alan E. *The Virus Invaders*. Watts, 1992. Answers questions regarding the nature of viruses and the role they play in disease.

Chapter 9

Morgan, Nina. *Louis Pasteur*. Bookwright Press, 1992. This biography details Pasteur's life, dreams, family, and accomplishments in chemistry and microbiology.

ELECTRONIC MEDIA

Chapter 8

Bacteria. Videocassette. National Geographic. Examines the forms of bacteria, the structure of a bacterial cell, its reproductive processes, and the importance of bacteria in the world.

Viruses: What They Are and How They Work. Videocassette. Britannica. Explains the structure and diversity of viruses, the viral life cycle, uses of viruses, and disease-producing viruses.

Chapter 9

Algae: The Food Chain. Videocassette. American School Publishers. This program looks at the food chain, classification, life cycle, and cell structure of algae.

Fungi: Activity and Structure. Videocassette. American School Publishers. This program explores life cycles, sexual and asexual reproduction, symbiotic relationships, and classification of fungi.

Protists: Form, Function, and Ecology. Videocassette. Britannica. The world of protists is explored using light electron microscopes.

UNIT 3

Discussion Explain to the students that Earvin "Magic" Johnson, Jr., was a professional basketball player whose career ended prematurely because he contracted the virus that causes AIDS, or Acquired Immune Deficiency Syndrome. Encourage the students to discuss the things they may have read or heard about the disease known as AIDS. Because many misconceptions and misunderstandings about AIDS exist, help the students separate fact from myth during the course of their discussion.

UNIT 3 SIMPLE LIVING THINGS

CHAPTERS

8 Viruses and Monerans 208

Despite your size and today's technology, your body may be no match against tiny foes such as viruses and monerans.

9 Protists and Fungi 228

The work of protists and fungi helps maintain the basic cycles of life.

Journal Activity Extend the discussion by explaining to the students that AIDS is an indiscriminate killer; anyone can acquire the deadly virus if he or she is exposed. Emphasize that AIDS is a disease that transcends all races, creeds, cultures, and classes of people. Ask the students to think of several questions they have about AIDS and record those questions in a journal entry. After the Unit has been completed, have the students review their questions and see whether they can answer any based on their studies. If the answer to any question is beyond the scope of the material covered in the Unit, challenge the students to use additional reference materials to find an answer and share those findings with their classmates.

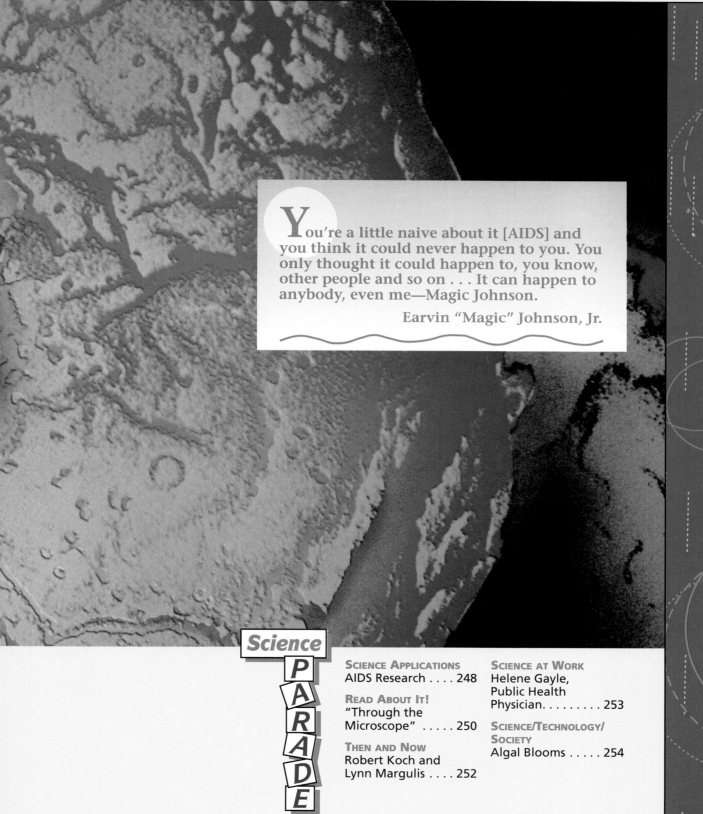

> You're a little naive about it [AIDS] and you think it could never happen to you. You only thought it could happen to, you know, other people and so on . . . It can happen to anybody, even me—Magic Johnson.
>
> Earvin "Magic" Johnson, Jr.

Science PARADE

SCIENCE APPLICATIONS
AIDS Research 248

READ ABOUT IT!
"Through the Microscope" 250

THEN AND NOW
Robert Koch and Lynn Margulis 252

SCIENCE AT WORK
Helene Gayle, Public Health Physician. 253

SCIENCE/TECHNOLOGY/ SOCIETY
Algal Blooms 254

UNIT 3

CHAPTER 8
VIRUSES AND MONERANS

PLANNING THE CHAPTER

Chapter Sections	Page	Chapter Features	Page	Program Resources	Source
Chapter Opener	208	*For Your Journal*	209		
Section 1: VIRUSES	210	Discover By Researching **(B)**	214	*Science Discovery**	SD
• The Microscopic World of Viruses **(B)**	210	Discover By Writing **(A)**	215	Investigation 8.1: Making Models of Viruses **(A)**	TR, LI
• Characteristics of Viruses **(B)**	211	Section 1 Review and Application	216	Reading Skills: Making a Chapter Outline **(A)**	TR
• Viruses and Diseases **(B)**	213	Skill: Identifying Viral Variables **(A)**	217	Connecting Other Disciplines: Science and Social Studies, Modern Medicine and Society **(A)**	TR
				Shapes of Viruses **(A)**	IT
				Reproduction of Viruses **(A)**	IT
				Study and Review Guide, Section 1 **(B)**	TR, SRG
Section 2: MONERANS	218	Discover By Calculating **(H)**	219	*Science Discovery**	SD
• Characteristics of Monerans **(B)**	218	Activity: What microorganisms cause various diseases? **(A)**	220	Investigation 8.2: Observing Cyanobacteria **(H)**	TR, LI
• Bacteria—Common Monerans **(B)**	219	Discover By Doing **(B)**	222	Thinking Critically **(H)**	TR
• Cyanobacteria **(A)**	222	Section 2 Review and Application	223	Extending Science Concepts: Calculating Bacterial Growth **(H)**	TR
		Investigation: Testing Disinfectants **(A)**	224	Record Sheets for Textbook Investigations **(A)**	TR
				Study and Review Guide, Section 2 **(B)**	TR, SRG
Chapter 8 HIGHLIGHTS	225	The Big Idea	225	Study and Review Guide, Chapter 8 Review **(B)**	TR, SRG
Chapter 8 Review	226	For Your Journal	225	Chapter 8 Test	TR
		Connecting Ideas	225	Test Generator	

B = Basic   **A** = Average   **H** = Honors
The coding Basic, Average, and Honors indicates subsections, features, and resources that might be appropriate for different levels of learners. For additional suggestions regarding choice of topic and depth of coverage, see the Pacing Chart on pages T26–T29.

*Frame numbers at point of use
(TR) Teaching Resources, Unit 3
(IT) Instructional Transparencies
(LI) Laboratory Investigations
(SD) *Science Discovery* Videodisc Correlations and Barcodes
(SRG) Study and Review Guide

CHAPTER MATERIALS

Title	Page	Materials
Discover By Researching	214	(per class) reference materials
Discover By Writing	215	(per class) reference materials (per individual) journal
Skill: Identifying Viral Variables	217	none
Discover By Calculating	219	(per individual) paper, pencil
Activity: What microorganisms cause various diseases?	220	(per group of 3 or 4) unlined paper, colored pencils
Teacher Demonstration	221	bouillon cubes (2), water, beakers (3), salt, sugar
Discover By Doing	222	(per group of 3 or 4) small amount of pond scum, microscope, paper, pencil
Investigation: Testing Disinfectants	224	(per group of 3 or 4) safety goggles, laboratory apron, agar-filled Petri dish, wax pencil, cotton swab, filter-paper disks (4), forceps, disinfectant solutions

ADVANCE PREPARATION

For the *Demonstration* on page 221, you will need to obtain bouillon cubes. For the *Discover By Doing* on page 222, samples of pond scum are needed for the activity. For the *Investigation* on page 224, obtain agar, Petri dishes, and filter-paper disks from a biological supply house.

TEACHING SUGGESTIONS

Field Trip
Plan a field trip to a local hospital. Ask whether various staff members of the hospital can show and explain procedures used in each area of the hospital to control diseases and infections and create a sterile hospital environment. Surgery rooms, communicable disease wards, and food-preparation areas provide excellent examples of such controls.

Outside Speaker
Invite the school nurse or other approved speaker to conduct a question-and-answer session about AIDS. Provide a question box for anonymous questions several days before the scheduled visit. Give the questions to the speaker in time to allow adequate preparation.

A speaker from the local health department could also be asked to speak to the class about how bacteria affect food. Topics might include spoilage or food poisoning.

CHAPTER 8
VIRUSES AND MONERANS

CHAPTER THEME—SYSTEMS AND STRUCTURES

In this chapter, the size and characteristics of viruses and monerans are discussed. The students will discover that, without the use of a microscope, viruses and most monerans are invisible because they are so tiny. The students will also learn that, despite their small size, viruses and monerans can have a huge impact on all organisms. **Systems and Structures** is also developed in Chapters 4, 5, 7, 9, 10, 12, 13, 16, 17, 18, 19, and 20. A supporting theme in this chapter is **Environmental Interactions**.

MULTICULTURAL CONNECTION

The growth of microbiology, specifically bacteriology and virology, as a science has been the result of the efforts of scientists from diverse cultural backgrounds. Suggest that the students report on the contributions of one or more of these or other scientists: Anton van Leeuwenhoek, Edward Jenner, Louis Pasteur, Robert Koch, Dmitri Ivanovski, Martinus Beijerinck, Giuseppe Sanarelli, Carlos Finlay, Shibasaburo Kitasato, Max Theiler, Masanori Ogata, Jonas Salk, Alexander Fleming, and so on. Suggest that the students place a flag on a world map to identify the nation from which the scientist they researched came. Following the students' reports, discuss the international character of scientific endeavor.

MEETING SPECIAL NEEDS

Second Language Support

Suggest that students with limited English proficiency keep a glossary of English terms with which they are unfamiliar. Each glossary entry might include the English term, the term's phonetic pronunciation, its definition, and a comparable term from the students' first language. The students can then use their glossary as a study tool.

CHAPTER 8
VIRUSES AND MONERANS

In the 1300s, about one-fourth of the European population died from the bubonic plague. No one knew at the time that the disease was caused by a moneran, a microscopic bacterium. Today, people have a better understanding of microscopic organisms, such as viruses and bacteria, and their impact on people.

▶ 208 CHAPTER 8

CHAPTER MOTIVATING ACTIVITY

Pour a small container of orange juice with pulp through a coffee filter into a glass. Ask the students to tell what they observed. (The juice went through the filter whereas the pulp remained in the filter.) Explain that using a similar technique, scientists first discovered that microbes smaller than bacteria existed. These scientists hypothesized that a plant disease was caused by bacteria and that by filtering out the bacteria and applying the remaining substance to a plant, they could prove their hypothesis if the plants remained healthy. When the plants became diseased even after the filtering process, the scientists determined that microbes smaller than bacteria existed.

For Your Journal

The students' answers to the journal questions should demonstrate their understanding of disease-causing microorganisms and health. After the students have completed their journal entries, encourage them to share their ideas with the class. The discussion may help you detect any misconceptions that the students may have about microorganisms and their impact on all other organisms. You may wish to address any misconceptions at appropriate times during the study of this chapter.

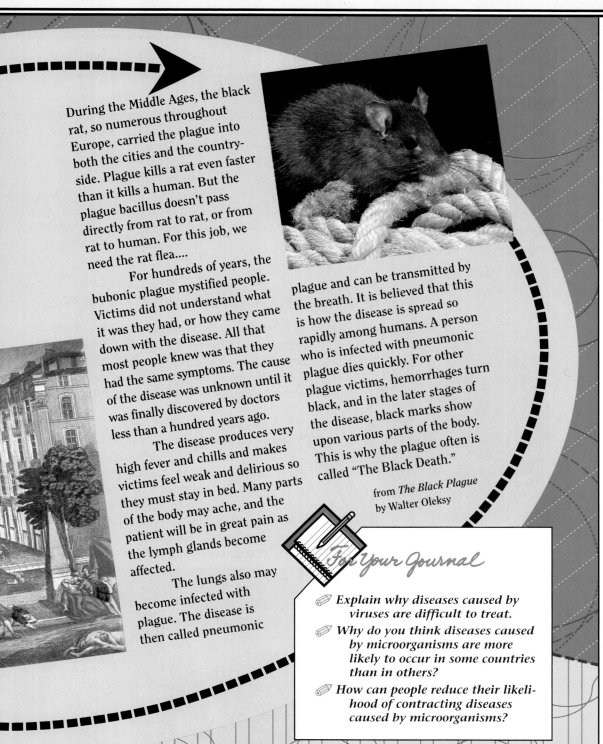

During the Middle Ages, the black rat, so numerous throughout Europe, carried the plague into both the cities and the countryside. Plague kills a rat even faster than it kills a human. But the plague bacillus doesn't pass directly from rat to rat, or from rat to human. For this job, we need the rat flea....

For hundreds of years, the bubonic plague mystified people. Victims did not understand what it was they had, or how they came down with the disease. All that most people knew was that they had the same symptoms. The cause of the disease was unknown until it was finally discovered by doctors less than a hundred years ago.

The disease produces very high fever and chills and makes victims feel weak and delirious so they must stay in bed. Many parts of the body may ache, and the patient will be in great pain as the lymph glands become affected.

The lungs also may become infected with plague. The disease is then called pneumonic plague and can be transmitted by the breath. It is believed that this is how the disease is spread so rapidly among humans. A person who is infected with pneumonic plague dies quickly. For other plague victims, hemorrhages turn black, and in the later stages of the disease, black marks show upon various parts of the body. This is why the plague often is called "The Black Death."

from *The Black Plague*
by Walter Oleksy

ABOUT THE LITERATURE

The Black Plague by Walter Oleksy is a nonfiction account of the history of the bubonic plague from ancient times through the 1980s. It focuses mainly on the European epidemic during the Middle Ages and its aftermath. The text explores how the Europeans reacted to outbreaks of the "Black Death." It also highlights the scientific endeavors that led to the discovery of the bacteria that cause plague and how the plague is spread.

For Your Journal

- Explain why diseases caused by viruses are difficult to treat.
- Why do you think diseases caused by microorganisms are more likely to occur in some countries than in others?
- How can people reduce their likelihood of contracting diseases caused by microorganisms?

Section 1: VIRUSES

FOCUS

This section discusses the characteristics of viruses and identifies how viruses resemble living and nonliving things. It includes information about the harmful effects that viruses can have on hosts and the role of vaccines in preventing viral diseases.

MOTIVATING ACTIVITY

Cooperative Learning Display a house plant and a nonliving object, such as a book. Have the class work in groups of three or four students to identify similarities and differences between the two items displayed. Ask the groups to focus their comparisons on the living and nonliving nature of the items. After the groups have completed their lists, have them share the comparisons with the class. Encourage the students to identify characteristics that distinguish living things from nonliving things.

PROCESS SKILLS
• Comparing • Synthesizing

POSITIVE ATTITUDES
• Enthusiasm for science and scientific endeavor • Curiosity

TERMS
• virus • host • epidemic
• immune system • vaccine

PRINT MEDIA
Viruses by Alan Nourse
(see p. 205b)

ELECTRONIC MEDIA
Viruses: What They Are and How They Work, Britannica
(see p. 205b)

Science Discovery
Thinking circle

BLACKLINE MASTERS
Study and Review Guide
Laboratory Investigation 8.1
Reading Skills
Connecting Other Disciplines

SCIENCE BACKGROUND

The common cold can be caused by dozens of different viruses. The chief viruses that cause colds are called *rhinoviruses*. Other viruses that can cause coldlike symptoms are *adenoviruses*. Although there is no cure for the common cold, modern medicines can treat the symptoms. A cold usually clears up after several days.

SECTION 1

Viruses

Objectives

Summarize the characteristics of viruses.

Compare viruses to living and nonliving things.

Identify five diseases caused by viruses.

Headaches, chills, fevers, aches, and pains—you think you have the flu. If you've ever had the flu, you've probably had viruses inside you similar to that pictured here. When compared to other viruses, influenza viruses are not particularly large, nor are they particularly small. To get some idea of what size an influenza virus really is, let's compare one to a penny. You would need about 950 000 of these viruses laid end to end just to reach across one side of a penny!

The Microscopic World of Viruses

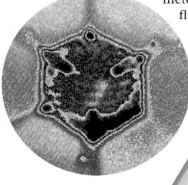

The size of viruses is measured in nanometers, or billionths of a meter. Suppose it were possible for you to lay 950 000 influenza viruses end to end across the surface of a penny at the rate of one virus each second. It would take you 950 000 seconds—about 11 straight days and nights—to complete the task!

If you think that 950 000 isn't a very large number, and that viruses really aren't very small, consider a different question: How many influenza viruses laid end to end could you fit along the length of a new pencil? This time the answer is a larger number—it would take about 9 500 000 viruses! Placing one virus per second, it would take you about 110 days, or almost four months, working day and night, to complete such a task!

▶ 210 CHAPTER 8

TEACHING STRATEGIES

● **Process Skills:** *Observing, Comparing*

Hold up a pencil with a sharp point. Tell the students that if the pencil point were increased in size 3400 times, its width would be about 3.4 m long. Cut a string or a strip of adding machine paper 3.4 m long. Then explain that if one of the largest viruses was increased in size 3400 times, its width would be about that of the pencil point. Have the students hold up the strip of paper and the pencil point. Encourage the students to discuss the relative sizes.

● **Process Skills:** *Classifying/Ordering, Inferring*

Have the students recall the names of the five kingdoms of classification. Write the names on the chalkboard. Ask the students in which kingdom viruses should be placed. (Viruses do not belong in any of the kingdoms.) Ask the students to explain why viruses are not included in the classification system. (Viruses are not living cells. Yet, once they enter a living cell, they can reproduce. As a result, they do not exhibit characteristics commonly associated with any of the kingdoms.)

Let's try one more question: How many influenza viruses laid end to end could you fit along Earth's equator? This time the answer truly is a large number—it would take about two quadrillion viruses. Quadrillion is a number containing 15 zeros. It would take you about one million lifetimes to finish laying two quadrillion viruses end to end!

The influenza virus is an average-sized virus. Other viruses are much smaller, like the virus that causes hoof-and-mouth disease in cattle. Still other viruses are much larger, like the human smallpox virus. Regardless of their size, viruses can pose a threat to you and your health.

Did you know that other animals and even plants can also become the victims of a virus attack? However, the viruses that attack plants do not attack people, and the viruses that attack people do not attack plants. You will never see a tree suffering from chicken pox!

▼ **ASK YOURSELF**

What size are viruses?

Characteristics of Viruses

A-a-a-choo! Your head aches, you have the sniffles and a cough, and your throat tickles. You have a cold, and you feel miserable. Did you know the common cold is caused by a virus? When it invades your body, the virus multiplies and soon you show all the symptoms of the common cold. A **virus** is a very small particle that can reproduce only inside a living cell. When a virus reproduces inside the living cells of your body, you don't feel too well.

If you could look at viruses using a very powerful microscope, you would see many different shapes. Some viruses are spherical, some are tube shaped, some look like miniature spaceships, and others look as though they are made from sticks and spools! Viruses do not look like any other organism. Besides unusual shapes, they have other unusual characteristics.

Dead or Alive?
Viruses are neither living things nor nonliving things, yet they have some characteristics of both. Viruses are like nonliving things in several ways. They do not grow or respond to outside stimuli, and they are not cells. Viruses cannot live on their own. They need to invade an organism in order to reproduce. Although viruses are able to reproduce, they can do so only inside a living cell.

Figure 8-1. The white fly carries a virus that attacks crop plants such as lettuce.

THE NATURE OF SCIENCE

Viruses were discovered in the 1890s by a Russian researcher and a Dutch botanist who were independently trying to isolate the cause of tobacco mosaic disease. Both scientists assumed the disease was caused by bacteria. However, plant fluid that they had poured through a filter with pores small enough to trap bacteria still caused the disease. The scientists theorized that the disease was caused by some infectious agent smaller than bacteria. The Dutch scientist, Martinus Willem Biejerinck, called the infectious agent a *virus*, which was taken from a Latin word meaning "poison." The term *virus* was once used to describe any disease-causing agent, but the term is now applied only to the microscopic agents that cause viral diseases.

 Although by the early 1930s scientists had discovered many different viruses, they were not able to observe the viruses directly. Most were much too small to be seen even with the most powerful light microscope. Then the electron microscope was developed in the 1930s. Improved electron microscopes could magnify objects more than 100 000 times. Finally, scientists had a tool that they could use to observe viruses.

ONGOING ASSESSMENT
▼ **ASK YOURSELF**

Viruses are very tiny and can only be seen through powerful microscopes.

TEACHING STRATEGIES, continued

- **Process Skills:** *Constructing/Interpreting Models, Applying*

Cooperative Learning Have the students work with partners to construct a model of a virus. Suggest that the partners find out information about a particular virus and construct a model of that virus. Have the partners display the models and discuss the variety of shapes displayed. Remind the students that the outer protein coating of viruses is responsible for the shape of all viruses.

- **Process Skills:** *Comparing, Evaluating*

Have the students look up the definition of the noun *host* in their dictionaries. Based on the definitions, the students should suggest the similarities and differences between a host to guests and a host to viruses. (A host entertaining guests usually welcomes them and enjoys visiting with them. A host to viruses does not want to have the virus and does not enjoy being sick.)

MEETING SPECIAL NEEDS

Gifted

Tell the students that a virus's nucleic acid is the part of the virus that takes over the host cell's activities and causes it to produce new viruses. Explain that the nucleic acid in most viruses is DNA, and the nucleic acid in others is RNA. Ask the students to research and write a report about the differences in the ways that these different types of viruses reproduce in the host cell. Have the students place their completed reports in their science portfolios.

ONGOING ASSESSMENT
ASK YOURSELF

The students are likely to suggest that since viruses cause diseases in organisms, organisms are usually not willing hosts. Some students may suggest that vaccine recipients are willing hosts to weakened forms of viruses.

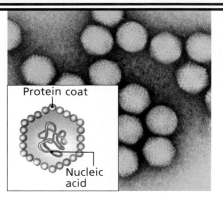

Figure 8–2. Although these viruses have different shapes, each virus is made of a core of nucleic acid surrounded by a protein coat.

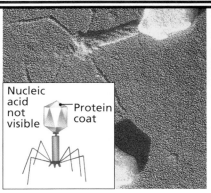

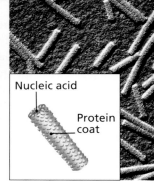

Viruses are similar to living things because they contain proteins and nucleic acids. The cells of your body also contain proteins and nucleic acids. *Nucleic acid* is the material that carries genetic, or hereditary, instructions. The nucleic acid is found in the center of the virus, and protein forms a coat around it. This protein coat is responsible for the many different shapes of the viruses shown in the illustrations.

Public Enemies Viruses are considered hostile because they invade and then attack the cells of living organisms. An organism that is invaded by a virus is called a **host**. Plants, animals, monerans, protists, and fungi all serve as hosts to different kinds of viruses. Once a virus invades a host cell, the virus can begin to produce new viruses. This occurs when the attacking virus releases its nucleic acid into the host cell. The host cell cannot tell the difference between its own nucleic-acid instructions and those of the virus. So the host cell begins following the virus's instructions to make new viruses.

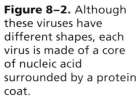

Figure 8–3. Viruses reproduce in a host cell until the cell can no longer hold them and bursts.

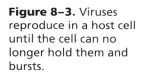

If you overfill a balloon with air, it bursts. The same thing happens to cells invaded by viruses. Eventually, the host cell fills with viruses and bursts, the new viruses are released, and they then attack other cells.

ASK YOURSELF

Is an organism ever a willing host to a virus? Why or why not?

- **Process Skills:** *Communicating, Interpreting Data*

Have the students raise their hands if they have had chicken pox. Ask the students to determine the percentage of class members who have had chicken pox. In their journals, the students should then plot that information on a bar graph. Follow the same procedure for students who have had common colds and influenza. Ask the students to use their graphs to help them make generalizations about these viral diseases. (The students are likely to determine that most, if not all, of the class members have suffered from these viral infections.) Ask the students why such a high percentage of the class has had these viral infections. (The viruses are easily transmitted from one person to another.)

- **Process Skills:** *Comparing, Inferring*

Ask the students why more people have probably suffered from the common cold than have had warts. (The students may suggest that viruses causing warts are not as infectious and cannot be transmitted as easily as those causing colds.)

Viruses and Disease

Viruses are present everywhere in the world around you. The fever blister that seems to develop just in time for school pictures is caused by a virus. Even a wart is the result of a virus invading your body.

Different viruses can invade the body in different ways. Many are airborne and are breathed in by a person. Once in the body, the viruses can begin to reproduce and to cause disease. Sometimes the effects of an invading virus create nothing more than a nuisance to the host. If you have a cold, your body is a host for invading viruses, but these viruses usually cause discomfort only for a short period of time. However, some invading viruses can cause much more serious illness. Did you have the chicken pox when you were younger? Remember how the pox itched? The chicken pox are caused by a virus. Other once-common childhood diseases, such as measles and mumps, are also caused by viruses.

Viruses also cause polio, a disease that can result in paralysis. Polio once swept through countries and communities in epidemics. An **epidemic** is the rapid spread of a disease through a large area. Older family members or friends may remember polio epidemics that swept the United States. Even a President of the United States was a victim of this disease. Franklin D. Roosevelt, the thirty-second President, spent most of his adult life in a wheelchair and wore leg braces because he had had polio. But he battled the disease and became a great success and an inspiration to others. Polio epidemics no longer occur because most people are vaccinated against the disease. In the next activity, you can find out more about how epidemics have been stopped.

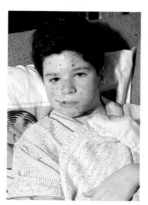

Figure 8–4. Chicken pox is an infectious disease that is caused by a virus.

Figure 8–5. Franklin D. Roosevelt, who served as President from 1933 to 1945, battled polio courageously.

> ✧ **Did You Know?**
> Chicken pox and flu viruses can trigger a rare childhood disease that kills from 3 to 5 percent of its victims. Reye's syndrome is a disease of the central nervous system and liver. Severe cases can result in brain damage or even death. Because many victims of Reye's syndrome were given aspirin to treat the symptoms of chicken pox or flu, doctors caution against the use of aspirin for treating viral diseases in children and young people.

INTEGRATION—Social Studies

Franklin Delano Roosevelt was stricken by polio in 1922 at the age of 39. He overcame the most severe effects of his disease and devoted himself to public service. As President, he led the country through the difficult years of the Great Depression and most of World War II until his death in 1945. Roosevelt was the only President to be elected to four terms in office. When he sought the presidency for the third time, he was breaking the two-term precedent established by George Washington. Following his death, the Twenty-Second Amendment to the Constitution was passed. It states that no person may be elected to the office of the presidency more than twice.

TEACHING STRATEGIES, continued

● **Process Skills:** *Applying, Expressing Ideas Effectively*

Have the students look up the definition of each word that comprises the acronym *AIDS*: acquired immune deficiency syndrome. (acquired: to get, to have; immune: protected from, free from; deficiency: a lack of something; syndrome: group of signs and symptoms) Ask the students to think about these separate definitions and then to describe AIDS in their own words.

● **Process Skills:** *Inferring, Analyzing*

Explain to the students that AIDS is caused by the HIV viruses. These viruses invade lymphocytes, which are cells that help the body fight disease. The viruses redirect the functions of the lymphocytes to produce new viruses. Ask the students to use this information to explain why AIDS victims cannot fight minor diseases, such as colds or flu. (The lymphocytes can no longer fight infections because they are now producing viruses rather than fighting diseases.)

DISCOVER BY *Researching*

To help the students begin their research, you may wish to direct them to look up smallpox and polio in their encyclopedias. The information provided can help the students outline their research and give clues to additional resources. Also help guide the students' research by suggesting that they look up the names Edward Jenner, Jonas E. Salk, and Albert B. Sabin in the encyclopedia.

After the students have completed their research for the *Discover by Researching* activity, have them write a report and place the report in their science portfolios.

SCIENCE BACKGROUND

The immune system protects the body from limitless numbers of foreign materials. Lymphocytes, or white blood cells, play key roles in the immune response. One type of lymphocyte produces antibodies when a foreign substance enters the body. After the antibodies inactivate a foreign substance, some of the lymphocytes remain in the body. These lymphocytes are called *memory cells*. If the same foreign substance enters the body again, the memory cells produce large amounts of the antibody against it.

 **DISCOVER BY** *Researching*

The stories that describe successful battles against epidemics can be exciting and inspiring. Use reference materials to discover more about our successes in halting the spread of epidemics, such as smallpox or polio epidemics, and share what you have learned with your classmates.

A Deadly Viral Disease In recent years, acquired immune deficiency syndrome (AIDS) has received a great deal of attention. AIDS is caused by a virus and is a serious infection that destroys the body's immune system. A healthy **immune system** provides the body with the ability to fight infection. The AIDS virus weakens the immune system so much that disease-causing organisms become life threatening.

People who have AIDS die from diseases that noninfected people can usually resist. A simple infection, for example, may prove deadly to an AIDS patient whose body can no longer stop the infection from spreading.

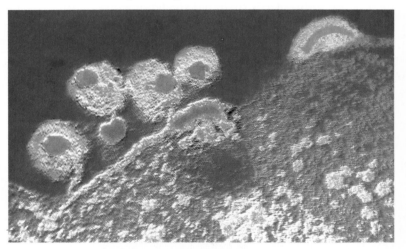

Figure 8-6. These AIDS viruses, shown as red and yellow spheres, create a disease that results in many different symptoms and is always fatal.

AIDS was first recognized in the early 1980s. The virus has been found in semen (the fluid carrying sperm), blood, saliva, and tears. However, the only known transmissions of the disease have been through semen and blood. Scientists know that AIDS is not transmitted by casual contact, such as shaking hands with an infected person or being in the same room with him or her. Sexual contact, contaminated blood products, and sharing of contaminated needles, however, can and do transmit the AIDS virus from person to person.

- **Process Skills:** *Inferring, Expressing Ideas Effectively*

In the past, some recipients of blood transfusions contracted AIDS. Today, blood is carefully screened to detect the presence of the AIDS antibodies. Have the students explain why the screening is necessary. (Protecting blood-transfusion recipients from the deadly virus makes screening blood supplies essential.)

- **Process Skills:** *Formulating Hypotheses, Inferring*

In most states, children are required to provide medical records that prove they have been vaccinated against such diseases as polio before they are permitted to enter school. Ask the students to hypothesize reasons for this requirement. (The requirements are designed to protect the students from infectious diseases.)

GUIDED PRACTICE

Write the following terms on the chalkboard: virus, host, epidemic, immune system, and vaccine. Ask the students to define the terms and to discuss how the terms are related to one another.

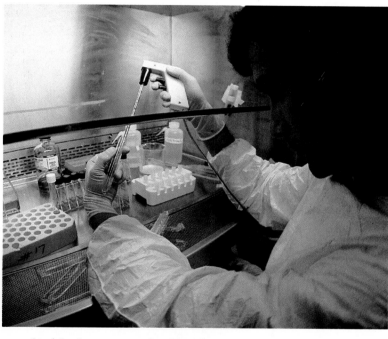

At this time, a cure for AIDS has not been discovered. Vaccines are not yet available to prevent people from getting this disease. However, great effort and millions of dollars are being spent to find a way to prevent this fatal disease. For more information about AIDS, see page 248 at the end of the unit.

Stopping Viral Infections Any illness caused by a virus is difficult to treat or cure. Because viruses are not living, they are very difficult to destroy. Since treatment of viral diseases is difficult, great energy is spent trying to prevent them. A **vaccine** is a substance given to people to keep them from getting a disease. You have probably had several vaccines against diseases such as measles, mumps, and polio. In the following activity, you can learn about vaccines and their makeup.

Discover By Writing

Have you ever wondered what a vaccine was made of or how it worked? When you receive a vaccination against polio, for example, you actually receive a dose of weakened viruses. Use reference materials to find out why people seldom contract a disease after receiving a vaccination against that disease. Summarize the information in your journal.

Figure 8–7. Research to find a cure for AIDS is ongoing.

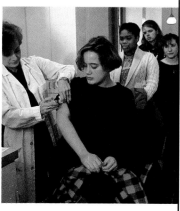

Figure 8–8. Today, vaccines are available to prevent diseases such as mumps and measles.

BACKGROUND INFORMATION

Louis Pasteur, a French scientist who had studied chemistry, made major contributions to medicine and industry. He developed the process of pasteurization, or the use of heat to kill microbes. Pasteur proved that many diseases were caused by microbes and that vaccinating animals with weakened microbes enabled the animals to develop immunity to the microbes. Pasteur first used his rabies vaccine on a small boy who had been bitten by a rabid animal. When the boy did not get rabies, Pasteur knew his new vaccine was effective.

LASER DISC
477

Thinking circle

DISCOVER BY *Writing*

To reinforce the students' research, you may wish to invite a health-care worker to speak to the class about vaccinations. Encourage the students to write questions in their journals that they would like to ask the speaker.

SECTION 1 215

TEACHING STRATEGIES, continued

● **Process Skills:** *Analyzing, Formulating Hypotheses*

Edward Jenner's discovery of the smallpox vaccination was the result of a scientific method in studying the smallpox disease. Ask the students to outline the method used by Jenner and to suggest the hypothesis Jenner might have formed. (If people are infected with the cowpox virus, they will not contract the smallpox virus.)

INDEPENDENT PRACTICE

Have the students answer the Section Review and Application questions in their journals. Then ask them to write a paragraph identifying the characteristics of viruses and relating those characteristics to the difficulty of curing diseases caused by viruses.

EVALUATION

Have the students discuss why vaccines against viral diseases are important. (The students should relate the need to prevent the viral diseases to the difficulty of curing these diseases.)

ONGOING ASSESSMENT
ASK YOURSELF

Unlike living things, viruses do not grow or respond to outside stimuli, they are not cells, and they cannot reproduce except in a living cell. Like living things, viruses are composed of proteins and nucleic acids.

SECTION 1 REVIEW AND APPLICATION

Reading Critically

1. The cells of living things contain proteins and nucleic acids, and so do viruses. But unlike living things, viruses are not cells, do not grow or respond to outside stimuli, and cannot reproduce except in living cells.

2. Viral diseases include chicken pox, measles, mumps, polio, and AIDS.

3. AIDS destroys the immune system and is always fatal.

Thinking Critically

4. Since viruses can only reproduce in a host, the chicken pox virus would not reproduce. The virus would be eliminated in that one generation.

5. It is now known that warts are caused by viruses. A toad's warts are a natural physical feature. Since they are not caused by viruses, people cannot get these warts by touching toads.

6. An epidemic would severely tax the health-care system, might disable or kill thousands of people, and might be difficult to contain.

▶ **216** CHAPTER 8

Table 8-1 Viral Diseases for Which Vaccines Exist

Hepatitis A and B	Measles
Influenza	Mumps
Polio	Rabies
Rubella (German measles)	

In 1796, Edward Jenner observed that people did not get smallpox, a deadly disease, if they first had cowpox, a minor disease. Jenner conducted an experiment in which he used material from a cowpox sore as a vaccine against smallpox. His experiment was successful, and since Jenner's time, billions of people worldwide have been vaccinated against smallpox. As a result, there have been no cases of smallpox anywhere in the world since 1981.

Figure 8-9. Since the discovery of a smallpox vaccine by Edward Jenner, vaccines against many other viruses have been developed.

 ASK YOURSELF

How are viruses like and unlike living organisms?

SECTION 1 REVIEW AND APPLICATION

Reading Critically
1. How are viruses like living things? How are they different?
2. List several diseases that are caused by viruses.
3. Why is AIDS such a serious disease?

Thinking Critically
4. What would happen if one generation of chicken pox viruses never found a host?
5. At one time, some people believed they could get warts by touching a toad. What is now known about viruses that indicates that this is probably not true?
6. Describe several ways in which an epidemic would pose serious health concerns.

RETEACHING

 Cooperative Learning Have the students work with partners to draw a small poster showing how a virus invades a cell, reproduces, and leaves a cell to invade others. To reinforce their visuals, the students should label each stage shown on their posters.

EXTENSION

Have the students find out information about how scientists use viruses for such purposes as developing vaccines and controlling insects.

CLOSURE

 Have the students write the key terms in their journals and then use these terms in a brief summary of the section's content.

SKILL

Identifying Viral Variables

Process Skills: Inferring, Analyzing

Grouping: Groups of 2 or 3

Objectives
- **Identify** the characteristics of viral infections.
- **Analyze** the characteristics of an infection to determine the viral disorder.

Discussion

The emphasis of this activity is on relating symptoms of several viral diseases to specific case studies. As a presentation alternative, you may wish to have several students role play the symptoms and have a panel of students diagnose the specific viral disease.

▶ **Application**

Knowledge of common health problems may enable parents to identify an illness in their children. People should not diagnose diseases without the help of healthcare practitioners because many diseases have similar symptoms and can be easily misdiagnosed. In some foreign countries, diseases may be common that are uncommon in the home country of a traveler.

✳ **Using What You Have Learned**

The students might suggest that the practices of veterinarians and physicians are similar. Each perform tests to isolate an illness. However, people, unlike animals, are able to tell a physician symptoms that are bothering them, thus helping to eliminate possibilities.

SKILL Identifying Viral Variables

▼ PROCEDURE

1. Study the table of various health problems and symptoms caused by viruses.
2. Examine each case study carefully, then infer from the symptoms which health problem might be involved.

Case Study 1: Patient felt nauseated, had a fever, felt weak, and ached all over. Several other family members felt the same way.

Case Study 2: Patient had a fever and painful facial swelling. Patient also had a headache and muscle aches.

Case Study 3: Patient had a headache and fever. Irritating spots, which turned into blisters and scabs, appeared on the patient's cheeks, arms, and legs.

Case Study 4: Patient suddenly experienced a fever, a splitting headache, and nausea. A student at the patient's school was hospitalized earlier that day with the same symptoms.

TABLE 1: SOME HEALTH PROBLEMS CAUSED BY VIRUSES

HEALTH PROBLEM	SYMPTOMS
Common cold	Runny nose; slight fever; chilly sensations; sore throat
Influenza	Headaches; nausea; contagious; similar to a very bad cold; fever; general weakness
Meningitis	Inflammation of membranes covering the brain and spinal cord; sudden fever; painful headache; vomiting; contagious; potentially fatal if untreated
Chicken pox	Crops of spots appearing on trunk, then on face and limbs; rash blisters and scabs; fever; headache
Mumps	Painful swelling of glands in cheeks; fever; headache; muscle aches

▶ **APPLICATION**

How can a knowledge of common health problems and their symptoms be helpful to a parent? Why is it important that people *not* make diagnoses of diseases without the help of medical professionals? Explain why a traveler to foreign countries is sometimes required to receive one or more immunization shots before departure.

 Using What You Have Learned

Just like people, pets sometimes do not feel well. When you don't feel well, a family member or doctor will usually ask you what's wrong or where it hurts. Pets don't have the ability to speak. How do you think the practices of a veterinarian would be similar to or different from the practices of your doctor?

SECTION 1 **217** ◀

Section 2:
MONERANS

FOCUS

This section discusses characteristics of monerans. The three general shapes of bacteria are described, and the conditions in which bacteria can survive are noted. The harmful and helpful actions of bacteria are also covered. The characteristics of cyanobacteria are introduced. The usefulness of cyanobacteria is also identified.

MOTIVATING ACTIVITY

Make a thin mixture of water and plain yogurt. Place a drop of the yogurt mixture on a slide and add a drop of methylene blue to the slide. Cover the slide with a coverslip, and have the students view it under a microscope. Ask the students what they see. Explain that the rod-shaped organisms they see are bacteria. These bacteria make yogurt by fermenting milk.

PROCESS SKILLS
• Comparing • Interpreting Data

POSITIVE ATTITUDES
• Enthusiasm for science and scientific endeavor
• Cooperativeness

TERMS
• monerans • antibiotics
• pigments

PRINT MEDIA
Louis Pasteur
by Ann Angel
(see p. 205b)

ELECTRONIC MEDIA
Bacteria,
National Geographic
(see p. 205b)

Science Discovery
Bacteria;
classification
Cyanobacterium; in geysers

BLACKLINE MASTERS
Study and Review Guide
Laboratory Investigation 8.2
Thinking Critically
Extending Science Concepts

✧ **Did You Know?**
A person's throat is sore when he or she has strep throat because the streptococci survive by decomposing the tissue in the throat.

REINFORCING THEMES—
Systems and Structures

When discussing the characteristics of bacteria, compare their features to those of viruses. Emphasize the relative differences in size, shape, structure, and reproductive methods of viruses and bacteria.

▶ 218   CHAPTER 8

SECTION 2

Monerans

Objectives

Compare and contrast viruses and bacteria.

Identify five diseases caused by bacteria.

Summarize the effects of bacteria on humans.

Although bacteria are microorganisms, they are enormous in size when compared to viruses. The size of viruses is measured in billionths of a meter; the size of bacteria is measured in thousandths of a meter. While they are much larger than viruses, bacteria are still too small to be seen without a microscope.

Recall that the conditions required for the reproduction of viruses are limited—they need to be inside living cells. Bacteria are much different—some bacteria can exist rather comfortably in a cup of steaming hot coffee!

Characteristics of Monerans

You just learned your best friend has strep throat. Now your throat is scratchy. You have a fever, and your tonsils are swollen. You may have strep throat, too. If you do, you have an infectious disease caused by bacteria. *Bacteria* are a common example of monerans. **Monerans** are organisms that do not have a nucleus but do have a cell wall. Bacteria are commonly found in three shapes: spheres, rods, and spirals. While some bacteria live singly, many are found in groups, pairs, chains, or clusters. The bacteria that cause strep throat are spheres and appear in chains.

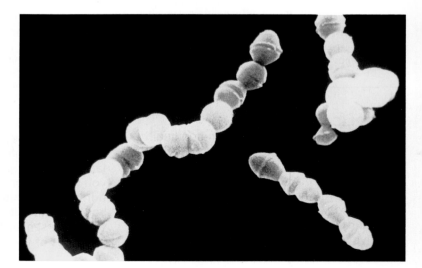

Figure 8–10. Strep throat is caused by bacteria that are spread from person to person through moisture sprayed by the nose or mouth.

TEACHING STRATEGIES

● **Process Skills:** *Comparing, Classifying/Ordering*

Have the students collect or identify objects to represent the three different shapes of bacteria. For example, the students might identify globes as spheres, chalk as rods, and rotini as spirals.

● **Process Skills:** *Observing, Applying*

Have the students examine Figure 8–10. Explain that it shows the bacteria that cause strep throat. Ask the students to identify the shape of the bacteria and to tell whether these bacteria are arranged in groups, pairs, chains, or clusters. (The bacteria that cause strep throat are spheres that appear in chains.) After the students have identified the shape and arrangement of the strep bacteria, point out that *strep* is short for *streptococci*. *Strepto-* means "twisted chain," and *cocci* means "spherical bacteria." Ask the students what the name of the strep bacteria tells us. (It tells the shape and arrangement of the bacteria.)

Figure 8–11. Paint pots, or hot springs that contain boiling water and mud, are often brightly colored by bacteria living in the boiling water.

Hot or Cold Bacteria have the ability to live in a variety of conditions. The illustration shows bacteria living successfully in boiling water. Bacteria have also been found in very cold environments and even buried 5 m deep in the soil! Because monerans have been found in some very unusual places, it's easy to see that they must adapt successfully to less-than-perfect environmental conditions.

Home Sweet Home Monerans can be found almost everywhere. While some monerans live freely, others live in plants or animals. Still others live in products made from plants and animals. Monerans are found in many foods, such as cheese and yogurt. Many monerans are helpful to plants and animals. Others, however, are harmful.

 ASK YOURSELF

Under what conditions can bacteria live?

Bacteria—Common Monerans

Bacteria reproduce quickly by splitting into two cells. Under proper conditions, reproduction can occur every 20 minutes. If conditions were perfect, one bacterium could produce so many others in 24 hours that the entire group would weigh 2 million kilograms, approximately the weight of a train locomotive!

 *Calculating*

The weight of the earth is estimated to be about 5.9 sextillion (10^{21}) tonnes. If one bacterium could produce a group of bacteria weighing 2 million kilograms in 24 hours, how long would it take that group to reproduce so that it weighed as much as the entire earth?

SCIENCE BACKGROUND

All monerans are prokaryotes. Prokaryotes are organisms that do not have a nuclear membrane separating their DNA from the cytoplasm, nor do they have any membrane-bound organelles. Prokaryotes do have a cell wall. Some monerans have a capsule outside the cell wall. Often this capsule protects disease-causing monerans from the immune system of their host.

ONGOING ASSESSMENT
ASK YOURSELF

Bacteria are able to exist under many different conditions, such as in high temperatures and in deep soil.

DISCOVER BY *Calculating*

Point out to the students that the term *tonnes* refers to metric tons, which are equal to 1000 kg. Using a scientific calculator and the equation $2(10^3) \times 2^x = 5.9(10^{21})$, the students will find that the group of 2 000 000 kg (or 2000 tonnes) will double about 61 times to equal the earth's weight. Assuming that the bacteria double every 20 minutes, it would take a little more than 20 hours.

 **LASER DISC**

2808

Bacteria; classification

SECTION 2

ACTIVITY

What microorganisms cause various diseases?

Process Skills: Communicating, Classifying/Ordering

Grouping: Groups of 3 or 4

Hints
Suggest that the students use microbiology textbooks to find the structures and shapes of the bacteria.

▶ Application
1. Boils, bacterial pneumonia, scarlet fever, tuberculosis, and whooping cough are caused by monerans.

2. Vaccines have been developed for chicken pox, rabies, whooping cough, yellow fever, pneumonia caused by bacteria, and tuberculosis.

3. The more a person knows about diseases, the better able that person is to avoid contracting a disease and to minimize symptoms of a disease if contracted.

★ PERFORMANCE ASSESSMENT
Ask the students to make a comparison-and-contrast chart for viruses and bacteria. Suggest that they begin with the diseases caused by viruses and then include viruses' sizes, shapes, structures, and so on. Have the students add to their charts as they finish the chapter. Then check their charts for accuracy.

TEACHING STRATEGIES, continued

● **Process Skills**: *Synthesizing, Classifying/Ordering*

Introduce these terms and their definitions: *cocci*, meaning "spherical bacteria"; *bacilli*, meaning "rod-shaped bacteria"; and *spirilla* meaning "spiral bacteria." Have the students research information about which of the bacterial diseases identified in Table 8–2 are caused by each of these different shapes of bacteria. Also have the students identify the symptoms of each of these diseases. (All of the diseases identified in the table are caused by bacilli except for strep throat, which is caused by cocci.)

● **Process Skills**: *Inferring, Applying*

Ask the students to explain why an antibiotic, such as penicillin, may cure one bacterial infection but not another. (The bacteria may have become resistant to the penicillin, or the penicillin is not effective on that particular type of bacteria.)

Fortunately, perfect conditions for bacteria reproduction are rare because the environment in which bacteria live does not have enough food to support so many organisms. The amount of waste products produced by the bacteria also would interfere with their continued growth and reproduction.

In Harm's Way Like viruses, some disease-causing bacteria invade other organisms. Since bacteria reproduce so quickly, the illnesses they cause can spread throughout the body rapidly. Bacteria cause disease in two ways. Some bacteria destroy cells. When cells in your body are destroyed, you feel weak and often have a fever. Other bacteria give off poisons, or toxins, that can damage your body. In the next activity you will identify some of the disease-causing bacteria and viruses.

ACTIVITY

What microorganisms cause various diseases?

MATERIALS
unlined paper, colored pencils

PROCEDURE
1. Find the name of the first disease listed in this activity (boils). Then note the type of organism that causes that disease (staphylococcus).
2. Use a reference book to find out what the microorganism looks like. Draw the microorganism and label it with its name and the disease it causes.
3. Repeat the procedure for the remaining diseases listed.

APPLICATION
1. Which diseases listed in this activity are caused by monerans?
2. For which diseases listed have vaccines been developed?
3. Why is it important for you to know about diseases, their causes, and their treatments?

SOME DISEASE-CAUSING MICROORGANISMS

Disease	Microorganism
Boils	Staphylococcus
Chicken pox	Virus
Pneumonia	Bacillus or virus
Rabies	Virus
Scarlet fever	Streptococcus
Tuberculosis	Bacillus (singly or in a V shape)
Whooping cough	Bacillus (singly or in pairs)
Yellow fever	Virus

▶ 220 CHAPTER 8

- **Process Skills:** *Comparing, Inferring*

Point out to the students that antibiotics are effective treatments against bacterial infections but not against viral infections. Ask the students to use the definition of *antibiotics* to explain why antibiotics can treat bacterial diseases but not viral diseases. (*Antibiotic* means "against life." Antibiotics can be used to destroy living organisms such as bacteria. Viruses do not have many of the characteristics of living things, and, therefore, antibiotics are not effective against them.)

EXTENSION

 List the following diseases on the chalkboard, and have the students research information about the causes, symptoms, and possible cures of each bacterial disease: cholera, fire blight, anthrax, botulism, leprosy, and yaws. Encourage the students to place their findings in their science portfolios.

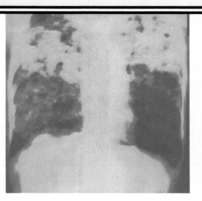
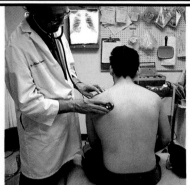

Figure 8–12. Improved preventive and treatment methods have reduced the number of people who die from tuberculosis. However, the disease remains a major concern, especially in developing countries.

Demonstration

You will need two bouillon cubes mixed in 500 mL of water, three beakers, salt, and sugar. Explain that preservatives are often used in food processing to prevent the growth of bacteria. Pour equal amounts of the bouillon solution into the three beakers. Put 1 teaspoon of salt in the first beaker, 1 teaspoon of sugar in the second, and add nothing to the third beaker. Label the beakers and set them in a warm place. After two days, examine the beakers. **CAUTION: Warn the students not to taste the liquids. Bacterial growth could result in food poisoning if the liquids are ingested.** Ask the students to describe the appearance of the liquids. (The "sugar" beaker should have the most bacterial growth and be very cloudy. The "salt" beaker should have the least amount of growth and be almost clear.) Explain that salt is a preservative that deters the growth of bacteria.

Table 8-2 Some Diseases Caused By Bacteria

Diphtheria	Strep throat	Tuberculosis
Leprosy	Tetanus	Whooping cough

Bacteria cause many illnesses, such as those listed in the activity. Fortunately, most bacterial diseases respond to treatment. If you've ever been sick with a bacterial infection and received medication for it, you probably received an antibiotic. **Antibiotics** are chemical substances used to kill or slow the growth of bacteria. The word *antibiotic* comes from two Greek words that mean "against life." Antibiotics interfere with the cell processes of bacteria and, as a result, kill them. Unfortunately, antibiotics may also kill some healthy cells. Occasionally, antibiotics may cause unusual side effects, such as rashes, diarrhea, or sensitivity to sunlight. These side effects, however, are tolerated by many patients because the side effects represent less of a threat to long-term health than the bacterial infection does.

With a Little Help

Even though it appears that many harmful bacteria inhabit our world, there are actually many more helpful bacteria than harmful bacteria. You may find this hard to believe after learning about some of the illnesses caused

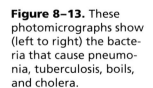

Figure 8–13. These photomicrographs show (left to right) the bacteria that cause pneumonia, tuberculosis, boils, and cholera.

MULTICULTURAL CONNECTION

The bacterial and viral diseases that the European explorers, conquerors, and settlers brought to the Americas caused far greater devastation to Native American populations than the Europeans themselves. The American Indians had never been exposed to and, as a result, had no resistance to diseases such as measles, diphtheria, and smallpox. Native communities lost 50 to 90 percent of their people from epidemics of diseases inadvertently brought to the Americas by the Europeans.

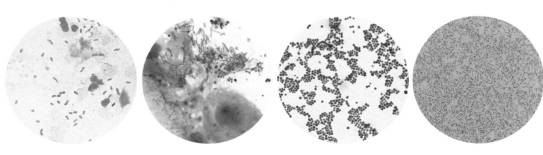

TEACHING STRATEGIES, continued

- **Process Skills:** *Inferring, Generating Ideas*

Explain that cyanobacteria, like plants, have chlorophyll, which enables them to make their own food. Ask the students to give reasons why cyanobacteria are classified as monerans rather than plants. (Cyanobacteria have cell walls and can make their own food just as plants do. However, unlike plants, they do not have distinct nuclei. As a result, they are classified as monerans rather than as plants.)

GUIDED PRACTICE

Write these pairs of terms on the chalkboard, and have the students state the relationship between the terms in each pair: monerans, bacteria; antibiotics, penicillin; pigment, chlorophyll. (In each pair, the second term is an example of the first term.)

INDEPENDENT PRACTICE

 Have the students answer the Section Review and Application questions in their journals. Then ask them to write two statements—one identifying the harmful effects of bacteria and one identifying the beneficial effects of bacteria on the earth.

ONGOING ASSESSMENT
ASK YOURSELF

The students may suggest that bacteria help them by decomposing materials and by helping to produce foods. The bacteria can harm them by causing diseases.

DISCOVER BY *Doing*

Make sure that the students understand that other organisms, such as protists, also inhabit ponds. If the students are unable to bring pond samples, have them look for "pond scum" in a home aquarium, or provide a prepared slide. **CAUTION: If the students are going to obtain pond-water samples, make sure they are aware of water safety rules. Caution them to go to the pond accompanied by a responsible adult.**

MEETING SPECIAL NEEDS
Mainstreamed

 Cooperative Learning Have students with learning disabilities work with partners to discuss and outline the major ideas in this section. Then have the partners write questions about the content on index cards. The partners can exchange questions with other partner teams and answer each other's questions.

Figure 8–14. Yogurt and cheese are produced through the action of bacteria on milk and cream. This photograph shows cheese being made.

---

by bacteria. However, many bacteria have useful functions. For example, bacteria are necessary for decay to take place. Imagine what life would be like if all of the dead plants and animals in the world around you never decayed! Garbage and waste would pile up and fill every city, state, and country with material that could not be used and would never go away. Bacteria decompose these materials and even help recycle valuable compounds contained in garbage and waste back into the environment.

Bacteria are also used in industry and in the production of food. In industry, for example, bacteria break down the fibers of plants used to make linen and rope. The string that you used to fly a kite was probably created with the help of bacteria. If you have ever cooked with vinegar or eaten cheese or sauerkraut, you have used products that were produced with the help of bacteria. Bacteria are even used to create antibiotics that kill other bacteria!

ASK YOURSELF

Name several ways that bacteria positively and negatively influence your life.

Cyanobacteria

Have you ever wanted to swim in a pond or lake but could not because green "scum" was covering the surface of the water? The "scum" is really a large group of organisms that may be cyanobacteria.

DISCOVER BY *Doing*

 If you live near a pond that contains a large amount of "scum," ask your teacher for permission to bring a small amount to class for observation under a microscope. Try to identify which organisms are cyanobacteria. Draw what you see under the microscope.

You may have seen some organisms in your sample that were not cyan (blue-green) in color and wondered what these organisms were. Cyanobacteria are not always blue-green, or even green. They can also be red, violet, or yellow, though they are always called cyanobacteria. Sometimes they are even called blue-green algae, but they are really monerans, not protists as true algae are.

▶ 222 CHAPTER 8

EVALUATION

 Encourage the students to write a paragraph explaining the impact of bacteria on their lives and how their lives might be different if bacteria did not exist.

RETEACHING

Draw a two-column chart on the chalkboard, labeling one column "viral diseases" and the other column "bacterial diseases." In a paper bag, place slips of paper with the names of various diseases written on them. Have the students draw a slip of paper from the bag and write the name of the disease under the appropriate column in the chart.

CLOSURE

 In their journals, the students can create idea webs to serve as summaries of the important ideas presented in this section.

Cyanobacterium; in geysers

① The students may suggest that black-and-white TV, for example, does not really project black-and-white images.

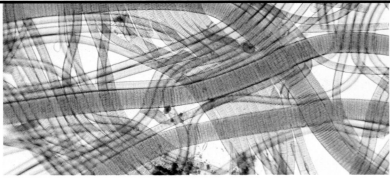
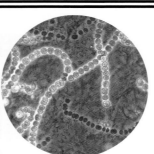

Figure 8–15. These moneran cyanobacteria are often mistaken for algae.

ONGOING ASSESSMENT
ASK YOURSELF

Cyanobacteria restore nitrogen to the soil, a necessary element for proper plant growth.

SECTION 2 REVIEW AND APPLICATION

Reading Critically

1. Bacteria are living things. It is not clear whether viruses are alive. Bacteria are one-celled organisms and are affected by antibiotics. Viruses are not cells and are not affected by antibiotics.

2. Cyanobacteria contain chlorophyll and can make their own food. Other types of bacteria do not contain chlorophyll and cannot make their own food.

3. Bacteria help humans by decomposing waste materials and recycling important chemicals back into the environment. Bacteria are also used in the production of food and antibiotics.

Thinking Critically

4. *Antibiotic* comes from Greek words meaning "against life." Bacteria are living organisms and thus are affected by antibiotics. Because viruses have characteristics of both living and nonliving things, they are not affected by antibiotics.

5. Adding chemicals to a pond or stream will upset the ecological balance of the pond or stream. Cyanobacteria restore nitrogen to the environment and serve as a food source for other organisms in the pond or stream.

If you think it's unusual that cyanobacteria are not always cyan in color, think of a violet. Violets may have flowers of a violet color, but the flowers can also be colored pink, white, or blue. Can you think of other instances in which the name given to something is somewhat misleading? ①

The various colors of the cyanobacteria represent different species. The colors come from the pigments in the cells. **Pigments** are chemicals that give color to the tissue of living organisms. *Chlorophyll*, the pigment that makes plants green, is the best-known pigment. Chlorophyll enables cyanobacteria to make their own food.

Cyanobacteria reproduce by simple cell division. They are so plentiful that they are considered a nuisance by many pool owners and by gardeners who try to scrape them off damp flowerpots. However, cyanobacteria also serve a useful purpose. They help restore nitrogen to the soil. Nitrogen is necessary for healthy plant growth and is often added to soil in fertilizers. The presence of cyanobacteria allows plants to grow with very little added fertilizer.

 **ASK YOURSELF**

How can cyanobacteria improve the soil?

SECTION 2 REVIEW AND APPLICATION

Reading Critically

1. Compare and contrast the characteristics of viruses with those of bacteria.
2. How are cyanobacteria different from other kinds of bacteria?
3. How are bacteria useful to humans?

Thinking Critically

4. Why are antibiotics effective against bacteria but not against viruses?
5. When certain chemicals are added to a swimming pool, the bacteria growing there die. Explain why it would not be good to add these chemicals to a pond or stream to kill the cyanobacteria growing there.

SECTION 2   **223**

INVESTIGATION

Testing Disinfectants

Process Skills: Observing, Identifying/Controlling Variables

Grouping: Groups of 3 or 4

Objectives
- **Identify** the variable(s) of an experiment.
- **Relate** the results of an experiment to everyday life.

Pre-Lab
Have the students read through the procedure. Ask them to predict (1) the effect of disinfectants on bacterial growth and (2) the disinfectant solution that will be the most effective.

Hints
Note that this activity is designed to show a disinfectant's effectiveness at room temperature. Petri dishes should be incubated upside down so that drops of condensation that form on the covers will not fall on the surfaces and ruin individual colonies. Have the students identify and record the ingredients that make up the disinfectants that they are using.

Analyses and Conclusions

1. Sections 1, 2, and 3 of the dish contained variables. Section 4 was the control.

2. The paper disc with the largest clear area (area of inhibition) around it had the most effective disinfectant.

3. Bacteria were already present on the unsterilized agar and Petri dish. If these bacteria were not destroyed by sterilization, they would grow on the agar plates, and the results from the swabbing would be inaccurate.

4. A control is used to compare the effects of a variable against a stable condition. The control of an experiment reveals what would have happened if nothing were changed.

Application
Because many bacteria can be harmful to people, hands should be washed frequently to prevent the spread of bacteria. It is especially important to wash your hands before eating.

✳ Discover More
The type and concentration of solution mixed by each student will determine the effectiveness of the homemade disinfectants.

Post-Lab
Have the students review their predictions. Did they choose the most effective disinfectant solution? If so, why did they choose that solution?

INVESTIGATION

Testing Disinfectants

▼ MATERIALS
- safety goggles • laboratory apron • agar-filled Petri dish • wax pencil
- cotton swab • filter-paper disks (4) • forceps • disinfectant solutions

▼ PROCEDURE

1. **CAUTION: Put on safety goggles and a laboratory apron and leave them on throughout this investigation.** Turn your Petri dish upside down and use a wax pencil to label the dish as shown.

2. Moisten a cotton swab with water and rub it lightly across a surface in the classroom, such as a desk, doorknob, or window.

3. Lift the top of the Petri dish just enough to insert the swab. Gently, but quickly, streak the swab in a zigzag motion over half the agar surface. Then turn the Petri dish one-half turn and streak the other half of the agar. Replace the cover.

4. Get four 5-mm disks of filter paper. Use your forceps to dip one in a disinfectant solution. Place the disk on section 1 of your Petri dish. Dip two more disks in different disinfectant solutions and place them on sections 2 and 3 of your Petri dish. Place the last disk without disinfectant on the fourth section.

5. Tape your Petri dish shut and incubate it upside down for 48 hours in a warm, dark place.

6. **CAUTION: Do not remove the cover of the Petri dish while observing the bacterial colonies. Wash your hands thoroughly after this investigation.** After incubation, compare the diameters of the clear areas around each disk.

7. Dispose of the Petri dish as directed by your teacher.

▶ ANALYSES AND CONCLUSIONS
1. Which section(s) of the dish contained the variable(s)? The control(s)?
2. Which disinfectant appears to be the most effective? How can you tell?
3. The agar and the Petri dish were sterilized before this investigation. Why do you think this was necessary?
4. What is the purpose of a control in an investigation?

▶ APPLICATION
Why do you think it is important to wash your hands thoroughly at the end of this investigation? List several ways in which this procedure applies to your everyday life.

✳ Discover More
People sometimes use homemade disinfectant solutions instead of solutions that are available commercially. **CAUTION: Do not mix any disinfectant with any other disinfectant. To make a disinfectant solution, just add water to a disinfectant.** Disinfectant solutions typically include water mixed with bleach, water mixed with ammonia, or water mixed with vinegar. With the approval and guidance of your teacher, create one of these homemade disinfectant solutions and repeat the steps of this investigation. Then compare and describe the results of each investigation.

CHAPTER 8 HIGHLIGHTS

The Big Idea—
SYSTEMS AND STRUCTURES

Lead the students to understand that the scale and structure of viruses enable them to invade living cells easily. Also point out to the students that because of monerans' structure and characteristics, monerans can easily be classified as living organisms, but viruses cannot so easily be classified. As the major theme is discussed, remind the students that despite their sizes, viruses and bacteria have a huge impact on all living organisms.

For Your Journal

The students' ideas should reflect the understanding that improved health care and cleanliness are important factors in inhibiting the spread of viral and bacterial diseases. The students should also consider the responsibility of people who are infected with certain viral and bacterial diseases to take precautions that will inhibit the spread of the diseases.

CONNECTING IDEAS

Have the students copy and complete the concept map in their journals. The concept map should be completed with words similar to those shown here.

Under *viruses*:
    common cold
    rabies
    influenza
    mumps
    measles
    polio
Under *monerans*:
    tuberculosis
    tetanus
    diphtheria
    whooping cough
    strep throat

CHAPTER 8 HIGHLIGHTS

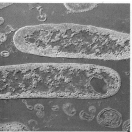

The Big Idea

People can see most organisms without the use of microscopes. However, they cannot see viruses and monerans. Both are tiny organisms that can be seen only by using a microscope. Although they are both microscopic organisms, viruses are much smaller than bacteria. Despite their size, viruses and bacteria have a huge impact on the environment.

Many viruses impact the environment in negative ways by causing diseases and other sicknesses in plants and animals. Monerans, such as bacteria, may benefit or harm organisms.

Review the journal entries that you wrote at the beginning of the chapter. Revise your entries to reflect what you have learned. Include information that explains why bacterial and viral diseases are more likely to break out in some areas than in others. Make recommendations for reducing the risk of contracting viral or bacterial diseases.

Connecting Ideas

Copy this unfinished concept map into your journal. Complete the concept map by writing the correct term in each blank.

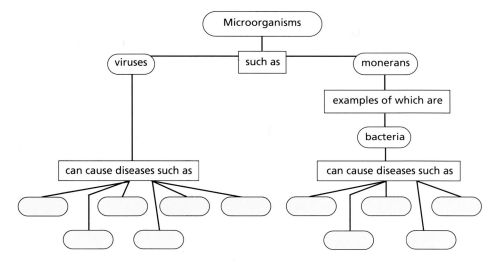

CHAPTER 8   **225**

CHAPTER 8 REVIEW

ANSWERS

Understanding Vocabulary

1. A virus requires a host to reproduce.

2. Antibiotics help the immune system deter and destroy bacteria.

3. Bacteria are a common example of monerans.

4. A vaccine helps control and confine the threat of a viral epidemic.

Understanding Concepts

Multiple Choice

5. d

6. a

7. c

8. c

9. b

Short Answer

10. An epidemic is the rapid spread of a disease through a large area.

11. Some scientists say that viruses are not living because viruses are not cells, they cannot grow, and they cannot respond to stimuli.

12. Antibiotics are effective against bacterial infections and not effective against viral infections. Vaccinations can prevent some viral infections.

Interpreting Graphics

13. Bacteria could reproduce in both environments; viruses could reproduce only in the human being.

Reviewing Themes

14. Viruses and monerans are relatively simple members of the microscopic world, but they influence the world around us in large and sometimes complex ways. For example, viruses and monerans can cause fatal and nonfatal illnesses.

15. Bacteria decompose waste materials and are used to create medicines and foods. Bacteria also cause diseases and illnesses in living things.

Thinking Critically

16. Antibiotics have an effect on living things such as bacteria. Because viruses do not have some of the characteristics of living things, antibiotics have no effect on viruses.

17. The students may identify problems associated with garbage

---

CHAPTER 8 REVIEW

Understanding Vocabulary

For each set of terms, explain the similarities and differences in their meanings.

1. virus (211), host (212)
2. immune system (214), antibiotics (221)
3. monerans (218), bacteria (218)
4. vaccine (215), epidemic (213)

Understanding Concepts

MULTIPLE CHOICE

5. Viruses are different from monerans because they
   a) are microorganisms.
   b) may cause diseases in humans.
   c) can be cured by antibiotics.
   d) can reproduce only in living cells.

6. All of the following are true about bacteria *except* that they
   a) have a distinct nucleus.
   b) can survive in very hot climates.
   c) can be treated with antibiotics.
   d) come in three different shapes.

7. Viral infections are difficult to treat because
   a) viruses are difficult to detect, even with microscopes.
   b) viruses have many different shapes.
   c) viruses do not appear to be living things.
   d) symptoms of the infections are difficult to diagnose.

8. Choose the location in which bacteria would be least likely to be found.
   a) in the atmosphere
   b) under water
   c) at the core of the earth
   d) on the earth's surface

9. Why is it not likely that people today will contract polio?
   a) The virus that causes polio no longer exists.
   b) Vaccines against polio have been developed.
   c) The polio virus exists today, but in a weakened form.
   d) Antibiotics have slowed the spread of the disease.

SHORT ANSWER

10. What is an epidemic?

11. Explain why some scientists say that viruses are not living things.

12. Compare and contrast antibiotics and vaccinations.

Interpreting Graphics

13. Look at the illustrations. In which environments could some bacteria reproduce? In which could some viruses reproduce?

226 CHAPTER 8

and waste buildup and the inability to manufacture different foods and medicines.

18. Diseases are controlled by the vaccines used against them. Because the diseases still exist on Earth, people should be vaccinated against them so that they do not contract the diseases.

19. The students might cite lack of research and development money, technology, or knowledge as reasons for our failure to create vaccines for all viral infections.

20. The students might suggest that AIDS is so dangerous because it is indiscriminate and always fatal. Also, because of its relatively rapid spread through different populations of the earth, it may be considered an epidemic.

Reviewing Themes

14. *Systems and Structures*
Compare the size and complexity of viruses and monerans to their role in the world around us.

15. *Environmental Interactions*
Explain how monerans influence you and your environment in positive and negative ways.

Thinking Critically

16. Why do bacteria respond to antibiotics, while viruses do not?

17. Describe some of the problems humans would face if there were no bacteria.

18. Polio and smallpox are now controlled. Why, then, are polio and smallpox vaccines still required in many states?

19. Scientists and researchers have created vaccines for various diseases such as polio and smallpox. List several reasons why you think scientists and researchers have not been able to create vaccines for other diseases such as AIDS.

20. This photograph shows AIDS viruses attacking a healthy cell. The AIDS viruses are shown in blue. Why is AIDS so dangerous? Could AIDS be considered an epidemic? Why?

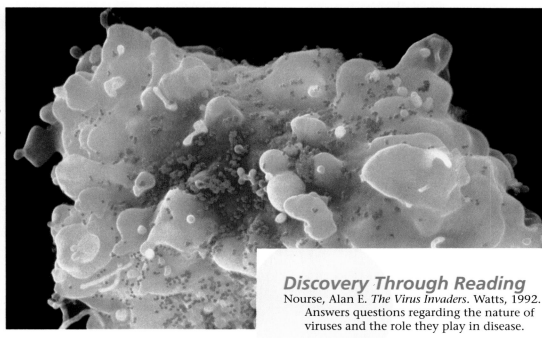

Discovery Through Reading

Nourse, Alan E. *The Virus Invaders*. Watts, 1992. Answers questions regarding the nature of viruses and the role they play in disease.

CHAPTER 9

PROTISTS AND FUNGI

PLANNING THE CHAPTER

Chapter Sections	Page	Chapter Features	Page	Program Resources	Source
Chapter Opener	228	For Your Journal	229		
Section 1: PROTOZOANS	230	Discover By Doing (A)	233	Science Discovery*	SD
• Characteristics of Protozoans (B)	230	Section 1 Review and Application	234	Reading Skills: Taking Notes (A)	TR
• Types of Protozoans (B)	231	Investigation: Observing Protozoans (A)	235	Protists (B)	IT
				Record Sheets for Textbook Investigations (A)	TR
				Study and Review Guide, Section 1 (B)	TR, SRG
Section 2: ALGAE	236	Discover By Researching (A)	236	Science Discovery*	SD
• Characteristics of Algae (B)	236	Discover By Observing (B)	237	Investigation 9.1: Observing and Classifying Algae (H)	TR, LI
• Types of Algae (B)	237	Section 2 Review and Application	238	Thinking Critically (H)	TR
		Skill: Organizing Data (B)	239	Study and Review Guide, Section 2 (B)	TR, SRG
Section 3: FUNGI	240	Discover By Doing (A)	241	Science Discovery*	SD
• Characteristics of Fungi (B)	240	Activity: What are the characteristics of fungi? (A)	243	Investigation 9.2: Observing the Growth of Yeast (A)	TR, LI
• Types of Fungi (B)	241	Discover By Writing (A)	244	Connecting Other Disciplines: Science and Language, Writing a Biographical Report (A)	TR
• Lichens and Slime Molds (A)	244	Section 3 Review and Application	244	Extending Science Concepts: Antibiotics (A)	TR
				Fungi (B)	IT
				Study and Review Guide, Section 3 (B)	TR, SRG
Chapter 9 HIGHLIGHTS	245	The Big Idea	245	Study and Review Guide, Chapter 9 Review (B)	TR, SRG
Chapter 9 Review	246	For Your Journal	245	Chapter 9 Test	TR
		Connecting Ideas	245	Test Generator	
				Unit 3 Test	TR

B = Basic   A = Average   H = Honors
The coding Basic, Average, and Honors indicates subsections, features, and resources that might be appropriate for different levels of learners. For additional suggestions regarding choice of topic and depth of coverage, see the Pacing Chart on pages T26–T29.

*Frame numbers at point of use
(TR) Teaching Resources, Unit 3
(IT) Instructional Transparencies
(LI) Laboratory Investigations
(SD) *Science Discovery* Videodisc Correlations and Barcodes
(SRG) Study and Review Guide

CHAPTER MATERIALS

Title	Page	Materials
Discover By Doing	233	(per individual) culture of paramecia, stained yeast cells, journal
Investigation: Observing Protozoans	235	(per group of 3 or 4) medicine dropper, slide, stock cultures or pond water, coverslip, compound light microscope
Discover By Researching	236	(per individual) labels of food and cosmetic containers
Discover By Observing	237	(per class) specimens of algae, microscope (per individual) journal
Skill: Organizing Data	239	(per group of 2) paper, pencil
Discover By Doing	241	(per individual) gilled mushroom cap, damp cotton ball, paper, glass or plastic cup
Teacher Demonstration	241	sugar, water, slides, microscope
Activity: What are the characteristics of fungi?	243	(per group of 3 or 4) medicine dropper, slides (2), forceps, fungi, coverslips (2), compound light microscope
Discover By Writing	244	(per individual) journal

ADVANCE PREPARATION

For the *Discover By Doing* on page 233, obtain cultures of paramecia and yeast cells that have been stained with congo red. For the *Investigation* on page 235, you will need to provide stock cultures of protozoans obtained from a biological supply house; or water obtained from rain, a pond, a mud puddle, a spring, or a stream. For the *Discover By Researching* on page 236, the students need to obtain labels from various food and cosmetic containers. For the *Discover By Observing* on page 237, you will need to obtain algae samples. For the *Discover By Doing* on page 241, you will need to provide gilled mushroom caps, cotton balls, and glass or plastic cups.

TEACHING SUGGESTIONS

Field Trip
Arrange a visit to a forest or wooded area with a lake or pond and try to locate samples of protists, algae, and fungi. The students can collect pond water (or any standing water), scrapings from rocks, tree bark, or bits of wooden fences and poles. If lakes and ponds are nearby, have the students look for evidence of *eutrophication* (excessive growth of algae that depletes the oxygen supply for animals).

Outside Speaker
Local physicians may be willing to speak about diseases of humans that are caused by protozoans and fungi. If specialists are available, invite a pathologist and a dermatologist. The pathologist can concentrate on internal diseases, whereas the dermatologist can concentrate on diseases of the skin, many of which are due to fungi.

CHAPTER 9
PROTISTS AND FUNGI

CHAPTER THEME—SYSTEMS AND STRUCTURES

This chapter introduces the students to the organisms called protozoans, algae, and fungi. The students will explore the characteristics used to identify the members of each group. The students will also study the interaction of these organisms with the environment and with humans. The theme of **Systems and Structures** is also developed through concepts in Chapters 4, 5, 7, 8, 10, 12, 13, 16, 17, 18, 19, and 20. A supporting theme of this chapter is **Energy**.

MULTICULTURAL CONNECTION

The kinds of food that people eat vary from country to country and even from region to region. People who live on seacoasts rely on food from the ocean, such as fish and seaweed, or algae. Algae may be used as fuel, fertilizer, livestock feed, and food. In Japan, a red alga called *nori* is dried, made into papery sheets, and used as a wrapper for seasoned rice and raw fish in a dish called *sushi*. So much algae is eaten in Japan that algae are grown commercially on special farms in the ocean. Algae are also an important part of the diets of people in the Pacific Islands, coastal China, Indonesia, and the Philippines.

MEETING SPECIAL NEEDS

Second Language Support

Fluent English-speaking students can be paired with students who have limited English proficiency to work together to draw and label a picture of a typical protozoan such as a sarcodine, ciliate, or flagellate. The pictures can then be displayed during the students' study of protozoans.

▶ 228   CHAPTER 9

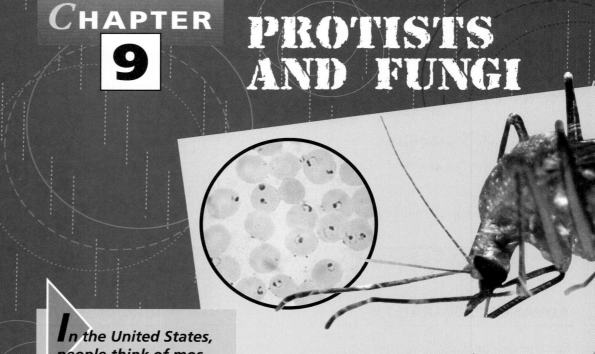

CHAPTER 9
PROTISTS AND FUNGI

In the United States, people think of mosquitoes mostly as a nuisance. In other parts of the world, however, mosquitoes can carry and transmit microscopic organisms that cause disease and sometimes death. Malaria, a disease transmitted by mosquitoes, is the cause of death for more than one million people each year.

People are sometimes afraid of things they cannot see. But when we think of things we cannot see, we usually think of large things. Seldom do we think of microscopic things.

The microscopic world of organisms contains a great variety of living things—all of which are too small for you to see. Some of these microscopic organisms are helpful, and others are harmful. Yet they are everywhere in the world around you.

The Anopheles mosquito shown in the photograph carries and transmits the disease malaria. Characterized by symptoms such as chills, fever, and weakness, malaria is a parasitic disease caused by microscopic organisms, which the mosquito injects into a person's bloodstream. Once in the bloodstream, the organisms invade the liver, causing the liver cells to produce more

228   Chapter 9

CHAPTER MOTIVATING ACTIVITY

Display a selection of familiar products that show the impact of fungi on human life. Products might include various cheeses (blue and gorgonzola show obvious mold activity), breads (with holes resulting from yeast activity), and different over-the-counter ointments (such as those for controlling athlete's foot). Ask the students why fungi are important and how fungi affect human health. (Responses should include the importance of fungi in the baking industry and in the production of cheese products. Responses should also indicate that fungi may cause disease.)

Answering the journal questions provides the students with an opportunity to demonstrate what they know about the benefits and harmful effects of microscopic organisms. Their answers will provide you with an opportunity to note any misconceptions they might have about microscopic organisms. You may choose to rectify those misconceptions immediately or at appropriate times throughout the chapter.

ABOUT THE PHOTOGRAPH

The saliva of the *Anopheles* mosquito carries a parasitic sporozoan that causes the disease malaria. When an infected *Anopheles* mosquito bites a person, spores are injected into the person's bloodstream, where they enter cells and live as parasites. The disease can also be transmitted to uninfected mosquitoes if those mosquitoes bite an infected person. Several hundred million people worldwide each year suffer from malaria. Although the disease is not commonly fatal, it can be seriously debilitating. Drugs such as chloroquinine can be used to treat some, but not all, strains of the disease. Consequently, the most effective means of stopping the spread of malaria is the elimination of areas in which *Anopheles* mosquitoes are known to breed, such as ponds, ditches, and other bodies of still water. The use of insect repellants, mosquito netting, and screens can also reduce people's contact with the mosquitoes. Eventually, scientists hope to develop a vaccine against malaria.

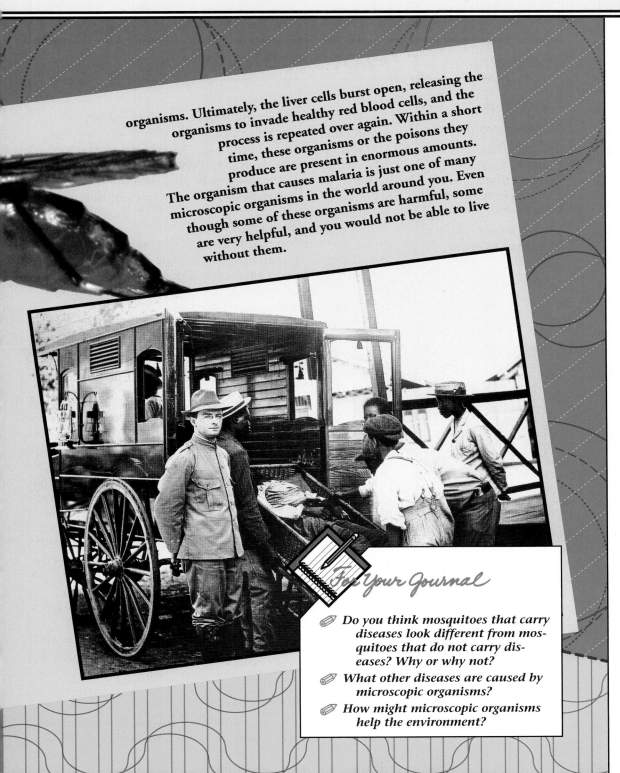

organisms. Ultimately, the liver cells burst open, releasing the organisms to invade healthy red blood cells, and the process is repeated over again. Within a short time, these organisms or the poisons they produce are present in enormous amounts.

The organism that causes malaria is just one of many microscopic organisms in the world around you. Even though some of these organisms are harmful, some are very helpful, and you would not be able to live without them.

For Your Journal

- Do you think mosquitoes that carry diseases look different from mosquitoes that do not carry diseases? Why or why not?
- What other diseases are caused by microscopic organisms?
- How might microscopic organisms help the environment?

Section 1: PROTOZOANS

FOCUS

In this section, the characteristics and structures of four groups of protozoans—sarcodines, ciliates, flagellates, and sporozoans—are described. Some of the harmful effects of protozoans are also presented.

MOTIVATING ACTIVITY

Cooperative Learning Some of the protists that the students will study have characteristics of both plants and animals. Working in small groups, the students can create a list of the characteristics of typical plants and a list of the characteristics of typical animals. Then ask the groups to discuss the characteristics on their lists and create two lists on the chalkboard that contain all of the most-agreed-upon plantlike and animal-like characteristics. Then, as the students study the chapter, they can refer to the chalkboard lists.

PROCESS SKILLS
• Observing • Comparing
• Generating Ideas

POSITIVE ATTITUDES
• Curiosity • Openness to new ideas

TERMS
• protozoans • pseudopods
• cilia • flagella

PRINT MEDIA
Simple Animals by John Stidworthy (see p. 205b)

ELECTRONIC MEDIA
Protists: Form, Function, and Ecology, Britannica (see p. 205b)

Science Discovery Ciliates Paramecium; movement

BLACKLINE MASTERS
Study and Review Guide
Reading Skills

ONGOING ASSESSMENT

ASK YOURSELF

Like animals, protozoans have structures that help them obtain food and react quickly to changes in the environment. Most protozoans also have the ability to move from place to place.

SECTION 1

Protozoans

Objectives

Compare protozoans and animals.

Identify several features that make protozoans different from each other.

Sketch a protozoan and **label** its major features.

Imagine that you are going on a two-week hiking trip in a wilderness area. You must carry with you everything you will need for the entire trip. One of your most important requirements is water. Because carrying enough water would be difficult, you hope to use the water you will find in the wilderness. You know you will find water in rivers and streams, or you can collect it by catching rain or perhaps by melting snow.

The water you find in the wilderness may look clean. But it probably contains microscopic organisms called *protozoans*. Drinking the water may make you ill. This is just one way in which microorganisms—living things you cannot see—can affect your life.

Figure 9–1. Although this stream looks clear, it may contain many protozoans.

Characteristics of Protozoans

Have you ever heard of African sleeping sickness or malaria? These diseases and many others are caused by protozoans (proht uh ZOH uhnz). **Protozoans** are microscopic organisms that are members of the protist kingdom. Like other protists, protozoans are single-celled organisms, but they usually do not have cell walls. Most protozoans are found in water.

Just as you have special structures in your body that help you move and eat, protozoans have special structures that help them get food and move from one place to another. Protozoans are able to take in oxygen and give off carbon dioxide directly through their cell membranes.

Protozoans are similar to animals in other ways. Just as animals can react quickly to changes in their environment, so too can many protozoans. Protozoans usually move toward food and better environmental conditions as animals often do. However, some protozoans cannot move on their own.

ASK YOURSELF

Although protozoans are not classified as animals, they are animal-like in some ways. Describe how protozoans are similar to animals.

TEACHING STRATEGIES

● **Process Skills:** *Comparing, Inferring*

Point out that the kingdom Protista includes both animal-like and plantlike single-celled organisms. The animal-like organisms are called *protozoans*. Explain that despite the fact that protists are single-celled, they carry out all of the basic life processes of living things. Ask the students to recall from their previous studies typical life processes of living things. (Responses should include growth and development, reproduction, energy consumption, and response to environmental stimuli. Remind the students that, because of the diversity of life, a list attempting to describe the characteristics of living things will always contain assumptions.) Ask the students whether they would expect to observe all the characteristics of life in protozoans. (Yes, although some of those characteristics may be difficult to observe in some protists.)

Types of Protozoans

Protozoans are classified, or grouped, according to their ability to move and the way in which they move. The four basic groups of protozoans are shown in the following table.

Table 9-1 Types of Protozoans

Protozoan	Locomotive Structure	Example
Sarcodines	Pseudopods	Amoeba
Ciliates	Cilia	Paramecium
Flagellates	Flagella	Trypanosome
Sporozoans	(none)	Plasmodium

Sarcodines *Sarcodines* are protozoans that move from one place to another by using **pseudopods** (SOO duh pahdz), or "false feet." Pseudopods are projections made of cytoplasm. Perhaps you have heard of or seen a picture of an amoeba (uh MEE buh). An example of a common sarcodine is the amoeba in the illustration. Though you might think that an amoeba looks something like uncolored gelatin, it is actually a small mass of cytoplasm surrounded by a cell membrane. Amoebas have no specific shape because they change shape almost constantly. Think of what your life would be like if you constantly changed shape! As the cytoplasm in an amoeba moves, new pseudopods are formed and the amoeba "creeps" along from place to place.

An amoeba also uses its pseudopods to capture food. When an amoeba finds a food particle, the pseudopods surround and close around the food particle, forming a food vacuole. Digestion takes place within the food vacuole. The digested food moves across the membrane of the vacuole and into the cytoplasm, where the nutrients are used. Food vacuoles can then move to the cell membrane and discharge undigested food or waste material from the cell.

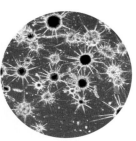

Figure 9-2. Found in untreated drinking water, these radiolarian protozoans can cause illness.

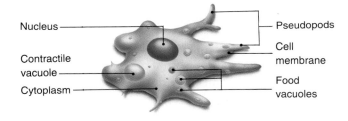

Figure 9-3. The movement of an amoeba is caused by the outward "flow" of cytoplasm to pseudopods.

✧ **Did You Know?**

The name *protozoan* comes from Greek words meaning "first animal." Originally, protozoans were classified as animals because of their animal-like characteristics.

THE NATURE OF SCIENCE

The kingdom Protista is made up of one-celled organisms that have distinct nuclei. Beyond that basic definition, however, great diversity exists among the more than 100 000 members of this kingdom. Some protists have characteristics they share with plants, others have animal-like characteristics, while still others resemble fungi in some ways. For these reasons, all scientists do not agree on which organisms should be classified as protists.

INTEGRATION—Language Arts

Point out that the students can use the meaning of the term *pseudopods* ("false feet") to help them remember this method of locomotion. Have the students investigate the derivations of the terms *cilia* and *flagella* (respectively, "eyelashes" and "whips") to help them envision the differences among the three types of locomotion used by protozoans.

TEACHING STRATEGIES, continued

● **Process Skills:** *Comparing, Applying*

Display pictures or drawings of a moneran cell and a protist cell. Ask the students to compare the cells. (Protist cells are usually larger and more sophisticated than moneran cells; moneran cells do not have nuclei, whereas protist cells do.)

● **Process Skills:** *Comparing, Expressing Ideas Effectively*

Remind the students that one of the life processes of living things is the response of a living thing to its environment. Paramecia are good examples to use in a discussion of the ways in which protozoans react to their environment. One reaction of paramecia occurs when they encounter danger: They can discharge spiny structures that can injure nearby cells. (These spiny structures can also anchor a paramecium during feeding.) Another reaction of paramecia is avoidance: They will back up if they encounter an obstacle. Then they will approach the same obstacle from a different angle and continue this behavior until they move past the obstacle. Point out that paramecia are classified as animal-like protists. Ask the students whether they agree or disagree with the classification and why.

BACKGROUND INFORMATION

Amoebic dysentery is a serious problem in many of the developing nations of the world. In those countries, some estimates show that dysentery-type diseases are the leading cause of death for children under the age of five. Dysentery can occur wherever people live in overcrowded conditions with poor sanitation. It was once common in hospitals, prisons, and army camps. Often more soldiers died of dysentery than were killed in battle.

MEETING SPECIAL NEEDS

Gifted

 Some protists live in *symbiotic relationships*. Ask the students to find two examples of symbiotic relationships mentioned in the chapter. Then have the students investigate and find examples of the different kinds of symbiotic relationships, focusing on those that involve protists. Have the students include their reports in their science portfolios.

LASER DISC
43820–44122
Ciliates

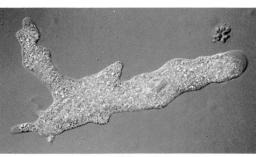

Figure 9–4. When an amoeba comes in contact with a food particle (top left), the pseudopod surrounds the food particle (top right) and slowly engulfs the particle (bottom left), forming a food vacuole (bottom right).

Figure 9–5. The thick protective covering, or cyst, surrounding this amoeba forms when the pond in which the amoeba lives dries up during a summer drought or freezes in the winter. When conditions are favorable, the amoeba will become active again.

Amoebas have other vacuoles called *contractile* (kuhn TRAK tuhl) *vacuoles*. These vacuoles collect extra water from the cytoplasm and release it through the cell membrane.

Most types of amoebas are harmless to you, but some cause disease. For example, one kind of amoeba causes amoebic dysentery (DIHS uhn tehr ee). Amoebic dysentery is a severe type of diarrhea. People may get the dysentery-causing amoeba by eating contaminated food or drinking contaminated water. Usually, such contamination occurs in areas with poor sanitation facilities. Another kind of amoeba lives in warm, still water. If swimmers inhale the amoeba, it causes a fatal inflammation of their brain and spinal cord.

Ciliates *Ciliates* (SIHL ee ihts) are the most complex protozoans. A ciliate has hundreds of short, hairlike structures called **cilia** that often cover the entire cell. These cilia are used to move the organisms through water in much the same way that your arms move you through water when you swim. In most ciliates, the beating cilia create a current of water that brings food to the organism.

The paramecium (par uh MEE shee uhm) [plural, *paramecia*], shown in the illustration on the next page, is an example of a common ciliate. Paramecia and other common ciliates are often found in pond water. You can see paramecia in action in the next activity.

▶ **232** CHAPTER 9

- **Process Skills:** *Comparing, Applying*

Ask the students to compare cilia and flagella. (Both enable an organism to move. Cilia are shorter than flagella.) Scientists now know that both have the same internal structure. But both terms remain in use.

- **Process Skills:** *Inferring, Generating Ideas*

Ask the students to think of possible reasons why sporozoans lack locomotive structures. (Sporozoans do not need to move about to find food or safety. All sporozoans are parasites that feed upon and reproduce in a host that can move.)

GUIDED PRACTICE

Write the following terms on the chalkboard: sarcodines, ciliates, flagellates, and sporozoans. Ask the students to explain what the terms have in common and how they are different.

INDEPENDENT PRACTICE

Have the students answer the Section Review and Application questions. Ask the students to describe how the relationship between a flagellate and a termite is beneficial to both.

Discover by Doing

Your teacher will prepare a culture of paramecia to which stained yeast cells have been added. Make a wet mount of the paramecia. After locating the red-stained food vacuoles, describe in your journal the process by which yeast cells are taken into the paramecia. Using what you have learned, predict how waste products are removed.

Most ciliates are free-living and do not cause disease in humans. Ciliates feed on dead plants and other small protozoans in the water, helping to continue the food chain. As cilia move ciliates through water, food enters the organisms. Food vacuoles, which form around the food particles, digest the food particles. Digested food is then absorbed by the cytoplasm through the membrane of a food vacuole, and waste material and undigested food are given off. As in the amoeba, contractile vacuoles remove extra water from the cell.

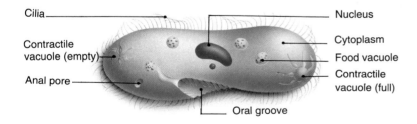

Figure 9–6. The cilia of a paramecium propel the organism quickly through the water. Cilia also help sweep food particles into the mouthlike oral groove.

DISCOVER BY *Doing*

This activity will work best if the yeast suspension is prepared the day before it is used. Just prior to making the wet mounts, boil the mixture and add red stain. The students' journal entries should describe paramecia's digestion of food and removal of waste products through food vacuoles.

PERFORMANCE ASSESSMENT

Ask each student to read his or her journal description to you. Use the description to evaluate whether the student understands the function of food vacuoles in protozoans.

LASER DISC

2814

Paramecium; movement

SCIENCE BACKGROUND

Plankton is a term used to describe the many small organisms that live near the surface of the ocean. Plankton can be divided into *phytoplankton*, which are simple, one-celled algae, and *zooplankton*, which are protozoans and tiny sea animals. Plankton is the main food source for many ocean-dwelling organisms.

Flagellates *Flagellates* (FLAJ uh layts) move by using long, whiplike structures called **flagella** (fluh JEHL uh). Flagellates are found in water and soil and even inside animals.

Have you ever wondered how termites eat wood? Within their digestive systems, they have a particular species of flagellate. These flagellates have enzymes that can digest wood. The digested wood is then used by the termites as food. Termites could not digest the wood without the flagellates.

Some flagellates are parasites. *Parasites* live in or on a host organism and cause harm to it. Some parasites cause diseases in humans or other organisms. The flagellate that causes African sleeping sickness is carried from one person to another by the tsetse (TSET see) fly. This flagellate gives off a toxin, or poison, that causes weakness in humans and can even cause death if the disease is not treated.

Figure 9–7. The flagellate that causes African sleeping sickness is shown above in human blood.

SECTION 1 233

EVALUATION

Ask the students to provide several examples of how protozoans can affect people in negative ways. (Examples might include diseases such as dysentery caused by amoebas, African sleeping sickness caused by the tsetse fly flagellate, or malaria caused by a sporozoan carried and transmitted by the *Anopheles* mosquito.)

RETEACHING

Ask four volunteers to give an oral description of each kind of protozoan without naming it. Have the remaining students listen to the description and decide whether a sarcodine, ciliate, flagellate, or sporozoan is being described.

EXTENSION

Point out that many sarcodines have shells. Ask the students to discuss how the presence of a shell might benefit or inhibit the feeding ability of a sarcodine. (Debates should conclude that the presence of a shell allows sarcodines to engulf only food small enough to fit through the opening in the shell.)

CLOSURE

Cooperative Learning Divide the class into four groups. Provide the groups with posterboard and ask them to create a sketch and description of a sarcodine, ciliate, flagellate, or sporozoan. Have the groups display the sketches as they describe the protozoan.

ONGOING ASSESSMENT
▼ **ASK YOURSELF**

Exposure to some protozoans can be reduced by cleaning and refrigerating food, decontaminating drinking water, and maintaining clean sanitation facilities. Exposure to other protozoans can be reduced by eliminating the conditions that are conducive to the reproduction of the insects that carry the disease-causing protozoans.

SECTION 1 REVIEW AND APPLICATION

Reading Critically

1. Three structures are cilia, flagella, and pseudopods.

2. Food vacuoles are organelles in which digestion occurs. Contractile vacuoles remove excess water.

3. A parasite is an organism that lives in or on a host organism and causes harm to it.

Thinking Critically

4. Sporozoans are the least animal-like because they do not move.

5. Flagellates and termites both benefit from their relationship. The sporozoan benefits and the red blood cell is harmed in their relationship.

Sporozoans All *sporozoans* (spawr uh ZOH uns) are parasites. They cannot move from one place to another on their own. Sporozoans may live in the muscles, the kidneys, or other organs of humans or other animals. They may also live in the blood. The best-known sporozoan lives part of its life in the red blood cells of humans, causing the disease malaria.

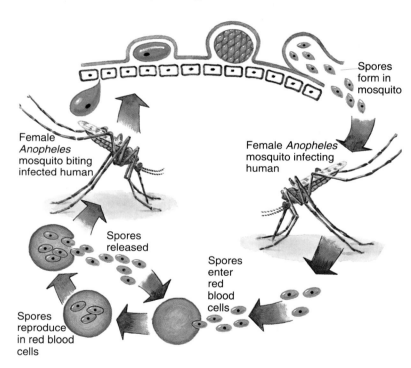

Figure 9–8. The *Anopheles* mosquito is responsible for transmitting the sporozoan that causes malaria.

 ASK YOURSELF

Some protozoans are harmful to humans. What steps might be taken to help reduce our exposure to these protozoans?

SECTION 1 REVIEW AND APPLICATION

Reading Critically
1. Name three structures that enable protozoans to move.
2. What is the function of food vacuoles? Of contractile vacuoles?
3. What is a parasite?

Thinking Critically
4. Which protozoan is the least animal-like? Explain.
5. How is the relationship between flagellates and termites different from the relationship between a sporozoan and a red blood cell?

INVESTIGATION

Observing Protozoans

Process Skills: Observing, Comparing, Applying

Grouping: Groups of 3 or 4

Objectives
- **Observe** and **compare** protozoans.
- **Communicate** ideas in the form of drawings.
- **Apply** knowledge of protozoans to the treatment of drinking water supplies.

Pre-Lab
Have the students read the procedure and study the illustrations. Ask them to predict the types of protozoans they will see.

Analyses and Conclusions

1. Ciliates are the protozoans most often observed. Some flagellates may also be seen. Amoebas may be observed if pond water is used along with some debris from the bottom of the container.

2. Drawings should include numerous ciliates and may include flagellates and amoebas. If the larvae of small crustaceans appear, have the students note their jointed appendages and multicellular nature. These structures will help the students distinguish crustaceans from protozoans.

3. The organisms in the water samples will depend on the source of the water. Pond water from a pond's edge is likely to show a variety of organisms.

Application
Protozoans can cause human diseases. Also protozoans in a water supply may indicate the unwanted presence of organic materials.

Discover More
Results and comparisons will vary depending on the new liquid used.

Post-Lab
Have the students compare their predictions with their results. How accurate were their predictions?

INVESTIGATION

Observing Protozoans

▶ MATERIALS
- medicine dropper
- microscope slide
- stock cultures or pond water
- coverslip
- compound light microscope

▼ PROCEDURE

1. Copy the table shown and complete it for each water sample you examine.
2. Using a medicine dropper, place one drop of water, provided by your teacher, on a microscope slide.
3. Cover the drop of water carefully with a coverslip.
4. Using low power on your microscope, observe the drop of water. Draw and label what you see.
5. Switch to high power. Again draw and label what you see.
6. Try to identify the types of organisms you find. Use the illustrations below to help you.
7. Repeat steps 2-6 with one or more additional water samples.

TABLE 1: OBSERVATION OF PROTOZOANS

Protist Name	Sketch	Description (Size/Shape/Color)	Type of Movement	Type of Feeding

  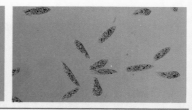

▶ ANALYSES AND CONCLUSIONS

1. What types of protozoans do you see? Flagellates? Ciliates? Amoebas? Use the photographs to help you identify the protozoans.
2. Compare your drawings with those of your classmates. How are the organisms in your samples similar to and different from the organisms in their drawings?
3. How do the organisms from the different water samples compare?

▶ APPLICATION
Why is it important for a scientist or a technician working in a water purification plant to be able to identify protozoans? Explain your reasoning.

Discover More
Ask your teacher to supply you with a different liquid and repeat the investigation. What microorganisms were present in the new liquid? Compare the results of your investigations and explain the similarities and differences.

★ PERFORMANCE ASSESSMENT

The students can demonstrate their ability to distinguish among protozoans by using their drawings as they describe the characteristics of each type of protozoan.

Section 2: ALGAE

FOCUS
Different types of algae and their characteristics are presented in this section. Some uses of algae and their role as a food source are also discussed.

MOTIVATING ACTIVITY

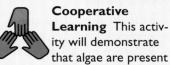

Cooperative Learning This activity will demonstrate that algae are present everywhere. Provide each group with a Petri dish. Ask the groups to pour tap water into the Petri dishes, then store the dishes in a warm location that will receive sunlight during the study of this section. Each day for three days, ask the groups to make slides of the water from their Petri dishes, then observe the slides using low and high power of a microscope. Have the students note and describe any presence of algae growth.

PROCESS SKILLS
- Observing • Comparing
- Applying

POSITIVE ATTITUDES
- Curiosity • Caring for the environment

TERMS
- algae

PRINT MEDIA
The First Plants by Theresa Greenaway (see p. 205b)

ELECTRONIC MEDIA
Algae: The Food Chain, American School Publishers (see p. 205b)

Science Discovery Algae

BLACKLINE MASTERS
Study and Review Guide
Laboratory Investigation 9.1
Thinking Critically

DISCOVER BY *Researching*

Carrageenin is also spelled *carrageenan*.

ONGOING ASSESSMENT

▼ **ASK YOURSELF**

Algae play an integral role in the ocean food chain. Without algae, other ocean organisms will eventually die.

 **LASER DISC**
2745
Algae

▶ **236** CHAPTER 9

SECTION 2

Algae

Objectives

Identify the general characteristics of algae.

Name and **describe** kinds of algae.

Summarize several ways in which algae are important to humans.

D o you know where you might find algae while hiking in the woods? Find some still water and look for the green stuff floating on the surface—that's algae. Believe it or not, you might be able to find algae more easily at home than in the woods! Here's a riddle: What might toothpaste, ice cream, mayonnaise, salad dressing, and instant pudding mix have in common? They may contain extracts of algae that are used to make them thick and "creamy." The next time you brush your teeth, read the ingredient label on the toothpaste while you brush. It's possible that your toothpaste contains algae!

Figure 9–9. These algae are sometimes called *sea lettuce* because they look like something you might find in a salad bowl.

DISCOVER BY *Researching*

Carrageenin, mannitol, and agar are some of the extracts of algae. Read the labels of several food and cosmetic containers. Identify and list products in your home that contain algae extracts.

Characteristics of Algae

Algae are plantlike organisms that contain chlorophyll and perform photosynthesis. They are characterized by several important features. Algae cells contain chloroplasts and have cell walls. Chloroplasts are the organelles in which food, in the form of sugars, is made in the process of photosynthesis. The sugars are then converted into starches, fats, and other compounds. These foods provide energy not only for the algae but also for the many types of organisms that eat algae.

▼ **ASK YOURSELF**

What might happen to other ocean organisms if all of the algae in the oceans suddenly disappeared?

TEACHING STRATEGIES

● **Process Skills:** *Inferring, Applying*

Point out that the kingdom Protista includes both animal-like and plantlike single-celled organisms. Protozoans are the animal-like organisms. The plantlike organisms are called *algae*. Ask the students what the presence of chloroplasts indicates about algae. (Algae are able to produce their own food through photosynthesis.)

● **Process Skills:** *Inferring, Applying*

Photosynthetic plants are not always green. Algae, for example, can be many colors. Ask the students how plants that are not green can perform photosynthesis. (Plant pigments capture and use light energy to make food. Different colors of pigments capture different colors of light. Each type of alga contains a different light-trapping pigment in addition to chlorophyll.)

Types of Algae

Not all scientists agree on which organisms are protists. Generally, though, all algae are classified as protists. Algae are grouped according to the different pigments they contain—green, red, or brown.

Euglenas One type of freshwater alga is the *Euglena* (yu GLEE nuh). *Euglenas* have characteristics of both animals and plants. They are animal-like because they are able to move from place to place. They are plantlike because they have chloroplasts that contain chlorophyll, giving them the ability to make their own food. When a *Euglena* is in an area that does not have enough light for photosynthesis to take place, it can absorb food from the water. Look for *Euglenas* when you do the next activity.

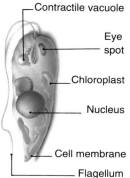

Figure 9–10. The presence of chloroplasts in *Euglena* makes the organism unique. It can move like an animal and make its own food like a plant.

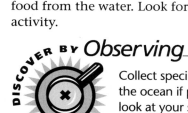 **Observing**

Collect specimens of algae from ponds, ditches, or the ocean if possible. Your teacher can help you look at your specimens under a microscope. In your journal, draw and describe what you see.

Dinoflagellates *Dinoflagellates* (dy noh FLAJ uh layts) are found in the ocean. A microscopic view of these organisms would reveal that each dinoflagellate moves using two flagella. The two flagella sometimes work together to spin the organism like a top!

One type of dinoflagellate is responsible for a situation known as *red tide*. Red tide occurs when dinoflagellates reproduce very rapidly. The toxins, or poisons, that they produce spread through the water, sometimes causing widespread fish kills. After an episode of red tide, beaches are often littered with dead fish. Humans can become ill if they eat shellfish that have absorbed the toxins produced by these dinoflagellates. You can read more about red tide in the article on page 254 at the end of this unit.

Some dinoflagellates have a very interesting characteristic. When the water in which they are living is disturbed by a ship or even a fish, these "fire algae" glow like tiny fireflies! Dinoflagellates contribute in a very helpful way to the balance within ecosystems. They serve as a major food source for animals in the ocean.

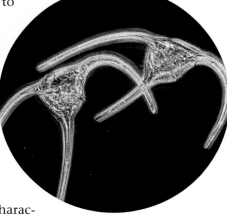

Figure 9–11. Although they are extremely small, dinoflagellates are a primary source of food for many fishes.

DISCOVER BY *Observing*

To help the students identify the different types of algae they may observe, have them compare their observations with the illustrations of a *Euglena*, dinoflagellates, and algae.

SCIENCE BACKGROUND

The production of light by living things is called *bioluminescence*. *Bio* means "living" and *luminescence* means "the emission of light." Like dinoflagellates, fireflies, fire beetles, squids, and lanternfish are bioluminescent organisms.

MEETING SPECIAL NEEDS

Second Language Support

Have the students compile a list of scientific words with irregular plural forms, writing the singular form in one column and the plural in another. (Words so far: cilium—cilia; flagellum—flagella; paramecium—paramecia; alga—algae) The students can add singular and plural forms of new words to their lists. Remind the students to use a dictionary to confirm the forms and spellings.

SECTION 2

GUIDED PRACTICE

 Have the students sketch *Euglena* anatomy using Figure 9–10. Then ask them to classify each labeled structure as animal-like, plantlike, or both. (animal-like: flagellum, eyespot, contractile vacuole; plantlike: chloroplast; both: nucleus, cell membrane) Have the students place their sketches in their science portfolios.

INDEPENDENT PRACTICE

 Have the students write answers to the Section Review and Application questions. In their journals, the students can write a paragraph describing the characteristics that are typical of all algae.

EVALUATION

Have the students explain the importance of algae in ocean food chains. Ask them to compare the role of algae to the role of plants in land-based food chains.

REINFORCING THEMES—Energy

When discussing algae as a food source, stress their contribution to the energy flow through the ocean ecosystem. Point out that tiny dinoflagellates serve as a primary food source for many fishes as well as for the largest organism to ever live on Earth—the blue whale.

ONGOING ASSESSMENT ASK YOURSELF

Toothpaste, ice cream, and pudding are examples of products that often contain extracts of algae.

SECTION 2 REVIEW AND APPLICATION

Reading Critically

1. Like plants, algae contain chlorophyll and perform photosynthesis. Algae cells have chloroplasts and cell walls.

2. Like protozoans, *Euglenas* use flagella for locomotion. Like algae, *Euglenas* have chloroplasts and can perform photosynthesis.

3. A diatom has a two-part shell-like cell wall made of a material similar to glass.

Thinking Critically

4. Algae can perform photosynthesis and serve as a food source for some aquarium organisms. Algae also add oxygen to the aquarium water.

5. Some land areas were once covered by oceans in which diatoms and other ocean organisms lived and died.

▶ **238** CHAPTER 9

Figure 9–12. Diatoms can be found in many different and beautiful shapes.

Diatoms *Diatoms* (DY uh tahmz) are golden brown algae and are the most common of all the single-celled organisms in the oceans. Diatoms are another important food source for animals that live in the water.

If you looked at a diatom with a microscope, you would see an organism that is very beautiful. The cell walls, or shells, of diatoms are made of a chemical similar to glass. The shells of diatoms are made up of two parts. The parts are of unequal size, with the larger part fitting over the smaller like the lid of a box.

When diatoms die, their empty shells collect on the ocean floor to form *diatomaceous* (dy uh tuh MAY shuhs) *earth*. Diatomaceous earth is a useful resource; it is collected, processed, and used in making insulation, toothpaste, and silver polish.

Green, Red, and Brown Algae

Algae live in many places, not just in water. You may have seen green algae in forests, growing on the trunks of trees. Green algae are also found in oceans, lakes, ponds, and swimming pools.

Although all algae are classified as protists, some are multicellular and grow to be quite large. Kelp, a type of brown algae that lives in the oceans, can grow to 50 m or more in length. Red algae also live in the oceans, attached to rocks on the ocean bottom. Red and brown algae are often used as thickening agents in ice cream and puddings.

Figure 9–13. Red algae (right), brown algae (center), and green algae (left) all contain chlorophyll. However, the red and brown algae also contain other pigments.

 ASK YOURSELF

What useful products contain algae?

SECTION 2 REVIEW AND APPLICATION

Reading Critically

1. How are algae like plants?
2. How are *Euglenas* like protozoans? Like algae?
3. Describe a diatom.

Thinking Critically

4. How can algae in an aquarium be helpful?
5. Why are large deposits of diatomaceous earth found inland despite the fact that diatoms live in the ocean?

RETEACHING

Display pictures of algae and green plants. Ask each student to give an oral explanation of how algae are like plants.

EXTENSION

The ability to emit light is an interesting property of many dinoflagellates. Have the students use reference materials to discover exactly how these organisms create light and then explain their findings to the class.

CLOSURE

Have the students write a paragraph in their journals that describes several ways in which algae positively and negatively affect other species.

SKILL

Organizing Data

Process Skills: Classifying/Ordering, Communicating, Evaluating

Grouping: Groups of 2

Objectives
- **Design** a method of organizing information.
- **Communicate** data using outlines and tables.
- **Evaluate** various methods of organizing data.

Discussion
The ability to organize data is an important skill. Although there are many ways to organize data, the students should understand how to organize data using outlines, tables, and graphs.

▶ **Application**

1. Table structure should demonstrate an understanding of how data can be organized clearly.

2. Outline structure should follow accepted formats.

✳ **Using What You Have Learned**

1. Some students may not have a preference. Whatever their preference, the students should support their choice with reasons.

2. Information can be easily located in organized graphs.

3. Each student should choose one method and describe the advantages and disadvantages of that method.

SKILL   Organizing Data

Sometimes it is easier to remember data if it is organized in some logical way. Scientists often have huge amounts of data that they need to use for their research. To organize the data you have read in a chapter, you can use some of the skills scientists use. Then, in addition to learning a skill, you will be studying at the same time.

▼ PROCEDURE

1. Go over the material you need to organize, and decide on the method you wish to use. You may wish to make an outline, a table, or a graph.
2. If you are going to use an outline, decide on the major ideas. Label the main ideas with Roman numerals.
3. Beneath each main idea, list topics related to the idea. Label these topics with capital letters.
4. Beneath each topic, list two concepts. Label each concept with an Arabic numeral.
5. If you are going to use a table, chart, or graph, sketch the way in which you are going to present the data. Decide on a title, headings, and the number of columns or sections.
6. Under the proper headings, list the material to be presented.
7. If measurements are needed, be sure to include the correct units.
8. Go through the material step by step to include all the important facts.
9. You may need to modify your organizational design as you actually add information.

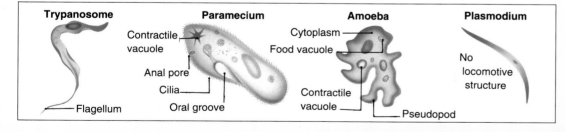

▶ **APPLICATION**
1. Organize the information about protozoans in a table. In your table, include shape, movement, and location of the organisms.
2. Organize the information about algae in an outline. In your data, include type of pigment, appearance, and uses.

✳ **Using What You Have Learned**
1. Which method of organization did you find most useful? Why?
2. Which method of organization made it easiest to locate information later?
3. Look at the diagrams and labels. Use any method described in this activity to organize the data in the diagrams. List the advantages and disadvantages of the method you chose.

Have the students place their tables and outlines from the *Skill* activity in their science portfolios.

Section 3: FUNGI

FOCUS

This section introduces the characteristics of fungi and examines the common groups of fungi, including sporangium fungi, club fungi, sac fungi, and imperfect fungi. Unusual protists such as lichens and slime molds are also presented.

MOTIVATING ACTIVITY

Display a variety of edible fungi for the students to examine. **CAUTION: Do not allow the students to taste any of the fungi. Some people may exhibit allergic reactions to them.** You could include button mushrooms, straw mushrooms, snow mushrooms (white fungus), cloud ears (brown fungus), tree fungus, wood ears, and morels. Make sure that the students understand that some fungi are extremely poisonous and others contain substances that alter the functions of the brain. Therefore, it is not advisable to eat wild mushrooms that have not been **positively** identified as safe.

PROCESS SKILLS
- Classifying/Ordering
- Observing • Comparing
- Evaluating

POSITIVE ATTITUDES
- Curiosity • Precision

TERMS
- fungi • lichen • slime molds

PRINT MEDIA
North American Mushrooms by Gary H. Lincoff (see p. 205b)

ELECTRONIC MEDIA
Fungi: Activity and Structure, American School Publishers (see p. 205b)

Science Discovery Fungi definition

BLACKLINE MASTERS
Study and Review Guide
Laboratory Investigation 9.2
Connecting Other Disciplines
Extending Science Concepts

SCIENCE TECHNOLOGY SOCIETY Often the immune system of the recipient of an organ transplant rejects a transplanted organ by treating the transplanted tissue as foreign and attacking it. Cyclosporine, a drug derived from fungi, suppresses the immune system of an organ recipient and increases the likelihood that the recipient's immune system will not reject the transplanted tissue.

▶ 240 CHAPTER 9

SECTION 3

Fungi

Objectives

Describe how fungi are different from other organisms.

Sketch the body structure of a typical fungus.

Summarize some of the ways in which fungi are important to humans.

What would you say if you went to the school cafeteria for lunch one day and you were given a plate of fungus? You'd probably say, "No, thank you!" The idea of eating fungus does not interest most people. However, some types of fungi can be very tasty. Have you ever eaten mushrooms? Mushrooms are a type of fungus. Have you ever eaten a slice of bread? Bread dough rises with the help of a fungus. Have you ever eaten cheese? Certain types of fungi help produce cheese. So the next time you are offered a plate of food that was produced by fungi, you might want to say, "Yes, thank you!"

Characteristics of Fungi

Fungi (FUHN jy) [singular, *fungus*] are organisms that decompose organic, or once-living, material. Sometimes called nature's recyclers, fungi cannot produce their own food the way plants can, and they do not have mouths to take in food as you and other animals do. Fungi must absorb their food through their cell walls.

Fungi may be *saprophytes* (SAP ruh fyts) or parasites. Saprophytes obtain food from dead organisms or from the waste products of living organisms. Parasites obtain food from a living host, usually causing injury or disease in the host.

The bodies of fungi are made of a network of threadlike structures called *hyphae* (HY fee). Fungi absorb food through their hyphae. Many hyphae together form a group of interlocking threads called the *mycelium* (my SEE lee uhm).

Most fungi reproduce by forming *spores*. The structures of a fungus that form spores are usually located above the surface on which the fungus grows. Because of this, the reproductive spores can be easily blown to different places by the wind. In the next activity, you can find out what some fungus spores look like.

TEACHING STRATEGIES

● **Process Skills:** *Applying, Inferring*

Display some fungi that form on food. (Bread without preservatives and orange rinds will readily form mold if left exposed to air and kept moist. It is advisable to cover such materials with plastic wrap because fungal spores may cause the students to sneeze, and the odor of decaying materials can be very unpleasant.) Explain that you did not add anything to these food items; they were simply left exposed to moist air. Ask the students what this reveals about the air we breathe. (The air carries spores of fungi that grow when they land on a suitable surface.)

● **Process Skills:** *Comparing, Inferring*

Remind the students that it is unsafe and extremely dangerous to eat wild mushrooms without expert knowledge of mushroom species. For example, one poisonous mushroom called the death cap mushroom can be fatal after only a few milligrams of the mushroom are eaten. Stress that expert knowledge about mushrooms is essential, because to the untrained eye, poisonous varieties of mushrooms can look very much like their nonpoisonous "relatives."

 DISCOVER BY *Doing*

You will need at least one cap from a gilled mushroom. (*Gilled* means the cap has ridges under it.) Place the cap gilled side down on a piece of paper. Put a damp cotton ball on top of the cap. Place a glass or plastic cup over the cap. Several hours later, remove the cap. The spores will have made a distinctive pattern on the paper. This pattern can be used to identify the mushroom.

▼ **ASK YOURSELF**

How do fungi obtain food?

Types of Fungi

The common groups of fungi are the sporangium (spaw RAN jee uhm) fungi, the club fungi, the sac fungi, and the imperfect fungi. Lichens and slime molds are grouped separately even though they are similar to fungi in many ways.

Sporangium Fungi Have you ever seen a neglected piece of bread become "fuzzy"? This fuzzy substance is a fungus called *bread mold*. Molds grow best in warm, moist, dark places. However, if you have ever seen spoiled food from a refrigerator, you know that molds can also grow in cold temperatures.

Sporangium fungi are named after the spore case, or *sporangium* [plural, *sporangia*], in which the spores are produced. As you can see in the illustration, the sporangia are held in the air by hyphae. When a spore case breaks open, the spores are released. These spores may remain inactive for months. When the conditions are right, the spores will produce new hyphae.

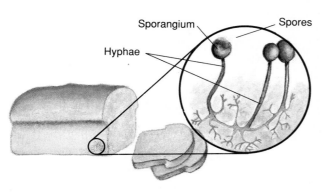

Figure 9–14. Magnified bread mold resembles tiny spheres sitting on top of thin stems. The sphere-shaped structures are the sporangia. The "stems" are the hyphae.

 DISCOVER BY *Doing*

Use only commercially produced mushrooms and white paper. If the caps do not produce a clear spore pattern after several hours, repeat the activity and allow the materials to stand for a longer period. Have the students place their spore prints in their science portfolios.

ONGOING ASSESSMENT
▼ **ASK YOURSELF**

Fungi obtain food by decomposing dead organisms or the waste products of living organisms. Fungi then absorb the matter through their cell walls.

Demonstration

Dissolve 25 g of sugar in 100 mL of water and store the solution in a dark place. Describe the ingredients of the solution to the students and ask them to predict what will happen. Allow the students to observe any changes in the solution. (They should observe white, cottonlike material growing in spots.) After 48 hours, have the students make slides of the growths in the water and observe the growths using a microscope. (The students should see the structures of mold.) **CAUTION: Students who are allergic to molds should not make their own slides.**

SECTION 3 **241** ◀

TEACHING STRATEGIES, continued

● **Process Skills:** *Inferring, Applying*

Discuss with the students the role of fungi in an ecosystem. Explain that many fungi act as decomposers; they break down matter from dead plants and animals and from the waste products of living organisms. Some nutrients from this organic, or once-living, matter are absorbed by the fungi while other compounds are returned to the soil. The decomposition of matter by fungi replenishes the soil by providing compounds that are needed for the growth, development, and reproduction of plants. Plants, in turn, use these compounds to manufacture nutrients for themselves. Some animals feed on plants, and the nutrients obtained from the plants cycle through their bodies. Ask the students to recall the name for fungi that obtain food from dead organisms or from the waste products of living organisms. (saprophytes)

● **Process Skills:** *Predicting, Generating Ideas*

The students may be interested to learn that some varieties of sac fungi are eaten as truffles—a costly gourmet food. Point out that sac fungi can be found in a variety of colors. The text discussion of sac fungi lists several benefits and harmful effects of

---

✧ **Did You Know?**
Fungi can grow in many unusual places. For example, some fungi can grow in jet fuel and can clog fuel lines.

MULTICULTURAL CONNECTION

Cooperative Learning Fungi, especially mushrooms, play an important role in ethnic cooking. Ask groups of students to visit the international foods section in a large supermarket or different ethnic food stores, or look through cookbooks or other reference materials to find out how different kinds of fungi are used in the cuisines of different countries. Have each group share its findings with the class.

SCIENCE BACKGROUND

Rusts and smuts cost money when they destroy crops. But ruined crops also mean a reduced food supply. In developing countries with little access to the chemicals or methods used to fight the diseases, rusts and smuts can mean famine.

LASER DISC
2829

Fungi definition

▶ 242 CHAPTER 9

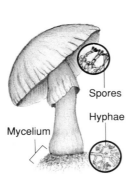
Figure 9–15. Mushrooms (left and right) and puffballs (center) are club fungi. Mushrooms produce spores under the cap.

Club Fungi The mushroom, the puffball, and the bracket fungus are all examples of club fungi. They are called *club fungi* because the structures in which spores are produced look like clubs, as shown in the illustration.

Sometimes fungi destroy agricultural crops. The rusts and smuts, shown in the photographs, are parasitic club fungi that destroy grain. Each year rusts and smuts cost the agricultural industry thousands of dollars in lost grains.

Figure 9–16. Corn smut (left) and wheat rust (right) destroy many crops. Scientists are working to develop types of corn and wheat that are resistant to these fungi.

Sac Fungi Sac fungi produce spores in sacs. Yeasts, one type of sac fungi, are important in the process of fermentation.

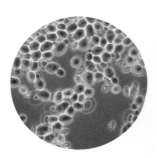

During the fermentation of bread dough, yeasts release carbon dioxide causing bread dough to rise.

Some sac fungi are parasites. Dutch elm disease and chestnut blight are tree diseases caused by powdery mildew, a sac fungus. In Dutch elm disease, the fungus is spread by bark beetles.

Figure 9–17. The morel (right) is an edible sac fungus. Yeasts (above) are used to make many food products.

sac fungi. Ask the students to predict other benefits and negative effects of sac fungi. (Responses should include brewing and crop destruction.)

GUIDED PRACTICE

Ask the students why saprophyte fungi are called "nature's recyclers." (These fungi decompose organic, or once-living, material, obtaining food from the material and returning compounds to the soil needed for plant growth.)

INDEPENDENT PRACTICE

 Have the students write answers to the Section Review and Application questions. Ask the students to write a paragraph summarizing the characteristics of fungi.

ACTIVITY

What are the characteristics of fungi?

Process Skills: Observing, Comparing

Grouping: Groups of 3 or 4

Hints

Obtain several mushrooms from a grocery store before the activity. Use a razor blade to slice thin sections of the mushrooms. Prepare both cross sections and longitudinal sections for the students to observe.

▶ **Application**

1. The students may be able to identify the cap, gills (the structures under the cap), stipe (the stemlike structure), hyphae, and spores.

2. Comparisons will depend on the observed structures.

3. Fungi have a simple reproductive process; the dead material from which they obtain food is usually abundant; and they can live in damp, dark conditions.

ACTIVITY

What are the characteristics of fungi?

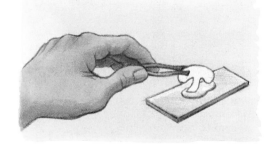

MATERIALS
medicine dropper, slides (2), forceps, samples of fungi, coverslips (2), compound light microscope

PROCEDURE
1. Refer to the Reference Section on page 574 to learn how to make a wet mount. Using a medicine dropper, place one drop of water in the middle of a slide. With forceps, place a thin piece of mushroom or other fungus on your slide.
2. Put a coverslip on the slide. Look at your slide under the microscope. Make a sketch of what you see. Try to identify the hyphae and spores in your sample.
3. Using an additional sample of fungus, make another wet mount.
4. Repeat step 2.

APPLICATION
1. What structures were you able to identify in your samples?
2. How do the two samples differ? How are they similar?
3. Why do fungi successfully grow and reproduce in so many different places?

PERFORMANCE ASSESSMENT

The *Activity* gives you an opportunity to evaluate the students' ability to make a wet mount and use a microscope correctly. Ask them to prepare and view another wet mount, describing the procedure as you listen.

Imperfect Fungi Imperfect fungi are called *imperfect* because their complete reproductive cycles are unknown. The fungi that cause ringworm and athlete's foot belong to this group.

Although many imperfect fungi cause disease, one type of imperfect fungus can help cure disease. Have you ever had an illness for which you took an antibiotic? Maybe you were given penicillin. The imperfect fungus *Penicillium* is the source of the antibiotic penicillin. The antibiotic properties of *Penicillium* were discovered by British scientist Alexander Fleming in 1928. Fleming noticed that a mold was stopping the growth of bacteria in laboratory cultures. After many tests, Fleming isolated penicillin, and its use since then has saved hundreds of thousands of lives.

Figure 9–18. *Penicillium* mold is shown under a microscope (left) and on fruit (above).

SECTION 3 **243**

EVALUATION

Ask the students to compare the reproductive structures of sporangium fungi, club fungi, sac fungi, and imperfect fungi. (sporangium fungi: spores produced in spore cases; club fungi: spore structures look like clubs; sac fungi: spores produced in sacs; imperfect fungi: reproductive cycle not thoroughly understood)

RETEACHING

Let each student examine a commercially purchased club mushroom. **CAUTION: The students should not eat any part of the mushroom.** Ask the students to sketch their mushroom in their journals and label as many structures as possible.

EXTENSION

Point out that people often associate agriculture with crops such as corn and wheat. However, crops of mushrooms are also grown. Ask the students to find out about commercial mushroom farming and share their findings with the class.

CLOSURE

Cooperative Learning Working in small groups, ask the students to create a list that describes two ways in which sporangium fungi, club fungi, sac fungi, imperfect fungi, lichens, and slime molds each might benefit or harm other organisms on Earth.

ONGOING ASSESSMENT
ASK YOURSELF

Responses might suggest that in a helpful way, fungi decompose organic materials and return essential compounds to the environment, and in a harmful way, some fungi destroy crops.

ONGOING ASSESSMENT
ASK YOURSELF

Protist; slime molds cannot be classified as fungi because when food is abundant, slime molds do not display characteristics that are typical of fungi.

SECTION 3 REVIEW AND APPLICATION

Reading Critically

1. The body of a common fungus is made up of threadlike hyphae. Hyphae are used for absorbing food.

2. Lichens are made up of organisms from two different kingdoms: algae (protists) and fungi. Slime molds creep, like amoebas (protists), and reproduce by forming spores, like fungi.

Thinking Critically

3. Because there is no shortage of waste products or dead organisms, saprophytes usually have abundant food supplies.

4. The alga in a lichen makes food for the organism, while the fungus retains water.

Figure 9–19. Lichens are often the first organisms to grow on bare ground.

Figure 9–20. In one stage of its life cycle, a slime mold is an amoebalike growth.

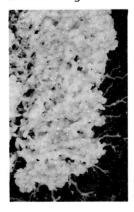

Discover by Writing

Choose the fungus that you find most interesting. Then in your journal draw and label the parts of that fungus. Describe the importance of the fungus.

▼ **ASK YOURSELF**

Are fungi helpful or harmful to the environment? Explain your answer.

Lichens and Slime Molds

A **lichen** is an organism that is part fungus and part alga. This combination produces an organism very different from either one. The alga in the lichen makes the food for the organism. How is this accomplished? The fungus gives the lichen support and helps it retain water. Lichens can be found in places where neither fungi nor algae can be found alone. Lichens can often be found attached to bare rocks.

Slime molds are very unusual organisms. They may look and reproduce like fungi, but they move like amoebas. Slime molds are often brightly colored and are found in damp soil and on rotting leaves or logs. The body of a slime mold behaves like an amoeba. It creeps along the ground or other surfaces and engulfs, or captures, food particles. When food is scarce, the slime mold becomes a spore-forming body, like a fungus. The spore-forming body releases spores that grow into new amoebalike organisms.

▼ **ASK YOURSELF**

If you had to identify a slime mold as either a fungus or a protist, which term would you choose? Why?

SECTION 3 REVIEW AND APPLICATION

Reading Critically
1. Describe the body of a common fungus.
2. Relate several reasons why lichens and slime molds are difficult to classify.

Thinking Critically
3. What advantages do fungi have as saprophytes?
4. How are lichens able to grow on the surfaces of bare rocks?

CHAPTER 9 HIGHLIGHTS

The Big Idea—SYSTEMS AND STRUCTURES

Help the students understand that although protozoans, algae, and fungi contribute in harmful and helpful ways to our environment, their overall effect on us is positive. Without these organisms, the world in which we live would change significantly. For example, without algae to eat, many ocean organisms would die. Without fungi to decompose organic matter, soil would not contain the nutrients that plants need to grow.

The ideas expressed by the students should reflect the understanding that microorganisms occupy an integral niche in the environment of the earth. You might wish to ask the students to describe the implications of a world without protozoans, algae, or fungi.

CONNECTING IDEAS

The concept map should be completed with words similar to those shown here.

Row 2: Algae

Row 4: the way they move; contain chlorophyll; perform photosynthesis; organic material

---

CHAPTER 9 HIGHLIGHTS

The Big Idea

Protozoans, algae, and fungi are very small organisms that make a large contribution to the world around us.

Protozoans comprise a diverse group of organisms that interact with the environment in helpful and harmful ways. While some protozoans cause disease, others feed on dead material, helping to replenish the supply of nutrients in the water environment.

Algae represent a primary food source for various aquatic organisms. Photosynthesis by algae also contributes tremendously to the earth's supply of oxygen. Some scientists estimate that algae release more oxygen into the atmosphere than all other land plants combined.

Without fungi, the speed with which organic materials decay would be decreased. Fungi also serve as a food source but are perhaps most noteworthy for their contribution to the world of germ-fighting antibiotics.

Think about what you wrote in your journal at the beginning of the chapter. What have you learned about microscopic organisms? Revise your journal entry to reflect this new information. Did you include your ideas on the importance of protists and fungi to the environment?

Connecting Ideas

Copy this unfinished concept map into your journal. Complete the concept map by writing the correct term in each blank.

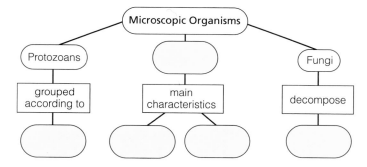

CHAPTER 9 **245**

CHAPTER 9 REVIEW

ANSWERS

Understanding Vocabulary

1. Protozoans and algae are generally classified as protists. Protozoans have some animal-like characteristics. Algae have some plantlike characteristics.

2. Algae are plantlike organisms that contain chlorophyll and perform photosynthesis. Fungi are organisms that cannot produce their own food; they decompose organic material and absorb food through their cell walls.

3. Most protozoans capture food and react to environmental stimuli, as animals do. Fungi cannot capture or produce their own food; they decompose organic material and absorb food through their cell walls.

4. Flagella and pseudopods are structures used for movement by two types of protozoans. Flagella are long, whiplike structures used by flagellates. Pseudopods are projections of cytoplasm used by sarcodines.

5. Ciliates and sporozoans are both protozoans. Ciliates use hairlike structures, or cilia, to move. Sporozoans are parasites that cannot move from place to place.

6. Lichens and slime molds are both hard to classify. Lichens are part fungus and part alga. Slime molds move like amoebas and reproduce like fungi.

Understanding Concepts

Multiple Choice

7. c
8. d
9. d
10. a
11. c

Short Answer

12. Like plants, amoebas can release excess water and become inactive during unfavorable conditions and active during favorable conditions. Like animals, amoebas can move and capture food.

13. Yes; fungi help break down organic material into components that are necessary for the survival of other organisms.

Interpreting Graphics

14. Water from the stream can contain microorganisms such as amoebas, which can cause serious illnesses in human beings.

15. A person should drink the water from the stream only if it has been treated to remove or kill the microorganisms.

CHAPTER 9 REVIEW

Understanding Vocabulary

For each set of terms, explain the similarities and differences in their meanings.

1. protozoans (230), algae (236)
2. algae (236), fungi (240)
3. fungi (240), protozoans (230)
4. flagella (233), pseudopods (231)
5. ciliates (232), sporozoans (234)
6. lichens (244), slime molds (244)

Understanding Concepts

MULTIPLE CHOICE

7. Which of the environmental conditions below is most likely to affect a protozoan?
   a) availability of food particles
   b) presence of light
   c) absence of water
   d) lack of oxygen

8. Choose the relationship that best describes the parasite/host relationship.
   a) A parasite seldom harms its host.
   b) A host seldom harms its parasite.
   c) A parasite frequently benefits its host.
   d) A host frequently benefits its parasite.

9. What advantage might a flagellate organism have when compared to an amoeba?
   a) efficiency of digestion
   b) efficiency of reproduction
   c) ease of respiration
   d) ease of movement

10. Which organism moves in a way that looks similar to a fish's movements?
    a) flagellate
    b) sarcodine
    c) sporozoan
    d) slime mold

11. Choose the way in which protists, algae, and fungi least benefit humans.
    a) They are a source of food.
    b) They replenish the supply of oxygen.
    c) They are a source of disease.
    d) They serve as a food source for other organisms we eat.

SHORT ANSWER

12. In what ways are amoebas similar to plants? In what ways are amoebas similar to animals?

13. Is it accurate to refer to fungi as decomposers? Explain your reasoning.

Interpreting Graphics

14. Why is it not a good idea to drink water from a mountain stream like the one shown in the picture?

15. Under what conditions would it be acceptable to drink water from a mountain stream?

246 CHAPTER 9

Reviewing Themes

16. Responses may describe how dinoflagellates serve as a major food source in the ocean environment, yet they can also cause red tide, which results in widespread fish kills.

17. Plants require nutrients for growth. They absorb these nutrients from the soil. Over time, the soil would become depleted if fungi did not decompose, or break down, organic material and return nutrients to the soil for plants to use.

Thinking Critically

18. Developing countries may lack sanitation facilities for disposing of human wastes. Human wastes can carry the amoeba that causes amoebic dysentery. If these wastes contaminate food or drinking water, people who eat the food or drink the water can contract the disease.

19. Diatomaceous earth contains the hard, glasslike shells of diatoms. These shells give silver polish a slightly abrasive quality suitable for polishing silver.

20. Responses should explain that because lichens can produce their own food, they can grow on bare rock, including cooled lava. Over time, the lichens break up the rock and produce a small amount of soil, then mosses will begin to grow. This is the second stage of primary succession.

21. Athlete's foot can be prevented by drying feet thoroughly after bathing or showering and by keeping feet, socks, and shoes clean and dry. This prevents the fungus spores, if they are present, from germinating. These measures are effective because the spores cannot germinate and grow unless there is moisture present.

22. The microscopic spores of these fungi are numerous and lightweight. They are carried everywhere by air currents and can circulate, for example, in closed air conditioning systems. The spores will germinate as soon as they land where a source of food is present and the environment is sufficiently moist and dark.

Reviewing Themes

16. *Systems and Structures*
Choose a specific type of alga and explain how that alga interacts positively and negatively with its environment.

17. *Energy*
Explain how forest fungi replenish the supply of energy to the forest community.

Thinking Critically

18. Why is amoebic dysentery a common problem in less developed countries?

19. Why is diatomaceous earth a good silver polish?

20. Lichens are often the first organisms to grow on cooled lava after a volcanic eruption. What characteristics of lichens enable this growth to occur? What organisms might grow and what processes might occur after the lichens are established on the lava?

21. You may have heard of ways to avoid getting athlete's foot. Based on what you know about how fungi grow, explain why these preventive measures are effective.

22. Why is it so difficult to control the spread of fungi such as mildew or bread mold?

Discovery Through Reading

Morgan, Nina. *Louis Pasteur*. Bookwright Press, 1992. This biography details Pasteur's life, dreams, family, and accomplishments in chemistry and microbiology.

Science PARADE

SCIENCE APPLICATIONS

Background Information

On the chalkboard, draw a horizontal line approximately one meter in length. Tell the students that the line represents a fraction bar. Above the line, write the number 10 000 000 and point out to the students that about 10 000 000 people in the world are infected with the AIDS virus. Below the line, write the number 5 000 000 000 and point out to the students that there are about 5 billion people in the world. Then simplify the fraction by erasing a zero from the number at the top of the fraction and erasing a zero from the number at the bottom of the fraction, continuing this pattern until there are no zeros left in the number at the top of the fraction. At this point, ask a volunteer to read the fraction (1/500) and explain to the students that worldwide, one out of every 500 people is infected with the AIDS virus.

SCIENCE APPLICATIONS

AIDS Research

Scientists have called it the modern plague. By the early 1990s, it had infected more than nine million people worldwide. This disease is AIDS. AIDS destroys the body's ability to fight infection. More than a decade after discovering it, scientists are still seeking ways to treat and prevent this deadly disease. Medical science still has not solved the mystery of AIDS.

Mystery Killer

The events surrounding the discovery of AIDS unfolded like a detective story. Physicians and scientists observed the first signs of the disease in the late 1970s. They noticed patients dying from a normally rare skin cancer and a usually mild form of pneumonia. Scientists suspected that something was weakening the patients' natural defenses against these infections. At the Centers for Disease Control in Atlanta, Georgia, physicians named the new disease AIDS and began looking for a cause.

A group of researchers was already investigating a family of viruses that attacked white blood cells, or lymphocytes. Lymphocytes control the body's immunity. Any virus that destroys lymphocytes leaves the body defenseless against pneumonia, cancer, and many other diseases. Researchers suspected that a virus of this type caused AIDS.

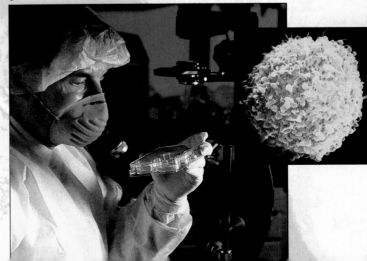

248 UNIT 3

Discussion

● **Process Skills:** *Inferring, Applying*

Lead the students in a discussion of the seriousness of the disease AIDS by asking the students to suggest reasons why AIDS should be considered a serious global health problem. Make sure the students understand that AIDS is transmitted primarily through sexual contact and intravenous drug use. It is essential that the students realize that they can significantly decrease their likelihood of contracting the disease by avoiding or protecting themselves against behaviors that may expose them to the virus. Encourage the students to guard against the naive attitude that "It will never happen to me."

Journal Activity
Many of the stories of people stricken with the AIDS virus are filled with courage and inspiration. Ask the students to use reference materials and write in their journals about a person such as Dawn Marcal, Ryan White, Mary Fisher, or Arthur Ashe.

Left: Dawn Marcal, infected as a teenager in high school, was a speaker for AIDS Education. She died in August, 1992.
Center: Magic Johnson, former basketball star, revealed that he is HIV positive in November of 1991.
Below: AIDS research

In 1984—less than three years after the first AIDS cases were identified—researchers successfully isolated the AIDS virus in the laboratory. They named it *Human Immunodeficiency Virus* (HIV). Scientists soon learned how the virus works. When HIV first enters the body, it stimulates the growth of antibodies, chemical substances that fight infection. However, the antibodies cannot stop the virus. Soon the virus attacks the body's lymphocytes, and the body cannot fight off infections and disease.

Possible Vaccines

The search for an AIDS vaccine illustrates the difficulties of AIDS research. One problem is making the vaccine itself. Scientists have tried deactivating the HIV virus and then injecting it into the body. They hope the body will respond by producing antibodies to fight HIV infection. In another approach, researchers alter the genetic makeup of living HIV. The altered virus does not cause AIDS but can reproduce in the patient's body, stimulating the immune system.

No Easy Task

Scientists may have to develop several AIDS vaccines. Recent findings indicate that a vaccine against intravenous infection may not guard against sexual transmission. A different vaccine may protect a fetus in the womb from infection by an HIV-positive mother. Doctors would like to vaccinate HIV-positive patients against the onset of AIDS. Others believe that an effective vaccine can only prevent an HIV carrier from passing the virus to others.

The nature of HIV complicates the search for a vaccine. First, the AIDS virus is highly variable. It can alter itself genetically within the person's body. In fact, the AIDS virus may have evolved genetically just in the years since its discovery.

Second, the human virus infects only humans, which complicates testing. Researchers can infect monkeys with a simian variety of HIV. However, a vaccine that is then found effective on monkeys may not work on humans.

Finally, reliable human testing requires that scientists give a vaccine to a test group and withhold it from a control group. If the vaccine fails, the test group might contract HIV. If it succeeds, AIDS patients might demand the vaccine before a reliable test is completed.

AIDS Drugs

The same problems apply to AIDS drug research. The antiviral drug AZT has become the standard treatment for HIV infection and AIDS. AZT apparently blocks the virus from taking over the DNA of white blood cells. Researchers are still testing AZT to determine its effectiveness. While the drug may slow AIDS onset, it does not necessarily extend the patient's life. Researchers are also seeking drugs to treat AIDS-related infections and diseases. ◆

READ ABOUT IT!

Discussion

Allow the students sufficient time to quietly read the selection, or encourage volunteers to read selected portions of the selection to the rest of the class.

- **Process Skills:** *Inferring, Expressing Ideas Effectively*

After the students have read about the discoveries of Anton van Leeuwenhoek, ask them to name some of the characteristics Leeuwenhoek displayed that are often shown by scientists and researchers. (He was extremely curious and asked many questions. Instead of leaving the questions unanswered, Leeuwenhoek was persistent and designed, conducted, and repeated experiments in the hope of answering the questions he had posed.)

- **Process Skills:** *Inferring, Expressing Ideas Effectively*

Explain to the students that during the course of a typical scientific investigation, scientists and researchers often create more questions than they answer. Ask the students to suggest reasons why questions are essential to

---

READ ABOUT IT!

Through the Microscope

by Arthur S. Gregor
from *Man's Conquest of Nature from Ancient Times to the Atomic Age*

"Gentlemen of the Royal Society, Honored Sirs. My name is Anton van Leeuwenhoek and I am the sheriff's deputy and janitor of the town hall here in the city of Delft in Holland, where I was born. I write to you because I have a hobby in which you might be interested. Much of my time I spend cleaning the town hall, but every moment I have to spare I arrange lenses in such a way that the tiniest things in the world are enlarged. In fact, I build microscopes, 247 of them so far.

Through my microscopes I look at hairs, at seeds, at the sting of bees and at the brains of flies. Gentlemen, I have seen things no man has ever seen before!"

At first [Anton van Leeuwenhoek] could not believe his eyes. He called out to his family to come and see the marvel. Under the lenses of his microscope, imprisoned in the water, were tiny creatures, turning, twisting, and leaping about.

Some swam with the help of tiny hairs. Some rolled over and over. Some actually changed their shape as they went lumbering from one side to the other. The drop of water was alive with thousands of little animals.

Leeuwenhoek had stumbled upon the world of microbes!

Here was a new world of which men had never dreamed, a world so small that a million of these creatures could fit into a thimble. Here was the cause—though Leeuwenhoek did not know it—of many of the diseases that have plagued mankind since the beginning of time.

He was too good a scientist to rest content with his discovery. Question after question rushed to his mind.

Where did these amazing little creatures come from? Could they have come out of the sky? How could he find out?

Well, he'd experiment as he always did when he came upon a new problem.

He cleaned and scoured a dish until it shone. Then

the work of scientists and researchers. (Lead the students to suggest that questions spark research. Without questions, the need for research would not exist. Without research, the body of scientific knowledge would not increase.)

Journal Activity
The development of the microscope by Leeuwenhoek is just one invention or discovery that was made possible by a curious scientist or investigator. Have the students describe other discoveries or inventions that were the result of scientific curiosity.

EXTENSION

Remind the students that Anton van Leeuwenhoek sought the acceptance of the Royal Society. Challenge interested students to discover more about the Royal Society and share their research with their classmates.

---

he caught a drop of rainwater directly in it. Now again he looked. But this time not a trace of the creatures!

He repeated the experiment over and over. Still no trace of them. So they did not come out of the sky. Well, then where did they come from?

He put the dish aside for several days and then looked again.

Ah, there they were, full of life, frolicking amidst bits of dirt and dust.

Now Leeuwenhoek turned detective and set to work to track down his creatures. He examined water from housetops, from mud puddles, from barrels, from wells, from the canals of Delft. And wherever there was dust and dirt he was sure to find his creatures.

Then he had a curious thought. "Why go out to find them. Can I have them come to me?"

He allowed a few grains of pepper to soak in water. And what he saw several days later sent him scribbling like mad to the Royal Society. The pepper water was alive with the creatures, millions of them, more than there were people in the whole of Holland.

Leeuwenhoek's simple microscope

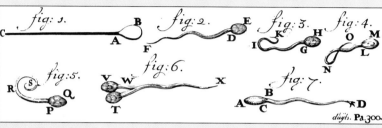

Anton van Leeuwenhoek had discovered one of the most important of all laboratory methods: a way of growing microbes!

"Nonsense!" cried some of the members of the Royal Society when Leeuwenhoek's letter came in 1680. "How much of this foolishness do we have to take?"

But Robert Hooke, a great experimenter himself and member of the Royal Society, did not shout "Nonsense!" at the man

Leeuwenhoek's drawings

who had seen the circulation of the blood with his own eyes.

Hooke built a new and powerful microscope and brewed pepper water himself. And what *he* saw sent him dashing back to the Royal Society.

"Gentlemen, our Dutch friend has not lied. Come and see for yourself."

The scientists and the professors crowded around Hooke's microscope. It was just as Leeuwenhoek had said; the water was teeming with microbes.

It was the greatest day in Anton van Leeuwenhoek's life when the reply from the Royal Society arrived. For along with it was a magnificent diploma in a beautiful silver case.

The sheriff's deputy and janitor of the town hall of the city of Delft was now an honored member of the Royal Society! On his death in 1723 he repaid the Society by leaving it his twenty-six most precious microscopes. ◆

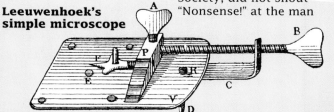

UNIT 3 251

THEN AND NOW

Discussion

● **Process Skills:** *Analyzing, Applying*

When discussing the life and work of Robert Koch, help the students recognize the persistence Dr. Koch displayed by pointing out that he incorrectly believed that he had found a cure for tuberculosis, yet he did not stop his work simply because he had made a mistake. Ask the students to explain how persistence or resiliency is an admirable characteristic in both scientists and nonscientists alike.

As the students read about and discuss the life and work of Lynn Margulis, point out that Margulis was one of the first researchers to suggest that cells with nuclei developed from cells without nuclei. Remind the students of the importance of symbiosis in the work of Dr. Margulis.

Journal Activity
Remind the students that the theories of Lynn Margulis were not always readily accepted by the scientific community. Ask the students to suggest several reasons why it is important that not all scientists agree when a theory is initially proposed.

THEN AND NOW

ROBERT KOCH
(1843—1910)

In 1890, Robert Koch (KAWK) thought he had a cure for tuberculosis in a bacterial extract called tuberculin. He was wrong, but he continued his research. Later, tuberculin was found useful for diagnosing tuberculosis.

Robert Koch was born in Clausthal, Germany, on December 11, 1843. He attended Göttingen University, where he studied medicine and the natural sciences. Koch became a physican but then devoted himself to research. He identified bacterial causes of diseases, such as anthrax, a deadly disease that occurs in sheep. Koch also discovered that bacteria were responsible for the major cause of death in his century, tuberculosis. A research institute for the study of tuberculosis was set up by the Prussian government and headed by Koch.

Koch developed methods that showed scientists how to search for and identify disease-causing bacteria. He also made contributions to the study of cholera, sleeping sickness, leprosy, cattle plague, and bubonic plague. For his work on tuberculosis and other diseases, the founder of bacteriology received many honors, including the Nobel Prize in Medicine in 1905. ◆

LYNN MARGULIS
(1938—)

In 1966, Lynn Margulis theorized that all multicellular organisms are the result of the joining together of monerans. Her ideas were not widely accepted then, but today, many scientists agree with her.

Lynn Alexander Margulis was born in Chicago on March 5, 1938. After receiving her bachelor's degree from the University of Chicago in 1957, she did graduate work at the University of Wisconsin and the University of California at Berkeley. Since receiving her Ph.D. in genetics in 1965, Margulis has taught at Boston University.

Margulis has written many articles on the classification and evolution of living organisms. She believes that many cellular organelles were originally bacterialike cells that over millions of years began to work together.

Not all scientists agree with Margulis's theories. But many biologists recognize that our understanding of the organization of cells has been greatly expanded through her work. In 1975, Margulis received the Dimond Award from the Botanical Society of America for her work on living bacteria, and in 1981, she received the Public Service Award from NASA. ◆

SCIENCE AT WORK

Discussion

● **Process Skills:** *Inferring, Applying*

Ask the students to suggest reasons why it is important to compile global and national health data in one easily accessible location such as the Centers for Disease Control in Atlanta, Georgia. (Compiling data allows researchers and other medical professionals an opportunity to observe trends and quickly react to those trends. Human suffering is then more likely to be limited or minimized.)

Journal Activity In their journals, have the students recall what AIDS is and define the ways in which it is transmitted. Also ask them to describe why it is extremely important for people to realize that AIDS is a preventable disease.

EXTENSION

Challenge interested students to use current reference materials to discover recent developments in the search for a cure for AIDS. Encourage the students to present their findings to their classmates.

SCIENCE AT WORK

HELENE GAYLE,
Public Health Physician

Dr. Helene Gayle provides information to people about how to prevent disease. She is a public health physician. Gayle began her career in disease control at the Centers for Disease Control in Atlanta, Georgia. The Centers for Disease Control, or CDC, is a part of the United States Public Health Service. Its main purpose is to prevent early death among the citizens of the United States. The CDC does this by finding ways to prevent and control diseases.

This work involves learning the ways a disease is spread. Gayle kept records of how many people had a disease, where the people lived, and what their lifestyles were like. Then, studying this information, Gayle looked for patterns to help explain how the disease was being spread and how to prevent it from spreading further.

Gayle is currently Chief of the AIDS Division for the United States Agency for International Development, a foreign assistance agency of the United States government. The AIDS Division is part of the agency program designed to improve health care in developing countries in Africa, Asia, Latin America, and the Caribbean. Through the agency,

Gayle provides health information and medical resources to assist in HIV and AIDS prevention.

Gayle believes that education is the best way to prevent the spread of AIDS. She wants young people to realize that the choices they make now may affect their entire lives.

It is Gayle's hope that her work will make a difference. The more people who know about AIDS and how it is spread, the better they can protect themselves. Gayle wants people everywhere to know that AIDS can be prevented. ◆

Discover More
For more information about careers in disease research, write to the
Association of Schools of Public Health
1015 15th Street, N.W. Suite 404
Washington, DC 20005

UNIT 3 253 ◀

SCIENCE/TECHNOLOGY/SOCIETY

Discussion

Ask the students whether they have ever seen a pond, lake, or other body of water that contained an unusually high concentration of algae. Encourage the students to describe their observations of the algae.

● **Process Skills:** *Comparing, Expressing Ideas Effectively*

Have the students recall the difficulty in classifying algae. (Algae can be unicellular or multicellular as well as protistlike in some ways and plantlike in other ways.) Ask the students to describe several plantlike and several protistlike characteristics of algae.

Point out to the students that decisions about the classification of living things are not always clear-cut. To help the students better understand this concept, remind them that libraries contain many different books, and those books are organized according to the content of each book. However, a given book might be classified as a mystery by one reader and as a romance by another. Yet the library must classify the book as

---

SCIENCE/TECHNOLOGY/SOCIETY

Algal Blooms

Algae can be found in all bodies of water. When conditions are favorable, algae reproduce rapidly and form colored patches called *blooms* on the surface of the water. Algal blooms can cause problems. They can damage aquatic plants and animals, certain industries that depend on marine life, and supplies of drinking water.

Algal blooms can result in massive fish kills.

Ecosystem Overload

Algae multiply rapidly and form blooms when nutrients enter fresh water. The nutrients may come from treated sewage or from the runoff of fertilized farm soil. When large numbers of algae die and decompose, they cause an oxygen shortage in the water. This condition is called *eutrophication*.

Algal blooms also limit the amount of sunlight that can reach the bottom of a lake or river. As a result, plants that need sunlight die, and the water ecosystem is damaged still further.

one or the other. Remind the students that similar decisions must be made by scientists who try to classify living things. For better or worse, these classification decisions are seldom unanimously agreed upon by the scientific community.

Journal Activity
Remind the students that algae occupy an important niche in the ecosystem of Earth. In a journal entry, have the students detail the implications for the ecosystem if all the algae on Earth were to disappear.

EXTENSION

Explain to the students that algae often feel "slimy" to the touch. Challenge interested students to discover more about this "slime" and share their findings with their classmates.

Deadly Consequences

A species of algae called *dinoflagellates* is responsible for toxic ocean blooms called *red tides*. These algae give the water a red color, and their toxin can contaminate marine animals such as mussels and clams. Eating this contaminated seafood can be fatal. Ocean spray that contains red-tide dinoflagellates can even cause respiratory problems in humans.

Search for Solutions

It is possible to control algal blooms in fresh water, but doing so is expensive. The water can be cleaned up by removing the algae and the bottom sediment. If the dumping of sewage into fresh water cannot be stopped, silica added to the sewage can help. Silica encourages the growth of diatoms, a type of algae that does not bloom.

The diatoms slow the growth of other algae and increase the food supply of fish.

So far no clear-cut control has been found effective against red tides in the oceans.

Now researchers are developing models for predicting red tides. Prediction rather than control may turn out to be the only good solution.

Scientists are also studying the possible benefits of red tides on the ocean ecosystem. There have been reports of larger-than-usual catches of shrimp in years following a major red tide. Scientists also hypothesize that dinoflagellate blooms are responsible for oil deposits in the North Sea. ◆

Bloom-causing dinoflagellates (top) can be offset by adding diatoms (center and bottom) or draining water.

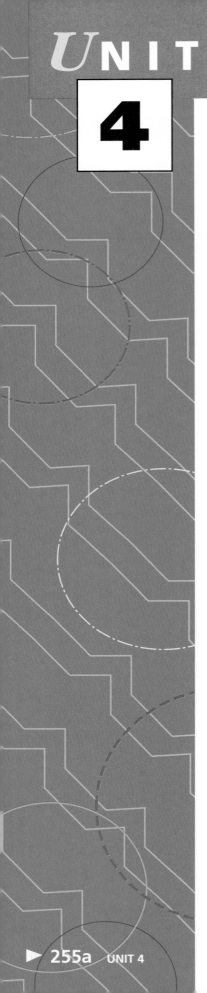

UNIT 4

PLANTS

UNIT OVERVIEW

Vascular and nonvascular plants are both introduced in this unit. Also discussed are the process of photosynthesis, a comparison of gymnosperms and angiosperms, and plant responses and adaptations.

Chapter 10: An Introduction to Plants, page 258

This chapter presents the different types of plants and the process of photosynthesis. First, the differences between vascular and nonvascular plants are explained. Both simple vascular plants and complex vascular plants are discussed, as well as the two main groups of complex vascular plants, gymnosperms and angiosperms. Nonvascular plants are also defined, and the process of photosynthesis is described. Then seed plant reproduction is described. The reproductive structures of gymnosperms and angiosperms are explained.

Chapter 11: Plant Growth and Adaptations, page 286

This chapter introduces the structures that develop during different stages in the life cycle of seed plants. The ability of plant structures to respond to light and gravity is presented. Various adaptations that help plants survive, including how they spread their seeds, live in harsh climates, and protect themselves against animals, are described and discussed.

Science Parade, pages 306-315

The articles in this unit's magazine explore the relationship of plants to hydroponics, to air pollution, and to life on a space station or base. The work of a plant scientist is featured in the career article. A famous plant scientist and a cellular biologist are the subjects of the biographical sketches.

UNIT RESOURCES

PRINT MEDIA FOR TEACHERS

Chapter 10

Asimov, Isaac. *How Did We Find Out About Photosynthesis?* Walker, 1989. Explains the process of photosynthesis and its importance to life on Earth.

Overbeck, Cynthia. *How Seeds Travel*. Lerner Publications, 1982. Describes seed function in plant reproduction and seed transportation by wind, water, and animals.

The Visual Dictionary of Plants. Dorling Kindersley, 1992. Text and illustrations describe various plants and their structures.

Chapter 11

Cooper, Jason. *Strange Plants*. Rourke Enterprises, 1991. Presents unusual plants and plantlike organisms, including carnivorous plants.

Gibbons, Gail. *From Seed to Plant*. Holiday House, 1991. Traces the development of a plant from a seed.

PRINT MEDIA FOR STUDENTS

Chapter 10

Facklam, Howard, and Margery Facklam. *Plants: Extinction or Survival?* Enslow, 1990. The importance of plants as a source of food, medicine, and other products is integrated with research contributions by scientists.

Chapter 11

Stidworthy, John. *Plants and Seeds*. Gloucester Press, 1990. Text and microscopic photographs introduce various forms of plant life, their methods of reproduction, and their assimilation of nutrients.

ELECTRONIC MEDIA

Chapter 10

Flowers: Structure and Function. Videocassette. Coronet. The parts of a flower are examined in relation to their functions in pollination and seed production.

How Green Plants Make and Use Food. Videocassette. Coronet. Shows how materials reach the leaves of plants and how light supplies energy for the manufacturing of food.

Pollination: The Insect Connection. Videocassette. Carolina. Examines the role of insects as pollinators. Discusses the interdependence of insects and flowering plants.

Chapter 11

Plant Tropisms and Other Movements. Videocassette. Coronet. Illustrates movements of plants in response to internal or external stimuli.

Seed Dispersal. Videocassette. Britannica. Reports on different types of seedcoats to show how external agents help each seed become dispersed.

UNIT 4

Discussion After reminding the students that plants are everywhere in the world around us, ask the students to describe ways in which people may tend to take plants for granted. Then point out to the students that most of the world's people use plants known as grasses as their primary source of food. Ask the students to name plants that might be included in the family of grasses. (The students should suggest corn, oats, wheat, rice, and barley.)

UNIT 4 PLANTS

CHAPTERS

10 *An Introduction to Plants* 258

No matter where you find yourself on planet Earth, there will be plants in one form or another.

11 *Plant Reproduction* 286

In the cycle of a plant's life, how does the blooming of a flower fit in? Is it the beginning or the end?

Journal Activity You can extend the discussion by informing the students that more than 10 000 different species of ferns are alive on Earth today. Given that small sample, ask the students to debate and estimate the total number of different plant species that are alive on Earth today. You might also ask the students to name the occasions during the course of a typical day during which a plant directly or indirectly influences their life. Ask the students to consider this question: If all plant life on Earth ceased to exist, why would all animal life also cease to exist?

Plants—without them you would have nothing. Just try to imagine your life without plants. You would have no food, clothing, shelter, or even oxygen—life would be impossible. Because of their importance to us, scientists study plants and go to great lengths and expense to find better ways to grow them.

Science Parade

SCIENCE APPLICATIONS
Salads Without Soil 306

READ ABOUT IT!
"Andy Lipkis and the Tree People" . . 308

THEN AND NOW
George Washington Carver and Elma Gonzalez 312

SCIENCE AT WORK
Shirley Mah Kooyman, Plant Scientist 313

SCIENCE/TECHNOLOGY/SOCIETY
Growing Plants in Space 315

UNIT 4   257

CHAPTER 10
AN INTRODUCTION TO PLANTS

PLANNING THE CHAPTER

Chapter Sections	Page	Chapter Features	Page	Program Resources	Source
Chapter Opener	258	*For Your Journal*	259		
Section 1: TYPES OF PLANTS	260	Discover By Observing (B)	261	*Science Discovery**	SD
• Vascular Plants (B)	260	Discover By Doing (B)	263	Investigation 10.1:	
• Complex Vascular Plants (A)	264	Discover By Doing (A)	267	Examining Mosses and	
• Simple Vascular Plants (A)	266	Section 1 Review and		Ferns (B)	TR, LI
• Nonvascular Plants (A)	268	Application	268	Reading Skills: Examining	
		Skill: Making a Bar Graph (B)	269	Similarities and Differences (B)	TR
				Root, Stems, and Leaves (A)	IT
				Monocots and Dicots (A)	IT
				Study and Review Guide, Section 1 (B)	TR, SRG
Section 2: PHOTOSYNTHESIS	270	Discover By Doing (B)	271	*Science Discovery**	SD
• How Plants Get Food (A)	270	Activity: What does a leaf look like when observed under a microscope? (A)	272	Thinking Critically (A)	TR
• How Photosynthesis Occurs (H)	273			Extending Science Concepts: Planning a Park (B)	TR
• Energy Transfer (B)	274	Discover By Doing (A)	273	Reactions of Photosynthesis (A)	IT
		Section 2 Review and Application	274	Study and Review Guide, Section 2 (B)	TR, SRG
Section 3: REPRODUCTION IN SEED PLANTS	275	Discover By Doing (A)	278	*Science Discovery**	SD
• Gymnosperms (B)	275	Discover By Researching (A)	280	Investigation 10.2:	
• Angiosperms (B)	277	Section 3 Review and Application	281	Comparing Seeds in Fruits and Cones (A)	TR, LI
		Investigation: Predicting the Effect of Colors of Light on Plant Growth (A)	282	Connecting Other Disciplines: Science and Health, Discovering Common Poisonous Plants (H)	TR
				Life Cycle of Gymnosperms (B)	IT
				Life Cycle of Angiosperms (B)	IT
				Record Sheets for Textbook Investigations (A)	TR
				Study and Review Guide, Section 3 (B)	TR, SRG
Chapter 10 HIGHLIGHTS	283	The Big Idea	283	Study and Review Guide,	
Chapter 10 Review	284	For Your Journal	283	Chapter 10 Review (B)	TR, SRG
		Connecting Ideas	283	Chapter 10 Test	TR
				Test Generator	

B = Basic   A = Average   H = Honors
The coding Basic, Average, and Honors indicates subsections, features, and resources that might be appropriate for different levels of learners. For additional suggestions regarding choice of topic and depth of coverage, see the Pacing Chart on pages T26–T29.

*Frame numbers at point of use
(TR) Teaching Resources, Unit 4
(IT) Instructional Transparencies
(LI) Laboratory Investigations
(SD) *Science Discovery* Videodisc Correlations and Barcodes
(SRG) Study and Review Guide

CHAPTER MATERIALS

Title	Page	Materials
Discover By Observing	261	(per group of 3 or 4) germinated seeds, hand lens, microscope
Discover By Doing	263	(per individual) white carnation (fresh), food coloring, scissors, water
Teacher Demonstration	265	dicot and monocot flowers, prepared slides of a cross section of a dicot and a monocot
Discover By Doing	267	(per group of 3 or 4) fern leaf, hand lens, toothpick, slide, microscope
Skill: Making a Bar Graph	269	(per group of 3 or 4) graph paper
Discover By Doing	271	(per group of 3 or 4) bean plants (2)
Activity: What does a leaf look like when observed under a microscope?	272	(per group of 3 or 4) prepared slide of leaf cross section, compound light microscope
Discover By Doing	273	(per individual) tall thin jar, small bowl, water, sprig of *Elodea*, desk lamp
Teacher Demonstration	276	mature male and female pine cones
Discover By Doing	278	(per individual) simple flower, forceps or toothpick, tape, paper, pencil
Discover By Researching	280	(per individual) journal
Investigation: Predicting the Effect of Colors of Light on Plant Growth	282	(per group of 3 or 4) healthy plants (4), cellophane (4 colors)

ADVANCE PREPARATION

For the *Discover By Observing* on page 261, obtain seeds that have germinated (such as beans or grasses). The students will need food coloring and fresh white carnations with long stems for the *Discover By Doing* on page 263. For the *Demonstration* on page 265, samples and slides of dicots and monocots will be used. For the *Discover By Doing* on page 267, the students will need a fern frond with sori on the underside. Bean plants are needed for the *Discover By Doing* on page 271. Obtain a prepared slide of a leaf cross section for the *Activity* on page 272. For the *Discover By Doing* on page 273, you will need *Elodea*. For the *Demonstration* on page 276, obtain a few mature male and female pine cones. Healthy plants from a nursery or greenhouse or class-grown plants are needed for the *Investigation* on page 282.

TEACHING SUGGESTIONS

Field Trip
A field trip to a botanic garden, nursery, or greenhouse would be helpful. The visit could be structured to parallel the story line in the chapter about a class visit to a botanic garden.

Outside Speaker
If you live near a major university or an agricultural experiment station, invite a plant physiologist to speak about the uses of plants. If no plant physiologist is available, a speaker from a plant nursery, florist's shop, or garden shop would be helpful. A pharmacist would be another good speaker. He or she could talk about how plants are used to make some medicines.

CHAPTER 10
AN INTRODUCTION TO PLANTS

CHAPTER THEME—SYSTEMS AND STRUCTURES

In this chapter, the students will be presented with the differences in structure among simple vascular, complex vascular, and nonvascular plants. The chapter will also describe the food-making process of plants known as photosynthesis and provide the students with an opportunity to explore the processes of reproduction in seed plants. The theme of **Systems and Structures** is also developed through concepts in Chapters 4, 5, 7, 8, 9, 12, 13, 16, 17, 18, 19, and 20. A supporting theme of this chapter is **Diversity**.

MULTICULTURAL CONNECTION

Cooperative Learning The environment will dictate the kinds of plants that grow in an area. For example, grasses, but few trees, grow in grasslands. Cacti and mesquite are common in some desert areas. Mosses and shrubs, but few trees, grow on the tundra near the Arctic Circle. Point out to the students that the available plant life will affect the people who live in the area and their way of life—what they eat, what they build, what they wear, and so on. Have the students work in small groups to investigate the influence of plants on the cultures of the Native Americans who lived on the Plains, in the Southwest, in the Far North, and in other areas.

MEETING SPECIAL NEEDS

Second Language Support

Fluent English-speaking students should be paired with students who have limited English proficiency to work together to design a poster. The poster should illustrate some of the many ways humans use plants. Encourage the students to label the poster in English and in any other language they know.

CHAPTER 10
An Introduction to Plants

Don't look now, but plants are everywhere around you. You know that the geraniums decorating the windowsill are plants. But what about the wood in your pencil, the cotton in your shirt, or the apple in your lunch? Wood, cotton, and apples all come from plants. And that's just three of the many ways people use plants.

CHAPTER MOTIVATING ACTIVITY

Prepare a display of plants grouped according to their uses. The display should include plants or plant products used for food, examples of cloth made from plant materials, plants used for medicine, and floral arrangements or decorative plants. If the actual plants or objects cannot be shown, use pictures instead. Examples: *Food*—greens (leaves); white potatoes (stems); turnips (roots); beans (seeds); apples (fruit); *Spices*— cloves, cinnamon; *Clothing* (cloth)—linen and cotton; *Medicine*— cinchona (source of quinine), foxglove (source of digitalis).

For Your Journal

Answering the journal questions gives the students the opportunity to demonstrate what they already know about plants and to describe the roles that plants play in their everyday lives. Their answers will also provide you with the opportunity to note any misconceptions they may have about plants and to correct those misconceptions at the appropriate times throughout the chapter.

Thousands of years ago, people discovered how to weave parts of plants to make baskets and coverings. In time, they learned to spin plant fibers into thread or yarn that could be woven into cloth. Cotton and linen are both materials that are woven from plant fibers.

People have also used plants and plant products for medicines for thousands of years. Today many medicines are still obtained from plants. Codeine in cough syrup is made from the sap of a type of poppy. Quinine, which is used to treat malaria, is produced from the bark of the cinchona tree.

People also use plants for decoration. These plants, called ornamentals, are used in outdoor parks and gardens; in bouquets, corsages, and arrangements; and inside houses and other buildings.

Nowhere is the long close relationship between people and plants more evident than in the many parks and gardens which people visit simply to enjoy looking at plants. At botanic gardens and arboretums, plants are displayed for people to see and enjoy. But these institutions have other important purposes. They collect, grow, and study plants to increase our knowledge about plants, to improve the varieties of plants, to conserve rare plants, and perhaps to discover new ways people can use plants in the future.

For Your Journal

- What do you visualize when you think about plants?
- How do you use plants in your daily activities?
- How do plants grow, make food, and reproduce?

ABOUT THE PHOTOGRAPH

Each year over a half million people visit Butchart Gardens near Victoria, British Columbia, Canada. Created in the early 1900s over part of a limestone quarry, the gardens are actually four separate gardens, one of which is the Sunken Garden. The other three are the English Rose Garden, the Japanese Garden, and the Italian Garden. Around the world botanical gardens, like Butchart Gardens, grow plants for scientific, educational, and artistic purposes. Many botanical gardens are open to the public. They offer tours, classes, and special displays as part of their objective to educate people. But botanical gardens are also research institutions. They work to improve plants through breeding and selection and to grow and preserve rare and exotic plants from all over the world. Some famous botanical gardens include the Royal Botanic Gardens in Kew, England, opened to the public in 1841, and the United States Botanic Garden in Washington, D.C., located near the Capitol.

CHAPTER 10   259

Section 1: TYPES OF PLANTS

FOCUS

The main focus of the section is the characteristics of vascular and nonvascular plants. Vascular plants are divided into complex and simple vascular plants, and then contrasted to nonvascular plants.

MOTIVATING ACTIVITY

Cooperative Learning Collect examples of seeds produced by vascular plants. These examples might include the seeds of different flowering plants, the fruits of several fruit trees, and the cones of varieties of coniferous trees. Divide the class into small groups, and allow each group sufficient time to study the different types of seeds. Then ask each group to describe the characteristics of the plants that produced those seeds. Write the groups' descriptions on the chalkboard and title the list *Vascular Plants*. Point out that the students have described some of the many characteristics of vascular plants.

PROCESS SKILLS
- Observing • Comparing
- Inferring

POSITIVE ATTITUDES
- Cooperativeness
- Openness to new ideas

TERMS
- xylem • phloem • vascular cambium • photosynthesis

PRINT MEDIA
The Visual Dictionary of Plants
(see p. 255b)

ELECTRONIC MEDIA
Pollination: The Insect Connection, Carolina
(see p. 255b)

Science Discovery
Tree rings
Plants growing

BLACKLINE MASTERS
Study and Review Guide
Laboratory Investigation 10.1
Reading Skills

---

① Xylem transports water and minerals; phloem transports food. Plants need both kinds of tissue to survive.

❖ **Did You Know?**
Some of our most important vegetables are fleshy taproots. These vegetables include beets, carrots, radishes, rutabagas, turnips, sweet potatoes, and cassavas.

SECTION 1 | Types of Plants

Objectives

Describe the structures of vascular plants.

Identify some of the groups of vascular plants.

Distinguish between vascular and nonvascular plants.

One warm spring day your class visits a botanic garden, where hundreds of varieties of plants are grown, studied, and displayed. Your guide through the garden is a botany student named Beth. She explains that you will visit greenhouses that recreate the kinds of environments that different plants need. Then Beth holds up a petunia plant. She points to the different parts of the plant as she talks about the functions of the parts.

Vascular Plants

Like the petunia, most of the plants you see every day are vascular plants, or plants with a system of vessels that carry materials throughout the plant. Inside the roots, stems, and leaves of these plants are two types of *vascular* tissues that transport nutrients and water. **Xylem** (ZY luhm) transports water and dissolved minerals from the roots to the leaves. **Phloem** (FLOH ehm) transports the food that is made in the leaves to all parts of the plant. Why is it necessary for plants to have both xylem and phloem? ①

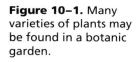

Figure 10–1. Many varieties of plants may be found in a botanic garden.

Anchors Away Beth gently removes the petunia from the soil. The many roots of the plant are thin and branching—soil particles cling to the roots.

Roots anchor many plants into the ground, so the plants are not blown away by wind or washed away by rain. Roots also absorb water and minerals from the soil. Thick roots, such as those of carrots and radishes, store food.

Plants have one of two types of root systems. Plants with taproots have one main root that grows almost straight down. *Taproots* can be long, like carrots, or rounded, like radishes. Plants with *fibrous roots* have many thin

TEACHING STRATEGIES

● **Process Skills:** *Observing, Inferring*

Ask the students to name the plants that they typically see in their community. (Most common plants are typically angiosperms—flowering plants with seeds.) Point out that these plants have vascular tissue, and that vascular tissue is used for carrying materials within the plants. Ask the students to name the system inside the human body that functions in a similar way to the vascular system of plants. (The human circulatory system carries nutrients and other materials through the human body.)

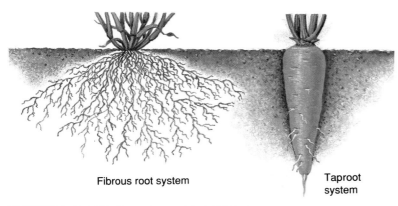

Fibrous root system

Taproot system

Figure 10-2. Fibrous roots spread out over a large area. Taproots are thick, single roots.

roots in a dense network and no one main root. Fibrous roots hold soil together and help keep it from eroding. Both types of roots have very fine, threadlike structures called *root hairs*. Root hairs absorb water and minerals from the soil by osmosis and diffusion. You can observe root hairs in the following activity.

DISCOVER BY *Observing*

Obtain some seeds that have germinated, or sprouted. Observe the roots with your unaided eye, with a hand lens, and under low power of a microscope. Describe your observations. How does magnification affect your observations?

Roots grow in both length and width. In plants, new cells are produced only in tissue called *meristem*. Cells produced in the root meristem begin to grow in the region of elongation. After the cells get longer, they develop into the other tissues of the root, including vascular tissue. The root cap protects the growing root tip as it pushes through the soil. Locate these root parts in the diagram.

Figure 10-3. Cell division occurs only in the meristem of the root tip. If a root tip is cut or broken, it cannot grow any longer.

DISCOVER BY *Observing*

This activity provides the students with an opportunity to closely observe the root system of a plant. The use of magnification will allow the students to examine the roots in greater detail and to see structures not visible to the unaided eye.

SCIENCE BACKGROUND

Roots have a zone of maturation in which cells have attained their full size and have become differentiated, or changed. Some of the cells become part of the epidermis, an outer covering that is one cell thick. The epidermis protects the inner cells of the root. Root hairs grow from epidermal cells. A single plant may contain millions of root hairs, which greatly increase the surface area of the root for absorbing water and minerals.

TEACHING STRATEGIES, continued

● **Process Skills:** *Comparing, Applying*

Ask the students to describe the similarities and differences between herbaceous and woody plant stems. Make sure that the students understand that although these two types of stems are different, they perform the same functions. Then ask the students to explain why the size of plants with herbaceous stems is limited. (The soft, flexible herbaceous stem cannot support much weight. Because herbaceous plant stems die periodically, they do not add woody tissue as woody plants do. As a result, herbaceous plants do not grow very tall.)

● **Process Skills:** *Observing, Comparing*

Refer the students to Figure 10–4 and have them compare the arrangement of xylem and phloem in a herbaceous plant to the arrangement of xylem and phloem in a woody plant. (In a herbaceous plant, the xylem and phloem appear in bundles; in woody plants, the xylem and phloem appear in rings, with the xylem found on the inside of the ring and the phloem found on the outside of the ring.) Remind the students that a woody plant contains a cambium layer; a herbaceous plant does not.

INTEGRATION—Language Arts

Have the students investigate the etymologies of the words *annual* (from Latin *annus*, meaning "year"), and *perennial* (from Latin *per-*, meaning "throughout" + *annus*). Ask the students how knowing the meanings of the words can help them remember the meanings of the concepts.

SCIENCE BACKGROUND

Annual growth rings have alternating bands of dark and light xylem because of different growing conditions in the summer and in the spring. The inner part of an annual ring is lighter because there is usually more moisture available in the spring, and new cells in the xylem produce rapidly and grow to a large size. The outer part of the annual ring grows in the summer when temperatures are greater and there is less moisture. The cells produced during this time are smaller. This "summer wood" appears darker because cell walls occupy a greater proportion of its area than in "spring wood."

LASER DISC

1490

Tree rings

Figure 10–4. In a herbaceous stem (left), xylem and phloem are grouped together in bundles. In a woody stem (right), the xylem is toward the inside and the phloem is toward the outside.

An Upright Job   Beth asks you to observe the petunia's stem. It is green and flexible. The functions of stems are to support the leaves and to connect the leaves to the roots. In some plants, such as cacti, stems produce and store nutrients.

Plant stems are sometimes divided into two types: *herbaceous* (hur BAY shuhs) and *woody*. Herbaceous stems are green and flexible. They depend on water pressure in their cells to hold them upright. Woody stems are harder and stiffer than herbaceous stems and contain wood cells that give them support.

Most plants with herbaceous stems grow, mature, reproduce, and die in one growing season and are called *annuals*. Many kinds of vegetables, including corn, peas, and radishes, are annuals. Many ornamentals, such as petunias, pansies, and marigolds, are also annuals.

Plants with woody stems and some plants with herbaceous stems live for more than one growing season. These plants are called *perennials* (puh REHN ee uhlz). Many grasses, shrubs, and trees such as oak and pine are perennials.

During each growing season, the stem, or trunk, of a tree gets thicker because new wood is deposited between the outer covering of the trunk, called *bark*, and the internal tissue. The new growth for each year forms a growth ring. The rings vary in thickness from year to year depending on the weather and other growing conditions. A rainy year may cause a tree to grow faster, producing a thick growth ring. Dry weather may cause a tree to grow more slowly, producing a thinner growth ring.

Stems contain both xylem and phloem. In a herbaceous plant, xylem and phloem are arranged in bundles, as shown in the following illustration. The long, stringy fibers you see when you bite a stalk of celery are the vascular bundles. In a woody stem, the xylem and phloem are arranged in a ring. The phloem is on the outside of the ring, and the xylem is on the inside.

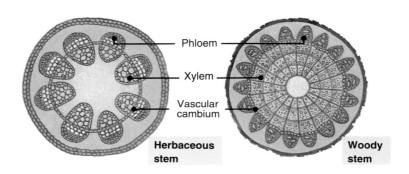

● **Process Skills:** *Comparing, Generating Ideas*

Ask the students if they have ever seen a cross section of a tree trunk, and have them describe what that cross section looked like. (The students should recall that there are distinct rings or circles in the cross section of a tree stump.) Ask the students if the rings of a tree are always the same size. (No, the rings can display different sizes, thicknesses, and colors.) Then have the students speculate why the rings of a tree stump may be different thicknesses. (Changes in growing conditions will result in different ring thicknesses.)

● **Process Skills:** *Interpreting Data, Expressing Ideas Effectively*

Ask the students to explain why it is possible to think of tree rings as a historical weather record. (The students might suggest that available moisture influences the size of the rings. If a tree shows limited ring growth during one or more seasons, it may have received less-than-average precipitation during that time.)

Between the xylem and the phloem is another tissue, called the *vascular cambium*. **Vascular cambium** is the growth tissue that produces new xylem and phloem cells in the stems of woody plants. As the xylem accumulates, the woody plant grows in diameter.

A stem, with its attached leaves and branches, grows from a region of dividing cells at its tip called the *apical meristem*. The apical meristem produces leaves that grow around it, forming the apical bud. Other buds form that become branches or flowers. In the mature region of a stem, the cells form into the vascular tissues and other tissues of the stem. You can observe the functioning of vascular tissue in the following activity.

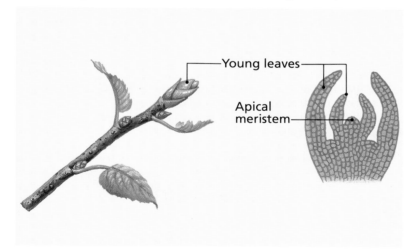

Figure 10–5. Just as roots grow at their tips, stems also grow at their tips.

DISCOVER BY *Doing*

The carnation used in this activity will change to a color corresponding to the food coloring in the water in which its stem was placed. The students should explain that the carnation changed color as the colored water was transported to the petals by the xylem.

PERFORMANCE ASSESSMENT

Evaluate the students' understanding of an experimental setup. Using the *Discover By Doing* activity, have the students formulate a hypothesis, describe the step-by-step procedure, record their observations, and summarize their results.

LASER DISC
36764–37213

Plants growing

Doing

You will need a fresh white carnation, food coloring, scissors, and a glass of water. Cut the stem of the carnation to about 10 cm in length. Add several drops of food coloring to the water in the glass. Place the carnation stem in the colored water and let it stand for 24 hours. What happens to the carnation? Explain.

Food Factories Leaves are the major food-making structures of a plant. They are the parts of the plant in which most photosynthesis takes place. **Photosynthesis** is the process by which green plants use chemicals from the environment and energy from the sun to make their own food. You will learn more about photosynthesis in the next section.

MEETING SPECIAL NEEDS

Second Language Support

Write a definition for each of these terms on the chalkboard: xylem, phloem, vascular cambium, photosynthesis. Have the students write the term for each definition. Expand the activity to include these terms: meristem, herbaceous, woody, annuals, perennials, epidermis, stomata.

SECTION 1 263

TEACHING STRATEGIES, continued

● **Process Skills:** *Comparing, Inferring*

Ask the students what single word best describes the function of vascular tissue. (transport) Then have the students examine a leaf and find the veins that contain the vascular tissue. Ask them to compare the veins with the streets in their community. (The students should notice that a leaf contains large veins that branch into smaller and smaller branches, similar to the appearance of a system of roads.) Point out that both roadways and veins transport things, and ask the students to name the specific things that leaf veins transport. (food, water, and minerals)

● **Process Skills:** *Observing, Inferring*

Have the students refer to Figure 10–6 and locate the stoma and its guard cells. Explain that excess water in the leaf causes the guard cells to swell and open the stomata. The excess water can then escape from the leaf and evaporate. Ask the students what happens when enough water has evaporated. (The guard cells shrink, and the stomata close.)

ONGOING ASSESSMENT
ASK YOURSELF

Root hairs absorb water from the soil into the root system. The water is then transported from the root system through the xylem of the stem, eventually reaching the leaves of the plant. Openings in the leaves called stomata can release the water into the atmosphere.

① Some plants that produce seeds include corn, hardwood trees, grasses, flowers, and coniferous trees.

② The pine tree is a gymnosperm; the petunia is an angiosperm.

INTEGRATION—*Language Arts*

Have the students look up the derivations of the terms *angiosperm* and *gymnosperm* in a dictionary. Point out that knowing the derivations of the two terms can help the students remember that whereas gymnosperms and angiosperms both have seeds, angiosperms have protective coverings around their seeds and gymnosperms do not. (*Gymnosperm*—from the Greek words *gymnos*, meaning "naked," and *sperma*, meaning "seed"; *angiosperm*—from the Greek words *angeion*, meaning "vessel," and *sperma*, meaning "seed")

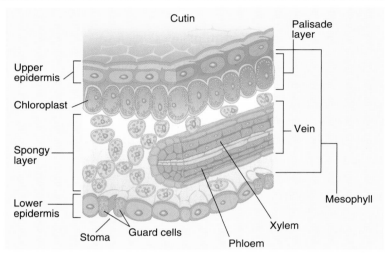

Figure 10–6. A leaf is many cell layers thick.

The illustration shows a cross section of a leaf. The outer layer of the leaf, or the *epidermis* (ehp uh DUR mihs), is a layer of thin, flat cells—like the cells of your skin. In some plants, the epidermis is covered with wax. The wax protects the inner parts of the leaf and reduces water loss. In the center of the leaf are the layers of cells in which photosynthesis takes place. Veins connect to the roots and stems. Openings in the epidermis, called *stomata*, allow air and water to pass into and out of the leaf. Each stoma's size is controlled by guard cells.

ASK YOURSELF

Trace the path of water from the root hairs to the leaf stomata.

Complex Vascular Plants

At the botanic garden, Beth walks over to a small pine tree. She points out that both the petunia and the pine tree are complex vascular plants. This means that they produce seeds. Many of the plants with which you are familiar are complex vascular plants. Some produce flowers and fruits, while others do not. But they all produce seeds. Name some plants that produce seeds. ①

There are two groups of complex vascular plants: the *gymnosperms* (JIHM nuh spurmz) and the *angiosperms* (AN jee uh spurmz). Gymnosperms lack a protective covering around their seeds, and they do not produce flowers. Plants that have a protective covering around their seeds and produce flowers are angiosperms. Which kind is the pine tree? Which kind is the petunia? ②

Gymnosperms There are over 500 species of gymnosperms, including cycads, ginkgoes, and conifers. Most gymnosperms keep their leaves throughout the year. As a result, they are often called *evergreens*. Gymnosperms produce seeds in cones.

- **Process Skills:** *Applying, Generating Ideas*

Point out that of all of the forests on Earth, one-third are coniferous (cone-bearing) forests. Then ask the students to speculate why one-third of Earth's forests are coniferous. (Some students might suggest that the climatic conditions that support the growth of coniferous forests are found in many places on Earth, resulting in a proliferation of these forests; others might suggest that coniferous trees have successfully adapted to specific climatic conditions found in many places on Earth.)

- **Process Skills:** *Inferring, Analyzing*

Point out that a common example of a gymnosperm—the ginkgo tree—is often found in cities planted along busy streets because ginkgoes survive better than other trees in high-traffic areas. Ask the students to speculate why ginkgo trees seem to tolerate the conditions found in high-traffic areas better than other species of trees. (The students might suggest that ginkgoes are better able to tolerate high levels of pollutants, including automobile exhaust and salt, than other species of trees.)

Angiosperms Most of the plants that live on Earth are angiosperms. Angiosperms produce seeds from flowers, and all produce some kind of fruit. Crop plants, hardwood trees, shrubs, grasses, and desert plants are all angiosperms. Angiosperms are divided into two groups: *dicots* (DY kahtz) and *monocots* (MAHN uh kahtz). Some of the differences between dicots and monocots can be seen in the diagram below.

The vascular tissue bundles are arranged differently in dicots and monocots. In dicots, they are arranged in a circle. In monocots, vascular bundles are found throughout the stem. The leaves of dicots are broad and have veins branching throughout them. Monocots have long leaves with veins that run from one end of the leaf to the other. Dicots include oaks, maples, elms, roses, beans, cabbages, tomatoes, and petunias. Monocots include grasses, palms, daffodils, irises, and lilies.

Figure 10–7. What angiosperms can you identify in this picture? ③

③ The angiosperms are tulips.

✧ **Did You Know?**
Today there are an estimated 250 000 species of living angiosperms—more than six times the number of species in all other plant groups combined.

Demonstration
Show the students a dicot flower and a monocot flower. Have the students notice the flower parts and the leaves and veins of each type of angiosperm. Also obtain a slide of a cross section of both a dicot and a monocot so that the students can see the different arrangements of the plants' vascular bundles.

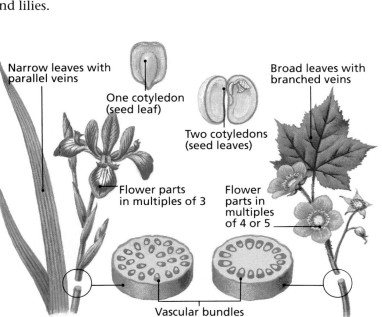

Figure 10–8. Monocots and dicots have characteristics that make them easily identifiable.

ONGOING ASSESSMENT
ASK YOURSELF

Gymnosperms are complex vascular plants that have seeds with no protective coverings. Angiosperms are complex vascular plants that have seeds with coverings and produce flowers.

ASK YOURSELF

What are the differences between gymnosperms and angiosperms?

TEACHING STRATEGIES, continued

● **Process Skills:** *Comparing, Observing*

Refer the students to Figure 10–9 and discuss the two types of simple vascular plants shown. Have the students who have seen ferns growing in nature describe their appearance. Emphasize that simple vascular plants have poorly developed vascular tissue, and they do not have seeds.

● **Process Skills:** *Comparing, Applying*

Point out that ferns are probably the most common examples of simple vascular plants and the ones that the students are most likely to be familiar with. Ask the students to recall their study of complex vascular plants and then identify the characteristics that separate complex vascular plants from simple vascular plants. (The xylem and phloem of simple vascular plants are very poorly developed compared to the xylem and phloem of complex vascular plants. Simple vascular plants do not have seeds; complex vascular plants do. Simple vascular plants require moisture for reproduction.)

BACKGROUND INFORMATION

Modern-day ferns are the descendants of some of the oldest plants on Earth; they first appeared over 350 million years ago. As tall as trees, they were some of the most abundant plants on Earth during the Carboniferous Period, about 250 million years ago. Most of the world's major coal deposits formed during this period. When the ancient plants died and fell into the swamps, they were buried by sand and mud. Over time, pressure and heat transformed the dead plants into coal.

THE NATURE OF SCIENCE

As new species are discovered and more information about the structure and chemistry of organisms is added, the system of classification used by scientists to organize living things has changed and expanded over the years. In fact, how species should be classified is constantly debated in the scientific community. For example, some scientists might classify some algae as plants, while other scientists classify algae as either protists or monerans.

Figure 10–9. Plants such as horsetails (inset) and ferns have poorly developed vascular tissues and do not produce seeds.

Figure 10–10. The life cycle of a fern

Simple Vascular Plants

Beth leads your class into a greenhouse where the air is moist and warm. Here some of the plants have leaves shaped like fiddleheads; others have thin, feathery leaves. Beth says that all of these plants are ferns. Plants such as ferns, horsetails, and club mosses are simple vascular plants. Simple vascular plants are seedless. Because many have simple, poorly developed xylem and phloem, these plants need a lot of moisture in their environment to survive and to reproduce.

Ferns If you have ever walked through a warm, damp forest, you have probably seen ferns. They are the most common seedless plants. Ferns have stems, roots, and leaves, but they do not make seeds. There are about 12 000 different kinds of ferns; they live in areas where the soil is moist and the climate is temperate. The life cycle of ferns includes both sexual and asexual phases.

During the asexual, or sporophyte, generation, small brown or orange structures can be seen on the bottom of the fern leaf. These groups of spore-bearing cases hold many spores. When the spores are released, they fall to the ground or are carried away by wind or water. When conditions for growth are right, a spore can grow into the sexual, or gametophyte, generation, which produces gametes—sperm and ova.

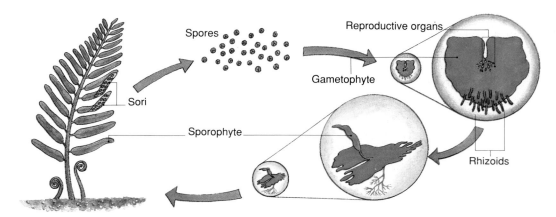

- **Process Skills:** *Inferring, Analyzing*

Remind the students that simple vascular plants can live only in a moist environment and that they are always seedless. Ask the students to consider these facts and then explain the relationship that exists between seeds and the ability to live in dry areas. (The students should infer that complex vascular plants can produce seeds and reproduce more successfully in dry environments than simple vascular plants. In a dry environment, simple vascular plants would have difficulty reproducing because their sperm cells require moisture to be able to swim to the ova. In other words, plants that reproduce by seeds are less dependent on moisture and can reproduce more successfully in dry conditions than seedless plants. Moisture is a necessary prerequisite for reproduction in simple vascular plants.)

Rootlike structures grow into the ground to anchor the tiny plant and supply it with minerals and water. On the bottom of the gametophyte, male and female reproductive organs form. When sperm cells are released, they must swim through water—dew or rain—to reach the female part of the plant. One sperm combines with an ovum to start a new sporophyte generation. In the following activity, you can observe the spores of ferns.

Discover by Doing

You will need a fern leaf, a hand lens, a toothpick, a slide, and a microscope. Using the hand lens, examine the underside of a fern leaf. Locate the spore cases and make a drawing of them. Scrape several spore cases onto a slide and make a wet mount of them. Examine them under the microscope and make a drawing of what you see. Try to germinate some spores in moist soil.

Other Seedless Plants Horsetails and club mosses are also seedless plants. Like ferns, these plants undergo sexual and asexual phases during their life cycles. Horsetails are found in areas where the weather is usually warm and moist. Their shoots look like the tails of horses. Only one genus of horsetails grows on Earth today. All of the others are extinct. Ancient horsetails, which lived over 300 million years ago, were as large as trees and lived in swampy areas.

The other seedless vascular plants are the club mosses. There are about 800 different species. The first club mosses lived about 350 million years ago. Club mosses are also found in moist areas.

ASK YOURSELF

How do ferns reproduce?

Figure 10–11. Horsetails (left) and club mosses (right) are among the most primitive vascular plants.

DISCOVER BY *Doing*

If fern leaves are not available in your supply of science materials, they can be ordered from scientific supply houses, obtained from a local plant shop, or found naturally in your area. If you collect fern leaves from ferns growing in your area, take only the number of leaves needed to complete the activity. Also it is better to remove one leaf from many different plants than it is to remove many leaves from one plant. Remind the students to focus their wet mounts first under low power and then under high power. Have the students place their drawings in their science portfolios.

ONGOING ASSESSMENT
ASK YOURSELF

Ferns reproduce in a cycle that includes a sexual phase involving sperm cells and ova, and an asexual phase involving spores.

SCIENCE BACKGROUND

Plant life cycles are marked by an alternation of generations. A sporophyte generation alternates with a gametophyte generation. The sporophyte generation produces spores after meiosis takes place in specialized cells. The spore divides by mitosis and produces a gametophyte. The gametophyte produces eggs and sperm, which fuse to form a zygote, the first cell of the next sporophyte generation.

GUIDED PRACTICE

Have each student write a list of the characteristics of simple vascular plants, complex vascular plants, and nonvascular plants. Then ask a student to read a characteristic from his or her list to another student who decides whether the given characteristic belongs to a simple vascular plant, a complex vascular plant, or a nonvascular plant.

INDEPENDENT PRACTICE

Have the students provide written answers to the Section Review and Application questions. Then ask the students to write a paragraph in their journals explaining how angiosperms and gymnosperms are different.

EVALUATION

Cooperative Learning Have the students work in small groups to construct a concept map, using the following terms: plants, vascular, nonvascular, simple vascular, complex vascular, gymnosperms, angiosperms, monocots, dicots.

RETEACHING

Provide the students with a simple drawing of a plant that includes the roots, stems, and leaves. Ask the students to label the parts of the plant and then explain the function of each part.

ONGOING ASSESSMENT
ASK YOURSELF

Because nonvascular plants do not have vascular tissue to transport water, all their parts must take in water directly from their environment.

SECTION 1 REVIEW AND APPLICATION

Reading Critically

1. Vascular tissue transports food that is made in the leaves to all parts of a plant, and transports water and dissolved minerals from the roots to the leaves of the plant.

2. Complex vascular plants have seeds. Simple vascular plants do not have seeds.

Thinking Critically

3. A fern is a vascular plant. It would have roots, leaves, and stems. A moss is a nonvascular plant. It would not have roots, stems, or leaves.

4. Responses might suggest that the spores produced in the asexual phase can wait to grow until environmental conditions are right, which increases the plant's chances of survival.

▶ **268**   CHAPTER 10

Figure 10–12.
Nonvascular plants such as mosses (left), hornworts (center), and liverworts (right) must grow in moist areas in order to survive.

Nonvascular Plants

In the next greenhouse, the air is very moist, the ground is very wet, and the plants are very small and low-growing. Beth points to what looks like a clump of green velvet and explains that it is a kind of moss. The moss is an example of a nonvascular plant. Nonvascular plants are simple plants that do not produce seeds and do not have vascular tissue. Some nonvascular plants have structures that look like but are not true roots, stems, or leaves.

Mosses, liverworts, and hornworts are simple land plants that need a moist environment. These plants have underground structures that anchor them in the soil and absorb water and minerals, but these structures are not roots. Other structures look like stems and leaves. However, the structures have no vascular tissue. Because food and water must pass from cell to cell, these plants don't grow very large.

 ASK YOURSELF

Why do nonvascular plants need a moist environment?

SECTION 1   REVIEW AND APPLICATION

Reading Critically

1. What is the function of vascular tissue in plants?
2. What is the difference between simple and complex vascular plants?

Thinking Critically

3. How would you distinguish between a moss and a fern?
4. In what ways might a sexual and an asexual phase of reproduction be an advantage to mosses and similar plants?

EXTENSION

Ask the students to investigate *bryophytes* and prepare a short oral report on these organisms to present to the class.

CLOSURE

Ask the students to locate and name a plant found on their school grounds and then classify that plant as a simple vascular plant, a complex vascular plant, or a nonvascular plant, providing reasons for their classification.

SKILL
Making a Bar Graph

Process Skills: Measuring, Analyzing

Grouping: Groups of 3 or 4

Objectives
- **Observe** and **measure** parts of an organism.
- **Communicate** using a bar graph.
- **Interpret** and **analyze** data from a bar graph.

Discussion

Before beginning the activity, distribute field guides and point out to the students the plants that are endangered or hazardous in your area. Remind the students that the leaves of these plants are not to be handled or collected.

Bar graphs are useful for showing relationships among subjects. Have the students discuss other contexts in which they have seen, used, or created bar graphs. Make sure that all students are proficient with a metric ruler before proceeding with the activity.

Application

4. The graphs created by each group will vary depending on the leaves and the graphing method they chose.

5. The students might determine that leaves from the same plant will most likely be about the same length.

6. A bar graph makes it easier to see how many leaves are in each length category.

7. Accept all responses that are logically supported.

Using What You Have Learned

Responses should reflect an understanding of how to compile the data and then organize the data into groups represented by bars.

SKILL Making A Bar Graph

▶ MATERIALS
graph paper

▼ PROCEDURE

1. Often it is necessary for scientists to collect and compile data. They may use the data to make a bar graph. Bar graphs can display a lot of information about the subject or process being studied in a very concise, efficient way.

2. To make a bar graph to show information, you must first decide on a title. Label the graph with the title and draw the vertical and horizontal axes so that they cover most of the paper.

3. Then you must decide how your axes should be labeled. Once you decide, make each square on the graph paper equal to a particular unit of measure. That unit could be 1 cm or 100 m. It does not matter, as long as each square stands for the same amount.

4. When you make a bar graph, the vertical axis usually indicates how many of something are the same. For instance, in this Skill activity, you might fill in three squares vertically if three of the plant leaves measured 10 mm in length.

▶ APPLICATION

1. **CAUTION: Do not attempt to collect leaves from any plants that are considered hazardous or endangered.** Choose a plant to study from an area near your home. Collect five leaves from this type of plant.

2. Decide on the unit that is best for the plant leaves you have chosen and measure the length of each leaf. Be sure that you are using an SI unit. Record this information in a table on the chalkboard. Copy the information for the leaves of the entire class.

3. Organize the leaves into categories according to length. For example, all leaves that are at least 6 cm long but less than 7 cm might be grouped in one category.

4. Make a bar graph of the class data. Compare your graph with those of your classmates. Did everyone in the class graph the information in the same way? Why or why not?

5. What general patterns do you see in the length of plant leaves for your class plant?

6. How does the bar graph help you see patterns and relationships in the data you collected?

7. After examining all the class graphs, which method used do you think was the best for representing the information?

✳ Using What You Have Learned
Select another characteristic of plants and create another bar graph. You may wish to consider height of plants, number of leaves on a stem, or number of petals on a flower.

SECTION 1 269 ◀

Section 2: PHOTOSYNTHESIS

FOCUS

This section presents the process of photosynthesis. The materials necessary for, and the products of, photosynthesis are described. The function of chlorophyll in photosynthesis is explained. The importance of this process to plants and animals is emphasized, as well as the importance of plants to life on Earth.

MOTIVATING ACTIVITY

If you live near a major university or an agricultural experiment station, invite a plant physiologist to speak to the students about the importance of photosynthesis as well as the importance and uses of plants. If a plant physiologist is not available, invite a speaker from a plant nursery, a florist's shop, or a garden shop, or the county extension agent. It is also likely that your local Department of Natural Resources will have classroom materials or other resources available for use.

PROCESS SKILLS
- Observing • Comparing
- Inferring • Evaluating

POSITIVE ATTITUDES
- Caring for the environment
- Cooperativeness

TERMS
- chlorophyll

PRINT MEDIA
How Did We Find Out About Photosynthesis? by Isaac Asimov (see p. 255b)

ELECTRONIC MEDIA
How Green Plants Make and Use Food, Coronet (see p. 255b)

Science Discovery Plant health Photosynthesis equation

BLACKLINE MASTERS
Study and Review Guide
Thinking Critically
Extending Science Concepts

SCIENCE BACKGROUND

The chlorophyll of a plant functions only in certain intensities of light. Therefore, there is a maximum rate of photosynthesis per unit of chlorophyll. Plants grow especially well in violet and blue intensities of light.

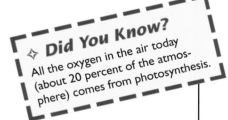

Did You Know? All the oxygen in the air today (about 20 percent of the atmosphere) comes from photosynthesis.

▶ 270 CHAPTER 10

SECTION 2 Photosynthesis

Objectives

Explain the process of photosynthesis.

Evaluate the importance of photosynthesis in energy transfer.

Standing in one of the largest greenhouses, you can see hundreds and hundreds of green plants. Overhead the sun streams through the glass panels. It is very quiet, but Beth says that actually there is a lot of activity going on. All the plants are busy, making, storing, and eating food! What does she mean?

Beth explains that unlike animals, plants make their own food. For a long time, people believed that plants ate soil just as animals ate food. In the 1600s, a man from Belgium, J. B. van Helmont, conducted an experiment to test the idea. First, some soil was dried and weighed. Then van Helmont planted a willow tree in a large tub filled with the soil. For five years, nothing was added to the soil but water. After five years, van Helmont removed the tree from the tub. Again he dried and weighed the soil. Although the tree had grown much larger, the soil weighed almost the same as it had five years earlier! What do you think van Helmont concluded? He concluded that plants don't eat soil but that they may eat water since that was all he had added to the tub.

Figure 10–13. The botanic garden in Oklahoma City

How Plants Get Food

Plants absorb sunlight. To understand how this works, Beth asks you to look at one green plant in the greenhouse. The sun shines on the plant. Energy from the sunlight is absorbed by chlorophyll in the cells of each leaf. **Chlorophyll** is the pigment that gives plants their green color. Chlorophyll traps solar energy and converts it into chemical energy. You can try the next activity to test the importance of sunlight for plants.

TEACHING STRATEGIES

- **Process Skills:** *Predicting, Applying*

The importance of photosynthesis cannot be overemphasized to the students, since photosynthesis not only provides food for organisms but also is the main source of oxygen in the atmosphere. Point out that the earth is said to be losing one hectare of tropical rain forest every two minutes. Ask the students why humans are destroying rain forests. (Responses should suggest that rain forests are being cleared so that the trees can be used for timber and the land can be used for farming and ranching.) Then ask the students to think about the process of photosynthesis and describe the implications associated with the elimination of tropical rain forests on Earth. (If the rain forests are destroyed, the available oxygen on our planet would significantly decrease, the carbon dioxide level in the atmosphere would significantly increase, and catastrophic damage to plant and animal food chains would occur.)

- **Process Skills:** *Comparing, Predicting*

Write the colors of the visible spectrum of sunlight on the board: red, orange, yellow, green, blue, indigo, violet. Ask the students to predict the colors in which plants grow best. (They grow best in colors toward the violet end of the spectrum.)

DISCOVER BY Doing

You will need two bean plants that are the same size. Put one in a dark place and the other in light. Wait for several days. Then bring out the bean plant that was in the dark place. Compare the two plants. What color is each plant? What other differences do you notice?

Plants need sunlight to stay green. Only green plants can make food. Therefore, plants can't make food without sunlight. The process that enables a plant to make its own food is called *photosynthesis*.

Figure 10–14. Whether green plants grow on land or under water, chlorophyll converts sunlight into chemical energy.

Besides making the food that plants use themselves, photosynthesis is important to all animals, including humans. Since most animals eat plants, or animals that have eaten plants, photosynthesis provides animals with food as well. In addition, during photosynthesis, plants remove carbon dioxide from the air and give off oxygen. Animals could not live without the oxygen that plants produce.

Photosynthesis takes place in organelles called *chloroplasts*. Chloroplasts contain chlorophyll. Most chloroplasts are located in leaves. In the next activity, you can see where most of the chloroplasts are located in a leaf.

DISCOVER BY Doing

When the students observe the plant that was placed in the dark for several days and compare it to a plant that received light, they will notice that the plant kept in the dark is less green and less healthy-looking. Use this opportunity to stress that light is a vital ingredient in the process of photosynthesis. Without the energy from light, photosynthesis could not take place.

 **LASER DISC**

3077

Plant health

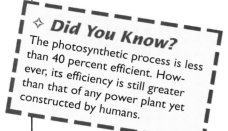

Did You Know?
The photosynthetic process is less than 40 percent efficient. However, its efficiency is still greater than that of any power plant yet constructed by humans.

MEETING SPECIAL NEEDS

Gifted

Organisms that make their own food are called *autotrophs*. Organisms that must obtain their food by eating other organisms are called *heterotrophs*. Have the students research the derivations of the words, find synonyms for the words, and list ten autotrophs and ten heterotrophs. Have the students share their findings.

SECTION 2

ACTIVITY

What does a leaf look like when observed under a microscope?

Process Skills: Observing, Communicating

Grouping: Groups of 3 or 4

Hints
This activity works best when the students observe dicot leaves. If prepared monocot and dicot slides are readily available, you might choose to have the students observe and compare these slides.

Procedure
3. Most chloroplasts are located in the palisade layer.

▶ **Application**
1. Epidermis—protects inside of leaf, contains stomata; vein—path for transport of materials; chloroplasts—contain chlorophyll, which traps energy needed for photosynthesis; stomata—allow water and gases to enter and exit leaf; guard cells—control stomata opening.

2. The structure of each part is adapted to its function.

① Responses might include plums, beets, and purple cabbage.

ONGOING ASSESSMENT
▼ **ASK YOURSELF**

Photosynthesis creates food for plants, and plants create food for animals that eat plants or other animals. Photosynthesis also uses carbon dioxide and gives off oxygen.

▶ 272 CHAPTER 10

TEACHING STRATEGIES, continued

● **Process Skills:** *Inferring, Applying*

Refer the students to the equation summarizing photosynthesis. Make sure they understand that this equation is a simplification of a process made up of many steps. Photosynthesis is a complex chemical process that involves many chemical reactions. (Several books would be needed to include everything that is known about the process. Entire college courses are devoted to photosynthesis.) Ask the students to explain why plants need food, such as glucose. (Glucose supplies the energy for the life activities of plants.) Then ask the students to explain how the energy stored in glucose is made available to plant cells. (Lead the students to infer that the process of respiration releases the energy stored in glucose.)

ACTIVITY
What does a leaf look like when observed under a microscope?

MATERIALS
prepared slide of leaf cross section, compound light microscope

PROCEDURE
1. Obtain a slide of a leaf cross section from your teacher.

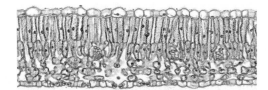

2. Observe the cross section under low and high power. Draw what you see.
3. Label the following plant tissues on your sketch: epidermis, vein, chloroplasts, stomata, and guard cells. Where are most of the chloroplasts located?

APPLICATION
1. What is the function of each part of the leaf that you labeled?
2. How is each of the tissues of the leaf especially well suited to the job it does?

Yellow, red, and orange pigments, in addition to chlorophyll, are also found in plants. However, in spring and summer, the leaves appear green because of the large amount of chlorophyll present. In the fall, the plant stops making chlorophyll and begins to use the existing chlorophyll as food, so the other colors become visible.

Other plant parts contain colored pigments, too. Carrots, for instance, have a great deal of orange pigment. Tomatoes and cherries have red pigment. What other examples of pigmented plant parts can you think of? ①

Figure 10–15. In the fall, plants stop making chlorophyll, so other colors show in the leaves.

 ASK YOURSELF

Why is photosynthesis important to plants and animals?

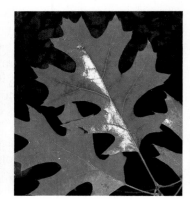

- **Process Skills:** *Observing, Generating Ideas*

Refer the students to Figure 10–16 and discuss the two stages of photosynthesis. Make sure the students understand that the energy-rich chemicals produced during the first stage of photosynthesis provide the energy with which the second stage of photosynthesis is performed. Ask the students to speculate how the energy-rich chemicals provide that energy. (Energy is stored in their chemical bonds and released when those high-energy bonds are broken.)

GUIDED PRACTICE

Write the simplified chemical equation for photosynthesis on the chalkboard and have the students explain the stages and components of the photosynthetic process.

INDEPENDENT PRACTICE

 Have the students provide written answers to the Section Review and Application questions. Then have them summarize the events that take place in the first and second stages of photosynthesis in their journals.

How Photosynthesis Occurs

The materials necessary for photosynthesis are water (H_2O) and carbon dioxide (CO_2). The most common products of photosynthesis are glucose ($C_6H_{12}O_6$) and oxygen (O_2). From glucose and other sugars, compounds such as starches and proteins are made. The chemical equation for photosynthesis is:

$$6CO_2 + 6H_2O \xrightarrow[\text{chlorophyll}]{\text{light}} C_6H_{12}O_6 + 6O_2$$

carbon dioxide   water   glucose   oxygen

Photosynthesis is a complicated process that occurs in two stages. During the first stage, light is absorbed by chlorophyll, and energy-rich chemicals are produced. These energy-rich chemicals are needed for the second stage of photosynthesis. Oxygen is released during the first stage, and glucose and water are produced during the second stage. In the following activity, you can learn more about photosynthesis.

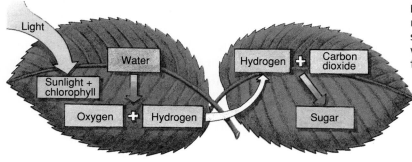

Figure 10–16. The reactions of photosynthesis provide plants with nutrients necessary for survival.

LASER DISC
3060
Photosynthesis equation

INTEGRATION—*Physical Science*

Electromagnetic waves are forms of energy that can be arranged in order of increasing frequency, or decreasing wavelength, to form the electromagnetic spectrum. Visible light makes up a narrow portion of the spectrum. Each of the colors of the visible spectrum has a particular wavelength and frequency. Chlorophyll is the primary photosynthetic pigment in plants. Chlorophyll absorbs primarily wavelengths of violet and blue light, and reflects, or transmits, wavelengths of green light. Accessory pigments are pigments in the chloroplast that can capture other wavelengths of light and transfer the energy to chlorophyll.

DISCOVER BY *Doing*

This activity works best if all of the air is removed from the inverted jar. The students should observe that the *Elodea* begins to perform photosynthesis in the presence of light.

DISCOVER BY *Doing*

You will need a tall thin jar (like an olive jar), a small bowl, some water, a sprig of *Elodea,* and a desk lamp. Put some water in the bowl. Fill the jar with water and drop the *Elodea* into it with the cut end at the top of the jar. Put your hand over the top of the jar and turn it over. Place it in the bowl. Try to keep as much of the water inside the jar as possible. Put the lamp close to the jar and turn the lamp on. Watch the *Elodea* for 20 minutes. What happens?

ONGOING ASSESSMENT
ASK YOURSELF

Light provides the energy used by chlorophyll to produce the energy-rich chemicals needed for the second stage of photosynthesis.

ASK YOURSELF
How is light used during photosynthesis?

SECTION 2   **273**

EVALUATION

Ask the students to describe the relationship of photosynthesis to Earth's supply of carbon dioxide and oxygen. (Photosynthesis uses carbon dioxide—a waste product of living things—and creates oxygen—a vital element for living things. Without photosynthesis, the oxygen in Earth's atmosphere would have been used up long ago.)

RETEACHING

Ask the students to design and draw a diagram that explains the process of photosynthesis. Have them imagine that they will use the diagram to explain the process to someone who has never heard of it. Remind them to use appropriate labels and to include explanations of the steps in their own words.

EXTENSION

Photosynthesis is sometimes said to occur in two distinct reactions—the *light reaction* and the *dark reaction*. Have the students research the meanings of these terms and share their discoveries with their classmates.

CLOSURE

Cooperative Learning Have the students work in small groups and write two questions based on the information in this section. Have each group exchange questions with another group and answer that group's questions.

ONGOING ASSESSMENT
ASK YOURSELF

Green plants capture some of the sun's energy and make it available to other organisms through photosynthesis. The energy is transferred when human beings eat plants or eat animals that eat plants or other animals.

SECTION 2 REVIEW AND APPLICATION

Reading Critically

1. The source of energy for photosynthesis is light.

2. Photosynthesis requires carbon dioxide, water, and chlorophyll.

Thinking Critically

3. The products of photosynthesis provide food and oxygen for animals.

4. Eventually, the colorful leaves fall off the tree. The tree no longer needs them, once their chlorophyll has been used up by the tree.

Figure 10–17. The energy stored by plants is transferred to humans and other animals when plants or animals are eaten.

Energy Transfer

Green plants are very important to humans as well as other animals; the survival of animals depends on green plants and other organisms that carry out photosynthesis. Most energy, except for geothermal and nuclear energy, comes either directly or indirectly from the sun. Green plants capture some of the sun's energy and make it available to other organisms through photosynthesis. When animals eat plants, the animals are getting this solar energy in the form of food. When, in turn, those animals are eaten by other animals, the energy is passed on through the food chain. Without plants, the energy from the sun could not be transferred from one organism to another.

You have probably eaten many different types of plants and plant parts. You eat roots when you eat carrots or radishes. You eat leaves when you eat lettuce. If you have white potatoes or asparagus with your dinner, you are eating stems. If you have cauliflower or broccoli, you are eating flower buds. If you have string beans or cucumbers, you are eating fruits. Peas and corn kernels are seeds.

 ASK YOURSELF

How does the sun supply energy to humans?

SECTION 2 REVIEW AND APPLICATION

Reading Critically
1. What is the source of energy for photosynthesis?
2. Name the materials needed for photosynthesis to occur.

Thinking Critically
3. In what way did photosynthesis make animal life on Earth possible?
4. What eventually happens to a tree's colorful fall leaves? Explain your answer.

Section 3: REPRODUCTION IN SEED PLANTS

FOCUS

This section presents seed plant reproduction. The reproductive structures of gymnosperms and the life cycle of angiosperms are examined. The parts of a flower are also presented, along with explanations of fruit and seed development, pollination and fertilization, and seed germination.

MOTIVATING ACTIVITY

Cut open and display the following: carrot, pepper, peas in pods, tomato, squash, celery, cucumber, raspberries, lettuce, peanut in shell, sweet potato, and sugar cane. Wrap each item in plastic wrap and label it. Write this definition of *fruit* on the chalkboard: a ripe ovary of a flower that contains seeds. Tell the students that vegetables do not contain seeds. Ask the students to list and classify each item as either a fruit or a vegetable, using the definition of *fruit*. They may be surprised to learn that many items commonly called *vegetables* or *nuts* are really fruits (pepper, peas, tomato, squash, cucumber, peanut).

Reproduction in Seed Plants

SECTION 3

At the botanic garden, Beth brings your class back to the pine tree and the petunia plant that she showed you when you first arrived. Do you remember what the pine tree and the petunia have in common? They are both complex vascular plants; they have roots, stems, and leaves, and they produce seeds.

The development of seeds is an adaptation that increased the ability of vascular plants to survive unfavorable environments. Seeds protect plant embryos from harsh conditions; seeds can remain dormant for years and still produce healthy plants.

Two groups of seed plants have evolved. Gymnosperms generally produce their seeds in cones and keep their leaves all year. Angiosperms produce flowers, protect their seeds in fruits, and usually lose their leaves in the fall. The pine tree is a gymnosperm, and the petunia is an angiosperm. How are their seeds different? ①

Objectives

Describe the life cycle of a conifer.

Name the parts of a flower.

Distinguish between pollination and fertilization.

PROCESS SKILLS
- Comparing • Observing
- Inferring

POSITIVE ATTITUDES
- Caring for the environment
- Cooperativeness

TERMS
- stamen • pistil

PRINT MEDIA
How Seeds Travel
by Cynthia Overbeck
(see p. 255b)

ELECTRONIC MEDIA
Flowers: Structure and Function, Coronet
(see p. 255b)

Science Discovery
Flower cross section
Flowers blooming

BLACKLINE MASTERS
Study and Review Guide
Laboratory Investigation 10.2
Connecting Other Disciplines

① The pine tree produces seeds in cones; the petunia produces seeds from flowers.

SCIENCE BACKGROUND
Many plants can reproduce by *vegetative propagation* in which a part of the plant, such as a root, stem, or leaf, grows into a whole new plant by regenerating the missing parts. For example, if a potato is cut into pieces, as long as each piece has one bud or eye, each piece will grow into a new potato plant.

Gymnosperms

Have you ever looked closely at a pine tree? If you have, you may have noticed that a pine tree has two different kinds of cones. It has male cones and female cones. The male cones, which contain pollen, are the smaller of the two.

Pollen grains produce sperm cells that fertilize the eggs within the larger female cones. Pollen is carried from the male cones to the female cones by wind. After fertilization, seeds mature and are released from the female cones. When conditions are right, the seeds grow into new gymnosperm plants.

Figure 10–18. The pine tree (above) is a gymnosperm. The petunias (left) are angiosperms.

SECTION 3　275

TEACHING STRATEGIES

● **Process Skills:** *Applying, Expressing Ideas Effectively*

Have the students recall what they learned about gymnosperms earlier in this chapter. (Gymnosperms are complex vascular plants that produce seeds in cones.) Ask the students to name three examples of gymnosperms. (Responses might include cycads, ginkgoes, and conifers.) Reminding the students that gymnosperms have seeds without protective coverings, ask the students to identify the two different kinds of cones produced by gymnosperms. (Gymnosperms produce male and female cones.)

Demonstration
Show the students a mature male pine cone and a mature female pine cone so that they can compare the two separate cones found on a pine tree. Pull off several bracts of the female cone to display the scars where the pine seeds developed.

✧ Did You Know?
Some bristlecone pine trees are almost 5000 years old, making them the oldest organisms on Earth.

INTEGRATION—*Social Studies*

Some of the tallest trees on Earth, redwoods and giant sequoias, are conifers. A giant sequoia called the General Sherman tree is over 80 m tall and 30 m in base circumference and is estimated to be 2200 to 2500 years old. The General Sherman tree lives in Sequoia National Park. Have the students use reference materials to discover the significance of the park and of the name *sequoia*. (The park was established in 1890 to protect some of the last groves of giant sequoias. The name *sequoia* honors a Cherokee chief named Sequoyah who developed a system of writing for the Cherokee language.)

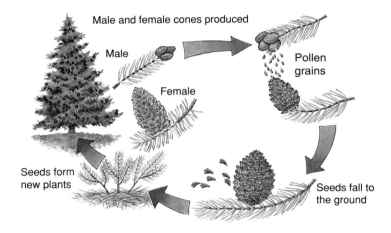

Figure 10–19. The life cycle of a typical gymnosperm

Conifers The word *conifer* means "cone-bearer." Pines, spruces, and firs have stiff seed cones and needlelike leaves. Forests of conifers are found mainly in northern climates and other regions with sandy soil and moderate rainfall.

Conifers survive in harsh conditions because their needles have little exposed surface area. Needles retain moisture through hot, dry summers and cold winters. Instead of losing their leaves all at once, conifers shed and replace needles throughout the year.

Other Gymnosperms Although they look nothing like the familiar conifers, cycads and ginkgoes are also gymnosperms. Cycads resemble palms but are unrelated. Their reproduction is similar to that of conifers, but instead of separate male and female cones on one tree, cycad trees are either male or female—that is, some cycad trees bear only male cones while others bear only female cones.

Figure 10–20. Conifer needles lose little moisture.

Figure 10–21. Cycads (left) are palmlike trees from the tropics, while ginkgoes (right) are native to China.

The ginkgo is the last living species of what was once a family of related trees. Ginkgoes have flat, fan-shaped leaves. Male ginkgoes produce pollen in small, conelike structures that hang down from the branches. After fertilization, female trees produce berrylike seeds. Ginkgoes are sometimes called *living fossils* because there are none left living in the wild.

 ASK YOURSELF

How does the life cycle of a gymnosperm compare with that of a fern?

Angiosperms

Angiosperms are flowering plants, like the petunia. All angiosperms have flowers at some time during their life cycles. Flowers are important because they contain the plant's reproductive organs. These organs produce the pollen and egg cells that will eventually develop into seeds. The seeds grow inside protective fruits.

Figure 10–22. The African violet is an angiosperm commonly found as a houseplant.

ONGOING ASSESSMENT
ASK YOURSELF

A gymnosperm reproduces when pollen containing sperm cells from a male cone is carried by the wind to the egg cells in a female cone. After fertilization, mature seeds are released from the female cone. A fern reproduces asexually when spore-bearing cases release spores and sexually when sperm cells use water to travel to the female reproductive organs.

MULTICULTURAL CONNECTION

Spices, which are seasonings used in food, are made from plants. But spices come from different parts of the plants, and they come from different parts of the world. For example, cloves come from buds, ginger from roots, mustard from seeds, and cinnamon from bark. Cinnamon comes from Sri Lanka, nutmeg from Indonesia, sesame from Africa, pepper from India, and cloves from Indonesia and Africa. Often certain spices are an important part of ethnic cuisines. Ask the students whether they can identify the spices commonly used in Mexican, Chinese, Greek, Italian, and Indian cuisines. Ask them to describe the spices used in ethnic dishes they are familiar with.

TEACHING STRATEGIES, continued

- **Process Skills:** *Observing, Applying*

Refer the students to Figure 10–23 and discuss the parts of a flower. Review the structure and function of each part. Use the diagram to trace the steps that occur in pollination and fertilization.

- **Process Skills:** *Applying, Inferring*

Many students have probably seen bees collecting nectar or pollen from a flower. Bees are the most frequent and characteristic pollinators of flowers. Discuss how flowers function to attract bees and other pollinators. Ask the students to think of other ways in which pollen is transferred. (Responses might suggest that birds and bats pollinate some species of flowers. Other angiosperms depend on the wind or water to carry pollen from one plant to another.) Then ask the students why angiosperms that depend on wind for pollination produce large amounts of pollen. (The chances of pollen reaching another flower by wind are greater if there is more pollen in the air.)

Did You Know? There are more than 20 000 different species of bees that serve as pollinators.

SCIENCE BACKGROUND

Angiosperms are flowering plants. They differ from other seed plants because their ovules are enclosed within highly modified leaves called *carpels*. Angiosperms are a very diverse group, ranging from tiny plants to some of the largest trees. They are the most abundant and dominant plants on Earth. Their reproductive features, which include the flower and the fruit, have contributed to their great success. Almost all of our plant sources of food are derived either directly or indirectly from angiosperms.

LASER DISC
3047

Flower cross section

DISCOVER BY Doing

Remind the students to handle forceps or toothpicks with care. Check the students' answers to ensure that they have labeled each flower part correctly. Their displays may differ from the diagram depending on the structures found in their chosen flower.

▶ **278** CHAPTER 10

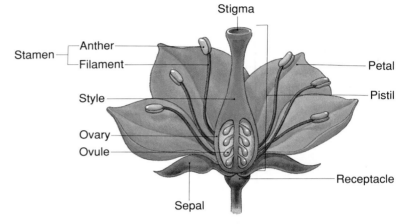

Figure 10–23. The parts of a typical flower are shown here. However, not all flowers contain all of these parts.

Parts of a Flower The parts of a typical flower are shown in the illustration. The male part of the flower is the **stamen.** The stemlike part of a stamen is the filament. At the top of a stamen is the anther, which contains the pollen grains. Most flowers have more than one stamen.

The female part of the flower is called the **pistil.** The slender part of the pistil is the style. At the tip of the style is the stigma. The stigma produces a sticky material that traps pollen grains. The base of the pistil is the ovary. The ovary contains ovules. Each ovule contains the female gamete, or ovum.

The base of the flower is the receptacle. Leaflike sepals grow from the receptacle and enclose the flower before it opens. If you look underneath the petals of a flower, you may be able to see the sepals. Petals grow above the sepals and protect the stamens and pistils. Petals usually have colors, shapes, or odors intended to attract certain insects or other animals.

The number of structures varies in different types of flowers. You may recall that monocots have flower parts in multiples of threes, and dicots have flower parts in multiples of four or five. In the following activity, you can observe the different parts of a flower.

DISCOVER BY Doing

You will need a simple flower, such as a tulip or a gladiolus, and a forceps or a toothpick. First use the diagram in your book to help you identify the flower's parts. Then carefully remove each part, tape it to a sheet of paper, and label it. How is your display different from the diagram in your book?

• **Process Skills:** *Inferring, Generating Ideas*

Discuss the process of pollination with the students. Help them relate the process of pollination and fertilization to their studies of sexual reproduction and genetics. Also remind them that cross-pollination occurs when the pollen of one flower is carried to another flower, and point out that most plants have physiological blocks to (or ways to prevent) self-fertilization. Ask the students to describe the genetic consequences of cross-pollination. (Cross-pollination creates a greater variation in offspring.) Ask the students if they can think of ways that flowers can ensure that they do not pollinate themselves and that cross-pollination occurs. (Responses might suggest that some plants develop their stamens and pistils at different times. A flower's pistil may develop only after the stamens have released their pollen.)

Figure 10–24. One function of a flower is to attract insects. When an insect lands on a flower, pollen sticks to the insect's body. When the insect moves to another flower, the pollen is transferred to that flower, where pollination may occur.

LASER DISC

42408–43042

Flowers blooming

 **SCIENCE TECHNOLOGY SOCIETY** The first pollen grains were not visible to humans until the development of the microscope in the seventeenth century. An English botanist, Nehemiah Grew (1641–1712), was one of the first people to describe and characterize the shapes of different pollen grains. Today scientists use their knowledge of pollen and genes to try to create specialized plants. Scientists have already been successful in creating plants that are "bigger and better." In the future, they hope to create, for example, plants that produce their own insecticides and plants that are immune to diseases.

Pollination and Fertilization The transfer of pollen from the anther to the stigma is called *pollination*. Pollen is moved from anther to stigma by wind, water, or insects or other small animals, depending on the kind of flower. When pollen is carried from the anther of one flower to the stigma of another flower, cross-pollination takes place. When a stigma receives pollen from the anther of the same flower, self-pollination occurs.

Immediately after pollination, a pollen tube grows from a cell in the pollen grain, through the stigma, to an opening in an ovule. Sperm cells move down the pollen tube and enter the ovule. Fertilization occurs when one sperm combines with each ovum to form a zygote. The zygote, or fertilized ovum, eventually develops into an embryo.

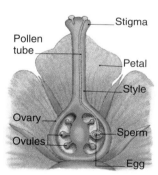

Figure 10–25. When pollen falls on the stigma, a pollen grain forms a pollen tube. The sperm inside the pollen grain moves down the pollen tube to fertilize the ovum.

REINFORCING THEMES— *Diversity*

When discussing pollination, encourage the students to recognize that all flowers display methods that ensure pollination, but the type of pollination that results in the greatest genetic diversity is cross-pollination. Cross-pollination mixes the characteristics of one plant with the characteristics of another plant, increasing the likelihood that the offspring will be different from the parents.

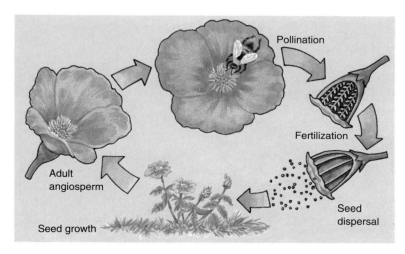

Figure 10–26. The life cycle of a typical angiosperm

SECTION 3

TEACHING STRATEGIES, continued

● **Process Skills:** *Inferring, Generating Ideas*

The students may not be aware that dry fruits such as walnuts, pecans, and acorns are actually fruits. Ask the students to explain how the simultaneous ripening of fruits and maturing of the seed are important in the dispersal of plants. (Fruits are usually not eaten by animals before the seed is developed because unripe fruit tastes sour. By the time the seed is fully developed, the fruit is ripe and therefore tasty to animals. The animals eat the fruit, and the seeds of the fruit pass undigested through their bodies and are typically deposited far away from the parent plant.)

GUIDED PRACTICE

Have the students choose a gymnosperm or an angiosperm seed plant and then take turns describing how their chosen plant reproduces.

INDEPENDENT PRACTICE

Have the students provide written answers to the Section Review and Application questions. Have the students summarize, in their own words, the function of each of the male and female parts of a typical flower.

SCIENCE BACKGROUND

Although fruits contribute to the health and well-being of people and other animals, a plant does not bear fruit for the enjoyment of other organisms. The function of a fruit is purely the dispersal of the plant's seeds. Fruit acts as a vehicle by which the seeds of the plant are transported from one place to another.

DISCOVER BY *Researching*

Encourage the students to use plant field guide books, botany textbooks, and encyclopedias to find the information needed to complete this activity.

MEETING SPECIAL NEEDS

Gifted

Have the students investigate the importance of plants with respect to economics (including food and textiles), medicine, aesthetics, and conservation, and as part of the oxygen-carbon dioxide cycle. Ask the students to present their research in any form they wish—oral report, chart, poster—and then place their work in their science portfolios.

Figure 10–27. Apples and tomatoes are fleshy fruits. Walnuts and pecans are dry fruits.

▶ **280**   CHAPTER 10

Fruit and Seed Development   A seed consists of three parts: the embryo, the stored food, and the seed coat. At the same time that the ovules are developing into seeds, the ovary is developing into a fruit. As the fruit swells and ripens, it supports and protects the developing seeds. The mature fruit will be either fleshy, like apples and tomatoes, or dry, like walnuts and pecans. Many fleshy fruits store sugars and, therefore, taste sweet. Dry fruits do not have sweet flesh; instead, oils and fats are stored in the seeds. In the following activity, you can make a list of the seeds, fruits, and other plant parts that people eat.

DISCOVER BY *Researching*

In your journal, make a table that has four columns with these headings: "Plant Name," "Plant Part Eaten," "Fruit or Vegetable," "Scientific Name." Under "Plant Name," list 20 plants that people eat. In the second column, list the part of the plant that is eaten. In the third column, indicate if the part eaten is commonly called a *fruit* or a *vegetable*. In the last column, write the scientific name of the part eaten.

Both dry and fleshy fruits play an important role in helping seeds disperse. The seeds in a fleshy fruit complete their development at about the same time that the fruit ripens and is ready to be eaten. Animals carry the fruits away, eat them, and leave the seeds in new locations. When the seeds are released from the fruit, they contain very little moisture and will not begin to grow until environmental conditions are favorable. Most dry fruits have a hard outer covering that protects the seeds until conditions are right for growth.

EVALUATION
Obtain several stalks of cut angiosperm flowers and ask the students to point to the major structures of the flowers as you name them.

RETEACHING
Have the students think of three different ways in which pollen is transferred from one plant to another and act out each method without using words. (The students' pantomimes should focus on water, wind, and insects or other small animals.)

EXTENSION
Point out that some fruits such as grapes have varieties with seeds and without seeds. Have the students discover how seedless fruits are created and share their findings with their classmates.

CLOSURE
Ask the students to draw two diagrams. One diagram should show the process of reproduction in a typical gymnosperm, and the other diagram should show the process of reproduction in a typical angiosperm.

Fruits can be classified as simple, aggregate, or multiple. A simple fruit is formed from a single ovary. Beans, peaches, tomatoes, and oranges are all simple fruits. The number of seeds in the fruit is the same as the number of ovules in the ovary.

Aggregate fruits form from several ovaries. All the ripe ovaries join to form a single fruit. Examples of aggregate fruits are blackberries, raspberries, and strawberries. Each ovary contains only one seed.

A multiple fruit has many single fruits growing close together. Pineapples and figs are examples of multiple fruits.

Figure 10–28. Apples are simple fruits, raspberries are aggregate fruits, and pineapples are multiple fruits.

Seed Germination After a period of inactivity, mature seeds germinate, or sprout. Seeds germinate when environmental conditions are good and the seeds have plenty of water. The seed coat breaks open, and the young seedling emerges. The stems and roots begin to grow, leaves and branches begin to form, and the plant tissues begin to develop. You will learn more about seed germination and plant growth in the next chapter.

▼ ASK YOURSELF
How does a sperm cell reach an ovum in an angiosperm?

SECTION 3 REVIEW AND APPLICATION

Reading Critically
1. How does reproduction occur in gymnosperms?
2. Describe the male and female parts of a flower.

Thinking Critically
3. How is pollination different from fertilization?
4. Why are angiosperms more successful at growing in a variety of environments than are gymnosperms?

ONGOING ASSESSMENT
▼ ASK YOURSELF
Immediately after pollination, a pollen tube grows from a pollen grain through the stigma to an opening in an ovule. Sperm cells then move down the pollen tube and enter the ovule to fertilize an ovum.

SECTION 3 REVIEW AND APPLICATION

Reading Critically

1. Reproduction in gymnosperms occurs after pollination. Fertilization occurs within the female cone. The ovule develops into a seed without a fruit.

2. The male part, the stamen, consists of a filament and an anther. The anther contains grains of pollen. The female part, the pistil, consists of a style and a stigma. The stigma produces a sticky material that traps pollen grains.

Thinking Critically

3. Pollination is the deposit of pollen on the stigma of a flower. Fertilization is the mating of sperm and egg.

4. The life cycle of gymnosperms is dependent upon the wind dispersing pollen from a male cone to a female cone. The life cycle of angiosperms is dependent upon wind, water, or insects or other small animals dispersing pollen from one plant to another. Pollination of gymnosperms can occur only by one method, while pollination of angiosperms can occur by a variety of methods in a variety of situations.

INVESTIGATION

Predicting the Effect of Colors of Light on Plant Growth

Process Skills: Observing, Inferring

Grouping: Groups of 3 or 4

Objectives
- **Observe** the effects that various wavelengths of light have on plant growth.
- **Measure** and **interpret** plant growth data.
- **Evaluate** the effectiveness of certain colors of light on plant growth.

Pre-Lab
Have the students read the procedure and study the illustration. Ask them to predict which of the plants will be the healthiest at the end of the experiment.

Analyses and Conclusions

1. The control plant and the plants wrapped in red and yellow cellophane should exhibit the healthiest appearances and the most growth because the process of photosynthesis reacts efficiently to the presence of blue light and to the presence of sunlight, which contains blue light.

2. The plant in blue cellophane should exhibit the least growth because blue cellophane blocks the passage of blue-violet light. Blue-violet light is required by chlorophyll to initiate the process of photosynthesis.

3. Colors toward the violet end of the visible spectrum encourage better plant growth than those colors toward the red end of the visible spectrum.

 Have the students place their designs and analyses from the *Investigation* in their science portfolios.

▶ **Application**
The light emitted by fluorescent lights is composed of different colors, or wavelengths, of light. Plants can use some of the colors contained in fluorescent light to perform photosynthesis.

✳ **Discover More**
Experiments might suggest that the soil containing lettuce seeds could be covered by various colors of cellophane. If all other conditions were identical, the differences in germination time could be determined.

Post-Lab
Have the students compare their predictions to the results of their experiment to see whether they were correct.

INVESTIGATION

Predicting the Effect of Colors of Light on Plant Growth

▶ **MATERIALS**
- four healthy plants (same kind, grown in the same conditions, each with several leaves)
- large sheets of colored cellophane (yellow, red, blue)

▼ **PROCEDURE**

1. Number the plants 1 through 4. Measure and record the height of each plant. Also record your observations about each plant's appearance. Use a table similar to the one shown below.

2. Use Plant 1 as the control. Place it in direct sunlight.
3. Design a way to test the effect of yellow, red, and blue light on plant growth. Remember to keep all other conditions the same.
4. Place Plants 2, 3, and 4 in the same sunlight as Plant

1. Remember to water the plants as necessary.
5. Allow the plants to grow for about a week.
6. Measure and record the heights of the plants on the table. Also record your observations about the plants' appearance.

TABLE 1: OBSERVATIONS OF PLANT GROWTH				
Date				
Plant 1 Control				
Plant 2 Blue Filter				
Plant 3 Red Filter				
Plant 4 Yellow Filter				

▶ **ANALYSES AND CONCLUSIONS**
1. Which plant grew the most? Which plant looks the healthiest?
2. Which plant grew the least? How does it look?
3. The cellophane filtered out the matching color of light. What can you conclude about the effects of certain colors of light on plants?

▶ **APPLICATION**
Fluorescent lights can be used to grow plants in places where there is no direct sunlight. How is that possible?

✳ **Discover More**
Try the experiment again using additional colors of cellophane to filter out different combinations of colors.

Design and carry out an experiment to find out what color of light speeds up the germination time of lettuce seeds.

CHAPTER 10 HIGHLIGHTS

The Big Idea—SYSTEMS AND STRUCTURES

Point out to the students that the organisms called plants are a vast and diverse group, varying widely in size and form. Scientists use the structures of plants to help them classify plants, or separate them into groups. This classification allows scientists from around the world to easily communicate and share information about plants with each other. Point out that whatever their differences, all plants have something that helps identify them as plants: They contain chlorophyll with which they can perform photosynthesis and thus make their own food.

CHAPTER 10 HIGHLIGHTS

The Big Idea

Plants come in many varieties. They range in size and appearance from the tallest redwood to a tiny green hornwort. They have adapted to live in practically every environment on Earth. Certain structural features can be used to divide plants into categories. Plants are classified as vascular or nonvascular, simple or complex vascular plants, or gymnosperms or angiosperms. Even within those divisions, plants can still vary significantly in form and appearance. Yet all plants, no matter what their structure, color, size, or shape, share two important characteristics: they contain chlorophyll and they can make their own food.

Look back at the answers you wrote in your journal before you began this chapter. Would you change your answers now? Revise your journal entry to reflect what you have learned about plants. Include information about the many kinds of plants and their differences and similarities.

For Your Journal

The ideas expressed by the students should reflect some of the concepts they have learned about plants while studying this chapter. You may wish to ask each student to think of a fact about plants that he or she did not know before and then have volunteers share that information with their classmates.

CONNECTING IDEAS

The concept map should be completed with words and phrases similar to those shown here.

Row 3: roots; stems
Row 5: anchor plant; connect leaves to roots; make food
Row 6: absorb materials from soil

Connecting Ideas

Copy this unfinished concept map into your journal. Complete the concept map by writing the correct term in each blank.

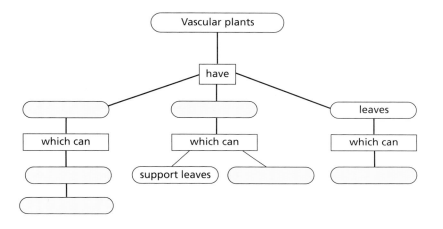

CHAPTER 10 REVIEW

ANSWERS

Understanding Vocabulary

1. Xylem and phloem are the two types of vascular tissue that transport water, dissolved minerals, and food in a plant.

2. Vascular cambium is the growth tissue, or layer of new xylem and phloem cells, in the stems of woody plants.

3. The chlorophyll of a plant absorbs light and produces energy-rich chemicals required during the process of photosynthesis.

4. The stamen and pistil are, respectively, the male and female reproductive organs of a flowering plant.

Understanding Concepts

Multiple Choice

5. d
6. b
7. c
8. a
9. c

Short Answer

10. A fruit usually forms from the ovary of a flower. An ovary gets fleshy and ripens as the seed develops inside. Other flower parts, along with the ovary, sometimes develop into parts of the fruit.

11. Pollination occurs when pollen is deposited on a stigma. A pollen grain develops a tube (pollen tube) that grows down the style of the pistil to an opening in the ovule. A cell in the pollen grain (which becomes the sperm cell) travels down the tube. Once inside the ovule, the sperm cell fuses with the egg cell, and fertilization occurs.

12. Plants produce organic compounds that contain energy and are used as food by humans and other animals.

Interpreting Graphics

13. The plant on the right, which has a fibrous root system, would be the best choice for a sloping field. A fibrous root system covers more area and more effectively holds soil to prevent erosion than a taproot system.

Reviewing Themes

14. Dispersal of pollen by wind is the primary method of gymnosperm reproduction. Because angiosperms disperse pollen by wind, water, and insects or other small animals, the probability is much greater that angiosperm cross-pollination will occur and produce a greater variety of plants.

15. The students' answers may suggest that all plants contain chlorophyll, they have the ability to make their own food through photosynthesis, they reproduce.

CHAPTER 10 REVIEW

Understanding Vocabulary

Explain how the term or terms in the first column play a role in the area of a plant's life listed in the second column.

1. xylem (260), phloem (260) ——— transport
2. vascular cambium (263) ——— growth
3. chlorophyll (270) ——— photosynthesis (263)
4. stamen (278), pistil (278) ——— reproduction

Understanding Concepts

MULTIPLE CHOICE

5. Complex vascular plants' distinguishing characteristic is their ability to
   a) grow roots.
   b) produce leaves.
   c) grow stems.
   d) produce seeds.

6. Which of the following is a vascular plant?
   a) moss
   b) fern
   c) hornwort
   d) liverwort

7. During photosynthesis, plants
   a) take in oxygen and give off carbon dioxide.
   b) take in glucose and give off carbon dioxide.
   c) take in carbon dioxide and give off oxygen.
   d) take in oxygen and give off water.

8. The parts of a flower's pistil include
   a) the stigma, style, and ovary.
   b) the stamen, filament, and anther.
   c) the receptacle, sepals, and petals.
   d) the anther, stigma, and pollen.

9. Which of the following is *not* generally a characteristic of an angiosperm?
   a) produces flowers
   b) loses its leaves in the fall
   c) produces cones
   d) protects its seed in a fruit

SHORT ANSWER

10. How is a fruit formed, and from what flower parts might it develop?

11. Describe pollination and fertilization in the angiosperm.

12. What is meant by the statement that plants produce food?

Interpreting Graphics

13. Look at the illustration below. Which plant would you choose to plant in a sloping field? Explain your answer.

284 CHAPTER 10

Thinking Critically

16. Fruits are considered to be a means of seed dispersal because fruits are eaten by an animal and pass through the animal's digestive tract. The seeds remain undigested, and by the time the seeds have been deposited along with the animal's waste, the animal has likely traveled far from the parent plant. Other types of seed dispersal, such as wind or water, are each effective means of dispersal in different conditions and in different environments, so it cannot be accurately said that one means of dispersal is more or less effective than another.

17. Because many angiosperms are pollinated by insects, the use of pesticides might kill the insects that pollinate the plants or disperse their seeds.

18. Simple vascular plants have poorly developed vascular systems. As a result, such plants must live near a constant source of water. Also they require moisture for reproduction.

19. Nonvascular plants do not have vascular tissue. Vascular tissue has great strength and can support a plant to great heights. Because nonvascular plants lack these strong supporting tissues, they cannot grow very tall.

20. Photographs: Virginia Creeper—herbaceous dicot, vascular, growing on woody dicot stem (tree); Butterfly weed—herbaceous dicot, vascular; Blue Flag or wild iris—herbaceous monocot, vascular.

Reviewing Themes

14. *Systems and Structures*
Compare and contrast the reproductive methods for gymnosperms and angiosperms. Use this information to explain why there are more types of angiosperms than there are gymnosperms.

15. *Diversity*
Dividing plants into categories, such as vascular and nonvascular, complex and simple, angiosperms and gymnosperms, and so on, is helpful in learning about the wide variety of plants. Now review what all plants have in common.

Thinking Critically

16. Why are fruits considered to be a means of seed dispersal? How effective are they in this role? Can you think of other types of seed dispersal that might be more effective or efficient? Explain.

17. Pesticides are sometimes used on crops to kill harmful insects. But pesticides may also kill helpful insects. What impact could this have on angiosperms?

18. Why is a moist environment important for the survival of simple vascular plants such as ferns?

19. Why can vascular plants grow much larger than nonvascular plants? Draw sketches of the plants to support your answer.

20. The photographs show three different kinds of plants. Determine whether each plant is woody or herbaceous, vascular or nonvascular, and a monocot or a dicot. Explain your choices.

Discovery Through Reading

Facklam, Howard, and Margery Facklam. *Plants: Extinction or Survival?* Enslow, 1990. The importance of plants as a source of food, medicine, and other products is integrated with research contributions by scientists.

CHAPTER 11

PLANT GROWTH AND ADAPTATIONS

PLANNING THE CHAPTER

Chapter Sections	Page	Chapter Features	Page	Program Resources	Source
Chapter Opener	286	*For Your Journal*	287		
Section 1: PLANT GROWTH	288	Discover By Doing **(B)**	288	*Science Discovery* *	SD
• Starting Out **(B)**	288	Discover By Doing **(B)**	289	Investigation 11.1:	
• Responding to the Environment **(A)**	290	Discover By Doing **(A)**	291	Investigating Geotropism **(A)**	TR, LI
• Reproduction **(A)**	292	Discover By Doing **(A)**	293	Connecting Other Disciplines: Science and Social Studies, Grafting **(B)**	TR
		Section 1 Review and Application	294	Extending Science Concepts: Tropisms **(A)**	TR
		Skill: Measuring Plant Growth **(H)**	295	Seed Germination **(B)**	IT
				Study and Review Guide, Section 1 **(B)**	TR, SRG
Section 2: PLANT ADAPTATIONS	296	Discover By Doing **(B)**	297	*Science Discovery* *	SD
• Spreading Seeds **(B)**	296	Activity: What can you learn from a collection of traveling seeds? **(A)**	298	Investigation 11.2: Observing Adaptations **(B)**	TR, LI
• Protection Against Animals **(A)**	298	Discover By Observing **(B)**	299	Reading Skills: Using Concept Maps **(B)**	TR
• Surviving Harsh Climates **(A)**	299	Section 2 Review and Application	301	Thinking Critically **(A)**	TR
		Investigation: Examining Seeds **(A)**	302	Record Sheets for Textbook Investigations **(A)**	TR
				Study and Review Guide, Section 2 **(B)**	TR, SRG
Chapter 11 HIGHLIGHTS	303	The Big Idea	303	Study and Review Guide, Chapter 11 Review **(B)**	TR, SRG
Chapter 11 Review	304	For Your Journal	303	Chapter 11 Test	TR
		Connecting Ideas	303	Test Generator	
				Unit 4 Test	TR

B = Basic   **A** = Average   **H** = Honors
The coding Basic, Average, and Honors indicates subsections, features, and resources that might be appropriate for different levels of learners. For additional suggestions regarding choice of topic and depth of coverage, see the Pacing Chart on pages T26–T29.

*Frame numbers at point of use
(TR) Teaching Resources, Unit 4
(IT) Instructional Transparencies
(LI) Laboratory Investigations
(SD) *Science Discovery* Videodisc Correlations and Barcodes
(SRG) Study and Review Guide

▶ 285A

CHAPTER MATERIALS

Title	Page	Materials
Discover By Doing	288	(per individual) raw peanuts, paper towel, hand lens, paper, pencil
Discover By Doing	289	(per individual) paper cups (4), potting soil, soaked bean seeds (4), pencil
Discover By Doing	291	(per group of 2) shoe box with lid, scissors, cardboard (2 pieces), potted bean seedling
Discover By Doing	293	(per group of 3 or 4) carrot tops, shallow dish, water, begonia cutting, spider plant "baby," pot, journal
Skill: Measuring Plant Growth	295	(per group of 2 or 3) bean plant, ruler, fine-point permanent marker, germinating bean seeds, straight pins, paper towel, jar with lid, cardboard
Discover By Doing	297	(per individual) maple seeds (or other winged seeds) or paper "helicopter" seeds
Teacher Demonstration	297	flowering impatiens plant
Activity: What can you learn from a collection of traveling seeds?	298	(per individual or pair) knee-high socks, envelopes, plant reference books, poster board, cellophane tape, marker, ruler
Discover By Observing	299	(per individual) ripe apple, cloth, water
Investigation: Examining Seeds	302	(per group of 2) bean seeds, scalpel, probes or toothpicks, hand lens, pencil, paper, corn seeds

ADVANCE PREPARATION

For the *Discover By Doing* on page 289, you will need to soak the bean seeds. Obtain potted bean seedlings for the *Discover By Doing* on page 291. Carrot tops, begonia cuttings, and spider plant "babies" are needed for the *Discover By Doing* on page 293. For the *Skill* on page 295, obtain bean plants and germinating bean seeds. Maple seeds or other winged seeds (paper models can be substituted) are needed for the *Discover By Doing* on page 297. For the *Activity* on page 298, the students will need to bring knee-high socks. The *Investigation* on page 302 needs bean seeds and corn seeds.

TEACHING SUGGESTIONS

Field Trip
Plan a neighborhood walking tour to locate, observe, and record information about local plant life. Have the students identify the type of plants, if possible, and note specific characteristics of each type. In the classroom, the students can research other information about the area's plants in reference books.

Outside Speaker
Check with local nurseries, greenhouses, or garden shops. People who work at these places are knowledgeable about plants. If a university is near, invite plant scientists to come and speak. Ask them to concentrate on discussing characteristics of plants that make them suitable for use around the home.

CHAPTER 11
PLANT GROWTH AND ADAPTATIONS

CHAPTER THEME—ENVIRONMENTAL INTERACTIONS

This chapter will introduce the students to the structures that develop during different stages in the life cycle of seed plants. The ability of plant structures to respond to light and gravity ensures that stems grow up, roots grow down, and leaves find light. Various adaptations that plants display, including how they spread their seeds, survive in harsh climates, and protect themselves against animals, help ensure plants' survival in their environments. The theme is also developed through concepts in Chapters 1, 2, 3, and 15.

MULTICULTURAL CONNECTION

Around the world, many varieties of plants are unique to the areas where they grow. The people who live in these areas often have developed interesting ways of using the native plants. Have each student investigate a plant native to a particular area, such as the baobab tree of Africa, the banyan tree of Asia, and the eucalyptus trees of Australia. Ask the students to draw a picture of the plant and write a brief description of the ways in which people use the plant. Then on an outline map of the world, have the students tape their pictures and descriptions in the appropriate locations.

MEETING SPECIAL NEEDS

Second Language Support

Fluent English-speaking students can be paired with students who have limited English proficiency to work together to tell the story of the life cycle of a seed plant. Encourage the partners to use any format they wish. For example, they could make a cartoon strip using pictures and text to describe the life of an oak tree, from acorn to mature tree.

CHAPTER 11
PLANT GROWTH AND ADAPTATIONS

From a stately oak to a colorful wildflower, all plants must reproduce, grow, and mature to survive. Plants depend on the environment for their livelihood. They need light to make food and water. They need to spread their seeds. They need to protect themselves from other organisms in their environment and often from the environment itself. Generation after generation of plants have survived because of the remarkable ways they have adapted.

In Florida the shore zone is usually dominated by red mangrove, chiefly young plants growing on ground that is periodically under water. . . . In meeting the ecologic challenge of life in the difficult, storm-wracked unstable zone where tidal salt water laps the edge of the land, the red mangrove has evolved into one of the most bizarre of all vegetables. Because the ecologic demands of the various kinds of seaside environments are very different, this species varies markedly in form. Some red mangroves are tall and straight-trunked; some are low-crowned domes standing on thin stilts; some spread out horizontally, and repeatedly reroot as they go, like vegetable centipedes.

Two of the fundamental adaptations that fit the red mangrove for its perilous life at the edge of the sea are salt tolerance and an ability to cling to unstable ground. A third adaptive achievement is a reproductive device that allows the

CHAPTER MOTIVATING ACTIVITY

Obtain pictures of individual trees of different kinds and sizes, growing in different environments. Point out to the students that the characteristics of a tree can reveal something about the history of that tree. For example, if its branches grow vertically, it may be that it had to compete with other plants for light, nourishment, and space. Have each student examine a tree in one of the pictures and write a paragraph describing what the history of that tree and its ecosystem might have been like. Encourage volunteers to share their paragraphs.

For Your Journal

By answering the journal questions, the students can explore their own knowledge of the development and adaptations of seed plants. Review their answers and note any misconceptions they might have about plant growth, development, and adaptations. Keep these misconceptions in mind and bring them up for discussion at the appropriate times during the study of the chapter.

species to disperse widely and colonize new habitat. Red mangroves have little yellow, waxy flowers; the seeds that these produce germinate before they leave the tree, generating a cigar-shaped seedling 6 to 12 inches long. When these seedlings fall, some may lodge and take root beneath the parent tree and become part of a growing forest there, but most of them drift away with tides and currents and may travel for hundreds or even thousands of miles before they strand or die. When a seedling strands in shallow water it quickly grows roots, and these pull the little stem erect. A few leaves appear, and in a short while a new little mangrove stands in the shallows. . . . A single, well-grown, tidal-zone red mangrove standing high on its thin legs is a strange-looking plant. A forest of such trees is one of the most offbeat kinds of vegetation to be found anywhere.

from *The Everglades: The American Wilderness* by Archie Carr and The Editors of Time-Life Books

For Your Journal

- How does a tiny seed become a giant tree?
- How are plants adapted to growing in a wide variety of climates?
- How do seeds disperse?

ABOUT THE PHOTOGRAPH

Mangrove forests dot the coast of Florida around its tip, including the area known as Ten Thousand Islands. Strictly speaking, the red mangrove is the only true member of the mangrove family to grow in Florida. However, the word *mangrove* is often applied to the other trees and vegetation that grow with the red mangrove in Florida's coastal swamps. The many roots that grow from the red mangrove's trunk and limbs bring up sap, raise the trunk above the water, and take in needed oxygen. The roots also hold the tree firmly in unstable ground and help accumulate mud and other materials, forming small islands where grasses and other vegetation can grow.

Mangroves are important to the ecology of the tidal zone environment. They protect the coastline against destruction by hurricanes. Insects and sea turtles feed on mangrove leaves. Manatees, alligators, crabs, and many fishes live in or near mangrove swamps. Mangroves provide resting places for many water birds, including herons, brown pelicans, and white ibises.

Section 1: PLANT GROWTH

FOCUS

In this section, the process by which a seed develops into a plant is presented. A plant's reactions to its environment are introduced. The ways in which a plant can reproduce, including grafting and vegetative propagation, are discussed.

MOTIVATING ACTIVITY

Cooperative Learning Obtain seeds from several trees that are native to your area. Give each group the seeds from one kind of tree. Have the group members design an environment in which they think their seeds will sprout. Then have them create that environment using containers, soil mixtures, and fertilizers and plant their seeds. The groups should periodically tend their seeds, and all the seeds that sprout into seedlings can then be transplanted into the local ecosystem.

PROCESS SKILLS
- Observing • Comparing
- Inferring

POSITIVE ATTITUDES
- Caring for the environment
- Curiosity • Enthusiasm for science and scientific endeavor

TERMS
- cotyledons • tropisms
- vegetative propagation
- regeneration

PRINT MEDIA
From Seed to Plant by Gail Gibbons (see p. 255b)

ELECTRONIC MEDIA
Plant Tropisms and Other Movements, Coronet (see p. 255b)

Science Discovery Sun tracking of flowers Plants from plant parts

BLACKLINE MASTERS
Study and Review Guide
Laboratory Investigation 11.1
Connecting Other Disciplines
Extending Science Concepts

DISCOVER BY Doing

Encourage the students to show and explain their diagrams. Point out that cotyledons vary in seeds. Some cotyledons are thick, some are thin. But in either case, they store food used by the seedling. Have the students save their diagrams in their science portfolios.

SECTION 1

Plant Growth

Objectives

Describe the internal structure of seeds.

Tell how plants respond to their environment.

Compare and contrast sexual and asexual reproduction in plants.

Some plants grow so quickly they almost appear to do so right before your eyes. For example, the red mangrove grows about 2.5 cm per hour. Red mangroves live in brackish waters along the coasts of southern Florida, Mexico, and the islands of the West Indies. Other plants hardly seem to grow at all. Bald cypress trees, which live in swampy areas of the Southeast from Delaware to Texas, may grow less than 2.5 cm in an entire year.

Starting Out

Plant growth begins with a seed. You already know something about seeds: they contain an embryo plant and they store food for the embryo's growth. Seeds also provide food for humans. Corn, peas, beans, and peanuts are all seeds. You can review the structure of a seed in the following activity.

 **Doing**

You will need several raw peanuts. Use your fingers to open the peanuts. Spread the parts out on a paper towel, and look at them with a hand lens. Make a diagram of a seed and label the parts.

Figure 11–1. A seed provides food for an embryo plant's growth.

288 CHAPTER 11

TEACHING STRATEGIES

● **Process Skills:** *Comparing, Applying*

Remind the students that the tiny plant inside a seed is called an *embryo*, and that this embryo is similar to the embryo inside of a human mother in that both embryos are created through sexual reproduction, and both have characteristics from their parents. Ask the students to describe the source of nutrients for each of these embryos. (The human embryo receives nutrients from its mother; the seed embryo receives nutrients from the contents of its seed. Much of the mass of a seed consists of stored food.) Then ask the students to describe why the development of a seed embryo into a seedling in terms of food supply may be much more tenuous than the development of a human embryo into an infant. (A human embryo is supplied with food throughout the period of its development; a seed embryo must develop into a seedling in a finite amount of time because its food source is limited.)

When a seed sprouts, or *germinates,* the first thing it does is take in water. This causes the seed to swell as the *embryo,* or developing plant, inside the seed starts to grow. The root tip is usually the first part of the embryo to emerge from the seed. Since seeds store little water, it is important for the water-absorbing part of the plant—the root—to begin growing first. As the root grows, it pushes the embryo, still mostly inside the seed, through the soil. The rapidly growing embryo, now called a *seedling,* faces many difficulties. If it fails to reach the surface and the light it needs for photosynthesis, it will die. The poem *Seeds* describes what can happen to seedlings.

The young stem and leaves are the next things to emerge in a seedling. These first leaves are called seed leaves, or **cotyledons** (kawt un LEE duhns). In the next activity, you can discover the importance of cotyledons.

Figure 11–2. The root is the first thing to sprout from the seed.

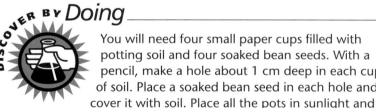

Discover by Doing

You will need four small paper cups filled with potting soil and four soaked bean seeds. With a pencil, make a hole about 1 cm deep in each cup of soil. Place a soaked bean seed in each hole and cover it with soil. Place all the pots in sunlight and keep them watered. After the plants have pushed their way through the soil, remove the cotyledons from two of the seedlings. During the next few days, compare the growth of the seedlings.

Cotyledons are different from the leaves that will form later. In beans, peas, and peanuts, the cotyledons are thick and store a lot of food. The cotyledons of squash and radishes are thinner—more like other leaves. They store less food.

After the cotyledons emerge, the seedling's first true leaves begin to grow. In some seeds, these leaves are already present inside the embryo. The leaves turn green in the sunlight and grow even larger. The seedling continues to use the food from the cotyledons until the true leaves can provide enough food for the seedling. Eventually, the cotyledons shrivel and die.

Seeds
by Walter de la Mare

The seeds sowed—
For weeks unseen—
Have pushed up pygmy
Shoots of green;
So frail you'd think
The tiniest stone
Would never let
A glimpse be shown.
But no; a pebble
Near them lies,
At least a cherry-stone
In size,
Which that mere sprout
Has heaved away,
To bask in sunshine,
See the Day.

▼ ASK YOURSELF

Why is it so important for a new seedling to reach sunlight quickly?

BACKGROUND INFORMATION

Walter de la Mare (1873–1956) was an English author who wrote for both children and adults. His poems, short stories, novels, and plays are often characterized by their blending of the real and the fanciful. *Bells and Grass* and *Peacock Pie* are collections of his poems for children. *Come Hither,* a compilation of rhymes and poems, is considered a children's classic.

DISCOVER BY *Doing*

The students should observe that removing the cotyledons removes the seedlings' initial food source and will significantly affect their development, unless the seedlings have developed leaves that can perform photosynthesis for food production.

PERFORMANCE ASSESSMENT

To check the students' understanding of the function of cotyledons as a food source, ask the students to explain the difference in the development of the seedlings with and without cotyledons.

ONGOING ASSESSMENT ▼ ASK YOURSELF

The amount of stored food in a seed is limited. The seedling must reach sunlight and begin photosynthesis before the food in the seed runs out.

TEACHING STRATEGIES, continued

● **Process Skills:** *Comparing, Evaluating*

Mention to the students that gardeners typically do one of two things with seeds: Either they plant the seeds in soil and then water the soil regularly, or they soak the seeds in water for a period of time, then plant the seeds in soil, and water the soil regularly. Also point out that the gardeners who soak seeds before planting think that soaking seeds provides an advantage. Ask the students to explain why soaking seeds before planting might be a good idea. (The students might suggest that seeds must absorb water to germinate. Soaking seeds allows the seeds to absorb water and split open, reducing the germination time required by the seeds once they are planted. The plants that develop from those seeds will reach maturity more quickly.)

● **Process Skills:** *Inferring, Applying*

Point out that plant seedlings often must struggle to survive in their environment. Factors such as diffused light caused by competing plants and damage caused by insects and other animals contribute to the difficulties seedlings experience as they

---

BACKGROUND INFORMATION

Throughout time, fires have periodically burned through forest environments. Despite popular fears about fire and the damage it can cause, a fire can actually benefit a forest by clearing away fallen trees, needles, leaves, and other debris on the forest floor; killing insect pests and unwanted plants; and allowing some seeds to germinate. In *prescribed burning*, small fires are sometimes intentionally set by forestry workers to remove the debris that could contribute to a larger, more devastating fire.

MULTICULTURAL CONNECTION

Corn is an important seed plant in many Native American cultures. It was first gathered in the wild and used as food by Native Americans in Mexico 10 000 years ago. They began to grow corn as a crop around 5000 B.C. By saving seeds from the best plants to use for their next crop, they created large, many-kerneled corncobs from the tiny wild cobs. Being able to cultivate corn, along with beans and squash, meant that the people could settle in villages and farm for food instead of traveling in search of food. Corn was grown by the Aztecs, the Mayas, and the Inca. Corn went to Europe when Christopher Columbus took corn seeds back to Spain. Eventually, corn was introduced into many other parts of the world.

Figure 11–3. The seedling (left) is no longer dependent on food stored in the seed. The sapling (right) grows rapidly.

Responding to the Environment

An embryo inside a seed grows by stretching, or elongating, its cells. However, for further growth to occur, new cells must be added by cell division. A seedling continues to grow in length at its stem tips and root tips. As the plant grows into a *sapling,* new cells elongate and mature. The processes of cell division and elongation continue as long as the plant is growing. For herbaceous plants, this is the main type of growth.

Woody plants, such as trees, shrubs, and many vines, have a lot of supporting tissue—xylem—in their stems and roots. In addition to growth at the tips of these organs, woody plants also have a layer of dividing cells under their bark that allows for growth in diameter.

If you have ever observed a plant growing on a window sill, you may have noticed that many of its leaves are turned toward the light. Why do the leaves of a plant grow toward light? Why do roots grow down and stems grow up? Why does a Venus' flytrap close when touched by a fly? All of these are examples of plants responding to the environment.

Figure 11–4. The Venus' flytrap responds to touch, quickly entrapping unsuspecting insects.

grow. Ask the students to list other factors that can negatively affect the growth of plant seedlings. (Responses might suggest that pollutants in rainfall and soil and unusual climatic conditions such as flooding and drought can create a hostile environment for developing seedlings.)

● **Process Skills:** *Predicting, Applying*

Make sure that the students understand that a plant's response to a stimulus in its environment—such as bending toward a light source—is a growth, not a behavioral, response. Point out that these responses to stimuli, or *tropisms*, can be considered positive or negative. Ask the students whether a plant growing toward a light source is displaying positive or negative phototropism. (positive phototropism because the growth is toward the stimulus) Then ask the students whether a plant's stem exhibits positive or negative geotropism. (negative geotropism because the growth is away from the stimulus)

When a Venus' flytrap closes around an insect, the plant is responding to the stimulus of touch. A *stimulus* is anything in the environment that causes a response from an organism. When a plant grows toward a window, it is responding to the stimulus of light. What is a plant responding to when its stems grow up and its roots grow down? In the next activity, you can observe firsthand a plant's response to a stimulus.

DISCOVER BY Doing

You will need a shoe box with a lid, a pair of scissors, two pieces of cardboard half as wide as the box, and a potted bean seedling. Cut a hole in one end of the shoe box. Then tape the cardboard strips to opposite sides of the box as shown. Put your plant in the end of the shoe box away from the hole. Put the lid on the box, and place it near a sunny window. Observe the growth of the plant for the next few days. Describe the growth of the plant.

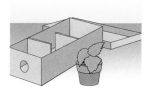

The plant in the activity bends because it needs light. It is responding to the light available. Plants' responses to the environment are called **tropisms** (TROH pihz uhmz). *Geotropism* is a plant's response to gravity. This tropism keeps the roots of a plant growing downward and the stems growing upward. *Phototropism*—response to light—is very important. A seedling must reach sunlight before its stored food runs out. It needs a stimulus so that it can grow in the right direction. That stimulus is the light itself.

Chemicals made in the tips of stems cause plants to respond to light. These chemicals make cells divide and lengthen, causing the stems to grow. Cells on the shady side lengthen more quickly, causing plants to bend toward the light. Even more of these chemicals are produced in the dark. Without these chemicals, a seedling would die underground.

Light is one of the most important stimuli for plants. When light and chemicals work together, interesting things sometimes occur. For example, you know that the number of daylight hours changes from season to season. Length of daylight affects the production of chemicals that control plant flowering.

Figure 11–5. When a plant bends toward a window, it is responding to light.

① The plant is responding to the stimulus of gravity.

DISCOVER BY Doing

This activity will enable the students to view phototropism—a plant's response to light. The students should observe that the bean seedling bends toward the hole and the source of light. You can extend the activity by having the students turn the plant away from the hole and watch it gradually move around toward the light again. Have the students place their descriptions from the activity in their science portfolios.

INTEGRATION—Language Arts

Point out that the term *tropism* is a word that has been adapted from the Greek word *tropos*. Ask the students to find out the meaning of *tropos* ("a turn, change") and explain why *tropism* is a good name for a plant's response to a stimulus.

◆ **Did You Know?**
The responses of plants to stimuli in their environment—tropisms—are caused by hormones.

SECTION 1 **291**

TEACHING STRATEGIES, continued

● **Process Skills:** *Evaluating, Applying*

Plants such as oak trees depend on the wind for pollination and reproduction. Ask the students to decide whether wind pollination is a relatively efficient or inefficient means of reproduction, and explain how an oak tree can greatly increase its chances of successful pollination. (The students might reason that pollination by wind is relatively inefficient because a tiny pollen grain must somehow land on a small female reproductive structure; to increase their chances, plants that rely on the wind for pollination typically release pollen in large quantities.)

● **Process Skills:** *Inferring, Expressing Ideas Effectively*

Ask volunteers to explain how vegetative propagation can economically benefit a commercial greenhouse. (Responses should reflect the idea that many stem cuttings or other parts can be taken from a plant. These cuttings cost nothing and can be grown into many new plants, which are then offered for sale.)

LASER DISC
43043–43336

Sun tracking of flowers

ONGOING ASSESSMENT
ASK YOURSELF

Plant roots must grow down so that they can anchor the plant in the soil and absorb water and minerals from the soil. Plant stems must grow up so that the leaves can reach the light that the plant needs in order to perform photosynthesis.

THE NATURE OF SCIENCE

One scientific study of plants made a major contribution to scientific knowledge. In the 1860s, an Austrian monk, Gregor Mendel, conducted experiments on inherited characteristics or traits using generations of pea plants. His studies became the foundation for *genetics*, the science of heredity. Using the principles that Mendel discovered, scientists went on to develop, for example, high-yield, disease-resistant varieties of corn, rice, and wheat that have helped to increase the world's food supply.

Figure 11–6. Violets and poinsettias are short-day plants.

Some plants, called *short-day plants,* will not flower if the number of daylight hours is too great. Short-day plants, such as poinsettias, flower in the early spring, late fall, or winter when the days are short. Other kinds of plants, such as spinach and clover, need twelve to sixteen hours of sunlight each day in order to flower. These are called *long-day plants.* Long-day plants flower during the longer days of early summer. Still other plants, such as corn, are *day-neutral plants.* These plants will flower in varying amounts of sunlight.

 ASK YOURSELF

Why is it important for a plant's roots to always grow down and its stems to always grow up?

Reproduction

Remember that flowers are the reproductive organs of angiosperms and cones are the reproductive organs of gymnosperms. Both angiosperms and gymnosperms produce seeds by sexual reproduction. Some plants can reproduce by asexual reproduction. Cuttings of roots, stems, or leaves can grow into new plants. This type of growth is called **vegetative propagation.**

Vegetative propagation is often used to grow plants faster and more reliably than with seeds. Many seedless varieties of flowering plants have been developed in this way. These plants, such as oranges and grapes, produce seedless fruits. They can be reproduced only by vegetative propagation.

A different kind of vegetative propagation takes place in the white potato plant. The potato is an underground stem. The "eyes" of the potato are tiny buds on the stem. Farmers cut up whole potatoes and plant the pieces that have eyes. A new plant quickly sprouts from each of these buds. Growing potatoes from seeds takes a much longer time.

Figure 11–7. Potatoes can be grown from seeds or from "eyes."

- **Process Skills:** *Interpreting Data, Analyzing*

Have the students imagine that a certain plant can reproduce only by vegetative propagation, and then ask them to explain how that plant might be at a disadvantage when compared to a plant that reproduces by seeds. (Responses might suggest that if the climate in which the plant was found significantly changed, and the plant could not adapt to the new climatic conditions, it would die. By reproducing only vegetatively, the plant would lack an efficient mechanism for transmitting its offspring to a different ecosystem with more suitable growth conditions.)

- **Process Skills:** *Comparing, Applying*

Remind the students that vegetative propagation is asexual reproduction. Then ask them to compare sexual and asexual reproduction and determine which type of reproduction contributes to the greatest genetic diversity of life. (Genetic diversity results from sexual reproduction because two parents are involved; asexual reproduction creates clones of the single parent organism.)

Bulbs are another example of vegetative propagation. A bulb is a short, underground stem. Planted in the fall, a plant, such as a lily, grown from a bulb will flower the following spring. A lily grown from a seed will not bloom until the second year. By the end of the growing season, bulbs will have divided into several pieces, which can be separated and planted—each piece producing a new plant the following year. Tulips, daffodils, onions, and garlic can also be grown from bulbs.

The strawberry plant and some grasses can reproduce by runners. A runner is a stem that "runs," or grows, along the ground. When one of its buds touches the ground, it forms roots and produces a new plant. Even pieces of stems and leaves can sometimes grow into new plants. The plant in the picture produces "babies," or tiny plants, at the edges of its leaves.

Plants that are grown through vegetative propagation have the same characteristics as the plant from which they were taken. For example, if you grow an African violet from a leaf cutting, that African violet will have the same kind of leaves, flowers, and other characteristics as the plant from which you cut the leaf. The new plant was once part of the old plant, so it has exactly the same traits as the old plant. You can find out more about vegetative propagation by doing the next activity.

Figure 11–8. Bulb-grown plants flower much sooner than those grown from seeds.

Figure 11–9. This kalanchoe is a "mother" plant with her "babies."

Discover by Doing

Place carrot tops in a shallow dish and water them regularly. Observe the growth of the roots. Make cuttings from a begonia and put them in water. Observe the development of new roots. Get a spider plant, cut off a "baby," and plant it in a pot. Observe how it grows. Record your observations in your journal. Find out how many ways you can grow plants by vegetative propagation.

LASER DISC

3082

Plants from plant parts

MEETING SPECIAL NEEDS

Gifted

Two other types of vegetative propagation—layering and budding—are relatively common practices. Ask the students to discover more about these practices (when are they used, what are their advantages and disadvantages, which foods result from layering and budding) and prepare a brief presentation for the class. Encourage the students to include appropriate pictures or diagrams.

DISCOVER BY *Doing*

In addition to the plants in the activity, many plants (including ivy, coleus, ferns, irises, blueberries, dandelions, apples, bananas, oranges, potatoes, gladioli, lilies, and tulips) reproduce from various plant parts. After the students have recorded their observations, have them compare their journal entries to those of other students to see how many different plants they were able to name.

GUIDED PRACTICE

Write the following terms on the chalkboard: cotyledons, tropisms, vegetative propagation, regeneration. Ask the students to define each term and explain how the terms are related to plant growth.

INDEPENDENT PRACTICE

Have the students provide written answers to the Section Review and Application questions. In their journals, have the students write a paragraph describing the struggle for survival from seed to sapling that a young oak tree experiences.

EVALUATION

Have the students describe the structure of a seed (a protective seed coat, stored food, a plant embryo, one or two cotyledons) and explain the importance of each part in the early growth of a plant.

RETEACHING

Ask the students to draw two pictures, one showing how a plant responds to light, and the other showing how a plant responds to gravity. Have the students label the pictures with the appropriate names. Encourage them to take turns explaining the concepts shown in their pictures.

SCIENCE BACKGROUND

Grafting does not create genetic diversity or variation; it simply combines the characteristics of two different plants.

ONGOING ASSESSMENT
ASK YOURSELF

In vegetative propagation, plants are grown from bulbs, stems, leaves, and roots.

SECTION 1 REVIEW AND APPLICATION

Reading Critically

1. Plant embryos obtain energy for growth from the food stored in their seeds.

2. Plant tropisms are plant growth responses to certain stimuli such as light and gravity.

3. A cotyledon is a seed leaf. Cotyledons contain the stored food used by a seedling until its true leaves can provide food.

Thinking Critically

4. Responses might suggest that rain could carry released pollen to the ground and not to female reproductive structures, or that wet foliage could cause pollen grains to adhere to the plant instead of being dispersed through the air.

5. Advantages include the need for only one plant in order to reproduce, and the transmission of the strengths of the parent to the offspring. Disadvantages include the transmission of the weaknesses of the parent to the offspring and a lack of genetic diversity in case of environmental changes.

▶ 294 CHAPTER 11

Figure 11–10. Grafting uses the root system of one plant and the branches of another.

Figure 11–11. This tree has five different varieties of apples. Grafting allowed one tree to represent an entire orchard.

Were you able to grow new plants from plant parts? If so, you were demonstrating regeneration. **Regeneration** is the ability to grow or replace missing parts. Many plants have this ability; some animals do, too. Other methods of asexual reproduction have been developed by scientists. These methods help farmers grow new plants without having to start them from seeds. *Grafting* is one of these methods.

Grafting enables farmers to make use of the best characteristics of two plants. For example, some types of orange trees are very hardy, but their fruit is extremely sour. Other types of orange trees are not as hardy, but their fruit is sweet and juicy. If the branches of the tree with sweet fruit are grafted to the roots of the sour orange tree, the hardy root system will support branches producing good fruit. It is also possible to graft branches from several different kinds of trees onto one root system, thereby producing a tree that bears several varieties of fruit.

 ASK YOURSELF

Describe several means of vegetative propagation.

SECTION 1 REVIEW AND APPLICATION

Reading Critically
1. How do plant embryos get energy for growth?
2. What are plant tropisms?
3. What is a cotyledon and why is it important to the growth of a new plant?

Thinking Critically
4. Wind pollinates many kinds of trees. How might wet weather conditions during flowering hinder the formation of seeds?
5. Any plant that reproduces asexually has exactly the same characteristics as the plant it grew from. What advantages and disadvantages could this have?

EXTENSION

Have the students explore the relationship between synthetic plant hormones and weed killers and share their findings with the class. (The synthetic hormones found in weed killers make weeds grow abnormally, a process that causes them to die.)

CLOSURE

 Cooperative Learning Have the students work in small groups to demonstrate their understanding of vegetative propagation by choosing a bulb, stem, root, or leaf to grow into a plant.

SKILL

Measuring Plant Growth

Process Skills: Observing, Comparing

Grouping: Groups of 2 or 3

Objectives
- **Compare** the growth characteristics of stems and roots.
- **Interpret** the data collected in an experiment.

Discussion
Encourage the students to recognize the value of making careful, precise measurements in an experiment. You might also wish to have the students read the procedure and predict the results of the experiment before they perform it. After they have collected the data, have the students compare their results to their predictions.

▶ **Application**
The students should observe that the spaces between the marks on the stem of the bean plant remained constant. Stem growth occurs only at the tip of the stem. Similarly, the students should also observe that the spaces between the marks on the root of the bean plant remained constant (although precise measurements will indicate that root cells tend to elongate slightly near the tip of the root). Root growth occurs only at the tip of the root.

Using What You Have Learned
In this activity, the students are able to observe that the production of new plant cells occurs only at the tips of a plant. (This production of new cells occurs in tissue called *meristem*.)

SKILL   Measuring Plant Growth

▶ MATERIALS

- a bean plant with at least an 8-cm stem • millimeter ruler • fine-point permanent marker • germinating bean seeds with roots 1 cm long • straight pins • paper towel • jar with lid • piece of cardboard

▼ PROCEDURE

1. Place the ruler next to the stem of the bean plant. Starting at the tip of the stem, make an ink mark every 5 mm along the stem. Be sure that you mark only the stem and not part of the youngest leaf.
2. Wait one or two days. Then complete step 3. While you are waiting, complete step 4.
3. Measure the spaces you marked the day or so before. Record your results.
4. Cut the piece of cardboard so that it will fit standing up inside the jar. It should be able to stand without falling over. Set this cardboard aside to use later.
5. Lay a germinated seed on the table. Starting at the tip of the root, make an ink mark every 2 mm. Work quickly so the root does not dry out.
6. Stick a pin through the center of the seed. Then pin the seed to the cardboard so that the root points downward. Place the cardboard in the jar.
7. Add enough water to the bottom of the jar until the root tip is 2.5 cm above the water.
8. Screw on the jar lid and keep the jar in a shady place. Wait one day.
9. The next day remove the seed from the jar. Measure the spaces you marked on the roots the day before. Record your results.

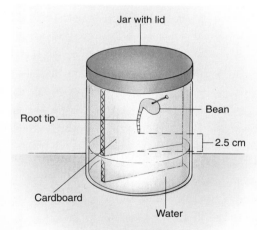

▶ APPLICATION
Did any of the spaces between marks on the stem get larger? Where does growth appear to occur on the stem? Did any of the spaces on the root increase in size? Where does growth appear to occur on the roots?

✳ Using What You Have Learned
What processes were you able to observe in this activity? What statement can you make about which parts of plants grow in length?

 Have the students place their tables and conclusions from the *Skill* activity in their science portfolios.

SECTION 1   **295** ◀

Section 2:
PLANT ADAPTATIONS

FOCUS
In this section, several plant adaptations, or ways that plants have developed to survive in their environments, are described. The different ways that plants spread their seeds, protect themselves against damage by animals, and survive harsh climates are discussed.

MOTIVATING ACTIVITY

Cooperative Learning Working in small groups, have the students consider this scenario: A seed plant native only to North Dakota begins to appear sporadically in Nebraska. Many years later, this seed plant begins to appear sporadically in Texas. Ask the students to explain how the plant was able to move a distance of 1500 km over land without any help from humans or their inventions.

PROCESS SKILLS
- Observing • Comparing
- Evaluating

POSITIVE ATTITUDES
- Caring for the environment
- Cooperativeness

TERMS
None

PRINT MEDIA
Strange Plants
by Jason Cooper
(see p. 255b)

ELECTRONIC MEDIA
Seed Dispersal,
Britannica
(see p. 255b)

Science Discovery
Palm, coconut
Plant adaptations
Cactus, barrel

BLACKLINE MASTERS
Study and Review Guide
Laboratory Investigation 11.2
Reading Skills
Thinking Critically

SCIENCE BACKGROUND
Many plants have coevolved with certain animal species. For example, a plant that depends on furry animals to disperse its seeds can find many candidates and would probably be a wide-ranging plant. But if all furry animal species became extinct, these plants would also be in danger of extinction. Only plants that did not depend on furry animals for seed dispersal would then be likely to survive.

▶ **296** CHAPTER 11

SECTION 2

Plant Adaptations

Objectives

Describe several plant adaptations.

Summarize how different plants survive harsh weather conditions.

Observe various types of seed dispersal.

Plants are found growing nearly everywhere on the earth. Not every kind of plant, however, is able to grow in every environment. Some plants grow only in wet environments. Other plants, such as those pictured, grow in the desert, where water is scarce. Desert plants have structures that help them conserve water. In fact, many plants have adaptations that enable them to survive in harsh environments.

Spreading Seeds

Figure 11–12. It's not easy being green.

One way plants can ensure their survival—at least for one more generation—is to successfully spread their seeds within an environment favorable for germination. Plants show remarkable adaptations for spreading seeds, or *seed dispersal*. Some plants enlist the aid of unsuspecting animals for this job. The burs of a burdock, for example, stick to the fur of many animals. The bur, which contains a seed, may be carried a great distance before it falls off.

Animals disperse seeds in other ways as well. Many fruits are eaten, seeds and all, by animals, especially birds. The seeds pass through their bodies undigested.

Humans, too, help with seed dispersal. Have you ever eaten watermelon outside and spit the seeds on the ground or tossed the core of an apple into an empty lot or field? An apple core contains many seeds, some of which might land in an environment suitable for germination. And the same seeds that stick to an animal's fur will also stick to your clothing.

Figure 11–13. Animals often provide "taxi service" for seed dispersal.

TEACHING STRATEGIES

● **Process Skills:** *Inferring, Applying*

Point out to the students that seed dispersal is the movement of one or more seeds away from a parent plant. Ask the students to explain why seed dispersal is important for the survival of a plant species. (A plant that successfully disperses its seeds increases its range, or the distance that it can be found from the parent plant; seed dispersal also provides a greater likelihood of survival because the dispersed seed is no longer in direct competition with the parent plant for space, light, water, and nutrients.)

● **Process Skills:** *Comparing, Applying*

Ask the students to name several fruits and describe what those fruits would taste like if they were not ripe. (Unripe fruits have a bitter or unpleasant taste.) Point out that the seeds inside unripe fruit are also unripe, or not mature. Emphasize that when the seeds of a fruit become ripe enough for dispersal and growth, a plant begins to pump sugars into the fruit, sweetening it and often changing its color. Animals are enticed to notice and eat the fruit, so that the animals will disperse the mature seeds.

Figure 11–14. Some seeds are adapted for flights of fancy.

Water and wind also aid in seed dispersal. A coconut can drop from a tree, travel on an ocean current, and germinate on a beach thousands of kilometers from where it fell. Red mangrove seeds bob along on tidal currents in a coastal marsh until they touch bottom. Then they sprout, growing prop roots to anchor themselves in the shallow water.

Many other seeds are fitted with "wings" and travel on currents of air. In the next activity, you can see how far winged seeds can fly.

DISCOVER BY Doing

Gather a cupful of maple seeds or other winged seeds, or make paper "helicopters" to represent seeds. Release some of the seeds or helicopters from a second-story window and observe where they land. Measure how far they fly. Do any of the seeds go farther than the spread of the branches of a large tree? Why do you think that would be important?

A few plants are adapted internally to ensure that their seeds are widely dispersed. The witch-hazel plant, shown in the photograph, launches its seeds from a miniature "catapult." Witch-hazel seeds can be propelled several meters from the parent plant.

When it comes to self-propulsion, however, the witch-hazel is a poor second to the dwarf mistletoe. Dwarf mistletoe fruits absorb water until internal pressure finally bursts the fruits open, throwing the seeds nearly 15 m from the "launch parent."

DISCOVER BY Doing

This activity will reinforce the students' understanding of the concept of seed dispersal. The data collected should lead the students to infer that some winged seeds travel farther than the spread of the branches of the tree from which they fall. Seed dispersal away from a parent plant is important because dispersal increases the likelihood of seed survival by reducing competition for vital space, light, water, and nutrients.

Demonstration

Some varieties of flowering plants disperse their seeds in a way that is similar to the way a cannonball is shot from a cannon. Obtain a flowering impatiens plant. (Impatiens are commonly used as house plants and landscape borders.) Have the students observe the fruit of an impatiens as it is touched. The fruit will burst open and fling out its seeds.

Figure 11–15. Witch-hazel seeds are catapulted into the air.

LASER DISC
2944
Palm, coconut

ACTIVITY

What can you learn from a collection of traveling seeds?

Process Skills: Comparing, Evaluating

Grouping: Individuals or pairs

Hints

Because some students may be susceptible to poison ivy and other such plants, check the area for these plants. Also ensure that the area is free of snakes and other organisms that are potentially dangerous.

If you or your students are able to identify various plants during your walk, share this information with each other.

▶ **Application**

Methods of seed dispersal and parent plant adaptations will vary depending on the location selected and the types of seeds collected during the activity.

★ PERFORMANCE ASSESSMENT

Review the students' posters and explanations to evaluate their understanding of the variety of seed dispersal methods as well as the function of seed dispersal.

ONGOING ASSESSMENT
▼ ASK YOURSELF

Successful seed dispersal greatly increases the probability that a plant species will survive by reducing potential competition for space, light, water, and nutrients and thus ensuring that offspring survive. Seeds that do not disperse are likely to end up in conditions that are not suitable for the proper growth and development of offspring.

TEACHING STRATEGIES, continued

● **Process Skills:** Observing, Applying

Obtain several pieces of Velcro, and distribute the pieces along with hand lenses to the students. Give them an opportunity to examine the structure of the material. Ask the students to explain, based on their observations, how Velcro works. (Responses should reflect the idea that the "hooks" on one face of the material stick to the "loops" on the other face of the material.) Then ask the students to explain how seeds such as burrs might attach themselves to animals' fur. (Responses should reflect the idea that burrs become hooked by an animal's fur and are detached from their parent plant. At some later time and at some distance from the parent plant, the burr becomes "unhooked" from the fur and falls to the ground.)

ACTIVITY
What can you learn from a collection of traveling seeds?

MATERIALS
old knee-high socks, envelopes, plant reference books, poster board, wide cellophane tape, marker, ruler

PROCEDURE
1. Pull an old pair of large knee-high socks over your shoes. Then walk through a field or overgrown lot. After your walk, check your socks and other clothing for seeds.
2. Collect as many different seeds as you can, placing each kind in a separate envelope.
3. At home or at school, use reference books to identify the seeds in your collection. The books will also tell you something about the methods of dispersal of the various seeds.
4. Attach the seeds to poster board with clear tape. Use a marker to label each kind of seed. Allow enough space to write a brief description of the method of dispersal for each kind of seed. Display your collection.

APPLICATION
What unusual methods of seed dispersal did you discover? How is the parent plant adapted for this dispersal method?

▼ **ASK YOURSELF**

How does successful seed dispersal help to ensure the survival of a plant species?

Figure 11–16. "Leaves of three—let it be."

Protection Against Animals

A species will soon become extinct if many of its individuals don't live long enough to reproduce. Plants are no exception. Some of the biggest dangers to the survival of plants come from grazing animals and insects. To defend against these attacks, some plants have adapted a kind of "chemical warfare." For example, the "milk" of the milkweed plant must have a bad taste because most insects avoid this plant.

Poison ivy, poison oak, and poison sumac use a slight variation of this adaptation. These plants secrete a chemical that is very irritating to the skin, keeping most animals away and allowing the plants to grow undisturbed.

- **Process Skills:** *Classifying, Evaluating*

Have the students suggest the names of ten different plants that reproduce using seeds, and record the names of these plants on the chalkboard. Remind the students that plants can disperse their seeds with the help of animals, humans, wind, and water. Write the four methods of seed dispersal on the chalkboard. Then ask the students to select a plant and debate the advantages and disadvantages of each method of seed dispersal for that particular plant.

- **Process Skills:** *Inferring, Generating Ideas*

Ask the students what might happen to a plant that develops without protection against animals, for example, a rose bush without thorns. (Responses should reflect the understanding that a rose bush without thorns is less likely to survive and reproduce than a rose bush with thorns.) Then ask the students how plants such as grasses do not appear to have any protection against animals but, nevertheless, survive. (Grasses can be cropped to the ground by grazing cattle, but as long as the roots remain, the grasses will grow again.)

Black walnuts seek to avoid a different kind of threat to survival—competition from other plants. The roots of a black walnut secrete a chemical into the soil that keeps other plants from growing too closely.

Some plants have a physical defense against animal agression. Roses, raspberries, black locusts, and wild citruses all have thorns to keep animals away. The spines of many desert plants, although not structurally the same, are equally effective adaptations that keep hungry animals away.

ASK YOURSELF

In what ways do plants protect themselves from animals?

Figure 11–17. Chemicals produced by the roots of the black walnut tree keep other plants from growing too closely.

Figure 11–18. Thorns among roses grow for protection.

 In genetic engineering, genes from the cells of one organism are transferred to the genes in the cells of another organism. By manipulating the genes of plants, scientists hope to create plants that can, for example, protect themselves against pests by producing their own insecticides. Potato plants could produce a chemical that warded off potato beetles. This idea is not so farfetched: In 1987, genes from bacteria were added to tomato plants to make the plants resistant to caterpillars. Gene manipulation, however, has its detractors as well as its supporters. Have interested students research and debate the issue.

ONGOING ASSESSMENT
ASK YOURSELF

Responses might suggest that plants protect themselves from animals by having an unpleasant taste, by releasing chemicals that blister skin or prevent other plants from growing nearby, or by having thorns.

 **LASER DISC**

3046

Plant adaptations

Surviving Harsh Climates

Remember, plants grow in almost every environment. However, no matter where a plant grows, one of the biggest dangers to its survival is loss of water. Like all living organisms, plants need water to survive, but it is not always readily available. So plants must conserve the water they do have. In the next activity, you can observe a water-conserving adaptation.

 **Observing**

Get a ripe apple and polish it with a cloth. Observe how the skin shines. Now put a drop or two of water on the apple's skin. What happens to the water?

TEACHING STRATEGIES, continued

● **Process Skills:** *Inferring, Synthesizing*

Make sure that the students understand the role of genetic variation in plant adaptations such as seed dispersal, protection against animals, and surviving harsh climates. Stress that all of the adaptations shown by plants today are a result of genetic variation. Over the course of time, variations have appeared in plants, and certain of these variations have enabled the plants of today to survive and reproduce. Without these genetic variations, plant adaptations would not occur.

GUIDED PRACTICE

Ask each student to describe an adaptation of plants related to seed dispersal, protection from animals, and survival in harsh climates.

INDEPENDENT PRACTICE

 Have the students provide written answers to the Section Review and Application questions. In their journals, ask the students to write a summary of the adaptations that enable cacti to survive in the desert.

---

① The wax helps the produce retain its moisture, thus keeping it fresher for the customers.

 LASER DISC
3003
Cactus, barrel

MEETING SPECIAL NEEDS

Mainstreamed

 Cooperative Learning Have the students make a game board on poster board. The name of the game is *Plant Adaptations*, and the game board is divided into three sections: *Seed Dispersal*, *Protection Against Animals*, and *Protection From Climate*. Help the students collect pictures of plants. Then have teams of students place the pictures in appropriate categories and explain their decisions.

REINFORCING THEMES— *Environmental Interactions*

Remind the students that individual plants do not choose the adaptations that will best suit them for survival. Rather, genetic diversity or variation provides plants with many adaptations, some of which are well suited to a particular environment, allowing the plants with those adaptations to survive and successfully reproduce.

▶ **300** CHAPTER 11

Figure 11–19. Plants have several adaptations to prevent water loss.

Figure 11–20. Cactus spines, which are modified leaves, help to reduce the loss of water.

Figure 11–21. Many desert plants have shallow roots to absorb the occasional heavy rains.

Apples are covered with a waxy coating. When you polish it, an apple gleams like a newly waxed car. On a car, wax causes water to bead and run off. But the waterproof skin of an apple keeps it from losing the water it stores inside. Many leaves also have waxy coats. This coating seals the leaves, preventing water from evaporating. Fruits and vegetables sold in supermarkets are sometimes coated with additional wax. How might this be an advantage to people who buy the produce? ①

There are many other ways in which plants are adapted to conserve water. For example, part of the bark of many trees contains a layer of cork cells. Commercial cork comes from certain species of oak trees. You probably know that cork floats, but did you know it does so because it does not absorb water? The cork cells in tree bark keep water from escaping from the tree's xylem.

Lack of water is the most limiting factor to desert life. However, many desert plants have adaptations to help them collect and store water. For example, the leaves of the cactus have become small, sharp spines. Photosynthesis occurs in the green stems, which are protected from drying by a thick layer of wax.

Desert shrubs have small, thick leaves with waxy coverings. During dry spells, these shrubs drop their leaves to conserve water. Photosynthesis continues in the stem cells. Desert plants also have large root systems. Sometimes the roots grow as deep as 30 m in search of water. Others have shallow, wide-spreading roots that absorb as much water as possible during heavy desert thunderstorms.

Shortly after a heavy rain, many of the desert's plants flower. The seeds that develop may lie dormant for many months, or even years, until the next rainy period. When enough moisture is present, they quickly germinate, grow, bloom, and produce new seeds.

EVALUATION

Ask the students to explain whether a seed with a tough, protective coat is an adaptation for protection from animals or an adaptation for survival in a harsh climate. (A protective seed coat prevents digestion of the seed by animals, *and* the coat allows the seed to remain dormant until climatic conditions for growth are favorable.)

RETEACHING

Show each student a picture of a plant. Ask the student to identify an adaptation that the plant has for the dispersal of its seeds, for protection from animals, or for survival in a harsh climate.

EXTENSION

Ask the students to use reference materials to research plant adaptations not mentioned in the chapter. Have the students find pictures, if possible, and share their discoveries with their classmates. Then have the students place their reports in their science portfolios.

CLOSURE

Cooperative Learning Ask the groups to discuss this statement: "The seed is the most important adaptation of plants." (Responses should reflect the understanding that a seed gives a plant the ability to reproduce without water and in a variety of environmental conditions.)

Plants of the tundra are adapted for survival in an equally harsh climate. These plants, too, must grow, flower, and produce seeds in a few short weeks. In addition, most tundra plants are small and are covered by winter snows. Snow actually insulates the plants from the freezing Arctic winds. That is why the Inuit, the natives of northern North America, used to make winter homes from blocks of snow.

Figure 11–22. The growing season is extremely short in the tundra.

It may seem strange, but tundra plants and desert plants face the same problem—lack of water. You might think cold would be the biggest problem for tundra plants. But the tundra is actually much like a desert, since it receives very little precipitation. As with desert shrubs, the leaves of many tundra plants have a thick coating of wax to keep them from drying out. Some tundra plants also have hairs on the leaves to keep water from evaporating.

 ASK YOURSELF

Why do both desert plants and tundra plants have thick, waxy coats on their leaves?

SECTION 2 REVIEW AND APPLICATION

Reading Critically
1. Give a specific example for each type of adaptation listed: seed dispersal, protection against animals, protection from climate.
2. Give examples of how seeds are spread by wind, water, and animals.

Thinking Critically
3. How do you think water birds, such as ducks, might spread seeds? What kinds of seeds would they spread?
4. Suppose you and your family moved from southern California to Michigan. Would you be able to grow the same kinds of trees in your Michigan yard as you had in your yard in California? Explain your answer.

ONGOING ASSESSMENT
ASK YOURSELF

Because desert and tundra areas receive only small amounts of precipitation annually, the plants that live in those areas have thick, waxy coatings on their leaves to conserve water by preventing evaporation.

SECTION 2 REVIEW AND APPLICATION

Reading Critically

1. Specific examples might include: seed dispersal—burrs becoming attached to an animal's fur; protection against animals—thorns; protection from climate—waxy coatings on leaves to prevent water evaporation.

2. Examples might include: wind—the winged seeds of maple trees are carried away from the parent tree; water—coconuts float in the oceans for many kilometers; animals—the seeds inside ripe fruits are undigested by animals and deposited in their waste products.

Thinking Critically

3. Responses might suggest that ducks could scatter the seeds of the aquatic weeds and grasses they have eaten everywhere they travel along their migratory route.

4. Responses might suggest that the climate in southern California is very different from the climate in Michigan. Trees adapted to live in warm, sunny southern California with its mild winters are not likely to survive through Michigan's cold, snowy winters.

INVESTIGATION
Examining Seeds

Process Skills: Observing, Comparing, Applying

Grouping: Groups of 2

Objectives
- **Observe** the structures of seeds.
- **Compare** the seeds of monocots to the seeds of dicots.
- **Infer** the advantage of soaking seeds before planting.

Pre-Lab
Have the students review the structures of seeds by examining the drawing of peanuts they made for the *Discover By Doing* activity at the beginning of the chapter.

Hints
Soak the bean and corn seeds overnight prior to performing the Investigation, if possible. It is critical for the students to understand that the structures that are easily identified in the soaked seeds do not appear because of the soaking; the structures are present in dry seeds also. Once the students can identify the structures in the soaked seeds, they can identify the structures in the dry seeds by comparison.
CAUTION: Scalpels are very sharp. Encourage the students to handle them with extreme care.

Analyses and Conclusions
1. A cross section and a longitudinal section allowed more structures within the seed to be observed.
2. Monocot seeds have one cotyledon, and dicot seeds have two cotyledons.
3. Tables should reflect accurate observations and good recording techniques.

▶ **302** CHAPTER 11

Application
The seeds were swollen and softer after soaking, which allowed parts of the seed to be more easily observed and identified. During soaking, the seed coat swells and ruptures due to the expansion of the seed. Immediately following the swelling, the seeds begin the process of germination. Gardeners often soak seeds to give the seeds a head start; the moisture causes germination to begin before the seeds are planted in soil.

✳ Discover More
Structures may vary depending on the seeds examined.

Post-Lab
Ask the students to compare their sketches of bean and corn seeds to their earlier drawing of peanuts. Do the structures in peanuts resemble those in beans or those in corn?

INVESTIGATION
Examining Seeds

▶ MATERIALS
- bean seeds, dry and soaked
- scalpel
- probes or toothpicks
- hand lens
- pencil
- paper
- corn seeds, dry and soaked

▼ PROCEDURE

1. **CAUTION: Scalpels are very sharp. Use extreme care.** Get two soaked bean seeds. Using the scalpel, cut one seed the long way. This makes a longitudinal section of the seed.
2. Cut another bean across the seed. This makes a cross section of the seed.
3. Use a probe or toothpick on both beans to find the structures of the bean seeds. A hand lens may make it easier for you to see some of the smaller parts.
4. Sketch what you see and label the parts.
5. Repeat steps 1-4 using dry bean seeds, soaked corn seeds, and dry corn seeds. Be careful when cutting the dry seeds.

▶ ANALYSES AND CONCLUSIONS
1. Why was it necessary to do a longitudinal section and a cross section of each type of seed?
2. List differences that you found between monocot (corn) and dicot (bean) seeds.
3. Make a table with a list of the parts that you were able to locate. Explain how each part formed.

▶ APPLICATION
What changes did soaking make in the seeds? Why do you think some gardeners soak seeds before planting them?

✳ Discover More
Find out what the inside structures of other seeds are like.

CHAPTER 11 HIGHLIGHTS

The Big Idea—
ENVIRONMENTAL INTERACTIONS

Lead the students to understand the relationship between adaptation and environment. Point out that genetic variations have provided plants with structural characteristics or adaptations that give them the ability to survive in harsh climates, protect themselves from animals, and disperse their seeds.

CHAPTER 11 HIGHLIGHTS

The Big Idea

Structural adaptations are important for plants to survive unfavorable conditions. Some adaptations help to ensure reproduction and seed dispersal. Other adaptations help protect plants from animals or harsh climates. Adaptations allow plants to live and grow in nearly every kind of environment on the earth.

Now how do you think tiny seeds grow into giant trees? How are plants adapted to unfavorable growing conditions? If your ideas have changed, revise your journal entries.

The ideas expressed by the students should reflect knowledge of the growth and development of a tree from a seed to a sapling. Their ideas should also reflect an understanding of how plants display characteristics, such as waxy leaf coverings and modified stem and root structures, that enable the plants to adapt to unfavorable growing conditions.

CONNECTING IDEAS

Each concept map should identify a plant and relate the adaptations of the plant to their specific functions.

Connecting Ideas

Think about a plant other than a cactus. In your journal, create a concept map similar to the one shown. Relate the plant's adaptations to their specific functions. Be sure to list the name of the plant you chose. Compare your concept map with those made by your classmates.

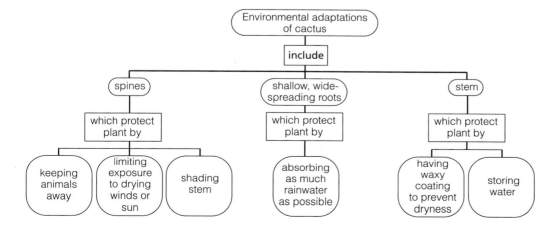

CHAPTER 11   303

CHAPTER 11 REVIEW

ANSWERS

Understanding Vocabulary

1. *Seeds* does not belong; stem cuttings and bulbs are two methods of vegetative propagation that create offspring with the same characteristics as the plant from which they were taken.

2. *Mosses* does not belong; a seed coat, an embryo, and cotyledons are structures of a seed.

3. *Seed dispersal* does not belong; regeneration, or the ability of one plant part to grow the other missing parts, is part of vegetative propagation, which is one type of asexual reproduction.

4. *Grafting* does not belong; phototropism and geotropism are two kinds of tropisms, or chemically induced plant reactions to stimuli.

Understanding Concepts

Multiple Choice

5. b
6. c
7. b
8. c
9. c

Short Answer

10. A plant's response to its environment is called a tropism. Examples of tropisms include phototropism, a plant's reaction to light, and geotropism, a plant's reaction to gravity.

11. Examples of ways in which seeds can be dispersed include: sticking to animals' fur or humans' clothing; being eaten but not digested by animals and eventually being eliminated in the waste products of those animals; being blown by the wind; and floating in water.

Interpreting Graphics

12. The roots are displaying geotropism (response to gravity), and the shoot is displaying phototropism (response to light).

13. The plant growing vertically is receiving equal amounts of light from all directions. The plant that is bent is exhibiting phototropism; it is bending toward a constant source of light.

Reviewing Themes

14. Tundra plants are small and easily covered by winter snows, which insulate them from freezing winds. The leaves of many tundra plants also have thick coatings and hairs that prevent the leaves from losing water by evaporation.

15. Explanations might describe how an animal can benefit a plant by dispersing its seeds, or harm a plant by eating it.

CHAPTER 11 REVIEW

Understanding Vocabulary

For each set of terms, identify the term that does not belong and explain the relationship of the remaining terms.

1. vegetative propagation (292), seeds, stem cuttings, bulbs
2. seed coat, cotyledons (289), mosses, embryos (289)
3. regeneration (294), asexual reproduction, missing parts, seed dispersal (296)
4. phototropism (291), grafting, geotropism (291), tropisms (291)

Understanding Concepts

MULTIPLE CHOICE

5. If a seedling is whitish rather than green, it probably needs
   a) less light.
   b) more light.
   c) less water.
   d) more water.

6. Which of the following plants was *not* grown through vegetative propagation?
   a) a begonia plant grown from a stem
   b) a strawberry plant that broke away from the "mother" plant
   c) a lily plant grown from a seed
   d) an iris plant grown from a bulb

7. The light source of a potted house plant that leans to the south is coming from the
   a) north.
   b) south.
   c) east.
   d) west.

8. A plant with leaves that are not eaten by insects when those of nearby plants are eaten may be protected by
   a) wind-dispersed seeds.
   b) thorns.
   c) chemicals.
   d) flat leaves.

9. Since onions are bulbs, which statement *cannot* be true about onions?
   a) Onions have meristems.
   b) Onions can reproduce asexually.
   c) Onions can only reproduce sexually.
   d) Onions have thick, fleshy leaves surrounding their meristems.

SHORT ANSWER

10. What are tropisms? Provide two examples of plant tropisms.

11. Describe the different ways that seeds can be dispersed.

Interpreting Graphics

12. Look at the photograph below. The corn kernel was planted upside down. What tropisms are the roots and the shoot displaying?

13. Look at the photographs below. Which plant is receiving light from only one direction? How can you tell?

304 CHAPTER 11

Thinking Critically

16. Plants that could not respond to stimuli could not grow and live. Response to stimuli is one of the fundamental characteristics of life. Without this characteristic, stems would not turn toward light sources or grow upward, and roots would not grow downward into the soil.

17. There is no obvious disadvantage to the plant giving off a foul odor. The advantages are that the odor offers protection from some animals and attracts certain flies and beetles that act as pollinators.

18. The tree would produce sweet apples because the grafted branch would receive only water and minerals from the tree. The branch would use these nutrients to perform photosynthesis and continue to create its own fruit.

19. Crocuses are day-neutral plants because different seasons of the year receive different amounts of sunlight.

20. No; the root system is not adapted to collecting water during sporadic, intense desert rains, and the water that was collected by the plant would soon be lost from evaporation through its unprotected leaves.

21. The plant that receives eight hours of light daily would grow faster because it has additional time to perform photosynthesis.

Reviewing Themes

14. *Environmental Interactions*
Structural adaptations have enabled plants to survive in harsh environments. Explain how the structure of tundra plants has enabled them to survive in their environment.

15. *Environmental Interactions*
Interactions between plant and animal life can be beneficial or harmful to plants. Explain how animal interactions can benefit plants and how they can harm the plants.

Thinking Critically

16. A plant's roots and stems respond to gravity. A plant's stems and leaves respond to light. What would happen to a plant that could *not* respond to the stimuli of gravity and light?

17. Some plants, such as the one shown below, give off a foul odor similar to that of rotting meat. Explain what advantages and disadvantages this type of odor might have.

18. If a sweet-apple tree branch is grafted to a sour-apple tree, what type of apple will the grafted branch produce? Why?

19. Different types of crocuses bloom in winter, early spring, or fall. Are crocuses long-day, short-day, or day-neutral plants? Explain your answer.

20. Imagine you have planted a plant that has deep roots and big leaves not covered by wax in a desert environment. Would you expect the plant to survive? Explain your answer.

21. Two plants are given the same amount of water and nutrients, but they are exposed to different amounts of light. One plant receives eight hours of light a day, and the other receives four hours of light. Which plant do you think would grow more rapidly? Why?

Discovery Through Reading

Stidworthy, John. *Plants and Seeds*. Gloucester Press, 1990. Text and microscopic photographs introduce various forms of plant life, their methods of reproduction, and their assimilation of nutrients.

Science PARADE

SCIENCE APPLICATIONS

Discussion

● **Process Skills:** *Comparing, Generating Ideas*

Have the students compare the similarities and differences between a traditional soil garden and a hydroponic garden. Then ask the students to describe how a hydroponic garden might look like a scene from a science-fiction movie.

● **Process Skills:** *Inferring, Applying*

Ask the students whether they think people would be more or less willing to purchase foods that have been grown hydroponically as compared to foods that have been grown traditionally. Encourage the students to support their reasoning.

● **Process Skills:** *Inferring, Expressing Ideas Effectively*

Ask the students whether they think traditional gardens that use soil will ever be completely

Science PARADE

SCIENCE APPLICATIONS

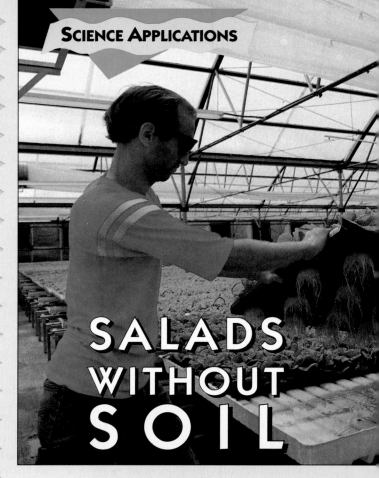

SALADS WITHOUT SOIL

An agricultural planner imagines city apartment dwellers growing fresh vegetables in rooftop gardens. NASA makes plans to grow vegetables in space-station gardens barely larger than closets. Farmers tend giant garden towers in which tons of vegetables are grown under plastic bubbles. All these visions of the future have one thing in common: a special kind of farming called *hydroponics*.

Gets the Dirt Out

On ordinary gardens and farms, plants are grown in soil. The plants' nutrients are provided by the soil or by fertilizers added to the soil. In hydroponic gardens, plants grow in water rather than soil. The plants' roots hang in water that is enriched with the chemicals necessary for growth. Researchers have learned exactly what ingredients must be added to the hydroponic "soup" for plants to thrive. Water, sunlight, and a mixture of about 15 chemical nutrients provide hydroponic crops with everything they need.

replaced by hydroponic gardens. Have them explain their reasoning. (The students might suggest that hydroponic gardens will never completely replace gardens that use soil as long as soil remains on the earth. However, the loss of soil would require food growers to find alternative ways to grow crops.)

● **Process Skills:** *Inferring, Generating Ideas*

Have the students describe the relationship that might exist between a hydroponic farm and the consumption of energy. (The students should infer that hydroponic equipment is likely to require electricity, as will any type of supplemental lighting.)

Journal Activity In their journals, ask the students to defend one of these views: Hydroponic farms are simply a fad. Hydroponic farms are here to stay, and their numbers will increase in the future.

EXTENSION

Challenge interested students to research the Biosphere 2 project and present their findings to the class.

Hydroponic gardens

Research on hydroponic lettuce farm

Space-Age Farming

Almost any crop can be grown hydroponically. Seeds are started in a spongelike material and then transplanted to a water tank. Often the base of the plant is anchored in sand. Plants do not grow bigger in hydroponic tanks or taste any different; they just grow faster. Hydroponics also eliminates many harmful pests that live in the soil.

A walk through a hydroponic farm may seem like a space-age dream. Rows of crops grow from giant tanks. Enriched water flows around the plants' roots or is dripped or sprayed from above. Most hydroponic farms are contained in greenhouses or plastic bubbles, which allow crops to be grown year-round.

Hydroponic gardens continue to pop up in the most unexpected places. In some places, restaurant patrons can pick their own fresh salad greens from a hydroponic tank, in the same way they choose a fresh fish or lobster. Rooftops are being transformed into productive gardens for the residents living in apartments below. In the future, fresh fruits might actually be grown inside supermarkets.

Hydroponic farming is not without its drawbacks. Because of the equipment required, hydroponic farms are very expensive to start. A farmer may have to spend $100,000 or more to start a small plot of only 1000m^2. However, a hydroponics farmer may quickly regain that investment by intensive production. Hydroponic farms can produce as many as five crops a year, whereas traditional farming methods yield only one or two. Hydroponics farmers also suffer fewer losses to insect pests and bad weather.

The Future Is Now

Hydroponics is becoming increasingly popular. More than 40 000 hectares of hydroponic crops are now grown in Europe, where land for farming is less plentiful than in North America. In Japan a national hydroponics research center has helped convert nearly a thousand farmers to hydroponic methods. Although there are presently few hydroponic farms in the United States, the science of hydroponics may some day put gardens in places where nothing could grow before. ◆

READ ABOUT IT!

Discussion

● **Process Skills:** *Inferring, Applying*

Remind the students of their earlier studies about the destruction of tropical rain forests. (The students should recall that slash-and-burn techniques are being used in some rain forests. After the trees have been cut and burned, the ashes are mixed into the soil, which is then used for agricultural purposes. After several years, the soil becomes depleted of nutrients and is often abandoned.) Write the term *reforestation* on the chalkboard and ask the students to infer the meaning of the term. (The students should infer that *reforestation* involves planting trees to compensate for or replace trees whose numbers have been affected by the actions of humans or the forces of nature.)

● **Process Skills:** *Inferring, Applying*

Have the students recall that Andy Lipkis became interested in reforestation after he discovered that smog was killing trees. Ask the students to name several

---

READ ABOUT IT!

Andy Lipkis and the TREE PEOPLE

by Mark Wexler
from *Ranger Rick*

"Take a good look now," said the forest ranger, "because in thirty years most of the pine trees in these mountains will be dead. There won't be anything left in some areas but bare hills."

The ranger was talking to a group of summer campers in the San Bernardino National Forest. The pine forest is located just east of Los Angeles in Southern California. "Smog is killing the trees," the ranger said. "Most of this pollution comes from cars and trucks."

The year was 1970. One of the campers was a tenth-grader from Los Angeles named Andy Lipkis. He was stunned by what he heard. "Are *all* of these trees dying?" he asked. "We've got to *do* something!"

"Until we get rid of the smog, there's not much we can do to save the trees," said the ranger. "There are some kinds of trees pollution won't kill. But not many of them grow here."

Andy thought about that, then asked, "Well, why don't we replace the dead trees with ones that can live with the smog?"

The ranger chuckled. He knew how difficult it would be to replace every dead tree.

But Andy wouldn't give up. If the forest couldn't be replanted, then at least he could plant some smog-proof seedlings around the summer camp. With permission from the camp director, Andy and some

▶ 308 UNIT 4

different sources of air pollutants. (The students might suggest motor vehicles and manufacturing or industrial factories.) Explain to the students that smog and other air pollutants can be traced directly to our use of fossil fuels. Point out that although some fossil fuels, such as natural gas, burn cleaner than others, such as coal, the burning of all fossil fuels produces substances that help make our atmosphere unhealthy.

● **Process Skills:** *Applying, Generating Ideas*

After reminding the students that the burning of fossil fuels contributes to our air pollution problem, have the students discuss and describe behaviors they could engage in that would help reduce air pollution.

friends dug up a baseball field and a parking lot. Then they bought two kinds of trees that are not harmed by pollution. They carefully planted them along with some mountain shrubs.

"It took us three weeks, and I really learned a lot," said Andy. "We all did."

That fall, Andy began studying how smog harms plant life. He learned that trees, like people, breathe in the air around them. In the forests of Southern California, however, trees also breathe in smog. The smog is loaded with a kind of gas known as ozone. Ozone destroys the green coloring called *chlorophyll* (KLOR-uh-fil) in the trees' needles. (All green plants need chlorophyll to make their food.) So when a tree breathes in ozone, it slowly loses its ability to make food for itself. It becomes weaker and weaker.

Trees weakened by smog become good targets for insects and disease. A healthy tree closes its wounds with sap and this helps keep pests away. A sick tree cannot defend itself very well. Pests work their way into its bark, and the tree dies.

Andy was angry about the air pollution. He knew it was the cause of the problem and that something would have to be done to stop it. But right now he figured he had to do something about replacing the dead trees. He thought it would be easy to get people to help. He decided to ask some large companies for money to buy seedlings. Then he would round up volunteers to plant the young trees.

In the next few months the teenager met with important people in several large California companies. He explained the problem smog was creating in the forests and how they could help. "But because I was just a kid," says Andy, "nobody would listen to me." He was so discouraged he gave up his plan—but only for

UNIT 4 309

Discussion

● **Process Skills:** *Comparing, Applying*

Remind the students that during the last twenty years, TreePeople have been responsible for planting more than 200 million trees. Ask the students to speculate whether that many trees have died during that time from air pollution. If the students feel that TreePeople are losing the battle because more than 200 million trees have died, ask the students if they believe the project still has merit and why. If the students believe that fewer than 200 million trees have died, ask them to discuss the merits of ceasing or scaling back the tree planting operation.

● **Process Skills:** *Comparing, Expressing Ideas Effectively*

Explain the meaning of the phrase "band-aid" solution. (A "band-aid" solution is a stop-gap or temporary measure that does not address the real underlying cause of a problem.) Ask the

a time. In 1972 Andy entered college and decided he would *not* abandon his tree planting project. "The forests were dying and I just couldn't let that happen," he said. Andy wrote letters to 25 summer camps in the San Bernardino Mountains. He told them about his idea and asked for their help. Soon he received letters from 20 camp directors saying they would like to help with the tree planting.

Then Andy called the California Division of Forestry, which grows seedlings to sell. He said he needed 20,000 seedlings for his project. "You're just in time," a man told him. "We're ready to plant some new trees. But to make room for them, we have to plow under last year's seedlings. Send us $600 and we'll save some of last year's trees for you." That made Andy really angry! Why couldn't the state just *give* him the trees instead of destroying them?

The more Andy thought about it, the madder he got. He began calling state officials. Finally, the state foresters agreed to donate 8000 smog-proof seedlings for Andy's project. But there was a hitch. The seedlings would be delivered late in the winter in several big boxes. That meant he would have to plant each tree in a pot by itself if it were to survive until summertime. So Andy got a local dairy to donate 8000 milk cartons for the seedlings. Then he and some friends worked day and night until the trees were safely potted.

When Andy learned smog was killing thousands of trees near Los Angeles, he acted! With friends he started planting trees that can live with smog. He planted this taller-than-Andy tree when he was still a teenager.

Andy's troubles, though, weren't over. He had his trees, but he needed money for tools and fertilizer. He turned to a Los Angeles newspaper for help. The newspaper ran a front-page story about Andy and the trees.

"I was amazed," Andy says. "All kinds of people sent money." Before long, Andy had more than $10,000 to use for his tree work.

In the summer of 1973, Andy worked with campers, Scouts, and other volunteers to plant the seedlings. At least 5000 of those 8000 trees are still growing, and Andy and his Tree People are still planting. So far they've planted

students to debate whether the work performed by TreePeople is a "band-aid" solution, or whether it truly addresses the environmental problem of air pollution. (The debate should conclude that the underlying cause of tree deaths is air pollution, and the planting of pollution-resistant trees in no way addresses that problem. Addressing air pollution would, for example, entail banning all energy sources that burn fossil fuels.)

Journal Activity In their journals, ask the students to name characteristics displayed by Andy Lipkis or his TreePeople that the students admire and to explain the reasons for their admiration.

EXTENSION

Have interested students research the similarities and differences between Andy Lipkis and the historical character known as Johnny Appleseed.

---

over 150,000 smog-proof seedlings!

"When we began our planting in 1973," notes Andy, "part of our goal was to save forests. But most important was to teach people that they must help cut down on smog." The Tree People know of a few good ways: You can use automobiles less and use bicycles and public transportation more. Also, you can use fewer electric gadgets. Making electricity makes pollution. "To have healthy forests we have to change some of the ways we live," Andy adds.

Young Tree People scramble along a California hillside planting and watering young trees. They hope a forest will someday stand where grass and weeds now grow.

Pollution is still killing thousands of trees in the Los Angeles area every year. But Andy and his friends are still working very hard to get people to stop smog and to help save the forests. "I get tired," he says," but I have a good time. And I have a good feeling inside, knowing that I'm doing something that counts." ◆

UPDATE

Today Andy Lipkis continues to plant trees. In more than 20 years, Andy Lipkis and his helpers have planted about 200 million trees. Andy's organization, called Tree People, has received attention all over the world for its replanting efforts.

Andy Lipkis and the Tree People are working on projects to teach communities about forest problems. They also have plans to plant 5 million more trees in the Los Angeles area!

Then and Now

Discussion

● **Process Skill:** *Comparing*

After the students have had an opportunity to read about George Washington Carver and Elma Gonzalez, ask them to identify similarities between the two people. (The students might suggest that they are persistent, determined, and curious in a scientific sense.)

Explain to the students that the humble beginnings of both Carver and Gonzalez serve as a reminder to other people—even though a person may not become a scientist or an inventor, persistence and desire can help that person make many accomplishments during his or her life.

Journal Activity In a journal paragraph, have the students explain how the work of George Washington Carver has influenced the people of his time and the people of today.

Then and Now

GEORGE WASHINGTON CARVER (1864—1943)

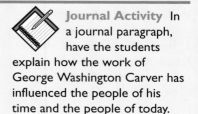

One of the greatest scientists of all times, George Washington Carver revitalized and revolutionized farming. He developed over 300 different products made from the oils and other chemicals found in peanuts. New industries sprang up around these products, and the economy of the war-ravaged South prospered. He produced similar results with sweet potatoes and pecans.

Born of slave parents on a plantation near Diamond Grove, Missouri, in 1864, Carver went out on his own at the age of ten to get a formal education. In 1891, after two years at Simpson College, he entered Iowa Agricultural College, now Iowa State University. Because of his outstanding work in botany and chemistry, Carver was appointed assistant instructor in botany and greenhouse director at the agricultural college.

After receiving his master's degree in 1896, Carver answered a call for teaching help from Booker T. Washington, head of Tuskegee Institute. Carver spent the rest of his life at Tuskegee Institute as a plant scientist. His research gradually won him the respect of Southern farmers by convincing them of the need for crop rotation. Years of cotton farming had depleted much of the soil of the South. Carver told the farmers that planting peanuts, clover, and peas would help restore the soil, since these crops release nitrogen as they grow. ◆

ELMA GONZALEZ (1942—)

The research of Elma Gonzalez may one day help scientists understand better how plant cells store energy and how seeds germinate.

Elma Gonzalez was born on June 6, 1942, in Mexico. Every spring Gonzalez would leave school to pick crops until late fall. She and her family were migrant farm workers. Through hard work and determination, Gonzalez graduated from high school. She went on to receive her bachelor of science degree in biology and chemistry at the Texas Woman's University in 1965 and then her doctorate in cellular biology at Rutgers University in 1972.

In 1974, Gonzalez became a professor of cellular biology at the University of California, Los Angeles. At the university, she has specialized in plant research. Gonzalez is now the head of her own research laboratory, where she studies the function of various internal bodies of plant cells. ◆

SCIENCE AT WORK

Discussion

Before reading the selection, ask the students to describe what they think the work of a plant scientist might involve.

● **Process Skills:** *Inferring, Applying*

Ask the students to explain why classroom teaching is an important component of the life and work of many researchers. (The classroom is one vehicle by which new information is presented and disseminated.)

● **Process Skills:** *Inferring, Expressing Ideas Effectively*

Help the students recall the reasons why Shirley Kooyman believes that plants are important. Then encourage the students to suggest additional reasons why plants are important. Have them identify several plants that are used for food, describe several products that are manufactured from plants or plant products, and name several components of people's homes that are constructed from plants or plant products.

SCIENCE AT WORK

Shirley Mah Kooyman, Plant Scientist

Shirley Mah Kooyman knows about the importance of raising healthy plants. She is a plant scientist in Chanhassen, Minnesota.

What do plant scientists do?

Plant scientists help ensure the survival of plants. They study ways to breed heartier and more productive plants. Our food, many of the things we wear, and even some parts of our homes come from plants. People also get enjoyment from looking at plants and I'm a botanist. I study plants in their natural state, in the wild. For example, I might do research to discover what makes wild wheat grow at the roadside. Then I could share my research results with other plant scientists.

Where do you work?

I work both indoors in a lab and outdoors in an arboretum. An arboretum is a place where many kinds of plants and trees are grown for study. I also work in the classroom, teaching plant identification and

Discussion

● **Process Skills:** *Inferring, Generating Ideas*

Remind the students that Shirley Kooyman believes that plants are sometimes taken for granted. Ask the students to describe different ways in which plants are taken for granted.

● **Process Skills:** *Comparing, Expressing Ideas Effectively*

Have the students suggest why it is important that plant scientists and researchers are able to create healthier, more efficient varieties of plants. (The students might suggest that the world's supply of usable farmland is diminishing, and the global population is increasing. Crops must remain vigorous and healthy in order to meet the global demand for food.)

 Journal Activity In their journals, ask the students to imagine they are a plant grower who uses artificial hormones and create an advertisement that is designed to relieve people's fears about eating plants that have been grown with the help of artificial hormones.

gardening. I enjoy teaching people about the plant kingdom.

Why do we need to know what makes plants grow?

They have many important roles. Unfortunately, plants are sometimes taken for granted, and people forget that they are living things. Plants need light, energy, water, and hormones. When the weather is too hot or cold, or if the soil is not properly nourished, plants can die. Research into the vital processes of plants can help make them healthier. This will help protect our food supply for years to come.

What are plant hormones?

Hormones are chemical messengers that tell a plant to produce other chemicals. This, in turn, stimulates plant growth. Hormones are produced in the stem and roots and travel to the rest of the plant. Scientists can now make plant growth regulators, or artificial hormones. If you place them on the root tip, artificial hormones make plants grow. Special types of botanists, called plant physiologists, work on the creation and production of artificial hormones.

Why do we need artificial hormones?

Artificial hormones are used to help produce bigger and better plants. In addition, artificial hormones can also prevent fruit from dropping off trees too early. These hormones are important because plants are used to feed more and more people.

What kind of education does a plant scientist need?

Most plant scientists need to have a four-year college degree in botany. After getting a college degree, plant scientists can work for private companies such as plant nurseries or seed companies. They might choose to work for federal, state, or local governments. For those who are interested in teaching at a college or university, more education is required. ◆

Discover More

For more information about careers involving botany, write to the

American Association of Botanical Gardens and Arboretums
786 Church Road
Wayne, PA 19087

▶ 314 UNIT 4

SCIENCE/TECHNOLOGY/SOCIETY

Discussion

Ask the students to discuss what they think an orbiting space station would be like. During the discussion, ask the students to consider if and when such a station might become a reality. Also encourage the students to discuss whether spending massive amounts of research money to bring such a project to fruition is a good idea.

Journal Activity

Ask the students to imagine they have been asked to work for an extended period aboard an orbiting space station. Ask the students to describe the type of job aboard the station that would most interest them.

EXTENSION

Challenge interested students to use reference materials and discover the plans of the United States and other countries for a permanent orbiting space station. Then ask volunteers to share their findings with their classmates.

SCIENCE/TECHNOLOGY/SOCIETY

GROWING PLANTS IN SPACE

During the twenty-first century, people may spend months and even years in space. Orbiting space stations and bases on moons and planets must be designed with balanced, self-sustaining life-support systems. These systems must be able to supply air, water, food, and warmth and must dispose of or recycle wastes. Green plants could serve several functions in life-support systems.

The Need for Plants

An adequate food supply for extended missions would take up a large amount of storage space, and its mass would require much fuel. Raising crops in space may help solve the food problem, and fresh food would be a welcome change from prepackaged, freeze-dried fare.

Plants could also recycle wastes, taking up carbon dioxide exhaled by astronauts and releasing oxygen back into the air. Digestive wastes could be used to fertilize the plants.

Plants could also provide a psychological benefit by surrounding people with living reminders of Earth. Plants could create a pleasant oasis in the high-tech environment of a space station or lunar base.

New Growing Techniques

To solve the problem of growing plants in the limited area of a space station, scientists are experimenting with water solutions as substitutes for soil. The solutions contain the elements essential for plant growth. This technology is a type of hydroponics, which you read about earlier.

Another problem with growing plants in space is the lack of gravity. Scientists are designing devices that put plant roots in contact with the hydroponic solutions without letting the solutions escape into the surrounding air.

Moon Dirt

Lunar soil is composed of minerals that are similar to those on Earth. However, they were formed in a different chemical environment. Lunar soil could supply plants with calcium, magnesium, iron, and small amounts of potassium and phosphorus. The two elements that plants require in the largest quantities—carbon from air and nitrogen from soil—are absent on the moon. Scientists must overcome such limitations to grow food away from the familiar soil of Earth. ◆

UNIT 4 315

UNIT 5

ANIMALS

UNIT OVERVIEW

The animal kingdom, from simple invertebrates, such as sponges and coelenterates, to complex warmblooded vertebrates, such as mammals, is discussed in this unit. Animal behavior is also presented.

Chapter 12: Types of Animals, page 318

This chapter presents the two main types of animals—vertebrates and invertebrates. The distinguishing characteristic of vertebrates, the backbone, is discussed. Also discussed are the different types of symmetry of body parts found among the vertebrates and invertebrates, as well as other distinguishing characteristics.

Chapter 13: Invertebrates, page 340

This chapter presents the major groups of invertebrates: sponges, coelenterates, worms, mollusks, echinoderms, and arthropods. The main distinguishing physical features of each group are described and compared. The life cycles of different members of each group are discussed. The distinguishing characteristics of arthropods are described. Insects are compared to other arthropods, and the stages of insect development are explained.

Chapter 14: Vertebrates, page 370

This chapter presents the groups of coldblooded vertebrates—fishes, amphibians, and reptiles—and the groups of warmblooded vertebrates—birds and mammals. The structures, characteristics, and behavior that distinguish these groups of vertebrates are discussed.

Chapter 15: Animal Behavior, page 398

In this chapter, animal behaviors, including defense mechanisms and methods of communication, are presented. Camouflage and mimicry are described. The chapter covers different ways that animals communicate, including the use of pheromones, and different ways that animal groups behave.

Science Parade, pages 420-429

The articles in the unit's magazine tie the theme of animals to the captive breeding of animals such as giant pandas and to the variety of animals in tropical rain forests. The work of a science teacher is discussed in the career feature, and the biographical sketches present an embryologist and an animal researcher.

UNIT RESOURCES

PRINT MEDIA FOR TEACHERS

Chapter 12

Cooke, John. *The Restless Kingdom: An Exploration of Animal Movement.* Facts on File, 1991. Text explains all types of animal movement and discusses how movement helps animals survive.

Loewer, Peter. *The Inside-Out Stomach: An Introduction to Animals Without Backbones.* Maxwell Macmillan, 1990. Lists characteristics, habits, and environments of many invertebrates.

Thomas, Warren D., and Daniel Kaufman. *Dolphin Conferences, Elephant Midwives, and Other Astonishing Facts About Animals.* J.P. Tarcher, 1990. A collection of fascinating facts about animals.

Chapter 13

Berger, Melvin. *Stranger Than Fiction: Killer Bugs.* Avon Books, 1990. Presents information about creatures such as army ants, killer bees, and black widow spiders.

Riefenstahl, Leni. *Coral Gardens.* Harper & Row, 1978. Photographs of corals in the Indian Ocean, Red Sea, and Caribbean.

Stidworthy, John. *Simple Animals.* Facts on File, 1990. Introduces invertebrate animals, including octopuses.

Wexo, John Bennett. *Insects.* Creative Education, 1991. Explores the world of insects.

Chapter 14

Allen, Missy, and Michel Peissel. *Dangerous Mammals.* Chelsea House, 1992. Describes various dangerous mammals from around the world, including the wolf and the African elephant.

Linley, Mike. *Reptiles.* Mallard Press, 1990. Introduces different kinds of reptiles, including snakes, crocodiles, turtles, and lizards.

Chapter 15

Bailey, Jill. *Mimicry and Camouflage.* Facts on File, 1988. A simple introduction to animal protection through camouflage and mimicry.

Nielsen, Nancy J. *Animal Migration.* Franklin Watts, 1991. Examines migratory patterns of different fishes, birds, and mammals.

PRINT MEDIA FOR STUDENTS

Chapter 12

Kerrod, Robin. *Mammals: Primates, Insect-Eaters and Baleen Whales.* Facts on File, 1989. Brief essays highlight the characteristics of the various mammals depicted to explain how each is unique.

Chapter 13

Wiessinger, John R. *Bugs, Slugs, and Crayfish—Right Before Your Eyes.* Enslow Publishers, Inc., 1989. Text and drawings describe the physical characteristics, habits, and habitats of insects and other animals without a backbone.

Chapter 14

Darling, Kathy. *Manatee On Location.* Lothrop, Lee & Shepard Books, 1991. Text and photographs describe the life history, physical characteristics, behavior, and underwater activities of the Florida manatee and how students and others are trying to save this endangered sea mammal.

Chapter 15

Patent, D. H. *How Smart Are Animals?* Harcourt Brace Jovanovich, 1990. Discusses recent research on levels of intelligence in both wild and domestic animals.

ELECTRONIC MEDIA

Chapter 12

Beginnings of Vertebrate Life. Videocassette. Britannica. Follows a vertebrate embryo from fertilization to hatching.

The Gray Squirrel's Neighborhood. Videocassette. Coronet. Describes the relationship between the gray squirrel and its environment.

How We Classify Animals. Videocassette. AIMS. Uses the diversity of animal life to illustrate classification.

Chapter 13

Echinoderms and Mollusks. Videocassette. Coronet. Views these animals in their natural habitats.

Insects. Videocassette. Britannica. Identifies characteristics of many insects.

Spiders. Videocassette. Britannica. Contrasts spiders to insects to reveal spiders' unique characteristics.

Sponges And Coelenterates: Porous And Sac-Like Animals. Videocassette. Coronet. Examines characteristics and behavior of sponges and coelenterates.

Chapter 14

Remarkable Reptiles. Videocassette. Marty Stouffer Productions. Presents reptiles and their adaptations.

The Vertebrates: Birds. Videocassette. Coronet. Explores the amazing variety of birds and their environments.

Chapter 15

Beyond Words: Animal Communication. Videocassette. National Geographic. Examines the communication skills of a variety of species.

Camouflage in Nature: Form, Color, Pattern Matching. Videocassette. Coronet. Photography reveals the ways in which animals conceal themselves by form, color, or pattern.

UNIT 5

Discussion Many people find wild animals such as pandas fascinating. When Huan Huan, a panda in a Tokyo zoo, gave birth, more than 125 000 people wrote to the zoo suggesting names for the baby panda. Several hundred thousand people called the zoo each day on a special telephone line to hear the baby panda make noises. Ask the students why they think people find pandas and other wild animals so interesting.

UNIT 5 ANIMALS

In recent years, many pandas have starved when all the bamboo in their area died. And an even bigger problem for the pandas is poachers. These are people who illegally set traps to catch or kill wild animals. When a panda gets caught in a trap, the poacher can sell its skin for a lot of money.

From "China's Precious Pandas," in *Ranger Rick* magazine

CHAPTERS

12 Types of Animals 318
The animal kingdom contains a wide variety of organisms. Many of them are common, but many more are unusual.

13 Invertebrates 340
There are far more invertebrates than any other type of animal on the Earth.

14 Vertebrates 370
Though outnumbered by invertebrates, the vertebrates are often more noticeable because of their size and color.

Journal Activity You can extend the discussion by asking the students to describe problems—other than poaching and periodic loss of bamboo—that pandas might face in their struggle for survival. (The students might suggest that pandas have lost habitat due to the expansion of agriculture and the encroachment of cities.) Remind the students that, despite their rarity, pandas are very popular. They have come to symbolize the plight of many endangered animals throughout the world. Ask the students to write a journal entry that describes their feelings about pandas and other wild animals and offers their opinions on whether and why these animals should be saved.

For millions of years, giant pandas have lived in China. So why are they in trouble now?

Science PARADE

15 *Animal Behavior* 398

The routines and habits of animals make up the fascinating array of animal behavior.

SCIENCE APPLICATIONS
Romancing the Panda 420

READ ABOUT IT!
"Watch Your Step" 424

THEN AND NOW
Ernest Just and Jane Goodall 426

SCIENCE AT WORK
Peter Aromando, Science Teacher ... 427

SCIENCE/TECHNOLOGY/ SOCIETY
Saving Endangered Wildlife 428

CHAPTER 12

TYPES OF ANIMALS

PLANNING THE CHAPTER

Chapter Sections	Page	Chapter Features	Page	Program Resources	Source
Chapter Opener	318	*For Your Journal*	319		
Section 1: INTRODUCTION TO ANIMALS	320	Discover By Observing **(B)**	320	*Science Discovery**	SD
• Increasing Awareness of Animals **(B)**	321	Activity: Can you match the animals with the facts? **(A)**	321	Extending Science Concepts: Animal Records **(A)**	TR
• Contributions of Animals **(B)**	323	Discover By Researching **(B)**	323	Study and Review Guide, Section 1 **(B)**	TR, SRG
• The Good, the Bad, and the Ugly? **(B)**	324	Section 1 Review and Application	324		
Section 2: CHARACTERISTICS OF ANIMALS	325	Discover By Writing **(B)**	326	*Science Discovery**	SD
• General Characteristics of Animals **(B)**	325	Discover By Doing **(B)**	328	Investigation 12.1: Animal Locomotion **(A)**	TR, LI
• Symmetry **(A)**	327	Section 2 Review and Application	330	Connecting Other Disciplines: Science and Mathematics, Identifying Patterns **(B)**	TR
		Skill: Identifying Symmetry **(A)**	331	Comparing Animal and Plant Cells **(A)**	IT
				Types of Symmetry **(B)**	IT
				Study and Review Guide, Section 2 **(B)**	TR, SRG
Section 3: ANIMAL CLASSIFICATION	332	Discover By Researching **(A)**	335	*Science Discovery**	SD
• Vertebrates **(B)**	332	Section 3 Review and Application	335	Investigation 12.2: Classifying Animals **(A)**	TR, LI
• Invertebrates **(B)**	334	Investigation: Comparing Vertebrates and Invertebrates **(A)**	336	Reading Skills: Determining Cause and Effect **(B)**	TR
				Thinking Critically **(H)**	TR
				Vertebrates and Invertebrates **(B)**	IT
				Record Sheets for Textbook Investigations **(A)**	TR
				Study and Review Guide, Section 3 **(B)**	TR, SRG
Chapter 12 HIGHLIGHTS	337	The Big Idea	337	Study and Review Guide, Chapter 12 Review **(B)**	TR, SRG
Chapter 12 Review	338	For Your Journal	337	Chapter 12 Test	TR
		Connecting Ideas	337	Test Generator	

B = Basic   **A** = Average   **H** = Honors
The coding Basic, Average, and Honors indicates subsections, features, and resources that might be appropriate for different levels of learners. For additional suggestions regarding choice of topic and depth of coverage, see the Pacing Chart on pages T26–T29.

*Frame numbers at point of use
(TR) Teaching Resources, Unit 5
(IT) Instructional Transparencies
(LI) Laboratory Investigations
(SD) *Science Discovery* Videodisc Correlations and Barcodes
(SRG) Study and Review Guide

▶ 317A

CHAPTER MATERIALS

Title	Page	Materials
Discover By Observing	320	(per individual) journal
Activity: Can you match the animals with the facts?	321	(per group of 3 or 4) paper, pencil
Discover By Researching	323	(per individual) journal
Discover By Writing	326	(per individual) journal
Teacher Demonstration	326	series of microscope stations, prepared slides of plant and animal cells
Discover By Doing	328	(per individual) paper, ink or paint
Skill: Identifying Symmetry	331	(per group of 2 or 3) pictures (10–15) of common objects found at home or school
Discover By Researching	335	(per individual) journal
Investigation: Comparing Vertebrates and Invertebrates	336	(per group of 3 or 4) preserved frog, preserved earthworm, paper towel

ADVANCE PREPARATION

For the *Demonstration* on page 326, you will need prepared slides of plant and animal cells. Gather pictures of common objects found at home or school for the *Skill* on page 331. For the *Investigation* on page 336, obtain preserved frogs and earthworms from a biological supply company.

TEACHING SUGGESTIONS

Field Trip
Take a field trip to a zoo, a large aquarium, a fish hatchery, or any other facility in your area that would provide an opportunity for animal observation. Prepare a list of questions that the students must answer while observing the animals on the field trip.

Outside Speaker
If field trips are not feasible, or as an added dimension, invite a speaker to discuss some aspect of animal life. Do not overlook elderly or retired persons who may have hobbies such as butterfly collecting, seashell collecting, bird-watching, beekeeping, or animal photography.

CHAPTER 12
TYPES OF ANIMALS

CHAPTER THEME—SYSTEMS AND STRUCTURES

This chapter introduces the students to the structural characteristics of animals. The students define animals as multicellular organisms that have cell membranes but no cell walls. They learn that most animals have lines of symmetry. The students discover that animals are classified as vertebrates or invertebrates based on the presence or absence of a backbone. **Systems and Structures** is also developed as a major theme in Chapters 4, 5, 7, 8, 9, 10, 13, 16, 17, 18, 19, and 20. A supporting theme of this chapter is **Energy**.

MULTICULTURAL CONNECTION

Cooperative Learning Animals have played prominent roles in the mythology, legends, and folklore of most cultures. Have the students choose a culture and investigate the role that animals played in its mythology, legends, or folklore. For example, the students might choose to investigate the Raven's role in creationist stories of the Northwest American Indians. They might investigate the wolf's role in nurturing Remus and Romulus in ancient Roman mythology. They might research the story of Arachne in ancient Greek mythology, or they might decide to look into the worship of cats by the ancient Egyptians.

MEETING SPECIAL NEEDS

Mainstreamed

Have students with learning disabilities work with student tutors. Together the students should take notes outlining important chapter information and then review the notes to ensure that the information is understood.

▶ 318 CHAPTER 12

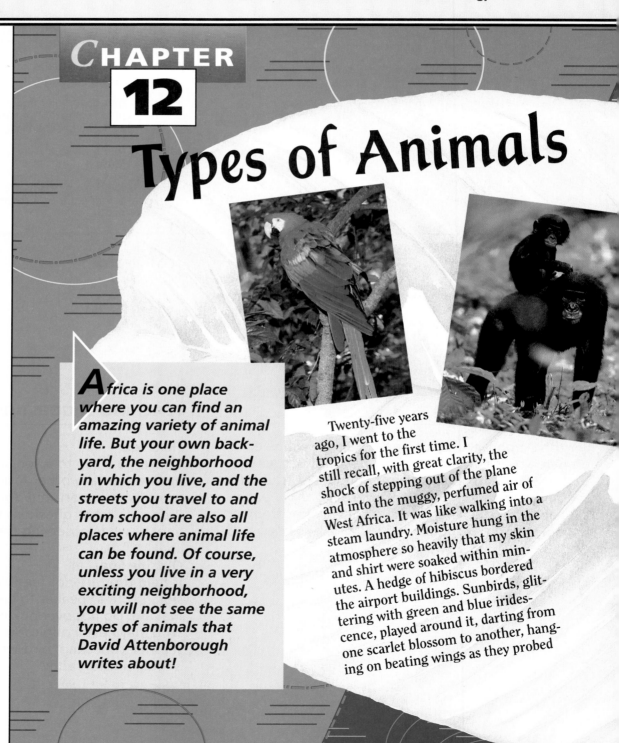

Chapter 12
Types of Animals

*A*frica is one place where you can find an amazing variety of animal life. But your own backyard, the neighborhood in which you live, and the streets you travel to and from school are also all places where animal life can be found. Of course, unless you live in a very exciting neighborhood, you will not see the same types of animals that David Attenborough writes about!

Twenty-five years ago, I went to the tropics for the first time. I still recall, with great clarity, the shock of stepping out of the plane and into the muggy, perfumed air of West Africa. It was like walking into a steam laundry. Moisture hung in the atmosphere so heavily that my skin and shirt were soaked within minutes. A hedge of hibiscus bordered the airport buildings. Sunbirds, glittering with green and blue iridescence, played around it, darting from one scarlet blossom to another, hanging on beating wings as they probed

CHAPTER MOTIVATING ACTIVITY

Cooperative Learning Have the students work in small groups to prepare an animal poster. The groups might draw animals, or they might cut pictures of animals out of old magazines. Encourage the groups to classify the animals in some way. For example, they might show land animals and sea animals, or they might choose to show large and small animals. After the posters have been completed, have the class view the posters and determine the categories used to classify the animals in each.

For Your Journal

By answering the journal questions, the students can demonstrate their knowledge of animals and animal characteristics. Throughout the chapter, when asking the students to consider the ways in which animals can help or harm humans, emphasize that every animal has a beneficial role in a balanced ecosystem, regardless of the nature of its interaction with humans.

---

for nectar. Only after I had watched them for some time did I notice, clasping a branch within the hedge, a chameleon, motionless except for its goggling eyes which swivelled to follow every passing insect. Beside the hedge, I trod on what appeared to be grass. To my astonishment, the leaflets immediately folded themselves flat against the stem, transforming green fronds into apparently bare twigs. It was sensitive mimosa. Beyond lay a ditch covered with floating plants. In the spaces between them, the black water wriggled with fish, and over the leaves walked a chestnut-coloured bird, lifting its long-toed feet with the exaggerated care of a man in snow-shoes. Wherever I looked, I found a prodigality of pattern and colour for which I was quite unprepared. It was a revelation of the splendour and fecundity of the natural world from which I have never recovered.

from *Life on Earth: A Natural History* by David Attenborough

ABOUT THE AUTHOR

Life on Earth by naturalist David Attenborough is based on a 13-episode nature series developed for the British Broadcasting Corporation. Like the television series, the book traces the history of animal life on the earth. Acclaiming the book, a reviewer noted, "With a pleasant enthusiasm [Attenborough] effortlessly encompasses 3000 million years in 300 pages, from the very start of life to its most bizarre and intricate manifestations." Attenborough has produced and hosted a number of other nature shows, such as *Zoo Quest*, *The Living Planet*, and *The First Eden: The Mediterranean World and Man*.

Hailing from Great Britain, Attenborough uses British, rather than American, spellings of words such as *splendor* and *color* throughout the text. He also uses some terms, such as *prodigality* (extravagance, wastefulness) and *fecundity* (reproductive fruitfulness) that may be unfamiliar to the students and should be defined.

For Your Journal

- List ten animals that have hair and ten animals that do not have hair.
- Write *helpful* or *harmful* after each animal, depending on whether you think it helps or harms humans.
- Think of another way to divide all animals into just two separate groups.

CHAPTER 12 **319**

Section 1:
INTRODUCTION TO ANIMALS

PROCESS SKILLS
- Observing • Inferring

POSITIVE ATTITUDES
- Curiosity • Caring for the environment

TERMS
None

PRINT MEDIA
Dolphin Conferences, Elephant Midwives, and Other Astonishing Facts About Animals by Warren D. Thomas and Daniel Kaufman
(see p. 315b)

ELECTRONIC MEDIA
The Gray Squirrel's Neighborhood, Coronet
(see p. 315b)

Science Discovery
Gecko; foot

BLACKLINE MASTERS
Study and Review Guide
Extending Science Concepts

DISCOVER BY *Observing*

 See the Motivating Activity for alternatives to the *Discover by Observing* activity. Another alternative would be to have the students leaf through a magazine and identify animals mentioned in the articles or shown in the pictures.

 **LASER DISC**
3693

Gecko; foot

▶ **320** CHAPTER 12

FOCUS
This section focuses on the vast diversity of animal life in the world. The section emphasizes that the animal kingdom includes insects, birds, and fishes as well as dogs, cats, and elephants. It also explores the impact that animals have on the environment and on humans.

MOTIVATING ACTIVITY
 As an alternative to the *Discover by Observing* activity, which may be used as a motivating activity, take the students on a field trip to a nearby nature center, forest preserve, park, farm, or botanical garden. Encourage the students to take notes as they observe the animal life and then to record their observations in their journals when they return to class. Remind the students to avoid contact with the animals and to disturb the environment as little as possible while making their observations.

SECTION 1

Introduction to Animals

Objectives

Describe the great variety of living animals.

Distinguish between helpful and harmful animals and give examples of each.

Debate the idea that an animal considered harmful might be helpful as well.

Do you suppose David Attenborough liked animals when he was young? How do people grow up to care so much about animals that they build a job, a career, and even a life around animals? Perhaps you feel as if you are one of those people. Perhaps you like some animals and dislike others but are curious about most animals.

How aware are you of the animals around you? You can increase your awareness of the animals in your environment by doing the following activity.

DISCOVER BY *Observing*

Spend about fifteen minutes outdoors looking for animals. Conduct your observations every day for five days. Try to observe at different times of the day or early evening. Write your observations in your journal. Depending on where you live, you may see very small animals or even animals that fly. If you aren't sure whether a living thing is an animal, list it with a question mark after it. Review your list after you have finished this chapter.

Figure 12–1. Marine biologists specialize in the study of plants and animals found in the world's oceans. They study organisms that live in shallow waters as well as those that live in the deep ocean.

TEACHING STRATEGIES

● **Process Skills:** *Evaluating, Expressing Ideas Effectively*

As a way of examining the students' perceptions about animals, have the students identify two animals that they fear or dislike and state their reasons. Then have the students identify two animals that they like and state their reasons.

● **Process Skills:** *Communicating, Interpreting Data*

Take a survey of the class to identify the number of students who have pets and the type of animals that they have as pets. Have the students create a bar graph identifying the number of students who own each type of pet. Then have the students identify the type of pet they would choose if they could have any pet they wished. Stress that wild animals do not make good pets. Have the students graph this information. Compare the two graphs for similarities and differences in the types of pets.

ACTIVITY

Can you match the animals with the facts?

Process Skills: Inferring, Comparing

Grouping: Groups of 3 or 4

Hints

Before the students begin to match animal names with the amazing facts, have them list characteristics that they associate with each of the animals. For example, the students may associate strength and industriousness with the ant. Have the students use the associations and the process of elimination to match the animals and facts. (Matches: 1, e; 2, h; 3, g; 4, k; 5, l; 6, b; 7, i; 8, f; 9, c; 10, a; 11, d; 12, j)

▶ **Application**

The students are likely to point out that factors such as length of life, speed, strength, size, and number of offspring all contribute to an organism's ability to survive.

Increasing Awareness of Animals

A biology teacher once explained that as a child she liked animals with fur, even mice. But she was frightened of cockroaches and lizards. Then she lived for two years in Asia where she learned to be grateful for lizards because they protected the household from insects whose bite or sting was painful and sometimes dangerous. She explained, "I never thought I would lie in bed at night and be glad to hear the clacking sound of geckos calling to each other. I knew if the geckos were there, other things were not." After a minute she added, "I still haven't found a way to like cockroaches, though."

The more we observe and read about animals, the more respect we have for them. How they move, what they eat, and how they go about their daily lives is a series of amazing facts. Check your animal knowledge by trying to match the facts with the appropriate animals in the following activity.

Figure 12–2. A typical gecko is small and nocturnal with large eyes and expanded toes. Often found in warm countries, it is harmless to humans and useful in destroying insects.

ACTIVITY

Can you match the animals with the facts?

MATERIALS
paper and pencil

PROCEDURE

1. Use the two lists provided. Match the animal from List #1 with the appropriate "amazing fact" from List #2.
2. You are not expected to know all of the facts listed in this activity. To match the lists, use logical reasoning and don't be afraid to guess.
3. Which of the interesting facts might help each animal survive? Give reasons for your answers.
4. What other animal facts do you know that might explain why certain animals are successful at getting food, escaping enemies, or just remaining healthy?

LIST #1 ANIMAL		
1. Ostrich	2. Flea	3. Emperor Penguin
4. Shrimp	5. Blue Whale	6. Elephant
7. Tuna	8. Giraffe	9. Mallard Duck
10. Box Turtle	11. Ant	12. Monarch Butterfly

LIST #2 AMAZING FACT
a. Has lived to be 129 years old
b. Snacks on 68 kg of hay a day
c. Lays 7-14 eggs at a time
d. Can pick up 50 times its own weight
e. One of its eggs scrambled would feed 12 people
f. Its tongue is 43 cm long
g. Can hold its breath underwater for 18 minutes
h. Can jump 33 cm
i. Travels 69 km per hour
j. Travels 16 km per hour
k. Travels 3 km per hour
l. Drinks 487 L of milk a day as a newborn

✧ **Did You Know?**
Cockroaches have inhabited the earth for about 250 million years. Only about 20 of the 3500 species of cockroaches are considered household pests.

▲ **MEETING SPECIAL NEEDS**

Second Language Support

Prepare flashcards with the pictures and English names of animals from the chapter. Have second-language students use the cards to become familiar with the English names of the animals.

TEACHING STRATEGIES, continued

● **Process Skills:** *Classifying/ Ordering, Comparing*

List the following sets of animals on the chalkboard and have the students identify something that the animals in each set have in common.
1. mosquito, robin, bat, chicken (wings)
2. whale, sponge, shark, lobster (sea animals)
3. turtle, snail, tortoise, clam (shells)
4. goldfish, cat, dog, parakeet (common pets)

● **Process Skills:** *Solving Problems/Making Decisions, Expressing Ideas Effectively*

Have the students imagine that they are going to a place where there are plenty of plants that can serve as food but no animals. Tell the students that they can take only one type of animal with them. Have the students name the type of animal they would choose and state reasons for their choices. (The students might choose dogs for companionship or cows for milk and a meat source.)

INTEGRATION—Language Arts

Many expressions in the English language refer to animals and their characteristics. Write these expressions on the chalkboard and encourage the students to give meanings for them: moving at a snail's pace (moving slowly); the early bird gets the worm (to achieve a goal, start working early). Ask the students to identify and define as many animal-related expressions as they can. Then have the students place their lists in their science portfolios.

SCIENCE BACKGROUND

Bats, the only flying mammals, live on all of the continents except Antarctica. Most of the 900 species of bats live in the tropics, however. Bats are nocturnal animals that sleep during the day and hunt at night. Some species rely on their vision and sense of smell to find food. Others, which do not see well at night, use echolocation to find food. Many bats prey on insects, and some eat small rodents or birds. Still others eat plants and seeds.

ONGOING ASSESSMENT
 ASK YOURSELF

The students might name yards, empty lots, sidewalks, trees, shrubs, and grasses.

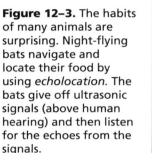

Figure 12–3. The habits of many animals are surprising. Night-flying bats navigate and locate their food by using *echolocation*. The bats give off ultrasonic signals (above human hearing) and then listen for the echoes from the signals.

Awareness of animals and a curiosity about them are what lead most people to pursue a career with animals. What animals are common where you live? Close your eyes and take an imaginary walk around your neighborhood. Are there birds perched on roofs, telephone wires, or tree branches? Is that a spider web attached to a fence? Look closely at an advertising sign. Have wasps built a nest behind it? Look for ants, bees, worms, cats, dogs, and butterflies. Maybe flies have found a food scrap that someone dropped or a garbage can that has not yet been emptied. You don't have to live on a farm or a ranch or in a jungle to find animals. Even large animals such as horses can be found in big cities. The horses of New York City's mounted police attract attention wherever they go in the city.

Sometimes we forget what the word *animal* includes. When we read about animals such as elephants and lions, it is easy to forget that the pigeons roosting on a nearby building are also part of the animal world. Thinking of butterflies and worms as animals may take a little practice. They don't bark, have fur, or walk on four legs like the animals that may share our homes.

 ASK YOURSELF

Name five places where animals may be found in your neighborhood.

Figure 12–4. Often the animals that share our environment go unnoticed. Sometimes we are just so used to seeing them that we don't pay attention.

- **Process Skills:**
Formulating Hypotheses, Identifying Variables

Cooperative Learning Organize the students into small groups and tell the groups to suppose that they are scientists designing an experiment to test the impact that spending time with a pet may have on a person with high blood pressure. Ask the students to state the hypothesis that they will test and to identify the variable and control for the experiment. (The students' hypothesis might be "Spending time with a pet will lower the blood pressure of a person with high blood pressure." The experimental variable would be the pet, and the control would be a group of people with high blood pressure who are not provided with a pet.)

- **Process Skills:** *Applying, Generating Ideas*

Write *earthworms, bats, spiders,* and *bees* on the chalkboard. Have the students identify ways in which each of these animals is important to the environment and helpful to humans. (Earthworms return nutrients to the soil and are often used as fishing bait, bats eat insects and bat guano is used as fertilizer, spiders eat insects, and bees pollinate flowers and make honey.)

Contributions of Animals

In at least some way, animals are a part of everyone's life. Many provide food for us and for other animals. Clothing and shoes are sometimes made from animal products. Some people may choose not to eat or wear animal products but are still influenced by animals in other ways. Research studies have shown that often people can be calmed and reassured by holding or petting a cat or dog. Heartbeat and blood pressure rates usually slow when a person is holding a soft, warm animal.

Some groups are experimenting with the idea of lending pets to elderly people or disabled people for a few hours a day. They may be unable to care for a pet 24 hours a day, but they can enjoy and benefit from the period of companionship. The animal is delivered and returned by the workers from the animal-lending group.

Have you ever seen an aquarium in a doctor's office? Physicians and dentists often have a large aquarium in the waiting room because the sight of fish swimming gracefully through the water can relax patients.

Figure 12–5. People of all ages enjoy animal companions. What types of animals would you choose for a Lend-a-Pet program? ①

① Chosen animals should be those capable of safe interaction with people.

MULTICULTURAL CONNECTION

Among the most beautiful contributions of animals to humans is silk. The silkworm's cocoon is spun into fibers and woven into cloth. According to Chinese legend, the empress Xilingshi discovered silk about 2600 B.C. when she investigated what was harming the mulberry trees in the emperor's garden. Although the legend may not be true, the Chinese did discover silk and held the secret to weaving it for about 3000 years.

DISCOVER BY *Researching*

As part of the activity, you may wish to have the students write to a nearby humane society, animal shelter, or animal protection organization for information about volunteer needs. As a class project, the students might do volunteer work for the organization.

ONGOING ASSESSMENT ASK YOURSELF

The students might describe a homeowner with a watchdog feeling protected and safe at night or a person who is not feeling well being cheered by a pet.

DISCOVER BY *Researching*

Use your library and phone directory to learn about safe ways student volunteers can be involved in organized activities with animals. Ask your parents and teachers for their opinions. Does your community have a rehabilitation area for injured or mistreated animals? Write your findings in your journal.

ASK YOURSELF

Give an example of how animals calm or relax humans.

SECTION 1 **323**

GUIDED PRACTICE

Help the students identify the ways in which different animals can help humans.

INDEPENDENT PRACTICE

 Have the students write answers to the Section Review and Application questions in their journals. Then ask the students to write a paragraph about an interaction between an animal and humans.

EVALUATION

Ask the students to list the names of three animals and then to identify the ways in which each of the animals is helpful or harmful to humans.

RETEACHING

Gather a set of pictures showing animals that may or may not be found in your environment. Have the students examine the pictures and tell if the animals shown are likely to be found in their neighborhoods.

EXTENSION

Have the students investigate ways in which some wild animals have adapted to urban and suburban living situations. The students can investigate coyotes, raccoons, or deer.

CLOSURE

 Cooperative Learning Have small groups write two statements that encompass the main ideas in this section.

BACKGROUND INFORMATION

Often referred to as POWs, prisoners of war are captured armed forces of warring nations. The first treaty that called for fair treatment of prisoners was signed by the United States and Prussia in 1785. Today, almost all nations have agreed to follow international rules established for the treatment of prisoners of war at the Hague (1899, 1907) and Geneva (1929, 1949) Conventions. However, during World War II, the Korean War, and the Vietnamese War, these rules were often ignored and prisoners were mistreated.

ONGOING ASSESSMENT
▼ **ASK YOURSELF**

Each animal listed should be supported with a description of how that animal is beneficial.

SECTION 1 REVIEW AND APPLICATION

Reading Critically

1. A gecko would eat insect pests that might invade the home.

2. Animals provide food and clothing for people, some eat insects or other pests, and some are domestic pets.

Thinking Critically

3. The students might suggest a number of animals; for example, cattle because they provide meat, milk, and materials for clothing.

4. The students might cite the fact that a lizard is a natural enemy of a cockroach.

▶ **324**   CHAPTER 12

Figure 12–6. You may think of stinging insects as harmful, but often they play an important role in pollinating plants.

The Good, the Bad, and the Ugly?

Are all animals helpful? What about the scary ones like rats that carry disease and termites that destroy houses? Many animals—particularly insects—are considered pests because they compete with humans for food and living space. Some animals, such as wolves or bears, can actually threaten human life.

All animals—even those that humans consider pests—have a role to play in their environment. For example, bees and other stinging insects are important in pollinating flowers. Removing a supposed "pest" can sometimes cause new problems. Killing the wolves that attack sheep and other livestock has, in some cases, caused deer populations to increase so much that the deer become the new "pests." Although it is important to control the numbers of animals that are destructive, the programs for control must be carefully thought out.

The biology teacher who still dislikes cockroaches is probably not too different from the rest of us. But she and others might be surprised to read some of the sad stories about people imprisoned during wartime. There have been many accounts of the loneliness and depression these prisoners experienced when kept completely alone in cells for months or even years. Several of these stories describe how grateful the prisoner was for the presence of crickets, cockroaches, or other insects in the cell. The prisoner studied the habits of the insects, thus finding a way to stay interested and alert because of the presence of another living thing—even a cockroach.

 ASK YOURSELF

List three animals you think of as pests. For each one, describe some activity of the animal that is actually beneficial.

SECTION 1   REVIEW AND APPLICATION

Reading Critically

1. Why could it be reassuring to have a gecko in your house?
2. List three ways animals are helpful to humans.

Thinking Critically

3. In the last two hundred years, what animal do you think has been the most useful to humans? Give reasons for your answer.
4. Would any of the animals mentioned in this section be natural enemies of each other? If so, give an example of such a pair and tell why they would be enemies.

Section 2:

CHARACTERISTICS OF ANIMALS

FOCUS

This section identifies the characteristics that distinguish animals from other organisms. It identifies symmetry as a feature common to most animals. The different types of body symmetry, such as bilateral symmetry and radial symmetry, are discussed, as is asymmetry.

MOTIVATING ACTIVITY

Give the students five minutes in which to list characteristics that distinguish animals from other organisms. Encourage the students to read their lists to the class and to discuss whether the characteristics listed actually distinguish animals from other organisms. For example, a student might list mobility as a distinguishing characteristic, but another student may point out that protists also have mobility.

Characteristics of Animals

SECTION 2

Objectives

Describe the basic characteristics of animals.

Determine whether or not a living thing is an animal.

Contrast radial and bilateral symmetry in animals.

If somewhere in this chapter you found a question asking you to explain the difference between a rose and a cow, you would probably laugh. The idea sounds funny because we don't usually have trouble seeing the difference between common plants and common animals. But David Attenborough in West Africa might easily have discovered a living thing that mystified him. Is the strange organism a plant or an animal?

Occasionally, we encounter a confusing organism that doesn't match our mind's picture of a common animal or plant. We may find it difficult to classify this organism. That's when it is helpful to think about the characteristics that all animals must have. Applying what we know about animals in general can help us decide whether a new organism is an animal or a plant.

PROCESS SKILLS
• Comparing • Observing

POSITIVE ATTITUDES
• Initiative and persistence
• Enthusiasm for science and scientific endeavor

TERMS
• symmetry
• radial symmetry
• bilateral symmetry
• asymmetry

PRINT MEDIA
The Restless Kingdom: An Exploration of Animal Movement by John Cooke (see p. 315b)

ELECTRONIC MEDIA
Beginnings of Vertebrate Life, Britannica (see p. 315b)

Science Discovery Anemone, colonial Planaria

BLACKLINE MASTERS
Study and Review Guide
Laboratory Investigation 12.1
Connecting Other Disciplines

General Characteristics of Animals

What do we know about animals in general? One characteristic of animals is that they must take their foods in from their environment. Animals cannot make their own food. They cannot perform the process of photosynthesis used by plants. If you have ever cared for pets, you know that they must be fed. Even you and your friends are good reminders of this characteristic of animals. A full day of sunshine and fresh air does not take the place of breakfast or supper, and your body complains to you if you delay eating for too long.

Figure 12–7. There are many ways of obtaining food, as can be seen by observing animals such as an American robin (below), a May beetle (bottom left), and a common snapping turtle (top left).

SECTION 2 **325**

TEACHING STRATEGIES

● **Process Skills:** *Comparing, Observing*

Display a picture of a plant and a picture of an invertebrate animal. Ask the students to list characteristics that the two organisms shown in the pictures have in common and characteristics that distinguish the two organisms. (The students are likely to note that both organisms are living things with the ability to grow. Structural characteristics might be listed as features that distinguish plants and animals.)

● **Process Skills:** *Classifying/Ordering, Comparing*

Have the students prepare a chart with the headings "Animal" and "Plant." Then have the students list cellular characteristics that distinguish each type of organism from the other. (Plants: presence of cell walls and chloroplasts, Animals: absence of cell walls and chloroplasts.)

SCIENCE BACKGROUND

All animals have a multicellular structure, and animal cells vary in size, although most are microscopic. The cells may be rectangular, spherical, disk-shaped, rod-shaped, or spiral. Yet, all the cells of an animal depend on one another for their own survival and the survival of the organism. In most animals, cells are organized into tissues, the tissues are organized into organs, and the organs into organ systems. Some simple animals do not have tissues or organs, but their cells have specific functions.

Demonstration
After reviewing the differences in plant and animal cells, set up a series of numbered microscope stations. At each station display prepared slides of plant or animal cells. Have the students move from station to station, identifying the magnified cells as either plant cells or animal cells. Have the students note the cellular features that they used to tell what type of cell is exhibited at each station.

① Chloroplasts are absent from animal cells because animal cells do not produce their own food by photosynthesis.

Figure 12-8. Some differences between animals and plants can be seen only on the microscopic level.

Figure 12-9. One difference between plant and animal cells is that animal cells do not have cell walls. Another difference is the absence of chloroplasts in animal cells. Why are chloroplasts missing in animal cells? ①

Another characteristic of animals is that they usually do not grow beyond a certain adult size. Imagine what would happen if humans, like trees, did not stop growing until they died!

A Closer Look The next two characteristics of animals are things you would not be able to tell if you just walked up to an animal such as the cow mentioned earlier. You would need to examine the animal with a microscope. In order to be called an *animal,* a living organism must be multicellular. Using a microscope, you would be able to see that some organisms have only one cell. Even though these one-celled organisms may look like animals, they are classified as *protists.*

If you used the microscope to look very closely at the cells of the cow, you would find that cow cells, like all animal cells, have cell membranes but no cell walls. Plant cells have both cell membranes and cell walls. So now you have a list of ways the cow and the rose are different, in case anyone ever asks! You can illustrate the differences in animal and plant cells by doing the following activity.

 **Writing**

In your journal, sketch a typical animal cell and plant cell as shown in the illustration. Label the cells to show the presence or absence of cell walls, cell membranes, and chloroplasts.

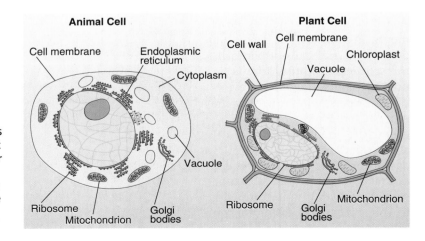

- **Process Skills:** *Observing, Comparing*

Have the students examine the diagrams of an animal and a plant cell on page 326. Ask the students to identify the animal tissue by distinguishing its cell structure from that of the plant tissue.

- **Process Skills:** *Generating Ideas, Analyzing*

Cooperative Learning Have the students work with partners to choose an animal and to write two clues about the animal based on its motility and two clues based on its appearance. Have partner teams exchange clues and try to name the animal described based on the clues given.

- **Process Skills:** *Classifying/Ordering, Applying*

Write the following verbs on the chalkboard: hop, slither, gallop, leap, and waddle. Ask the students to name animals that they associate with each type of movement. (The students might associate kangaroos and rabbits with hop, snakes and worms with slither, horses and zebras with gallop, frogs and tigers with leap, and ducks and penguins with waddle.)

Moving Along

A final characteristic of most animals is that they move from place to place. This type of movement is called *locomotion*. Most of the animals we encounter fly, run, swim, crawl, hop, or slither. Then we come to the animals that confuse us. Some animals move only during early stages of their lives. As adults they anchor onto some solid object and remain there the rest of their lives. Barnacles that attach themselves to piers and boats are an example of animals whose movement is limited. Sponges, which don't seem very animal-like at first glance, also have limited movement. We are so accustomed to watching animals move that it is usually their movement that makes us recognize them as animals. When animals seem to stay in one place most of their life, it can confuse observers.

Figure 12–10. Most animals are *motile*, which means that they move around during their lives. However, this fur seal does not seem to be in a hurry to go anywhere!

 ASK YOURSELF

Animals cannot make their own food. How do animals get the food they need?

Symmetry

All animals except for sponges have some arrangement of their body parts that enables us to describe them in terms of symmetry. **Symmetry** is the balanced arrangement of body parts around a center point or center line. Look at the picture of the starfish on the next page. The starfish has neither a head nor a tail. Its body parts are arranged in a circle around a center point. If you drew imaginary lines through the center of the starfish, they would resemble the spokes of a wheel. This circular body plan is called **radial symmetry.** The jellyfish is another example of radial symmetry.

SCIENCE BACKGROUND

At one time, sponges were considered plants because adult sponges do not move and are attached permanently to a base. However, sponges have been classified as animals because they do not produce their own foods as plants do and they do not have cell walls as plants do. Sponges belong to the phylum of animals known as Poriferans. The name *Poriferans* comes from a Latin word meaning "pore bearer." Most species of sponges live in saltwater environments, but a few live in freshwater environments. Sponges vary in size, shape, and color. Some are asymmetrical in shape, whereas others have radial symmetry.

ONGOING ASSESSMENT
ASK YOURSELF

Animals obtain the food they need by eating plants and other animals in their enviroment.

LASER DISC
3108

Anemone, colonial

SECTION 2 327

TEACHING STRATEGIES, continued

● **Process Skills:** *Applying, Experimenting*

Give each student a piece of paper with an octagon and a kite shape drawn on it. Have the students draw a dot in the center of the octagon. Then ask the students to draw as many lines of symmetry as they can. Ask the students what type of symmetry the octagon has. (The octagon has radial symmetry.) Then have the students draw a dot in the center of the kite shape and draw as many lines of symmetry as they can. Ask the students what type of symmetry a kite shape has. (A kite shape has bilateral symmetry.)

● **Process Skills:** *Applying, Constructing/Interpreting Models*

Reinforce the concepts of *anterior* and *posterior* by pointing out that in animals with bilateral symmetry the head is the anterior end and the tail is the posterior end. Reinforce *dorsal* and *ventral* by reminding the students that the belly is the ventral side and the back is the dorsal side. Then have the students cut pictures of animals out of old newspapers and magazines and label the anterior, posterior, ventral, and dorsal parts of each animal's body. Have the students place their work in their science portfolios.

REINFORCING THEMES— Systems and Structures

Although the animal kingdom is comprised of diverse species of organisms with many distinct characteristics, most of the organisms exhibit either radial symmetry or bilateral symmetry. As you discuss animal characteristics, point out that all vertebrates exhibit bilateral symmetry as do most invertebrates. However, some invertebrates, particularly certain sea creatures, have radial symmetry. Still other animals are asymmetric. Yet, all animals are multicellular organisms with structures that enable them to ingest foods.

MEETING SPECIAL NEEDS

Gifted

Seven essential functions are carried out by animals to ensure their survival. Assign each student a specific animal and have the student investigate how the animal carries out each of the seven functions. The functions include feeding, respiration, internal transport, elimination of waste products, response to environmental conditions, reproduction, and movement. Have the students place their research notes in their science portfolios.

Discover by Doing

Create "inkblot animals." First, fold a sheet of paper in half. Then open the sheet and flatten it out. Put ink or paint along the center fold line. Fold the paper in half again. Open the paper. What do you see? What type of symmetry does your inkblot animal have? Make a variety of "inkblot animals."

Figure 12–11. Radial symmetry can be seen in this starfish (left) and in this jellyfish (right).

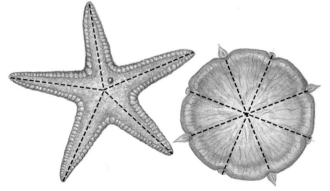

Most animals, including humans, have bilateral symmetry. An animal with **bilateral symmetry** has its exterior body parts arranged the same way on both sides of its body. A dog has two legs on each side, a butterfly has one wing on each side, and humans have one arm and one leg on each side. Nearly every animal with bilateral symmetry has a definite front end and a definite back end. The front end of the animal is called the *anterior*. The back end is called the *posterior*.

● **Process Skills:** *Observing, Inferring*

Use a toy animal as a prop. Label the anterior and posterior ends and the left, right, dorsal, and ventral sides of the animal. Use sheets of paper to show how the left side can be divided from the right side. Ask the students if this demonstrates bilateral symmetry. (Yes, because the two sides mirror each other.) Ask the students if you would get a line of symmetry if you divided the dorsal side from the ventral side. (No, because the ventral and dorsal sides are not mirror images of one another.) Use sheets of paper to show such a division. Then ask the students if you would get a line of symmetry if you divided the anterior end from the posterior end. (No, because the two ends do not mirror each other.) Again use paper to show the division. Ask the students to state how an animal would have to be divided to show bilateral symmetry. (The left side of the animal would have to be divided from the right to show bilateral symmetry.)

To picture bilateral symmetry in your mind, draw an imaginary line from the anterior end of an animal to the posterior end. Then imagine that the animal is divided in half along that line. As you can see from the illustration of the butterfly, the two long halves, or the left and right sides, are mirror images of each other.

Animals with bilateral symmetry also have an upper side and an underside. The upper side is called the *dorsal* side, and the underside is called the *ventral* side. When you rub a puppy's tummy, you are rubbing the ventral side. Petting its back is petting the dorsal side. When the puppy licks your hand and wags its tail, it is using anterior and posterior body parts to tell you that it likes to be petted. The anterior and posterior ends and the dorsal and ventral sides of a dog and a chimpanzee are shown in the illustrations.

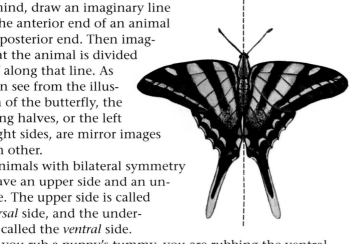

Figure 12–12. An animal with bilateral symmetry can be divided by an imaginary line into two equal right and left halves. The halves are mirror images of each other.

LASER DISC
3126

Planaria

SCIENCE BACKGROUND

Radial symmetry is a characteristic exhibited by some of the simplest animals, such as sea anemones and starfish. These animals have body parts that repeat around an imaginary line that is drawn through the center of their bodies. Animals with radial symmetry do not have a defined head per se. Also animals with radial symmetry usually are unable to move or move about in a somewhat unpredictable fashion.

In biology, the study of animals is known as *zoology*. At first, zoologists classified animals according to physical resemblances, habitats, and economic use. With the technological advances of microscopy, zoologists have been able to study animals' cellular structure, function, and genetic makeup. These improvements have led to more refined categories of classification.

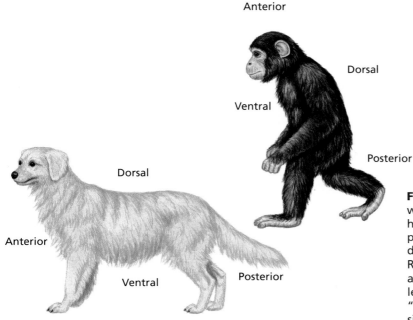

Figure 12–13. Animals with bilateral symmetry have anterior and posterior ends as well as dorsal and ventral sides. Regardless of whether an animal walks on four legs or two legs, the "belly" is the ventral side and the "back" is the dorsal side.

SECTION 2 329 ◀

TEACHING STRATEGIES, continued

● **Process Skills:**
Constructing/Interpreting Models, Applying

Give the students a sheet of paper with an asymmetrical figure drawn on it. Ask the students to draw as many lines of symmetry on the figure as possible. Ask the students to tell whether the figure is symmetrical or asymmetrical based on their experience. (The figure is asymmetrical because it does not have any lines of symmetry.)

GUIDED PRACTICE

Write the following terms on the chalkboard: symmetry, bilateral symmetry, radial symmetry, asymmetry, anterior, posterior, dorsal, and ventral. Ask the students to define the terms and to identify the terms with opposite meanings.

INDEPENDENT PRACTICE

 After they write the answers to the Section Review and Application questions in their journals, ask the students to sketch examples of animals or objects that have asymmetry, bilateral symmetry, and radial symmetry.

---

① The skeletons of some sponges are soft, and they have the ability to absorb water.

ONGOING ASSESSMENT
ASK YOURSELF

Radial symmetry is exhibited by starfish and jellyfish, and bilateral symmetry is exhibited by humans and butterflies.

SECTION 2 REVIEW AND APPLICATION

Reading Critically

1. Characteristics might include presence of cell membranes, lack of cell walls, method of locomotion, or the way in which food is obtained.

2. Some sponges have no symmetry. This type of body arrangement is known as asymmetry.

Thinking Critically

3. Some students might suggest that internal organs are located closer to the ventral side than the dorsal side of the human body. Other students might suggest that it is a natural reflex to protect one's face by turning away, and in so doing, the ventral side of the body is also protected.

4. An animal with bilateral symmetry can move effectively. An animal with radial symmetry moves somewhat randomly or has limited movement.

▶ **330**   CHAPTER 12

Figure 12–14. The soft skeletons of some types of sponges are often used for bathing. What characteristic of the sponges makes them a good choice for this purpose? ①

No Front End   A few animals are asymmetrical, or show asymmetry. The prefix *a-* in front of a word sometimes means "not." For example, *atypical* means "not typical." **Asymmetry** means "lacking symmetry"; *asymmetrical* means "not symmetrical." Asymmetrical animals have their body parts arranged in such a way that you cannot divide them into equal halves. Some sponges are examples of asymmetrical animals.

Uses of Symmetry   When artists and manufacturers design homes, automobiles, swimming pools, jewelry, or clothing, they often use symmetry. Cars with sleek lines can be divided down the middle to create mirror images. Domes on buildings and pieces of pottery have a radial design. The word *radial* is even applied to automobile tires. In the Skill activity on the next page, you will have the opportunity to find symmetry in other nonliving objects.

 ASK YOURSELF

Name two types of symmetry, and give examples of animals that exhibit each type.

SECTION 2   REVIEW AND APPLICATION

Reading Critically
1. List three characteristics of animals other than symmetry.
2. Name an animal that has no symmetry. What is its type of body arrangement called?

Thinking Critically
3. When humans sense they may be hit by an oncoming object, such as a thrown ball, they usually turn in such a way as to protect their ventral side. Why do you think this is so?
4. Compare the movement of an animal that has bilateral symmetry with the movement of an animal that has radial symmetry.

EVALUATION

Have the students describe the cellular characteristics that distinguish plants and animals.

RETEACHING

Give the students the following statements and ask them to identify the statements as *true* or *false*. The students should rewrite false statements so that they are true. 1. Animal cells have cell membranes and cell walls. (False, animal cells have cell membranes but do not have cell walls.)
2. Humans have radial symmetry. (False, humans have bilateral symmetry.)
3. The dorsal side of an animal is its back. (True)

EXTENSION

Have the students investigate characteristics that could be used to distinguish animals from monerans, fungi, and protists.

CLOSURE

Ask the students to create concept maps that distinguish plants from animals.

SKILL

Identifying Symmetry

Process Skills: Observing, Classifying

Grouping: Groups of 2 or 3

Objectives
- **Organize** objects into groups by symmetry.
- **Summarize** the methods used to categorize objects.

Discussion

You might wish to choose an object for each group and have that group debate the maximum number of symmetrical lines that could be drawn through that object.

▶ **Application**

1. The students should have drawn as many lines of symmetry as possible through each drawing to help them group the objects.

2. Although some objects exhibit only one type of symmetry, the students might suggest that a paper tube has both bilateral symmetry and radial symmetry.

3. Some students may believe that the bilateral symmetry of their face is perfect (although it is not), while others may argue that differences in hair style or in eyebrow arches make the bilateral symmetry imperfect.

✷ **Using What You Have Learned**

Animal groups will depend on the animals selected. Most pictures are likely to show animals with bilateral symmetry.

SKILL   Identifying Symmetry

▶ **MATERIALS**
- ten to fifteen pictures of common objects found at home or school

▼ **PROCEDURE**

1. Sort the objects into three groups: those with radial symmetry, those with bilateral symmetry, and those with asymmetry.
2. Sketch each object.
3. Draw a line or lines through your sketch of each object to show the type of symmetry.

▶ **APPLICATION**
1. What test did you use to determine the type of symmetry for each object?
2. Do any of the objects show more than one type of symmetry?
3. Hold two mirrors at right angles and move so that you can see one image of your face in the mirrors. The two mirrors are reversing your right and left sides. (This is actually how you look to other people!) Is the bilateral symmetry of your face and hair perfect? Explain your answer.

✷ **Using What You Have Learned**
Select pictures of ten different animals. Divide the animals into groups according to their symmetry.

Section 3: ANIMAL CLASSIFICATION

FOCUS
This section introduces and differentiates the vertebrates and invertebrates. The structure of the backbone, the feature that distinguishes vertebrates from invertebrates, is discussed.

MOTIVATING ACTIVITY
Have the students run their fingers down the center of their back. Tell them that what they feel is their backbone. Then display pictures of animals with and without a backbone. Ask the students to identify which animals they think have a backbone.

PROCESS SKILLS
• Comparing • Analyzing
• Classifying/Ordering

POSITIVE ATTITUDES
• Curiosity • Enthusiasm for science and scientific endeavor

TERMS
• vertebrate • invertebrate

PRINT MEDIA
The Inside-Out Stomach by Peter H. Loewer (see p. 315b)

ELECTRONIC MEDIA
How We Classify Animals, AIMS (see p. 315b)

Science Discovery Classification; The five kingdoms

BLACKLINE MASTERS
Study and Review Guide
Laboratory Investigation 12.2
Reading Skills
Thinking Critically

THE NATURE OF SCIENCE
Taxonomists, biologists who classify living organisms, attempt to group organisms in ways that show their evolutionary relationships. The taxonomists do this by studying homologous structures in fully grown organisms. Homologous structures are similar in shape and developmental pattern, but they may differ in function. For example, the wings of a bat and the arms of a human are homologous because they have similar structural patterns of development. Yet, each performs a different function.

SECTION 3

Animal Classification

Objectives

Compare and contrast vertebrates and invertebrates.

List examples of vertebrates and invertebrates.

Analyze the advantage of one type of animal over another.

If you could be another animal for a day, which would you choose to be? Would you choose one of the strange and colorful animals David Attenborough found in West Africa? Would you pick a more familiar animal—a household pet or an animal that you have seen living in your neighborhood?

Before you answer, think about what your day might be like as the animal you choose. If you choose a bird or a butterfly, you might spend a lot of the day flying over the countryside enjoying the view. Of course, as a bird, you would also have to eat worms! You might think that the cat next door has an easy, pleasant life. Would you choose to be a cat? Since this is turning out to be a little more involved than it sounded at first, maybe we should look closely at the main groups of animals. Finding out more about animal classification might help you in making your decision.

One decision you would have to make is whether or not you wanted a backbone. Any animal that has a backbone is a **vertebrate.** The group gets its name from the small bones called *vertebrae* that make up the spine or backbone. An animal that does not have a backbone is called an **invertebrate.**

Figure 12–15. What type of animal would you choose to be for a day? If you enjoy swimming, you might choose to be a fish.

Vertebrates

In some ways, vertebrates seem to have an advantage over animals without a backbone. The bones of a vertebrate are part of the living tissue of the animal and grow as the animal grows. The bones support the body from within and also protect vital organs. Being squished is not as likely for a vertebrate as it appears to be for the soft, less-protected worm.

TEACHING STRATEGIES

● **Process Skills:** *Comparing, Synthesizing*

Point out that the prefix *in-* means "not." Ask the students to tell the meaning of the following words based on their knowledge of the meaning of the prefix *in-*: inability (not having the ability), indirect (not direct), and informal (not formal). Then tell the students that the word *vertebrates* means "animals that have a backbone." Ask them to define *invertebrates* based on the definitions of the prefix *in-* and the word *vertebrates*. (*Invertebrates* means "animals that do not have a backbone.")

● **Process Skills:** *Applying, Solving Problems/Making Decisions*

Cooperative Learning Emphasize the fact that the 41 700 species of vertebrates make up only about 3 percent of the species in the animal kingdom. Ask the students to work with partners to determine the approximate total number of species in the animal kingdom and the approximate number of invertebrates in the kingdom. (The students should use the equation .03X = 41 700 to determine that there are about 1 390 000 species in the animal kingdom. Of these, about 1 348 300 are invertebrate species.)

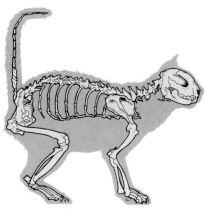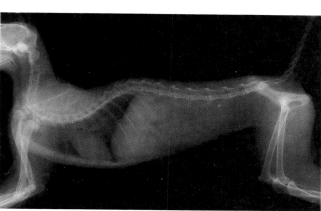

Part of what the vertebrae are supporting and protecting is the *spinal cord*. The spinal cord is a bundle of nerves that is surrounded by the vertebrae. The spinal cord is connected to the brain and carries electrical messages from the brain to all other parts of the body. The messages can also travel the other way.

If you choose to continue being a vertebrate—remember, as a human, you already are a vertebrate—you could choose to be one of 41 700 species of vertebrates. In spite of that large number, vertebrates make up only about 3 percent of the animal kingdom. The other 97 percent of animals are invertebrates.

The following tables give some examples of vertebrates and invertebrates. You may recall that *phylum* is the name given to a large group of animals. *Class* is the term used to further subdivide a phylum. In some cases, a descriptive name is used instead of the more accurate Latin name for a phylum or class. For example, in Table 12–2, instead of using the name *Platyhelminthes*, we will use the descriptive name *Flatworms*.

Figure 12–16. The backbone of a vertebrate forms the central portion of the animal's skeleton. The rest of the skeleton is attached to the backbone.

Table 12-1 Vertebrates

Class Name	Example of Animal
Fishes	Trout
Amphibians	Frog
Reptiles	Alligator
Birds	Hummingbird
Mammals	Elephant

▼ **ASK YOURSELF**

How did vertebrates get their name?

SCIENCE BACKGROUND

Vertebrates belong to the Chordate phylum. At some time in their lives, all chordates have a notochord that runs down the back of their bodies. In the vertebrates, the notochord is replaced or surrounded by the vertebral column. All chordates have bilateral symmetry and gills. Gills, however, may appear only in the undeveloped stages of growth in many vertebrates.

MEETING SPECIAL NEEDS

Mainstreamed

Reinforce the concept of vertebrates by discussing pictures of the endoskeletons of vertebrates and exoskeletons of some invertebrates. To demonstrate the importance of the endoskeleton, wrap a piece of paper around a pencil and try to crush the paper. Then take another piece of paper and crush it. Ask the students which piece of paper is like a vertebrate and why and which is like an invertebrate and why. (The paper around the pencil is like a vertebrate because it is protected by an inner structure. The other paper is like an invertebrate because it needs an outside structure to protect it.)

ONGOING ASSESSMENT
▼ **ASK YOURSELF**

They got their name from the small bones, called *vertebrae*, that make up the backbone.

TEACHING STRATEGIES, continued

● **Process Skills:** *Inferring, Analyzing*

Have the students examine Table 12–2. Point out that the phylum Arthropods has been divided into four classes. Have the students identify the classes. Then ask the students whether a mosquito would be related more closely to a spider or to a grasshopper. Have the students explain their answers. (A mosquito is related more closely to a grasshopper than to a spider because mosquitoes and grasshoppers are in the same class of animals.)

GUIDED PRACTICE

Have the students tell how vertebrates and invertebrates are alike and how they are different. (Both are members of the animal kingdom. However, vertebrates have a backbone and invertebrates do not.)

INDEPENDENT PRACTICE

Have the students write answers to the Section Review and Application questions in their journals. Then have the students write a paragraph identifying the most important concepts presented in the section.

 Scientists are able to determine the approximate age of fossils through radiocarbon dating. In the 1940s, an American scientist Willard Libby developed this technique of measuring a fossil's radiocarbon content to determine the fossil's age. All living things contain radiocarbon, which begins to decay after the organisms die. By comparing the amount of the radiocarbon in a fossil to the amount of material produced by the decaying radiocarbon, scientists can determine the age of the fossil. Using this method, scientists have identified invertebrate fossils that are estimated to be about 700 million years old. The oldest vertebrate fossils are about 500 million years old.

SCIENCE BACKGROUND

Coelenterates, which are also called cnidarians, are soft-bodied animals with stinging tentacles. Coelenterates include hydras, sea anemones, jellyfish, and corals. There are about 9000 species of coelenterates, and most live in the sea.

 **LASER DISC**

2807

Classification; The five kingdoms

Invertebrates

If you would like to be part of the animal group that was on Earth before animals had backbones, you would choose to be an invertebrate for a day. Fossil evidence indicates that invertebrates appeared on Earth long before vertebrates. Invertebrates may have been able to survive so long because their relatively simple structures enable them to readily adapt to new living conditions.

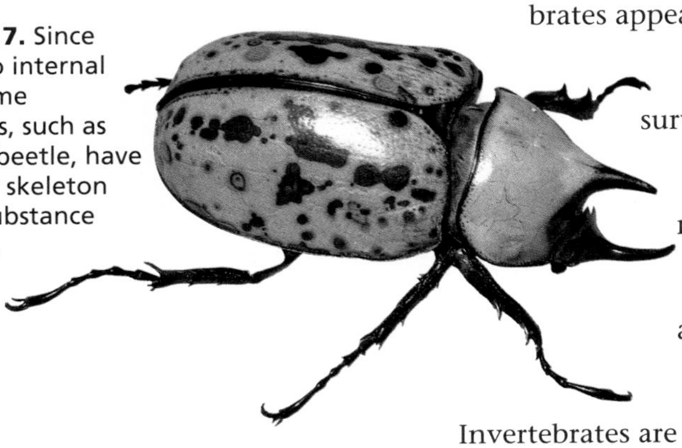

Many invertebrates are well protected by coverings of shell, by spines, or by outside skeletons. Invertebrates are extremely varied, ranging from butterflies and corals to worms and spiders. They include starfish, jellyfish, grasshoppers, clams, and even octopuses.

Certain invertebrates that live in water have radial symmetry. The jellyfish and the starfish are invertebrates with radial symmetry. Most invertebrates and all vertebrates have bilateral symmetry. Remember the asymmetrical sponge? That sponge is also an invertebrate.

Figure 12–17. Since they have no internal skeleton, some invertebrates, such as the unicorn beetle, have a hard outer skeleton made of a substance called *chitin*.

Table 12-2 Invertebrates

Phylum Name	Example of Animal
Poriferans	Sponge
Coelenterates	Jellyfish
Flatworms	Tapeworm
Roundworms	Hookworm
Segmented Worms	Earthworm
Mollusks	Snail
Echinoderms	Starfish
Arthropods	
Arachnids	Spider
Crustaceans	Lobster
Myriapods	Centipede
Insects	Grasshopper

EVALUATION

Have the students make a two-column chart with the columns labeled "Vertebrate" and "Invertebrate." Then have the students list as many animals as they can under the appropriate columns. The students should place their completed charts in their science portfolios.

RETEACHING

Cooperative Learning Have partners write the class names of vertebrates and the phylum names of invertebrates on index cards. One partner draws a card, tells whether it names a class of vertebrates or phylum of invertebrates, and names a representative animal.

EXTENSION

Have the students choose one of the invertebrate phyla and research information about the characteristics of organisms in the chosen phylum.

CLOSURE

Have the students write the section subheads in their journals and summarize the information provided in the text following the subheads.

DISCOVER BY Researching

List twenty-five animals that are as different from one another as possible. Using your observational skills as well as reference books, list two advantages and two disadvantages of being each of the animals. Record the information in your journal. Then choose the animal you would most like to be for a day.

Figure 12–18. Invertebrates come in many different forms. The tiger swallowtail butterfly (left), the cowrie (top right), and the marbled orb-weaver spider (bottom right) are all invertebrates.

What factors did you use in making your decision? In the next two chapters you will read more about invertebrates and vertebrates. You may change your mind when you know more about both types of animals!

 ASK YOURSELF

List three invertebrates and describe any protective covering each has.

SECTION 3 REVIEW AND APPLICATION

Reading Critically
1. How are vertebrates and invertebrates different?
2. Name two functions of vertebrae.
3. Why is the spinal cord important enough to be enclosed by protective bone?

Thinking Critically
4. Why are vertebrates considered more complex animals than invertebrates?
5. Grasshoppers and turtles both have hard outer coverings. Can you conclude that both are invertebrates? Explain.

DISCOVER BY Researching

The students may wish to limit the number of animals they research. Remind the students to use references such as encyclopedias and biology texts to help them identify advantages and disadvantages of being a particular type of animal.

ONGOING ASSESSMENT
ASK YOURSELF

A starfish has a spiny skin, a grasshopper has an outside skeleton, and a clam has a hard shell.

SECTION 3 REVIEW AND APPLICATION

Reading Critically

1. Vertebrates have a backbone; invertebrates do not.

2. Vertebrae support the animal and protect the spinal cord.

3. The spinal cord is the connection between the brain and all other parts of the body. The spinal cord carries messages to and from the brain.

Thinking Critically

4. The students might suggest that vertebrates have more sophisticated body structures and characteristics than invertebrates have.

5. Hard outer coverings do not indicate the presence or absence of a backbone. Grasshoppers are invertebrates, and turtles are vertebrates.

SECTION 3   **335**

INVESTIGATION

Comparing Vertebrates and Invertebrates

Process Skills: Observing, Comparing, Inferring

Grouping: Groups of 3 or 4

Objectives
- **Identify** structures of vertebrates and invertebrates.
- **Compare** structures of vertebrates and invertebrates.

Pre-Lab
Ask the students how they would identify a vertebrate and an invertebrate.

Hints
Rinse off the specimens in order to remove excess preservatives as well as odor. **CAUTION: Some preservatives may irritate eyes and skin.** Be sure that the classroom is well ventilated and that the students keep their hands away from their faces and wash their hands immediately after the Investigation. Do not use formaldehyde-preserved specimens.

Analyses and Conclusions

1. The frog is a vertebrate because it has a backbone.

2. The earthworm is an invertebrate because it does not have a backbone.

★ PERFORMANCE ASSESSMENT

Have the student groups research information about how the animals in the Discover More list differ in symmetry, vertebrae, sense organs, coloration, general shape, locomotive structures, and food-getting structures. Evaluate the students' findings for appropriate differentiations.

▶ Application
The students should conclude that the frog is far more complex than the earthworm. The frog has observable characteristics, such as eyes, jointed appendages, and a backbone. The frog can move in more ways than the earthworm.

✳ Discover More
Rankings should list asymmetrical organisms as the least complex animals and organisms with bilateral symmetry as the most complex.

Post-Lab
Have the students compare the results of their investigation to their original ideas.

INVESTIGATION

Comparing Vertebrates and Invertebrates

▶ MATERIALS
- preserved frog
- preserved earthworm
- paper towel

▼ PROCEDURE

Part A
1. Place the frog and the earthworm side by side on the paper towel so that you can observe them at the same time.
2. List five ways in which the frog is like the earthworm and five ways in which the two animals differ.
3. Which animal seems more complex to you? Give reasons for your answer.

Part B
1. Run your finger down the middle of the frog's dorsal side. (The dorsal side is the one with the frog's eyes.) Move your fingers from the anterior end to the posterior end of the frog's body. Can you feel a backbone? Record your answer.
2. On a sheet of paper draw the outline of the frog's body. Draw a dotted line from anterior to posterior, dividing the frog's body down the middle. Do the two halves resemble each other? What kind of body symmetry does a frog have? Record your answers.

Part C
1. Now examine the earthworm. The dorsal side of the earthworm is darker in color than the ventral side. Be sure the earthworm is lying on its ventral side. The anterior end of the earthworm is larger than the posterior end.
2. Run your finger down the middle of the dorsal side of the earthworm. Move your fingers from the anterior end to the posterior end of the earthworm. Can you feel a backbone? Record your answer.
3. On a sheet of paper draw the outline of the earthworm's body. Draw a dotted line from anterior to posterior, dividing the earthworm's body down the middle. Do the two halves resemble each other? What kind of body symmetry does an earthworm have? Record your answers.

▶ ANALYSES AND CONCLUSIONS
1. Is the frog a vertebrate or an invertebrate? Explain.
2. Is the earthworm a vertebrate or an invertebrate? Explain.

▶ APPLICATION
Have you changed your mind about which animal is more complex? If so, explain why. If not, give new reasons to support your opinion.

✳ Discover More
Make a list that includes one animal with radial symmetry, one with bilateral symmetry, and one with asymmetry. Rank the animals in order from the least advanced to the most advanced. Explain your reasons for ranking the animals as you did.

CHAPTER 12 HIGHLIGHTS

The Big Idea—
SYSTEMS AND STRUCTURES

As you discuss the chapter, lead the students to understand that all animals exhibit certain characteristics despite their size. All animals are multicellular organisms with structures that enable them to ingest food. Most have either bilateral or radial symmetry, and most have structures that enable them to move. Emphasize that cellular structure is a criterion used by scientists to distinguish plants and animals.

As the students revise their journal entries, they should be able to identify easily the main feature that distinguishes vertebrates from invertebrates. In addition, they should have an increasing awareness of the beneficial role all animals play within an environment. The students' views should reflect their increased awareness of the importance of animals to the environment and to humans.

CONNECTING IDEAS

Each diagram should contain a description of how that animal can be helpful and harmful. For example, the students might identify the bee's ability to sting as a harmful characteristic and the bee's ability to produce honey and pollinate flowers as beneficial characteristics. Emphasize again that although an animal's behavior might be harmful to humans, the behavior is always beneficial to the animal.

CHAPTER 12 HIGHLIGHTS

The Big Idea

Animals are found everywhere, and they vary greatly in size. They range from tiny insects to enormous elephants and whales. But although they come in many different sizes, all animals have common structural traits.

Some animals have a backbone that protects the bundle of nerves called the spinal cord. These animals are called *vertebrates*. Animals without a backbone are called *invertebrates*.

All animals share the same general characteristics. Animals take in food from their environment, they are multicellular, they have cells without cell walls, and most have some pattern of body symmetry.

Add to the journal entries you made before you read the chapter. Write the word vertebrate *or* invertebrate *next to each animal you listed. Were most of your animals vertebrates? If so, add some more invertebrates to your list. For those animals you labeled as* harmful, *add one positive trait for each animal.*

Connecting Ideas

Animals make many contributions to humans and to the environment. Sometimes, however, animals are considered pests, or even dangerous. The confusing thing is that many animals are considered both helpful *and* harmful. Make a list of animals you consider to be both helpful and harmful. Try to show why your opinions could both be true. For each animal, illustrate your point with a diagram like the one shown.

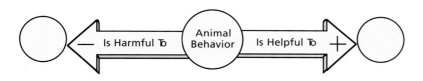

CHAPTER 12 REVIEW

ANSWERS

Understanding Vocabulary

1. A vertebrate is different from an invertebrate because a vertebrate has a backbone whereas an invertebrate does not.

2. Asymmetry is different from symmetry because asymmetrical shapes cannot be divided into mirror images whereas symmetrical shapes can be divided into mirror images.

3. Bilateral symmetry is different from radial symmetry because shapes with bilateral symmetry can be divided into mirror-image halves whereas shapes with radial symmetry can be divided equally around a center point.

Understanding Concepts

Multiple Choice

4. a
5. d
6. b
7. b

Short Answer

8. A mosquito can be classified as an animal because it is multicellular, it does not have cell walls, it does not make its own food, it does not have chloroplasts in its cells, and it can move from place to place.

9. The spinal cord functions as a connection between the brain and other parts of the body. It carries electrochemical messages to and from the brain.

10. Suggestions may include shell homes, stingers, and spiny or hard outer coverings.

Interpreting Graphics

11. The animal has bilateral symmetry. Check each diagram to ensure that the students have correctly labeled the anterior, posterior, dorsal, and ventral portions of the animal's body.

12. In the animal kingdom, 97 percent of animals are invertebrates and 3 percent are vertebrates. In the diagram, 97 of the 100 units are unshaded to represent the invertebrate animals, and 3 of the 100 units are shaded to represent the vertebrate animals.

Reviewing Themes

13. The students might suggest that the brown squirrel is a vertebrate and the brown tarantula is an invertebrate.

14. Animals must obtain energy by eating food from the environment in which they exist. Plants obtain energy by using light to perform photosynthesis to make their own food.

Thinking Critically

15. Environmental groups might believe that many people feel comfortable interacting with "cute" animals but not with less attractive organisms such as snakes.

CHAPTER 12 REVIEW

Understanding Vocabulary

For each pair of terms, complete this statement:
_____ is different from _____ because _____.

1. A vertebrate (332); an invertebrate (332)
2. Asymmetry (330); symmetry (327)
3. Bilateral symmetry (328); radial symmetry (327)

Understanding Concepts

MULTIPLE CHOICE

4. A characteristic of animals is that
   a) they take in food from their environment.
   b) they grow throughout their lifetimes.
   c) their cells are protected by a cell wall.
   d) their cells have no nuclei.

5. Some sponges are asymmetrical because
   a) they live in the water.
   b) they are invertebrates.
   c) their body parts are mirror images of each other.
   d) their body parts do not have a balanced arrangement around a center point or line.

6. All animals are
   a) symmetrical.
   b) multicellular.
   c) vertebrates.
   d) invertebrates.

7. An animal with radial symmetry
   a) has anterior and posterior ends.
   b) has a circular arrangement of body parts.
   c) has difficulty moving quickly.
   d) has more than one spinal cord.

SHORT ANSWER

8. Justify the classification of a mosquito as an animal.

9. Explain the function of the spinal cord.

10. Give two examples of ways in which invertebrates protect themselves.

Interpreting Graphics

11. Copy the "animal" below. What type of symmetry does it have? Use the labels *anterior, posterior, dorsal,* and *ventral* to mark the different parts of the animal.

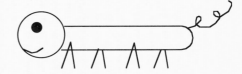

12. Explain how the diagram below shows the ratio of vertebrates to invertebrates in the animal kingdom. Hint: Recall that *percent* is a ratio of 100.

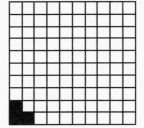

338 CHAPTER 12

16. Broad-spectrum pesticides eliminate many insects that feed on the farmer's crops as well as insects that feed on the crop-eating insects. Other crop-eating animals, such as rabbits, will not have to compete for food and their populations may increase, becoming a new pest problem.

17. Most animals have growth controls that restrict the size to which a healthy animal will grow. Some plants, however, continue to grow throughout their lifetimes. Plants with restricted space, light, or soil minerals will not grow as large as other plants with fewer restrictions.

18. Some animals in a home may remain hidden during the day and become active at night when most people are sleeping. Other animals in a home may be active during the day but are too small to be easily observed.

19. An advantage of an internal skeleton is that it grows as the animal grows. A disadvantage is that some portions of the torso are not protected by the skeleton. An advantage of an external skeleton is that the entire torso is protected by the skeleton. A disadvantage is that the skeleton does not grow and must be shed and replaced as the animal grows, leaving the animal unprotected for a time.

Reviewing Themes

13. **Systems and Structures**
    A brown tarantula (a very large spider) and a brown squirrel might both be about 25 cm long. In what ways would the animals be different?

14. **Energy**
    How do animals differ from plants in the way they get the energy they need to live?

Thinking Critically

15. Why do environmental groups working to save an endangered habitat find it necessary to choose a "cute" animal as their mascot? For example, why might a small, fuzzy monkey be chosen instead of a colorful snake?

16. A "broad-spectrum" pesticide is one that kills a wide variety of insects. Explain why these pesticides have, in some cases, created new pest problems for farmers.

17. Discuss why growing to a predetermined size is a characteristic of animals rather than a characteristic of living things in general.

18. Your friend claims there are no animals living in her house because her family does not have house pets. Explain why she might be wrong in her claim.

19. In this chapter you read about both vertebrates and invertebrates. A vertebrate, like the cat shown here, has a skeleton inside of its body. Since invertebrates do not have a backbone, they must have some other means of body support. The grasshopper shown here has its skeleton outside its body. What are the advantages and disadvantages of an internal skeleton? What are the advantages and disadvantages of an external skeleton?

Discovery Through Reading

Kerrod, Robin. *Mammals: Primates, Insect-Eaters and Baleen Whales.* Facts on File, 1989. Brief essays highlight the characteristics of the various mammals depicted to explain how each is unique.

CHAPTER 13

INVERTEBRATES

PLANNING THE CHAPTER

Chapter Sections	Page	Chapter Features	Page	Program Resources	Source
Chapter Opener	340	For Your Journal	341		
Section 1: SPONGES, COELENTERATES, AND WORMS	342	Discover By Problem Solving (A)	343	Science Discovery*	SD
		Discover By Doing (A)	345	Investigation 13.1: Studying the	
• Characteristics of Sponges (B)	342	Discover By Observing (A)	346	Body Parts of an Earthworm (A)	TR, LI
• The Life Cycle of Sponges (A)	343	Discover By Writing (B)	346	Connecting Other Disciplines:	
• Characteristics of Coelenterates (B)	344	Activity: How can you directly observe the characteristics of a typical coelenterate? (A)	347	Science and Social Studies, Exploring Sponge Harvesting (A)	TR
• The Life Cycle of Coelenterates (A)	346	Discover By Observing (A)	351	Extending Science Concepts: Regeneration of Planarians (H)	TR
• Flatworms (B)	348	Section 1 Review and Application	352	Earthworm Anatomy (B)	IT
• Roundworms (B)	350	Investigation: Tracing the Life Cycle of a Liver Fluke (A)	353	Record Sheets for Textbook Investigations (A)	TR
• Segmented Worms (B)	350			Study and Review Guide, Section 1 (B)	TR, SRG
Section 2: MOLLUSKS AND ECHINODERMS	354	Discover By Observing (B)	356	Science Discovery*	SD
• Types of Mollusks (B)	354	Discover By Calculating (A)	357	Internal Anatomy of Mollusks (B)	IT
• Animals with Spiny Skin (B)	356	Section 2 Review and Application	357	Study and Review Guide, Section 2 (B)	TR, SRG
Section 3: INSECTS—THE MOST COMMON ARTHROPODS	358	Discover By Researching (A)	361	Science Discovery*	SD
• Characteristics of Arthropods (B)	358	Section 3 Review and Application	361	Investigation 13.2: Dissecting a Grasshopper (A)	TR, LI
• Characteristics of Insects (B)	359			Thinking Critically (A)	TR
• Classification of Insects (A)	359			Arthropods (B)	IT
• Insect Development (A)	360			Study and Review Guide, Section 3 (B)	TR, SRG
• Economic Importance of Insects (A)	361				
Section 4: OTHER ARTHROPODS	362	Discover By Doing (B)	363	Science Discovery*	SD
• Arachnids (B)	362	Section 4 Review and Application	365	Reading Skills: Making an Outline for a Report (A)	TR
• Crustaceans (B)	364	Skill: Classifying Insects (A)	366	Arthropods (B)	IT
• Myriapods (B)	365			Study and Review Guide, Section 4 (B)	TR, SRG
Chapter 13 HIGHLIGHTS	367	The Big Idea	367	Study and Review Guide, Chapter 13 Review (B)	TR, SRG
Chapter 13 Review	368	For Your Journal	367	Chapter 13 Test	TR
		Connecting Ideas	367	Test Generator	

B = Basic   A = Average   H = Honors
The coding Basic, Average, and Honors indicates subsections, features, and resources that might be appropriate for different levels of learners. For additional suggestions regarding choice of topic and depth of coverage, see the Pacing Chart on pages T26–T29.

*Frame numbers at point of use
(TR) Teaching Resources, Unit 5
(IT) Instructional Transparencies
(LI) Laboratory Investigations
(SD) *Science Discovery* Videodisc Correlations and Barcodes
(SRG) Study and Review Guide

CHAPTER MATERIALS

Title	Page	Materials
Discover By Problem Solving	343	(per individual) journal
Discover By Doing	345	(per individual) poster board, pencil
Discover By Observing	346	(per class) coral jewelry, journal
Discover By Writing	346	(per individual) journal
Activity: How can you directly observe the characteristics of a typical coelenterate?	347	(per group of 3 or 4) living hydras, Petri dish, water, microscope or magnifying glass, ruler, pencil, medicine dropper, *Daphnia*
Discover by Observing	351	(per individual) earthworm, glass jar, soil, journal
Teacher Demonstration	351	worm farm
Investigation: Tracing the Life Cycle of a Liver Fluke	353	(per group of 2 or 3) die, pencil, paper
Discover By Observing	356	(per individual) live snail, piece of glass or ceramic tile, sandpaper
Discover By Calculating	357	(per individual) paper, pencil
Discover By Researching	361	(per individual) journal
Discover By Doing	363	(per class) spider web location, journal
Skill: Classifying Insects	366	(per group of 2 or 3) journal

ADVANCE PREPARATION

For the *Discover By Observing* on page 346, you will need coral jewelry (or coral samples). Obtain living hydra and *Daphnia* from a biological supply house for the *Activity* on page 347. For the *Discover By Observing* on page 351, earthworms and glass jars filled with soil are needed. The *Demonstration* on page 351 needs a worm farm, which can be obtained from a biological supply house. For the *Discover By Doing* on page 356, the students will need live snails.

TEACHING SUGGESTIONS

Field Trip
If there is a natural or artificial site in the area where crustaceans can be observed, plan a field trip to the site. If field trips are not possible, try to arrange to have a speaker bring samples of crustaceans to the class. Sometimes owners of seafood restaurants or fish markets will share samples and can often provide a substantial amount of information. Visits by such a speaker are often very exciting because the visitor may not be accustomed to such invitations and will take extra care to plan for the occasion. When requesting such a speaker, do not ask for information about the biology of crustaceans. Instead, ask for information on the commercial aspects of crustaceans and how they are processed.

Outside Speaker
Invite a veterinarian or a veterinarian's assistant to speak to the class about heartworms. Ask the speaker to bring preserved specimens to show to the class.

CHAPTER 13 INVERTEBRATES

CHAPTER THEME—SYSTEMS AND STRUCTURES

This chapter presents many kinds of invertebrate animals, including some that the students may not have studied before. The students will explore various groups of invertebrate animals, including sponges, coelenterates, worms, mollusks, echinoderms, and arthropods. The students will also examine some of the positive and negative ways in which these organisms interact with humans and the environment. The theme of **Systems and Structures** is also developed through concepts in Chapters 4, 5, 7, 8, 9, 10, 12, 16, 17, 18, 19, and 20. A supporting theme of this chapter is **Diversity**.

MULTICULTURAL CONNECTION

Although the students may be familiar with edible invertebrates such as clams, oysters, and crabs, insects are an important food source in many cultures. For example, the aborigines of central Australia eat an insect larva called a witchetty grub. This grub, which is 50 percent protein and 50 percent fat, is highly prized by people who must search for food in a harsh, arid environment. The Bantu in South Africa and the Shona in Zimbabwe stew and dry locusts to use as a relish. Crickets are eaten wrapped in leaves in Indonesia and fried in Myanmar. South American Indians eat termites, as do many Hindu groups in India. Beetles of various kinds are eaten in Japan and China. Native Americans in California gathered and roasted grasshoppers or ground them into flour. They also ate dried caterpillars.

MEETING SPECIAL NEEDS

Second Language Support

Fluent English-speaking students should be paired with students who have limited English proficiency to draw a background of ocean water and collect pictures of invertebrate animals that are commonly found in the ocean. Ask the students to label the picture of each organism with its name, then display the pictures by attaching them to the ocean background.

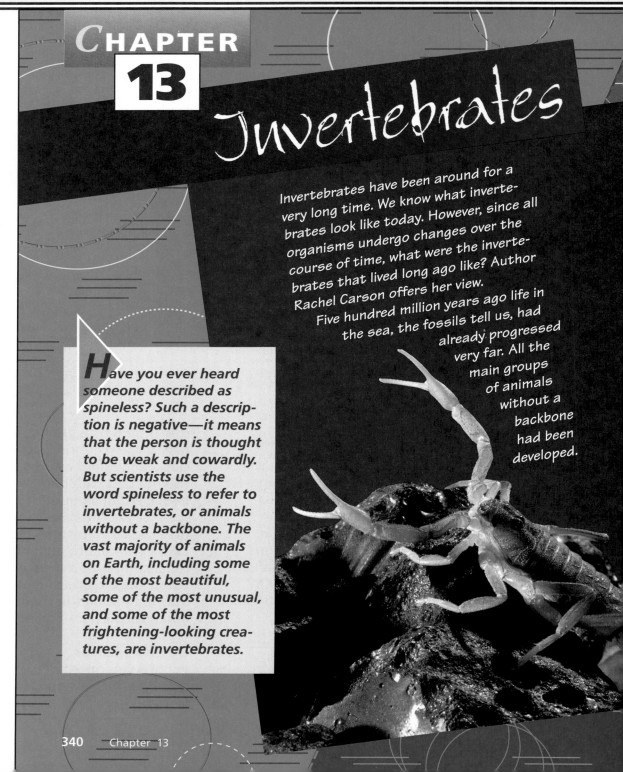

CHAPTER 13 Invertebrates

*H*ave you ever heard someone described as spineless? Such a description is negative—it means that the person is thought to be weak and cowardly. But scientists use the word spineless to refer to invertebrates, or animals without a backbone. The vast majority of animals on Earth, including some of the most beautiful, some of the most unusual, and some of the most frightening-looking creatures, are invertebrates.

Invertebrates have been around for a very long time. We know what invertebrates look like today. However, since all organisms undergo changes over the course of time, what were the invertebrates that lived long ago like? Author Rachel Carson offers her view.

Five hundred million years ago life in the sea, the fossils tell us, had already progressed very far. All the main groups of animals without a backbone had been developed.

CHAPTER MOTIVATING ACTIVITY

Cooperative Learning Obtain a selection of natural sponges and distribute them throughout the classroom. Using hand lenses or microscopes, the students can examine the structure of a sponge. Remind the students that they are looking at the skeleton of an animal. Then ask them to make a detailed sketch of a small portion of what they observe, and display the sketches throughout the classroom while the students learn about invertebrates.

Answering the journal questions provides the students with an opportunity to demonstrate their knowledge of invertebrates. Their answers will also provide you with an opportunity to note any misconceptions they might have about invertebrates.

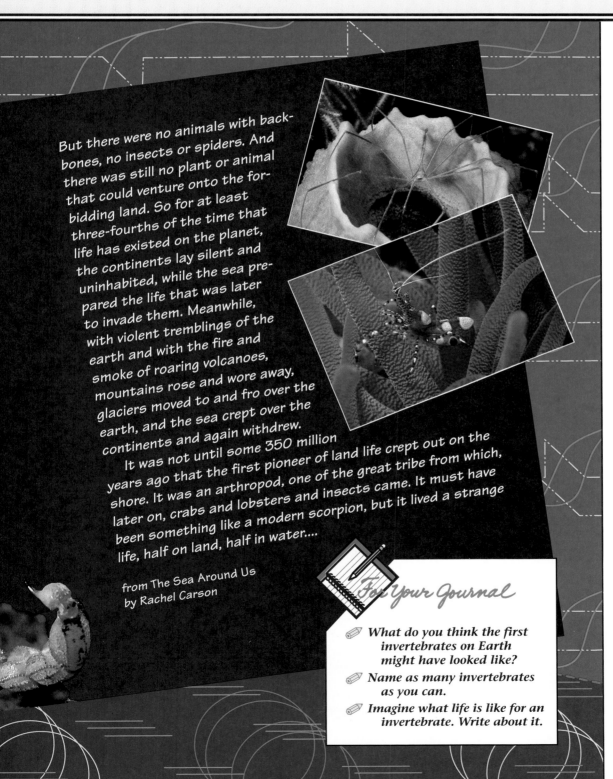

But there were no animals with backbones, no insects or spiders. And there was still no plant or animal that could venture onto the forbidding land. So for at least three-fourths of the time that life has existed on the planet, the continents lay silent and uninhabited, while the sea prepared the life that was later to invade them. Meanwhile, with violent tremblings of the earth and with the fire and smoke of roaring volcanoes, mountains rose and wore away, glaciers moved to and fro over the earth, and the sea crept over the continents and again withdrew.

It was not until some 350 million years ago that the first pioneer of land life crept out on the shore. It was an arthropod, one of the great tribe from which, later on, crabs and lobsters and insects came. It must have been something like a modern scorpion, but it lived a strange life, half on land, half in water....

from The Sea Around Us by Rachel Carson

For Your Journal

- What do you think the first invertebrates on Earth might have looked like?
- Name as many invertebrates as you can.
- Imagine what life is like for an invertebrate. Write about it.

ABOUT THE AUTHOR

Rachel Carson (1907–1964) brought the then-new science of ecology to public attention in the 1950s and 60s. Trained as a marine biologist, Carson began her career as a science writer for the United States Fish and Wildlife Service in 1935. Her articles and reports led to her writing her first book, *Under the Sea-Wind* (1941). Her second book, *The Sea Around Us*, explored the history, geography, and biology of the ocean. The book was a best-seller in 1951, as was Carson's third book, *The Edge of the Sea*, in 1955. However, Carson is probably best known for her fourth and last book, *Silent Spring* (1962), which describes the destructive effects of pesticides. Carson warned that these chemicals killed birds and fish and destroyed humans and other animals' food supplies. Although Carson stressed the interrelatedness of all living things and human dependence on nature in all of her books, *Silent Spring* became the focus of a national debate. Her book helped lead to restrictions on the use of pesticides and other federal environment protection laws.

CHAPTER 13 **341**

Section 1:
SPONGES, COELENTERATES, AND WORMS

FOCUS

This section describes the characteristics of the simplest invertebrates—sponges, coelenterates, and worms. The skeletal types and reproductive methods of sponges, the polyp and medusa life-cycle stages of coelenterates, and the three major groups of worms—flatworms, roundworms, and segmented worms—are presented.

MOTIVATING ACTIVITY

Invite a local veterinarian or veterinarian's assistant to speak to the class about heartworms. Ask the speaker to bring preserved specimens or other display materials to share with the class.

PROCESS SKILLS
- Comparing • Observing
- Evaluating • Solving Problems/Making Decisions

POSITIVE ATTITUDES
- Caring for the environment
- Curiosity

TERMS
regeneration

PRINT MEDIA
Coral Gardens by Leni Riefenstahl (see p. 315b)

ELECTRONIC MEDIA
Sponges and Coelenterates: Porous and Sac-Like Animals, Coronet (see p. 315b)

Science Discovery Portuguese man-o-war Planaria Stimulus; on a worm

BLACKLINE MASTERS
Study and Review Guide
Laboratory Investigation 13.1
Connecting Other Disciplines
Extending Science Concepts

✧ Did You Know?
A sponge 10 cm tall and 1 cm in diameter pumps approximately 22.5 L of water through its body each day.

Sponges, Coelenterates, and Worms

SECTION 1

Objectives

Explain regeneration.

Distinguish between the sexual and asexual stages of coelenterate reproduction.

Summarize the effects that different worms have on humans.

The underwater environment contains many fantastic sights and sounds. Imagine scuba diving, not in Earth's ancient ocean, but in the crystal-clear water of a modern-day tropical paradise, and seeing beds of gigantic seaweed and many beautiful fish. It would be obvious to you that the seaweeds are plants and the fish are animals. Then you might see sponges like the ones in the illustration and ask yourself, "Are they plants or are they animals?" Other people have asked the same question. In fact, scientists debated the answer to that question for more than 2000 years!

Characteristics of Sponges

Today scientists know that sponges are animals. Sponges are considered to be the simplest invertebrates because their bodies are only two cell layers thick. They have no tissues and, therefore, no organs or systems.

Sponges can be found in salt water and in fresh water. In its adult form, a sponge is attached to an object in the water and does not move from that spot. Animals that do not move from place to place are called *sessile* (SEHS ihl).

The body of a sponge is covered with many small openings called *pores*. Water enters the pores of a sponge, carrying algae, protozoa, and other one-celled organisms. Inside the sponge, these organisms are filtered out of the water and digested as food. The filtered water is then pushed out of the sponge through an opening in the top part of its body.

The body of a sponge is supported by a type of "skeleton" made of either spongin or spicules. *Spongin* is a flexible protein. *Spicules* (SPIHK yoolz) are very sharp spikes that look like crystals when viewed under a microscope. Which type of sponge would you rather bathe with? You will discover more about sponges in the next activity.

Figure 13–1. Pores allow water and nutrients to flow constantly into and out of a sponge.

TEACHING STRATEGIES

● **Process Skills:** *Observing, Comparing*

Have the students examine Figure 13–1. Point out that the classification of sponges was difficult for early taxonomists because sponges did not resemble typical animals or plants. It was not until the nineteenth century that sponges were definitively classified as animals. Ask the students to describe the characteristics of sponges that would help them determine that sponges are not plants. (Sponges do not have chloroplasts, stems, roots, or leaves; they do not have reproductive structures like plants'; they do not perform photosynthesis.)

● **Process Skills:** *Interpreting Data, Expressing Ideas Effectively*

Ask the students to discuss why a sponge, regardless of its size, must filter as much as one ton of water to gain one ounce of body weight. (To obtain enough food, sponges must filter great quantities of water because they can feed only on the tiny material that is able to enter their pores.)

DISCOVER BY *Problem Solving*

Artificial sponges are less expensive to buy than natural sponges. Since natural sponges are available in great numbers in salt water and in fresh water, why would a sponge manufactured in a factory be less expensive? Develop a hypothesis to explain this difference in cost. Write the hypothesis in your journal.

DISCOVER BY *Problem Solving*

Encourage the students to discuss what the sponge manufacturing process and the methods of harvesting natural sponges might involve before they develop their hypotheses. To extend the activity, you may wish to have the students research information to support their hypotheses.

Figure 13–2. Can you tell which sponges are natural and which are synthetic? ①

① Natural sponges are on the left, and synthetic sponges are on the right.

Synthetic sponges are made in factories and are packaged for use in household cleaning. You've probably seen many of these sponges, which can be purchased in a wide array of colors. Occasionally, natural sponges may also be brightly colored. The natural coloring is a result of algae and bacteria that live with the sponges.

ONGOING ASSESSMENT
ASK YOURSELF

A sponge is a simple invertebrate because it has no tissues, organs, or systems.

 ASK YOURSELF

Why is a sponge classified as a simple invertebrate?

The Life Cycle of Sponges

Sponges can reproduce both sexually and asexually. A sponge that reproduces sexually produces both sperm cells and egg cells. A sperm and egg combine to produce a fertilized cell. The fertilized cell then develops into a *larva* (plural, *larvae*) or young sponge. The sponge larva shown in the illustration has many flagella. The motion of the flagella allows the larva to swim around in the water. After about 24 hours, the larva settles on a rock or another object and attaches itself. This location becomes its permanent home. The larva then grows into the mature form we recognize as a sponge.

Figure 13–3. Although adult sponges are sessile, larvae, like this one, are free swimming.

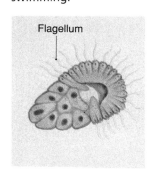

SECTION 1 **343**

TEACHING STRATEGIES, continued

• **Process Skills:** *Predicting, Applying*

Remind the students that sponges can reproduce both sexually and asexually. During sexual reproduction, a sponge produces sperm cells and egg cells, which combine to produce fertilized cells. Explain to the students that during asexual reproduction, sponges can produce gemmules, or they can grow new sponges by means of fragmentation, or budding. In budding, a piece of a mature sponge breaks off and forms a completely new sponge. Ask the students to explain why it is so easy for new sponges to grow by means of regeneration or fragmentation. (It is easy for sponges to regenerate from just a few cells because sponges do not have specialized organs.) Ask the students to predict whether sponges produced asexually would tend to be like or unlike the parent sponge from which they came. (The sponges would be like their parent.)

• **Process Skills:** *Inferring, Applying*

Point out that soft, natural bath sponges have been used for many centuries. In many parts of the world, several species of sponges are harvested by professional sponge divers for commercial use. The sponges are cured by allowing them to dry

---

ONGOING ASSESSMENT
ASK YOURSELF

During regeneration, a piece of mature sponge forms a completely new sponge. During gemmule production, a small group of sponge cells with a supply of food and a hard protective covering is manufactured and released.

REINFORCING THEMES— *Diversity*

Remind the students that freshwater sponges may produce gemmules during harsh conditions. Point out that the gemmules of some sponges display adaptations for periods of extreme cold. These gemmules remain within the body of a sponge, which dies during the cold season. When conditions again become favorable, the gemmules are released after the body of the sponge decays. This adaptation ensures that the population of sponges does not become extinct, since the parent sponges all die each winter.

MEETING SPECIAL NEEDS

Second Language Support

Have pairs of students look up the word *coelenterate* in a dictionary. (from Greek *koilos*, meaning "hollow," and from Greek *enteron*, meaning "intestine.") Ask the students how an understanding of the word's parts will help them remember what the word means and how it describes the organisms it names.

▶ **344** CHAPTER 13

Figure 13–4. A Portuguese man-of-war may look beautiful, but its tentacles can be deadly to small animals and sometimes even to humans.

Sponges also have the ability to reproduce asexually. This can occur in two ways. *Fragmentation* occurs when a small branch or piece of a mature sponge breaks off and forms a completely new sponge. This process is called **regeneration.** You would have the ability to regenerate if, for example, your ear could break off and then grow into a whole brand-new you!

A less common form of asexual reproduction may occur in sponges that live in fresh water. During harsh conditions, a sponge may produce gemmules (JEHM yoolz). A *gemmule* is a small group of sponge cells with a supply of food and a hard protective covering. Gemmules can live through unfavorable conditions, such as the drying up of a lake or stream. When the conditions are again right for survival, the gemmules open, and the cells escape to form new sponges.

  ASK YOURSELF

Describe how sponges reproduce.

Characteristics of Coelenterates

On your scuba-diving expedition, you see an unusual animal called a jellyfish. Despite its name, a jellyfish is not a fish but a *coelenterate* (sih LEHN tuh rayt), a baglike animal that has tentacles around its mouth. Other examples of coelenterates are the hydra, Portuguese man-of-war, sea anemone, and coral. Although coelenterates are more complex than sponges, they are still considered "simple" because even though they have tissues, they have no organs.

The tentacles around the mouth of a coelenterate have stinging cells, which paralyze or kill prey. These tentacles also hold prey and carry it into the coelenterate's mouth, where the prey enters the body cavity. Any waste material from the digestive process leaves the body cavity through the animal's mouth.

The tentacles and stinging cells of a coelenterate can cause serious allergic reactions and pain in humans. Any swimmer or person walking on a beach who is stung by a coelenterate should contact a physician immediately. The following activity will help you compare the body plans of coelenterates and sponges.

out in the sun. After the soft tissues rot and are washed away, the spongy skeleton remains. Remind the students that sponges are a living resource, and that overharvesting of any one type could lead to a depletion of that population of sponges. Ask the students to describe what could be done to avoid such a loss. (Sponges can regenerate from small pieces. By cutting up some sponges and "reseeding" them into their habitats, sponge divers could prevent depletion of their sponge populations.)

● **Process Skills:** *Inferring, Generating Ideas*

Point out that the coelenterates or cnidarians are animals that attach themselves permanently to either a substrate or float, with minimal control over their movements through the water. Most coelenterates are carnivores, yet they cannot actively pursue their prey. Ask the students how having tentacles with stinging cells compensates for not being able to move about to obtain prey. (The tentacles can reach out and feel for passing prey; the stingers can immobilize the prey; and the tentacles can carry the prey to the coelenterate's mouth.)

Make a poster showing the body plan of a sponge and that of a coelenterate. Identify all the parts you can. Below each diagram, list the differences between the two animals.

The Portuguese man-of-war shown in the photograph on page 344 is a type of coelenterate that is made up of several animals living together in a group called a *colony*. These animals function in a way similar to workers in a factory—everyone works at different tasks to reach a common goal. Some animals in a colony gather food, others are stingers, and still others are used during reproduction.

Another type of coelenterate is the jellyfish. While some jellyfish are small, some have tentacles that may be 9–12 m long and a bell 2 m across. One jellyfish was reported to have tentacles so long that they would touch the ground if the jellyfish were placed on the roof of a 20-story building!

Figure 13–5. Sea anemones use stinging cells in their tentacles to capture small animals for food.

Sea anemones and corals are often compared to flowers. The photographs show that these coelenterates can be very beautiful. However, the brightly colored tentacles that look like harmless flower petals are used for stinging and capturing crabs and small fish for food.

Corals also live in colonies. Unlike the Portuguese man-of-war, however, each coral is a single animal. Corals have hard skeletons that are often made of limestone. Coral reefs are created over hundreds of years as living corals build their skeletons on top of the skeletons of dead corals. The following activity offers you the opportunity to think about a certain use of coral.

LASER DISC
49630–50081

Portuguese man-o-war

DISCOVER BY *Doing*

As they are designing their posters, encourage the students to include the number of cell layers, the presence (or lack) of tissues, a body cavity, and any specialized cells in both coelenterates and sponges. You might have the students display their posters in the classroom while studying this section. Then have the students save their posters in their science portfolios.

INTEGRATION—*Language Arts*

The stinging cells of coelenterates are called *nematocysts*. The derivation of this word is from Greek words: *nemato-* refers to thread; *-cyst* refers to a bladder or pouch. The nematocyst is actually a hollow sac to which a hollow thread is attached.

TEACHING STRATEGIES, continued

● **Process Skills:** *Comparing, Applying*

Direct the students' attention to Figure 13-6, which shows the life cycle of a jellyfish. Ask the students to identify the free-swimming stage of a jellyfish's life. (medusa) Explain to the students that most coelenterates experience the polyp and medusa stages in their life cycle. However, the hydra exists only in the polyp stage. Point out that hydras can creep from place to place by means of a disk, float upside down by secreting bubbles, or move by "somersaulting" in the water.

Ask the students why these forms of hydra locomotion are rather unusual. (The polyp stage of coelenterates is usually sessile and attached; the medusa stage of coelenterates, which is not present in hydras, is typically the mobile floating or swimming stage.)

● **Process Skills:** *Inferring, Generating Ideas*

Have the students recall that sponges can reproduce asexually by fragmentation, or budding. Point out that hydras can also reproduce asexually by developing buds that fall off the parent and then grow into adult hydras. Ask the students to explain how

---

DISCOVER BY *Observing*

The students' journal entries might suggest that removing coral will damage coral reefs, destroy breeding places for some ocean organisms, and reduce protective cover for ocean fishes. Help the students realize that coral reefs develop very slowly over a long period of time, while damage to the reefs can occur very quickly.

ONGOING ASSESSMENT
ASK YOURSELF

Coelenterates have tissues; sponges do not.

DISCOVER BY *Writing*

The journal entry of each student should identify significant stages in the development of each student.

PERFORMANCE ASSESSMENT

Remind the students that the different stages of an animal's life make up its life cycle. Ask the students to draw and label a poster that compares and contrasts the life cycles of a jellyfish and a sponge. Evaluate the accuracy of each poster.

▶ **346** CHAPTER 13

Figure 13-6. The life cycle of a jellyfish is shown on the right. Although the adult jellyfish (left) is free swimming, the larval stage is sessile.

DISCOVER BY *Observing*

If possible, examine pieces of coral jewelry. Research the use of coral in jewelry-making. In your journal, explain how this use of coral might negatively impact the ocean environment. Find out whether there is such a thing as "fake" coral jewelry for the environmentally conscious consumer.

ASK YOURSELF

Explain how coelenterates are more complex than sponges.

The Life Cycle of Coelenterates

If you look at photographs of yourself taken at different times of your life, what changes would you notice? Even though you have gone through many changes, there has never been a time in your life when you did not have the form that we recognize as that of a human being. However, some animals look very different from one stage of their growth to the next. In either case, the different stages of an animal's life make up its *life cycle*. In the following activity, you can reflect on the changes in an animal's development.

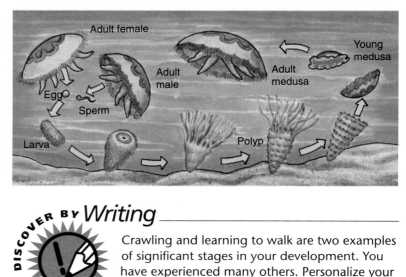

DISCOVER BY *Writing*

Crawling and learning to walk are two examples of significant stages in your development. You have experienced many others. Personalize your journal by recording the significant stages in your development. You might wish to illustrate the various stages you have experienced.

the budding process of hydras might be different from the budding process of sponges. (The hydra body form is more complex and sophisticated than that of the sponge; the sponge's bud is essentially a number of sponge cells, whereas the hydra's bud looks like a small hydra, complete with tentacles.)

● **Process Skills: Constructing/Interpreting Models, Applying**

It may be difficult for some students to differentiate between the polyp and medusa stages of the coelenterate life cycle. Have the students make models of a polyp and a medusa from modeling clay to reinforce their understanding of these two stages in the life cycle of a typical coelenterate.

ACTIVITY

How can you directly observe the characteristics of a typical coelenterate?

Process Skills: Observing, Evaluating

Grouping: Groups of 3 or 4

Hints
Place the Petri dish on a piece of paper that contrasts with the color of the hydra. Encourage the students to show concern for the fragile animals under study. Hydras should be touched gently. Also remind students that predator-prey relationships, such as that of the hydra and *Daphnia*, are a part of nature.

▶ **Application**
1. The ability of hydras to move indicates that they possess cells that can contract. Their ability to immobilize prey (*Daphnia*) shows that they have stinging cells.

2. Green hydras look like plants, and they often remain in one place.

3. Hydras do move and eat. They cannot make their own food, as plants do; they must catch food.

4. Responses might suggest that the hydra has a hollow, saclike body with radial symmetry. It has tentacles, stinging cells, and a single "mouth" opening and is very small in size.

Most coelenterates are unlike you—their life cycles occur in two stages, the polyp and the medusa. In the *polyp* stage, the animal's body is vase-shaped. In the *medusa* stage, it is bell-shaped. The illustration on page 346 shows how the jellyfish experiences both the polyp stage and the medusa stage in its life cycle.

Just as the life cycles of different coelenterates vary, the ways in which coelenterates reproduce also vary. The hydra can reproduce sexually, with sperm cells swimming to nearby egg cells that are attached to the hydra for fertilization. The hydra can also reproduce asexually by developing buds that grow into new hydras. The following activity will help you learn more about hydras.

ACTIVITY How can you directly observe the characteristics of a typical coelenterate?

MATERIALS
living hydras, Petri dish, water, microscope or magnifying glass, millimeter ruler, pencil, medicine dropper, *Daphnia*

PROCEDURE
1. Place one or two hydras in a Petri dish and add enough water to cover them. Using a microscope or a magnifying glass, observe a hydra carefully.
2. Measure the length of the hydra when it is stretched out. What happens when you touch the hydra gently with a pencil point?
3. Use a medicine dropper to create a current around the hydra. How does it react?
4. Add two or three live *Daphnia* to the Petri dish. If the hydra is hungry, it will swallow and digest the *Daphnia*.

APPLICATION
1. What evidence indicates that a hydra has cells that contract and cells that sting?
2. Why might early observers have thought that the hydra was a plant?
3. What does careful observation reveal that indicates the hydra is an animal?
4. Imagine that you are the first person ever to see a hydra. How would you describe it to everyone else?

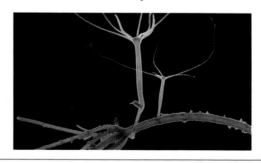

It is easy to see that hydras have a fairly simple life. Not all coelenterates, however, lead such a simple life, especially in terms of reproduction. Jellyfish reproduction is more sophisticated than that of the hydra. In jellyfish reproduction, the female medusa releases egg cells into the water, and the male

★ PERFORMANCE ASSESSMENT

Use the *Activity* as an opportunity to observe the students' ability to handle equipment correctly, follow directions accurately, and work together cooperatively.

SECTION 1   **347** ◀

TEACHING STRATEGIES, continued

● **Process Skills:** *Comparing, Generating Ideas*

Explain to the students that the nervous system of the flatworm is more developed than the nervous system of coelenterates. Flatworms have two longitudinal nerves running the length of each side of their bodies and a series of transverse nerves connecting them. They also have a simple "brain." Ask the students what aspect of the planarian's body form could indicate the presence of such a nervous system. (A planarian has symmetry (bilateral) and has definite head and tail regions. In addition, a flatworm has eyespots in its head region.)

● **Process Skills:** *Predicting, Applying*

Have the students examine the life cycle of the liver fluke shown in Figure 13–8. Encourage a volunteer to paraphrase the liver fluke life cycle. Then ask the students to predict what would happen if sheep were removed from the life cycle of liver flukes. (Liver flukes require several different hosts during their life cycles. Without host organisms such as sheep, a generation of liver flukes would die.)

ONGOING ASSESSMENT
▼ **ASK YOURSELF**

During the polyp stage, the animal's body is vase-shaped. During the medusa stage, the animal's body is bell-shaped. The polyp stage occurs before the medusa stage.

BACKGROUND INFORMATION

American zoologist Libbie Henrietta Hyman (1888–1969) performed many experiments on free-living flatworms. After writing two successful laboratory manuals, Hyman began a six-volume encyclopedia of the invertebrates in which she drew all the illustrations herself. After she died, the encyclopedia was completed by other scientists, and it made important contributions to the study of invertebrates. For her work, Hyman received honors from several scientific societies.

LASER DISC
3126

Planaria

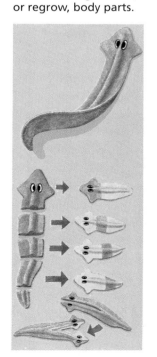

Figure 13–7.
Planarians are sometimes called "cross-eyed worms." An injured planarian can regenerate, or regrow, body parts.

medusa releases sperm cells near the egg cells. The larvae that develop after fertilization swim around until they attach themselves somewhere on the bottom of the ocean. After the larvae develop into polyps and then mature, they release themselves to become free-swimming young jellyfish.

 ASK YOURSELF

How do the polyp and the medusa stages of development differ?

Flatworms

The name *flatworm* gives a very accurate description of the animals in this group—all have bodies that are completely flat. Even though flatworms are considered to be "simple," they are more complex than sponges and coelenterates. The bodies of flatworms have three tissue layers; coelenterates have two. Unlike sponges and coelenterates, flatworms have definite head and tail regions. Some flatworms are parasites.

Flat as a Plane The planarian shown in the illustration is an example of a flatworm that is not a parasite. Most planarians live in water, but some live on moist land. They are usually about 1 cm long, although some planarians grow to about 60 cm long! A planarian's head is shaped like a triangle, and it has two *eyespots*. The eyespots are not really eyes at all; they are tissue that can only sense changes in light. The eyespots help the planarian move about its environment.

The planarian has a mouth on its bottom side. From its mouth it can extend a feeding tube called a *pharynx* (FAR inks). The pharynx draws in food in the form of microscopic organisms. Waste matter also passes out of the mouth, so the mouth is similar in function to that of coelenterates.

Planarians reproduce both sexually and asexually. However, as is true of most flatworms, there are no male or female planarians. Each individual worm produces both egg cells and sperm cells. Also, a small body part can form a completely new organism through the process of regeneration. The illustration shows how planarians regenerate.

It's a Fluke Flukes are parasitic flatworms that must find several different hosts to continue their life cycle. The liver fluke, shown in the next illustration, must live in the digestive tract of a sheep during one stage of its development and in a snail during another stage!

▶ **348** CHAPTER 13

- **Process Skills:** *Inferring, Analyzing*

Remind the students that while worms such as liver flukes and tapeworms are parasites, worms are not the only animals that exhibit parasitic lifestyles.

Ask the students to explain why it is not advantageous for a parasite to harm its host so severely that it kills the host. (Killing a host would destroy a parasite's source of nourishment and, in some cases, shelter.)

- **Process Skills:** *Inferring, Solving Problems/Making Decisions*

Explain to the students that although sheep liver fluke infections are rare in humans, other parasitic worm infections can cause harm to people. Ask the students to describe ways in which exposure to parasitic worms could be reduced or eliminated. (Suggestions might include eliminating various hosts in the life cycles of the parasitic worms, cooking food properly, and not drinking or bathing in unclean or untreated water.)

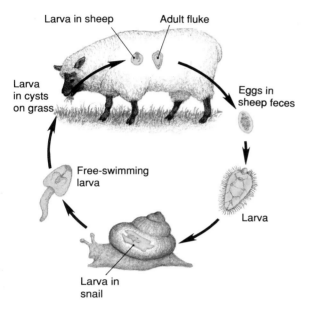

Figure 13–8. Liver flukes require a different host for each part of their life cycle.

SCIENCE BACKGROUND

Most flatworms are not distinctly male or distinctly female; each individual worm produces both sperm cells and egg cells. An organism that has both male and female reproductive organs is known as a *hermaphrodite*.

> ✦ **Did You Know?**
> Pearls are formed when an irritation causes an oyster to produce a secretion that forms a protective covering around the object causing the irritation. Often this object is a piece of debris such as a grain of sand. However, the irritation may also be caused by a parasitic worm.

Sticks Like Tape, Too Tapeworms are probably the best known of the flatworms. You may have heard about tapeworms. They look like long tapes or ribbons. These parasites enter hosts that graze on grasses containing tapeworm eggs. Once inside the host, the tapeworm egg develops into a larva. The larva burrows into the flesh of the host and lives there. Humans can get tapeworms if they eat raw or undercooked meat that contains tapeworm larvae. Once a tapeworm larva is in the body, it attaches itself to the intestine and lives on the digested food of its host. Its eggs may then pass out with the feces of the host. A tapeworm is dangerous because it uses the nutrients that should feed the host. Tapeworms, like other parasites that are found in food, can be killed by the proper cooking of food before it is eaten and by the proper disposal of human waste. Careful inspection of meat has made tapeworm infections in humans rare today.

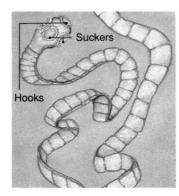

Figure 13–9. The hooks on the head of a tapeworm are used by the animal to attach itself to the intestine of its host.

ONGOING ASSESSMENT
ASK YOURSELF

Parasitic worms are dangerous to a host organism because they use the nutrients the host needs for itself.

Explain how parasitic worms pose a threat to their host organisms.

TEACHING STRATEGIES, continued

● **Process Skills:** *Inferring, Generating Ideas*

Roundworms are more advanced in their body structures than flatworms. Roundworms have separate sexes, rather than being hermaphroditic like many flatworms. Ask the students what the advantage may be to having separate sexes. (The female can produce and lay a great number of eggs, thus producing many offspring. Roundworms are, in fact, among the most numerous of multicellular animals.)

● **Process Skills:** *Inferring, Expressing Ideas Effectively*

Discuss with the students the serious side effects of being infected with a parasitic roundworm. Then ask the students why they should wear a foot covering. (The larva of the parasites often enters a human host through the soles of the feet; wearing shoes or sneakers prevents the larva from touching the skin and entering it.)

ONGOING ASSESSMENT
ASK YOURSELF

Roundworms can be male or female. Flatworms are not distinctly male or female. Roundworms have a more complex digestive system than flatworms.

LASER DISC

37215–37898

Stimulus; on a worm

SCIENCE BACKGROUND

The name for the phylum of segmented worms is *Annelida*, which means "little ring" and refers to the ringlike segments into which the annelid body is divided.

INTEGRATION—
Mathematics

To help the students better understand the sizes of various worms relative to one another, have each student draw a line the length of a notebook page and mark the line in millimeter and centimeter divisions. Then have the students draw a line for the average length of each worm they have studied. Provide these measurements: planarian, 3-5 mm; tapeworm, 1-6 m (the students can double the line back on itself); roundworm, 30 cm; earthworm, 20 cm. Ask the students to label each line and add their diagram to their science portfolios.

Roundworms

Roundworms are more advanced than all of the other worms you have met so far. They have a fairly complex digestive tract and separate sexes.

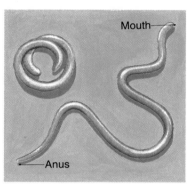

While roundworms do have a more complex digestive system than flatworms, the system is still considered to be "simple." The roundworm's digestive tract consists of a long tube with two openings: a mouth and an anus. Food enters through the mouth, and waste is removed through the anus.

The occurrence of two sexes in roundworms is a great advance over flatworms. Roundworms only reproduce sexually. During reproduction, the sperm cell and egg cell combine to form a larva. Since many roundworms are parasites, a host is often required to complete the development of the larva.

Figure 13–10. Most roundworms are parasites and cause serious illness in humans and other animals. The anatomy of a roundworm (top) is very simple.

The hookworm is an example of a typical parasitic roundworm. During the larval stage, the hookworm enters a human host. The larva can enter a human being when the person walks barefoot in areas where hookworm larvae live. The larva is so small that it can pass painlessly through the skin on the soles of the feet. Once in the body, the larva makes its way into a blood vessel and is carried into the lungs. The human host then coughs up the larva and swallows it into the digestive system. In the intestine, the worm continues its development and feeds on the blood of the human host. This blood loss can be very dangerous to the host. A hookworm can also damage the host's small intestine. Hookworms are a problem in most warm, damp areas of the world, including the southeastern United States.

 ASK YOURSELF

How are roundworms more advanced than flatworms?

Segmented Worms

Segmented worms are the most advanced variety of worms. Like other types of worms, they have three tissue layers. In addition, segmented worms have a "tube-within-a-tube" body plan. This means that their internal organs are separated from the other parts of their body. Segmented worms also have simple circulatory, muscular, and nervous systems. The bodies of these worms have segments that look like a series of rings running the length of the body.

- **Process Skills:** *Comparing, Applying*

Point out to the students that the phylum name for the roundworms is Nematode, and the word *nematode* is derived from Greek words meaning "threadlike." Ask the students to describe how roundworms are like threads. (Roundworms are long, slender, smooth animals, tapered at both the anterior and posterior ends.) Have the students look at the illustration on page 350 and note the body form of nematodes, or roundworms.

- **Process Skills:** *Inferring, Applying*

Ask the students to explain why the most important worm to humans is the earthworm. (Without earthworms, it would probably be impossible for people to farm crops. Earthworms loosen soil, allowing air, water, and minerals from earthworm waste products to penetrate the soil.)

Earthworms
by Valerie Worth
Garden soil,
Spaded up,
Gleams with
Gravel-glints,
Mica-sparks, and
Bright wet
Glimpses of
Earthworms
Stirring beneath:
Put on the palm,
Still rough
With crumbs,
They roll and
Glisten in the sun
As fresh
As new rubies
Dug out of
Deepest earth

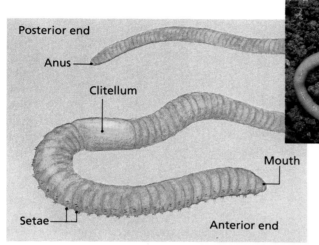

Figure 13–11. The best-known segmented worm is the earthworm. Segmented worms are more complex than flatworms or roundworms.

DISCOVER BY *Observing*

The students should observe that the earthworm moves through the soil, loosening and breaking it apart. They should conclude that the action of earthworms benefits gardeners by allowing air and water, and the roots of plants, to reach into the spaces in the loosened soil.

PERFORMANCE ASSESSMENT

Have the students compare the features of their earthworm with those in Figure 13–11. Then evaluate the students' knowledge of earthworms by asking them to identify the clitellum, setae, segments, and anterior and posterior ends of the earthworm without looking at the diagram.

Gone Fishing Earthworms are probably the best known of the segmented worms. You may have seen many earthworms. If you look closely at the outside of an earthworm, you can see a distinct feature—a large band of tissue that surrounds several of the segments near the anterior end of the worm. The band, called the *clitellum* (klih TEHL uhm), is involved in the reproduction of the worm.

If you gently rubbed your finger over an earthworm, you would find that it is not completely smooth. *Setae* (SEET ee), or tiny bristles, are located on the ventral side of the worm. The setae are attached to muscles that help the earthworm move efficiently through soil. Does the poem help you picture earthworms in a garden? Use your observations from the next activity to help you write your own poem about earthworms.

DISCOVER BY *Observing*

Carefully examine an earthworm. Feel its setae, count its segments, and find its mouth and anus. Then place it in a glass jar partially filled with soil. Observe the earthworm for a few days. How has it affected the soil? How could this kind of activity help gardeners? Write your conclusions in your journal. Then compose your own poem about earthworms. Release the earthworm in the soil outside.

Demonstration

Obtain a "worm farm" from a biological supply house and assemble the kit for the students. The kit will typically include a polystyrene box, bedding, food, worms, and instructions for maintenance. A "worm farm" will allow the students to observe earthworms in their habitat without harming or intruding on the animals. Encourage responsible students to perform the periodic maintenance of the "farm."

GUIDED PRACTICE

Cooperative Learning Have the groups discuss the three main types of worms—flatworms, roundworms, and segmented worms. Ask the students to summarize the basic body structures of each type of worm and describe the life cycles of the parasitic and nonparasitic worms.

INDEPENDENT PRACTICE

Have the students provide written answers to the Section Review and Application questions. Ask the students to compare and contrast sponges and coelenterates in their journals.

EVALUATION

Ask the students to describe several ways that some coelenterates and worms can be harmful to people. Then ask the students to describe several ways that exposure to such organisms could be reduced or eliminated.

RETEACHING

Distribute several natural sponges throughout the classroom. Ask the students to handle and study the sponges, paying particular attention to how the sponges feel. Then have the students record their reactions in a journal paragraph.

---

① Earthworm tunnels allow air and moisture to penetrate the soil.

ONGOING ASSESSMENT
ASK YOURSELF

Unlike roundworms and flatworms, earthworms have digestive, muscular, circulatory, nervous, respiratory, excretory, and reproductive systems; they also have a "tube-within-a-tube" body plan.

SECTION 1 REVIEW AND APPLICATION

Reading Critically

1. Liver flukes and tapeworms are examples of parasitic worms.

2. The stinging cells of some coelenterates can cause serious allergic reactions and pain in humans.

3. The hookworm larva passes through the skin on the soles of feet, into a blood vessel, and then into the lungs. The human host coughs up the larva and swallows it into the digestive system.

Thinking Critically

4. A snail host is needed for the liver fluke's life cycle to continue. Without a snail host, there will not be a next generation for that liver fluke.

5. In rural areas, where there is more dirt and people are more likely to walk barefoot, hookworm larvae can enter the human body through the soles of the feet.

6. Sponges are mobile in their larval stage; the larva swims freely until it attaches to an object.

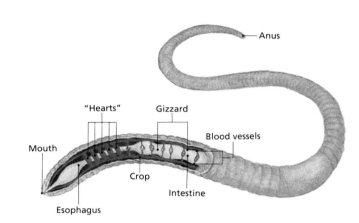

Figure 13–12. The body plan of an earthworm is a tube within a tube. Segmented worms have digestive, muscular, circulatory, respiratory, nervous, excretory, and reproductive systems.

The illustration shows that the digestive system of segmented worms consists of a tubelike structure called the *esophagus*, a thin-walled crop for storage, and a thick, muscular gizzard for grinding. The rest of the digestive system is an intestine that runs the length of the worm and ends with the anus. Solid wastes, called *castings,* pass through the anus. Castings are rich in minerals and improve the soil in which they are deposited. In what other ways are earthworms good for the soil? ①

Have a Heart (Hearts?) Segmented worms are the first animals you have met in this chapter that have blood vessels. Earthworms have five pairs of "hearts." Actually, the "hearts" are very simple muscles that pump blood into the blood vessels.

Earthworms reproduce sexually, but each earthworm produces both sperm and eggs. Although both sexes are found in the same individual, the worms do not fertilize themselves. During reproduction, two worms exchange sperm cells, and both are fertilized. The fertilized eggs are covered with a sticky substance secreted by the clitellum and then deposited. The sticky substance forms a kind of cocoon, which protects the eggs.

 ASK YOURSELF

In what ways are earthworms more advanced than roundworms or flatworms?

SECTION 1 REVIEW AND APPLICATION

Reading Critically
1. Give two examples of worms that are parasites.
2. Why should people be cautious when they are near some coelenterates?
3. Describe how hookworms enter the human digestive system.

Thinking Critically
4. If a liver fluke cannot find a snail host, what will happen to the next generation of liver flukes?
5. Hookworm infections occur more frequently in rural areas than in cities. Explain why.
6. Knowing that adult sponges are sessile, how would you answer scientists who say locomotion is a necessary characteristic of animals?

EXTENSION

It takes millions of coral skeletons, and sometimes even millions of years, to build coral reefs. The Great Barrier Reef off the coast of Australia supports a great variety of marine life. It is also one of the most beautiful natural areas in the world. Ask the students to find out more about the Great Barrier Reef.

CLOSURE

Cooperative Learning Ask each group to create an activity based on the information in the section. The activity might be a crossword puzzle using important terms or flash cards with questions and answers. Ask the groups to exchange and try out their activities.

INVESTIGATION

Tracing the Life Cycle of a Liver Fluke

Process Skills: Observing, Inferring

Grouping: Groups of 2 or 3

Objectives
- **Observe** the life cycle sequence of liver flukes.
- **Infer** reasons for the presence or absence of liver flukes in various geographical areas.

Pre-Lab
Review with the students the stages in the life cycle of a liver fluke using Figure 13–8. Ask the students to draw their own diagram of the life cycle to use during the Investigation.

▶ **Analyses and Conclusions**

1. Some groups may complete the life cycle in as few as six turns. Others will need more turns. Some groups may not complete the life cycle within the scheduled activity period.

2. The students should give logical reasons to support their response.

▶ **Application**

A sheep liver fluke infection of humans is rare; however, the Chinese liver fluke infection affects more than 200 million people worldwide. The fluke lays eggs inside the liver and bile ducts. The adult flukes often block these passages. Diseases caused by liver flukes are uncommon in the United States because of generally careful sanitation practices. Liver flukes may be a problem in less developed areas.

✳ **Discover More**
Responses should explain any similarities or differences that the students may find in their results.

Post-Lab
Ask the students to write the results of the investigation (i.e., the number of attempts) on their life-cycle drawing.

INVESTIGATION

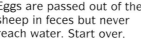

Tracing the Life Cycle of a Liver Fluke

▶ **MATERIALS**
- die • pencil • paper

▼ **PROCEDURE**

The procedure in this investigation is similar to a classification key. At each step, you will throw the die and follow the directions for an odd number or an even number. Record your results for each try. Continue the investigation until you have completed the life cycle.

1. **Odd number**
   The sheep eats grass on which a cyst has been deposited. Move to step 2.
   Even number
   The grass is mowed, and the cyst is destroyed before the sheep eats it. Start over.

2. **Even number**
   The cyst dissolves in the sheep, and the larva develops into an adult fluke. Move to step 3.
   Odd number
   The cyst dissolves in the sheep, but the larva dies. Start over.

3. **Odd number**
   Eggs are passed out of the sheep in feces and fall into water. Move to step 4.
   Even number
   Eggs are passed out of the sheep in feces but never reach water. Start over.

4. **Even number**
   Eggs develop into larvae, and the larvae enter a snail host. Move to step 5.
   Odd number
   Eggs develop into larvae, but the larvae cannot find a snail host. Start over.

5. **Odd number**
   Larvae reproduce asexually in the snail and then leave the snail host. Move to step 6.
   Even number
   The snail dies before the larvae reproduce asexually. Start over.

6. **Even number**
   The larvae swim to grass and form cysts. Life cycle is completed.
   Odd number
   The larvae have been carried too far away by the snail and cannot reach a grassy area on which to settle. Start over.

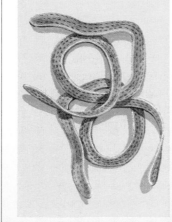

▶ **ANALYSES AND CONCLUSIONS**

1. How many attempts did it take for your liver fluke to complete its life cycle?
2. Do you think there would be this many problems in real life? Explain.

▶ **APPLICATION**

Liver flukes are commonly found in Asia, Africa, and South America. How does the presence of liver flukes affect humans? Why are diseases caused by liver flukes uncommon in the United States? Where in the United States might diseases caused by liver flukes be a problem?

✳ **Discover More**
Change the outcomes for each roll of the die to reflect the life cycle of a tapeworm. Then repeat the investigation. Compare your results from each investigation. How do you account for the similarities and differences in your results?

SECTION 1 353

Section 2:
MOLLUSKS AND ECHINODERMS

FOCUS
The section presents the great variety of animals that are known as mollusks and echinoderms. Mollusks—soft-bodied animals with and without shells—are described. Various echinoderms—animals with hardened bumpy or spiny plates under their skin—are also described.

MOTIVATING ACTIVITY

Cooperative Learning Distribute a clam, an oyster, or a scallop to each group of students. (These may be purchased fresh or frozen in supermarkets.) **CAUTION: Remind the students not to eat the shellfish and to wash their hands after handling the shellfish.** Ask each group to pry open the shell and examine the animal inside. Point out to the students how vulnerable the animal would be to enemies if it did not have a shell. Ask the students to write a description of the animal and its shell and explain why a shell is necessary to the animal. Ask the students to include a drawing with their descriptions.

PROCESS SKILLS
- Comparing • Measuring
- Observing

POSITIVE ATTITUDES
- Caring for the environment
- Curiosity

TERMS
None

PRINT MEDIA
Simple Animals by John Stidworthy (see p. 315b)

ELECTRONIC MEDIA
Echinoderms and Mollusks, Coronet (see p. 315b)

Science Discovery Octopus, blue-ring Regeneration; sea star arm

BLACKLINE MASTERS
Study and Review Guide

MULTICULTURAL CONNECTION

Many products from mollusks have been used by people for centuries. Pearls, which form inside oysters, are greatly valued. Parts of shells, such as mother-of-pearl, are used for jewelry. The cuttlefish produces cuttlebone and a brown fluid from which the brown ink, sepia, was derived. The octopus releases a blackish fluid that was once used for ink. Have the students collect more information about mollusk products and save their findings in their science portfolios.

▶ 354 CHAPTER 13

SECTION 2

Mollusks and Echinoderms

Objectives

List the major characteristics of mollusks and echinoderms.

Compare the movement of various mollusks.

Evaluate the comparative strength of mollusks and echinoderms.

"Happy as a clam!" Have you ever heard this expression? What does it mean—can clams be happy or sad? Clams are one example of mollusks. Were clams among the first organisms to emerge from the sea? Perhaps they were, but we are not sure. Regardless of which organisms first crawled from the ancient ocean, mollusks have evolved into many diverse life forms. If you like eating clams, oysters, scallops, or squid, your taste buds relish the mollusk phylum!

Types of Mollusks

Mollusks (MAHL uhsks) are soft-bodied animals that may have shells to protect them. Some mollusks live in the ocean and in fresh water, while others live on land. You probably have seen different kinds of mollusks—oysters, clams, and snails are mollusks with shells; slugs and octopuses are mollusks without shells. Does it surprise you that these animals are in the same phylum? The differences between them remind us that animals don't have to look alike for their tissue and organ development to be similar.

Mollusks are more complex than any of the worms. Mollusks have all of the advanced characteristics of the segmented worms, and they also have a true heart and are either male or female. They also have body parts—three main parts, in fact: the head-foot, the visceral mass, and the mantle.

The *head-foot* is the part of the mollusk that contains the mouth and sensory organs. It is also the body part that helps the animal move from place to place. The *visceral mass* contains all the other organs of the body, including the heart. Digestion occurs within the visceral mass. The *mantle* is a thin membrane covering the visceral mass. The mantle produces the shell of the mollusk.

TEACHING STRATEGIES

- **Process Skills:** *Inferring, Generating Ideas*

Have the students examine the mollusks shown in Figure 13–13. Explain that the shell of a mollusk is secreted by the mantle and protects the soft body of the mollusk. Ask the students what means of protection or defense mollusks without external shells, such as an octopus and a squid, may have developed. (An octopus and a squid can squirt an inklike fluid to confuse and escape from predators; they move very quickly by means of propulsion; they have tentacles with suckers.)

- **Process Skills:** *Inferring, Applying*

Clams, oysters, mussels, and scallops are all mollusks that are considered edible. They feed by passing water through their bodies and filtering out microscopic organisms to eat. Explain that coastal waters are sometimes closed to mollusk harvesting for health reasons, then ask the students why it would be unsafe to eat mollusks collected from these waters. (Mollusks in polluted water filter contaminated food particles out of the water, and the contaminants concentrate in their tissues.)

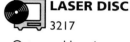

LASER DISC
3217
Octopus, blue-ring

SCIENCE BACKGROUND

Scallops are mollusks that have a row of bright blue eyes. These eyes only sort out light and dark; they do not provide images like human eyes do. The octopus has the most complex eyes of all invertebrate animals; they are comparable in structure and function to human eyes.

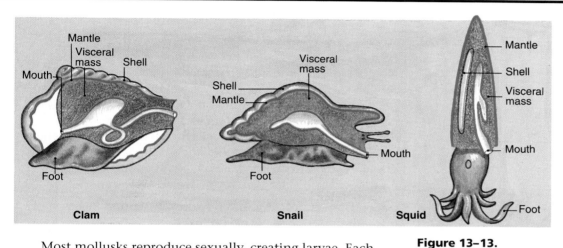

Most mollusks reproduce sexually, creating larvae. Each larva swims freely until it becomes an adult.

Some mollusks such as clams and oysters have a shell that is made of two parts, or valves. A clam, an oyster, or a mussel has two valves and moves from place to place on a hatchet-shaped foot extended from its shell. For that reason, such mollusks are called hatchet-footed mollusks.

The snail is an example of a mollusk with a shell consisting of only one valve. Snails are found both in water and on land. On land, a snail moves by gliding across a layer of mucus that is secreted by its foot. Because the snail appears to glide on its stomach, it is called a stomach-footed mollusk. The following activity will provide you with the opportunity to directly observe how a snail moves.

Figure 13–13. Although these animals look very different from one another, they are all mollusks. Note that all have the body parts of a typical mollusk.

MEETING SPECIAL NEEDS

Second Language Support

Write the term *bivalve* on the chalkboard. Help the students find out the derivation of the term. (*bi-*, two; *-valva*, leaf of a folding door) Point out that bivalve mollusks have two parts or shells that are hinged together at one edge, like a folding door, and that they open and close using one or two strong muscles.

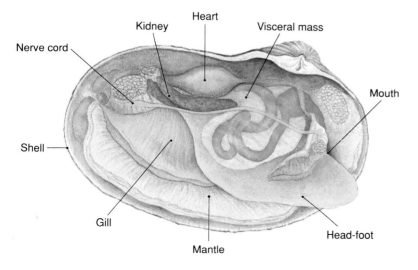

Figure 13–14. As water flows through the body of a clam, nutrients and oxygen are removed from the water, and waste and carbon dioxide are removed from the animal.

SECTION 2

GUIDED PRACTICE

Have the students identify and describe the three main body parts of a mollusk. (head-foot—contains mouth and sensory organs and allows movement; visceral mass—contains all of the other organs of the body, including the heart; mantle—covers the visceral mass and produces a shell)

INDEPENDENT PRACTICE

Have the students provide written answers to the Section Review and Application questions. In their journals, ask the students to describe how a starfish is able to eat a mollusk with a shell, such as a clam or an oyster.

EVALUATION

Pose this question to the students: "Would a hatchet-footed, stomach-footed, or head-footed mollusk be more likely to win a race?" Have the students explain their reasoning. (A head-footed mollusk would be most likely to win a race because it moves using a set of tentacles.)

DISCOVER BY Observing

The students should observe that the snail moves by sliding on a layer of mucus. The mucus is the shiny trail the snail leaves. The students should note that the snail moves more smoothly and quickly over the smooth surface than over the rough surface. The snail hesitates in its movement when its soft tissues come in contact with the rough sandpaper.

PERFORMANCE ASSESSMENT

As part of the above activity, ask each student to identify the parts of the actual snail. Evaluate the student's knowledge of mollusks.

ONGOING ASSESSMENT
ASK YOURSELF

Mollusks are male or female soft-bodied animals that have a heart and may or may not have a shell. The main body parts of mollusks include the head-foot, mantle, and visceral mass.

① The rays of a starfish and the arms and legs of a human being help the respective animal move and obtain food.

> ✧ **Did You Know?**
> The female of one species of starfish lays eggs on the bottom of the ocean and covers them with her body until they hatch.

Figure 13–15. The octopus may look like a frightening animal, but it is really rather timid.

DISCOVER BY Observing

Your teacher will supply you with a live snail. Put the snail on a smooth surface, such as a piece of glass or a ceramic tile. How does the snail move? Describe the secretion left by the snail. Now put the snail on a rough surface such as sandpaper. Compare the movement of the snail on the smooth surface and the rough surface. Was the reaction of the snail to sandpaper the same as your reaction to walking barefoot across gravel? Explain.

Ogden Nash, a poet who often wrote nonsensical verses, had this question for one of the mollusks.

> *Tell me, O Octopus, I begs,*
> *Is those things arms, or is they legs?*
> *I marvel at thee, Octopus;*
> *If I were thou, I'd call me us.*

This silly rhyme focuses on the octopus—an advanced mollusk. The octopus and the squid are the most advanced mollusks. Both are male or female head-footed mollusks. A head-footed mollusk has a very distinct head, and its foot is a set of tentacles with suckers (the arms or legs of the octopus, for example). An advanced mollusk may have an external or internal shell or no shell at all.

 ASK YOURSELF

Identify some of the characteristics of mollusks.

Animals with Spiny Skin

An *echinoderm* (ih KY nuh durm) is an animal with hardened bumpy or spiny plates under its skin. Starfish, sea urchins, and sand dollars are common echinoderms. The body of an echinoderm usually has five parts that look like rays coming from a center point.

Coats of Many Colors Starfish are found in colorful shades of deep purple and red as well as in the yellowish sand color of most dried starfish. Starfish are quite strong. Their rays can pry open the shells of clams and oysters. Starfish sometimes use their rays for climbing. For example, they may climb up the support poles of piers and docks. How are the rays of a starfish similar to your arms and legs? ①

RETEACHING

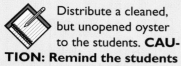 Distribute a cleaned, but unopened oyster to the students. **CAUTION: Remind the students not to eat the oyster and to wash their hands after handling the oyster.** Reminding them that a starfish can open mollusks such as an oyster, have the students attempt to open the oyster using only one hand. After the activity, ask the students to describe their experience in a journal entry.

EXTENSION

Mollusks have a circulation system that is described as an "open system." Challenge the students to discover more about the circulation system of mollusks and share their findings with the class. (In mollusks, the blood bathes organs directly rather than moving through blood vessels.)

CLOSURE

 Cooperative Learning Ask each group to use materials typically found in the classroom to create a model of a clam. Each model should include a head-foot, visceral mass, and mantle.

Can You Stomach This? Starfish are meat eaters that use their great strength to pry open mollusk shells. Starfish can be very persistent and patient—some will struggle for as long as two days and nights to open the shell of an oyster or a clam! Would you be able to wait that long to eat? Once the starfish has pried a shell open, it inserts its stomach into the shell. Its stomach secretes digestive juices that slowly turn the soft part of the mollusk into a liquid, which the starfish draws up through a tube. Then it moves its stomach back into its own body.

Even though starfish eat meat, they would never want you for dinner—you're much too large! Besides, you're able to move too fast. A starfish moves only about 15 cm per minute.

DISCOVER BY Calculating

Using meters or kilometers, estimate the distance from your home to your school. Then estimate the length of time it would take you to walk that distance. Given the fact that starfish move at a rate of about 15 cm per minute, determine how long it would take a starfish to travel from your home to your school.

And Baby Makes Three or More Starfish reproduce sexually. A female starfish may release as many as 2.5 million eggs at one time. She expels the eggs into the water at the same time that the male releases sperm. Starfish can also reproduce asexually by regeneration. Some starfish can grow and develop a whole new individual from one detached ray.

▼ ASK YOURSELF

List characteristics of starfish.

Figure 13–16. Mollusks provide food for starfish.

SECTION 2 REVIEW AND APPLICATION

Reading Critically
1. How do mollusks move?
2. Is a clam stronger than a starfish, or is a starfish stronger than a clam? Explain.
3. How does a starfish digest a meal?

Thinking Critically
4. Since mollusks have rather heavy shells to drag around with them, how have they managed to become so widespread in the ocean?
5. Gatherers of clams and oysters used to destroy starfish by cutting them up and throwing the pieces back into the water. Was this a good way to keep the starfish from damaging clam and oyster beds? Explain.

LASER DISC
4151

Regeneration; sea star arm

ONGOING ASSESSMENT
▼ **ASK YOURSELF**

Starfish are invertebrate animals with hardened bumpy or spiny plates under their skin. Found in various colors, starfish exhibit radial symmetry, which typically consists of five rays originating from a center point.

SECTION 2 REVIEW AND APPLICATION

Reading Critically

1. The head-foot of a mollusk allows movement from place to place.

2. A starfish can pry open the shell of a clam. If it does, it can be said that the starfish is stronger than the clam.

3. A starfish inserts its stomach into the soft mass of its prey. Digestive juices from the stomach liquefy the mass, which the starfish draws up through a tube.

Thinking Critically

4. The larvae of mollusks swim freely and distribute themselves throughout the oceans before settling down and developing shells.

5. No; starfish can regenerate from a single ray. By cutting the starfish up, the gatherers actually increased the population of starfish.

Section 3:
INSECTS—THE MOST COMMON ARTHROPODS

FOCUS

This section introduces arthropods. The most common arthropods are insects. The general characteristics of insects, the development of insects through a process known as metamorphosis, and the benefits that insects provide to the environment of Earth are also explored.

MOTIVATING ACTIVITY

Obtain a crab shell or lobster claw from a fish market or from the frozen food section of a supermarket. Allow the students to feel and tap on the chitinous covering. Point out that *chitin* is the material that makes up the outer skeleton of an arthropod, which protects the soft parts of the body and prevents the loss of water. Discuss with the students how this type of outer skeleton also retains the body shape of the animal. Then explain to the students that there are more arthropods in the world than any other type of animal.

PROCESS SKILLS
- Classifying/Ordering
- Comparing • Applying

POSITIVE ATTITUDES
- Cooperativeness
- Curiosity

TERMS
- exoskeleton • complete metamorphosis
- incomplete metamorphosis

PRINT MEDIA
Insects by John Bennett Wexo (see p. 315b)

ELECTRONIC MEDIA
Insects, Britannica (see p. 315b)

Science Discovery
Arthropod identification
Insect; life cycle terms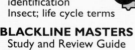

BLACKLINE MASTERS
Study and Review Guide
Laboratory Investigation 13.2
Thinking Critically

MEETING SPECIAL NEEDS

Gifted

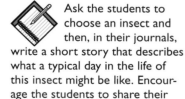
Ask the students to choose an insect and then, in their journals, write a short story that describes what a typical day in the life of this insect might be like. Encourage the students to share their stories with the class.

▶ 358 CHAPTER 13

SECTION 3

Insects—The Most Common Arthropods

Objectives

List characteristics of arthropods.

Identify the economic importance of certain insects.

Distinguish between complete and incomplete metamorphosis.

What were the first insects on Earth like? Were they anything like the insects that we have today? Were the pesky mosquitoes or ugly earwigs around then? We can't be sure, but we do know that today there are nearly one million different insects. When people talk about insects, they often refer to insects as "bugs." Correctly speaking, bugs are insects, but not all insects are bugs. This harmless mistake doesn't seem to bother most people—does it "bug" you?

Characteristics of Arthropods

Insects belong to the phylum of arthropods. The word *arthropod* means "jointed leg." Arthropods have several pairs of jointed legs, or appendages, which extend out from their bodies. Arthropods may also have other jointed appendages such as claws, jaws, fangs, egg depositors, and pincers.

The body of an arthropod is divided into body sections, or segments, and is covered by an outer skeleton called an **exoskeleton.** The exoskeleton protects soft body parts like a suit of armor. Because the exoskeleton does not grow, the arthropod sheds it occasionally and forms another. This shedding process is called *molting*. After molting, an arthropod is defenseless until the new exoskeleton forms.

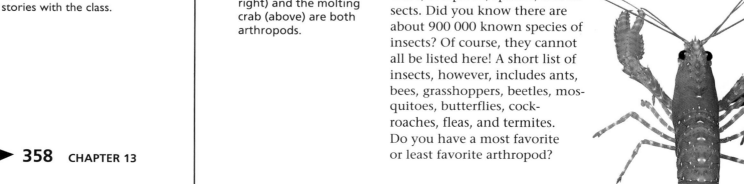

Figure 13–17. The red Hawaiian lobster (far right) and the molting crab (above) are both arthropods.

Examples of arthropods include lobsters, crabs, scorpions, spiders, and insects. Did you know there are about 900 000 known species of insects? Of course, they cannot all be listed here! A short list of insects, however, includes ants, bees, grasshoppers, beetles, mosquitoes, butterflies, cockroaches, fleas, and termites. Do you have a most favorite or least favorite arthropod?

Arthropods have complex circulatory and nervous systems. Those that live in water use gills to breathe, and those that live on land breathe through air ducts or through structures called *book lungs*. Spiders are arthropods with book lungs—their lung tissue is layered and looks something like the pages of a book.

ASK YOURSELF
What are the characteristics of arthropods?

Characteristics of Insects

You are probably very familiar with many different kinds of insects. If you looked closely at an insect, you would see three distinct body sections—the head, the thorax, and the abdomen. In addition, you would see three pairs of legs. Although most insects have wings, some are wingless. Insects also have antennae and compound eyes. An insect seeing through compound eyes will see many images of an object instead of just a single image.

Insects, as well as other arthropods, have specialized mouthparts. Each part performs a different function, depending on the needs of the animal. As the illustrations show, the mouthparts of insects are adapted to their feeding habits.

ASK YOURSELF
What are the three body parts of insects?

Classification of Insects

You have probably classified, or ordered, various things in your life. Maybe you've ordered your music collection alphabetically.

One of the many characteristics used to classify insects is wings. A close examination of insects shows that they display a great variety of wings. Some insects have one pair of wings, others have two pairs, while still others have no wings at all. Wings may be thin and delicate, covered with scales, straight, or curved. Wings may be out to the side, straight down the back, or overlapping each other.

ASK YOURSELF
How do insect wings differ?

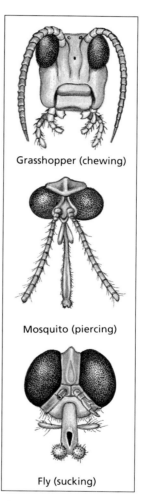

Figure 13–18. The structure of a compound eye causes an insect to see multiple images. Insects also have various types of mouthparts.

 LASER DISC
3474
Arthropod identification

ONGOING ASSESSMENT
ASK YOURSELF

Arthropods have jointed appendages, body segments, exoskeletons, and complex circulatory and nervous systems.

ONGOING ASSESSMENT
ASK YOURSELF

The three body parts are the head, thorax, and abdomen.

ONGOING ASSESSMENT
ASK YOURSELF

Insects may have one or two pairs of wings or no wings at all. The wings may be thin and delicate, covered with scales, straight, or curved, depending on the particular insect.

THE NATURE OF SCIENCE

Approximately one million different insects are known to scientists today. Because many new insects are discovered each year, the classification system used by taxonomists to group and identify insects is likely to change. The body of scientific knowledge of living things is always changing and growing, and as a result, the ways in which scientists classify and organize their knowledge of living things also change.

SECTION 3

GUIDED PRACTICE

Ask the students to describe the complete metamorphosis of a butterfly. Their descriptions should include the name of each stage and a detailed description of what occurs during each stage.

INDEPENDENT PRACTICE

 Have the students provide written answers to the Section Review and Application questions. In their journals, ask the students to summarize the characteristics of typical arthropods.

EVALUATION

Have the students compare and contrast complete and incomplete metamorphosis and provide the name of an insect that experiences each process.

 **SCIENCE TECHNOLOGY SOCIETY** Many insects are considered to be destructive because they feed on crops cultivated for human use. Scientists have developed ways to control the number of these kinds of unwanted insects. One way is through the use of chemicals called *insecticides*. Unfortunately, while these chemicals do kill insects, they also harm other organisms in the environment. As a result, biologists have begun to develop safer ways to protect food crops. One method is to introduce into the environment an insect that is a natural predator of the insect pests. Another method is to create genetically altered crops that are more resistant to the destructive effects of insects.

 **LASER DISC**
3472

Insect; life cycle terms

ONGOING ASSESSMENT
ASK YOURSELF

The pupa stage is present in complete metamorphosis but absent from incomplete metamorphosis.

Figure 13–19. The Palamedes swallowtail butterfly experiences complete metamorphosis during its life cycle.

Insect Development

Many insects change form through the various stages of their life cycle. Often the immature form does not resemble the adult form. *Metamorphosis* is a series of changes that an insect experiences during its life cycle. Metamorphosis can be complete or incomplete.

What Am I—Identity Crisis An insect that goes through **complete metamorphosis** has four stages of development—egg, larva, pupa, and adult. The butterfly experiences complete metamorphosis. When a butterfly egg hatches, a very young caterpillar emerges. A caterpillar is the larval stage of a butterfly. After a time, the caterpillar enters an inactive pupa stage. During this stage, the caterpillar forms a chrysalis (KRIHS uh lihs), a protective covering, around itself. While in this stage, the caterpillar gradually changes into a butterfly. When the butterfly is completely developed, it leaves the chrysalis as an adult.

Child or Adult? The grasshopper experiences **incomplete metamorphosis,** which consists of only three stages of development—egg, nymph, and adult. A young grasshopper, called a *nymph,* looks much like an adult grasshopper. Although a nymph looks like an adult, the nymph does not have wings. The wings develop as the insect matures. Since the grasshopper does not go through an inactive pupa stage, its metamorphosis is incomplete.

Some insects, such as the silverfish, do not experience metamorphosis. Like you, these insects just grow bigger as they become adults.

 ASK YOURSELF

Which stage of development is not present in incomplete metamorphosis but is present in complete metamorphosis?

RETEACHING

Distribute several different prepared insect specimens throughout the classroom. Ask the students to describe the differences in the wing structure of the insects and debate the value of using wings as a classification criterion for insects.

EXTENSION

Camouflage is the ability of insects to blend into their surroundings; many insects are the color of leaves, bark, or grass. This coloration protects them from predators by making them more difficult to see. Ask the students to research this topic and collect pictures of camouflaged insects.

CLOSURE

Cooperative Learning Have each group of students bring in an insect from outside. (Each insect should be placed in a jar with holes in the top and leaves and several drops of water inside.) Ask each group to create a detailed sketch of their insect. When the sketches are complete, ask the students to release the insects into their natural environment.

Economic Importance of Insects

Insects are everywhere in the world around you, and they affect many things that you do. Have ants ever ruined a picnic for you? Have mosquitoes ever made your camping trip miserable? For better or for worse, insects have an impact on the environment of Earth.

The destruction caused by an insect, while a serious problem, is told with humor in the following poem.

The Termite
by Ogden Nash

Some primal termite knocked on wood
And tasted it and found it good,
And that is why your Cousin May
Fell through the parlor floor today.

Although some insects cause damage, many others can benefit the environment. Some insects pollinate crops. Bees provide honey, butterflies provide beauty, and most insects are a food source for other animals. The following activity will help you discover some positive and negative ways that insects are economically important.

DISCOVER BY Researching

After thinking over what you have learned about insects and after interviewing friends and relatives, compile lists of both positive and negative ways insects are economically important. Record the lists in your journal.

ASK YOURSELF

What benefits do insects provide?

SECTION 3 REVIEW AND APPLICATION

Reading Critically
1. Describe the characteristics of insects.
2. List four characteristics of arthropods.
3. Describe the four stages of complete metamorphosis.

Thinking Critically
4. A grasshopper nymph has a tough exoskeleton. What happens to the exoskeleton as the nymph becomes an adult?
5. In what ways do you think the behavior of an arthropod might change when it is molting?

SCIENCE BACKGROUND
Many insects live together in large communities; these insects are called *social insects*. Social insects display division of labor; that is, different individuals perform specific tasks that benefit the community as a whole.

ONGOING ASSESSMENT
ASK YOURSELF

Some insects are beneficial because they pollinate crops and are a food source for other animals.

SECTION 3 REVIEW AND APPLICATION

Reading Critically

1. Insects have three distinct body sections—the head, thorax, and abdomen. Insects also have antennae, compound eyes, specialized mouthparts, and sometimes wings.

2. Arthropods have jointed appendages, exoskeletons, and complex circulatory and nervous systems.

3. The four stages are egg, larva, pupa, and adult.

Thinking Critically

4. The nymph molts, or sheds, its exoskeleton and grows a new, larger exoskeleton to fit its adult size.

5. The students might suggest that since an arthropod is vulnerable during molting, it will likely hide.

Section 4: OTHER ARTHROPODS

FOCUS

This section explores arthropods such as arachnids, crustaceans, and myriapods. (Millipedes and centepedes are collectively called *myriapods*. They are classified into two classes, Chilopoda and Diplopoda.)

MOTIVATING ACTIVITY

Cooperative Learning Explain to the students that spiders and other arachnids sometimes suffer from unfair stereotypes. These stereotypes, for example, might foster the idea that all spiders are to be feared or disliked.

Working in groups, have the students think of situations in their lives in which they have been exposed to unfair images of spiders and other arachnids. (Movies sometimes portray spiders as villains.) Ask the groups to share their ideas and record their thoughts.

PROCESS SKILLS
- Classifying/Ordering
- Comparing • Observing

POSITIVE ATTITUDES
- Curiosity

TERMS
None

PRINT MEDIA
Stranger Than Fiction: Killer Bugs by Melvin Berger (see p. 315b)

ELECTRONIC MEDIA
Spiders, Britannica (see p. 315b)

Science Discovery Scorpion; desert

BLACKLINE MASTERS
Study and Review Guide
Reading Skills

SCIENCE BACKGROUND

Spider silk is a protein secreted as a liquid by several large glands located in the central portion of a spider's abdomen. Each species of spider has several types of glands, each secreting a different kind of silk. As the liquid silk is drawn out through the spinnerets, it hardens into a strong, elastic solid. Some silks are used to build webs, while others are used to build cocoons.

SECTION 4

Other Arthropods

Objectives

Organize the large number of arthropods and **identify** criteria for their classification.

Compare and contrast arthropod classes.

Explain why spiders are not insects.

Imagine what it must have been like for the first arthropod that emerged from the sea. What did this creature find? What did it look like? Why was it drawn to the land? Rachel Carson thought that the first arthropod might have looked something like a modern scorpion and that it probably lived its life half on land and half in water. If they were similar to scorpions, then these arthropods were ancestors of present-day arachnids.

Arachnids

Spiders
by Mary Ann Hoberman

Spiders seldom see too well.
Spiders have no sense of smell.
Spiders spin out silken threads.
Spiders don't have separate heads.
Spider bodies are two-part.
Spider webs are works of art.
Spiders don't have any wings.
Spiders live on living things.
Spiders always have eight legs.
Spiders hatch straight out of eggs.
Since all these facts are surely so,
Spiders are not insects, no!

The poem says spiders are not insects. Many people, however, incorrectly think of them as insects. As you read, you will discover differences between spiders, one type of arachnid (uh RAK nihd), and insects.

What's the Difference? Arachnids have eight legs, while insects have only six. Insects have three body segments, but arachnids have only two—the *cephalothorax* (sef uh luh THAWR aks) and the abdomen. The cephalothorax is the arachnid's head and chest region. The abdomen contains most of its internal organs. An arachnid can have as many as 12 simple eyes. These eyes can

TEACHING STRATEGIES

● **Process Skills:** *Observing, Generating Ideas*

Have the students examine the anatomy of a spider as shown in Figure 13–20. Explain that, in addition to four pairs of walking legs, spiders have other specialized appendages. Ask the students to locate the cephalothorax in the diagram. Point out that the first two pairs of appendages on the cephalothorax are specialized for functions other than walking. The first pair have fangs connected to poison glands; the second pair are sensory. Have the students locate the abdomen in the diagram. The abdomen has specialized organs called *spinnerets*, which are used to spin silk into webs or cocoons. Ask the students to describe how the location of the specialized appendages of the cephalothorax helps the spider with feeding. (These appendages are used to inactivate prey and hold it in place while the spider sucks out the soft tissues.)

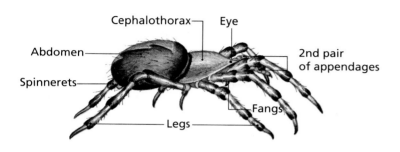

Figure 13–20. Spiders can be identified by their eight legs and two body segments.

DISCOVER BY *Doing*

After the students locate a spider web, ask them to note the location and orientation (vertical or horizontal) of the web and any other design characteristics that might vary from one type of web to another.

sense light and dark, but they cannot form images. You can see arachnids much better than they can see you!

In addition to having an extra pair of legs, arachnids have more appendages than insects. Arachnids have sharp, fanglike structures that are used to inject poison into their prey. The poison creates a kind of "milkshake" by liquefying the soft inner part of the prey. The arachnid then sucks out this soft material and swallows it.

Oh, the Web You Weave Spiders have structures that enable them to make webs. At the rear of the spider's abdomen are several small structures called *spinnerets*. Spinnerets are tiny tubes through which a liquid protein passes. This protein hardens into the familiar silk of spider webs. In the following activity, you will observe a spider web.

 *Doing*

Locate and observe a spider web. In your journal, draw what you see. If the spider is in the web, observe it for several minutes. Write a description of the spider and sketch it. Describe its behavior. Use a classification guide to help you identify the type of spider that made the web.

PERFORMANCE ASSESSMENT

Distribute a photograph or sketch of a spider web to each group of students. Ask each group to use a classification guide to identify the type of spider that created the web. Evaluate each group on its ability to use a classification guide correctly.

① The survival rate of the eggs of some spiders is very low.

MEETING SPECIAL NEEDS

Mainstreamed

To help the students remember why spiders are called *arachnids*, tell them the story of Arachne. She was a Greek girl who was so skilled at weaving that she challenged the goddess Athena to a weaving contest. This was not a good idea. Athena was so angry that she turned Arachne into a creature that could weave beautiful designs—the spider.

How Many in a Batch? Reproduction in spiders begins when the male deposits sperm cells into the female. The female spider can store sperm cells for as long as 18 months. As a result, she can fertilize several batches of eggs with sperm from one deposit. Some spiders can lay as many as 2000 eggs in a batch, depending upon the size of the spider. The average female lays about 100 eggs. Why do you think it is an advantage for some spiders to lay so many eggs? ①

GUIDED PRACTICE

Ask the students to compare and contrast the structures of arachnids and crustaceans. (Similarities: Their bodies are divided into a cephalothorax and an abdomen; the cephalothorax has specialized appendages for protection and catching food as well as four pairs of walking legs. Differences: Crustaceans have antennae, mandibles, compound eyes located on stalks, and more appendages.)

INDEPENDENT PRACTICE

Have the students provide written answers to the Section Review and Application questions. In their journals, have the students describe something that they previously did not know about arachnids, crustaceans, or myriapods.

EVALUATION

Have the students construct a table that can be used to classify arachnids, crustaceans, and myriapods. Each table should include headings for characteristic structures and common habitats.

LASER DISC
3344

Scorpion; desert

ONGOING ASSESSMENT
 ASK YOURSELF

Insects have six legs and three distinct body segments—the head, thorax, and abdomen. Arachnids have eight legs and two distinct body sections—the cephalothorax and abdomen.

THE NATURE OF SCIENCE

 Cooperative Learning Ask the students to work in groups and debate the value of the research performed by scientists on organisms such as invertebrates. (The investigations performed by scientists and researchers are valuable to us in many ways. Knowledge of the structures and lifestyles of invertebrates allows scientists not only to appreciate the great diversity of living things, but also to determine the environmental conditions that threaten the existence of the invertebrates and design ways to save threatened organisms from extinction.)

Spider Relatives Spiders and scorpions are the two best-known arachnids. A scorpion has a large stinger at the end of its abdomen. The scorpion is able to whip the stinger over its back and stab its prey, injecting poison into the victim. Ticks and mites are also arachnids—the tiniest ones. Usually less than one millimeter in length, many ticks and mites are parasites and carry disease-causing bacteria.

 ASK YOURSELF

Compare the characteristics of insects and arachnids.

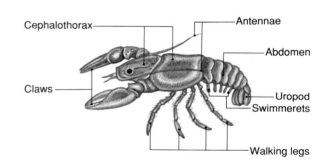

Figure 13–21. Crayfish live in fresh water and look like lobsters.

Crustaceans

Crustaceans have more appendages than arachnids. Crustaceans have two pairs of appendages called *antennae* (an TEHN ee). The antennae are sense organs that provide information about the environment. Crustaceans also have structures called *mandibles*, or jaws, which are used for chewing, crushing, and grinding food.

Most crustaceans live in water. Examples include lobsters and crayfish—lobsters live in salt water and crayfish live in fresh water. There are a few crustaceans, however, that live in damp places on land. The pill bug is one such crustacean.

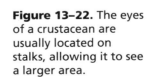

Figure 13–22. The eyes of a crustacean are usually located on stalks, allowing it to see a larger area.

A crustacean such as a crayfish uses its small claws to help hold food and its large claws to protect itself and to capture food. The crayfish's next four pairs of appendages are "walking legs." Behind these are five pairs of swimmerets. Swimmerets are much shorter than the other appendages and are used in swimming. The uropod, a flipperlike structure at the posterior of the animal, is also used for swimming. A crayfish can move backward very quickly by snapping its tail forward.

▶ **364** CHAPTER 13

RETEACHING

Ask the students to share an experience that they have had with an arachnid, a crustacean, or a myriapod. Encourage them to make their oral story as funny or as serious as they wish.

EXTENSION

Challenge the students to research the effects of the bite of arachnids such as the scorpion and black widow spider. Encourage the students to share their findings with their classmates.

CLOSURE

 Cooperative Learning Remind the students of the discussion they had about arachnid stereotypes during the Motivating Activity. Ask the groups to review the subject and try to validate or invalidate any of the stereotypes using information they learned in the section. (The students should conclude that, in most cases, such stereotypes of arachnids and their relatives are unfair.)

Some crustaceans are male; others are female. In most cases, fertilized eggs are attached to the swimmerets and carried by the female until they hatch.

 ASK YOURSELF

Which animal has more appendages—a spider or a crayfish?

Myriapods

Centipedes and millipedes are myriapods. *Myriapod* means "having many legs." If you ever see millipedes or centipedes, you might be able to tell them apart by watching how they move. Centipedes wriggle because the legs on either side of the body move alternately. Millipedes have a smoother motion because the legs on either side of the body move at the same time. The reason for this difference in motion is the number of legs each myriapod has. Centipedes have one leg on each side of every body segment. Millipedes have two legs on each side of every body segment.

All myriapods need a moist environment in which to live. They may be found in moist soil, in decaying logs, and under objects such as rocks.

Millipedes have rounded, tube-shaped bodies. They are vegetarians that live in damp, dark places and eat dead plants. Millipedes often give off an unpleasant odor when they are frightened. Centipedes have bodies that look like flattened tubes. They often have fangs that they use to attack other invertebrates, including earthworms. Both of these arthropods care for their eggs.

Figure 13–23. A centipede (top) may have from 15 to 170 pairs of legs. A millipede (bottom) may have up to 115 pairs of legs.

 ASK YOURSELF

How is a centipede different from a millipede?

SECTION 4 REVIEW AND APPLICATION

Reading Critically

1. Name four types of arthropods and give an example of each.
2. State two ways arachnids are different from crustaceans.
3. List several ways in which centipedes and millipedes differ from each other.

Thinking Critically

4. Could millipedes and centipedes live in the same environment? Explain.
5. Why don't spiders have mouthparts?
6. What characteristics of arthropods point to the supposition that these animals may have been the first to adapt to a land-based lifestyle?

ONGOING ASSESSMENT ASK YOURSELF

Crustaceans (crayfish) have more appendages than arachnids (spider).

ONGOING ASSESSMENT ASK YOURSELF

Centipede—preys on animals, flatter bodies, fangs for protection. Millipede—eats plant material, sometimes uses unpleasant odor for defense, moves more smoothly.

SECTION 4 REVIEW AND APPLICATION

Reading Critically

1. Crustaceans (crayfish), arachnids (spiders), myriapods (centipedes), insects (grasshoppers)

2. Crustaceans have more appendages than arachnids; crustaceans have chewing jaws, or mandibles, while arachnids have fangs for holding and eating prey.

3. Centipedes—carnivores, flat bodies, fangs. Millipedes—herbivores, tube-shaped bodies, sometimes use unpleasant odor for defense.

Thinking Critically

4. Yes; both prefer damp, dark environments, but they eat different foods.

5. Spiders have no need to chew. Their injected venom liquefies their prey, which they can then swallow.

6. The number and variety of their jointed appendages and the presence of either gills or book lungs for breathing suit a land-based lifestyle.

SKILL

Classifying Insects

Process Skills: Comparing, Classifying/Ordering

Grouping: Groups of 2 or 3

Objectives
- **Compare** the characteristics of various insects.
- **Classify** insects according to observed characteristics.

Discussion
Have the class study the characteristics given in the table before beginning the activity. Also review the order names so that the students are confident of the pronunciations.

Application
1. Homoptera (aphid); Orthoptera (grasshopper); Diptera (mosquito)

2. The students are likely to mention either the wing description or the piercing, sucking method of eating.

3. Responses might include mosquitoes, flies, and gnats.

4. The order would be Hymenoptera. The mouthparts of the Hymenoptera are not piercing; the sting would have been from the abdomen.

Using What You Have Learned
Responses should reflect an understanding of the classification table used in the activity.

 Have the students place their work in their science portfolios.

▶ **366** CHAPTER 13

SKILL   Classifying Insects

▶ **BACKGROUND**

All insects are classified as follows:

Kingdom: Animal; **Phylum:** Arthropod; **Class:** Insect

The next level of classification is **Order**. To determine order, it is necessary to examine the insect's wings, legs, and mouth parts.

▼ **PROCEDURE**

Imagine you have returned from a field trip on which you observed various insects. Back in your laboratory, you read the characteristics recorded in your journal. Use the table to help you identify the insects in these three entries.

1. "Small insect resting on green leaves, feeding on plant with a piercing, sucking motion, has two pairs of wings that make a rooflike structure over the insect's body when it is at rest."

2. "Insect about as long as the first joint of my forefinger. Brownish color, two pairs of wings, legs powerful enough for the insect to jump a meter."

3. "Small insect with one pair of thin wings. Landed on my wrist. Bit or stung me with a piercing needlelike structure."

TABLE 1: INSECT ORDERS AND CHARACTERISTICS

Order		Examples	Characteristics
Orthoptera ("straight-winged")		Grasshoppers, crickets	Two pairs of straight wings; legs modified for jumping
Homoptera ("uniformly winged")		Cicadas, aphids, scale insects	One or two pairs of wings that form a roof over body when at rest (some species are wingless); mouthparts adapted for piercing-sucking; feed on plants
Diptera ("two-winged")		Flies, gnats, mosquitoes	One pair of wings; mouthparts adapted for lapping or piercing-sucking
Coleoptera ("shield-winged")		Beetles (includes fireflies, ladybugs)	Two pairs of wings; hard forewings join to form a straight line down the back
Hymenoptera ("membrane-winged")		Ants, bees, wasps	Two pairs of wings; chewing mouthparts, social insect

▶ **APPLICATION**

1. To which order does each recorded insect belong?
2. In the first journal entry, what characteristic helped you decide on a certain order?
3. Name three members of the order you chose for the third journal entry.
4. If the insect described in the third journal entry had two pairs of wings instead of one, what order would you have thought about? What other characteristic might have caused you difficulty in classifying this insect?

✴ **Using What You Have Learned**

Observe some insects in their natural habitat. To avoid bites and stings, do not touch or otherwise disturb the insects. Using the procedure outlined in this activity, classify the insects into their proper orders.

Chapter 13 HIGHLIGHTS

The Big Idea—
SYSTEMS AND STRUCTURES

Lead the students to understand that even though people do not always cherish pleasant images of invertebrates such as worms, mollusks, insects, and arachnids, these animals nonetheless occupy important places in the ecosystem of the earth. These and other invertebrates contribute positively in many ways to the world in which we live.

Chapter 13 HIGHLIGHTS

The Big Idea

Invertebrates have evolved and successfully adapted to become the largest group of animals in our world. Invertebrates are also diverse. Land and marine organisms with a diameter of several millimeters and giant squids with lengths of almost 20 m are invertebrates. All invertebrates occupy a significant role in their surrounding habitat. The food chain, for example, would be affected if various orders of invertebrates suddenly disappeared. Invertebrates also make important economic contributions to the lives of people.

For Your Journal

You have just finished studying a group of animals that make up almost all of the animals in the world. Look back at the ideas you wrote at the beginning of the chapter. Thinking of the different animals you studied, add to your list of beautiful and frightening-looking invertebrates. Also include the name of your favorite invertebrate and a brief explanation of why it is your favorite.

For Your Journal

The ideas expressed by the students should reflect the understanding that the diverse group of organisms known as invertebrates benefits other organisms in many ways. You might encourage interested students to share their explanations of their favorite invertebrates with each other.

CONNECTING IDEAS

The concept map should be completed with words similar to those shown here.

Under <u>mollusks</u>:
   snails
   clams
   oysters
Under <u>worms</u>:
   flatworms
   segmented worms
   roundworms
Under <u>arthropods</u>:
   insects
   arachnids
   crustaceans
   myriapods
Additional categories:
coelenterates
   jellyfish
   hydras
echinoderms
   starfish
   sea urchins

Connecting Ideas

Copy the following concept map into your journal. Complete the concept map by adding appropriate terms in the blanks.

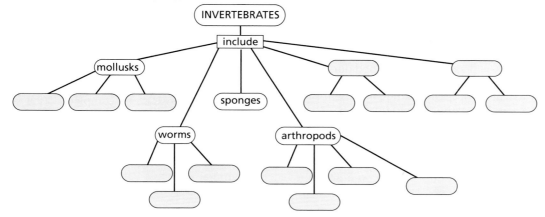

CHAPTER 13 REVIEW ANSWERS

Understanding Vocabulary

1. Both are structures of worms. A pharynx is the feeding tube of a flatworm. A clitellum is involved in the reproduction of a segmented worm.

2. Setae are bristlelike structures that help a segmented worm move from place to place. Sessile animals do not move from place to place.

3. Spinnerets are the structures that release the material a spider uses to make webs.

4. The outer skeleton that protects the body of an arthropod is called an exoskeleton. Regeneration is an asexual reproductive method in which a fragment of an organism grows into a whole organism.

5. During complete metamorphosis, an organism experiences four stages of development—egg, larva, pupa, and adult. During incomplete metamorphosis, an organism experiences only three stages of development—egg, nymph, and adult.

Understanding Concepts

Multiple Choice

6. a
7. d
8. a
9. c
10. d
11. d

Short Answer

12. A sessile animal is one that remains stationary. Corals and sponges are sessile animals.

13. A starfish typically eats mollusks, which it forces open using the pressure of its tube feet. Once open, the starfish pushes its stomach into the shell of the mollusk and injects digestive juices that dissolve the mollusk's soft body. The starfish then draws the liquid up through a tube.

14. Complete metamorphosis involves four stages during which an animal changes form completely. The butterfly is an example of an animal that develops through complete metamorphosis. During the caterpillar (larva) stage, the animal has a completely different form than when it is in the butterfly (adult) stage. Incomplete metamorphosis is a simpler process in which the young animal looks very much like it will look as an adult. The grasshopper finishes incomplete metamorphosis by adding wings.

15. The Portuguese man-of-war is a colony made up of many animals, each performing specific tasks that contribute to the survival of the whole colony. A colony, such as the one in which the corals live, is a group of totally independent animals that cluster together but do not depend on each other for the tasks that are necessary for survival.

CHAPTER 13 REVIEW

Understanding Vocabulary

For each set of terms, explain the similarities and differences in their meanings.

1. pharynx (348), clitellum (351)
2. setae (351), sessile (342)
3. webs, spinnerets (363)
4. exoskeleton (358), regeneration (344)
5. incomplete metamorphosis (360), complete metamorphosis (360)

Understanding Concepts

MULTIPLE CHOICE

6. Why is it important for sponges to live in water that moves?
   a) Moving water offers sponges a greater and more constant variety of nutrients.
   b) Moving water gives sponges the ability to move from a less desirable location to a more desirable location.
   c) Moving water helps protect sponges by keeping predators away.
   d) Moving water is not an advantage to sponges.

7. How do earthworms benefit soil?
   a) Earthworm castings provide nutrients for soil.
   b) Earthworm pathways allow oxygen to penetrate soil.
   c) Earthworm pathways help rainfall penetrate soil.
   d) All of these choices are correct.

8. When an arthropod molts, it takes between several hours and several days for the new exoskeleton to harden. What is the most likely behavior of an arthropod during this time?
   a) The arthropod will hide because it is especially vulnerable to predators.
   b) The arthropod will become more active than usual.
   c) The arthropod will lie in the sun to help speed the hardening process.
   d) The arthropod will carry on normal life activities.

9. Which type of mollusk could move great distances in the shortest period of time?
   a) hatchet-footed
   b) stomach-footed
   c) head-footed
   d) All mollusks move at the same speed.

10. Which of the following invertebrates is least likely to be eaten by another animal?
    a) clam
    b) starfish
    c) worm
    d) sponge

11. The setae of an earthworm are similar to what part of your body?
    a) hair
    b) heart
    c) eyes
    d) legs

SHORT ANSWER

12. Explain the term *sessile* and give two examples of a sessile animal.

13. Explain how a starfish obtains and digests food.

14. Explain the difference between complete metamorphosis and incomplete metamorphosis. Give an example of an insect that experiences each type of metamorphosis.

15. Explain the difference between a colonial animal, such as a Portuguese man-of-war, and animals that live in colonies, such as corals.

Reviewing Themes

16. Responses might include the following: Invertebrates occupy important niches in all food chains. Some insects eat other destructive insects. Some invertebrates are the sources of antibotics and other drugs. Mollusks are an important food source for people. Beetles and termites aid in the process of decomposition. Bivalves produce pearls. Corals provide places for many marine organisms to live. Silkworms spin cocoons from which silk is made.

17. Responses might describe a sponge developing the ability to produce gemmules or an earthworm developing the ability to sense changes in its environment.

Interpreting Graphics

18. The sea urchin's spiny skin makes it an unpalatable meal for most predators.

19. Responses might suggest that either the spines of a sea urchin do not bother the parrot fish, or the parrot fish has found a way to avoid the urchin's spines and attack its soft underside.

Thinking Critically

21. Each member of a colony performs a task that is essential for the survival of the colonial organism.

22. Responses might suggest that birds and other animals that depend on insects for food would die; animals that depend on other insect-eating animals would also die; plants that depend on insects for pollination would not be pollinated and would not reproduce; fewer plants and animals would cause serious food shortages; and decaying matter that is usually eaten by some insects would build up, increasing the incidence of disease and disrupting the cycle of nature.

23. Some responses might suggest that while an exoskeleton protects the soft body parts of an arthropod, the exoskeleton must also be shed periodically, leaving the arthropod vulnerable to attack.

24. Generally speaking, it is an evolutionary fact that regeneration is lost in higher-life forms. Specifically, regeneration is lost because of the change in the volume of nerve cells to body mass at the point of damage or loss.

25. In certain animals, some larvae are free swimming. Such larvae usually move by means of a flagellum or cilia, which would be of little value to them if they tried to move on land. Since many of the animals that produce this type of larvae are sessile in the adult stage, larvae that have the ability to move allow a wider distribution of the animals through their water environment.

26. It is advantageous for a flatworm parasite to have both sexes in the same animal because, once the flatworm is inside its host, it is restricted to that location. Having both sexes in the same animal allows it to mate with any other member of its species, increasing its chance of finding a mate.

Reviewing Themes

16. *Systems and Structures*
Using examples other than those you read about, list five ways that invertebrates benefit the world economically.

17. *Diversity*
Choose an invertebrate and describe a situation that shows the invertebrate adapting to its environment.

Interpreting Graphics

18. How does the skin of a sea urchin help protect it from predators?

19. A parrot fish will eat a sea urchin. Explain how you think this might occur.

Thinking Critically

20. A female starfish will lay about 2.5 million eggs at one time. Why aren't the world's oceans overcrowded with starfish?

21. How do the members of the colony in a colonial organism benefit one another?

22. Make a list of the problems that might occur if all of the insects of the world were to die simultaneously.

23. Discuss the advantages and disadvantages of having an exoskeleton.

24. Why do you suppose that regeneration of an entire organism from a single part is not a characteristic of higher-level animals?

25. Many animals that begin as larvae live in water. Using what you know about larvae, explain why this is true.

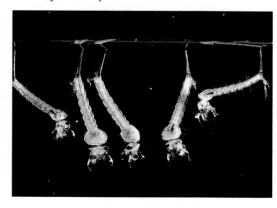

26. Why is it advantageous for parasitic flatworms to have both sexes in the same animal? Explain in detail.

Discovery Through Reading

Wiessinger, John R. *Bugs, Slugs, and Crayfish—Right Before Your Eyes.* Enslow Publishers, Inc., 1989. Text and drawings describe the physical characteristics, habits, and habitats of insects and other animals without a backbone.

CHAPTER 14

VERTEBRATES

PLANNING THE CHAPTER

Chapter Sections	Page	Chapter Features	Page	Program Resources	Source
Chapter Opener	370	For Your Journal	371		
Section 1: FISHES, AMPHIBIANS, AND REPTILES	372	Discover By Observing (B)	373	Science Discovery*	SD
• Fishes (B)	372	Activity: How would you redesign a fish for life on land? (A)	377	Investigation 14.1: Dissecting a Frog (H)	TR, LI
• Amphibians (B)	376	Discover By Writing (A)	379	Reading Skills: Finding the Main Idea (B)	TR
• Reptiles (B)	380	Section 1 Review and Application	382	Connecting Other Disciplines: Science and Health, Discovering the Uses of Snakes (A)	TR
		Investigation: Identifying Foods That Tadpoles Need (A)	383	Extending Science Concepts: Comparing Cartilaginous and Bony Fishes (A)	TR
				Fish Anatomy (B)	IT
				Frog Anatomy and Metamorphosis (B)	IT
				Record Sheets for Textbook Investigations (A)	TR
				Study and Review Guide, Section 1 (B)	TR, SRG
Section 2: BIRDS AND MAMMALS	384	Discover By Researching (B)	385	Science Discovery*	SD
• Birds (B)	384	Discover By Observing (A)	388	Investigation 14.2: Bird Reproduction (A)	TR, LI
• Structure and Reproduction of Birds (A)	386	Discover By Writing (B)	390	Thinking Critically (H)	TR
• Mammals (B)	389	Discover By Doing (B)	390	Bird Anatomy and Egg Structure (A)	IT
• Types of Mammals (A)	392	Section 2 Review and Application	393	Study and Review Guide, Section 2 (B)	TR, SRG
		Skill: Making and Interpreting a Graph (A)	394		
Chapter 14 HIGHLIGHTS	395	The Big Idea	395	Study and Review Guide, Chapter 14 Review (B)	TR, SRG
Chapter 14 Review	396	For Your Journal	395	Chapter 14 Test	TR
		Connecting Ideas	395	Test Generator	

B = Basic   A = Average   H = Honors
The coding Basic, Average, and Honors indicates subsections, features, and resources that might be appropriate for different levels of learners. For additional suggestions regarding choice of topic and depth of coverage, see the Pacing Chart on pages T26–T29.

*Frame numbers at point of use
(TR) Teaching Resources, Unit 5
(IT) Instructional Transparencies
(LI) Laboratory Investigations
(SD) *Science Discovery* Videodisc Correlations and Barcodes
(SRG) Study and Review Guide

▶ 369A

CHAPTER MATERIALS

Title	Page	Materials
Discover By Observing	373	(per class) fish in fishbowl or aquarium, journal
Activity: How would you redesign a fish for life on land?	377	(per group of 2) paper, pencil, aquarium, fishes
Discover By Writing	379	(per individual) journal
Investigation: Identifying Foods That Tadpoles Need	383	(per group of 3 or 4) aquaria (2), pond water or dechlorinated water, rocks, tadpoles (4–6 per liter of water), aquarium plants or boiled lettuce, air pump (optional), ground beef or mealworms
Discover By Researching	385	(per individual) journal
Discover By Observing	388	(per class) vinegar, eggshell
Teacher Demonstration	389	food coloring, water, hard-boiled egg
Discover By Writing	390	(per individual) journal
Discover By Doing	390	(per individual) journal
Skill: Making and Interpreting a Graph	394	(per group of 2) graph paper, pencils (2 colors)

ADVANCE PREPARATION

For the *Discover By Observing* on page 373, you will need fish in a fishbowl or an aquarium. The same fish may be observed for the *Activity* on page 377. The *Investigation* on page 383 requires aquaria, pond water, plants or boiled lettuce, tadpoles, and ground beef or mealworms. For the *Discover By Observing* on page 388, obtain vinegar and eggshells. A hard-boiled egg is needed for the *Demonstration* on page 389.

TEACHING SUGGESTIONS

Field Trip
Arrange a field trip to a local zoo. Contact the curator or the education director and ask him or her to meet with the students (or to provide a knowledgeable employee to meet with them) to share some behind-the-scenes information about mammals. Diet, care of newborn animals, and other special concerns of the zoo management should be discussed.

Outside Speaker
Invite someone from a local pet shop to discuss aquaria and terrariums. The speaker should touch on such topics as which fish can live in the same tank, whether turtles can be added to aquaria and terrariums, which natural enemies one might accidentally combine, and what animals have unusual food needs. How to enjoy these hobbies on limited budgets should also be discussed.

Arrange for an expert on birds or bird care to visit the class and talk to the students. Veterinarians who specialize in birds, or pet shops, are good sources for speakers. In addition to health care, it would be interesting to hear about the import/export laws involving birds.

CHAPTER 14 VERTEBRATES

CHAPTER THEME—CHANGES OVER TIME

This chapter introduces the students to the groups of cold-blooded vertebrates—fishes, amphibians, and reptiles—and to the groups of warmblooded vertebrates—birds and mammals. While studying the chapter, the students will learn about the structures, characteristics, and behavior that have developed in these groups of vertebrates. The theme of **Changes Over Time** is also developed through concepts in Chapter 6. A supporting theme of this chapter is **Systems and Structures**.

MULTICULTURAL CONNECTION

Fishes are an important food source for people from many cultures around the world. Ask each student to choose a different location on Earth and then use reference materials to discover the kinds of fishes that are used as food and other ways in which fishes are an important resource in that area. Encourage the students to share their findings with their classmates. Then ask them to summarize their information on a sheet of paper and display it in the appropriate location on a world map.

MEETING SPECIAL NEEDS

Mainstreamed

Point out to the students that the skins of reptiles, including snakes, alligators, and crocodiles, are sometimes obtained illegally and used to make items such as belts, purses, and shoes. Assign each mainstreamed student a student tutor, and have the pairs work together to write a letter to the United States Fish and Wildlife Service asking for information about the penalties that can be imposed on a person caught illegally catching reptiles.

CHAPTER 14 VERTEBRATES

Fishers are fur-bearing animals that can reach a length of three feet. They are found in the forests of eastern North America and are extremely rare. In his novel The Winter of the Fisher, *Cameron Langford describes one year in the life of a fisher. Notice in this part of the story how the fisher learns from and about other animals. All of them are vertebrates—animals with a backbone.*

He followed and spied on many animals in the triangle, unconsciously filling in the gaps in his education left by his mother's death. From the weasels he learned the use of cover, from the foxes the way to circle a rabbit until it ran itself out; the deer showed him how to blend motionlessly with any background, and the squirrels and jays taught him the meaning of their danger calls....

He learned the language of the tracks, written fleetingly across the impermanent face of the snow. The tiny marks of deer mice crisscrossed everywhere, small and bird-like, the thin streak of the pendant tail cutting like a knife edge between each set of tightly grouped prints.

CHAPTER MOTIVATING ACTIVITY

Obtain several temperature strips. Use a strip to take your own temperature and record the data on the chalkboard. Ask a volunteer to use another strip to take his or her temperature, and record that data on the chalkboard. Point out that because humans are warm-blooded vertebrates, their temperature always remains about the same, regardless of the temperature of the room. Explain that coldblooded vertebrates, such as fishes, amphibians, and reptiles, do not have a constant body temperature. Their temperature changes according to the temperature of their environment.

For Your Journal

The journal questions provide the students with an opportunity to demonstrate their knowledge of vertebrates. Their answers to the questions will provide you with an opportunity to note any misconceptions they might have about vertebrates. You might choose to focus on those misconceptions at appropriate times during the study of the chapter.

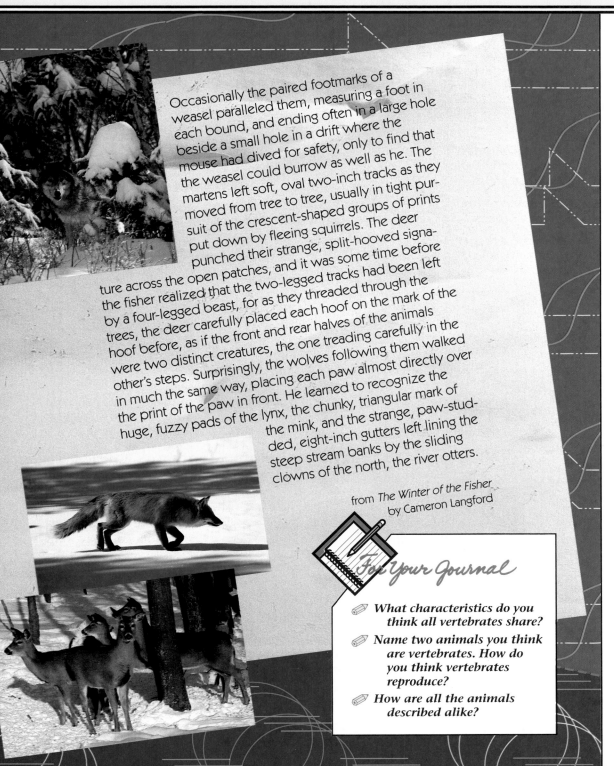

Occasionally the paired footmarks of a weasel paralleled them, measuring a foot in each bound, and ending often in a large hole beside a small hole in a drift where the mouse had dived for safety, only to find that the weasel could burrow as well as he. The martens left soft, oval two-inch tracks as they moved from tree to tree, usually in tight pursuit of the crescent-shaped groups of prints put down by fleeing squirrels. The deer punched their strange, split-hooved signature across the open patches, and it was some time before the fisher realized that the two-legged tracks had been left by a four-legged beast, for as they threaded through the trees, the deer carefully placed each hoof on the mark of the hoof before, as if the front and rear halves of the animals were two distinct creatures, the one treading carefully in the other's steps. Surprisingly, the wolves following them walked in much the same way, placing each paw almost directly over the print of the paw in front. He learned to recognize the huge, fuzzy pads of the lynx, the chunky, triangular mark of the mink, and the strange, paw-studded, eight-inch gutters left lining the steep stream banks by the sliding clowns of the north, the river otters.

from *The Winter of the Fisher* by Cameron Langford

For Your Journal

- What characteristics do you think all vertebrates share?
- Name two animals you think are vertebrates. How do you think vertebrates reproduce?
- How are all the animals described alike?

ABOUT THE BOOK

Cameron Langford's book is a novel, or a work of fiction, but it is based on fact. The author tells the story of a year in the life of a fisher, from his birth one spring to his first mating the next spring. During the course of the year, the fisher claims a territory, fights other predators, and survives a harsh winter. The fisher learns to fear the trapper, who sets metal traps to catch animals, and to trust an old Ojibway Indian, who comes to the fisher's rescue several times. The book also describes many other animals that share the fisher's world—porcupines, foxes, bears, skunks, rabbits, wolves.

The fisher is known by many names, including fisher marten, fisher cat, black fox, and black cat, but it is neither a fox nor a cat. A member of the weasel family, the fisher has a body like a weasel's, a bushy tail, a slender muzzle, and rounded ears. Trappers have long valued its thick, brownish black fur, so much so that the fisher is now rare in its northern forest habitat.

Despite its name, the fisher does not fish. It hunts and eats small rodents, particularly porcupines and snowshoe hares. Sometimes the fisher eats fruits and nuts.

Section 1:
FISHES, AMPHIBIANS, AND REPTILES

FOCUS
This section introduces fishes, amphibians, and reptiles—the groups of coldblooded vertebrates. The body temperature of coldblooded vertebrates changes as the temperature of their environment changes. The important structures and unique characteristics that differentiate fishes, amphibians, and reptiles are examined in detail.

MOTIVATING ACTIVITY

Cooperative Learning To help the students better understand the external anatomy of a bony fish, obtain a preserved specimen or a bony fish from a supermarket. Show the students the seven fins of the fish. Then write the number and the name of the fins on the chalkboard: one dorsal, one anal, one caudal, two pectoral, and two pelvic. Allow small groups of students an opportunity to closely examine the fish. Ask the groups to prepare a diagram of the fish, use the chalkboard list to predict the name of each fin, and label their diagram accordingly.

PROCESS SKILLS
- Observing • Comparing
- Inferring

POSITIVE ATTITUDES
- Caring for the environment
- Curiosity

TERMS
coldblooded

PRINT MEDIA
Reptiles by Mike Linley (see p. 315b)

ELECTRONIC MEDIA
Remarkable Reptiles, Marty Stouffer Productions (see p. 315b)

Science Discovery
Grunt, blue striped; scales
Frog life cycle
Snake, body temperature; comparison with humans

BLACKLINE MASTERS
Study and Review Guide
Laboratory Investigation 14.1
Reading Skills
Connecting Other Disciplines
Extending Science Concepts

 **SCIENCE TECHNOLOGY SOCIETY** Lampreys from the Atlantic Ocean were able to penetrate into all of the Great Lakes when a canal was dug around Niagara Falls to connect Lake Ontario and Lake Erie. Since then, lampreys have become a problem in all of the Great Lakes. One of the methods currently used to reduce the lamprey population is the chemical treatment of lamprey-spawning areas.

SECTION 1
Fishes, Amphibians, and Reptiles

Objectives

Organize fishes according to major characteristics.

Summarize frog metamorphosis.

Describe the characteristics of reptiles.

Imagine being a fish for a day! Where would you most like to live—in a stream, a lake, or an ocean? Or, would you like to be a fish in a large aquarium where you would not have to search for food? Look at the two fishes in the illustration. Would you prefer to be a very large fish like the shark or a colorful fish like the betta? Perhaps you would have enemies to worry about. Which of the animals described at the beginning of the chapter might be your enemy?

Fishes

As a fish, you would notice several things—first of all, you are very wet! Second, your body temperature is constantly changing. All fishes, as well as amphibians and reptiles, are coldblooded vertebrates. **Coldblooded** animals do not maintain a constant temperature. Their temperature changes when the temperature of the environment changes.

With or Without Bones If you decided to be a fish for a day, you would have another decision to make. Would you want to be a jawless fish, a cartilaginous fish, or a bony fish? Before you make such a decision, you may want to find out more about each type of fish.

Figure 14–1. The seas abound in strange and beautiful creatures, such as the hammerhead shark (above) and the betta fish (right).

TEACHING STRATEGIES

● **Process Skills:** *Comparing, Applying*

Explain to the students that jawless fishes and cartilaginous fishes both have skeletons made of cartilage; they do not have bones. Then ask the students to explain why, if both fishes have skeletons made of cartilage, they are not both classified as cartilaginous fishes. (Cartilaginous fishes have jaws; jawless fishes do not.) Ask the students to predict which group of these fishes is most primitive. (jawless) Point out to the students that the development of a jaw marked a great evolutionary advancement in vertebrates.

DISCOVER BY Observing

Observe a fish in a fishbowl or an aquarium. Describe how the fish swims. How many fins does the fish have? How does it use its fins in swimming? What other characteristics or behaviors can you identify? Sketch the fish in your journal, and write notes based on your observations of the fish.

Jawless Fishes

Jawless fishes are the least developed vertebrates living today. This group once had many members but now includes only two types of fishes—lampreys and hagfish.

Jawless fishes could also be called *boneless fishes* because they have no real bones. Their skeletons are made of cartilage, and as a result, they are very flexible. Both the lamprey and the hagfish have long, round, tubelike bodies. Each fish attaches itself to another fish and sucks out the blood and body fluids of the host fish. Hagfish also eat the tissue of the host. While lampreys attack healthy, living fishes, hagfish usually feed on dead or dying animals.

Figure 14–2. The mouth of the lamprey (left) is perfectly suited to its feeding behavior. The hagfish (right) has no bony skeleton.

Cartilaginous Fishes

Like jawless fishes, *cartilaginous fishes* have skeletons made of cartilage. Unlike jawless fishes, however, they have jaws, which are also made of cartilage. Instead of the small, round scales that cover most fishes, cartilaginous fishes have sharply pointed scales. These scales make the fishes' skin feel rough, like sandpaper.

Sharks, rays, and skates are examples of cartilaginous fishes. Sharks have a fish shape with pectoral, pelvic, dorsal, and caudal fins. Skates and rays do not look like fish at all. Instead, their bodies are flattened. Their shape allows skates and rays to glide through the water in a graceful manner. They almost look as if they are flying through the water.

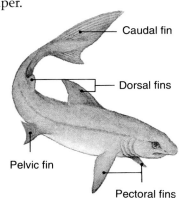

Figure 14–3. The positions of its fins enable a shark to change directions quickly as it pursues its prey.

DISCOVER BY Observing

Suggest that the students try to include a feeding period in their observations. Remind the students not to disturb the fish in any way. The students are to observe the normal behavior of the fish. Encourage them to share their observations with their classmates.

PERFORMANCE ASSESSMENT

After the students complete the *Discover By Observing* activity, ask them what structural characteristics they observed that allow the fish to live in water. Use their responses to evaluate their ability to observe, record, and infer from data.

INTEGRATION— Language Arts

Point out to the students that the scientific name for the class of jawless fishes is *Agnatha*. Write *Agnatha* on the chalkboard. Ask the students to use a dictionary to find out the meanings of the prefix *a* and the word root *gnath*. Then ask the students how the word parts describe the class of jawless fishes. (The prefix *a* means "without"; *gnath* means "jaws.")

TEACHING STRATEGIES, continued

● **Process Skills:** *Inferring, Generating Ideas*

As the students examine the photographs of a shark, point out these interesting facts about sharks: Sharks are cartilaginous fishes, but their skin and teeth are made of bonelike material; the skin of a shark is covered by scales, and each scale has a backward-pointing spine covered by enamel (this enamel is similar to the enamel that covers teeth); even though the skin of a shark looks smooth, it is rough to the touch; a shark may have up to several thousand teeth in its mouth, arranged in rows; as a shark's tooth wears out, a new tooth moves forward. Ask the students why a constant supply of new, sharp teeth is an adaptive feature of sharks. (Most sharks are predators; it is to their advantage to have a constant supply of sharp teeth for catching and cutting prey.)

● **Process Skills:** *Inferring, Applying*

Point out to the students that, although the sense of smell may be the strongest sense for fishes, eyesight is important as well. Most fishes are nearsighted because of the shape of their eyes, although the sight of some sharks is as good as that of

---

INTEGRATION—
Economics

Refer to the map and information that the students compiled during the Multicultural Connection activity in the chapter opener. Point out that millions of people depend on fishes for their basic nourishment and enjoy fishing as a livelihood and sport. Explain to the students that it is known that fishes have been kept in captivity since ancient times; for example, fishes were kept in ponds for a food source. Encourage the students to list ways in which fishes have affected their lives. Have the students place the lists in their science portfolios.

BACKGROUND INFORMATION

Emile Victor Rieu (1887–1972) established the Oxford University Press in India and published classical works for Penguin Books, including his own translation of *The Odyssey*. His book of children's poems, *The Flattered Flying Fish and Other Poems*, was published in 1962.

① The students should compare the size of the ray to the size of the diver to make their estimate.

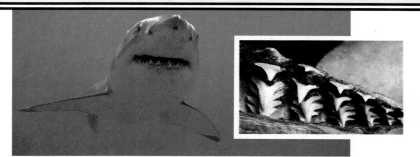

Figure 14–4. A shark's powerful jaws contain several rows of razor-sharp teeth.

Figure 14–5. Despite a threatening appearance, the manta ray is actually a rather shy and gentle animal. Estimate the size of the manta ray shown. ①

Guess Who's Coming to Dinner? Have you ever seen a movie about shark attacks on humans? At first, all seems fine as swimmers enjoy the water. Then threatening music builds. Suddenly, a shark attacks! Contrary to popular belief, however, most sharks avoid humans and are not a threat to them. In fact, the chance of a person being attacked by a shark is less than the chance of a person being struck by a car while crossing a street. Even so, swimmers should leave the water if a shark is spotted. The diet of most sharks includes crustaceans, mollusks, and other fishes. In the following poem, E. V. Rieu takes a light-hearted look at the feeding habits of a shark.

The Flattered Flying Fish

Said the shark to the Flying Fish over the phone:
"Will you join me tonight? I am dining alone.
Let me order a nice little dinner for two!
And come as you are, in your shimmering blue."

Said the Flying Fish: "Fancy remembering me,
And the dress that I wore at the Porpoises' tea!"
"How could I forget?" said the shark in his guile:
"I expect you at eight!" and rang off with a smile.

She has powdered her nose; she has put on her things;
She is off with one flap of her luminous wings.
O little one, lovely, light-hearted and vain,
The Moon will not shine on your beauty again!

Rays of the Sea The flying fish became a meal for the shark, but would she have been a likely meal for a ray? Rays usually feed on creatures such as clams, fishes, and even plankton. Like most sharks, most rays are harmless to humans. For example, the large manta ray shown in the picture is not dangerous to humans even though it weighs as much as 1360 kg. The stingray, however, produces painful wounds if its spiny tail strikes a swimmer. The spine at the base of the tail gives off a venom that can be fatal to humans. Other rays can produce an electric shock of as much as 220 volts.

humans. Some fishes have color vision. Although fishes do not need eyelids and tear ducts because they live in water, some fishes do venture onto dry land. Many fishes have developed land adaptations such as a thick layer of clear skin over the eyes, a layer of pigment on the eyes, and the ability to roll their eyes back into moist sockets. Ask the students to think of the environmental conditions on land from which fishes that move onto land have had to protect their eyes. (Fishes have had to protect their eyes from drying out by the air, from being injured by objects, and from sunlight that is more intense than the sunlight that penetrates the water.)

- **Process Skills:** *Inferring, Generating Ideas*

Ask the students to examine the internal and external structures of the bony fish shown in the diagram. Point out that, unlike the bony fishes, cartilaginous fishes do not have a swim (or gas) bladder, and as a result, cartilaginous fishes do not have an internal means of adjusting their depth in the water. Ask the students to infer how a cartilaginous fish such as a shark avoids sinking. (Most sharks have to keep moving forward all the time to avoid sinking.)

Bony Fishes When you hear the word *fish*, you probably do not first think of lampreys, sharks, or rays. You are more likely to think of trout, perch, salmon, codfish, catfish, and flounder. These are examples of *bony fishes*.

Most bony fishes are ray-finned fishes. *Ray-finned* simply means that the paired fins of the fish have long, bony rays. These rays allow the fins to fan out and to hold their shape. If you have ever held a bony fish, you may have noticed that you could lift the fins up and fold them back down, much as you might open and close a fan.

The illustration shows the external features of bony fishes, including the fins. Find the *operculum*. This is a bony plate that covers and protects the gills. Note that there are no eyelids. Since fishes are in water, they have no need for eyelids and tear ducts, which protect and moisten the eyes of land animals. Also notice the openings on the snout of the fish. These openings are like your nostrils. The sense of smell seems to be a fish's strongest sense.

A fish can control its movements and receive messages about its environment by using two unusual structures. A fish has a swim bladder that fills with or loses air. This structure allows the fish to rise and sink in the water. A fish also has a lateral line extending the length of the side of the body. The lateral line is sensitive to vibrations that occur in the water and permits the fish to sense nearby movement and the presence of food.

When a fish swallows a worm, an insect, or some other piece of food, the food goes through a digestive system. The system consists of an esophagus, a stomach, and an intestine. In addition, there is a liver that helps digest fats and produce proteins. Waste matter from digestion passes out of the intestine through the anus.

Figure 14–6. Although bony fishes vary greatly in appearance, they all have the same basic internal and external structures.

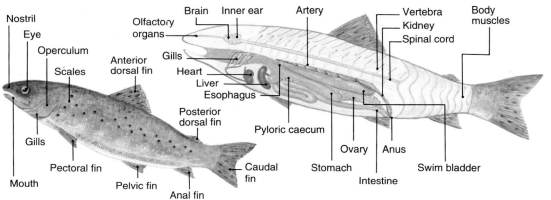

REINFORCING THEMES— *Changes Over Time*

A coelacanth (SEE luh kanth) is a lobe-finned fish that at one time was thought to have been extinct for approximately 60 million years. Scientists think that four-legged animals evolved from lobe-finned fishes. Ask interested students to research this topic and report their findings to the class. Ask them to include a sketch of a fossil coelacanth or a picture of a living coelacanth.

SCIENCE BACKGROUND

Cartilaginous fishes, as well as bony fishes, have a lateral line system extending along both sides of the body that is filled with sensory receptors. These sensory receptors detect pressure waves caused by the movements of other organisms in the water. Sharks also have pits on their snouts that are sensitive to the minute amounts of electricity that are produced by living things. This sense is used by sharks to locate prey. In fact, some sharks have difficulty locating baited pieces of dead meat because the dead meat does not give off any electrical impulses.

 **LASER DISC**

3542

Grunt, blue striped; scales

SECTION 1 **375** ◀

TEACHING STRATEGIES, continued

● **Process Skills:** *Predicting, Generating Ideas*

Remind the students that the gills of fishes function not only to absorb oxygen from the water, but also to eliminate carbon dioxide (the waste product of breathing) by means of diffusion. Explain that the gills of ocean fishes also excrete the excess salt that is taken in with salt water. Ask the students to predict why it is important for ocean fishes to be able to excrete excess salt. (Because they live in a saltwater environment, these fishes take in more salt than they need to survive. Yet, the cells of their tissues still require water for survival. By excreting excess salt, ocean fishes are able to obtain the water their cells need and not accumulate a high salt content in their tissues.)

● **Process Skills:** *Observing, Inferring*

Ask the students to examine the frogs shown in Figure 14–8. Point out that frogs exist in a wide variety of colors and sizes. Have the students look at the extended bubblelike throat sac of one of the frogs, then ask the students what role this feature

---

ONGOING ASSESSMENT

▼ **ASK YOURSELF**

Cartilaginous fishes have skeletons made of cartilage; bony fishes have skeletons made of bone.

MEETING SPECIAL NEEDS

Second Language Support

Have the students look up the derivation of the word *amphibian* in a dictionary. Ask them how an understanding of the meaning of *amphibian* can help them remember why the word names this group of vertebrates. (*Amphi-* means "double, on both sides" and *bios* means "life"; amphibians live a "double life." The name alludes to the fact that amphibians live on land and in water.)

✧ **Did You Know?**
The longest recorded jump for a frog is 6.5 m.

① The color of the poison-arrow frog warns predators that the frog is poisonous; the color of the wood frog enables the frog to blend in with tree bark; the color of the green tree frog camouflages the frog among the leaves of trees. The colors of wood frogs and green tree frogs help protect them from predators.

▶ **376** CHAPTER 14

Figure 14–7. Fish eggs are often fertilized externally. As this male grunion swims over the eggs, he sprays them with milt.

Figure 14–8. Frogs, such as the poison-arrow frog, the wood frog, and the green tree frog, are found in many different parts of the world. What adaptations does each frog display? ①

A fish obtains oxygen from the water that continually passes over its *gills*. The gills contain many blood vessels that absorb the oxygen from the water. The oxygen passes through the thin cell membranes of the blood vessels and is carried by the blood to all parts of the fish's body. The blood is pumped by a simple two-chambered heart.

Bony fishes reproduce sexually. The female lays eggs, and the male fish sprays them with a fluid called *milt*. Milt contains sperm. Fishes lay many more eggs than most other vertebrates because fishes' eggs are usually unprotected. Some fishes do guard their eggs, but most do not protect them. The only other vertebrates that lay such large numbers of unprotected eggs are the amphibians.

▼ **ASK YOURSELF**

How do bony fishes differ from cartilaginous fishes?

Amphibians

Frogs are probably the best-known *amphibians*. Like all amphibians, frogs have successfully made the adjustment to living on land, but they still need a water environment.

Although adult frogs have lungs and are able to breathe oxygen from the air, they spend a great deal of time submerged in water. The skin of a frog is very thin, and oxygen from water can pass directly through the skin cells. Frogs and other amphibians can also absorb oxygen through the linings of their mouths.

You have seen that fishes are specifically designed to function in water. Over time, some sea creatures developed characteristics that enabled them to survive partly on land. What changes were needed for them to move from the water to the land and back again? The ancestors of today's amphibians developed some, but not all, of the adaptations needed to survive partly on land. What kind of adaptations do you think these animals had to make? The following activity may help you decide on the types of changes that were necessary.

may play in the mating of frogs. (The male uses this throat sac to make a mating sound to attract females.) Also mention to the students that the sound made by one species of frog is different from the sound made by a frog of another species.

• **Process Skills:** *Inferring, Generating Ideas, Comparing*

Point out to the students that some frogs are poisonous, and poisonous frogs typically are brightly colored in red or blue or have other bold coloration patterns. Ask the students to explain why a typical nonpoisonous frog will be green, gray, or brown, while a poisonous frog displays bright colors. (A frog that does not have a toxic form of defense often is camouflaged for protection and thus its colors match the color of leaves, dirt, or its surroundings. The bright colors of a poisonous frog warn predators to avoid it.)

ACTIVITY

How would you redesign a fish for life on land?

Process Skills: Comparing, Inferring

Grouping: Groups of 2

Hints

The use of an aquarium and fishes is optional.

Procedure

1. The students should explain how each structure of a fish is adapted for underwater survival. For example, gills allow fishes to extract oxygen from water.

2. The students might suggest that a fish would need a life-support system that recreates the conditions of its natural environment.

▶ **Application**

1. Humans can explore the underwater environment by using snorkeling equipment, scuba gear, or submersible vessels.

2. The temperature of coldblooded organisms such as fishes changes to match the environment's. Adapting to rapid temperature changes is easier for warmblooded organisms such as humans.

ACTIVITY

How would you redesign a fish for life on land?

MATERIALS
paper, pencil, aquarium, fishes

PROCEDURE

1. List the ways that fishes are specifically designed for life in the water.
2. Imagine that you must create a method for a fish to enter the air-breathing world but still spend most of its life in the water. Design ways to overcome problems resulting from fish structure. For example, fishes have no eyelids or tear glands because these structures are not needed in an underwater environment. Fishes have a protective covering of slime on their scales that would quickly dry out in the air. Fishes are also coldblooded.

APPLICATION

1. List some ways in which humans have prepared themselves for entering and exploring the underwater environment.

2. Humans can move from a very hot environment to a very cold one without experiencing much discomfort. Why do fishes have problems with rapid changes in temperature?

What changes did you make in your redesigned fish? Your fish needed a way to take in oxygen from the air, but it still had to be able to take in oxygen when it returned to the water. After the changes you made, could your fish return to water without drowning? Your fish would probably need legs or some sort of muscle development so that it could move on land. Your redesigned fish may resemble amphibians in some ways. Amphibians are adapted for life on land and in water.

Leapfrog You are probably familiar with many of the external features of frogs. A frog has four legs, the back legs being longer and stronger than the front legs. Its large, round eyes bulge from the top of its head. Because its nostrils are also on the top of its head, a frog can breathe even when almost totally submerged in water. Look at the three frogs in the photographs on the previous page. How are they alike and how are they different? ②

Have the students place their designs in their science portfolios.

② The frogs are similar in structure and different in color.

SECTION 1   **377** ◀

TEACHING STRATEGIES, continued

● **Process Skills:** *Observing, Inferring, Applying*

Have the students observe the internal anatomy of the frog shown in the diagram. Ask them to identify an internal feature of the frog that reflects an important adaptation to life on land. (Responses might suggest that the development of lungs made the breathing of air possible.) Then ask the students to identify how the legs of the frog may also reflect an important adaptation to life on land. (Responses might suggest that the hind legs are long and strong, making movement on land possible.)

● **Process Skills:** *Inferring, Applying*

Point out that toads tend to be more terrestrial than frogs, and that some toads even burrow into the ground during the day. Ask the students why toads have drier, bumpier skin than frogs and why toads exhibit such behavior as burrowing in the ground. (Frogs have a thin, moist skin that must be in close contact with water in order to avoid loss of fluid. Toads have a thicker skin to help retain body fluid. By burrowing into the ground during the heat of the day, toads avoid excessive water loss from their skin.)

INTEGRATION—Language Arts

Ask the students to use reference materials to discover interesting statistics and facts about frogs. For example, the students might find the names of the largest frog, the smallest frog, and the female frog that lays more than 25 000 eggs. Ask the students to organize their findings in a chart in their journal. You might have the class use the collected information to create a newsletter called *Frog Facts*.

> ✦ **Did You Know?**
> Touching the dry, warty skin of a toad does not cause a person to get warts. However, some toads do give off a poison that can irritate or cause illness. The unpleasant taste of this poison is the toads' protection against predators. But the poison has no effect on one predator, the hognose snake.

LASER DISC
3638
Frog life cycle

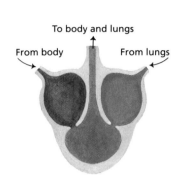

Figure 14–9. The internal anatomy of the frog is shown. Note the three-chambered heart, which is more sophisticated than the two-chambered heart of the fish.

Figure 14–10. Toads such as this *Bufo* do not live in the water, but they must return to the water to breed.

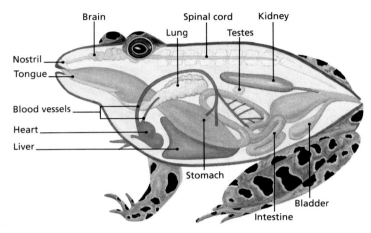

Frogs and other amphibians have skeletons of bone. Because they are vertebrates, they have a backbone. Their small brain is protected by a bony skull, and their limbs are supported by bones. Hinged joints enable frogs to bend and move their limbs.

The internal organs of the frog are similar to those of more complex vertebrates. The liver is the largest of these organs. It produces bile and aids in digestion. Blood is pumped through the frog's body by a three-chambered heart, which is more advanced than the two-chambered heart of the fishes, but not as advanced as the hearts of birds and mammals.

Frogs and other amphibians reproduce sexually. The eggs are fertilized outside the female's body. The female produces large numbers of eggs that fill her body cavity before they are released. During mating season, the male frog grasps the female from the back. The female releases the eggs, and the male sprays them with sperm.

A Bumpy Frog? Toads, while they appear similar to frogs, are quite different. Toads are a little fatter, and their skin is bumpier and drier. Toads do not spend as much time in water as frogs do. In fact, many toads have adaptations that allow them to live in dry areas where a frog would die. Some toads even live in the desert. However, toads must always return to water to reproduce.

Frogs are the subject of many stories, poems, and cartoons. Kermit the Frog and the Frog Prince may have been childhood favorites of yours. In the following activity you can create some of your own frog characters.

- **Process Skills:** *Inferring, Applying*

Remind the students that in Chapter 13 they learned about metamorphosis. Have the students recall the stages in the life cycle of a butterfly and a grasshopper. Ask the students to examine the life cycle of a frog and compare the metamorphosis experienced by frogs to that experienced by butterflies and grasshoppers. Point out that not all frogs and toads pass through an aquatic larva stage in their metamorphosis to the adult stage. Some frogs undergo their larval stage in or on the body of one of their parents. Ask the students to explain why this variation in development may exist. (Some species live in a moist habitat such as a rain forest, but not near a body of water.)

Discover by Writing

Make a list in your journal of frog characters you remember. Tell some fact about each frog that made it special. For example, Kermit's eyes were once made of Ping-Pong balls. Create a frog character of your own. Name your frog and tell facts that make it special. Draw a picture of your frog character or write a nonsensical poem about it.

Just a Tad Different Recall that a butterfly undergoes great changes on its way to the adult stage. From egg, then caterpillar, then pupa, the butterfly emerges. This type of dramatic change is called *metamorphosis*. Frogs and toads also develop through metamorphosis. Frog and toad eggs hatch at various times from a few days to several weeks after fertilization. Water temperature has some influence on the length of time—warmer water apparently speeds up the hatching process somewhat.

The young frogs and toads that emerge from the eggs do not look at all like adults. They look a great deal like small fish. The young are called *tadpoles*. A tadpole has gills and must stay in the water. Gradually, a tadpole develops legs, its tail is absorbed, its lungs develop, and the animal becomes an adult frog or toad. In some cases, metamorphosis can take several years to complete. As the tadpole develops into the adult form, it changes its diet from plant life to small animals, such as insects. Using the illustration below as a guide, describe the life cycle of the frog. ①

Figure 14–11. A tadpole may look like a small fish, but it will soon develop legs and lose its tail.

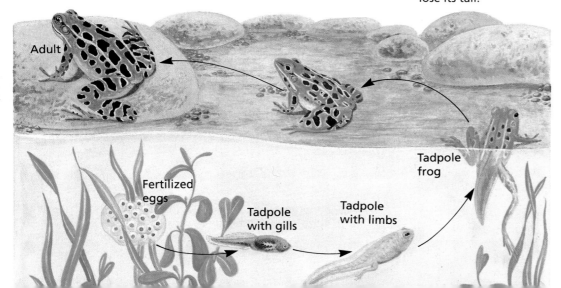

DISCOVER BY Writing

Encourage the students to share their lists, drawings, and poems with their classmates. Have a volunteer compile a class list of all the frog characters that the students remembered.

SCIENCE BACKGROUND

There is a great variety in the means of reproduction and the development of different species of frogs. Some frogs lay their eggs directly in water. Others place their fertilized eggs in foamy nests located in vegetation that overhangs water, thus ensuring that the tadpoles will fall into the water when they emerge. Some frogs carry their eggs in pouches on their backs or in their mouths.

✧ Did You Know?

Frog eggs are black and white. The white portion of a frog egg is the yolk, or food, for the developing embryo, and the black portion is the developing embryo.

① The students' descriptions should begin with the fertilized eggs and proceed through each stage to the adult frog.

SECTION 1 379

TEACHING STRATEGIES, continued

● **Process Skills:** *Comparing, Classifying/Ordering*

Have the students examine the salamander in the photograph and note the caption. Ask the students to describe what features they would look for to determine whether a cold-blooded vertebrate was a salamander (amphibian) or a lizard (reptile). (Salamanders are amphibians; thus, they have a smooth, moist skin without scales, do not have claws on their feet, and are usually found near water. Lizards have dry, scaly skin and claws on their feet; they are found in a wide variety of habitats, including dry areas such as deserts.)

● **Process Skills:** *Inferring, Generating Ideas*

Point out to the students that reptiles were the first vertebrates that were truly terrestrial because they did not require water for reproduction; nevertheless, many species of reptiles evolved to live in a water environment. Ask the students to explain why reptiles, which were able to live independent of water, evolved many aquatic species. (Responses might suggest that many reptiles began eating animals that lived near or in the water just as seals, which eat fish, have become aquatic mammals; because so many kinds of reptiles evolved on

SCIENCE BACKGROUND

Caecilians are long-bodied amphibians without legs and tails. Although they resemble earthworms, caecilians are not related to the segmented worms. One unique feature of caecilians is their inwardly curving teeth that are modified for holding and cutting such prey as earthworms and insects.

ONGOING ASSESSMENT
ASK YOURSELF

An adult frog has lungs and legs; a tadpole has neither.

> ✦ **Did You Know?**
> Some scientists hypothesize that dinosaurs were warmblooded animals. In addition, recent fossil evidence found in nests indicates that at least some dinosaurs cared for their young, unlike most cold-blooded animals.

LASER DISC
3638

Snake, body temperature; comparison with humans

Figure 14–12. A salamander (top) and a caecilian (bottom) are amphibians. Yet salamanders are often mistaken for lizards, and caecilians are mistaken for earthworms.

Cousins to the Frog Salamanders are quite different from frogs and toads. In fact, a salamander looks more like a lizard than a frog or a toad. Sometimes it is difficult to remember that these lizardlike creatures are amphibians. Unlike lizards, salamanders have smooth skin with no scales, and they have no claws on their feet. As the photograph shows, salamanders have a tail, and their legs grow from the sides of the body. Most salamanders live near water, and most return to the water to reproduce. Some species do lay their eggs on the ground, but the eggs must always be kept moist. Unlike the eggs of frogs and toads, salamander eggs are fertilized in the body of the female.

Caecilians are even less like frogs and toads than salamanders are. In fact, caecilians resemble earthworms more than they do frogs, toads, or salamanders. Compare the photograph of a caecilian with those of a salamander and a frog.

ASK YOURSELF

Name two structures found in an adult frog but not found in a tadpole.

Reptiles

The fossil record suggests that the earliest organisms on the earth lived in water. Later, amphibians appeared, spending part of their lives in water and part on land. The first vertebrates to be true land dwellers were the *reptiles*.

Reptiles do not have to protect their skin from drying out, and they do not have to return to water to reproduce. Many reptiles, such as lizards, snakes, and tortoises, live in dry areas. However, crocodiles, alligators, and many turtles prefer watery environments.

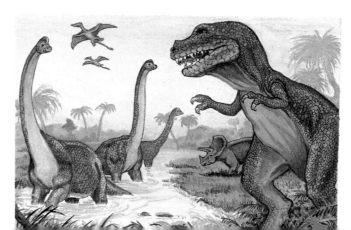

Figure 14–13. The name *dinosaur* means "terrible lizard," yet many dinosaurs were small plant eaters that ran from danger.

land, competition for food drove some species to eventually adapt to life in the water; as amphibian species became more scarce over time, the aquatic habitat may have become more available to reptiles.)

● **Process Skills:** *Inferring, Analyzing*

Explain that reptiles that lay eggs with shells perform internal fertilization and do not have to lay their eggs in water. Ask the students what feature of a reptile egg necessitates internal fertilization and frees reptiles from needing water for the development of their eggs. (When the eggs are laid, they are covered with a leathery shell. Thus, the sperm has to reach the egg before the shell has formed. This shell affords the developing embryo some protection from drying out in a land environment.)

EXTENSION

Explain to the students that if fishes that live in the depths of the sea are brought to the surface, they will literally explode. Challenge the students to discover why this reaction occurs and to share their findings with the class.

In a While, Crocodile
Reptiles have hard, leathery scales covering their bodies. The scales of alligators and crocodiles are so hard that they are more like plates. If you have ever spent time looking at alligators and crocodiles at the zoo, you have probably heard other visitors trying to decide which of the animals were alligators and which were crocodiles. Experts say you can tell the difference by observing the snouts or the teeth. Observing snouts seems like the safer method. Alligators have broad, somewhat flat snouts. The snout of a crocodile is much narrower. Sometimes you can see two bottom teeth of the crocodile even when the mouth is closed.

In a poem by Ogden Nash, Professor Twist insists on distinguishing alligators and crocodiles despite unusual conditions.

> **The Purist**
>
> I give you now Professor Twist,
> A conscientious scientist.
> Trustees exclaimed, "He never bungles!"
> And sent him off to distant jungles.
> Camped on a tropic riverside,
> One day he missed his loving bride.
> She had, the guide informed him later,
> Been eaten by an alligator.
> Professor Twist could not but smile.
> "You mean," he said, "a crocodile."

Eggs, Leathery Side Up
Many reptiles, including turtles, alligators, and crocodiles, lay eggs covered with shells. Snakes are unusual reptiles because, although some lay eggs, others bear live young. Reptile eggs are fertilized inside the female's body. The shells are usually not as breakable as those of bird eggs. Since the shells of most reptile eggs are somewhat leathery, they bend and tear rather than shatter during hatching.

Get That Bug
Of the many species of lizards, most are harmless, and, in fact, do a good job of controlling small, pesky insects. The Gila (HEE luh) monster and the beaded lizard are the only venomous lizards. *Venomous* animals are able to inject a poison into victims by biting them.

Alligator

Crocodile

Figure 14–14. Describe the observable differences between the alligator and the crocodile. ①

Figure 14–15. Komodo dragons are the largest living lizards. They are native to islands of Indonesia. They are aggressive animals and can be dangerous to humans.

① The alligator's snout is broad and somewhat flat; the crocodile's snout is much narrower. Two teeth of the crocodile may be visible when its mouth is closed.

SCIENCE BACKGROUND
Crocodiles may look primitive, but they are actually the most advanced of the reptiles in both physical structure and behavior. Crocodiles have a complete division between the ventricles of their hearts. As a result, oxygenated and deoxygenated blood do not mix in the heart, and more oxygenated blood can reach the brain and the body. Crocodiles exhibit such advanced behavior as parental care of the young; in some species, both parents care for the young.

BACKGROUND INFORMATION
Ogden Nash (1902–1971) sold his first poem to *The New Yorker* magazine in 1930 and went on to write twenty books of humorous verse. He is famous for his unusual rhymes; he often made up words to fit the rhyme pattern.

 The bite of venomous snakes and other reptiles may require an injection of antivenin. Antivenin is a substance used to counter the effects of animal venom. Many antivenins are manufactured in interesting ways. Challenge the students to discover more about antivenins.

GUIDED PRACTICE

Ask the students to name the three classes of fishes and describe the distinguishing features of each class.

INDEPENDENT PRACTICE

 Have the students answer the Section Review and Application questions. Ask the students to name the coldblooded vertebrate that they found most interesting during their study of this section and to list several reasons why they chose that vertebrate.

EVALUATION

 Ask the students to make a chart that lists and compares the features that are common to fishes, amphibians, and reptiles, and the features that differentiate fishes, amphibians, and reptiles. Have the students place their charts in their science portfolios.

RETEACHING

Ask the students to recall the difference between the skeletons of cartilaginous fishes and those of bony fishes. Point out that humans also have cartilage and bones in their bodies. Have the students feel cartilage by gently feeling their outer ear structures. Have them feel bone by gently squeezing their fingers.

---

① The students should note the differences in the color patterns.

ONGOING ASSESSMENT
▼ ASK YOURSELF

Reptiles do not have protect their skin from drying out; they do not have to return to water to reproduce. Reptiles can live entirely in a land environment.

SECTION 1 REVIEW AND APPLICATION

Reading Critically

1. Lamprey: Jawless fishes have skeletons of cartilage and no jaws. Shark: Cartilaginous fishes have skeletons of cartilage and jaws. Perch: Bony fishes have skeletons of bone.

2. Amphibians' skin dries out quickly without water, and they need water for reproduction.

3. The students might describe lizards, snakes, tortoises, crocodiles, alligators, or turtles.

Thinking Critically

4. The bright color of poisonous amphibians warns would-be predators.

5. Reptile eggs are larger than amphibian eggs because they contain more stored food. As a result, the female reptile cannot lay as many eggs at one time as a female amphibian, whose eggs are much smaller.

6. The students might relate the development of fins and gills in fishes, lungs and legs in amphibians, and shell-covered eggs in reptiles to conditions in the animals' environments.

▶ **382** CHAPTER 14

Figure 14–16. The nonvenomous king snake (top) looks much like the venomous coral snake (bottom). What differences can you observe? ①

No Need for Stockings Snakes are reptiles without legs. The jaws of most snakes can unhinge, or separate, from one another. An unhinged jaw allows the snake to swallow animals that are actually larger around than itself. The photograph below shows a snake eating an egg with the roof of its mouth. The skin of a snake is not wet and slimy; it is dry and patterned with scales.

Snakes frighten many people—possibly because there is a lot of false information about snakes. Although there are several venomous snakes in the United States, most snakes are harmless. The water moccasin, copperhead, rattlesnake, and coral snake are the best-known venomous snakes in the United States.

Figure 14–17. An egg-eating snake does not actually swallow an egg whole. It cracks the egg open with its upper jaw and swallows the contents.

 ASK YOURSELF

How are reptiles different from amphibians?

SECTION 1 REVIEW AND APPLICATION

Reading Critically

1. Give an example of a jawless fish, a cartilaginous fish, and a bony fish. Tell how they are alike and how they are different.
2. Give two reasons why amphibians need to live near water.
3. Describe four different types of reptiles.

Thinking Critically

4. Poisonous amphibians are usually brightly colored. How does this coloration benefit the amphibian and other animals?
5. Explain why reptiles do not produce as many eggs as amphibians do.
6. Describe the adaptations that probably occurred in the evolution of fishes, amphibians, and reptiles.

CLOSURE

Cooperative Learning Ask one member in each group to identify an important term or concept from the section. Have the other members discuss that term or concept. Have the members exchange roles until all the material in the section has been discussed.

INVESTIGATION

Identifying Foods That Tadpoles Need

Process Skills: Experimenting, Inferring, Identifying/Controlling Variables

Grouping: Groups of 3 or 4

Objectives
- **Observe** the effect of eating meat on the development of tadpoles.
- **Interpret** and **record data** from an experiment.
- **Infer** the importance of a balanced diet.

Pre-Lab
Ask the students to read the procedure and then predict how the tadpoles in the two containers will look after the investigation.

Hints
Be sure to use only pond water or dechlorinated water in both containers. Do not place the containers near cool drafts or windows during cold weather.

Analyses and Conclusions
1. Yes, all of the tadpoles grow in size, and some of the tadpoles change shape.
2. Each journal entry should state that the tadpoles in container A developed into frogs, while the tadpoles in container B became larger tadpoles but never developed into frogs.
3. The container with meat—container A—is the variable.

Application
1. Without meat in their diet, the tadpoles did not develop normally into adult frogs.
2. The students might answer yes based on an adult frog's diet of insects.

Discover More
A well-balanced diet provides the nutrients for proper cell growth, repair, and development, and also supplies the energy necessary for the maintenance of life processes. A well-balanced diet should include foods from all the major food groups.

Post-Lab
Ask the students to review the predictions they made before they began the investigation. Encourage them to explain any differences that may exist between their predictions and their results.

INVESTIGATION

Identifying Foods That Tadpoles Need

► MATERIALS
- aquaria (2) • pond water or dechlorinated water • rocks • tadpoles (4–6 per liter of water)
- aquarium plants or boiled lettuce • air pump (optional) • ground beef or mealworms

▼ PROCEDURE

1. Pour pond water or dechlorinated water into both containers until the water is about 10 cm deep. Tap water may be used if it is allowed to stand at least 24 hours. Make sure both containers are kept in exactly the same type of environment.
2. Place several rocks in each container so that part of each rock is above the water level.
3. Place an equal number of tadpoles in each container.
4. To both containers, add three or four sprigs of an aquarium plant, such as *Elodea*, or a few lettuce leaves.
5. If an air pump is used, force air into the water at a slow pace. If an air pump is not used, change the water every few days by removing 1/4 to 1/3 of the water and replacing it with fresh pond water or dechlorinated water.
6. Mark one container "A" and the other one "B." To container A, add a few pinches of raw hamburger meat or a few mealworms each day for 10 days. Do not add any meat or mealworms to container B. Before adding fresh food to container A, remove any uneaten food left from the day before.
7. Check the containers each day. Observe the tadpoles closely. Record your observations in your journal.

► ANALYSES AND CONCLUSIONS
1. Do any of the tadpoles grow in size or change their shape?
2. Do you find any differences between the tadpoles in container A and those in container B? Describe the differences in your journal.
3. What is the variable in the experimental setup?

► APPLICATION
1. What happens to the tadpoles that do not have any meat in their diets?
2. Would you describe a frog as a meat-eating animal? Explain your answer.

✳ Discover More
Why is it important for human beings to eat a well-balanced diet? Give examples of foods that should be included in a person's daily diet.

SECTION 1 **383** ◄

Section 2:
BIRDS AND MAMMALS

PROCESS SKILLS
- Communicating
- Interpreting Data
- Evaluating

POSITIVE ATTITUDES
- Caring for the environment
- Curiosity • Enthusiasm for science and scientific endeavor

TERMS
- warmblooded • incubation
- gestation

PRINT MEDIA
Dangerous Mammals by Missy Allen and Michel Peissel (see p. 315b)

ELECTRONIC MEDIA
The Vertebrates: Birds, Coronet (see p. 315b)

Science Discovery Skimmer; feeding Horse; mare and colt Grebes; courting Platypus, duckbilled

BLACKLINE MASTERS
Study and Review Guide
Laboratory Investigation 14.2
Thinking Critically

MEETING SPECIAL NEEDS

Mainstreamed

The students may have difficulty remembering the difference between warmblooded and coldblooded animals. Suggest that they can think of a warmblooded animal as having its own heating and cooling mechanisms to keep its temperature constant, so it is always just "warm enough." A coldblooded animal has no such internal control.

FOCUS

This section presents the concept of warmblooded vertebrates and identifies birds and mammals as the two groups of warmblooded vertebrates. The characteristics, structures, and development of birds are examined. The characteristics and classification of mammals are also presented.

MOTIVATING ACTIVITY

Ask the students to predict whether their heartbeat rate will be faster or slower than that of a bird. Have the students record their own pulse by placing their index and middle fingers on the opposite wrist just below the fleshy part of the thumb and counting the number of pulses in one minute. Point out that this pulse rate is equal to the number of heartbeats per minute. Ask the students to record their prediction and their pulse rate on a sheet of paper. Create a chart on the chalkboard that has two columns, one labeled "Type of Bird" and the other labeled "Heartbeats/Minute." Then add the following information to the chart: chicken, 300; canary, 514;

SECTION 2

Objectives

Summarize how birds are adapted for flight.

Recognize the characteristics of mammals.

Differentiate between birds and mammals in regard to reproduction.

Birds and Mammals

Imagine hiking in a forested area on a warm spring morning. You come to a clearing, and you see a doe and her fawn grazing quietly. You hear birds calling to one another. You see chipmunks racing across the path ahead of you. As you reach a pond, you notice a duck nesting in the reeds. You marvel at the beauty of spring as you walk quietly down the path. You wonder what the lives of the animals you see must be like, and you want to know more about them.

You know that birds and mammals are warmblooded animals. A **warmblooded** animal keeps a nearly constant body temperature regardless of the temperature of its environment. You also know that you are a mammal, and as a mammal, your body temperature stays approximately 37°C. This is true whether you are playing volleyball outdoors or shopping in an air-conditioned store. Because birds use a large amount of energy for flight, their body temperature is slightly higher than yours. A sparrow's average body temperature is 42°C, while that of a thrush is as high as 45°C.

Figure 14–18. Bald eagle and mallard duck in flight

Birds

Birds are easy to identify. They have wings, a bill or beak, two legs, and a covering of feathers all over their bodies. Feathers, besides being pretty, help birds in many ways. They provide birds with a covering that is strong, lightweight, and flexible. The movement of some feathers, called *flight feathers,* helps birds fly. Feathers also help hold in body heat and, in some cases, repel water.

Light as a Feather Feathers grow from the outer layer of a bird's skin and can be compared to the scales on a reptile. A feather has three parts: the vane, the rachis, and the quill. The *vane* is the flat part of a feather, made up of many threadlike structures that are zipped together. The *rachis,* or shaft, supports the vane. The *quill* is the bottom section of the rachis and is attached to the skin of a bird.

TEACHING STRATEGIES

hummingbird, 1000; sparrow, 800. Ask the students why the heartbeat rates of these birds are so much higher than those of humans. (Responses should reflect an understanding of the rapid activities of birds as compared to the activities of humans.)

● **Process Skills:** *Classifying/Ordering, Comparing*

Have the students name as many birds as they can, and ask a volunteer to record the names given by the students. Ask the students to identify characteristics that are common to all the birds listed on the chalkboard.

(Responses might include wings, feathers, and a beak.) Ask the students to identify characteristics that might change from bird to bird. (Responses might suggest size, color, type of feather, type of beak, body shape, or claw type.)

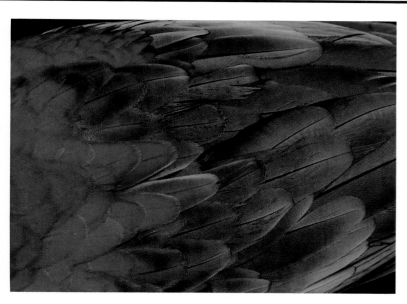

Bird for a Day If you were a bird for a day, what color feathers would you choose? Would bright feathers attract enemies? Would dull feathers be boring? Feather color is the result of *pigment*, the coloring matter in cells and tissues. Each species of bird has characteristic colors. Although it is rare for a bird's color to be affected by its diet, flamingo feathers are an exception. Zoo keepers have discovered that flamingos kept in captivity and away from their natural diets lose their pink color. However, adding shrimp and carrot juice to their diet restores their bright color.

Colored feathers are legally used in craft projects as well as in the fashion industry. However, to protect certain birds, especially songbirds, laws have been passed limiting the right to use or even to own certain feathers.

DISCOVER BY Researching

Research the laws that control what feathers can be used and under what conditions. Check which feathers individuals are able to possess legally. Were special exceptions made for Native Americans who use certain feathers in cultural activities? Write your findings in your journal.

ASK YOURSELF
What are the functions of feathers?

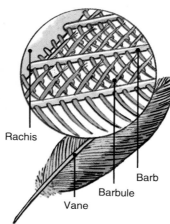

Figure 14–19. Depending on the species, a bird may have up to 25 000 feathers. A bird's feathers help it fly and keep warm.

Rachis
Barb
Barbule
Vane

REINFORCING THEMES— Changes Over Time

Remind the students that birds and other warmblooded animals have developed methods or adaptations for retaining body heat. Feathers, fur or other body hair, and layers of fat all serve to insulate a warmblooded animal's body and help the animal maintain a constant body temperature, regardless of the temperature of its environment.

DISCOVER BY Researching

Help the students find reference materials containing information about endangered and threatened bird species, if they have difficulty locating information about the uses and regulations of feathers.

ONGOING ASSESSMENT ASK YOURSELF

The feathers of birds help retain body heat and repel moisture; provide a lightweight, strong, flexible covering; and help birds fly.

◆ **Did You Know?**
Birds shed their feathers and replace them with new ones once a year. This process, called *molting*, usually takes place after the breeding season. Birds often alternate between a colorful set of feathers and a dull set.

SECTION 2 **385**

TEACHING STRATEGIES, continued

● **Process Skills:** *Comparing, Generating Ideas*

Explain to the students that many birds have the ability to live in a variety of environments. Ask the students to describe the various adaptations birds have developed that enable them to live in many different environments. (The responses might, for example, describe how the feet of birds are adapted for different lifestyles: swimming birds have webbed feet; walking or stalking birds have long legs with toes that can be spread apart; other birds have feet with claws that are adapted for grasping and balancing.)

● **Process Skills:** *Comparing, Expressing Ideas Effectively*

Have the students recall what they learned about reptiles in the preceding section. Then have the students recall the names of five common reptiles. (lizards, turtles, snakes, alligators, and crocodiles, for example) Ask the students to compare birds with reptiles, and discuss the ways in which birds are similar to reptiles. (Birds and reptiles—except for snakes—have legs and feet that are covered with scales. Both birds and most reptiles lay eggs that are covered with shells.)

INTEGRATION—
Mathematics

Some birds eat their own mass in food each day. A human with a mass of 76.5 kg would need to eat 153 kg of food a day, if he or she used energy at the same rate as a hummingbird!

BACKGROUND INFORMATION

John James Audubon (1785–1851) had a lifelong interest in birds and traveled through middle and southern America sketching species of birds in their natural environments. Like many other naturalists of his time, Audubon was interested in discovering new species of birds and learning about their behavior. Though untrained as an artist, Audubon created detailed drawings and paintings of birds at a time when cameras and binoculars did not exist. Unable to publish his work in the United States, Audubon traveled to England and began publishing his master work, *The Birds of America*. His work is still acclaimed today.

LASER DISC

47833–48053

Skimmer; feeding

Figure 14–20. The bones of a bird's wings are similar to those of the human arm. However, the shape and function of the wings are modified for flying.

Figure 14–21. The digestive system of a bird is extremely efficient. Almost all of the nutritious components that can be obtained from food are absorbed into the blood.

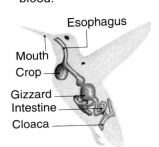

Structure and Reproduction of Birds

All birds have very efficient bodies, and most are adapted to flying. Even a bird's skeleton has special features to help it fly. For example, most birds have lightweight bones that contain air spaces.

Wonderful Wings The most specialized parts of a bird's body are its wings. The wings are also the strongest parts of a bird's body. Although a bird's wing and a human arm and hand are used for entirely different purposes, the bones in them are very much alike. Use the illustration to compare the structure of a bird's wing and the structure of a human arm.

Let's Eat The digestive system of a bird is similar to that of other vertebrates. It includes an esophagus, a stomach, and an intestine. However, birds have unique structures—a crop and a gizzard—that other vertebrates do not have. The esophagus connects the bird's mouth to its crop, which is used to store food. The gizzard grinds the food. This organ is especially important because birds do not have teeth. Waste materials travel through the intestine and leave the bird's body through an opening called the *cloaca* (kloh AY kuh). Use the illustration to trace the passage of food through a bird's digestive system.

The diets of birds vary greatly. Many birds are *herbivores*, eating seeds and other plant parts. However, some are *carnivores*, or meat eaters, and still others are *omnivores*, which eat both plants and animals. Regardless of what a bird eats, it eats large amounts for its size. People who nibble or pick at their food are often accused of "eating like a bird," while people who eat large amounts of food are often accused of "eating like a horse." Actually, in relation to the size of each animal, a bird eats far more food than a horse does. Birds must eat large amounts of food to provide the energy needed for flying.

- **Process Skills:** *Comparing, Evaluating*

Explain to the students that even though different birds eat different things (some are herbivores, some are carnivores, some are omnivores), the digestive systems of all birds are similar and extremely efficient. Ask the students to evaluate the advantages of a bird's crop and gizzard. (The crop is used to store and moisten food that is digested or regurgitated at a later time; the gizzard is a muscle (often containing small rocks swallowed by a bird) that crushes food, making digestion easier for the bird.)

- **Process Skills:** *Observing, Inferring*

Direct the students' attention to the respiratory system of a bird shown in the illustration. Have the students note the many air sacs. Ask the students to infer how the sacs aid a bird's breathing and respiration. (Air sacs increase the supply of air available to the lungs, which in turn increase the supply of oxygen available to the cells.) Explain that flight requires a tremendous amount of energy, and when energy is being consumed, heat is being produced. The air sacs also help cool the bird by allowing cool external air near warm internal organs.

With Every Breath

The illustration shows the respiratory system and heart of a bird. The respiratory system takes in oxygen through the nostrils. The oxygen passes down the trachea to the branched bronchi. Located in front of the bronchi is the syrinx (SIHR ihngks), or voice box, of the bird. From the bronchi, the air enters small spongy lungs. Branching off from the lungs are air sacs that fill with air and make the bird lighter.

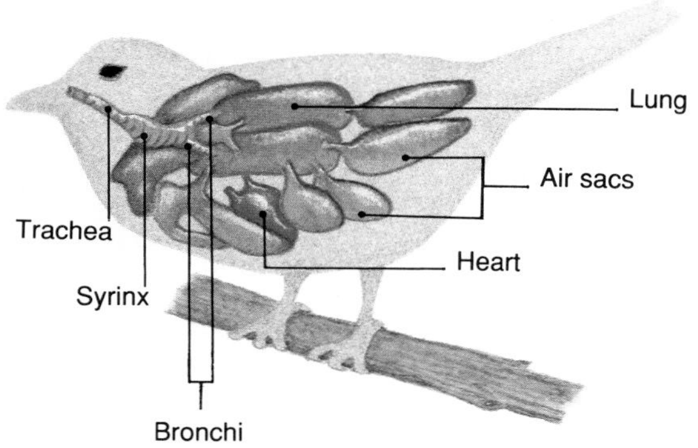

And the Beat Goes On

A bird's heart has four chambers and therefore is the most advanced of any vertebrate heart that you have studied so far. The four-chambered heart allows blood to flow through the lungs for a fresh supply of oxygen before it circulates through the body again.

Figure 14–22. A bird's respiratory system is designed to help the bird in flight. Air sacs attached to the lungs make the bird lighter.

Reproduction

Even the reproductive organs of birds help make the body lighter for flying. In females, a single *ovary* develops as the bird matures. This ovary does not always stay the same size; it gets larger at mating time. In this way, the weight of the female bird is kept to a minimum most of the time.

The male bird produces sperm in reproductive organs called *testes*. After they are fertilized in the female, the eggs move down a tube called the *oviduct*. Albumen, which is the white part of the egg, serves as food for the developing bird. The yolk is also a food source. The egg passes through a shell gland where it is covered with a shell. The egg with its shell leaves the female's body through the cloaca. Egg production is the same whether or not the egg is fertilized. Most eggs sold in stores are not fertilized.

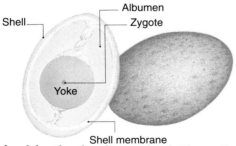

Figure 14–23. The bird's egg is attached to the inside of the shell. The yolk and albumen feed the developing bird until it hatches.

Eggs of Many Colors and Shapes

Bird eggs are found in many different colors. Generally, dark eggs or eggs with spots and blotches are laid by birds whose nests may be easily seen. The colors and spotting are a type of camouflage. White eggs are usually laid by birds whose nests are well hidden or by birds whose mates share the responsibility of staying with the eggs.

REINFORCING THEMES— *Changes Over Time*

In the scientific community, there has been speculation that dinosaurs and birds have a common ancestry, and that birds are more closely related to the dinosaurs than to reptiles. A cross section of dinosaur bones reveals air spaces and pockets very similar to those of birds, while other reptile bones appear solid, very much like human bones. Some dinosaur skeletons have been found with polished rocks in the vicinity of their gizzards. This evidence has led to the speculation that dinosaurs were also warmblooded.

MULTICULTURAL CONNECTION

Louis Agassíz Fuertes (1874–1927) found a way to combine his two loves: nature and drawing. At 9 he discovered John James Audubon's *The Birds of America*, and at 15 Fuertes decided he too wanted to study and draw birds. While attending Cornell University, Fuertes showed his bird drawings to Eliot Coues, a noted ornithologist (a scientist who studies birds), who was impressed by the young man's talent. After college, Fuertes traveled widely to increase his knowledge of birds. He was known for his photographic memory and attention to detail. He illustrated most bird books published between 1896 and 1927.

TEACHING STRATEGIES, continued

● **Process Skills:** *Inferring, Expressing Ideas Effectively*

Remind the students that birds reproduce sexually. Ask the students if they think fertilization precedes or follows the formation of the shell, and have them explain their reasoning. (The shell is formed after fertilization. The sperm could not penetrate the eggshell to fertilize the egg if the shell formed first.)

● **Process Skills:** *Predicting, Applying*

Remind the students that the color of bird eggs can vary from species to species and sometimes includes spots. Ask the students to explain the function of color and spotting on bird eggs. (Colors and spotting may be a type of camouflage.) The students might be interested to learn that spotted or unevenly colored eggs usually have more spots and deeper color at their larger end. This occurs because the egg passes through the oviduct large end first, picking up most of the available pigment as it passes through the tube.

● **Process Skills:** *Inferring, Generating Ideas*

Remind the students that although the eggs produced by different bird species have the same general shape, they do not always have the same specific

---

✧ Did You Know?
The eye structure of birds is different from that of humans. The eye structure of humans allows them to see only one view at a time. The eye structure of birds enables them to see two, sometimes three, views at one time.

★ DISCOVER BY *Observing*
The students should observe that the eggshell quickly begins to deteriorate in the acidic environment of the vinegar solution. They should conclude that diet and environmental conditions can significantly affect the development of eggshells.

★ PERFORMANCE ASSESSMENT
Working in small groups, ask each group to identify and discuss several human actions that are likely to negatively affect the development of bird eggs. Use the discussion to evaluate the students' ability to work cooperatively and generate ideas.

① Both male and female Canada geese parents take turns incubating their eggs.

Figure 14–24. You are probably most familiar with the oval, white eggs of chickens. Yet bird eggs come in a variety of colors and shapes.

Bird eggs come in many shapes and sizes. Birds that have nests on bare rocks or cliffs produce eggs that are very pointed at one end. This shape seems to protect the eggs if they are disturbed. Instead of falling off the rock or cliff, the eggs roll in small circles. If these eggs were shaped like chicken eggs, they would fall as easily as a chicken egg can roll off your kitchen counter. In the next activity, you will find out about other dangers to bird eggs.

Figure 14–25. Both Canada goose parents—male and female—care for the eggs. As one parent keeps the eggs warm, the other searches for food.

DISCOVER BY *Observing*

In addition to natural hazards, such as enemies and weather, bird eggs are also threatened by human-caused pollution. Your teacher will demonstrate the effect of vinegar on an eggshell. If an eggshell can be destroyed so easily through contact with a common household substance, what effect do you think environmental conditions and a bird's diet might have on its shell? ✎

Warm Up Unlike fishes, amphibians, or reptiles, birds are warmblooded. As a result, it is necessary for bird eggs to be kept warm. Keeping eggs warm until they hatch is called **incubation.** Depending on the type of bird, incubating eggs is the job of the female, the male, or both parents. Which sex incubates the eggs of Canada geese? ①

shape. Have the class discuss how some eggs are shaped to protect them from being disturbed. Then ask the class to explain how diet and environmental conditions can influence the development of eggshells. (Brittle, malformed, or otherwise imperfect shells could result from a poor diet or poor environmental conditions.)

● **Process Skills:** *Comparing, Applying*

Ask the students to compare the amount of parental care reptiles give to their eggs with the amount of parental care birds give to their eggs. (Reptiles generally do not care for their eggs; birds tend and incubate their eggs.)

● **Process Skills:** *Predicting, Applying*

The students may be interested to learn that some birds, such as the cowbird, lay their eggs in the nests of other birds. Ask the students how cowbirds benefit from this practice. (By laying their eggs in the nests of other birds, cowbirds do not have to incubate their own eggs. In addition, the other birds take care of the young cowbirds.)

While in the egg, the developing bird is nourished by the yolk and albumen and receives oxygen through the pores in the eggshell. Incubation times vary and may be as long as 80 days for the large white egg of the royal albatross or as short as 12 days for the yellow-and-brown egg of the lark. The green egg of the mallard duck must incubate approximately 26 days, whereas the chalky white egg of the flamingo needs about 32 days. A chicken egg develops in 21 days.

ASK YOURSELF
Describe the reproduction of birds.

Mammals

Spending a day at a zoo, such as the San Diego zoo, shows that humans are interested in other mammals. Which of the animals in the photographs would you expect to see in a zoo?

Who's Who at the Zoo Even though other classes of vertebrates also live in zoos, mammals are often the leading attractions. Close your eyes for a moment and take an imaginary walk around a zoo. Which mammals would you probably stop to look at for the longest time?

Figure 14–26. Mammals include a wide variety of animals, including humans. Shown here are the tiger, opossum, humpback whale, and giraffe.

ONGOING ASSESSMENT
ASK YOURSELF

Birds reproduce sexually. The male bird produces sperm. After fertilization in the female, the eggs move through the oviduct, then through a shell gland, and out through the cloaca. After a period of incubation, the eggs hatch.

Demonstration
Create a solution of food coloring and water that is mixed to the ratio 1 mL food coloring to 40 mL water. Place a hard-boiled egg in the solution, and ask the students to predict how the food coloring will affect the egg white of the hard-boiled egg. After 24 hours, peel the egg and provide the students with an opportunity to examine the appearance of the egg white. The students should note that the splotching of the egg white with color occurred because the food coloring passed through the pores of the eggshell and the shell membrane.

Ask the students to write a brief explanation of the setup and results of the Demonstration, along with their predictions and observations, and place the explanation in their science portfolios.

SECTION 2   **389**

TEACHING STRATEGIES, continued

● **Process Skills:** *Inferring, Analyzing*

Point out to the students that the two characteristics used to classify mammals are the presence of hair somewhere on the body at some time during the animal's life and the presence of mammary glands that produce milk for the animal's young. Ask the students to use these characteristics to help them name five mammals: one that eats insects, one that flies, one that lives in water, one that has a trunk, and one that eats only meat. Then ask the students to describe how each mammal they named is adapted to both its diet and its environment. (Responses might suggest that the insect-eating anteater has a long, pointed snout; the wings of a bat are formed from thin, smooth skin supported by long arms; the limbs of a seal are modified as flippers; an elephant eats plants with a long trunk that has the ability to grasp and tear; a meat-eating lion has sharp teeth for biting and long canine teeth for tearing.)

DISCOVER BY *Writing*

Review the classification criteria for mammals with the students before having them complete the activity. Remind the students that the conditions in their zoo should duplicate as closely as possible the natural environment of each of the animals they choose.

DISCOVER BY *Doing*

You might choose to extend the activity by having the students debate how the tracks left by a running animal might differ from the tracks left by that animal when walking.

★ PERFORMANCE ASSESSMENT

Cooperative Learning Hand out a copy of drawings of the tracks of several different animals that can be found in the students' outdoor environment. Have small groups of students work together and use a classification key to identify the tracks. Use the performance of each group to evaluate its members' ability to use a classification key correctly.

▶ **390** CHAPTER 14

DISCOVER BY *Writing*

If you could safely and appropriately care for mammals in your own private zoo, which ones would you select? In your journal, list 15 mammals you would allow to live in your zoo. Tell why you selected each of the animals. Did the possibility of extinction influence your selection? Draw a layout of your zoo. Write a description of the conditions in your zoo.

Figure 14–27. Experienced trackers can identify what tracks belong to which animal.

In the Human World All animals contribute to the environment in some way. Mammals, perhaps because they are so visible, seem to make major contributions. In some cases, animals greatly improve the quality of human life. Guide dogs for the visually impaired and dogs trained to assist hearing-impaired people provide their owners with the opportunity to enjoy independent living.

Mammals are often present in our lives without letting us see them. Have you ever wandered into wooded areas and seen the tracks left by animals? The tracks tell stories without words about the activities of the animals. In the passage at the beginning of the chapter, the fisher learned much from observing animal tracks. Give some examples of what the fisher learned. Then do the following activity to become more of an expert yourself!

DISCOVER BY *Doing*

Using a book such as *Field Guide to Animal Tracks*, draw and label 20 animal tracks in your journal. Study the tracks often until you become an expert at recognizing them.

Sometimes, when you are riding in a car at night, you might catch the reflected glow of a mammal's eyes. Many mammalian eyes reflect light. The eyes of bears and opossums reflect red light. The eyes of house cats, cougars, and other cats reflect green light. Deer eyes reflect amber light.

Mammal or Not? Two main characteristics are used to classify animals as mammals. One of these is the presence of hair somewhere on the body at some time during the animal's life. The other characteristic is the presence of *mammary glands* that produce milk for the young. The mammary glands give this class of vertebrates its name. Most young mammals need parental care for longer periods of time than other vertebrate young do. Feeding the young from milk-producing glands is part of that parental care. The milk contains water; butterfat; lactose, a sugar; proteins; and various salts.

In some cases, the amount of hair a mammal has on its body has been reduced to only a few bristles. Whales and dolphins are mammals that have just a few bristles of hair around the chin and snout early in their lives. Dolphins actually have no more than eight of these bristles.

Mammals whose hair covering has been reduced usually have thickened skin or even armored plates to protect them. Elephants, for example, have little hair but very thick skin. Armadillos have armorlike skin that protects them from the weather and from enemies. In some animals, hair is associated with the senses, especially with the sense of touch. Whiskers help animals recognize the presence of objects, especially in burrows or other dark places.

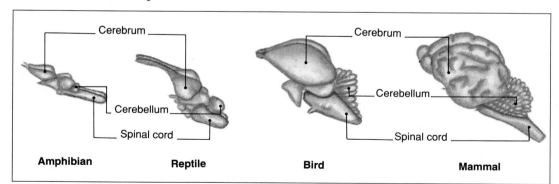

Mammals have a very well-developed brain, which is the center of their nervous system. The *cerebrum*, which is the largest part of the mammalian brain, is involved with intelligence and the organizing of information. The illustration shows amphibian, reptile, bird, and mammal brains. As animals become more complex, the cerebrum becomes more developed. As a result, the human cerebrum is the most developed of all animals.

Figure 14–28. The brain of mammals is much more advanced than that of birds, amphibians, and reptiles.

▼ **ASK YOURSELF**

By what characteristics are mammals classified?

TEACHING STRATEGIES, continued

● **Process Skills:** *Predicting, Applying*

Ask the students to consider the developing embryo of a monotreme, a marsupial, and a placental mammal, and then list an advantage or a disadvantage of monotreme, marsupial, and placental mammal development. (The students might suggest that monotreme eggs are vulnerable to predators when left unattended; marsupials are born underdeveloped and must find their own way to their mother's pouch; placental mammals are more protected during embryonic development and more fully developed at birth.)

GUIDED PRACTICE

Ask the students to describe the characteristics that birds and mammals have in common and the characteristics that differentiate birds and mammals.

INDEPENDENT PRACTICE

Have the students provide written answers to the Section Review and Application questions. Ask the students to make a sequence chart of the events in the pregnancy of placental mammals.

---

① The spiny anteater

SCIENCE BACKGROUND

The only North American marsupial is the opossum. It is about the size of a cat and has a long snout, big ears, and a long, almost hairless tail. An opossum can use its tail to hang upside down from a tree branch. Opossums carry their young in a stomach pouch for two months, then the young ride on their mother's back for several more weeks. Opossums are night-hunting omnivores. When threatened, they lie very still and act as if they are dead; hence, the expression "to play possum."

LASER DISC
3914
Platypus, duckbilled

Did You Know?
Both kinds of monotremes and all but two groups of marsupials live only in Australia and New Guinea and on nearby islands.

LASER DISC
3978
Horse; mare and colt

Figure 14–29. The duck-billed platypus is one of only two mammals that lay eggs. What is the other mammal? ①

Figure 14–30. Marsupials, such as this kangaroo, give birth to premature young. The young climb into the mother's pouch, where they attach themselves to the mammary glands and remain until fully developed.

Types of Mammals

There are three main types of mammals. They are the monotremes (MAHN uh treems), the marsupials (mahr SOO pee uhlz), and the placentals (pluh SEHN tuhlz). These animals are grouped according to how their young develop.

Mammal Eggs? We don't usually think about mammals laying eggs. Most bear live young. If someone tried to sell us an elephant egg, we would know right away it was a joke. But there really are two mammals that lay eggs. The duck-billed platypus and the spiny anteater are members of a primitive group of mammals called *monotremes*. The platypus actually lays eggs in a nest. Can you imagine finding a nest of platypus eggs? Unless you live in Australia or Tasmania, you will never find such a nest. The spiny anteater keeps her eggs in a pouch on the side of her body. After monotreme eggs are hatched, the newborns feed on milk from their mother's mammary glands.

In the Pocket *Marsupials* are mammals with pouches, such as the kangaroo and the opossum. The marsupials are the main group of mammals found in Australia. Marsupials give birth to their young long before the young are fully developed. The tiny, underdeveloped animals must then find their way to the mother's pouch, where the mammary glands are located. The young continue their development in the pouch, nourished by the mother's milk.

On the Life Line *Placental* mammals are those in which the young develop completely within the mother's body. More than 95 percent of all mammals are placental mammals. This group gets its name from the placenta. The *placenta* is an organ that attaches the young unborn animal to the mother. The young mammal is attached to the placenta by the umbilical cord. During development of the young, nutrients and oxygen pass from the mother to the young. At the same time, carbon dioxide and wastes move from the young to the mother.

EVALUATION

 Ask the students to make a comparison-and-contrast chart listing the major characteristics of fishes, amphibians, reptiles, birds, and mammals. Have the students place their completed charts in their science portfolios.

RETEACHING

Write the names *monotreme*, *marsupial*, and *placental mammal* on the chalkboard. Display pictures of a kangaroo, a platypus, and a dog. Have the students match the pictures with the names and explain the differences among the three kinds of mammals.

EXTENSION

Many bird species display fascinating courtship behaviors. Use the Laser Disc to encourage the students to discover more about these behaviors.

 LASER DISC
50536–51140
Grebes; courting

CLOSURE

Cooperative Learning Have the students work in small groups and develop a diagram displaying the characteristics of mammals. Each diagram should identify the three major groups of mammals, include two examples for each group, and describe the development of each group's young.

Figure 14–31. The well-developed young of placental mammals can feed on the milk provided by the mother. Some placental animals, such as humans, are dependent on the mother for a longer time than others, such as dogs.

Placental mammals give birth to relatively well-developed young. Once an egg is fertilized by a sperm in the body of a female, a zygote is formed. The zygote attaches itself to the wall of an organ called the *uterus*. This attachment begins the period of **gestation** (jehs TAY shuhn), or the time when a young mammal is developing within the mother's body. As the zygote grows, it is called an *embryo*.

The young grows inside the mother's body until it is time for birth. The muscular uterus begins to contract and relax, firmly pushing the new mammal out of the mother's body. This process is the same for all placental mammals, including bats, whales, deer, gorillas, and humans.

Another word for gestation is *pregnancy*. The period of gestation varies for different mammals, just as the incubation time of birds does. The gestation period of a mouse is as short as three weeks, while that of an elephant is nearly two years! Gestation time for humans is approximately 40 weeks.

▼ ASK YOURSELF

How are development and birth different in the three types of mammals?

SECTION 2 REVIEW AND APPLICATION

Reading Critically
1. Describe how birds are adapted for flight.
2. Describe the relationships between humans and other mammals.
3. What is the purpose of the placenta?

Thinking Critically
4. What are the advantages of bird parents sharing duties during the incubation period?
5. Most mammal eggs contain a small amount of yolk in comparison to bird eggs. Why do you think this is so?

ONGOING ASSESSMENT
▼ ASK YOURSELF

Monotreme mammals lay eggs; marsupial mammals give birth to young that are not fully developed; placental mammals give birth to young that develop completely within the mother's body.

SECTION 2 REVIEW AND APPLICATION

Reading Critically

1. They have wings, lightweight bones with air spaces, feathers, air sacs around the lungs, and minimum weight in all systems.

2. The relationships between humans and mammals can be beneficial or negative. For example, humans benefit from mammals' companionship; overgrazing by mammals can destroy valuable cropland.

3. A placenta allows a mammal embryo to obtain food and oxygen and discard waste products during its development.

Thinking Critically

4. If bird parents take turns incubating eggs, the eggs are never left unattended and vulnerable to predators.

5. A mammal embryo obtains food through the placenta during its development. The only food supply for a bird embryo is the food stored in the yolk. Therefore, a bird egg must contain a much larger supply of yolk than a mammal egg.

SKILL

Making and Interpreting a Graph

Process Skills: Interpreting Data, Communicating

Grouping: Groups of 2

Objectives
- **Communicate** using a line graph.
- **Interpret** data displayed on a line graph.

Discussion
Emphasize that a graph can be used to combine two sets of related data. A graph shows the relationship between different data and makes interpretation of that data less difficult.

Application
1. At 10:29
2. At 10:10, 10:15, 10:20, and 10:22
3. Responses should reflect an understanding that a line graph makes numerical data easier to compare.

Using What You Have Learned
The students' graphs should show their understanding of correct graph construction.

PERFORMANCE ASSESSMENT
As a means of evaluating the students' ability to construct and interpret a line graph, ask questions that the student can answer by referring to the line graph he or she made for the *Skill* activity.

Have the students place their line graphs in their science portfolios.

SKILL Making and Interpreting a Graph

▶ **MATERIALS**
- graph paper
- pencils in 2 different colors

▼ **PROCEDURE**

A science student observed that the number of birds at a bird feeder was constantly changing. As part of his observations, he listed the times and numbers of birds at the feeder in a table like the one shown.

TABLE 1: BIRD-FEEDING TIMES

Times	Number of Birds	Times	Number of Birds
10:02	3	10:18	7
10:08	5	10:20	0
10:10	0	10:22	0
10:14	2	10:23	1
10:15	0	10:29	9

1. Use a ruler to draw a vertical line down the left side of a sheet of graph paper. Leave space to the left of the line to write a title and numbers. This line is called the *Y axis*.
2. At the bottom of the graph paper, draw a horizontal line that intersects with the bottom of the vertical line. Make sure that there is enough space under the line to write a title and numbers. This line is called the *X axis*.
3. Starting at the bottom of the *Y axis*, number from 1 to 10 going up the axis. There should be an equal amount of space between each two numbers. This is the *range* of the *Y axis*—in this case, the number of birds.
4. Starting at the left of the *X axis*, write the times in 5-minute intervals starting with 10:00. There should be an equal amount of space in each interval. This is the *range* of the *X axis*—in this case, time.
5. Plot the data from the table on the graph. For example, the first data point is directly over the time 10:02 and directly to the right of the number 3 (for 3 birds).

▶ **APPLICATION**
1. During what time period was there the greatest number of birds at the feeder?
2. During what time period was there the least number of birds at the feeder?
3. How is a line graph an advantage over a table when you are reading certain types of information?

✴ **Using What You Have Learned**
You can compare data on a line graph by using different colors. Suppose you wanted to see which of two bird feeders was used most during specific times of the day. Make up data giving the number of bird visitors for a second bird feeder. Using a different colored pencil, chart the information with a second line on the same graph.

CHAPTER 14 HIGHLIGHTS

The Big Idea—
CHANGES OVER TIME

Reinforce the idea that a classification system can help identify the evolutionary adaptations displayed by the organisms it classifies. In this chapter, the students examined the characteristics of vertebrates that enable scientists and taxonomists to classify the vertebrates into groups. Also remind the students that the adaptations displayed by modern-day organisms are likely to change in the future. As a result, the criteria used to classify organisms in the future are likely to be different from the criteria used today. Because all living things evolve, or change over time, the system and criteria used to classify those organisms must also change.

CHAPTER 14 HIGHLIGHTS

The Big Idea

Classification schemes for living organisms help organize the great quantity of information collected about life. In this chapter, the vertebrates are classified into five groups: fishes, amphibians, reptiles, birds, and mammals. The groups help to emphasize common characteristics and also point to evolutionary links among the species.

Early fossil records show the appearance of ocean-dwelling creatures. Gradually, over long periods of time, some of the fishlike creatures began to adapt for life on land. The five groups of vertebrates parallel the development of life in Earth's history. Each class of vertebrates is more advanced than the previous one. Mammals are the most advanced. Brain and nervous system advances as well as lengthy and responsible care of young are seen in human beings, the most advanced of all the mammals.

Review the questions you answered in your journal before you read the chapter. Add to your journal entries by describing ten more vertebrates. Include two animals from each of the five classes covered in this chapter: fishes, amphibians, reptiles, birds, and mammals. Sketch each animal and describe a few of its characteristics. How are these animals alike? How are they different?

The journal entries of the students should reflect an appreciation of the many different kinds of vertebrates, including their similarities and their differences. You might ask volunteers to describe what a world without fishes, amphibians, reptiles, birds, and mammals would be like.

CONNECTING IDEAS

 The concept map should be completed with words similar to those shown here.

Row 3: coldblooded animals

Row 5: fishes, amphibians, reptiles, birds, mammals

Connecting Ideas

Copy this unfinished concept map into your journal. Complete the concept map by writing in the appropriate terms.

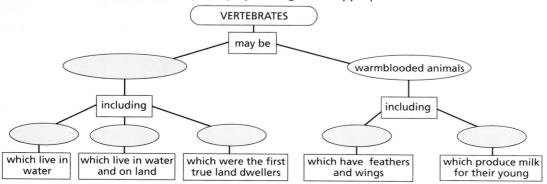

CHAPTER 14 **395**

CHAPTER 14 REVIEW ANSWERS

Understanding Vocabulary

1. Coldblooded animals do not maintain a constant body temperature; their temperature changes to match the temperature of the environment. Warmblooded animals maintain a constant body temperature, regardless of the temperature of the environment.

2. Amphibians and reptiles are both vertebrate animals. Amphibians must spend part of their lives in water and part of their lives on land. Reptiles can spend all of their lives on land.

3. Both cartilaginous and jawless fishes have skeletons made of cartilage. Cartilaginous fishes have jaws and are more developed than jawless fishes.

4. Incubation and gestation are both terms related to the development of animal offspring. The process of keeping bird eggs warm until they hatch is called incubation. Gestation is the period of time a young mammal develops within its mother's body.

Understanding Concepts

Multiple Choice

5. d
6. c
7. c
8. d
9. a

Short Answer

10. Their body temperature changes as the temperature of their environment changes.

11. Male fishes fertilize eggs by spraying them with milt after the female fishes have deposited the eggs. The eggs could not be fertilized if they were covered with shells.

12. The students should name birds.

13. The colors of some bird eggs help camouflage the eggs from predators.

14. Flight consumes tremendous amounts of energy. Birds have to eat large quantities of food to provide the energy needed for flight.

15. Both foods contain the essential nutrients for proper growth and development.

Interpreting Graphics

16. Responses might suggest that because many other amphibians have lungs and can breathe oxygen from air, the presence of external gills indicates that mud puppies can obtain oxygen only from water and must remain in a water environment. Therefore, mud puppies are less advanced than other amphibians.

▶ 396 CHAPTER 14

CHAPTER 14 REVIEW

Understanding Vocabulary

For each pair of terms, explain how they are alike and how they differ.

1. coldblooded (372) and warmblooded (384)
2. amphibians (376) and reptiles (380)
3. jawless fishes (373) and cartilaginous fishes (373)
4. incubation (388) and gestation (393)

Understanding Concepts

MULTIPLE CHOICE

5. The first vertebrates to be true land dwellers were the
   a) fishes.   b) mammals.
   c) amphibians.   d) reptiles.

6. Tadpoles develop into frogs through a process called
   a) incubation.   b) gestation.
   c) metamorphosis.   d) reproduction.

7. Birds that nest on cliffs may lay eggs that are
   a) round.   b) very large.
   c) pointed at one end.   d) colored like rocks.

8. One characteristic that distinguishes mammals from other vertebrates is that
   a) they are warmblooded.
   b) their young are fully developed when born.
   c) they do not lay eggs.
   d) they have body hair at some time during their lives.

9. Which process is described by the term *gestation*?
   a) A young mammal is developing within its mother's body.
   b) Eggs are kept warm until they hatch.
   c) A new mammal is created by means of internal fertilization.
   d) Young mammals continue their development by getting nourishment from the mother's milk.

SHORT ANSWER

10. How are coldblooded animals adapted to their environment?

11. What might happen if fish eggs had shells?

12. One example of a warmblooded vertebrate is the mammal. Give a second example.

13. Why are some bird eggs colored?

14. Explain why birds eat large quantities of food.

15. What similarity exists between the food provided for developing birds and that provided for young mammals?

Interpreting Graphics

16. The mud puppy shown here is a type of amphibian that resembles a tadpole. Note the mud puppy's external gills, which are never lost, even as an adult. Are mud puppies more or less advanced than other amphibians? Justify your answer.

Reviewing Themes

17. Possible answers include: for fish—gills for obtaining oxygen from water, fins to move through water; for amphibians—the ability to live in a water environment during reproduction; for reptiles—the protective shells of eggs, lungs for breathing oxygen, scaly skin that does not dry out.

18. Similarities include: both birds and mammals care for their offspring until such time as the offspring can live independently; differences include: a developing bird embryo receives its nourishment entirely from an egg, while a developing mammal embryo receives its nourishment from its mother.

Thinking Critically

19. The reproductive process of amphibians makes their offspring extremely susceptible to predation because the eggs and the young are unprotected. Also the need for water limits the areas in which amphibians are able to live successfully.

20. Bird feeders containing foods favored by birds will encourage birds to visit a yard. Clean, accessible water and the absence of pets will also help to attract birds. The noise and droppings produced by large numbers of birds can become a nuisance or a health hazard.

21. Responses might suggest that placental development is more successful for mammals than marsupial development. Marsupials are most successful in Australia where there is little competition from other mammals for niches.

22. Conservation laws were probably passed to protect the birds that were killed to obtain their feathers. At one time, large numbers of feathers were used by the fashion industry. The feathers of many songbirds were sought because of their interesting patterns and bright colors.

23. Fishes produce the most eggs; mammals produce the fewest offspring. As the animal type becomes more advanced, the probability for the successful survival and development of the offspring increases because the degree of parental care also increases. However, the greater the amount of parental care, the fewer the number of offspring a parent can successfully care for.

24. Modern reptiles often seek shade during the heat of the day and bask in the sun during cool weather. Some reptiles (such as turtles) spend time in water to help maintain their body temperature. Since most reptiles can only survive within a limited temperature range, temperatures above and below that range can be dangerous or fatal to the reptile.

Reviewing Themes

17. *Changes Over Time*
List the characteristics of fishes, amphibians, and reptiles that you think are the most valuable for their survival.

18. *Systems and Structures*
Compare and contrast how birds and humans care for their young. Include as many different facts as you can.

Thinking Critically

19. There are fewer amphibians on Earth than any other type of vertebrate. Explain why you think this is so.

20. If you wanted to encourage birds to visit your yard, what steps would you take? Why are birds discouraged from visiting some places?

21. Why do you think placental mammals are more numerous than marsupial mammals?

22. There are several laws that prohibit owning certain types of feathers. Why do you think these laws were passed?

23. Fishes produce many young. They actually produce even more eggs than those that hatch. Is this true of amphibians? What about reptiles? Do birds lay as many eggs as frogs do? Do elephants have as many baby elephants as ducks have baby ducks? Over a lifetime, what animal probably has the fewest offspring? Develop a hypothesis to explain why numbers of eggs and numbers of young are reduced as the animal type becomes more advanced.

24. If a coldblooded animal is to remain living, it must learn to adapt itself to changes in the temperature of its environment. Dinosaurs had some very unusual adaptations, such as huge fins, that helped keep their bodies at a constant temperature. What are some of the things that modern reptiles do to keep from becoming too hot or too cold? Do you think being too hot would be more dangerous than being too cold? Explain.

Discovery Through Reading

Darling, Kathy. *Manatee On Location*. Lothrop, Lee & Shepard Books, 1991. Text and photographs describe the life history, physical characteristics, behavior, and underwater activities of the Florida manatee and how scientists and others are trying to save this endangered sea mammal.

CHAPTER 15

ANIMAL BEHAVIOR

PLANNING THE CHAPTER

Chapter Sections	Page	Chapter Features	Page	Program Resources	Source
Chapter Opener	398	*For Your Journal*	399		
Section 1: DEFENSE	400	Discover By Researching **(B)**	401	*Science Discovery**	SD
• Camouflage **(B)**	401	Activity: How effective is protective coloration? **(B)**	404	Investigation 15.1: Observing Insect Adaptations **(A)**	TR, LI
• Mimicry **(B)**	403	Section 1 Review and Application	405	Reading Skills: Predicting Outcomes **(A)**	TR
• Other Means of Defense **(B)**	404	Skill: Interpreting Data **(A)**	406	Extending Science Concepts: Animal Camouflage **(H)**	TR
				Protective Coloration in Animals **(B)**	IT
				Group Behavior in Animals **(B)**	IT
				Study and Review Guide, Section 1 **(B)**	TR, SRG
Section 2: GROUP BEHAVIOR	407	Discover By Researching **(A)**	409	*Science Discovery**	SD
• Social Behavior **(H)**	407	Activity: What behavior can you observe in an ant colony? **(A)**	411	Investigation 15.2: Water Temperature and Its Effect on Fish Respiration **(H)**	TR, LI
• Communication **(H)**	410	Discover By Researching **(A)**	412	Connecting Other Disciplines: Science and Social Studies, Exploring Methods of Communication **(A)**	TR
• Seasonal Behavior **(A)**	413	Section 2 Review and Application	415		
• Parental Behavior **(A)**	414	Investigation: Predicting Animal Responses **(A)**	416		
				Thinking Critically **(H)**	TR
				Group Behavior in Animals **(B)**	IT
				Record Sheets for Textbook Investigations **(A)**	TR
				Study and Review Guide, Section 2 **(B)**	TR, SRG
Chapter 15 HIGHLIGHTS	417	The Big Idea	417	Study and Review Guide, Chapter 15 Review **(B)**	TR, SRG
Chapter 15 Review	418	For Your Journal	417	Chapter 15 Test	TR
		Connecting Ideas	417	Test Generator	
				Unit 5 Test	TR

B = Basic **A** = Average **H** = Honors
The coding Basic, Average, and Honors indicates subsections, features, and resources that might be appropriate for different levels of learners. For additional suggestions regarding choice of topic and depth of coverage, see the Pacing Chart on pages T26–T29.

*Frame numbers at point of use
(TR) Teaching Resources, Unit 5
(IT) Instructional Transparencies
(LI) Laboratory Investigations
(SD) *Science Discovery* Videodisc Correlations and Barcodes
(SRG) Study and Review Guide

▶ 397A

CHAPTER MATERIALS

Title	Page	Materials
Discover By Researching	401	(per individual) journal
Teacher Demonstration	401	terrarium, chameleon, objects for terrarium (brown to green in color)
Activity: How effective is protective coloration?	404	(per group of 2 or 3) yellow pencil, material for camouflaging pencil
Skill: Interpreting Data	406	(per group of 2 or more) pencil, graph paper
Discover By Researching	409	none
Teacher Demonstration	410	tape recorder, cassette tape
Activity: What behavior can you observe in an ant colony?	411	(per group of 3 or 4) ant colony, shovel, newspaper, small trowel, plastic bag, paper tube, glass jar (large), cornmeal, grass seeds or sugar, sponge, tape, black paper
Discover By Researching	412	(per individual) journal
Investigation: Predicting Animal Responses	416	(per group of 3 or 4) scissors, shoe box with lid, sow bugs (10), sponge, water

ADVANCE PREPARATION

For the *Demonstration* on page 401, obtain a terrarium, a chameleon, and objects ranging from brown to green in color for the terrarium. For the *Activity* on page 411, you will need to locate an ant colony and obtain sugar and cornmeal. The *Investigation* on page 416 requires sow bugs and shoe boxes.

TEACHING SUGGESTIONS

Field Trip
Arrange a field trip to an area with natural wildlife. Have the students observe the animals and note examples of mimicry and camouflage. Also have them note animals with special defense strategies or group living patterns. A tape recorder can be used to save the communication sounds of the animals.

Outside Speaker
Invite someone from the zoo or other source to talk about camouflage and mimicry in animals. Encourage the speaker to bring slides to illustrate both camouflage and mimicry.

CHAPTER 15
ANIMAL BEHAVIOR

CHAPTER THEME—ENVIRONMENTAL INTERACTIONS

This chapter describes how change helps ensure an organism's survival in its environment. The students will observe changes in physical features that, over time, have increased an organism's chance of survival. They will also learn how animals have acquired certain characteristics of behavior to help ensure their survival. This theme is also developed through concepts in Chapters 1, 2, 3, and 11. A supporting theme of this chapter is **Nature of Science**.

MULTICULTURAL CONNECTION

Humans have learned from other animals about how to adapt to different environmental conditions in order to survive. Many of the customs of dress and behavior that characterize different cultures around the world have been acquired over the years in order to ensure the survival of that group of people. This is evident, for example, in the way people may dress to hunt for wild animals. In the African kingdom of Benin, the Edo hunters wore large, bushy helmets to disguise themselves in the forest. The Inuit and other people in the Arctic region wear white clothing in order to blend in with their snowy surroundings when hunting for polar bears or seals. Have the students research the dress of other cultures that depend on hunting for survival.

MEETING SPECIAL NEEDS

Second Language Support

Students with limited English proficiency can use observation to better understand the concept of animal behavior. As the students progress through the chapter, have them create a collage of pictures that show animal defenses and behavior they have read about. Filmstrips and videocassettes on this topic can also be used to enhance understanding.

CHAPTER 15
Animal Behavior

My stop! I whistled to the driver. The air brakes wheezed and puffed as the bus slowed and stopped on a steep hill. I hauled my backpack down the bus steps and jumped out into brilliant sunshine.

If you have a pet, you can probably think of quite a few animal behaviors. But wild animals and those in zoos have specific behaviors, too. Most behaviors help ensure an animal's survival. Now travel through the rain forest of Costa Rica with Dr. Adrian Forsyth, studying the behaviors of some of the animals that live there.

398 CHAPTER 15

CHAPTER MOTIVATING ACTIVITY

Use felt-tip markers to color sixty toothpicks red, sixty green, and sixty brown. Randomly scatter thirty of each color in an area that has dirt, and thirty of each in an area of grass. (You can also color two of the sets of toothpicks to match two different colored floors.) Give the students 20-second time periods to collect all the toothpicks they can from each area. Have the students make a table and record the number of each color toothpick they gathered in each 20-second time period. Discuss the students' results, particularly the effect of color on their ability to find the toothpicks.

For Your Journal

Review the students' journal entries and be prepared to discuss their answers to the questions. You will want to focus on their previous experiences with animals and reveal possible misconceptions they may have about animal behavior. Have the students share their ideas about the importance of animal behavior to scientists.

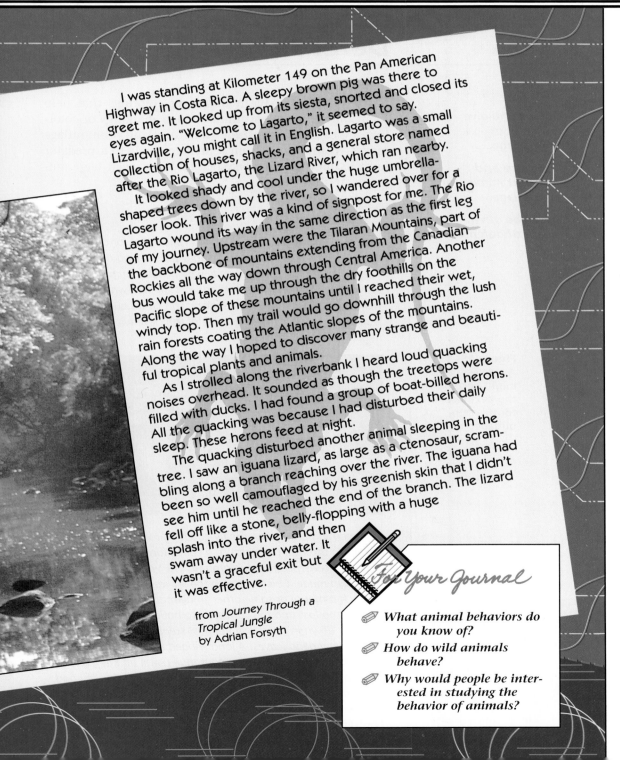

I was standing at Kilometer 149 on the Pan American Highway in Costa Rica. A sleepy brown pig was there to greet me. It looked up from its siesta, snorted and closed its eyes again. "Welcome to Lagarto," it seemed to say. Lizardville, you might call it in English. Lagarto was a small collection of houses, shacks, and a general store named after the Rio Lagarto, the Lizard River, which ran nearby.

It looked shady and cool under the huge umbrella-shaped trees down by the river, so I wandered over for a closer look. This river was a kind of signpost for me. The Rio Lagarto wound its way in the same direction as the first leg of my journey. Upstream were the Tilaran Mountains, part of the backbone of mountains extending from the Canadian Rockies all the way down through Central America. Another bus would take me up through the dry foothills on the Pacific slope of these mountains until I reached their wet, windy top. Then my trail would go downhill through the lush rain forests coating the Atlantic slopes of the mountains. Along the way I hoped to discover many strange and beautiful tropical plants and animals.

As I strolled along the riverbank I heard loud quacking noises overhead. It sounded as though the treetops were filled with ducks. I had found a group of boat-billed herons. All the quacking was because I had disturbed their daily sleep. These herons feed at night.

The quacking disturbed another animal sleeping in the tree. I saw an iguana lizard, as large as a ctenosaur, scrambling along a branch reaching over the river. The iguana had been so well camouflaged by his greenish skin that I didn't see him until he reached the end of the branch. The lizard fell off like a stone, belly-flopping with a huge splash into the river, and then swam away under water. It wasn't a graceful exit but it was effective.

from *Journey Through a Tropical Jungle* by Adrian Forsyth

For Your Journal

- What animal behaviors do you know of?
- How do wild animals behave?
- Why would people be interested in studying the behavior of animals?

ABOUT THE AUTHOR

Adrian Forsyth, one of North America's leading natural science writers, specializes in animal behavior and rain forest ecology. After receiving his doctorate in biology from Harvard University, Dr. Forsyth taught for several years before devoting himself to his writing and biological fieldwork.

Dr. Forsyth, a native of Ottawa, Ontario, has won three gold awards for science writing from the Canadian National Magazine Awards Foundation. His book *Journey Through a Tropical Jungle* uses text and pictures to describe the animals, plants, and people of the Costa Rican rain forest. He is also the coauthor of *Tropical Nature*, which has been described as an excellent introduction to tropical biology.

Section 1: DEFENSE

FOCUS

This section covers some of the ways in which animals are able to protect themselves. Types of camouflage that help animals blend in with their environment are discussed, as is mimicry, an adaptation whereby organisms resemble other organisms or objects. Other means of defense such as speed, size, and odor are also presented.

MOTIVATING ACTIVITY

Cooperative Learning Have the students work in groups of three or four to create lists of animals from their neighborhood. Then ask the students to describe how each animal defends itself. Also ask them to think about and make a list of ways in which humans have defended themselves by using defense mechanisms or behaviors similar to those of animals. Responses might include protective armor, electric fences, mace, tear gas, poison darts, running, hiding, playing dead, and so on. Have each group share its lists with the other groups.

PROCESS SKILLS
- Observing • Constructing/Interpreting Models

POSITIVE ATTITUDES
- Curiosity
- Cooperativeness
- Creativity

TERMS
- camouflage • mimicry

PRINT MEDIA
Mimicry and Camouflage by Jill Bailey (see p. 315b)

ELECTRONIC MEDIA
Camouflage in Nature: Form, Color, Pattern Matching, Coronet (see p. 315b)

Science Discovery Ptarmigan, willow Frog, Tinctorius poison dart

BLACKLINE MASTERS
Study and Review Guide
Laboratory Investigation 15.1
Reading Skills
Extending Science Concepts

REINFORCING THEMES—
Environmental Interactions

Most students are aware of the changes that plant life undergoes with the changing seasons. But many animals also undergo seasonal changes. Point out to the students that the arctic fox, the snowshoe hare, and a bird called a ptarmigan go through an annual cycle of color changes that helps them survive in their environment. Ask the students to describe seasonal changes in how they dress and to explain how their comfort and even their survival are affected by these changes.

SECTION 1

Defense

Objectives

Define and give examples of camouflage.

Compare and contrast mimicry and camouflage.

Describe other means of defense.

Many animals have markings or colorings that help them hide from other animals. Such markings or colorings are called **camouflage** (KAM uh flahj). Camouflage may save an animal from a predator or help a predator hunt its prey. As Dr. Forsyth found out, even large animals can hide.

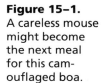

Figure 15–1. A careless mouse might become the next meal for this camouflaged boa.

One of my goals on this trip was to collect flies for a friend of mine, a scientist who was studying a special kind of fly. These flies are fast, and the only way I could catch them was to walk slowly along the trails and net the flies when they were sunning themselves on leaves. This calm, quiet way of walking helped me discover some of the cleverly camouflaged animals of the jungle.

One morning on my favorite fly-catching route, a path that ran down through the forest to a coffee farm, I spotted a large boa constrictor on the ground. The wavy pattern of the boa's skin made it blend in with the leaves, helping the boa to catch a mouse or some other small mammal that was moving too fast and too carelessly to notice it.

Lower down in the valley I almost stepped on an even better mimic of dead leaves. A long winged bird fluttered up from under my foot. It was a pauraque (PAHR uh kay), a kind of whippoorwill that nests on the ground during the day and hunts at night. Any bird that nests on the ground has to be very skilled at hiding from its enemies. I had seen for myself how difficult it is to spot one of these birds standing still.

Figure 15–2. This bird nests on the ground, but it is well hidden by its protective coloration.

TEACHING STRATEGIES

● **Process Skills:** *Observing, Communicating*

Many books with pictures showing examples of animal camouflage are available from libraries. Have the students observe pictures that show examples of animal camouflage. Ask the students to share their own observations about animals that use camouflage for protection.

● **Process Skills:** *Predicting, Generating Ideas*

Have the students predict what would happen to the population of white-furred arctic foxes and snowshoe hares if the snowfall during the winter season were very sparse. (These animals depend on snow so that they can blend in with their surroundings. The lack of snow would make them easier to spot and, therefore, more vulnerable to enemies.)

Figure 15–3. The arctic fox blends in with the snow of winter and the bare ground of summer.

Camouflage

Camouflage is important for the survival of many animals. An animal that lives in tall grasses might have stripes to confuse a predator. The stripes of zebras and certain antelopes, for example, help them to blend into their surroundings. Tigers also have stripes, but for slightly different reasons. Its stripes help a tiger to sneak up on its prey without being seen.

The best-camouflaged animals are those whose colors match their surroundings. Its white fur helps a polar bear blend into its icy environment.

The coats of some animals even change color to keep up with the changing seasons. An arctic fox's gray fur blends in with the bare ground of the summer tundra. In the fall, its fur changes to white, blending in with the snow of the long tundra winter.

Animals such as certain lizards, fishes, and invertebrates have the ability to change color as needed. Chameleons are so good at changing color that the word *chameleon* has a second meaning in the dictionary. In addition to a type of lizard, *chameleon* is also defined as "a person who is changeable or moody."

DISCOVER BY *Researching*

Use your library to learn more about the lizards called *chameleons*. Find out about their environment and the causes of their color changes. Write your findings in your journal and report them to your classmates.

DISCOVER BY *Researching*

Remind the students that chameleons are reptiles and that information on them can be found in articles and books about reptiles. You may want to extend the *Discover By Researching* by asking the students to find out about other organisms that are able to change color like the chameleon. (For example, the mollusk, octopus, and squid are adept at color change. An octopus can change to all colors ranging from black to white in order to match rocks, seaweed, and other objects in the ocean.)

Demonstration

Set up a terrarium in the classroom and purchase a chameleon for the students to observe. Place a variety of objects in the terrarium ranging from brown to green in color so that the students are able to observe the chameleon's changes in color.

LASER DISC
3792, 3793

Ptarmigan, willow

SECTION 1 **401**

TEACHING STRATEGIES, continued

- **Process Skills:** *Comparing, Generating Ideas*

Ask the students to think of examples of how humans use bright colors to warn or to make something easier to see. (Answers should include objects such as road signs, safety vests, fire trucks, tennis balls, golf balls, and so on.)

- **Process Skills:** *Inferring, Applying*

Review natural selection with the students by reminding them it is the process by which organisms pass an inherited characteristic that helps them survive from one generation to another. Using the idea of natural selection, have the students infer why cheetahs, as a group, are faster than other animals. (Weaker, slower cheetahs were not able to survive as well as the stronger, faster cheetahs. Those that survived passed their characteristics along to their offspring. The weaker, slower cheetahs were not likely to survive to pass their characteristics on to their offspring. Since only the fast and strong survived to reproduce, cheetahs, as a group, evolved to become very speedy animals.)

LASER DISC

3620

Frog, Tinctorius poison dart

MEETING SPECIAL NEEDS

Mainstreamed

Any student who needs additional help with the content of this section can be paired with a student tutor to review the main points of the section. The tutor can help the student write a summary or design a diagram that focuses on the main points.

ONGOING ASSESSMENT
ASK YOURSELF

Camouflage may protect an animal from a predator or help a predator hunt its prey.

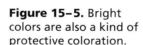

Figure 15–4. The dark top and light bottom of killer whales help them to blend in with their ocean surroundings.

Figure 15–5. Bright colors are also a kind of protective coloration.

Another type of camouflage, seen often in fishes and other marine animals, is countershading. Countershaded animals are darker on top than they are on their undersides. The killer whale, for example, has a black upper surface and a white underside. The underside blends with the water surface when seen from below, while the topside blends with the ocean depths when seen from above. This camouflage allows the killer whale to hunt with better results.

Some animals use bright colors instead of camouflage for protection. Bright colors "advertise" the danger of close encounters. Dr. Forsyth describes one close encounter.

One day I was walking along, looking for flies, when I suddenly stopped and jumped back without really knowing why. I looked down, and at my feet was an unusually beautiful snake. Its markings were bold and bright, the exact opposite of camouflage. When I realized the snake was banded with shiny rings of black and red, I jumped back even farther. I was glad I was wearing my tall rubber boots. It was a dazzling but deadly coral snake, with venom strong enough to paralyze and kill a human.

But the snake had no interest in harming me. Snakes, after all, have no arms or legs to help them capture food, so this coral snake couldn't have been considering anything as big as me for dinner. Their venom is simply a special saliva used to subdue more suitable prey and then help digest it. But they also use their poisonous bite to defend themselves from animals that step on them. So the bright colors of the coral snake had done us both a favor. They provided a warning that had stopped me from injuring the snake, and my retreat had stopped the snake from having to bite me in self-defense.

ASK YOURSELF

In what ways is camouflage useful for an animal?

- **Process Skills:** *Observing, Comparing*

Obtain two or three types of fish from a local market so that the students may observe the characteristic of countershading. If possible, obtain a leopard frog or other amphibian that exhibits countershading. Have the students compare the different organisms and draw conclusions about how countershading aids survival.

- **Process Skills:** *Expressing Ideas Effectively, Observing*

Introduce the students to the topic of mimicry by asking them to explain what it means to *mimic* someone. (to copy or imitate the person's appearance, behavior, and speech) Then explain how mimicry helps protect animals from predators. Have the students look at Figure 15–7 and ask them whether they would be able to tell the two butterflies apart without the labels. Relate the butterflies' similarity to the protection it affords the viceroy butterfly.

Mimicry

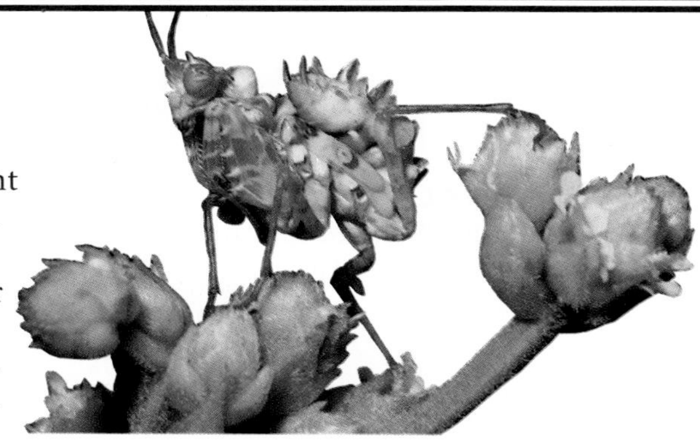

Mimicry (MIHM ihk ree) is another type of protective coloration. But mimicry is different from camouflage. **Mimicry** makes an animal look like something else. That is, the animal mimics, or copies, the appearance of another organism.

Some animals mimic parts of their environment. For example, the pink flower mantis, shown in the photograph, looks like a beautiful orchid. Certain other insects look so much like small tree branches that they are called *walking sticks*. Dr. Forsyth found animals that even copy the behavior of the object they are mimicking.

> There were moths that looked just like dead sticks and old leaves. Their wings were gray and brown, with fine lines and spots looking like tree bark or leaf veins, and they were twisted in crinkled, wrinkled shapes. They even had blotchy patches on them that looked like mold and fungus. The moths that looked like dead twigs stayed completely still, and the ones that looked like dead leaves let go of the branches they were resting on and fluttered slowly to the ground.

Animals can also mimic other animals. For example, certain nonpoisonous frogs have the same bright colors as those frogs that do have poison glands. Predators cannot tell the nonpoisonous frogs from the poisonous ones. As a result, predators usually pass both types of frogs by for a safer meal. The viceroy butterfly, which birds find tasty, has the same colors and markings as the monarch butterfly, which will actually make birds sick if eaten. Once a bird eats a monarch, it avoids eating another one. This benefits the viceroy, because birds will not eat it either. In the next activity, you can test the effectiveness of some types of camouflage and mimicry.

Figure 15–6. The coloration of the pink flower mantis is an example of mimicry.

Figure 15–7. The monarch butterfly (left) and the viceroy butterfly (right) look alike to birds.

SCIENCE BACKGROUND

The harmless puff adder or hognose snake is an example of mimicry because it looks like the deadly rattlesnake. But the adder takes things a step or two farther. It puts on quite a bluff when cornered. The adder flattens out and widens its head, rearing up into a striking pose and hissing. If this doesn't scare off potential enemies, it will act as if it is going into convulsions and then fake its own death.

THE NATURE OF SCIENCE

The idea of animal mimicry was the result of work by the English naturalist Henry Walter Bates. In the mid-nineteenth century, Bates spent several years studying animals living along the Amazon River. He discovered butterfly specimens that looked alike but were members of different species. He learned that one of the species had an unpleasant odor and taste. The other species, however, did not exhibit this unpleasant odor and taste. Bates's discovery was the first known example of animal mimicry. Today, instances of mimicry in which one species resembles another species with an unpleasant odor and taste are referred to as Batesian mimicry.

ACTIVITY

How effective is protective coloration?

Process Skills: Comparing, Constructing/Interpreting Models

Grouping: Groups of 2 or 3

Hints
Have each group place its pencil in the classroom without the others observing the location.

▶ **Application**
Responses should note that the most successful example of protective coloration was the pencil that blended in the best with its classroom location. It was the most successful because it was the most difficult for observers—who in the wild might be predators—to find.

ONGOING ASSESSMENT
ASK YOURSELF

With protective coloration, the color(s) of an organism match or blend in with the color(s) of its environment, preventing the organism from being seen. In mimicry, an organism copies the appearance of another organism or an object, making no attempt to remain unseen.

INTEGRATION—Language Arts

Ask the students to think of phrases that refer to animal behavior, particularly survival behavior. (Examples: to play possum, to run like a rabbit, to follow the herd, to lock horns, to hightail it) Have the students investigate the meaning of each phrase and explain whether it describes animal behavior accurately.

GUIDED PRACTICE

Review the terms *camouflage* and *mimicry* with the students. Have them cite examples of each to demonstrate their understanding of the terms.

INDEPENDENT PRACTICE

 Have the students complete the Section Review and Application questions. Then have them describe in their journals at least two examples of animal defense not mentioned in their books.

EVALUATION

Have the students describe the difference between a chameleon's and a viceroy butterfly's means of defense.

ACTIVITY

How effective is protective coloration?

MATERIALS
yellow pencil, materials for camouflaging pencil

PROCEDURE
1. Using whatever materials are available, camouflage or disguise your pencil so that it can be hidden in the open somewhere in your room.
2. Have another team try to find your pencil while you try to find their "hidden" pencil.

APPLICATION
What do you think was the most successful example of protective coloration of a yellow pencil in your class? Explain why you think it was successful.

ASK YOURSELF

What is the difference between protective coloration and mimicry?

Other Means of Defense

Camouflage and mimicry are often used as methods of defense. These methods involve tricking predators or sneaking up on prey. Animals also have other adaptations for protection. Some animals are able to swim, fly, or run faster than most of their predators. The Thompson's gazelle, for example, has been clocked at 50 km/h, while one of its predators, the lion, can run only about 35 km/h.

Have you ever heard the saying "There is safety in numbers"? Animals often travel in large groups for protection. Predators have a hard time attacking large numbers of animals. However, any animal that becomes separated from the herd is in danger of attack. In the activity at the end of this section, you can see how large numbers of offspring help to ensure a species' survival.

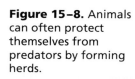

Figure 15–8. Animals can often protect themselves from predators by forming herds.

RETEACHING

 Have the students create a three-column chart with the headings *camouflage*, *mimicry*, and *other*. Under each heading, direct the students to write the names or draw pictures of animals that exhibit these types of defense. Have the students place their charts in their science portfolios.

EXTENSION

In nature, plants also have structural and chemical adaptations to decrease the likelihood of being eaten. Ask the students to think of examples of such adaptations in plants. (Examples include thorns, spines, "hairy" leaves, and chemicals such as those given off by poison ivy, poison oak, and poison sumac.) Interested students should be encouraged to find out more about plant defenses and report their findings to the class.

CLOSURE

 Cooperative Learning Have the students work in small groups to prepare concept maps or diagrams for the section. Have the groups exchange and complete the maps.

Sometimes size is the best defense. Very few predators will attack an elephant simply because elephants are so large. However, young or sick elephants sometimes become prey. Stronger and larger elephants try to protect weaker elephants whenever possible. Elephants have been known to free other elephants from traps.

Figure 15–9. The belly of the seemingly well-protected porcupine is open to attack by the fisher.

Outer coverings can also offer protection for animals. Armadillos have tough, leathery plates that protect their internal organs by making it difficult for predators to bite or claw them. Armadillos try to flee from attack, but if given no choice, they can roll themselves into balls, making it even harder for predators to find an unprotected place to attack. However, no animal is immune to attack. For example, the fisher, a relative of the weasel, preys on well-armored porcupines by rolling them over and attacking their soft bellies.

Skunks and other animals with scent glands protect themselves by spraying attacking animals with an offensive-smelling liquid. Skunks are usually left alone by would-be predators that have had unpleasant experiences with this little striped animal.

 ASK YOURSELF

Explain how traveling in large groups can be a benefit to both predators and prey.

SECTION 1 REVIEW AND APPLICATION

Reading Critically

1. Which types of defense depend on an animal deceiving its predator?
2. Give three examples of animal defenses that do not depend on deceiving a predator.

Thinking Critically

3. Some toads blend so well with the ground that they are almost invisible. Is this camouflage or mimicry? Explain your answer.
4. Some fish are able to puff themselves up to look much larger than they really are. Would this be an example of camouflage, mimicry, or another defense? Explain your answer.

✦ **Did You Know?**
The opossum and the hognose snake play dead to trick predators. The hognose snake tries so hard to convince its enemy it is dead that if flipped over, it will flip itself over on its back again, thus ruining its ruse!

ONGOING ASSESSMENT
ASK YOURSELF

A predator may be unwilling to attack a large group of prey, fearing it might be overwhelmed by the numbers. Predators in a group can work together to attack and outrun prey.

SECTION 1 REVIEW AND APPLICATION

Reading Critically

1. Camouflage and mimicry are methods of defense that deceive predators.

2. Examples include speed, agility, size, outer coverings, and scents.

Thinking Critically

3. The toads make no attempt to look like other organisms, so they are not defending themselves using mimicry. Because the toads blend in with their surroundings, they are using camouflage.

4. Puffing up to a larger size is neither camouflage nor mimicry. The puffed-up fish is not hidden, and it does not resemble another organism. It is trying to trick a predator by suddenly becoming much larger.

SECTION 1   **405**

SKILL
Interpreting Data

Process Skills: Interpreting Data, Generating Ideas

Grouping: Groups of 2 or more

Objectives
- **Interpret** data from tables.
- **Record** data from tables.
- **Generate** ideas about the value of scientific data.

Discussion
Remind the students that when creating a graph, they must choose a standard unit for the horizontal axis and a standard unit for the vertical axis. In other words, if a student decides to list the number of months on the horizontal axis, the unit chosen to represent those months must, for example, be 2, 4, 6, etc., but not a combination of those numbers.

Also help the students interpret the given graph and answer the questions posed in the procedure.

Application
After one year, 19 percent of the remaining hatchlings have survived. After two years, 41 percent of the remaining hatchlings have survived. After three years, 59 percent of the remaining hatchlings have survived.

Using What You Have Learned
The students might suggest that interpreting data helps scientists answer some questions and ask other questions and thus adds to scientific knowledge.

PERFORMANCE ASSESSMENT
Have the students explain how they constructed their graphs, including what they chose for standard units and how they plotted the data. Use their explanations to evaluate their ability to construct and interpret graphs.

SKILL: Interpreting Data

▼ MATERIALS
- pencil
- graph paper

▼ PROCEDURE

1. Loggerhead sea turtles lay great numbers of eggs about three times each year. They lay their eggs in nests dug into sandy beaches. The baby turtles, called *hatchlings,* crawl from their nest and scurry down the beach toward the sea. If all the hatchlings survived, the population of loggerhead sea turtles would be very large.

2. Study the graph showing the survival rate of loggerhead hatchlings. The graph shows what would happen to the population of turtles after three years if all the eggs hatched and all the hatchlings survived. Interpret the graph. How many turtles would there be after one year? After two years? After three years?

3. Hungry predators, such as gulls, ospreys, and crabs, grab many hatchlings from the sand. Since not all hatchlings survive, the population of sea turtles does not change much. Use this data table to construct your own graph of hatchling survival rates.

TABLE 1: SURVIVAL RATE OF LOGGERHEAD TURTLE HATCHLINGS	
Number of Months	Percent of Remaining Hatchlings Surviving
4	6
8	11
12	19
16	28
20	35
24	41
28	51
32	57
36	59

▶ APPLICATION
Interpret the data on your graph. How many turtles would there be after one year? After two years? After three years?

✳ Using What You Have Learned
Get together with a group of classmates. List your ideas about the ways in which interpreting data might be important to scientists in their work.

Section 2:
GROUP BEHAVIOR

FOCUS

In this section, the behavior of animals in groups is investigated. The focus is on the benefits animals derive from living in groups. Communication between animals within a group is discussed. The effects of seasonal changes on group behavior are studied. Also the topic of parental care is presented.

MOTIVATING ACTIVITY

Cooperative Learning Have the students work in groups of three or four. Ask them to think about social behavior and communication as exhibited in their classroom or school. Have them write short reports describing the behavior of students as a group. You may want the groups to spend one session making observations and a second session planning and writing their reports. Caution the students to avoid making derogatory remarks about any individual or group.

PROCESS SKILLS
- Observing • Comparing

POSITIVE ATTITUDES
- Curiosity
- Cooperativeness
- Enthusiasm for science and scientific endeavor

TERMS
- pheromones • torpor

PRINT MEDIA
Animal Migration by Nancy J. Nielsen (see p. 315b)

ELECTRONIC MEDIA
Beyond Words: Animal Communication, National Geographic (see p. 315b)

Science Discovery Geese flying Whale, beluga; sonogram of song Bear, black

BLACKLINE MASTERS
Study and Review Guide
Laboratory Investigation 15.2
Connecting Other Disciplines
Thinking Critically

Group Behavior

SECTION 2

"*Aaaaarrrroooooo-oooo-oooo-gaaaahhhh!*"
I almost jumped out of my rubber boots. It sounded as if there was a huge and fierce creature right above me. Branches were rattling, sticks and leaves rained down, along with an avalanche of noise.

I had startled a large male howler monkey feeding at the end of a branch not far above my head. And he had startled me! The howler glared down, shaking the shaggy mane and beard around his face. He thrust out his strong lower jaw and began bellowing out more threats at me.

Objectives

Describe four types of social behaviors among animals.

Explain several ways in which animals communicate.

Compare and contrast seasonal behaviors.

Social Behavior

Their behavior is one way the howler monkeys defend their territory. Dr. Forsyth was interested in territorial behaviors. His story continues:

I wasn't worried. Howlers are big monkeys, but they are harmless in spite of their fearsome calls. In fact the animal's annoyance was my good luck. I had hoped to get a good look at the special equipment the male howler has for producing his deep roars. I could see clearly that his throat was massive, with a giant voice box. He was using this sound system for producing deep gravelly bass notes.

Figure 15–10. A troop of howler monkeys claims a territory.

INTEGRATION—
Geography

Point out that the area being explored by Dr. Forsyth is a tropical rain forest. Rain forests are in countries that lie near the equator, including Brazil, Peru, and Surinam in South America; Panama, Nicaragua, and Honduras in Central America; Indonesia, Malaysia, and Thailand in Asia; and Zaire, Gabon, and Cameroon in Africa. Have the students locate these countries on a world map.

TEACHING STRATEGIES

- **Process Skills:** *Analyzing, Formulating Hypotheses*

Discuss the howling behavior of the howler monkeys with the students. Ask them to analyze the timing of the howling sessions and offer hypotheses about why the howling occurs just before the monkeys bed down at night and shortly after they wake in the morning. (Answers might include that the howlers are re-establishing their territory at the beginning and end of each day.)

- **Process Skill:** *Comparing*

Have the students compare the group behavior of howler monkeys and the white pelicans with that of other more familiar animals they have observed. (The students might mention that if one dog barks in the neighborhood, others begin barking, or if one crow squawks, others will. They may even suggest their own behavior in the classroom (for example, if two students begin to talk, others soon join in) or other human group behavior, such as everyone looking up if one person looks up.)

✧ **Did You Know?**
When in danger, a herd of musk oxen will form a circle with their horned heads pointing out and their young calves in the center. This circle presents a formidable barrier to any enemy and as long as the circle remains unbroken, it is an effective defense.

INTEGRATION—
Social Studies

Some of the most fascinating social behavior can be observed among insects such as termites, bees, wasps, and ants. Have the students study the roles of the members of an insect colony and then compare the roles to those in a human community, using a chart to organize the information.

 Have the students place their *Integration* charts in their science portfolios.

SCIENCE BACKGROUND

The killer whale is not a whale; it is a dolphin. Unlike other dolphins, it hunts in groups of up to 20 or more. Killer whales are ferocious and will attack almost anything. They are also known as *sea wolves* because they hunt in packs like wolves. They will attack a huge whale by ramming it with their heads. Once the whale is stunned, they will literally eat it alive.

Figure 15–11. Male howler monkeys have a massive voice box. Howler monkeys spend the night in the treetops.

The howler's roaring is an important social behavior directed at other howler monkeys. A troop normally roars together, led by the largest dominant male. These howling sessions begin each morning as the monkeys wake up. Often they repeat the chorus just before bedding down in the treetops for the night. You can hear the calls echoing down the vast valleys. The clamor advertises their location and discourages other troops from using the same area of forest and fighting over the same feeding trees.

The roaring of the howler monkeys is an example of group behavior. Animals often live or work together in groups. For example, white pelicans fish in groups. All the pelicans in the group dip their bills into the water at the same time. Any fish that escapes the bill of one pelican is likely to swim right into the open bill of another pelican. Few fish escape this cooperative effort, and all the pelicans get enough to eat.

Some animals, such as dolphins, form social groups rather than work groups. During the day, dolphins swim and play in groups that may number nearly 100. At night they split up and feed on their own.

Figure 15–12. Animals often feed together.

- **Process Skills:** *Analyzing, Formulating Hypotheses*

Predatory fish such as sharks tend to be loners, seeking companionship only at times of mating. Have the students analyze this behavior and offer hypotheses to explain it. (The students should recognize that sharks don't need the help of others for protection or for securing food, since one shark is capable of protecting itself and catching its own food. Sharks do, however, need a mate in order to ensure the survival of their species.)

- **Process Skills:** *Comparing, Inferring*

Explain that the number of fish in a school varies. For example, tuna live in schools of less than 25, whereas herring may form schools of over 100 000 000. Ask the students to infer reasons for these variations in number. (Answers might include reasons such as the larger schools exist because there are more of that type of fish, or smaller fish tend to form larger groups for increased protection.) Point out that predators may be discouraged by the sheer size of a school, and that many fish swimming together may appear as a few large fish to a predator.

Figure 15–13. Dolphins socialize only part of the time.

Protection is another benefit of group living. Fish, for example, often live in large groups called *schools*. An attacking predator is likely to catch only the fish from the fringe of the school. The fish in the center are protected.

Did you ever wonder why the word *school* is used for a group of fish? All kinds of names are used for different animal groups. For instance, a group of whales is a *pod*, while a group of baboons is a *troop*. What other unusual names do you know? In the next activity, you can test your knowledge of animal group names.

Figure 15–14. Schooling is a form of group behavior for some fish.

LASER DISC
45002–45513

Geese flying

DISCOVER BY *Researching*

Answers: an army of frogs, a pack of wolves, a pride of lions, a bed of oysters, a murder of crows, a flock of sheep, a gaggle of geese, a town of prairie dogs. Extend the activity by having the students find more unusual group names, such as a sloth of bears, a muster of peacocks, a knot of toads, and a smack of jellyfish. The book *Animal Crackers* by Robert Hendrickson contains an extensive list.

MEETING SPECIAL NEEDS

Second Language Support

Have students who speak another language share their knowledge of the names given to certain groups of animals in their native language. Encourage the students to compare these names with English names to see whether there are any similarities.

ONGOING ASSESSMENT
ASK YOURSELF

Group behavior may help animals protect their territory or protect themselves.

Here is a list of animals. Next to the list are terms that refer to groups of these animals. See how many animals you can match with their group names. Add any other names for groups of animals that you know.

a _____ of frogs	bed
a _____ of wolves	army
a _____ of lions	flock
a _____ of oysters	pack
a _____ of crows	pride
a _____ of sheep	murder
a _____ of geese	town
a _____ of prairie dogs	gaggle

What are some of the reasons for group behaviors among animals?

TEACHING STRATEGIES, continued

● **Process Skills:** *Inferring, Applying*

Ask the students what they think birds are doing when they chirp or sing. (The students should suggest that the birds are communicating with each other.) Ask the students to give other examples of animal communication. Encourage them to speculate on what the animals are communicating to each other.

● **Process Skills:** *Observing, Evaluating*

Ask any of the students who have been in a back yard, a park, or other outdoor location on a summer night to describe one of the predominant sounds they heard. (the sound made by crickets) Ask the students what they think the sound a cricket makes communicates. (Crickets make sound by crossing their forewings. The sound acts as either an aggressive message or, in most cases, identifies a possible mate.)

Demonstration

Use a tape recorder to record forms of animal communication. For example, tape a dog barking, growling, yipping, whining, or making other sounds under different circumstances—while begging for food, playing, greeting its master, reacting to a stranger. Play the tape for the students and ask them how the different sounds might communicate different messages. You could also tape birds at different times of the day and under different circumstances. Then play the tape for the students and have them try to detect differences in the sounds and analyze what the differences mean.

MULTICULTURAL CONNECTION

Communication behavior varies among cultures. In some, it is important to look directly at people when talking to them. Other cultures frown on this behavior. The acceptable distance between two people who are talking differs among cultures. If the class represents a diversity of cultures, have student volunteers discuss ways people behave when communicating in their ethnic groups. Other investigations could focus on greetings in different cultures and other means of communicating, such as sign language.

Figure 15–15. Pigeons make use of many structures built by humans.

Communication

Animals communicate with one another in many ways. Perhaps you have watched birds and listened to them sing or call to one another. Although animal communications can sometimes be seen or heard, humans often fail to understand them. The most interesting communication occurs as animals make one another aware of boundaries, predators, food, or nesting sites.

For example, when they are ready to mate, pigeons signal by clapping their wings, bowing, and sometimes dragging their tail feathers. Nest sites are almost always chosen by the male. He communicates his choice to his mate by nodding repeatedly toward the ground. The female joins him and also makes nodding motions if she agrees with his selection.

Although they do protect their nests, pigeons do not seem concerned about territory. Perhaps this is because pigeons often live in crowded, busy places.

Many interesting behaviors can also be observed among animals that almost never come in contact with humans. Dr. Forsyth has observed and photographed the fascinating behavior of rain-forest lizards as they communicate about territory.

I saw something strange by the river. A huge lizard, as big as a large dog, waddled along the ground. I thought, "Perhaps this is how the river (Lagarto) got its name." It was a male ctenosaur (TEHN uh sahr) lizard looking like something out of the Dinosaur Age. He had scaly skin banded with stripes of blue-gray and black and a huge spiny crested head with a very wide mouth. He was tilting his head up and down again and again. The cause of all this head-bobbing was another ctenosaur nearby, who was also head-bobbing. Each was signaling that this was his territory: "Keep out, no trespassing." But it soon became clear that both lizards would rather feed than fight. As I watched, they put their heads down and began to pull up large mouthfuls of weeds.

Figure 15–16. Aggression between lizards is a form of communication.

- **Process Skill:** *Expressing Ideas Effectively*

Ask volunteers to act out or demonstrate bee dances. Encourage them to research the topic in order to present the dances more accurately. Point out to the students that the dances tell both the direction and the distance of the food source from the bee hive. A bee dance gives the direction in relation to the sun. The speed of the dance tells the distance—the faster the dance, the closer the food source. A good source of information is Karl von Frisch's *The Dance Language & Orientation of Bees*.

Complex communication occurs among bees. Bees use "dances" to tell other bees where food is located. If the food is closer than 90 m, a bee does a "round dance." If the food is farther away, the dance becomes a more complicated "waggle dance." In the next activity, you can observe other insects.

Figure 15–17. Communication among animals can help to ensure the survival of the entire group.

ACTIVITY

What behavior can you observe in an ant colony?

Process Skills: Observing, Comparing

Grouping: Groups of 3 or 4

Hints

You may want the students to wear gloves or put plastic bags over their hands when digging up ants. Be careful about the kinds of ants that the students dig up. Those living in the southern United States should be particularly careful to avoid digging up fire ants.

After the jars are covered and the ants are put away for the first day, ask the students to predict the kinds of behavior the ants might display during the course of the activity.

▶ **Application**

The students may compare home building and home preparation skills: Both ants and people build and prepare homes. Group behavior may also be compared: Both ants and people can accomplish more work when they divide tasks and work together.

 Have the students place their observations of ant behavior in their science portfolios.

 **LASER DISC**
52189–52304

Whale, beluga; sonogram of song

ACTIVITY

What behavior can you observe in an ant colony?

MATERIALS

ant colony; shovel; newspaper; small trowel; plastic bag; small paper tube; large, clear glass jar with tiny holes in the lid; cornmeal, grass seeds, or sugar; moist sponge; tape; black paper

PROCEDURE

1. **CAUTION: Some ants may bite. Do this activity only under the supervision of your teacher.** Look for an ant colony. Dig down with a shovel, and place the soil on the newspaper.
2. Use the trowel to sort through the soil. Tiny white objects that look like grains of rice are ant cocoons and larvae. Place ants, cocoons, and larvae—about 30 in all—in a plastic bag along with some soil. Seal the plastic bag, and place it in the refrigerator for 5 to 10 minutes to slow down the ants.
3. Stand the paper tube in the center of the jar. Then transfer the ants and the soil into the jar around the outside of the tube. Cover the jar with the lid.
4. Feed the ants by placing a pinch of cornmeal, grass seeds, or sugar on top of the soil. Place a small piece of moist sponge on the top of the soil, too. Keep the sponge moist, and give your ants fresh food every few days.
5. Tape the black paper around the jar to keep out the light. Remove the paper to observe the ants every day for two weeks. Each time, re-cover the jar with the paper after making your observations.
6. When you have finished your observations, return the ants to their colony by carefully emptying the jar near the place where you found them.

APPLICATION

How might you compare the ants' behavior to the behavior of people?

SECTION 2   **411**

TEACHING STRATEGIES, continued

● **Process Skills:** *Observing, Inferring*

Introduce the topic of pheromones by asking the students if they have ever watched a moth flying at night. Have the students describe the moth's flight. (Moths fly back and forth in a zigzag fashion.) Ask the students if they can think of any way this flight pattern is of value to a moth. (Accept reasonable answers.) Tell the students that female moths give off pheromones that attract mates. Male moths detect the pheromones with chemoreceptors in their antennae. The males follow the trail of the female's pheromone. The zigzagging path of the male moth indicates that it is staying within the range of the female's pheromone.

● **Process Skills:** *Inferring, Synthesizing*

Review with the students the social behavior of bees. Ask them if they think pheromones play any part in bee society. (Pheromones are released by the queen bee and inhibit any other queen bees from being produced. Pheromones also attract a following of worker bees to the queen when she leaves the hive.)

✧ Did You Know?

The honeybee depends on group behavior to help it survive cold weather. The bees form a ball in the hive. Those in the center continually vibrate their wings to generate heat. The bees on the outside change places with those on the inside at regular intervals so that no one bee is exposed to the cold for very long.

 Pheromones have been artificially produced as a way to control some insect populations. Artificial female pheromones are used to attract male insects into traps where they are destroyed to prevent mating. Another method uses fibers soaked with artificial pheromones spread over crops. The pheromones give off an odor that attracts male insects and prevents them from finding females to mate with. Since the insects don't mate, the insect population is eventually decimated. The advantage of using pheromones is that they are not harmful to the environment, as pesticides are.

ONGOING ASSESSMENT
▼ **ASK YOURSELF**

The females send out pheromones, and the males pick up the scent, thus allowing the insects to identify and locate each other.

▶ 412 CHAPTER 15

Figure 15–18. Communication by pheromones enables ants to create and follow trails.

The types of communications seen so far have involved sight or sound. Animals also communicate using chemicals called **pheromones** (FEHR uh mohnz). These chemicals serve as signals that are picked up by animals of the same species. Depending on the animal, the signals are picked up by antennae, nostrils, or tongues.

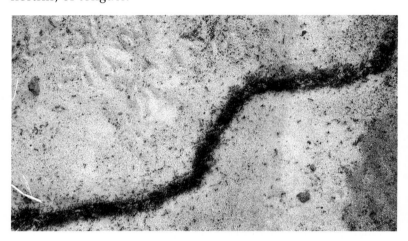

You may have seen ants trailing after one another on the way to some food scraps found by ant scouts. Because of the pheromones being deposited by the ant scouts, the rest of the colony knows where to gather and can easily follow the trail.

Attracting mates is another use for chemical communications. Moths use pheromones in this way. The female usually sends out the chemical, and the male picks up the scent with his antennae. Butterflies also attract mates, but they appear to rely on their bright colors and patterns more than on scents. This may be due to the fact that butterflies mate in the daytime, while moths mate at night.

⭐ DISCOVER BY Researching

Scientists are investigating the use of pheromones as a way to control insect pests. Use your library resources to learn more about the biological control of insects. Write your findings in your journal and be sure to present the pros and cons of this method of pest control.

▼ **ASK YOURSELF**

How might animals use pheromones to find mates?

- **Process Skills:** *Comparing, Generating Ideas*

Discuss how humans use chemical odors as signals. Perfume, cologne, and after-shave lotion are examples of odor enhancers that people apply to their bodies. Also potpourri and scented air fresheners are used to make places smell more inviting. Ask the students to compare these behaviors with those of other animals. Ask the students to cite other examples of odor uses. (breath mints, mouthwash, body powder, rug cleaners)

- **Process Skills:** *Analyzing, Generating Ideas*

Ask the students why they think a frog must hibernate during the winter. Remind the students that a frog is a coldblooded animal. (Because a frog is cold-blooded, its body temperature changes as the temperature of its environment changes. Therefore, if the temperature of the environment decreases, the frog's temperature decreases. To survive the cold, the frog must hibernate to maintain its body temperature.)

Seasonal Behavior

Animals often change their behaviors as the seasons change. Animals that could be in danger of freezing or starving during a harsh winter sometimes survive by going into *hibernation*. Actually, the term for this inactive period is torpor. **Torpor** is a state in which an animal's body functions, or metabolism, actually slows down. Body temperature drops and breathing and heartbeat rates slow considerably.

Contrary to what most people think, bears do not hibernate. Although they sleep long and often during cold months, bears wake periodically and even venture outside their dens on warm days. More importantly, their metabolism does not slow down.

The ground squirrel, a small North American rodent, is an animal that truly hibernates. Before its winter torpor, a ground squirrel eats extra food, which is stored as fat, to supply the energy its body will need to survive. Then it finds a cozy place to sleep for the entire winter. The warm days of spring awaken the squirrel. By this time a new supply of food will await the hungry animal, who may have lost 50 percent of its normal body mass.

Some desert animals go into a kind of torpor, called *estivation*, to escape the heat of summer. The spadefoot toad, for example, digs deep into the bottom of a desert water hole. Even if the water hole dries up, the toad is protected by cool mud until the late summer rains and the cooler days of fall return.

Figure 15–19. The ground squirrel hibernates in winter, while the spadefoot toad estivates in summer.

 **LASER DISC**
3996
Bear, black

 Both hibernation and estivation result in the slowing of body metabolism. Cryogenics is a science that deals with extremely low temperatures. Physicians have been interested in cryogenics as a method for slowing down or halting metabolism in living tissue, thus preserving living body tissue, such as blood, as well as body organs for future use. Using cryogenics would enable doctors to save healthy organs from accident victims for use in future transplants. Sperm and eggs from animals can be stored using cryogenics and later used to conceive new organisms. This use is of particular interest to biologists who are trying to save certain species of animals from extinction.

GUIDED PRACTICE

Write the categories *social behavior, communication, seasonal behavior,* and *parental behavior* on the board. Have the students discuss the categories and provide examples that can be listed in each one.

INDEPENDENT PRACTICE

 Have the students complete the Section Review and Application questions. In their journals, the students can create and answer two additional questions about the material in the section. Encourage the students to ask their questions in class.

EVALUATION

Have the students select three animals and describe one behavior each uses to ensure its survival and one example of how it communicates.

RETEACHING

Have the students create a short oral story that describes the social behavior of an animal of their choice. The story should be written as though the animal were telling it. Invite the students to take turns telling their stories to the class.

SCIENCE BACKGROUND

The Arctic tern migrates the greatest distance of any animal. Each year it journeys from the Arctic to the Antarctic and back. The round-trip distance it travels is equivalent to a trip around the world (40,000 km). A species of snipe has been known to fly non-stop from Japan to Tasmania (an island south of mainland Australia) —a distance of about 8000 km.

BACKGROUND INFORMATION

People have long been fascinated by the homing instincts of birds. One well-known example involves some cliff swallows and a mission church in the town of San Juan Capistrano in California. Every year these swallows leave the mission about October 23 to winter in South America and then return to California on March 19. Long ago, the Romans made use of the swallows' homing instinct. The birds were used to carry messages during wars. They were taken from their nests to outlying forts and then released to return to their nests with messages tied to their legs.

ONGOING ASSESSMENT
ASK YOURSELF

Both hibernation and estivation are inactive states in which organisms' body functions are slowed. Hibernation occurs in the winter and estivation in the summer.

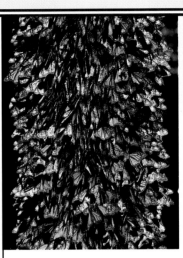

Figure 15–20. Monarch butterflies fly to Mexico in the fall to avoid the harsh winter weather of Canada.

Not all animals escape unfavorable conditions by hibernating or estivating. Some simply move to different locations. This seasonal movement of animals from one location to another is called *migration*.

Monarch butterflies, for example, spend the summer in Canada, then fly to Mexico for the winter. Humpback whales mate and feed off the coast of Alaska. In winter they migrate to the waters off Baja California, where the females give birth before heading north again in the spring. Many African animals, such as zebras, antelopes, and wildebeests, migrate across the veldt, not to seek warmer climates, but to find food and water. Even fishes, such as eels, leave the open ocean to lay their eggs in coastal rivers and streams.

 ASK YOURSELF

How are hibernation and estivation similar? How are they different?

Parental Behavior

In the animal kingdom, there are few examples of parents caring for their young. Most animals simply mate, lay their eggs, and leave their young to fend for themselves. Mammals are the notable exception. A young mammal is unable to survive in the world on its own. Mammals need a certain amount of parental care while learning to find food and defend themselves from predators. Dr. Forsyth observed that howler monkeys spend a great deal of time with their young. Notice the similarities in behavior between howler monkeys and human children.

> *Soon I came upon the rest of the howler troop, quietly feeding on leaf buds. Small babies were riding on their mothers' backs or clinging to their bellies as the females wandered slowly from one branch to another, looking for fresh new growth. They seemed slow and sleepy. Sometimes they would stop feeding to scratch, groom each other or do a little sunbathing, dozing peacefully with their legs draped over the tree limbs. One of their common gestures was to stick their tongues out at one another. It was just like watching a family on a picnic.*

EXTENSION

 Cooperative Learning Working in groups of two or three, the students can research the migration of three or four species of animals. Encourage the students to select at least one animal from outside of North America. Using a globe or world map, have the students trace the migratory paths of their animals. Provide each group with an outline map of the world to plot the migratory paths of the animals. Have the students label the point of origin and the destination as well as major countries and bodies of water that lie along each migratory path.

CLOSURE

Have each student prepare an outline that presents the main points of the section and two or three supporting details for each main point.

Figure 15–21. Howler monkey families are similar to human families.

Although parental care is most common among mammals, a few other animals, even some invertebrates, care for their young. Dr. Forsyth describes an encounter he had with a scorpion mother:

> I knew it was a mother because her back was covered with baby scorpions. The mother scorpion carries her young around so that she can keep them out of harm's reach. She is better at catching food than her youngsters, so they stay with her until they are large enough to catch supper by themselves. Until then the mother grabs large insects such as cockroaches with her strong pincers, then prepares a meal by chewing open the skin of the roach and adding her saliva, which is full of digestive enzymes. These enzymes begin to pre-digest the food. The young scorpions can then climb down and drink their dinner of liquid cockroach.

Figure 15–22. This mother scorpion provides supper as well as transportation for her young.

ASK YOURSELF

Why do mammals take good care of their young?

SECTION 2 REVIEW AND APPLICATION

Reading Critically
1. Describe three reasons why animals need to communicate with each other.
2. What are pheromones? Explain how they are used by ants to find food.

Thinking Critically
3. Explain how an understanding of animal behavior might help a performer who works with animals.
4. When might humans migrate?
5. Why is parenting more common among mammals than among other animal groups?

ONGOING ASSESSMENT
ASK YOURSELF

A young mammal would die if its parents did not care for it until it learned to find food and defend itself.

SECTION 2 REVIEW AND APPLICATION

Reading Critically

1. Animals communicate to establish territorial boundaries, alert others in their group to danger, attract mates, and reveal the location of food sources.

2. Pheromones are chemicals used by animals to communicate with each other. Ant pheromones mark a trail to food. The ant "scout" that finds food deposits pheromones at the spot and along the trail back to the colony. Then other ants can follow the trail the scout marked back to the food source.

Thinking Critically

3. Understanding animal behavior allows the performer to avoid types of behavior that the animals might find threatening, thus keeping the performer out of danger.

4. Humans might migrate to find better living or employment conditions.

5. Parenting is necessary for the survival of the mammal young who, unlike other animal young, are unable to fend for themselves.

SECTION 2

INVESTIGATION

Predicting Animal Responses

Process Skills: Formulating Hypotheses, Predicting

Grouping: Groups of 3 or 4

Objectives
- **Design** an experiment.
- **Predict** sow bug responses.
- **Hypothesize** the location of sow bugs based on their behavior.

Pre-Lab
Discuss with the students the likely responses of insects to conditions of light versus darkness and dampness versus dryness. How will the students be able to tell the insects' preferences?

▶ Analyses and Conclusions
The students should note that the sow bugs tended to move to a damp, dark environment.

▶ Application
Sow bugs are likely to be found in relatively damp and dark environments.

✳ Discover More
Environmental preferences will vary depending on the type of animal chosen.

Post-Lab
Ask the students whether the insects reacted in the manner that they predicted. How did the students determine the insects' environmental preferences?

PERFORMANCE ASSESSMENT
Ask the students to explain how and why they set up their experiment as they did. Ask them to state their hypothesis, give their data, and summarize their results. Use the students' responses to evaluate their knowledge of experimental design.

INVESTIGATION

Predicting Animal Responses

▶ **MATERIALS**
- scissors • shoe box with lid • 10 sow bugs • sponge • water

▼ PROCEDURE

1. Copy this table for your data.

SOW BUG RESPONSES				
Sow Bugs	Number in Light	Number in Dark	Number with No Sponge	Number with Wet Sponge
Predicted				
Observed				

2. Predict how many sow bugs will prefer a light environment and how many will prefer a dark one. Write your predictions in the table.
3. Design a simple experiment using the shoe box to see if your prediction is correct.
4. Gently remove the sow bugs from the shoe box. Now predict how many sow bugs will prefer a damp environment and how many will prefer a dry one. Record your predictions.
5. Design an experiment to see if your prediction about a damp or dry environment is correct.
6. When you are finished with them, return the sow bugs to their natural environment.

▶ **ANALYSES AND CONCLUSIONS**
Were your predictions correct? How did the sow bugs respond to the changes in their environment?

▶ **APPLICATION**
Using your data, hypothesize where you might find sow bugs.

✳ **Discover More**
Find out the kind of environment another animal, such as an earthworm, prefers.

CHAPTER 15 HIGHLIGHTS

The Big Idea—ENVIRONMENTAL INTERACTIONS

The students should be led to the understanding that animal behavior changes as conditions change and that the ability to change plays an important role in an animal's survival. Be sure that the students understand the importance of the connection between the environment and an animal's behavior and appearance.

The students' lists of behaviors should indicate whether they have an appreciation of the variety of animal behaviors. Their lists should now demonstrate a broader understanding than that shown in their earlier lists. Have the students compare their lists with those of other students.

CONNECTING IDEAS

Possible answers are:

1. white pelicans
2. ants
3. zebras
4. ground squirrels
5. spadefoot toads
6. monarch butterflies
7. howler monkeys

CHAPTER 15 HIGHLIGHTS

The Big Idea

Scientists observe animal behavior and how it changes as conditions change. Analysis of observations helps scientists make generalizations about relationships in nature and patterns of change. These generalizations in turn help scientists predict what will happen under different circumstances.

Look back at the list of animal behaviors you wrote in your journal. What new behaviors can you add to your list now that you have completed this chapter? You may wish to do more research and continue to add to your list.

Connecting Ideas

The boxes below contain generalizations about animal behavior. Write the number of each generalization in your journal. After each number, name an animal whose behavior is described by the generalization.

1 Some animals live together because they feed together in the same area.

2 Some animals live in colonies and work together for survival.

3 Some animals live in large groups for protection.

4 Some animals enter an inactive state to survive cold winters.

5 Some animals enter an inactive state to escape summer heat.

6 Some animals move from one location to another to avoid harsh conditions.

7 Some animals spend a lot of time caring for their young.

CHAPTER 15 REVIEW ANSWERS

Understanding Vocabulary

1. Mimicry and camouflage are two methods of defense used by animals.

2. Sounds, movements, and pheromones are examples of ways in which animals communicate with each other.

3. Torpor is an inactive state in which an organism's body functions slow down. Hibernation is winter torpor, and estivation is summer torpor.

4. By forming a herd, a school, or other grouping, animals can often protect themselves from predators.

Understanding Concepts

Multiple Choice
5. c
6. b
7. a
8. b

9. The concept map should be completed with these words.

	torpor	
hibernation		estivation
winter		summer

Short Answer
10. Pheromones are probably more important to moths than to butterflies because moths mate at night when scent would play a more important role. Butterflies mate during the day. They use their bright colors and patterns to attract other butterflies.

11. The offspring of mammals are usually helpless at birth and would not be able to survive without parental care.

Interpreting Graphics

12. The deer is not making any attempt to hide its identity; because it is blending in with its surroundings, it is an example of camouflage. The walking stick is attempting to hide its identity and make predators believe that it is a twig or stick; it is an example of mimicry.

Reviewing Themes

13. Animals migrate to avoid harsh conditions, to find food, or to breed. Monarch butterflies migrate to Mexico to avoid the harsh winter weather of Canada. Zebras, wildebeests, and antelopes migrate across the African plains as the amount of available food changes.

14. Dr. Forsyth is recording numerous and detailed observations of animal behavior in the tropical rain forest. His collected data may support a hypothesis he has already made, or it may cause him to revise that hypothesis and collect additional data. The students may choose any destination, but they should suggest making observations, collecting data, forming hypotheses, conducting experiments, and drawing conclusions as possible methods to use.

CHAPTER 15 REVIEW

Understanding Vocabulary

Explain how the terms in each set are related to each other.
1. mimicry (403), camouflage (400)
2. sounds, movements, pheromones (412)
3. torpor (413), hibernation (413), estivation (413)
4. herd, school

Understanding Concepts

MULTIPLE CHOICE

5. An example of mimicry is
   a) the stripes of a tiger.
   b) the white coat of a polar bear.
   c) the markings on a viceroy butterfly.
   d) the light underbelly of a killer whale.

6. Pigeons use body movements to communicate with each other about
   a) when to migrate.
   b) where to build their nest.
   c) how many eggs to lay.
   d) how to find food sources.

7. Ants know when a food supply is gone because
   a) the other ants are no longer leaving pheromones.
   b) they see other ants returning without food.
   c) they no longer smell the food.
   d) other ants signal to them with their antennae.

8. Caribou herds moving south in the winter is an example of
   a) hibernation.
   b) migration.
   c) estivation.
   d) communication.

9. Copy and complete the concept map.

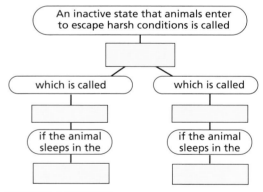

SHORT ANSWER

10. What reason have scientists given to explain why moths use pheromones more than butterflies do?

11. Why do mammal parents have to take good care of their offspring?

Interpreting Graphics

12. Look at the photographs below. Which shows an example of camouflage? Which shows an example of mimicry? Explain the difference.

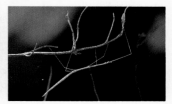

418 CHAPTER 15

Thinking Critically

15. The students might suggest body language including smiles, frowns, crossed arms, posture, nods, and eye contact. It has been estimated that almost 90 percent of human communication is conveyed by nonverbal means.

16. Accept logical responses. The students might suggest it is possible that pheromones act as attracting agents between humans, just as they do for other animals.

17. The students might suggest that military personnel wear clothing designed to blend with the environment so that they cannot be seen by the enemy. Their equipment is similarly disguised. Burglars who work at night may wear dark clothing because they want to blend in with the darkness.

18. The stripes of zebras help them blend together as a group, making it difficult for a predator to pick out and attack a single zebra.

19. The top of the airplane is painted to blend in with the ground so that an enemy looking down on the plane cannot see it. The bottom of the airplane is painted to blend in with the sky so that an enemy looking up cannot see the plane.

20. The patterns are so similar that predators and humans learn to react quickly to the bright colors, which are a more obvious warning, and thus avoid both kinds of snakes.

Reviewing Themes

13. *Environmental Interactions*
Explain why some animals migrate each year. Give at least two examples.

14. *Nature of Science*
How does the journey of Dr. Forsyth demonstrate a scientific method? If you were to make a similar journey, where would you go? What methods would you use in your studies?

Thinking Critically

15. Make a list of the nonverbal communications exchanged by human beings. Explain what you think each communicates.

16. Scientists have hypothesized that humans use pheromones, just as other animals do. In fact, humans often use the pheromones of other animals as perfume. Musk is an example of this. If humans do have pheromones, what do you think their functions are?

17. Do humans ever use camouflage? List examples and explain the purpose of each kind of camouflage.

18. Zebras' black-and-white stripes help them blend in with each other. How might this blending help them avoid predators?

19. A military aircraft may use countershading as camouflage. What would the airplane look like? How would countershading give the airplane an advantage when flying?

20. The venomous coral snake has red, yellow, and black bands. The nonvenomous scarlet king snake also has red, yellow, and black bands, but they are arranged in a different pattern from the coral snake's. Yet the scarlet king snake is still avoided by predators and humans. Why?

Discovery Through Reading

Patent, D. H. *How Smart Are Animals?* Harcourt Brace Jovanovich, 1990. Discusses recent research on levels of intelligence in both wild and domestic animals.

Science PARADE

SCIENCE APPLICATIONS

Discussion

● **Process Skills:** *Inferring, Applying*

Point out to the students that zoo animals such as pandas are kept in captivity for reasons other than the enjoyment and education of the people who visit the zoos. Ask the students to give other reasons for keeping animals in captivity. (The students might suggest that captive animals are shielded against the harmful actions of humans that occur both in and to the natural environment of the animals, and in the case of endangered or threatened animal species, the breeding of such species in a controlled environment might ensure that the species will not become extinct.)

Science PARADE

SCIENCE APPLICATIONS

Romancing the Panda

Probably no pair of animals had ever been so closely watched as Ling-Ling and Hsing-Hsing, the giant pandas at the National Zoo in Washington, D.C. Human observers and television cameras recorded the couple's every move, waiting for the pandas to mate and give the American people one of the world's rarest zoo animals—a panda born in captivity.

Ling-Ling and Hsing-Hsing at the National Zoo

Bear Went a Courtin'

Getting pandas to reproduce outside their natural habitat is not easy. The reproductive cycle of a female panda is, at best, unpredictable. Panda watchers keep an eye on the female for signs that she is in heat, or ready to mate. Then zookeepers must decide how long to keep the male and female together. Pandas are solitary creatures and will fight with each other if left together too long. Even in their native mountains, female pandas often go through heat without meeting a likely male.

Every time Ling-Ling went into heat, panda fans hoped Hsing-Hsing would show some interest in fatherhood. Observers were ecstatic when the pair successfully mated for the first time in 1983, after a two-day courtship. However, biologists were not sure that Ling-Ling was pregnant, so they arranged for her to be artificially inseminated.

In artificial insemination, sperm from a male is injected into the female. The procedure is commonly used in farm animals but is often unsuccessful with wild animals.

After the mating and insemination, zoo biologists kept an eye on Ling-Ling for any sign that she might be pregnant. With pandas it is not easy to tell. A panda's pregnancy is impossible to detect until very near birth. One problem is the difference in size

- **Process Skills:** *Inferring, Applying*

Explain to the students that conservation efforts such as captive breeding of pandas are sometimes met with resistance; not everyone favors such programs. Ask the students to think of several reasons why people or governments may not favor animal conservation programs. (The students might suggest that some people feel the money for such programs could be better spent elsewhere, and governments might not want to fund programs that show little immediate, as opposed to long-term, results.)

EXTENSION

The article mentions that tamarin monkeys have been successfully bred in zoos. Encourage the students to find out about other successful zoo breeding programs, focusing particularly on what made the program work. Point out that breeding animals in captivity is the subject of the article on page 428.

---

between mother and fetus. Ling-Ling weighed as much as a football lineman, while a panda cub at birth weighs only as much as a stick of butter. Another problem is that the time from fertilization to birth can vary from 97 to 175 days. Biologists hypothesize that a female panda's reproductive system actually times the birth to occur in early fall. That way the cub will be weaned in early spring, when the weather is good and food is plentiful.

To the joy of panda watchers, Ling-Ling had cubs several times. But each time, the joy was short-lived because the cubs died within days of birth.

The Need to Breed

Breeding pandas in captivity is important because so few pandas are left in the wild. Pandas live only in remote mountain regions of southern China. Their numbers have dwindled to less than 1000 because humans have destroyed their natural habitat.

Pandas have also been the victims of natural disasters. Pandas feed almost exclusively on bamboo. Once every 100 years or so, bamboo plants flower and die off in great numbers. This happened in the 1970s and left wild pandas with little to eat. In the past, they might have moved to other locations and eaten bamboo that was on a different cycle. But today much of China's bamboo has been cut down to provide farmland for its growing population.

Pandas use a sixth finger to grasp bamboo.

Another reason for their decreasing numbers is that pandas are not very adaptable. They have the simple stomachs and the short intestines typical of meat-eating animals, but they are too slow to catch prey. So they live by eating up to 24 kg of bamboo shoots a day. They have large grinding teeth and a sixth "finger" for grasping bamboo. Because they are such specialized feeders, pandas cannot easily switch to another food supply. When the bamboo dies off, pandas starve. Nearly 140 pandas died during the most recent bamboo die-off.

The Only Hope

Captive breeding may be the only way to save the giant panda. Biologists hope that one day they will have as much success at breeding pandas as they have had with other endangered species, such as tamarin monkeys. These monkeys live in the rapidly disappearing rain forests of coastal Brazil. By studying the few remaining wild tamarin monkeys, biologists learned that they are monogamous—that is, each has only one mate. When zoo biologists imitated this arrangement, the monkeys bred successfully.

Ling-Ling died unexpectedly on December 23, 1992. She was 23 years old. Scientists continue to attempt captive breeding with mating pairs in other locations. Will scientists ever be as successful at breeding pandas? So far, only a handful of cubs have survived in zoos outside China. Without giant pandas, the world would be a poorer place. ◆

Giant panda at the Metro Zoo in Toronto, Canada

READ ABOUT IT!

Discussion

● **Process Skills:** *Inferring, Predicting*

Ask the students to describe the climatic conditions that exist in a rain forest. (Rain forests typically receive a lot of precipitation and have high temperatures throughout the year.) Ask the students to describe where rain forests might be located on the earth. (Generally speaking, rain forests are located near the equator; they can be found in Central America, north central South America, central Africa, Southeast Asia, northern Australia, and the Indonesian islands in the Pacific Ocean.)

● **Process Skills:** *Inferring, Expressing Ideas Effectively*

Explain that the tropical rain forests of the earth contain a rich diversity of animal life; about one-half of all living things on Earth can be found in tropical rain forests. Ask the students to describe the different kinds of animal life that a visitor to a rain forest might encounter. (The students might suggest that animals such as gorillas, monkeys, lemurs, colorful birds, lizards, snakes, and giant millipedes live in the rain forest.)

READ ABOUT IT!

Watch Your Step

by Adrian Forsyth
from *Journey Through a Tropical Jungle*

Michael and Patricia Fogden

A rooster that wouldn't quit crowing had us moving as soon as the first light hit the treetops. It was one of those rare mornings with just a few drifting clouds and the promise of sunshine—not to be wasted. After a quick breakfast of bananas, I reorganized my insect-collecting gear. Eladio worked on his machete, filing a fresh, razor-sharp edge on it, and we headed off into the forest with Eladio leading the way.

Eladio was the perfect person to walk with in the jungle. He was silent, alert and skillful. He was able to find and cut a path with no apparent effort. His machete strokes were neat and economical, disposing of a vine here or a fallen branch there. It seemed as easy as walking. But I knew that his skill was based on years of practice and that it would take me twice as much time to cut a trail half as long.

Eladio opened up a trail into some of the wettest rain forest I had ever seen. Small streams poured down the steep slopes. Mushrooms and molds were sprouting on every fallen log. I even found insects coated with fungus. One of them was a large speckled weevil with a long snout. The weevil had actually been invaded by a fungus that spreads through its body, feeding

- **Process Skills:** *Inferring, Generating Ideas*

Point out that people often have misconceptions about the rain forest. Ask the students whether they believe rain forests are so dense that a visitor must use a machete to make a path. (Accept all responses. In reality, the floor of a mature rain forest is often bare, except for a layer of fallen leaves and other decaying foliage. Less mature rain forests may have thick vegetation on the ground.) Ask the students to describe the conditions that would create bare ground on the floor of a rain forest. (Accept logical responses. Explain that a mature rain forest is covered by a canopy of leaves at a level about 30 m from the ground. Because of this dense canopy, sunlight is diffused, and few plants can grow on the ground beneath the canopy.)

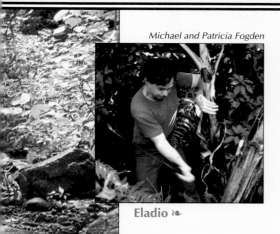

Michael and Patricia Fogden

Eladio

on its tissues. Before the weevil dies this remarkable fungus somehow causes the weevil, normally a low-lying creature, to crawl to the very top of a palm leaf and clamp on in a tight death grip. Then the fungus bursts right out of the weevil, growing into a mushroom-like structure that opens to spread its infective spores on the breeze.

When we reached the Peñas Blancas River, Eladio headed upriver to cut more trail for another day's hike. Before I turned back I sat down for a rest on the river edge. The river was swollen with yesterday's rain and ripping along at high speed. I could hear the groaning of large rocks and boulders grinding and rolling under water along the riverbed. The water and gravity were working at the slow and endless task of wearing the mountain down, turning rock into gravel and gravel into sand, and carrying the sand to the lowlands and the ocean. I looked at the sand and gravel bar and thought how long it had taken to turn a huge boulder into fine sand. I thought about how long it had taken the river to carve the valley out of the mountain and how much time had gone into growing the great forest that blanketed the slopes.

And I had so little time to see it all. I splashed cool clean river water in my face and headed back along the trail alone.

Along the streams, where the sun broke through, sun-loving plants like the heliconia grew in large clumps and produced tremendous stalks of flowers. I stood under one that reached from just above my head all the way down to my ankles. It was easy to be entranced by the almost magical vegetation. I kept staring up, guessing at the height of tremendous trees, and I had to remind myself that this was a place where one had to walk very, very carefully.

The warm lowland rain forest is home to a great variety of venomous snakes and stinging insects. Pit vipers, for example, thrive in warm wet jungles, and I soon saw a mottled green one coiled and camouflaged in some vegetation along the trail. It was warm enough here for the fer-de-lance, a large camouflaged snake that waits coiled beside game trails to ambush small mammals. One thing I wanted to avoid was a close encounter with one of these snakes.

Actually, I had to watch out for ants even more than for snakes. I knew that Eladio had lived down here for many years without once being bitten by a snake. But he had told me that he was always getting stung by ants and wasps.

Some of these ants showed up in strange places. Along one section of trail near the river there were many cecropia trees with hollow stems that looked a bit like bamboo. When I tapped one stem, small azteca ants came boiling out of holes all along the trunk. They ran around furiously with their pointed back ends held straight up. When they reached my finger they swarmed all over it and began biting. Luckily their bite was irritating only on the soft skin between

Fer-de-lance, 2 m (6 ft.) long

Adrian Forsyth

Background Information

In 1950, about 15 percent of the earth's surface was covered by rain forests. Today, less than half of that area remains, and it is likely that only about 5 percent of those rain forests will still exist by the year 2000.

Discussion

● **Process Skills:** *Inferring, Generating Ideas*

Remind the students that along with plants and animals, some people also call the rain forests home. Ask the students to describe some of the hardships that people of the rain forest might face in their everyday lives. (The students might suggest that foods such as fishes and berries are seasonal in nature and require the inhabitants of the area to move periodically from areas of scarce food to areas where food is more plentiful. The students might also suggest that the high temperatures and humidity of the rain forest make the preservation of foods difficult, especially with a lack of amenities such as electricity and refrigeration.)

Cecropia tree swarming with ants

my fingers. But I could see that climbing a cecropia tree would be almost impossible because hundreds of these ants would swarm over my face and body.

The cecropia tree provides a home for the aztecas, and they aggressively attack any insect or animal they encounter on their tree. With the ants on duty, the tree stays free from grazing and leaf damage. To encourage this relationship, cecropia trees have hollow stems where the ants live. They also have special glandular areas that produce tiny white bits of a special protein-rich food for the ants.

For some reason large cecropias have few ants or no ants at all. Perhaps that is why the three-toed sloths that love to eat cecropia leaves are usually seen in the largest cecropia trees, the ones that have lost the ant colonies.

At one point the trail skirted around a large fallen tree. The fallen tree had opened up a gap in the foliage, allowing sunlight to reach the forest floor, encouraging a dense growth of weedy plants. On these plants I found some green treehopper insects and some black and yellow polka-dotted ones. They had also developed a relationship with the ants. The hoppers were sucking sap out of plant stems, and the ants were walking all over the hoppers, milking the sweet sap from them. But the ants were working for their supper: when I poked at the treehoppers the ants rushed at my finger, biting wildly. The arrangement was something like dairy farming, with the treehoppers getting protection as they fed on the plant, and the ants getting sugary rewards for tending the hoppers.

But while I was watching all this, I found that I had stepped where I shouldn't have. Swarming up my pant leg, both inside and out, was a cloud of huge ants.

I was standing in the midst of an army ant raid! I jumped back along the trail, kicking off my boot and hiking up my pants. Only one large soldier ant, as big as a paperclip, had made its way onto my skin, but that was enough. It locked its great sickle-shaped mandibles into my calf and

Journal Activity
Have the students recall what they have read and discussed about tropical rain forests. After reminding the students that tropical rain forests are disappearing at a rapid pace, ask the students to write a journal entry that describes actions that they could perform to help prevent or slow the loss of the earth's tropical rain forests.

EXTENSION

Studies performed by the United Nations and other organizations have estimated the rate at which the rain forests are being destroyed. Challenge interested students to use reference materials and discover how many hectares of rain forest are lost every minute of every day. Encourage the students to share their findings with the class. (Typical estimates range from 10 to 40 hectares of tropical rain forest being permanently destroyed or seriously degraded each minute.)

Army ants carry off a wasp larva.

Adrian Forsyth

plunged its sting repeatedly into the skin. The fiery hot sting made me jump even higher.

When I yanked the army ant off my leg, I saw how determined its grip was. Its head remained firmly attached to my skin even when the rest of its body was broken off. I could see why South American Indians use these gripping army ant heads to stitch up open wounds.

I hastily brushed the remaining ants off my pants and boots and stepped back carefully to take a closer look at the raid. It was easy to do as long as I kept an eye on the shifting columns of ants spreading through the forest. At the head of the raid, scouts were dragging their abdomens along the ground and along branches, laying down a chemical signal to direct the other army ants to new areas and food sources. Sometimes to cross a gap they strung their bodies together in a chain so that other ants could run across from tree to tree. The endless columns of ants divided and criss-crossed and blended together again, always advancing, weaving their way up and across the forest.

Careful to keep out of their path (one army ant sting was more than enough), I gazed at the spectacle unfolding before me. The army of ants surged forward. It was a tremendous feat of chemical coordination and communication among tens of thousands of individuals. Somehow all those ants, working without language or leaders, had managed to organize an effective hunt. ◆

THEN AND NOW

Discussion

● **Process Skills:** *Inferring, Applying*

Ask the students how the work of scientists such as Ernest Just influences the lives and health of people today and in the future. (Just's research provided scientists and medical researchers with a better understanding of how organisms develop. This improved knowledge allows medical professionals to be better trained and able to increase the quality and length of human life.)

Ask the students why the word *courageous* has often been used to describe the life and work of Jane Goodall. (The students might infer that Goodall conducts her research for long periods of time far from civilization in an environment some might consider dangerous.)

Journal Activity
Have the students describe why research—such as that performed by Goodall, Just, and others—is important to everyone. Encourage the students to cite particular examples of research efforts that have benefited people.

THEN AND NOW

Ernest Just (1883—1941)

In 1908 Ernest Just began his career at the Marine Biology Laboratory in Woods Hole, Massachusetts. For 20 years he studied embryology, or the development of organisms. While researching marine invertebrates, he also served as head of the Department of Physiology at Howard University.

Earnest Just was born in Charleston, South Carolina, in 1883. After high school, he worked his way north and enrolled at Kimball Union Academy in New Hampshire. Just continued his education at Dartmouth College, where he graduated with honors in 1907. In 1916 he received his Ph.D. in zoology from the University of Chicago.

Just spent many years in Europe, teaching and doing research. He wrote several books and numerous articles on cell physiology. His work was respected by eminent biologists of his day, and in 1915, the NAACP awarded Just the first Spingarn Medal for his efforts to improve the quality of medical education at Howard. ◆

Jane Goodall (1934—)

Jane Goodall is best known for her study of the behavior of wild chimpanzees. In June 1960 she set up camp in the Gombe Stream Game Preserve in Tanzania. At first she had trouble just finding chimpanzees. When she did find them, she established a routine for observing them at a distance, using binoculars. After nearly a year, the wild chimpanzees became accustomed to Goodall's presence. She was then able to record firsthand observations of nearly every part of their behavior. She found that chimpanzees make and use simple tools and that they hunt for and kill other animals.

Jane Goodall was born in London, England, in 1934. As a child, she developed a strong interest in animals. She studied them and recorded observations of their behavior in her notebooks. At the age of 18, she took a job at Oxford University to earn money for a trip to Africa. Once there, she met paleontologist Louis S. B. Leakey and became his assistant at the Coryndon Museum of Natural History in Nairobi, Kenya.

Although she began her studies of wild chimpanzees without formal education, she later earned her doctorate from Cambridge University. Today Goodall continues her research. One of her concerns is the increasing pressure on wild chimpanzee communities from the interference of humans. ◆

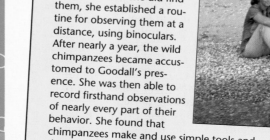

426 UNIT 5

SCIENCE AT WORK

Discussion

● **Process Skills:** *Applying, Generating Ideas*

Point out to the students that Peter Aromando is an enthusiastic science teacher. Ask the students why teaching science might be an enjoyable job to have. (The students might suggest that being with people and performing demonstrations are enjoyable activities.) Then have the students recall that Aromando believes science is important to everyone. Ask the students to describe several ways in which science has affected their lives. (Encourage creative responses. Science affects people's lives in countless ways.) Then have the students create, without using a dictionary or other reference book, a definition of the word *science*.

SCIENCE AT WORK

Peter Aromando, SCIENCE TEACHER

After placing the last plant on the table, Peter Aromando steps back and makes a mental note. There are slides, forceps, a scalpel, a microscope, and a plant at each lab station. Everything is set up. The next day's laboratory is ready. He then looks at the clock and heads out the door for a conference. Aromando teaches science at Deland Middle School in Deland, Florida.

Before becoming a teacher, Aromando was a medical technologist. He worked at a blood bank and in hospital laboratories. "What has helped me a great deal in my career as a science teacher," says Aromando, "is my practical work experience in the science field. This work experience helps me bring real-life lab skills into the classroom. I try to teach students safety in the lab, accuracy in experiments and in gathering data, and the proper care of equipment."

Aromando is the science teacher on the seventh-grade teaching team. He and the other teachers on the team meet daily to plan activities and field trips and to review their students' progress. Using the team concept, Aromando is able to combine the skills he teaches in his science class. Aromando feels the science classroom is an excellent place to demonstrate the use of all of a student's skills. "I believe the science classroom should be an active classroom with stimulating lab activities. Students should view the science classroom as a dynamic place where they can learn about the environment," says Aromando.

In the science lab, students get to learn how science works. Using tools such as the microscope, balance, and models, Aromando tries to take science from the textbook and bring it to life for students. "Science is everywhere," explains Aromando. He adds, "Students learn best by hands-on experiences. By using their senses, students can learn about the world around them."

Aromando believes that science also helps all students develop analytic and critical-thinking skills. They can learn how to analyze facts, gather data, and solve problems. He feels science is an excellent foundation for many careers. All of the health fields—medicine, nursing, lab technology, physical therapy, nuclear medicine, and so on—require a basic science education. Other science-related career opportunities are environmental studies, research, space exploration—and, of course, teaching. ◆

Discover More
For more information about careers in teaching science, write to the
National Science Teachers Association
Attn: Public Information Office
1742 Connecticut Avenue, N.W.
Washington, DC 20009

SCIENCE/TECHNOLOGY/SOCIETY

Discussion

● **Process Skills:** *Inferring, Applying*

Ask the students to name some actions of humans that have contributed to the endangerment of animal species. (The students might name the pursuit of natural resources, the development of farmland, and the growth of cities.) Ask the students whether these human actions intentionally or unintentionally affect animal species in negative ways. (Accept logical responses. Human actions can both intentionally and unintentionally harm wildlife. For example, the harvesting of forest for paper products intentionally decreases the amount of available habitat for tree-dwelling creatures, whereas the pollutants emitted by motor vehicles unintentionally harm all of the wildlife on the earth.) Have the students describe a scenario in which something other than human actions causes animals to become endangered or extinct. (The students might recall the extinction of dinosaurs due perhaps to catastrophic changes in the climate of the earth.)

SCIENCE/TECHNOLOGY/SOCIETY

Saving Endangered Wildlife

The population of the giant pandas of China is dwindling as human settlements destroy their natural territory. The okapi, a giraffelike animal that lives in the rain forests of Zaire, is losing its home to farming and lumbering. As their habitats disappear, countless species are threatened with extinction.

The Role of Zoos

Over the last few decades, zoos have become increasingly involved in the breeding of endangered species. Unfortunately, saving animals from extinction is more complex than simply supplying

Embryo transfer allowed this eland to give birth to a bongo calf.

▶ **428** UNIT 5

Journal Activity Explain to the students that because of the stress of captivity, captive animals generally do not live as long as their relatives in the wild. Ask the students to debate whether the benefits outweigh the disadvantages of captivity. Then encourage volunteers to share their journal entries with their classmates.

EXTENSION

You might choose to extend the discussion by challenging the students to describe how artificial insemination can be used to create offspring that are genetically superior to those created by more natural methods of reproduction.

This horse was a surrogate mother for the baby zebra.

them with mates. Animals in captivity often do not get along. The stresses that go with living in small areas often keep animals from reproducing. In addition, zoos cannot presently house the number and variety of individual animals needed for long-term survival of a species.

To counter these problems, researchers have created a computer database of animals in captivity. It is called the International Species Inventory System (ISIS). ISIS helps identify likely candidates for breeding programs. A British database, the *National Online Animal History,* offers information that indicates how closely animals may be related to each other. This information is needed because breeding close relatives limits the genes available for the next generation.

Technology to the Rescue

The most commonly used breeding method is *artificial insemination*. It involves the collection of sperm from a male and its placement into the reproductive tract of a female. Another technique, *in vitro fertilization,* involves placing an ovum and sperm together in a glass dish. After fertilization, the resulting embryo is transferred to a selected female.

Embryo transfer, another breeding method, consists of placing the embryo of one female into the reproductive tract of another female, called the *surrogate mother.* Scientists have even done cross-species embryo transfers between species that are closely related. For example, biologists placed the embryo of a bongo antelope into a surrogate eland antelope, which later gave birth to the bongo calf.

Because the reproductive systems of exotic animals differ from one species to another, researchers have successfully adapted artificial breeding methods to only about 20 species. Perhaps advances in reproductive technology will make artificial breeding possible with a greater variety of endangered animals. ◆

UNIT 5   **429**

UNIT 6

THE HUMAN BODY

UNIT OVERVIEW

This unit discusses the systems of the human body, how they function, how they work together, and what is needed to maintain their health. The unit also deals with substances that are often responsible for the destruction of a person's body—drugs, alcohol, and tobacco.

Chapter 16: Support and Movement, page 432

This chapter presents the characteristics and functions of the skeletal system. The formation of bone and muscle tissues is described, as well as how the tissues work together to supply the body with a supportive structure that enables movement.

Chapter 17: Digestion and Circulation, page 456

A description of the digestive system is presented in this chapter, along with an explanation of the processes involved in digestion. Carbohydrates, lipids, proteins, vitamins, water, and minerals, and how the body utilizes them are discussed. The chapter also discusses blood and the human circulatory system. The components of blood and their functions are described. This chapter presents the structure of the heart and a description of blood vessels. Finally, the path of blood through the body is explained.

Chapter 18: Respiration and Excretion, page 488

This chapter describes the structures and functions of the respiratory and excretory systems. Breathing and gas exchange are explained. The structures and functions of the skin are also presented.

Chapter 19: Coordination and Control, page 510

This chapter discusses the major components of the central nervous system and explains how these components coordinate all body activities. The major categories of drugs are discussed, with emphasis on such drugs as alcohol and tobacco.

Chapter 20: Reproduction and Development, page 542

This chapter presents the endocrine and reproductive systems of the human body. The male and female reproductive systems are described. Fertilization of an ovum and development of an embryo are explained, as are the processes of prenatal development and birth.

Science Parade, pages 564-573

The articles in this unit's magazine relate the topic of the human body to nutrition and physical fitness, to sleep patterns, and to the aging process. The career feature presents the work of a pharmacist, and the biographical sketches feature a blood researcher and an aerospace physician.

UNIT RESOURCES

PRINT MEDIA FOR TEACHERS

Chapter 16
Klein, Madelyn. *Arthritis*. Franklin Watts, 1989. Discusses types, causes, and treatments of arthritis.

Selim, Robert D. *Muscles: The Magic of Motion*. U.S. News Books, 1982. Examines the different types of muscles and their ability to make the human body move.

Chapter 17
Asimov, Isaac. *How Did We Find Out About Blood?* Walker, 1986. Follows the history of human knowledge about blood and its functions.

Avraham, Regina. *The Digestive System*. Chelsea House, 1989. A comprehensive explanation of the structures and functions of the digestive system.

McGowen, Tom. *The Circulatory System: From Harvey to the Artificial Heart*. Franklin Watts, 1988. Traces the development of theories about circulation.

Chapter 18
Parker, Steve. *The Lungs and Breathing*. Franklin Watts, 1989. Describes the function of the respiratory system.

Walzer, Richard A. *Healthy Skin: A Guide to Lifelong Skin Care*. Consumers Union, 1989. Discusses recent developments in the field of dermatology.

Young, Alida E. *Is My Sister Dying?* Willowisp Press, 1991. A girl donates a kidney to help keep her sister alive.

Chapter 19
Kittredge, Mary. *The Senses*. Chelsea House, 1990. Describes the function of sensory organs and explains their relationship with the brain and nervous system.

Silverstein, Alvin, and Virginia Silverstein. *World of the Brain*. Morrow, 1986. Describes the structures and functions of the brain and nervous system.

Chapter 20
Growth and Development: The Span of Life. Torstar Books, 1985. Describes the physical, mental, and emotional development of a human throughout a lifetime.

Little, Marjorie. *The Endocrine System*. Chelsea House, 1990. Presents the structure and function of endocrine glands and hormones.

Nourse, Alan E. *Menstruation*. Franklin Watts, 1987. Discusses the menstrual cycle and its significance.

PRINT MEDIA FOR STUDENTS

Chapter 16
Parker, Steve. *The Skeleton and Movement*. Watts, 1989. Presents human anatomy and the skeletal system.

Chapter 17
Lambourne, Mike. *Down the Hatch: Find Out about Your Food*. Millbrook Press, 1992. This book is about food and digestion.

Chapter 18
Lambourne, Mike. *Inside Story: The Latest News about Your Body*. Millbrook Press, 1992. This book discusses human organ systems.

Chapter 19
Parker, Steve. *The Brain and Nervous System*. Watts, 1990. Describes the control system of the body.

Chapter 20
Avraham, Regina. *The Reproductive System*. Chelsea House Publishers, 1991. Discusses aspects of reproduction and explains the stages of growth from fertilization to birth.

ELECTRONIC MEDIA

Chapter 16

Muscular System. Videocassette. Coronet. Explains how muscles move the body.

The Skeleton. Videocassette. Britannica. Identifies the structure of the skeleton.

Chapter 17

Circulatory System. Videocassette. Coronet. Discusses the components of the circulatory system.

Nutrition: Eating Well. Videocassette. National Geographic. Investigates nutritional needs.

The River of Life. Videocassette. Britannica. Examines the structure and function of blood.

Chapter 18

The Lungs and Respiratory System. Videocassette. Britannica. Shows how oxygen gets from lungs to blood to cells.

The Skin: Its Structure and Function. Videocassette. Britannica. Examines structures and functions of skin.

Work of the Kidneys. Videocassette. Britannica. Illustrates structures of the kidneys.

Chapter 19

Drug Abuse: Sorting It Out. Videocassette. Britannica. Discusses substance abuse.

The Nervous System. Videocassette. Britannica. Shows how the nervous system controls body activities.

Sense Organs. Videocassette. Coronet. An overview of sensory receptors in the skin, ears, eyes, nose, and mouth.

Chapter 20

The Endocrine System. Videocassette. Britannica. Describes the endocrine system as a chemical control mechanism.

Human Reproduction. Videocassette. Britannica. Illustrates aspects of reproduction.

Reproductive System. Videocassette. Coronet. Explains how the endocrine and reproductive systems make reproduction possible.

UNIT 6

Discussion Remind the students that they each possess genes that determine every trait they currently display or will develop during their lives. Point out that each student has, for example, genes for height. Ask the students to describe several behaviors that can help them grow to their maximum height and develop to their maximum physical potential. (Accept logical responses. Quality of diet and frequency of exercise are two of the most important factors that contribute to a person's maximum height and physical potential.)

UNIT 6 — THE HUMAN BODY

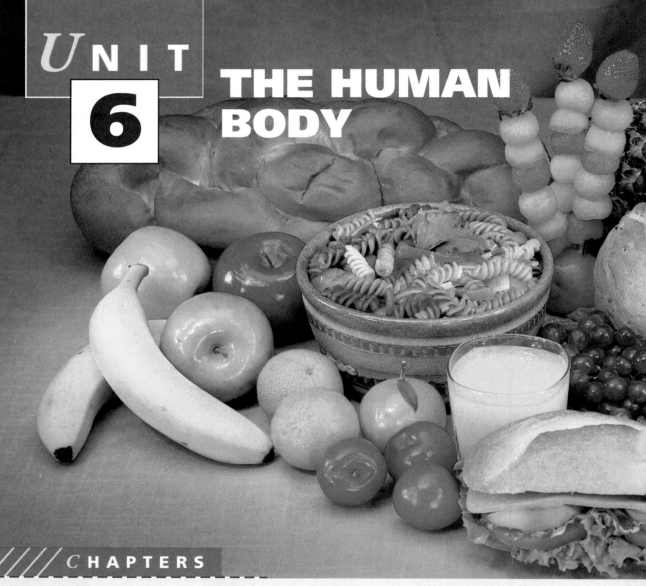

CHAPTERS

16 Support and Movement 432

Bones and muscles work together to make the coordinated movements of your body possible.

17 Digestion and Circulation 456

Each time you eat, nutrients are released and then circulated throughout your body to meet the energy needs of every cell.

18 Respiration and Excretion 488

Every breath you take is part of a complex process that helps fuel and cleanse your body. How does your skin aid this process?

19 Coordination and Control 510

The movement of your eyes along these lines of type and the resulting words and thoughts in your mind are parts of your body's complex coordination and control systems.

Journal Activity You can extend the discussion by focusing on eating trends. Point out, for example, that sometimes people who lead very busy lives feel they have little time to eat as they should. They skip meals or rely too much on fast foods or frozen dinners. Ask the students to describe situations in which they did not eat as well as they should have. What was wrong with those meals? Then ask the students to describe in a journal entry the importance of eating well and to suggest what they could do to ensure that they eat better balanced meals more of the time.

Mmmmm. Faced with all this food, what would you choose? Every day you make choices about the food you eat. Do you make the best choices?

Scientists keep learning about how the body works and what is necessary to maintain fitness. The type and amount of food you eat can have an impact on the health of your body. More than ever the daily choices you make can shape the future of your body's fitness.

20 *Reproduction and Development* 542

You began as new life and developed into who you are today.

Science PARADE

SCIENCE APPLICATIONS
Daily Decisions About Your Body 564

READ ABOUT IT!
"I Slept for Science: A Study of Sleep Problems" 567

THEN AND NOW
Charles Drew and Irene Duhart Long 571

SCIENCE AT WORK
Mamie Lou, Pharmacist 572

SCIENCE/TECHNOLOGY/ SOCIETY
Aging 573

UNIT 6 **431**

CHAPTER 16

SUPPORT AND MOVEMENT

PLANNING THE CHAPTER

Chapter Sections	Page	Chapter Features	Page	Program Resources	Source
Chapter Opener	432	*For Your Journal*	433		
Section 1: THE SKELETAL SYSTEM	434	Discover By Doing (B)	434	*Science Discovery**	SD
• What Bones Do (B)	434	Discover By Observing (A)	437	Investigation 16.1:	
• Types of Bones (B)	436	Discover By Calculating (A)	439	Human Movement (A)	TR, LI
• The Composition and Structure of Bones (B)	436	Discover By Observing (B)	441	Reading Skills: Determining Cause and Effect (B)	TR
• Growth and Formation of Bones (A)	438	Discover By Observing (A)	442	Connecting Other Disciplines: Science and Health, Coordinating the Muscular and Skeletal Systems (H)	TR
• The Skeleton (H)	439	Section 1 Review and Application	443	The Human Skeletal System (A)	IT
• Joints (A)	442	Skill: Organizing Information (B)	444	Joints (B)	IT
				Study and Review Guide, Section 1 (B)	TR, SRG
Section 2: THE MUSCULAR SYSTEM	445	Discover By Doing (A)	447	*Science Discovery**	SD
• Types of Muscles (B)	445	Discover By Calculating (A)	449	Investigation 16.2: Muscle Pairs (H)	TR, LI
• How Muscles Work (A)	449	Discover By Doing (A)	449	Thinking Critically (H)	TR
		Activity: How do the muscles, bones, tendons, and ligaments in your hand work? (B)	450	Extending Science Concepts: Muscle Fatigue (A)	TR
		Discover By Observing (B)	451	The Human Muscular System (A)	IT
		Section 2 Review and Application	451	Record Sheets for Textbook Investigations (A)	TR
		Investigation: Observing Chicken Wing Muscles and Bones (A)	452	Study and Review Guide, Section 2 (B)	TR, SRG
Chapter 16 HIGHLIGHTS	453	The Big Idea	453	Study and Review Guide, Chapter 16 Review (B)	TR, SRG
Chapter 16 Review	454	For Your Journal	453	Chapter 16 Test	TR
		Connecting Ideas	453	Test Generator	

B = Basic   **A** = Average   **H** = Honors
The coding Basic, Average, and Honors indicates subsections, features, and resources that might be appropriate for different levels of learners. For additional suggestions regarding choice of topic and depth of coverage, see the Pacing Chart on pages T26–T29.

*Frame numbers at point of use
(TR) Teaching Resources, Unit 6
(IT) Instructional Transparencies
(LI) Laboratory Investigations
(SD) *Science Discovery* Videodisc Correlations and Barcodes
(SRG) Study and Review Guide

▶ 431A

Journal Activity You can extend the discussion by focusing on eating trends. Point out, for example, that sometimes people who lead very busy lives feel they have little time to eat as they should. They skip meals or rely too much on fast foods or frozen dinners. Ask the students to describe situations in which they did not eat as well as they should have. What was wrong with those meals? Then ask the students to describe in a journal entry the importance of eating well and to suggest what they could do to ensure that they eat better balanced meals more of the time.

Mmmmmm. Faced with all this food, what would you choose? Every day you make choices about the food you eat. Do you make the best choices?

Scientists keep learning about how the body works and what is necessary to maintain fitness. The type and amount of food you eat can have an impact on the health of your body. More than ever the daily choices you make can shape the future of your body's fitness.

20 *Reproduction and Development* 542

You began as new life and developed into who you are today.

Science PARADE

SCIENCE APPLICATIONS
Daily Decisions About Your Body 564

READ ABOUT IT!
"I Slept for Science: A Study of Sleep Problems" 567

THEN AND NOW
Charles Drew and Irene Duhart Long 571

SCIENCE AT WORK
Mamie Lou, Pharmacist 572

SCIENCE/TECHNOLOGY/ SOCIETY
Aging 573

CHAPTER 16

SUPPORT AND MOVEMENT

PLANNING THE CHAPTER

Chapter Sections	Page	Chapter Features	Page	Program Resources	Source
Chapter Opener	432	*For Your Journal*	433		
Section 1: THE SKELETAL SYSTEM	434	Discover By Doing (B)	434	*Science Discovery* *	SD
• What Bones Do (B)	434	Discover By Observing (A)	437	Investigation 16.1:	
• Types of Bones (B)	436	Discover By Calculating (A)	439	Human Movement (A)	TR, LI
• The Composition and Structure of Bones (B)	436	Discover By Observing (B)	441	Reading Skills: Determining Cause and Effect (B)	TR
• Growth and Formation of Bones (A)	438	Discover By Observing (A)	442	Connecting Other Disciplines: Science and Health,	
• The Skeleton (H)	439	Section 1 Review and Application	443	Coordinating the Muscular and Skeletal Systems (H)	TR
• Joints (A)	442	Skill: Organizing Information (B)	444	The Human Skeletal System (A)	IT
				Joints (B)	IT
				Study and Review Guide, Section 1 (B)	TR, SRG
Section 2: THE MUSCULAR SYSTEM	445	Discover By Doing (A)	447	*Science Discovery* *	SD
• Types of Muscles (B)	445	Discover By Calculating (A)	449	Investigation 16.2:	
• How Muscles Work (A)	449	Discover By Doing (A)	449	Muscle Pairs (H)	TR, LI
		Activity: How do the muscles, bones, tendons, and ligaments in your hand work? (B)	450	Thinking Critically (H)	TR
				Extending Science Concepts: Muscle Fatigue (A)	TR
		Discover By Observing (B)	451	The Human Muscular System (A)	IT
		Section 2 Review and Application	451	Record Sheets for Textbook Investigations (A)	TR
		Investigation: Observing Chicken Wing Muscles and Bones (A)	452	Study and Review Guide, Section 2 (B)	TR, SRG
Chapter 16 HIGHLIGHTS	453	The Big Idea	453	Study and Review Guide, Chapter 16 Review (B)	TR, SRG
Chapter 16 Review	454	For Your Journal	453	Chapter 16 Test	TR
		Connecting Ideas	453	Test Generator	

B = Basic   A = Average   H = Honors
The coding Basic, Average, and Honors indicates subsections, features, and resources that might be appropriate for different levels of learners. For additional suggestions regarding choice of topic and depth of coverage, see the Pacing Chart on pages T26–T29.

*Frame numbers at point of use
(TR) Teaching Resources, Unit 6
(IT) Instructional Transparencies
(LI) Laboratory Investigations
(SD) *Science Discovery* Videodisc Correlations and Barcodes
(SRG) Study and Review Guide

▶ 431A

CHAPTER MATERIALS

Title	Page	Materials
Discover By Doing	434	none
Discover By Observing	437	(per group of 3 or 4) raw chicken leg bone, vinegar, journal
Discover By Calculating	439	(per group of 3 or 4) measuring tape
Discover By Observing	441	none
Discover By Observing	442	(per individual) journal
Skill: Organizing Information	444	(per group of 2 or 3) paper, pencil
Discover By Doing	447	none
Teacher Demonstration	448	small mirrors
Discover By Calculating	449	(per individual) paper, pencil
Discover By Doing	449	none
Activity: How do the muscles, bones, tendons, and ligaments in your hand work?	450	(per group of 3 or 4) blank white paper, pencil, colored pencils, crumpled paper, tape, coin
Discover By Observing	451	(per individual) wire (9-cm), dinner knife, heavy book
Investigation: Observing Chicken Wing Muscles and Bones	452	(per group of 3 or 4) raw chicken wing, dissecting pan, scissors, forceps, paper towels

ADVANCE PREPARATION

For the *Discover By Observing* on page 437, the students will need vinegar and a raw chicken leg bone. Raw chicken wings are needed for the *Investigation* on page 452.

TEACHING SUGGESTIONS

Field Trip
Visit an athletic training and rehabilitation facility associated with a local university, hospital, or professional sports team. The facility's trainer can describe the equipment and programs used to rehabilitate damaged limbs.

Outside Speaker
Have a doctor, or other health professional who is involved with sports and fitness-related injuries, visit your class. Ask the speaker to concentrate on skeletal and muscular problems that can be avoided by proper health habits and fitness training.

CHAPTER 16
SUPPORT AND MOVEMENT

CHAPTER THEME—SYSTEMS AND STRUCTURES

This chapter introduces the students to the skeletal and muscular systems of the human body. The students learn that through the interaction of the skeletal and muscular systems, a human is able to move. The structures and functions of bones and muscles are also discussed. The major theme of this chapter **Systems and Structures** is also developed in Chapters 4, 5, 7, 8, 9, 10, 12, 13, 16, 17, 18, 19, and 20.

MULTICULTURAL CONNECTION

Any type of physical activity requires the coordination of bone and muscle movement. One of the most popular and graceful forms of physical activity is dance. Dance is a form of artistic and personal expression through rhythmic movements, which is often accompanied by music. Most cultures have developed traditional folk dances. Have the students learn about the folk dances of a particular culture. They may want to investigate a particular dance, such as the Irish jig, the Spanish flamenco, the Hopi rain dance, the Yoruba Apala dance, or the Japanese bugaku. Suggest that the students investigate a dance's origins and traditional meaning. As the students investigate the dance, they might learn a few basic steps and demonstrate these for the class.

MEETING SPECIAL NEEDS

Mainstreamed

During the study of this chapter, remember to be particularly sensitive to the feelings and reactions of students who have disabilities involving bones and muscles. Also be sure to adapt any activities that they may have difficulty performing.

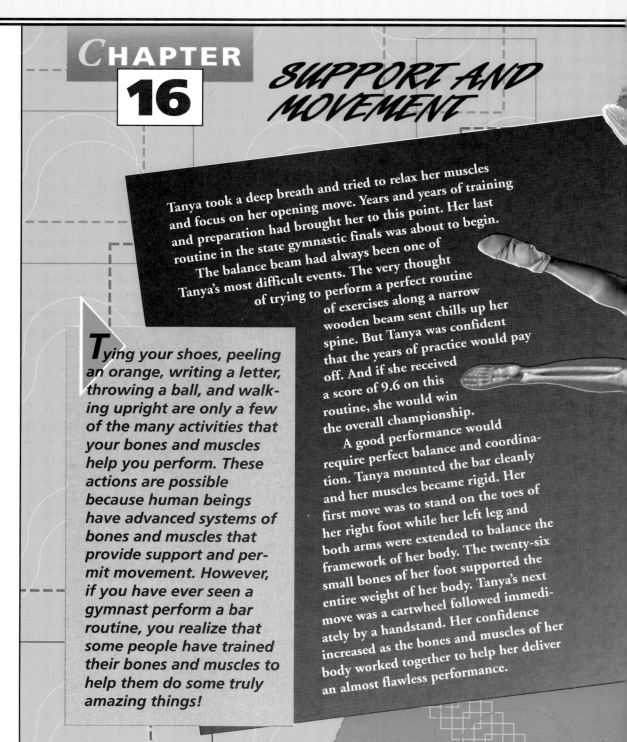

CHAPTER 16
SUPPORT AND MOVEMENT

*T*ying your shoes, peeling an orange, writing a letter, throwing a ball, and walking upright are only a few of the many activities that your bones and muscles help you perform. These actions are possible because human beings have advanced systems of bones and muscles that provide support and permit movement. However, if you have ever seen a gymnast perform a bar routine, you realize that some people have trained their bones and muscles to help them do some truly amazing things!

Tanya took a deep breath and tried to relax her muscles and focus on her opening move. Years and years of training and preparation had brought her to this point. Her last routine in the state gymnastic finals was about to begin.

The balance beam had always been one of Tanya's most difficult events. The very thought of trying to perform a perfect routine of exercises along a narrow wooden beam sent chills up her spine. But Tanya was confident that the years of practice would pay off. And if she received a score of 9.6 on this routine, she would win the overall championship.

A good performance would require perfect balance and coordination. Tanya mounted the bar cleanly and her muscles became rigid. Her first move was to stand on the toes of her right foot while her left leg and both arms were extended to balance the framework of her body. The twenty-six small bones of her foot supported the entire weight of her body. Tanya's next move was a cartwheel followed immediately by a handstand. Her confidence increased as the bones and muscles of her body worked together to help her deliver an almost flawless performance.

CHAPTER MOTIVATING ACTIVITY

In a straight line along the length of the classroom, place two parallel strips of masking tape 10 cm apart. Ask the students to remove their shoes and walk between the two strips. Then have each student stand on one foot between the strips as long as he or she can. Ask the students to repeat the activity several times throughout the study of the chapter. Have the students record the experiences in their journals. Was it easy or difficult to walk between the strips? Was standing in place on one foot easy? Why or why not? Does repeated practice make the exercise easier to perform?

For Your Journal

Through their answers to the journal questions, the students will reveal what they know about the skeletal and muscular systems. Although most students will have some understanding of how the human body moves through interactions of these systems, they may not know how these interactions take place. They may also have some misconceptions about the structures and functions of bones and muscles. Address any misconceptions you note during the discussion of the chapter's content.

ABOUT THE PHOTOGRAPH

This gymnast is in the midst of a routine on a balance beam, which is a wooden beam about 10 cm wide. The routine, which must last at least 70 seconds, will include somersaults, handsprings, and cartwheels. The gymnast will be given a score based on her routine's difficulty, her execution of the routine, and her poise and grace. The balance beam event is just one of four events in which the gymnast will compete. She will also perform in the side horse vault, uneven parallel bars, and floor routine events. In national and international competitions, the gymnast must perform compulsory routines, in which certain exercises are required, and in optional routines, in which exercises are tailored to the strengths of the gymnast.

After several more moves, all that remained was the ever-important dismount. Tanya's arms swept forward over her head and down toward the beam. Her back and leg muscles flexed to propel her strong body off the beam. Flipping through the air, she locked her knee joints, and her leg muscles became rigid. As her feet hit the mat, shock waves rippled upward through her bones and muscles. Her leg muscles immediately tightened, leaving her feet securely planted on the mat exactly where they had first touched. With her arms stretched out, Tanya stood tall and erect, knowing that she had given the performance of her life.

For Your Journal

- What purpose do bones and muscles serve?
- What are bones and muscles made of?
- How do muscles work? Illustrate your answer.

CHAPTER 16 433

Section 1: THE SKELETAL SYSTEM

FOCUS

This section describes the functions and structures of bone and cartilage and the formation and growth of bone. Components of major bone groups are discussed. This section also identifies the different kinds of joints at bone articulation sites and describes the type of movement that each type of joint allows.

PROCESS SKILLS
• Observing • Measuring

POSITIVE ATTITUDES
• Curiosity • Precision

TERMS
• skeletal system • marrow
• cartilage • joints
• ligaments

PRINT MEDIA
Arthritis by Madelyn Klein
(see p. 429b)

ELECTRONIC MEDIA
The Skeleton, Britannica
(see p. 429b)

Science Discovery
Fetus, human; with dyed skeleton
Skeletal system; human

BLACKLINE MASTERS
Study and Review Guide
Laboratory Investigation 16.1
Reading Skills
Connecting Other Disciplines

INTEGRATION—
Language Arts

Explain that the English language has idioms and other expressions that use the word *bone*. Ask the students what the expression "tickled my funny bone" means. ("made me laugh" or "amused me") Challenge the students to think of additional expressions that use the word *bone* and to define those expressions.

MOTIVATING ACTIVITY

Cooperative Learning Divide the class into several small groups. Give each group the following materials: paste, straws or toothpicks, construction paper. Ask the groups to use the materials to construct a paper house or other building. Have the groups display their constructions and explain how they used the materials to make their paper buildings. Encourage the students to conclude that the straws or toothpicks were needed to provide support and structure for the paper sides of their buildings.

SECTION 1

The Skeletal System

Objectives

Summarize the functions and structure of bones.

Distinguish among the different parts of the skeleton.

Compare the types of joints and **contrast** their functions.

As a gymnast, Tanya needs to know about bones and how they work because such knowledge is important to her sport. But bones and their functions are important to you, too. Beneath your skin lies a complex system of 206 bones known as your **skeletal system.** These bones are much like the steel beams that make up the framework of a skyscraper or the wood that forms the frame of a house. As your body's framework, bones give you shape and help you move. Take a moment to feel the parts of your skeletal system.

Discover by Doing

Use your thumb and index finger to feel the small bones of your hands and feet. Probe along your spine at the base of your neck and below your waistline. Feel the large bones of your legs and arms. Examine your ribs and shoulders. Check out the bones of your jaw and skull. Feel around your joints. Your skeletal system extends from the top of your head to the tips of your toes and fingers. As you can feel, the bones that make up this system are different shapes and sizes.

What Bones Do

When you think about what your bones do, you probably think about support. That seems to be the most obvious thing that bones do. But they do more than just provide support.

Bones perform five major functions. First, bones do support other parts of the body. In doing so, they also give shape to your body. Without a skeletal system, you might look like the character in this cartoon!

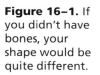

Figure 16–1. If you didn't have bones, your shape would be quite different.

TEACHING STRATEGIES

- **Process Skills:** *Evaluating, Inferring*

Have the students list the five functions of the bones. Ask the students to evaluate the importance of each function in terms of the body's ability to operate properly. Have them consider whether any of the functions are unessential. (The students might conclude that all functions are essential to the proper operation of the human body. Some students might suggest that the bones' role in supplying emergency minerals could be eliminated if adequate supplies of the minerals were taken in.)

The second function of bones is to help parts of your body move. Look at your hand while you move your fingers back and forth. Feel the bones move as you wiggle your fingers. These movements are accomplished not only by bones. The bones must team up with muscles to perform the movements.

The third job of bones is protection. Sturdy bones protect many of the delicate organs in your body. Remember when you felt your skull and ribs? Your skull is like a helmet surrounding perhaps the most important organ in your body—the brain. Your ribs protect your heart, lungs, kidneys, liver, and other important organs. Why do you think it is called a rib cage? ①

Other functions performed by the bones are less obvious than the first three. Their fourth function is to provide an important source of minerals for emergency use by your body. For example, calcium is necessary for muscle contraction and blood clotting. When your diet does not contain enough calcium, your body uses its emergency supply stored in the bones.

Figure 16-2. Bone marrow, found in the center of many large bones, performs many vital life processes.

The fifth and final function is performed inside some of the larger bones of your body. In the center of these bones is a soft substance called **marrow** (MAR oh), which makes blood cells. Every minute, marrow produces millions of blood cells. These new cells move out of the bones and into the bloodstream and spread throughout your body. Since blood cells live only a few weeks and since you need billions of them, the marrow is busy making new blood cells 24 hours a day.

ASK YOURSELF

What are the five major functions that bones perform?

① It is called a rib cage because the ribs enclose the chest to protect vital organs in much the same way as a cage encloses an area.

SCIENCE BACKGROUND

The bone marrow is the manufacturer of red blood cells, white blood cells, and platelets. One disease that affects the marrow's ability to produce these blood cells is leukemia. Leukemia is a kind of cancer in which the bone marrow produces a large quantity of abnormal white blood cells and loses its ability to produce red blood cells, normal white blood cells, and platelets. As a result, a person with leukemia becomes anemic, a condition in which there are insufficient numbers of red blood cells. The person lacks the ability to fight off infection because normal white blood cells are not produced. In addition, because a person with leukemia does not produce platelets, which aid blood clotting, bleeding is extremely dangerous to leukemia patients.

ONGOING ASSESSMENT
ASK YOURSELF

Bones support other parts of the body, help parts of the body move, protect organs, provide a source of minerals, and produce blood cells.

TEACHING STRATEGIES, continued

- **Process Skills:** *Observing, Comparing*

If possible, display samples of long, short, and flat bones in the classroom. Have the students examine the samples and suggest reasons for the shapes.

- **Process Skills:** *Observing, Constructing/Interpreting Models*

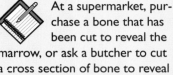

At a supermarket, purchase a bone that has been cut to reveal the marrow, or ask a butcher to cut a cross section of bone to reveal the bone marrow. Also ask the butcher to cut away a small section of the periosteum, or outer protective layer of the bone. Display the bone in the classroom and have the students draw the bone and label the parts in their journals. Then weigh the bone sample. Ask the students whether removing the marrow from the bone would greatly reduce the weight of the bone. (No, it will not greatly affect the weight of the bone because the marrow is not as heavy as other parts of the bone.) Remove the marrow and reweigh the bone to check the students' predictions. **CAUTION: If the students handle the bone, remind them to wash their hands afterwards.**

ONGOING ASSESSMENT
▼ ASK YOURSELF

Differences in the shapes and sizes of bones are related to the functions of bones. Long bones help support weight and assist in movement. The ends of short bones slide over one another to give maximum movement. Flat bones protect organs.

SCIENCE BACKGROUND

The bones are a network of living cells and fibers that are supported by mineral, mostly calcium and phosphorus, deposits. The protective membrane around bone is the periosteum. This layer is responsible for repairing damaged bone. Beneath it is a layer of hard bone. This layer is made up of compact bone. The ends of some bones and the entire composition of other bones are made up of spongy bone, which gives strength and resiliency. Spongy bone is covered by a thin layer of compact bone. The tunnels carrying blood vessels and nerves through compact bone are called the Haversian canals. The interior of the bone contains bone marrow. There are two types of bone marrow—yellow marrow and red marrow. Yellow marrow is mostly fat and serves as a fat storage area. Red marrow manufactures red blood cells, special white blood cells, and platelets.

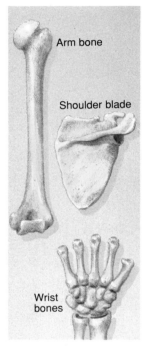

Figure 16–3. Long bones (top), flat bones (middle), and short bones (bottom) are found in different parts of the body. However, the primary functions of all bones are support, movement, and protection.

Types of Bones

If you have ever looked closely at the lumber used to build a house, you may have noticed that it comes in many different shapes and sizes. Long, thick pieces are used to support the floor and roof. Long, thinner pieces form the framework of the walls. The framework is then covered with flat plywood boards.

Like the lumber used to build a house, the bones of your skeletal system are different sizes and shapes. The size and shape of the bones in each group are tied to the bones' functions. For example, the long bones of your arms and legs make up one group of bones. Because of their shape and size, these bones can help support your weight and assist with movement. The smaller long bones of your fingers and toes are also part of this group.

Some bones are flat to help protect organs. The ribs, skull bones, and hip bones are examples of flat bones. Flat bones also contain marrow. Yet another group is the short bones. Most of the short bones are in the wrists and ankles. The shape of these bones helps them to slide easily over one another, giving these areas maximum movement.

▼ ASK YOURSELF

How are the three types of bones different?

The Composition and Structure of Bones

Bones are amazingly strong. In fact, they are often compared to concrete, granite, or cast iron. Some bones can withstand pressure of over 220 kg per square cm. While bones are as strong as concrete and iron, they are much more flexible and weigh much less. Unlike concrete and iron, bones are living tissue capable of changing and growing.

About 45 percent of bone is made of nonliving minerals. These minerals are mostly calcium and phosphorus, which give bones their strength. Living bone cells and blood make up 30 percent of bone tissue, and the remaining 25 percent is water.

Look at the cutaway drawing of a long bone on the next page. You can see that the bone has three layers. The first layer is a very thin covering of cells that forms a tough, protective membrane around the bone. This membrane produces the new bone cells needed for the growth and repair of bone tissue. If you break a bone, this membrane is responsible for mending the damaged area.

- **Process Skills:** *Generating Questions, Applying*

  Cooperative Learning Divide the class into several small groups, and ask each group to write a set of six "trivia" questions about bone structure and composition. For example, the students might ask what part of the bone is responsible for repairing damaged bone tissue. After the groups have written their questions, have them challenge other groups to answer the questions.

- **Process Skills:** *Applying, Analyzing*

  Point out to the students that although minerals make up about 45 percent of the composition of bones, they account for about 66 percent of the mass of the bones. Ask the students to determine how much mass the minerals in a 2-kg bone have. (about 1.32 kg)

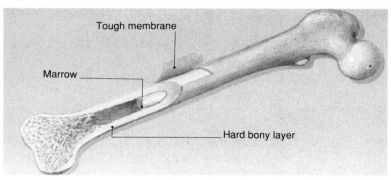

Figure 16–4. Many bones consist of three distinct layers. What would happen if your bones could not make any new bone cells? ①

① Broken bones could not heal.

② Since one of the major functions of the long bones of the arms and legs is to provide mobility, it is essential that they weigh less than other bones. If long bones weighed more, it would be more difficult to move them.

Beneath the protective membrane is a layer of hard, white, rocklike minerals. This layer is what most people picture when they think of bones. Tiny chambers spread through the mineral layer like caves in a mountain. Each microscopic chamber contains a single living bone cell. This hard layer of minerals and bone cells surrounds a soft core of marrow. Passageways spread through the hard layer of the bone like a maze of interconnecting tunnels. These tiny tunnels connect the bone cell chambers and marrow with the surface of the bone. The tunnels are filled with blood vessels. Blood carries the nutrients and oxygen needed to keep the cells alive.

The center or core of the long bone is filled with marrow. Marrow is lighter than the other material making up bones and, therefore, helps reduce the weight of long bones. Why is it important for the long bones of the arms and legs to weigh less than other bones? ②

All bones do not have the same three-layer structure. Bone structure varies depending on the type of bone and its function. For example, the short bones and many of the flat bones are mostly spongy bone surrounded by a thin layer of hard bone. The following activity focuses on the composition of a bone.

DISCOVER BY *Observing*

 Because vinegar is an acid, the hard minerals in the leg bone dissolve in the vinegar. As a result, the bone becomes less rigid and more flexible.

PERFORMANCE ASSESSMENT

Have the students place another leg bone in distilled water. Ask them to compare what happens to the bone in water to what happened to the bone in vinegar. Evaluate the students' responses to determine whether they understand that water is not an acid, and so it does not dissolve the minerals in the bone.

ONGOING ASSESSMENT

ASK YOURSELF

The protective membrane surrounds the bone and manufactures new bone cells; the hard bony layer contains nonliving minerals and living bone cells; the marrow is the soft inner core of the bone, which manufactures blood cells.

DISCOVER BY *Observing*

Get a clean leg bone from an uncooked chicken and place it in vinegar. Let it set for five to seven days. Then remove the bone. Feel it and bend it. Write your observations in your journal. How did the bone change? Why did this happen?

ASK YOURSELF

Describe the three layers of a long bone.

TEACHING STRATEGIES, continued

● **Process Skills:** *Comparing, Applying*

Point out that bone and cartilage are both connective tissues, which bind body parts together and provide structural support. Ask the students how bone and cartilage are different. (Bone is stiff and inflexible, whereas cartilage is softer and more flexible.)

● **Process Skills:** *Inferring, Expressing Ideas Effectively*

Tell the students that the region at the end of the long bones, the epiphysis, produces a large amount of cartilage that is then replaced by bone until a person is about 20 years old. Ask the students to infer why the epiphysis produces the cartilage and why it stops producing cartilage. (Long bones continue to grow by adding bone tissue at the ends of the bones in order to add height throughout the growing years. Cartilage is no longer produced after full growth has taken place.)

● **Process Skills:** *Interpreting Data, Analyzing*

Tell the students that together an arm and a hand have 30 bones and a leg and a foot have 30 bones. Ask the students to determine what percentage of the 206 bones in an adult human are bones in the arms and legs. (58.24 percent) Then

---

LASER DISC
4221, 4222

Fetus, human; with dyed skeleton

MULTICULTURAL CONNECTION

In recorded history, the first physician identified by name was Imhotep, an ancient Egyptian who lived in about 2650 B.C. An architect as well as a physician, Imhotep treated injuries of workers building pyramids that he designed. A papyrus published at a later date describes 48 injuries and wounds to the bones and joints and how to treat them. The papyrus identifies procedures for setting broken bones and realigning dislocated joints. The information provided in the papyrus is believed to reflect the teachings of Imhotep.

ONGOING ASSESSMENT
▼ **ASK YOURSELF**

Before birth, the cartilaginous skeleton begins to form bone. New bone replaces cartilage after birth as well. Throughout the growing years, cartilage is formed and ossifies to become bone. Cartilage is no longer manufactured by the bone after adult height is reached.

▶ **438** CHAPTER 16

Figure 16-5. Human bone cells (top) develop from cells of cartilage (bottom).

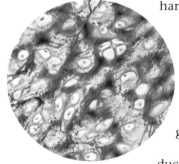

Figure 16-6. As a child grows and develops into an adult, the long bones gradually grow in length. This growth takes place in growth areas made of cartilage.

Growth and Formation of Bones

The bones of your body did not begin to form until about seven months before your birth. At that time, your skeleton was made of a softer and more flexible substance called **cartilage**. Your nose and ears are made of cartilage. If you hold the tip of your nose or ear and wiggle it, you will know what cartilage feels like.

Slowly, the cells that make up cartilage begin producing minerals. The minerals are deposited around the cells, and the cartilage gradually hardens and turns to bone. It is not until you enter your late teens that most of the cartilage is replaced by bone. At the time you were born, your body contained about 300 individual bones. After your birth, some of these bones joined when the cartilage that separated them hardened. So instead of 300 bones, an adult has only 206 bones.

Once your bones are formed, they continue to grow until you are in your mid-twenties. This growth takes place in an area toward the ends of the growing bones. During your growing years, cartilage is constantly being made in this area. The cartilage then hardens into bone, resulting in longer bones. In time the whole bone hardens, leaving no cartilage at the ends of the bones. This signals the end of growth for bones.

Even after your body stops growing, you continue to produce new bone tissue. The new bone tissue is formed when existing bone cells divide. This new tissue is needed to repair and replace damaged or worn-out cells.

▼ **ASK YOURSELF**

How do your bones change from before you are born to the time you stop growing?

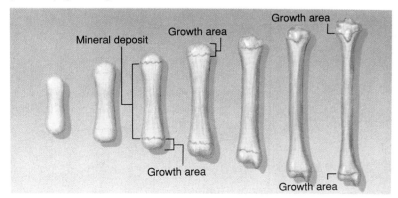

tell the students that each shoulder has two bones and the hips are made up of three bones. Have the students determine the percentage of bones that all the appendages make up. (61.65 percent) Point out that if the fused bones in the spine are included, there are 33, rather than 24, vertebrae in the spine.

● **Process Skills:** *Experimenting, Inferring*

Cooperative Learning Divide the class into small groups, and give each group two blocks of wood and some petroleum jelly. Have the students take turns rubbing the blocks of wood together.

Afterwards, have the students feel the wood where it was rubbed together. Ask them to note what they feel. Then have the students liberally apply petroleum jelly to one block and rub the blocks together again. Ask them to note how rubbing wood against wood differed from rubbing wood with a layer of petroleum jelly against wood. (More friction was produced when the wood was rubbed against wood. Evidence of the friction was the heat generated. The petroleum jelly reduced the friction.) Explain that the petroleum jelly provides the same kind of lubrication and protection as the fluid within the joints.

The Skeleton

The word *skeleton* comes from a Greek word that means "dried up." The skeleton does look like a completely dried-up human being! In fact, the skeleton alone can give us a good idea of what an organism looked like. For example, although no one has ever seen a dinosaur, scientists have determined what dinosaurs looked like by putting together the remains of their skeletons.

In some ways your skeleton is like a tree. A tree has two main parts—the trunk and the branches. Your skeleton can also be divided into two parts. One part, the *central skeleton,* is made up of the bones that form the trunk of your body. These include the skull, spine, and rib cage. The other part is like the branches of a tree and is called the *appendages.* The appendages include the bones of your arms and legs and the bones of your hips and shoulders to which your arms and legs are attached.

The Central Skeleton The spine is the core of the central skeleton. The spine consists of 24 small, irregularly shaped bones, called *vertebrae,* which are stacked one on top of the other. The spine is the only support for the upper body. Without your spine, you would not be able to hold yourself upright. The bones of the spine also protect the soft nerve tissue that carries messages between your brain and other parts of your body. The illustration on the following page shows some of the major bones of the human skeleton.

The vertebrae are separated from each other by disks of cartilage. These disks act like cushions and prevent the vertebrae from grinding against each other. Sometimes a disk can rupture or slip out of place. This can be a very painful condition because of the many sensitive nerves that run through your spine. Use the following activity to find out another interesting fact about the spine.

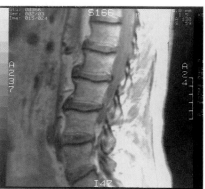

Figure 16–7. Disks of cartilage between the vertebrae can be seen in this photograph. The disks absorb shock when you run or jump. In addition, the whole spinal column is shaped like an S. This shape acts like a spring to absorb shock.

Discover by Calculating

Your height changes during the day. You can check to see how much it changes by measuring your height when you first get up in the morning. Later in the day, measure your height again. How much did you shrink? Check with your classmates. Do tall people shrink more than short people? What is the average shrinkage for all the people you checked?

✧ **Did You Know?**
As people age, they actually get shorter. Between the ages of 27 and 77, a person may shrink 7 cm in height because the spine bends and the disks contract.

DISCOVER BY Calculating

Remind the students that they should carefully measure themselves so that small differences in height can be measured. The students should determine that they do shrink a little during the course of the day.

PERFORMANCE ASSESSMENT

Ask the students to repeat the activity, this time measuring an adult. Have them note any differences in the amount of shrinkage and draw conclusions about the shrinkage rates of adults and themselves based on their findings. Evaluate the logic of the students' conclusions based on their recorded measurements.

TEACHING STRATEGIES, continued

● **Process Skills:** *Observing, Classifying/Ordering*

Have the students carefully examine the skeletal system shown on the page. Then ask them to give the scientific names for the following bones: shoulder blade, kneecap, breastbone, collarbone, backbone, shinbone, and thigh bone.

(scapula, patella, sternum, clavicle, vertebral column, tibia, and femur)

● **Process Skills:** *Applying, Classifying/Ordering*

Cooperative Learning Divide the class into small groups, and ask each group to set up a two-column chart with one column labeled "Central Skeleton" and the other labeled "Appendages." Have one member of each group randomly call out the names of bones. The other group members should work together to place the name of the bone in the appropriate column.

● **Process Skills:** *Comparing, Observing*

Have the students study the bones in the two main parts of the skeletal system. Ask the students to compare the function of the central skeleton to that of the appendages. (The central skeleton protects vital organs, including the brain, the heart,

---

SCIENCE BACKGROUND

The bones serve as a storehouse of minerals for the body. Bones store about 99 percent of the body's calcium and about 85 percent of the body's phosphorus. Calcium is a mineral needed by muscles to function and by blood to clot. If a person does not take in enough calcium for the body to use, calcium is supplied by the bones. If the amount of calcium intake remains too low, the body uses the calcium in the bones, and the bones deteriorate.

LASER DISC
4154

Skeletal system; human

SCIENCE BACKGROUND

The skeletons of all vertebrates are structurally similar. However, certain adaptations in bone structures reflect differences in methods of locomotion and body shape. In humans, the last few vertebrae of the spine, the coccyx, form a vestigial tail. In other mammals, additional bones at the base of the vertebral column form the structural support of the tail.

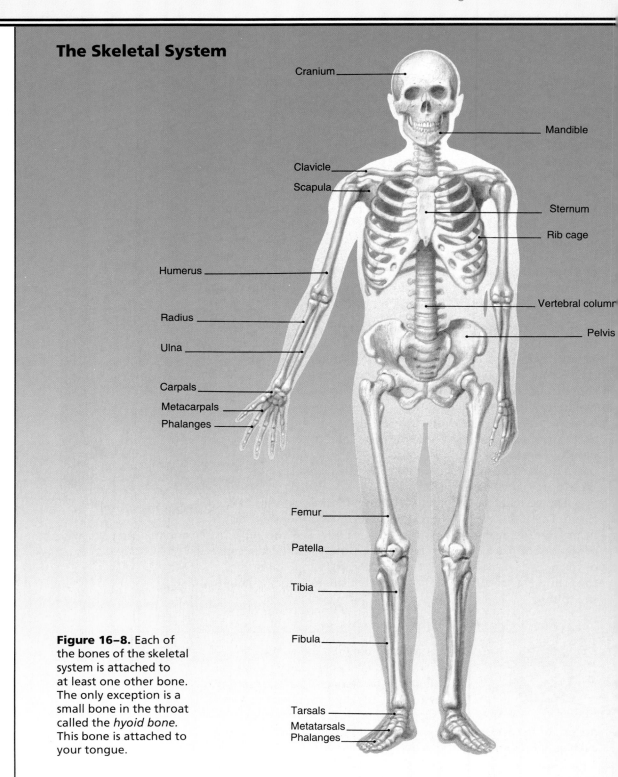

The Skeletal System

Figure 16–8. Each of the bones of the skeletal system is attached to at least one other bone. The only exception is a small bone in the throat called the *hyoid bone*. This bone is attached to your tongue.

and the lungs. The appendages provide structural support and movement.)

● **Process Skills:** *Classifying/ Ordering, Comparing*

Remind the students that there are three different types of bones—long bones, short bones, and flat bones. Help the students to identify the various bones of the body by their type. For example, the carpals and tarsals are short bones; the metacarpals, phalanges, ulna, radius, and humerus are long bones; and the skull, ribs, vertebrae, and shoulder blades are flat bones. Ask the students to tell whether the central system or the appendages have more flat bones. (The central system has more flat bones because one of its primary functions is to protect vital organs.)

● **Process Skills:** *Inferring, Applying*

Point out that the top 10 pairs of ribs are connected to the sternum by cartilage. Ask the students why the bones are connected by cartilage rather than by bone. (Bone is rigid, whereas cartilage is flexible. The chest needs to expand as a person breathes. If the ribs were connected by bone to the sternum, the chest could not expand.)

Connected to the spine are 12 pairs of ribs. They combine to form the rib cage. Each pair of ribs is attached to a different vertebra of the spine. The top 10 pairs of ribs are also attached to the breastbone, forming your chest. The lowest two pairs of ribs attach to vertebrae, but their front tips are not connected to any other bone. Why do you think these last two pairs of ribs are often called the floating ribs? ①

Attached loosely by muscle to the top of the spine is the skull. The skull consists of 22 bones. All but one of these bones are locked in place and cannot move. The one bone that can move is the lower jaw, which is used in chewing food.

Discover BY Observing

Place your thumb against your upper teeth and perform a chewing motion. Now place your thumb against your lower teeth and do the same chewing motion. What did you observe?

In a newborn, the bones of the upper skull are not joined. This is important because the head is the largest part of a newborn baby, and the head has to be a little flexible to get through the birth canal. The gaps in the skull slowly fill in during a child's first two years of life.

The Appendages Hanging from the central skeleton are the bones of the shoulders, hips, arms, and legs. These bones form the appendages of the human body. The arms are attached to the central skeleton by two sets of shoulder bones. The arm is divided into three parts—the upper arm, the lower arm, and the hand. These bones are the upper appendages.

The largest bone in the body, the upper leg bone, is attached directly to the hip bone. The hips are fused to the lower spine forming one solid mass of bone. The lower appendages continue below the knee with two long bones that make up the lower leg. The feet and ankles complete the bones of the lower appendages.

 ASK YOURSELF

Trace an imaginary line from your skull to your foot. Describe the bones along this line.

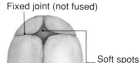
Fixed joint (not fused)
Soft spots
Newborn

Fixed joint (fused)
Adult

Figure 16–9. The soft areas of a baby's skull (top) are gradually filled with bone as the child grows and develops. In the adult skull (bottom), these spaces are completely closed.

---

① They are called floating ribs because they are not attached to the sternum.

DISCOVER BY *Observing*

The students should observe that when they chew, the upper jaw does not move, whereas the lower jaw does move.

SCIENCE BACKGROUND

A newborn baby's skull has six soft spots where the bones have not joined together yet. After a few years, the skull bones are completely fused in a zigzag fashion by fixed joints called sutures.

ONGOING ASSESSMENT ASK YOURSELF

Descriptions should include bones of the skull, spine, ribs, pelvis, legs, and feet. The students should identify the type of bones and the part of the skeleton the bones belong to.

SECTION 1 **441**

TEACHING STRATEGIES, continued

● **Process Skills:** *Observing, Comparing*

Have the students feel their wrists, knees, shoulders, and the back of the neck near the skull as they move these body parts. Have them tell how the movements are different and identify the type of joint involved. (Wrists are gliding joints, knees are hinge joints, shoulders are ball-and-socket joints, and the back of the neck is a pivot joint.)

GUIDED PRACTICE

Have the students define each of the following key terms and explain its function in the human body: skeletal system, marrow, cartilage, joints, ligaments.

INDEPENDENT PRACTICE

 Have the students write their answers to the Section Review and Application questions in their journals. Then ask the students to list the different types of joints and identify at least one place in the body where each type of joint is found.

MULTICULTURAL CONNECTION

Since it was first introduced by Kenji Takagi, a Japanese surgeon, the arthroscope has revolutionized surgery for torn cartilage at the joints. Today's refined arthroscope is a medical instrument with multiple lenses and optic fibers. It is used by doctors to examine and treat joint problems. By inserting the arthroscope through a small incision, a doctor can view the damaged joint and guide other instruments precisely to remove torn cartilage. The use of the arthroscope to remove torn cartilage has reduced hospitalization and recovery time for patients. Patients also have less pain from arthroscopic surgery than from traditional surgery.

MEETING SPECIAL NEEDS

Second Language Support

To help students with limited English proficiency name the bones identified, photocopy the illustration of the skeletal system. Have the students write the name of each bone in their first language. Then suggest that the students use both their first language and English to identify the body parts.

 Encourage the students to place their labeled illustrations in their science portfolios.

▶ **442** CHAPTER 16

Joints

Try to walk without bending your knees. Try to write without bending your fingers, wrist, or arm. Bones are tough, rigid structures that do not bend. Yet in order to move and perform tasks, the parts of our body must bend, twist, and turn. Fortunately, where bones join, we have structures called **joints.** Joints make different kinds of movement possible.

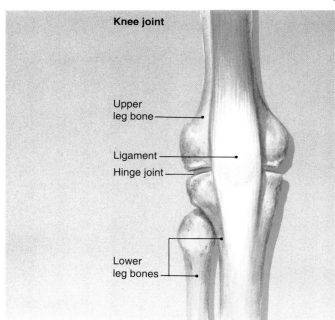

Figure 16–10. The bones of the upper and lower leg are held together at the knee by a ligament.

When two adjoining bones move, they create friction. To help reduce friction, the ends of bones are covered with a smooth layer of cartilage. The cartilage lets bones slide rather than grind against each other. In addition to cartilage, the moving joints produce a thick fluid that acts like oil to lubricate the joints.

How are bones that meet at a joint held in place? Muscles alone cannot do the job. It takes strands of a very strong, tough tissue, called **ligaments,** to hold the bones in place. The ligaments are attached to each of the bones at a joint.

There are 65 joints that join the bones in your body. Not all of these joints allow movement. One type of joint, called a *fixed joint,* locks the joining bones in place. The bones of your skull, except for your lower jaw, are fixed joints. Most of the remaining joints can be divided into four kinds of movable joints. Each type of joint allows a particular kind of movement. Study the illustrations on the next page to better understand how the joints work.

DISCOVER BY Observing

Look around your home, neighborhood, and school for things with moving parts that are joined together. Make a list of at least five things in your journal. Describe how their parts are joined together. Next to each item, write the type of skeletal joint that the connection most resembles.

EVALUATION

Display an unlabeled transparency of the skeletal system. Point to a bone and ask the students to name the bone and identify the type of bone it is.

RETEACHING

Discuss the five functions of the bones. Then ask the students to identify bones that are designed primarily for protecting vital organs and those that are designed for movement.

EXTENSION

Have the students research information about sprains. Ask the students to write short paragraphs in their journals describing what sprains are and how they should be treated.

CLOSURE

Cooperative Learning Divide the class into small groups and have the groups list five important facts about the skeletal system. Ask the groups to share their lists with the class and explain their choice of the particular facts.

A *pivot joint* allows a rotating or rolling motion. The two top vertebrae form this type of joint, which allows you to roll and rotate your head.

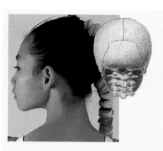

A *fixed joint* does not move. The bones of your skull are joined by fixed joints.

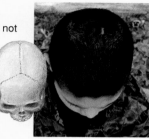

A *gliding joint* allows the bones to slide forward, backward, and sideways. The bones of your ankles and wrists are joined by gliding joints.

A *hinge joint* allows only a back-and-forth movement. It is much like the hinge on a door. Your elbows and knees are hinge joints.

A *ball-and-socket joint* allows movement in nearly all directions. It gets its name because the ball-like end of one bone fits into the socketlike end of another bone. Your hips and shoulders are ball-and-socket joints.

ASK YOURSELF

Describe each type of joint that you find along an imaginary line from your fingertip to the top of your head.

SECTION 1 REVIEW AND APPLICATION

Reading Critically
1. What is the function of bone marrow?
2. How does cartilage differ from bone?
3. How is a hinge joint different from a ball-and-socket joint?

Thinking Critically
4. What problems might be experienced by a person whose bones are weakened by disease?
5. Why do you think broken bones heal faster for people under the age of 27 than for people over the age of 27?
6. How are the joints of the skull related to its major function?

ONGOING ASSESSMENT ASK YOURSELF

The students should identify the joints of the hand as hinge joints, of the wrist as gliding joints, of the elbow as a hinge joint, of the shoulder as a ball-and-socket joint, of the top two vertebrae as a pivot joint, and of the skull as fixed joints.

SECTION 1 REVIEW AND APPLICATION

Reading Critically

1. Bone marrow produces blood cells.

2. Cartilage is a softer and more flexible substance than bone.

3. A hinge joint allows a back-and-forth movement, whereas a ball-and-socket joint allows movement in all directions.

Thinking Critically

4. A person with weakened bones is more likely to break his or her bones than other people are.

5. The students might suggest that younger people produce bone cells more rapidly than older people so a fracture would heal faster in younger people.

6. The skull's main function is to protect the brain. Its fixed joints do not move and thus provide more protection for the brain than any other type of joint could.

SKILL

Organizing Information

Process Skills: Communicating, Classifying/Ordering

Grouping: Groups of 3 or 4

Objectives
- **Locate** specific bones.
- **Organize** information into a usable format.
- **Appraise** the effectiveness of organizational tools.

Discussion

Because the information about the human body is so vast, using a system of organization to remember all of the information is essential. Suggest that the students use mnemonic devices to help them remember all of the bones of the body. For example, the students might use the following mnemonic device to remember the bones of the hand and arm: *People may carry red umbrellas home.* The initial letters represent the terms *phalanges, metacarpals, carpals, radius, ulna, humerus.*

▶ Application

1. The bones are located in the following places: lower jaw (mandible), upper arm (humerus), thigh (femur), lower leg (tibia), shoulder (clavicle), shoulder (scapula), lower arm (radius), chest (sternum), knee (patella), lower arm (ulna), head (cranium), chest (rib), lower leg (fibula), fingers (phalanges), hips (pelvis), and back (vertebrae).

2. The students should categorize the bones using each method.

3. The students should choose the method that works best for them.

4. You may wish to play a game to test the students' recall.

✳ Using What You Have Learned

The organization system used should be similar to the one used to group the bones.

▶ **444** CHAPTER 16

SKILL *Organizing Information*

▼ PROCEDURE

1. One way to organize information is to put the list of topics in alphabetical order.

2. Another way to organize information is to make an outline of the topics. For example, you might begin outlining bones in the following manner:

  I. Central framework
    A. Skull
      1. mandible

3. Clustering is a grouping technique used to organize information. For example, make a diagram like the one below that groups the bones of the body in circles by location. The distance between circles indicates the closeness of the relationship.

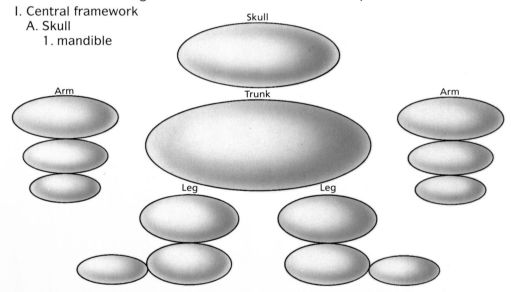

▶ APPLICATION

1. Use Figure 16-8 on page 440 and a model human skeleton, if available, to find the location of each of the following bones:

mandible	pelvis	cranium	ulna
phalanges	fibula	sternum	tibia
vertebrae	patella	scapula	rib
humerus	radius	clavicle	femur

2. Use one of the three methods of organizing information described above. Organize the same information using each of the other methods.

3. Select the method of organization that seems to work best for you. Explain why you like this method.

4. Study the information and have a partner test your memory when you feel ready.

✳ Using What You Have Learned

Use the diagram of the muscular system on page 446, and organize the muscles as you did the bones.

Section 2:
THE MUSCULAR SYSTEM

FOCUS

This section describes the three types of muscles and how they work. The different types of fiber that skeletal, smooth, and cardiac muscles have are described. The section also explains how muscles are attached to bone and how muscles and bones work together to move body parts.

MOTIVATING ACTIVITY

After showing the students how to take their pulse, have student partners take and record each other's pulse. As a class, perform simple exercises for five minutes. Then have the partners once again take and record each other's pulse. Have the students compare the two pulse rates for change.

PROCESS SKILLS
- Comparing • Observing

POSITIVE ATTITUDES
- Curiosity • Enthusiasm for science and scientific endeavor

TERMS
- muscular system • tendons
- voluntary muscles
- involuntary muscles

PRINT MEDIA
Muscles: The Magic of Motion by Robert D. Selim (see p. 429b)

ELECTRONIC MEDIA
Muscular System, Coronet (see p. 429b)

Science Discovery Running; person

BLACKLINE MASTERS
Study and Review Guide
Laboratory Investigation 16.2
Thinking Critically
Extending Science Concepts

The Muscular System

SECTION 2

As amazing as your skeletal system is, it does not and cannot move itself. For every motion, from a heartbeat to the wink of an eye to a handstand on a balance beam, your body depends on its muscular system. Tanya could never have competed in or won the state gymnastics meet had she not trained her muscles to work smoothly with each other and with her skeletal system to perform the moves of her routines.

Your body has about 650 separate muscles. These muscles, working together to coordinate the movements of your body, form your **muscular system.** Not all the muscles in your muscular system are the same. Different muscles are needed to do the many different jobs required of them. But what is incredible is that muscles themselves, no matter what type they are, can perform only one motion. They can only *contract*, or make themselves shorter. Yet muscles work to help you do all the things you do in a typical day.

Objectives

Distinguish among the different types of muscles.

Demonstrate how a muscle works.

Types of Muscles

Muscles move your bones, churn the food in your stomach, fill your lungs with fresh air and push the old air out, and pump blood through your blood vessels. Because muscles do a number of different kinds of work, there are different kinds of muscles. Scientists have identified three basic kinds of muscle—*skeletal, smooth,* and *cardiac*.

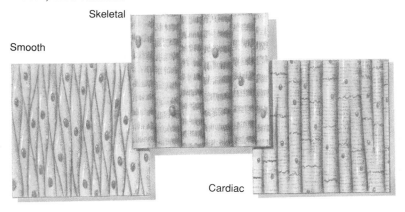

Figure 16–11. The three types of muscle in your body are shown here—smooth muscle (left), skeletal muscle (center), and cardiac muscle (right).

MEETING SPECIAL NEEDS

Mainstreamed

 Cooperative Learning Assign the students partners and have the partners read the section together. Encourage the partners to discuss each paragraph and decide on the main idea of the paragraph. The students should write the main ideas and use them as a study aid.

TEACHING STRATEGIES

● **Process Skills:** *Comparing, Observing*

Have the students examine prepared slides of the three types of muscle tissue under a microscope. Ask them to identify similarities and differences. (All three have nuclei, cell membranes, and other features present in animal cells. They differ in shape, the presence or absence of striations, and arrangement.)

● **Process Skills:** *Inferring, Applying*

Point out to the students that only skeletal muscles are shown in the diagram. Ask the students what body systems would have to be shown to illustrate smooth and cardiac muscles. (The students might name the circulatory system and the digestive system.)

● **Process Skills:** *Observing, Comparing*

Remind the students that humans have bilateral symmetry. Ask the students what bilateral symmetry is. (The body has left and right sides that are mirror images of each other.) Then ask the students to use the diagram to point out and identify the skeletal muscles that appear on the right and left sides.

REINFORCING THEMES—
Systems and Structures

When discussing the interaction of the skeletal system and the muscular system, stress the importance of connecting tissues such as cartilage, ligaments, and tendons. In addition, help the students understand that these systems interact not only with one another but also with other systems, such as the nervous and circulatory systems.

The Muscular System

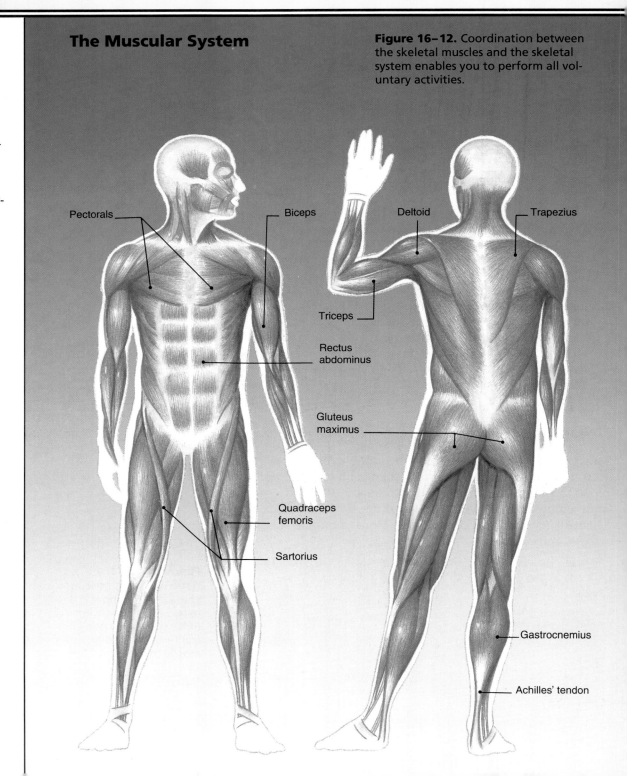

Figure 16–12. Coordination between the skeletal muscles and the skeletal system enables you to perform all voluntary activities.

- **Process Skills:** *Comparing, Applying*

After the students have read about tendons, ask them to compare and contrast tendons with ligaments. (Both are strong connecting tissues. The tendons connect muscle to bone, and the ligaments connect bone to bone.)

- **Process Skills:** *Expressing Ideas Effectively, Inferring*

Cooperative Learning Divide the class into groups. Remind the groups that the skeletal muscles are called voluntary muscles because a person can move these muscles whenever he or she wants to. Ask the students to discuss if they have ever experienced involuntary movement of skeletal muscles. If necessary, help the students relate information about reflex actions (such as the hand moving quickly away from a hot stove), tics, and cramps. All three of these actions are involuntary movements of the muscles. Have the groups find out more information about these three types of actions and report their findings to the class.

Skeletal Muscle For the most part, skeletal muscles are muscles that are attached to bones. But skeletal muscles are not directly attached to bones. Instead, tough bands of connecting tissue, called **tendons,** attach the muscles to bones. Tendons are very strong and do not stretch. In one instance, scientists testing the strength of tendons found that a two-cm thick tendon was able to support over 8000 kg! In the next activity you can observe the tendons in your ankle.

Discover by Doing

Hold your ankle and move your foot. Do you feel the tendons? Put your fingers on the front of your ankle and your thumb on the back. Point your toes up, then down. You should feel two strong cords of tissue moving. What do those cords do? The tendon behind your ankle connects a muscle in the back of your leg to the bones of your toes. When this muscle shortens, it pulls the toes down. A muscle in the front of your leg pulls on another tendon, raising the toes.

Tendons are also present in your fingers. In fact, there are no muscles at all in your fingers. Instead, skeletal muscles in your forearm pull on the tendons to control the movements of your fingers.

Skeletal muscles are also known as **voluntary muscles** because you have control over their movement. When you want to wiggle your toes, throw a ball, smile, or frown, you can start and stop the action by sending signals through your brain to skeletal muscles.

Muscles cannot push; they can only pull. Therefore, skeletal muscles must work in pairs. One muscle contracts to pull a bone in one direction, and the other contracts to pull it in the opposite direction. In order for a body part to move, the muscles in a pair must relax and contract at opposite times.

Figure 16–13. The skeletal muscles move when you want them to. You can control and coordinate their action.

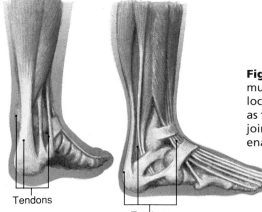

Figure 16–14. The muscles and tendons located near joints such as the ankle hold the joints together and enable movement.

BACKGROUND INFORMATION

The large tendon at the back of the ankle is called the Achilles' tendon. It was named after Achilles, a great Trojan War hero in Greek mythology. According to legend, Achilles's mother, a sea nymph, dipped him in the River Styx to protect his body from harm. Unfortunately, the water did not reach the heel by which his mother held him. During the Trojan War, Achilles killed Hector, who was the brother of Paris, a prince of Troy. Paris vowed to avenge his brother's death and did so by shooting an arrow into Achilles's unprotected heel. Achilles died of the wound.

THE NATURE OF SCIENCE

Of the muscular dystrophies, or diseases that weaken skeletal muscles, the most common is Duchenne dystrophy. It is named after a French neurologist Guillaume Duchenne (1806–1875), who first described the disorder in 1861. He was the first to describe several nervous and muscular disorders and to develop treatments for them. Duchenne studied the effects of electrical stimulation on nerves and muscles. Using electrodes on the skin to send electric current into the body, Duchenne discovered that stimulating nerves caused muscular responses. His book *Physiologie des Mouvements* (1867) described the functions of all the muscles in the muscular system.

SECTION 2

TEACHING STRATEGIES, continued

● **Process Skills:** *Applying, Observing*

Explain to the students that the muscles work in pairs called *antagonistic pairs*. Demonstrate the actions resulting from the contractions of the biceps and triceps, a pair of antagonistic muscles in the upper arm. Point out that the biceps is a flexor, or a muscle that bends an appendage. The contraction of the biceps allows the arm to bend at the elbow. When the biceps contracts, the triceps relaxes. The triceps is an extensor, or a muscle that straightens some part of the body. When it contracts, the triceps causes the arm to straighten at the elbow and the biceps relaxes. Have the students stand up and alternately bend (flex) and straighten (extend) one leg. Ask the students where the flexor and the extensor muscles are located in the upper leg. (The flexor is located in the back of the leg, and the extensor in the front of the leg.)

● **Process Skills:** *Comparing, Inferring*

 Have the students set up a three-column chart in their journals. Then have them chart the characteristics of the skeletal, smooth, and cardiac muscles. Ask the students to use the chart to identify characteristics

---

Demonstration

Make the classroom as bright as possible. Have several small mirrors available so that the students can view their pupils. Have the students note the size of the pupils after exposure to light. Then darken the classroom as much as possible. After several minutes, have the students once again look at their pupils and note their size. Explain that two muscles control the size of the pupils. Ask the students whether these muscles are voluntary or involuntary.

 Before technological advances such as the electron microscope and X-ray diffraction were developed, most scientists theorized that filaments in muscle fiber contracted by folding or coiling. Then in the 1950s, two research teams, coincidentally headed by scientists with the same last name, drew almost identical conclusions about the way muscle fibers contract. Using modern technology, Hugh Huxley and his colleague Jean Hansen and Andrew Huxley and his colleague Rolf Niedergerke concluded that the filaments in muscle fiber do not contract but actually slide over one another. Today, their sliding filament theory is the most widely accepted theory of muscle contraction.

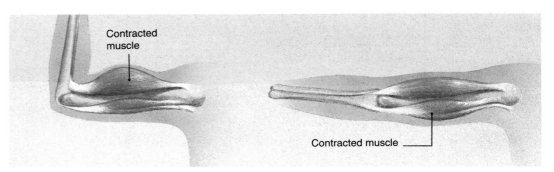

Figure 16–15. As the biceps muscle contracts to lift the arm, the triceps muscle relaxes. In order to straighten the arm, the biceps must relax and the triceps must contract.

Figure 16–16. You have no control over smooth muscle.

Look at the illustration of the muscles in the upper arm. The muscle on top, the *biceps*, pulls the lower arm up by contracting. At the same time, the muscle on the bottom, the *triceps*, is relaxing. When you want to lower your arm, the triceps contracts while the biceps relaxes. This kind of action controls the movement of all appendages in your body.

Remember, each muscle is attached to bones by tendons. One end of the muscle is attached to a movable bone while the other end is attached to a bone that does not move. It is a bit like a spring attached to a screen door. One end is connected to the movable door, and the other end is connected to the immovable door frame. When the spring contracts, like a muscle, it pulls the door closed.

The cells of skeletal muscle are long and thin. They form long fibers that can be as long as 50 cm. The fibers lie side by side in long, narrow, untwisted strips. If you look at these fibers under a microscope, you will see that they have light and dark bands that make them look striped.

Smooth Muscle The muscles of internal organs such as the stomach and intestines are smooth muscle. This type of muscle moves food through your digestive tract. Smooth muscles also control the size of blood vessel openings and change the size of the pupils in your eyes.

Unlike skeletal muscles, the smooth muscles are **involuntary muscles.** This means that they work without signals from your brain. Usually you are not aware of the movements of smooth muscles and couldn't stop them even if you wanted to.

Smooth muscles also differ from skeletal muscles in that their cells are smaller and not as long as the cells of skeletal muscles. In addition, there are no light and dark bands across the fibers of smooth muscle. Instead of being bundled together like skeletal muscles, smooth muscle fibers are arranged in compact layers.

that cardiac muscle has in common with skeletal muscle and characteristics that cardiac muscle has in common with smooth muscle. (Cardiac muscle cells are striated and arranged in bundles as skeletal muscle cells are. Like smooth muscle cells, cardiac muscle cells are short, and they work automatically.)

Ask the students why it is important that the cardiac muscle is an involuntary muscle. (If a person had to consciously control the pumping action of the heart, the person could never sleep or rest because the heart must work continuously to pump blood throughout the body.)

● **Process Skills:** *Applying, Inferring*

Explain to the students that when smooth muscles in the arteries contract, blood is forced into other parts of the body. Tell the students that in cold weather, the small arteries in the skin contract supplying less blood to the surface of the body. Ask the students how the contraction of the small skin arteries helps protect the individual in cold weather. (Less blood at the surface of the body results in less heat loss, thus helping to keep the body's temperature stable.)

Cardiac Muscle The third type of muscle, cardiac muscle, is found only in the heart. Cardiac muscle has characteristics of both of the other types of muscle. Like skeletal muscle, it is striped and arranged in bundles. Like smooth muscle, its cells are shorter than those in skeletal muscle, and they work involuntarily. The fibers are joined in a continuous network that makes cardiac muscle very strong. Why is it important that cardiac muscle be strong? ①

Figure 16–17. Like smooth muscle, cardiac muscle is an involuntary muscle.

① Cardiac muscle must have enough strength to move blood through thousands of kilometers of blood vessels in the human body for the duration of the human life.

 **Calculating**

In most people, the heart contracts about 70 times a minute during normal activity. It continues contracting at this rate for a lifetime. How many times does it contract in an hour? A day? A year?

DISCOVER BY *Calculating*

 Encourage the students to write the equations for solving the problems and the answers in their journals. If the heart contracts 70 times in a minute, it will contract 4200 times in an hour, 100 800 times in a day, and 36 792 000 times in a year.

▼ **ASK YOURSELF**
Describe the three types of muscles.

How Muscles Work

Even though your body has different kinds of muscles, they all work in the same way, by contracting and relaxing. In fact, contracting is the only activity a muscle can do. If it is not contracting, it is doing nothing, just relaxing. Yet this one simple activity enables you to do an amazing amount of work.

Muscle fibers can contract many times in a single second. All of this contracting takes a lot of energy. This is why the muscle tissue receives a rich supply of blood. The blood is continually bringing nutrients and oxygen to the cells, while at the same time, the blood removes waste products from the cells.

When muscle fibers contract, they contract completely. After contracting, the fibers must relax so that a fresh supply of nutrients and oxygen can enter, and waste products can leave. These cycles of contraction and relaxation happen very quickly. In fact, they can occur as rapidly as 10 per second.

 **Doing**

Slowly draw your fingers into a loose fist. Now squeeze your fist more tightly and hold the squeeze for 30 seconds. You control the speed and force of contraction. Can you continue squeezing even as the muscles tire? Why?

Figure 16–18. Coordination between muscles and groups of muscles is necessary for any physical activity. Athletics requires coordination, strength, and endurance.

★ **PERFORMANCE ASSESSMENT**

Ask the students to explain how they solved the problems in the *Discover By Calculating* activity. Evaluate the accuracy of the students' calculations.

ONGOING ASSESSMENT
▼ **ASK YOURSELF**

Skeletal, or voluntary, muscle contains cylindrical cells that form long, parallel fibers arranged in bundles. Smooth, or involuntary, muscle contains oval-shaped fibers that are tightly packed end to end in layers. Cardiac muscle contains striated fibers that are arranged end to end in overlapping bundles. The cardiac muscle is an involuntary muscle.

SECTION 2   **449** ◀

ACTIVITY

How do the muscles, bones, tendons, and ligaments in your hand work?

Process Skills: Observing, Applying

Grouping: Groups of 3 or 4

Hints

Before beginning this activity, review with the students the information they have studied about the skeletal and muscular systems. The goal of this activity is to develop an appreciation for the range of motions and activities the human hand can perform. By observing the smooth operation of the bones, muscles, tendons, and ligaments, the students can better understand the processes that occur when they use their hands.

▶ Application

1. The hand has 27 bones.

2. Fingers have hinge joints, the thumb has a hinge joint and a gliding joint, and the wrist has a gliding joint.

3. Hand movements are controlled by muscles that originate in the forearm and attach to the bones of the hand by tendons.

4. The students might note that grasping things, picking things up, or using tools would be difficult or impossible without thumbs.

LASER DISC

1812

Running; person

▶ 450   CHAPTER 16

TEACHING STRATEGIES, continued

● **Process Skills:** *Applying, Inferring*

Point out that when a muscle cell receives a nerve impulse, it contracts. The number of nerve cells that transmit impulses to muscle cells affects the degree of contraction of the whole muscle. Ask the students if the number of nerve impulses to the muscle would be greater when a person is lifting a heavy load than when a person is lifting a lighter load. (More work is needed to pick up a heavy load than a lighter load, so a muscle must receive more nerve impulses for greater contraction when a person is lifting a heavy load.)

GUIDED PRACTICE

Write the following terms on the chalkboard: muscular system, tendons, voluntary muscles, involuntary muscles. Ask the students to define the terms and to tell how the terms relate to one another.

ACTIVITY

How do the muscles, bones, tendons, and ligaments in your hand work?

MATERIALS
blank white paper, pencil, colored pencils, crumpled paper, tape, coin

PROCEDURE

1. Place your nonwriting hand palm down on a sheet of white paper. Spread your fingers and thumb as far apart as you can and carefully trace around your hand and wrist with a pencil.
2. Examine the bones, muscles, joints, and tendons of your nonwriting hand and wrist. On your drawing, sketch in the bones of the hand and wrist. Try to determine the length, thickness, and shape of the bones.
3. Using a colored pencil, put a capital *L* on your sketch wherever you think a ligament is located.
4. Move your fingers, thumb, and wrist in all directions and squeeze a crumpled piece of paper in your hand repeatedly so you can see the tendons. Use a different colored pencil to draw lines where you think these tendons are located.
5. Tape the thumb of your writing hand securely to the side of the palm and try either to pick up a coin or to button a shirt. Using your thumb-restricted hand, write your initials on your sketch.

APPLICATION

1. How many bones do you think you have in your fingers, thumb, palm, and wrist?
2. What types of joints do you have in your fingers, thumb, and wrist?
3. Where are the muscles that pull the tendons of the hand and wrist when you clench your fist?
4. What purpose does your thumb serve? How would your life be different without your thumbs?

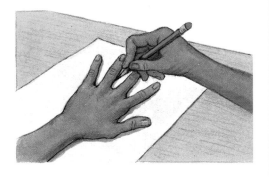

All this contracting of muscle fibers produces the heat your body needs to stay warm. The more active muscles are, the more fuel they burn and, therefore, the more heat they produce. This explains why you feel much warmer when you exercise.

Before a muscle can contract, it needs a signal from the nervous system. The skeletal muscles get their signals from the brain. During your waking hours, these muscles are being bombarded with messages at the rate of about 50 per second. The smooth and cardiac muscles get their messages more directly from the spinal cord, and the messages arrive at a slower rate.

INDEPENDENT PRACTICE

 Have the students write their answers to the Section Review and Application questions in their journal. Then ask the students to explain in a journal entry how muscles work.

RETEACHING

 Cooperative Learning Organize the students into small groups. Have each group orally read the section titled "The Muscular System" together and write a brief summary of the section. Then have the groups share their summaries with the class.

EXTENSION

Have the students research information about the human heart and how it operates.

EVALUATION

Have the students compare and contrast the three types of muscle cells.

CLOSURE

 Have the students write a journal entry that explains the importance of the muscular system to the human body.

Most skeletal muscles remain partially contracted at all times by alternately contracting individual muscle fibers. For this reason, a healthy muscle still feels firm even though it is relaxed. We say that such muscles have good *muscle tone*. Regular exercise develops good muscle tone. Without regular exercise, fat or other connective tissue replaces muscle or muscle loses its tone. For instance, when a cast is removed from a leg that was broken, the muscles are smaller. The fibers have decreased in both size and number. Exercise will make the muscles stronger by increasing the size of the fibers.

While a muscle fiber must totally contract, a complete muscle does not. For example, when you pick up your textbook, that movement requires fewer fibers of your biceps to contract than when you pick up a large pail of water. But in both situations, those fibers in the muscle that do contract must fully contract. You can observe muscle contractions by doing the following activity.

DISCOVER BY Observing

Bend a 9-cm piece of wire into a U-shape. Balance the wire on a dinner knife. Hold the knife with your arm extended and place the ends of the wire so that they barely touch your desktop. Keep your arm steady, but don't brace it against your body or on your desk. Can you see the tiny contractions of your arm muscles? Now try holding a heavy book with your arm extended until your arm becomes very tired. Then repeat the balanced-wire exercise. What difference did you observe?

▼ ASK YOURSELF

What are the two benefits of muscle contraction?

SECTION 2 REVIEW AND APPLICATION

Reading Critically
1. How do muscles work together to produce movement of parts of the body?
2. Why are tendons needed?
3. Describe the difference between voluntary and involuntary muscles.

Thinking Critically
4. Mitochondria are organelles that release energy in body cells. How do you think the number of mitochondria in muscle cells compares to the number in other kinds of cells? Why?
5. What would happen if a muscle constantly contracted and never relaxed?
6. Why is it important that cardiac muscles are not voluntary muscles?

ONGOING ASSESSMENT
▼ **ASK YOURSELF**

Muscle contraction allows movement of and support for the body.

SECTION 2 REVIEW AND APPLICATION

Reading Critically

1. Skeletal muscles work in pairs. One muscle of the pair contracts, causing movement of a bone in one direction. The other muscle of the pair then contracts, returning the bone to its original position.

2. Tendons connect muscles to bones. Contraction of a muscle pulls on a tendon, which in turn pulls on a bone.

3. Involuntary muscles have the ability to perform without a person thinking about them. Voluntary muscles must be consciously instructed to perform.

Thinking Critically

4. The muscular system provides the body with the ability to do work. As the amount of work performed increases, the amount of energy required also increases. As a result, the number of mitochondria will be greater in muscle cells than in other cells.

5. A muscle in a constant contracted state would cause serious, painful cramping in that area of the human body.

6. If cardiac muscles were voluntary, human beings would spend all of their time concentrating on making the muscles perform work. If humans forgot about their voluntary cardiac muscles, they would die.

SECTION 2   451 ◀

INVESTIGATION

Observing Chicken Wing Muscles and Bones

Process Skills: Observing, Comparing

Grouping: Groups of 3 or 4

Objectives
- **Inspect** the muscles, tendons, and bones of a chicken wing.
- **Relate** structure to function.

Pre-Lab
Have the students read the steps of the procedure. Then ask them to discuss the purpose and function of skeletal muscles, tendons, and bones.

Hints
CAUTION: Remind the students to be careful when using sharp instruments and to wash their hands thoroughly. Thoroughly disinfect each surface and tool that comes in contact with the chicken. The chicken wings should be kept refrigerated until they are used. Tell the students to be careful not to cut into the muscle tissue as they remove the skin. Encourage the students to gently use forceps to push aside the muscles of the chicken in order to view the underlying bones.

Analyses and Conclusions

1. The muscles are long and thin, thicker at their centers, and tapered at their ends.

2. A tendon is similar in appearance to white rope and extends from the ends of the muscle. A tendon is thinner than the muscle from which it extends.

3. Tendons join muscles to bones. Because tendons are thinner than the muscles from which they extend, several muscles can be attached to the same bone by tendons.

4. A muscle is attached to two bones near a joint. Contraction of the muscle pulls the bones closer together or pushes the bones farther apart.

5. Each drawing should display two muscles that work as a pair to move a bone.

Application
Knowing how the muscles work would help a body builder select exercises that use the muscles that he or she wants to develop.

✷ Discover More
The joints of the wing are similar to those of the human arm. The wing has ball-and-socket joints and pivot joints.

Post-Lab
Encourage the students to relate the shape and structure of the tissues that they observed to the functions of the bone, tendons, and muscles.

INVESTIGATION

Observing Chicken Wing Muscles and Bones

▶ **MATERIALS**
- chicken wing, raw • dissecting pan • scissors • forceps • paper towels

▼ **PROCEDURE**

1. **CAUTION: Sharp instruments may cause injury. Use them with care and follow the directions of your teacher.**
2. Rinse the chicken wing and dry it with a paper towel. Place the wing in the dissecting pan.
3. With the forceps, carefully lift the skin at the cut end of the wing and insert the scissors.
4. Cut the skin to the tip of the wing and carefully peel the skin away from the muscle.
5. Observe the muscle structure of the wing. Note where the muscles attach to the various bones.
6. Cut away the muscles.
7. Observe how bones fit together. Move the bones and observe how the joints work.
8. **CAUTION: Wash your hands carefully with soap and water after handling the chicken wing. Clean your work area with a disinfectant.**
9. Make a drawing of the muscles, tendons, and bones that you were able to observe.

▶ **ANALYSES AND CONCLUSIONS**
1. Describe the shapes of the muscles.
2. How do tendons and muscles differ in appearance?
3. How does the structure of a tendon relate to its function?
4. To how many bones is each muscle attached? Why?
5. On your drawing, indicate two muscles that work as a pair. Describe the movement that they would cause.

▶ **APPLICATION**
How would knowing about the way muscles work help a body builder develop his or her muscles?

✷ *Discover More*
Reexamine the joints of the chicken wing. What types of joints are found in the wing? How do they compare to the types of joints in human appendages?

CHAPTER 16 HIGHLIGHTS

The Big Idea—SYSTEMS AND STRUCTURES

Point out to the students that the structures of the skeletal and muscular systems give form to the body and enable the body to perform a variety of tasks. Encourage the students to identify the ways in which the muscular and skeletal systems interact. In addition, help the students relate the shape and structure of different types of bones to their functions and identify differences in the three types of muscular cells.

For Your Journal

The students' ideas should reflect an increased awareness of the many purposes of bones and muscles. Their entries in their journals should also indicate that the structures of bones and muscles are related to their functions and purposes.

CONNECTING IDEAS

The concept map should be completed with words similar to those shown here.

Row 2: Muscular
Row 4: support, movement, movement, digestion
Row 5: mineral storage, respiration

CHAPTER 16 HIGHLIGHTS

The Big Idea

Our bones and muscles form systems that interact with each other to perform important functions for us; in particular, these systems provide support and movement. But the skeletal and muscular systems do more than that. The bones protect important organs, produce blood cells, and serve as an emergency mineral reserve. The muscles are an important source of heat needed to keep the body warm, and they are part of the processes of digestion, respiration, and circulation.

All muscles can perform only one action: they contract. But this one action allows muscles to perform an amazing amount of work. There are three types of muscles, each with its own particular structure and function. Some permit voluntary movement, while others function involuntarily. Skeletal, smooth, and cardiac muscles make up your muscular system.

Do you think you have learned a lot about bones and muscles from this chapter? Look back at what you wrote in your journal before you began the chapter. How has your understanding of these important body systems changed since then? How would you revise your ideas on the purpose, structure, and function of bones and muscles?

Connecting Ideas

Copy the unfinished concept map into your journal and fill in the blank boxes.

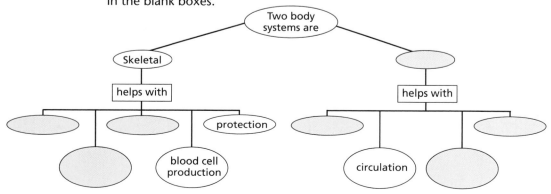

CHAPTER 16 REVIEW

ANSWERS

Understanding Vocabulary

1. The skeletal muscular system permits the movement of the skeletal system.

2. A smooth layer of cartilage at the ends of bones lets the joints slide over one another.

3. Marrow is the soft inner core of a bone that produces blood cells.

4. Both ligaments and tendons are connective tissues that bind other tissues together.

5. The muscular system of the human body is controlled by voluntary muscles, which can be moved at will by a human, and involuntary muscles, which operate without conscious thought by a human.

Understanding Concepts

Multiple Choice

6. a
7. d
8. b
9. c
10. c

Short Answer

11. The various shapes and compositions of bones allow bones to perform different functions within the human body.

12. Skeletal muscle contains cylindrical cells that form long, parallel fibers arranged in bundles and provides voluntary movement of bones and other tissues. Smooth muscle contains oval-shaped fibers that are tightly packed end to end in layers, provides involuntary movement of digestive organs, and allows the constriction of some blood vessels. Cardiac muscle contains fibers that are arranged end to end in overlapping bundles and allows regular, continued beating of the heart.

Interpreting Graphics

13. Muscle A contracts as muscle B relaxes.

14. Muscle B contracts as muscle A relaxes.

Reviewing Themes

15. The bones form the skeletal system, which provides support and movement in the body. The muscles form the muscular system with voluntary and involuntary muscles that make movement possible and generate heat. The interaction of the two systems enable humans to move.

16. A long bone consists of a tough, protective membrane; a hard, nonliving bony layer; and marrow. Skeletal muscles contain cylindrical cells that form long, parallel fibers arranged in bundles.

CHAPTER 16 REVIEW

Understanding Vocabulary

Explain the connection between the terms in each pair.

1. skeletal system (434), muscular system (445)
2. cartilage (438), joints (442)
3. marrow (435), bone
4. tendons (447), ligaments (442)
5. involuntary muscles (448), voluntary muscles (447)

Understanding Concepts

MULTIPLE CHOICE

6. Which of the following is *not* a function of the skeleton?
   a) carrying nerve messages
   b) giving support
   c) providing a mineral reserve
   d) producing blood cells

7. Which of the following words most accurately describes what muscles can do?
   a) expand   b) move
   c) push   d) contract

8. Which of the following is *not* part of the central skeleton?
   a) skull   b) spinal cord
   c) ribs   d) vertebrae

9. Two functions of muscles are movement and
   a) waste removal.
   b) blood supply.
   c) heat production.
   d) protection of body parts.

10. Cartilage is important to the skeletal system because it
    a) joins muscle to bone.
    b) joins bone to bone.
    c) reduces friction.
    d) strengthens bone tissue.

SHORT ANSWER

11. Why do bones have different shapes and compositions?

12. Describe the differences in form and function among the three types of muscles.

Interpreting Graphics

13. Describe the action of muscles A and B when the bone in the following diagram moves from position 1 to position 2.

14. Describe the action of muscles A and B when the bone in the following diagram moves from position 1 to position 2.

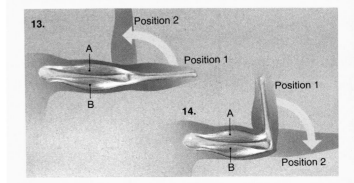

Thinking Critically

17. Hollow bones weigh less than other bones. A bird with hollow bones requires less energy for flight.

18. A single backbone would restrict movement of the human torso.

19. Because muscle cells perform work, they require an increased amount of blood to supply oxygen and remove carbon dioxide.

20. The hardness of bones helps protect and support the tissues of the human body. The flexibility and strength of muscle cells allow the human body to perform a wide range of movements.

21. Cold temperatures reduce muscle contraction and impair coordination. The signatures made after cooling the hand will reflect impaired coordination when compared to those made during normal conditions.

22. The skull consists primarily of flat bones. Thin flat bones protect the brain and other tissues by covering large surface areas without becoming too heavy. The bones of the leg are thick and hollow, providing ease of movement and support for the entire body.

23. The appendages correspond in structure because they originated embryonically from the same tissue. The bones that comprise the appendages have become modified to adapt to flying, swimming, and walking.

Reviewing Themes

15. *Systems and Structures*
Describe bones and muscles in terms of systems and explain the interaction between the two systems.

16. *Systems and Structures*
Describe the structure of a long bone and the structure of skeletal muscles.

Thinking Critically

17. Unlike human bones, many bird bones are hollow. What advantage do hollow bones have for birds?

18. How would your life be different if your spinal column were made of a single, hollow bone rather than being made of separate vertebrae?

19. Why do muscle cells get a richer supply of blood than many other types of cells?

20. Skeletal muscle cells are long and skinny. Bone cells are surrounded by hard mineral deposits. How does the structure of these cells help each cell perform its job?

21. Determine the effect of temperature on muscle contraction by doing the following experiment: Write your name 20 times on a piece of paper. Soak your hand and forearm in a pan of ice water for 30 seconds and then write your name 20 more times. Compare the two sets of signatures. What is your conclusion?

22. Compare the bone structure of the human leg to that of the human skull. How are the differences related to the functions of these bones?

23. Study the diagrams of the bone structures of the human arm, a bird's wing, a whale's flipper, and a dog's front leg. How are these structures similar and how are they different? How have the bones been adapted for different purposes?

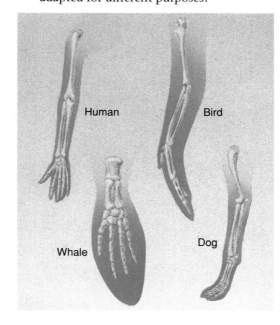

Discovery Through Reading

Parker, Steve. *The Skeleton and Movement.* Watts, 1989. Presents human anatomy and the skeletal system in an up-to-date, easy-to-read format.

CHAPTER 17
DIGESTION AND CIRCULATION

PLANNING THE CHAPTER

Chapter Sections	Page	Chapter Features	Page	Program Resources	Source
Chapter Opener	456	*For Your Journal*	457		
Section 1: DIGESTION AND NUTRITION	458	Discover By Observing **(B)**	459	*Science Discovery* *	SD
• The Digestive System **(B)**	458	Discover By Doing **(A)**	460	Investigation 17.1: Digestion of Carbohydrates **(H)**	TR, LI
• Shooting the Food Tube **(A)**	460	Discover By Researching **(A)**	462	Reading Skills: Mapping Concepts **(A)**	TR
• Nutrition **(B)**	464	Discover By Doing **(A)**	462	Connecting Other Disciplines: Science and Geography, Reading a Map **(A)**	TR
		Activity: How is starch digested? **(H)**	465	Thinking Critically **(A)**	TR
		Discover By Doing **(A)**	466	Extending Science Concepts: Food Additives **(A)**	TR
		Discover By Calculating **(H)**	467	The Human Digestive System **(A)**	IT
		Discover By Researching **(A)**	469	Study and Review Guide, Section 1 **(B)**	TR, SRG
		Discover By Problem Solving **(B)**	470		
		Section 1 Review and Application	470		
		Skill: Classifying Nutritional Information **(A)**	471		
Section 2: BLOOD	472	Discover By Calculating **(A)**	474	Record Sheets for Textbook Investigations **(A)**	TR
• The Functions of Blood **(B)**	472	Discover By Doing **(A)**	476	Study and Review Guide, Section 2 **(B)**	TR, SRG
• Components of Blood **(B)**	473	Section 2 Review and Application	476		
		Investigation: Observing Hemoglobin **(A)**	477		
Section 3: CIRCULATION	478	Discover By Calculating **(A)**	478	*Science Discovery* *	SD
• A Story with Heart **(B)**	478	Discover By Researching **(A)**	479	Investigation 17.2: Circulation in Vertebrates **(A)**	TR, LI
• Blood Vessels **(B)**	480	Discover By Calculating **(A)**	484	The Human Heart and Blood Vessels **(A)**	IT
• Path of Blood **(B)**	482	Section 3 Review and Application	484	Study and Review Guide, Section 3 **(B)**	TR, SRG
Chapter 17 HIGHLIGHTS	485	The Big Idea	485	Study and Review Guide, Chapter 17 Review **(B)**	TR, SRG
Chapter 17 Review	486	For Your Journal	485	Chapter 17 Test	TR
		Connecting Ideas	485	Test Generator	

B = Basic **A** = Average **H** = Honors
The coding Basic, Average, and Honors indicates subsections, features, and resources that might be appropriate for different levels of learners. For additional suggestions regarding choice of topic and depth of coverage, see the Pacing Chart on pages T26–T29.

*Frame numbers at point of use
(TR) Teaching Resources, Unit 6
(IT) Instructional Transparencies
(LI) Laboratory Investigations
(SD) *Science Discovery* Videodisc Correlations and Barcodes
(SRG) Study and Review Guide

▶ 455A

CHAPTER MATERIALS

Title	Page	Materials
Discover By Observing	459	(per group of 3 or 4) hand mirror, encyclopedias or reference books
Discover By Doing	460	(per individual) unsalted soda cracker
Teacher Demonstration	461	person trained in first aid
Discover By Researching	462	(per class) reference materials
Discover By Doing	462	(per individual) vegetable oil, dish, liquid dish detergent
Activity: How is starch digested?	465	(per group of 3 or 4) soda crackers (2), paper towels, one dropping bottle containing amylase solution and one containing iodine solution, white potato, cooked egg white, carrot slice
Discover By Doing	466	(per individual) brown paper bag, shortening, egg white, raw potato
Discover By Calculating	467	(per individual) paper, pencil
Discover By Researching	469	(per individual) paper, pencil
Discover By Problem Solving	470	(per individual) paper, pencil
Skill: Classifying Nutritional Information	471	(per group of 3 or 4) food products (2 or 3), paper, pencil
Discover By Calculating	474	(per individual) paper, pencil
Discover By Doing	476	(per group of 3 or 4) prepared slide of human blood, microscope, journal
Investigation: Observing Hemoglobin	477	(per group of 3 or 4) beakers (2), dechlorinated water, *Daphnia*, medicine dropper, yeast suspension, plastic food wrap
Discover By Calculating	478	(per individual) watch with a second hand, journal
Discover By Researching	479	(per class) reference materials
Discover By Calculating	484	(per individual) journal

ADVANCE PREPARATION

For the *Discover By Doing* on page 460, you will need unsalted soda crackers. A trained first-aid person is needed for the *Demonstration* on page 461. For the *Activity* on page 465, obtain soda crackers, amylase solution, iodine solution, a white potato, cooked egg white, and carrots. A raw potato and egg white are needed for the *Discover By Doing* on page 466. For the *Skill* on page 471, obtain several food product packages with nutrition information. For the *Discover By Doing* on page 476, obtain slides of human blood cells only from a biological supply house. The *Investigation* on page 477 will need *Daphnia*, dechlorinated water, and a yeast suspension.

TEACHING SUGGESTIONS

Field Trip
Visit a facility that performs stress testing for cardiac patients. The students can see how the tests are performed and how an electrocardiograph is traced.

Outside Speaker
Have a nutritionist from a local hospital visit your class. In addition to a discussion of good basic nutrition, your guest might explain how nutrition and diet relate to certain illnesses.

Chapter 17: Digestion and Circulation

CHAPTER THEME—SYSTEMS AND STRUCTURES

This chapter introduces the students to the digestive and circulatory systems of the human body. While studying the chapter, the students will explore how the digestive and circulatory systems work together to supply the entire human body with nutrients and oxygen. The students will also examine the importance of a balanced diet in their lives. The theme of **Systems and Structures** is also developed through concepts in Chapters 4, 5, 7, 8, 9, 10, 12, 13, 16, 18, 19, and 20. A supporting theme of this chapter is **Energy**.

MULTICULTURAL CONNECTION

Dr. Daniel Hale Williams (1856–1931) was the first surgeon to perform a successful open-heart operation on a human. He opened a man's chest to repair the damage caused by a knife wound. Williams accomplished this feat without blood transfusions, X-rays, or antibiotics, none of which existed in 1893.

As a boy, Williams was first apprenticed to a shoemaker and then to a barber. He finished high school while working part time. At 22, Williams became an apprentice to a doctor, a common form of medical training in those days. But he wanted more training, so he enrolled in Chicago Medical College (later Northwestern University Medical School). He graduated in 1883 and began a practice on Chicago's South Side. Williams was a very successful surgeon. But opportunities in hospitals were limited for African American doctors, so in 1891, Williams opened the first interracial hospital in the United States, Provident Hospital and Training School Association. The hospital also offered a training program for nurses, open to students of all races. In 1912, Williams was appointed as a surgeon to prestigious St. Luke's Hospital in Chicago. He was a founding member of the American College of Surgeons in 1913.

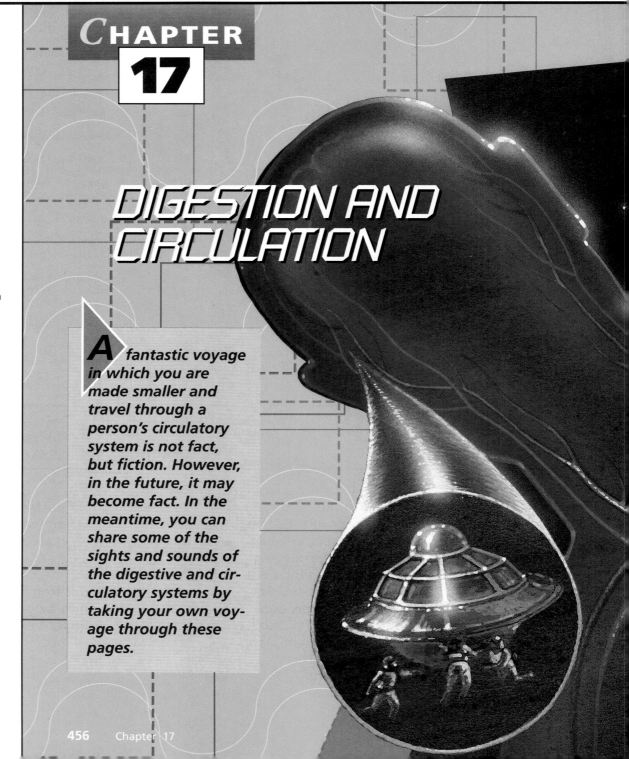

CHAPTER 17

DIGESTION AND CIRCULATION

A fantastic voyage in which you are made smaller and travel through a person's circulatory system is not fact, but fiction. However, in the future, it may become fact. In the meantime, you can share some of the sights and sounds of the digestive and circulatory systems by taking your own voyage through these pages.

CHAPTER MOTIVATING ACTIVITY

Ask the students to explain why a balanced diet is important. (A balanced diet is essential for the proper growth and development of the human body.) Encourage volunteers to suggest different menus that would comprise a healthy breakfast. List the menus on the chalkboard, then focus attention on the components of each menu. Ask the students how the individual components of each menu may contribute to the health of the human body. Use the students' responses to develop the generalization that a balanced meal or diet includes foods that contain a variety of essential nutrients.

For Your Journal

Encourage the students to use the journal questions to explore their knowledge of the human circulatory and digestive systems. Their answers to the questions should give you an idea of the depth as well as the accuracy of the students' knowledge about these systems. At suitable moments during the chapter, you can address the students' ideas and misconceptions about digestion and circulation.

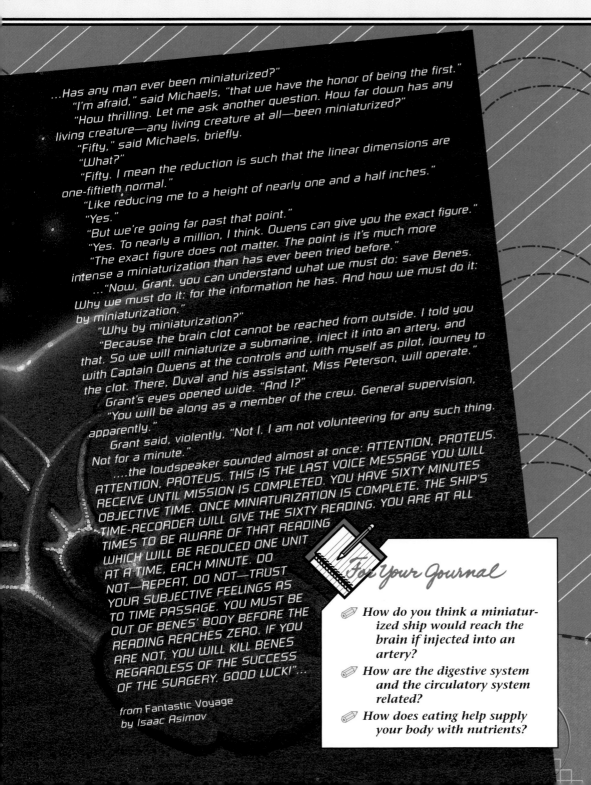

"...Has any man ever been miniaturized?"
"I'm afraid," said Michaels, "that we have the honor of being the first."
"How thrilling. Let me ask another question. How far down has any living creature—any living creature at all—been miniaturized?"
"Fifty," said Michaels, briefly.
"What?"
"Fifty. I mean the reduction is such that the linear dimensions are one-fiftieth normal."
"Like reducing me to a height of nearly one and a half inches."
"Yes."
"But we're going far past that point."
"Yes. To nearly a million, I think. Owens can give you the exact figure."
"The exact figure does not matter. The point is it's much more intense a miniaturization than has ever been tried before."
"...Now, Grant, you can understand what we must do: save Benes. Why we must do it: for the information he has. And how we must do it: by miniaturization."
"Why by miniaturization?"
"Because the brain clot cannot be reached from outside. I told you that. So we will miniaturize a submarine, inject it into an artery, and with Captain Owens at the controls and with myself as pilot, journey to the clot. There, Duval and his assistant, Miss Peterson, will operate."
Grant's eyes opened wide. "And I?"
"You will be along as a member of the crew. General supervision, apparently."
Grant said, violently. "Not I. I am not volunteering for any such thing. Not for a minute."
...the loudspeaker sounded almost at once: ATTENTION, PROTEUS. ATTENTION, PROTEUS. THIS IS THE LAST VOICE MESSAGE YOU WILL RECEIVE UNTIL MISSION IS COMPLETED. YOU HAVE SIXTY MINUTES OBJECTIVE TIME. ONCE MINIATURIZATION IS COMPLETE, THE SHIP'S TIME-RECORDER WILL GIVE THE SIXTY READING. YOU ARE AT ALL TIMES TO BE AWARE OF THAT READING WHICH WILL BE REDUCED ONE UNIT AT A TIME, EACH MINUTE. DO NOT—REPEAT, DO NOT—TRUST YOUR SUBJECTIVE FEELINGS AS TO TIME PASSAGE. YOU MUST BE OUT OF BENES' BODY BEFORE THE READING REACHES ZERO. IF YOU ARE NOT, YOU WILL KILL BENES REGARDLESS OF THE SUCCESS OF THE SURGERY. GOOD LUCK!"...

from *Fantastic Voyage* by Isaac Asimov

For Your Journal

- How do you think a miniaturized ship would reach the brain if injected into an artery?
- How are the digestive system and the circulatory system related?
- How does eating help supply your body with nutrients?

ABOUT THE AUTHOR

Isaac Asimov (1920–1992) was born in Russia and spent his early years studying and working in his father's candy store. At nine, he became an ardent reader of science fiction, and by eleven, he was writing his own science-fiction stories. From that time on, writing was an important part of Asimov's life. School was also important; he graduated from high school and continued on, earning a bachelor's degree, a master's degree, and a Ph.D. in chemistry. He was a college professor for ten years before he began to write full time.

Although Asimov was known for his science fiction, most of his books were nonfiction about science and technology. He was able to make complex information accessible to the nonscientist. A very prolific writer, Asimov wrote 400 books and innumerable short stories, articles, and essays.

MEETING SPECIAL NEEDS

Second Language Support

Fluent English-speaking students should be paired with students who have limited English proficiency to work together and create a drawing of what they believe the miniature ship used by the scientists in *Fantastic Voyage* might look like. Ask the students to display their drawings in the classroom for the duration of their studies of the digestive and circulatory systems.

Section 1:
DIGESTION AND NUTRITION

FOCUS
In this section, the events that occur during the digestion of a typical meal are traced. The relationship of the nutrients carbohydrates, lipids, proteins, vitamins, water, and minerals to human health are examined.

MOTIVATING ACTIVITY
Arrange for a nutritionist or registered dietitian from a local hospital or wellness center to visit the class. (Many organizations in the dietetics community have existing educational programs designed for presentation to schools.) In addition to a discussion of good basic nutrition, you might ask your guest speaker to explain how nutrition and diet are related to certain illnesses and discuss the inherently unhealthy nature of most fad diets for weight loss.

PROCESS SKILLS
- Comparing • Inferring
- Observing • Experimenting

POSITIVE ATTITUDES
- Curiosity

TERMS
- nutrient • digestion
- absorption

PRINT MEDIA
The Digestive System by Regina Avraham (see p. 429b)

ELECTRONIC MEDIA
Nutrition: Eating Well, National Geographic (see p. 429b)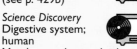

Science Discovery
Digestive system; human
Membranes; human body
Food • Pasta • Spinach
Milk • Ingredient panel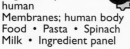

BLACKLINE MASTERS
Study and Review Guide
Laboratory Investigation 17.1
Reading Skills
Connecting Other Disciplines
Thinking Critically
Extending Science Concepts

MEETING SPECIAL NEEDS

Second Language Support

 To help the students remember the organs of the digestive system, ask them to make a simple "flow chart" with labels that name and describe each organ and its function. Have the students place the charts in their science portfolios.

▶ 458 CHAPTER 17

SECTION 1

Digestion and Nutrition

Objectives

Compare and contrast mechanical digestion and chemical digestion.

Diagram the events that occur during the digestion of a typical meal.

Evaluate the nutritional content of foods.

Has your body ever told you it was hungry by making your stomach rumble and your mouth water? These are ways your body reminds you to replenish your energy supply by eating. People tend to treat a plate full of delicious food in an interesting way. Even though great effort was taken to prepare it and care was taken to arrange it on a plate in an appealing way, what do people do? They respond by chewing, grinding, tearing, shredding, churning, pummeling, and mashing this mouth-watering meal into a thin paste!

Let's imagine taking a field trip through the digestion system. Since a big yellow school bus could not take you on this trip, you must travel inside an inner-space capsule that is about the size of a tiny pea. Climb in and let's follow a meal as it moves through the digestive system. All Aboard!

Figure 17–1. The organs of the digestive system

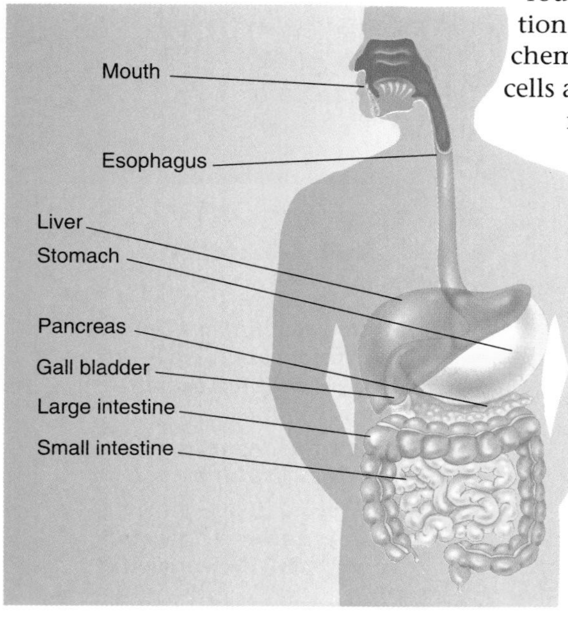

The Digestive System

Your trip begins with some background information about the digestive system. A **nutrient** is a chemical substance that your body needs to build cells and to keep those cells alive. Foods contain nutrients, and eating balanced meals is the best way that you can supply your body with the nutrients you need for good health. Nutrients are obtained from the foods you eat by the process of digestion. **Digestion** is the process that changes food into a form that the body can use.

Your digestive system changes large, complex substances, like many different foods on a plate, into nutrients that can pass through the cell membranes of your digestive system and into your bloodstream. The digestive system is a group of organs that work together to digest food for use by the body.

TEACHING STRATEGIES

● **Process Skills:** *Inferring, Applying*

Ask the students to explain why the digestive system must break down foods into very simple, very small molecules. (In order to be used by cells, nutrient molecules must be small enough to pass through cell membranes into cells.)

● **Process Skills:** *Observing, Generating Ideas*

Direct the students' attention to the digestive system shown in Figure 17–1. Have the students note the name of each organ involved in the process of digestion. Ask the students to infer why the digestive tract of the human body is sometimes called a canal. (The digestive tract is actually a long tube through which food passes during the digestive process.)

● **Process Skills:** *Comparing, Expressing Ideas Effectively*

Ask the students to explain the difference between mechanical and chemical digestion. (Mechanical—the physical breaking of large pieces of food into smaller pieces in the mouth; chemical—the breaking of complex molecules into smaller and simpler molecules.) Ask the students if the use of a knife during a meal is more like mechanical or chemical digestion. (mechanical)

Me and My Big Mouth The digestive process begins with *ingestion,* or eating. The mouth prepares food for travel down a nine-meter tube called the *digestive tract.* The digestive tract winds its way through the human body like a long tunnel. Each part of the tube performs a different function that helps digest food.

The first step in the digestive process is called mechanical digestion. *Mechanical digestion* is the physical breaking of large pieces of food into smaller pieces in the mouth. Hold on tightly as the tongue and cheeks move your capsule and food around the mouth. As you watch, giant, gleaming white teeth cut, tear, and grind food. Run your tongue over your own teeth. How many different kinds of teeth can you feel? The following activity will help you identify the different types of teeth in your mouth.

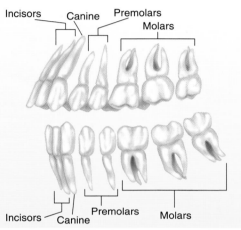

Figure 17–2. Notice the shapes of the teeth. Each shape is related to the tooth's function.

Using a hand mirror, examine both your upper and lower teeth. How many teeth do you have? Draw a diagram that illustrates the shape and number of your teeth. Identify the purpose of each different type of tooth. Use encyclopedias or other reference books if necessary.

From the activity you learned that front teeth have a narrow, knifelike edge. This makes them well suited for slicing and cutting. On each side of the mouth, next to the front teeth, is a large, pointed canine tooth. In other animals, canine teeth are used for capturing and holding live prey. Why do you think canine teeth are found in a human's jaw? The teeth at the rear of the mouth are broad and flat. They grind and mash food.

Chemical digestion, which breaks large molecules into smaller ones, also begins in the mouth. Glands release *saliva,* a mixture of mucus, water, and enzymes. About 500 mL of saliva flow through the mouth each day. Saliva moistens food, making it easier to swallow. The enzymes in saliva begin breaking down complex carbohydrate molecules into smaller and simpler molecules. The following activity will help you better understand chemical digestion.

DISCOVER BY *Observing*

The students should be able to identify at least three kinds of teeth: front teeth, or incisors, to cut food; canine teeth to tear food; and back teeth, or molars, to grind food. In an adult set of 32 teeth, eight are incisors, four are canines, and 20 are premolars and molars.

PERFORMANCE ASSESSMENT

Cooperative Learning After the students have correctly related the structure of human teeth to their function, show the jaw of another animal to groups of students and evaluate their ability to generalize by having them determine from the structure of the animal's teeth whether it was a herbivore or a carnivore.

① Canine teeth are named for their resemblance to a dog's fangs. Humans use their canine teeth to bite and tear their food.

SCIENCE BACKGROUND

Animals that have different kinds of teeth for different uses, as humans do, have *heterodont* teeth. Heterodont teeth are common among mammals. Animals with teeth that are all the same size and shape and are used for the same purpose have *homodont* teeth. Most fish and reptiles have homodont teeth.

TEACHING STRATEGIES, continued

● **Process Skills:** *Predicting, Applying*

Ask the students to predict in what part of the digestive system food is first acted on by an enzyme. (Saliva released in the mouth contains enzymes that act upon food in the mouth.) Remind the students of the cracker they chewed in the *Discover By Doing* activity. Then ask them to explain the function of saliva enzymes. (The enzymes present in saliva begin to break complex carbohydrate molecules into sugar molecules.)

● **Process Skills:** *Inferring, Applying*

Point out that the mouth cavity and the nasal passages of the human body share a common chamber, the pharynx. Liquids occasionally splash back into the nasal chamber when you cough or laugh while swallowing. Ask the students to explain why the respiratory and digestive systems do not permit simultaneous swallowing and breathing. (If simultaneous swallowing and breathing could occur, accidental inhalation of food particles would block the trachea and cause choking or suffocation.)

DISCOVER BY *Doing*

The students should experience a sweet taste as the starch in the cracker is broken down into sugar by digestive enzymes present in saliva.

ONGOING ASSESSMENT
ASK YOURSELF

Mechanical digestion is the physical breaking of large pieces of food into smaller pieces by the teeth. Chemical digestion is the breaking of large molecules into smaller ones by enzymes.

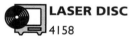

LASER DISC
4158

Digestive system; human

① Food may get into the windpipe, and you may cough or choke.

SCIENCE BACKGROUND

In the human stomach, hydrochloric acids cause milk and milk products to curdle. In the manufacture of cheeses, the enzyme rennin is obtained from the lining of a calf's stomach and used to curdle milk. The curds (solid part) are then separated from the whey (liquid part) and pressed into blocks of cheese.

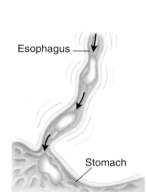

Figure 17–3. Waves of muscular contractions move food down the esophagus during peristalsis.

 **Doing**

Put an unsalted soda cracker in your mouth. Without swallowing, chew the cracker until you notice a change in taste. Describe what happens to the cracker and why the change occurs.

 ASK YOURSELF

What is the difference between mechanical and chemical digestion?

Shooting the Food Tube

Close your eyes and concentrate on what happens when you swallow. Can you describe what happens? Hold on tight, because suddenly the tongue presses against the roof of the mouth, forcing food and your capsule down the throat into a narrow tube called the *esophagus* (ih SAHF uh guhs). This action sends you on a 25-cm journey to the stomach.

Your windpipe moves upward as you swallow. This movement closes the entrance to the windpipe, preventing food from entering. Sometimes the opening to the windpipe does not close fast enough. What happens when you try to eat and talk or laugh at the same time? ①

Your capsule and food do not fall down the esophagus. A series of muscular contractions called *peristalsis* (pehr uh STAWL sihs) pushes material through the tube. The illustration shows how food moves through the esophagus. Contractions squeeze part of the tube shut to force the food farther along. You might feel something like toothpaste being squeezed through a tube.

Fill It Up Your trip down the esophagus to the stomach takes about five seconds. When you first enter the empty stomach, it looks like a cylinder. As it fills with food, it stretches into a J-shaped sac slightly larger than your closed fist. When completely expanded, the stomach holds about 1.5 L of food.

Bottoms Up If you thought being squeezed through the esophagus resulted in a rough ride, hold on because the stomach provides an even rougher ride. Three layers of muscle form the outer walls of the stomach. Each layer moves the stomach in a different way. Thus, the stomach acts like a mixer, throwing you first in one direction and then in another. Waves of contractions occur about every 20 seconds. As you tumble over and over, you hear the sounds of food and liquids gurgling

● **Process Skills:** *Comparing, Expressing Ideas Effectively*

Remind the students that by the time food reaches the stomach, it has been chewed and mixed with saliva. Ask the students to describe what form food is in when it reaches the stomach. (Food has been ground into a pastelike mixture.)

● **Process Skills:** *Predicting, Applying*

Explain to the students that the chemical composition and amount of partially digested food are the most important factors controlling the time it takes for the stomach to empty. Point out that, generally speaking, the more complex the food, the more time it takes for that food to be digested. Ask the students to predict which food, cereal or a steak, would take longer to leave the stomach and to justify their predictions. (Steak is primarily made up of protein and would remain in the stomach longer than cereal, which is a complex carbohydrate.) Have the students recall the "heavy feeling" they may have had in their stomach after eating a large portion of food such as steak, then describe other foods that do not seem to produce a "heavy feeling." Ask the students to debate the relationship between a heavy feeling in the stomach and digestion time.

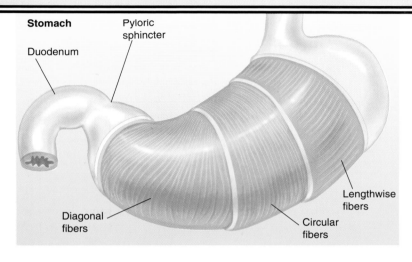

Figure 17–4. The stomach has three layers of smooth muscle, each running in a different direction. These muscles squeeze and churn the food to help break it apart.

Demonstration
A technique called *abdominal thrusts* can be used to dislodge material from the trachea of a choking victim. Ask a person trained in first aid to demonstrate the technique on you while the students watch.

✦ **Did You Know?**
The capacity of the human stomach is approximately 1.4 L. In contrast, a cow's stomach can hold more than 150 L.

and sloshing as the stomach churns its contents and your capsule.

During a pause in the mixing, you shine a light on the inner walls of the stomach. *Gastric juice,* a clear, colorless liquid containing enzymes and acid, drips from 35 million tiny pits in the walls. These gastric pits produce about 3 L of gastric juice each day. Thick, slimy mucus glistens on the stomach walls, protecting them from damage by the acidic gastric juice. Fortunately, your capsule protects you from exposure to this harsh environment, although gastric juices and mucus have smeared your capsule.

Even with a protective coating of mucus, the surface layer of stomach cells is attacked by the harsh chemicals inside the stomach. In a healthy person, this layer of cells is replaced every three days. Sometimes, gastric juice or irritating medicines, such as aspirin, eat away at the cells faster than they can be replaced. When this happens, gastric juice burns the stomach wall and causes a painful sore, called an *ulcer.*

SCIENCE BACKGROUND

If the mucus that lines the walls of the stomach does not adequately protect the stomach, the hydrochloric acid and pepsin in the digestive juices can actually begin to digest part of the stomach lining, causing an ulcer. An ulcer that forms in the stomach is called a *gastric ulcer.* An ulcer in the duodenum, or upper part of the small intestine, is called a *duodenal ulcer.* Digestive system ulcers may be the best known, but ulcers can also occur on the legs and eyes and in the mouth and bladder.

Figure 17–5. This electron micrograph shows the pits that line the inner walls of the stomach. Glands within these pits release digestive enzymes and acid.

TEACHING STRATEGIES, continued

● **Process Skills:** *Observing, Applying*

Point out that the rings of muscles at the entrance to and exit from the stomach are called *sphincters*. Help the students visualize the opening and closing of sphincters by using an inflated balloon. Hold the balloon closed by wrapping your forefinger and thumb around the open end. Then push in the sides of the balloon, open your fingers, and let the air rush out. Ask the students what purpose is served by the contraction of stomach muscles. (The movement of stomach muscles mixes the food with the digestive juices found in the stomach.)

● **Process Skills:** *Inferring, Applying*

Ask the students to describe two structural features of the small intestine that increase the surface area available for the absorption of food molecules into the bloodstream. (The inner wall of the small intestine is folded, and the folded surfaces are covered with microscopic fingerlike villi—singular: villus.) Explain to the students that each villus of the small intestine receives blood from a dense network of capillaries. Ask the students to name the process by which nutrient molecules are absorbed into these capillaries. (absorption)

DISCOVER BY *Researching*

The research should suggest that there are no *known* links between stress and ulcers. However, stress is *believed* to be a contributing factor to ulcers, as well as heart disease, hypertension, and stroke. Behavioral changes to relieve stress include exercise, a well-balanced diet, and stress management skills.

> ✧ **Did You Know?**
> A 100-kg person has as much as 400 g of living bacteria in the large intestine.

SCIENCE BACKGROUND

The digestive juices released by the pancreas are basic (high pH) and include sodium bicarbonate. Sodium bicarbonate is a basic compound that helps neutralize stomach acids (low pH).

DISCOVER BY *Doing*

After the detergent has been added to the vegetable oil and stirred, allow the dish to stand for 15 to 30 minutes. The students should then observe that the detergent begins to break, or separate, the oil into small droplets and infer that bile breaks fatty substances into smaller molecules.

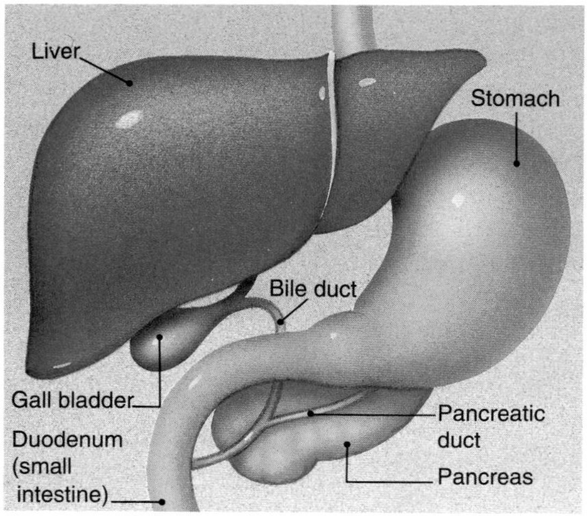

Figure 17-6. Bile is produced by the liver but concentrated and stored in the gall bladder.

DISCOVER BY *Researching*

Excessive stress is thought to be linked to various health problems. Using reference materials, explore the connection between ulcers and stress. If you discover a link, describe changes that people might make to reduce stress in their lives. List some effective treatments for ulcers.

Rings of muscles guard the entrance and exit of the stomach like gates. The entrance gate is a valve that keeps food from being pushed back into the esophagus. The exit valve keeps the food in the stomach until it becomes a liquid. A typical meal spends from two to six hours in the stomach. Eventually, the exit valve opens, and your capsule moves through along with some liquid food.

An Absorbing Story with a Lot of Twists Although the small intestine is more than 6 m long, most digestion occurs within the first meter. In the small intestine, peristalsis pushes food and your capsule along at a much slower rate than in the esophagus. After your turbulent time in the stomach, the small intestine seems almost still and quiet.

Glands in the walls of the small intestine release additional digestive juices into the liquid food, continuing the breakdown of nutrients into smaller molecules. The liver and pancreas also share a relationship with the small intestine. The liver produces bile, and the pancreas produces enzymes, both of which are used in the small intestine. The following activity will help you understand how bile aids digestion.

DISCOVER BY *Doing*

Put a spoonful of vegetable oil in a dish. Pour some liquid dish detergent on top of the oil. Stir the mixture. What happens to the oil? Why do you think this happens? If bile works the same way on fatty substances in the small intestine, what happens to the fatty substances?

● **Process Skills:** *Inferring, Applying*

Explain to the students that by the time undigested food reaches the large intestine, most of the essential nutrients that were available in that food were absorbed as it passed through the small intestine. Much of the remaining material in the large intestine is indigestible plant fiber called *cellulose*. Point out to the students that a diet high in fiber, or cellulose, is known to have beneficial effects on human health. For example, fiber decreases the risk or incidence of colon cancer by increasing peristalsis through the gastrointestinal tract. Water-soluble fiber helps to lower cholesterol. Fiber appears to prevent hyperglycemia—high blood sugar—in diabetics. Fiber also relieves constipation and promotes regularity. Stress the general idea that the undigested fibrous material that passes through the large intestine provides healthy benefits to the digestive system and to overall human health. The students may be interested to learn that foods with a high fiber content include dried beans and peas, fresh fruits and vegetables, whole grains such as whole wheat breads and bran cereals, and nuts and seeds.

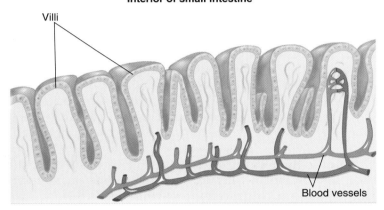

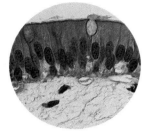

Figure 17–7. Villi increase the surface area of the small intestine and improve the body's ability to absorb nutrients into the bloodstream.

✧ **Did You Know?**
Depending on the kinds of foods eaten, the human body can complete the process of digestion in as few as 12 hours or as many as 24 hours.

 A proctoscope is an illuminated optical device that is used to inspect the rectum and the colon. Polyps, or growths, that can form in the large intestine of some people are sometimes cancerous, and early detection of these growths by a proctoscope will significantly increase the chance of survival for people who develop this type of cancer. A proctoscopic examination should be a regular part of routine health care for many older people.

Bile works in much the same way as the detergent worked on the oil in the dish—breaking large clumps of fat into tiny droplets. Enzymes from the pancreas help to neutralize stomach acid so that it does not damage the intestine.

The rippled walls of the small intestine look much different than the walls of the stomach. As you look closely, you see thousands of fingerlike projections sticking out of the walls. These projections, or *villi,* contain many blood vessels and greatly increase the working surface area of the small intestine. As you observe the action, you activate the shrinking mechanism in your capsule. Your capsule becomes smaller, and it and nutrient particles are absorbed by the blood vessels in the villi. **Absorption** is the movement of nutrient molecules into these blood vessels.

Large Intestine: A Wasteful Organ
As your capsule is absorbed by the villi, material that the body cannot digest moves into the large intestine. The large intestine is named for its diameter, not its length—it is only about 1.5 m long. The walls of the large intestine absorb extra water from the undigested material. The entire process of digestion releases about 7 L of fluid into the digestive tract each day. All but about 500 mL of this fluid return to the bloodstream before the waste leaves the body.

Bacteria in the large intestine break down much of the remaining waste material. Then the large intestine forms the waste into a soft, solid mass called *feces*.

▼ ASK YOURSELF
Describe the digestive process.

ONGOING ASSESSMENT
▼ **ASK YOURSELF**

Digestion is the process during which food is changed into a form that can be used by the body. The digestive system includes the mouth, esophagus, stomach, liver, pancreas, small intestine, and large intestine. The process includes ingestion, peristalsis, mechanical digestion, chemical digestion, absorption, and elimination.

TEACHING STRATEGIES, continued

● **Process Skills:** *Inferring, Applying*

Remind the students that life processes are essentially a series of chemical reactions. For life processes to occur, the body must be provided with the chemical substances that are used to provide energy and manufacture the building blocks of cells. Ask the students to name some life processes carried out by the human body that require nutrients. (All life processes require energy; suggested responses might include respiration, growth, metabolism, and movement.)

● **Process Skills:** *Inferring, Expressing Ideas Effectively*

Ask the students to recall and discuss what they previously learned about cellular respiration. (Cellular respiration is the aerobic mechanism by which cells use oxygen and break down nutrient molecules to release energy. The chemical processes of cellular respiration occur in the mitochondria of a cell.) Ask the students to explain the role of carbohydrates in cellular respiration. (Carbohydrates include groups of sugars such as glucose. In respiration, glucose and oxygen are

---

LASER DISC
4159

Membranes; human body

SCIENCE BACKGROUND

In 1774 Antoine Lavoisier performed experimental work demonstrating that combustion, or burning of a material, results from chemicals combining with oxygen. This research led Lavoisier to believe that the breaking down of food by the human body results in the release of the energy used to maintain life processes. As a result of his work, Lavoisier is sometimes called the "father of the science of nutrition."

MULTICULTURAL CONNECTION

Throughout human history, carbohydrate starches have been and continue to be an important dietary component. Many Native Americans obtained starch in their diets from the consumption of corn. People in Asia obtain starch in their diets from the consumption of rice. People in Caribbean and African countries obtain starch from cassavas and yams. Many other cultures around the world depend on corn, wheat, rice, beans, and tubers in varying forms as basic foods in their daily diets.

Figure 17–8. Honey is a sugar that is an animal product. Most sugars, however, come from plants such as sugar cane.

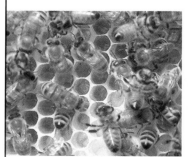

Nutrition

Your body needs about 45 different nutrients to remain healthy. Although you may not have known this, you probably know or have a good idea about other aspects of food, such as which foods are healthy, which are not so healthy, and why a balanced diet is important. The more you know about foods—which foods are healthy, which foods are not; which foods you need, which foods you don't—the greater your likelihood of growing to your greatest physical potential and living a long, healthy life.

Your trip through the digestive system has made you hungry—it is mealtime for you and your crew. As you eat, you think about the nutrition you receive from your food.

Nutrients give cells the energy they need to work properly and supply the building materials needed to grow new tissue and repair damaged cells. The six major groups of essential nutrients include carbohydrates, lipids, proteins, vitamins, minerals, and water.

Fuel for Energy Sugars, starches, and cellulose are examples of *carbohydrates*. Most animals cannot produce carbohydrates. Animals must rely on plants to produce carbohydrates. Sugars are the simplest group of carbohydrates. The cells of your body rapidly break down simple sugars to release energy. The simple sugar fructose gives apples, oranges, and other fruits their sweet taste.

converted to carbon dioxide and water with the release of energy needed by cells to complete the life processes of the human body.)

● **Process Skills:** *Inferring, Applying*

To help the students better understand the relationship of carbohydrates to human health, point out that carbohydrates should be the body's main source of energy. Carbohydrates are relatively low in calories and are the easiest of the six essential nutrients for the human body to break down. A lack of carbohydrates in the human body can be a serious problem because that lack causes the human body to break down first, fat and then protein stores for energy, and these protein stores are typically muscle tissue.

ACTIVITY

How is starch digested?

Process Skills: Experimenting, Inferring

Grouping: Groups of 3 or 4

Hints

Prior to the activity, prepare the egg white by hardboiling several eggs. Small cubes of egg white can then be cut from the hard-boiled eggs. White potatoes and carrots should also be cut up before the activity. Obtain several dropping bottles for use with the iodine and amylase.

▶ **Application**

1. Yes; the digestion of starch by amylase was demonstrated when iodine was dropped on the crackers and potato after they had been treated with amylase. In each case, starch was no longer present.

2. The crackers and potato cubes contained starch.

3. Amylase is produced by the salivary glands and is found in saliva.

Sometimes, two simple sugar molecules join to make a double sugar. Being larger, double sugars have more energy than simple sugars. Because of its size, a cell needs more time to break down a double sugar and release its energy. Table sugar is an example of a double sugar.

The Chain Gang Long chains of simple sugars form very large, complex carbohydrate molecules. Complex carbohydrates, such as starches, store large amounts of energy. Pasta, breads, rice, whole-grain cereals, and potatoes contain large amounts of starch. Some athletes, especially marathon runners, often eat these foods the evening before an event. The following activity will give you a hands-on opportunity to explore complex carbohydrates.

Figure 17–9. These foods contain starch. The process of digestion changes starches into sugars.

★ **PERFORMANCE ASSESSMENT**

Ask the students to name other foods that they predict contain starch. After collecting samples of those foods, have groups of students repeat the *Activity* to test their predictions. Evaluate the ability of each group to follow the established procedure of an experiment.

 **LASER DISC**
1814

Food

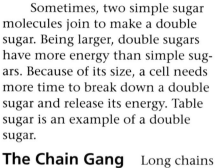

How is starch digested?

MATERIALS
soda crackers (2), paper towels, dropping bottle containing amylase solution, dropping bottle containing iodine solution, white potato, cooked egg white, carrot slice

PROCEDURE
1. Lay two soda crackers on a paper towel. Put five drops of amylase solution on one cracker and wait several minutes.

2. Test both crackers for the presence of starch by dropping two drops of iodine on each cracker. When testing the cracker that contains the drops of amylase, be sure to put the iodine on the same spot.

3. If starch is present, the iodine will turn the starch dark purple to black. Record your results.

4. Test all other foods in the same way you tested the crackers.

APPLICATION
1. Does amylase digest starch? Explain.

2. Which foods contained starch?

3. Where do you think amylase is found in your body?

SECTION 1 **465** ◀

TEACHING STRATEGIES, continued

● **Process Skills:** *Inferring, Applying*

Explain to the students that the term *oil* refers to lipids that are in a liquid state at room temperature. Point out that many lipids, including cholesterol, can be synthesized, or manufactured, by the body from carbohydrates. For this reason, only a small amount of dietary fat is needed by the human body, and excess intake of carbohydrates results in the body storing some of the carbohydrates in the form of fat.

● **Process Skills:** *Inferring, Expressing Ideas Effectively*

Help the students understand the relationship between proteins and cells by asking the students to explain why proteins are often called "the building blocks of cells." (Much of the material in cells is made up of protein.)

● **Process Skills:** *Inferring, Applying*

The students may be interested to learn that fats can be described as "saturated" or "unsaturated." Unsaturated fats are more readily metabolized by the human body than saturated fats. One type of saturated fat, cholesterol, has been linked to

---

LASER DISC

2352

Pasta

DISCOVER BY *Doing*

The students should discover that light penetrates the paper smeared with shortening and conclude that shortening contains lipids. Egg whites and potatoes do not contain lipids.

✧ **Did You Know?**
Fats, which have 9 Calories per gram, are the most dense of the energy nutrients. Carbohydrates and proteins have 4 Calories per gram.

The starches you discovered were large, complex carbohydrate molecules. Another large carbohydrate molecule is cellulose, or fiber, which is produced by plants. Humans cannot digest cellulose, yet cellulose is an important part of a healthy diet because it adds bulk to food. Bulk helps your body move waste through the digestive system. Grains and raw vegetables give your body the cellulose it needs.

Energy Warehouses Fats and oils are examples of *lipids*. Lipids store large amounts of energy. A molecule of lipid stores more than twice as much energy as a molecule of most sugars.

You may already be familiar with the lipid products shown in the photograph. Lard, butter, margarine, red meats, and corn oil are examples of foods that contain lipids. Fats and oils carry important vitamins, so they should be part of your diet but only about 30 percent of your daily intake. The following activity will help you identify a food that contains lipids.

Figure 17–10. A healthy diet includes small amounts of lipids, such as these.

DISCOVER BY *Doing*

Cut three squares of brown paper from a bag. Rub shortening on one square, egg white on the second square, and a slice of raw potato on the third square. Let the pieces of paper dry. Hold each square up to a bright light. Fat allows light to pass through the paper. Which of the foods that you tested contained lipids? Which did not? Test some additional foods for lipids.

arteriosclerosis and heart disease. It has also been shown that eating unsaturated fats may lower blood cholesterol. For this reason, it is recommended that health-conscious consumers read the labels of oils and products that contain oil. Consumers can then choose foods containing unsaturated fats. Coconut and palm oils, which are used in such foods as crackers and cookies, contain large amounts of saturated fats. Olive oil, corn oil, and other vegetable oils are recommended as substitutes for saturated plant oils and animal fats. Have the students identify which foods in Figure 17–10 contain "good" fats and which contain "bad" fats.

● **Process Skills:** *Inferring, Generating Ideas*

Explain to the students that fats are the most calorically dense of the energy nutrients. Along with carbohydrates and proteins, fats provide energy for the human body. Ask the students to describe how fats can protect the internal organs of the human body from injury. (Fatty tissues cushion organs from impact.) Also ask the students to explain how fatty tissue can help regulate the temperature of the human body. (Fatty tissue acts as insulation for the body.)

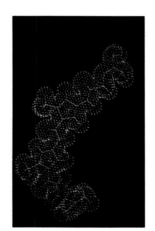

Figure 17–11. These foods are high in protein.

Figure 17–12. This computer-generated image shows the complexity of a protein molecule.

Chains of Amino Acids *Proteins* are very large molecules made of chains of many smaller molecules called *amino acids*. Common sources of proteins include eggs, meat, fish, cheese, milk, and peanut butter. Your hair, your fingernails, and your toenails are made of proteins. Muscles and skin also contain large amounts of protein.

Your body arranges various combinations of 20 amino acids to create thousands of different proteins. How can only 20 kinds of amino acids build so many proteins? They do so in much the same way as the 26 letters of the alphabet are used to make up tens of thousands of words. The words are made up by arranging different letters in different numbers and orders. Proteins are made up by different arrangements of the various amino acids. By changing the number, kind, and order of amino acids in the chain, the body makes thousands of different proteins.

DISCOVER BY *Calculating*

Twenty different amino acids are used by plants and animals to create proteins. Although not all combinations of amino acids create proteins, how many different combinations of 20 amino acids would be possible?

DISCOVER BY *Calculating*

Twenty different amino acids can be combined $2.432902008 \times 10^{18}$ different ways.

PERFORMANCE ASSESSMENT

Ask groups of students to label four separate scraps of paper with the letters A, B, C, and D. Explain that the four pieces of paper represent four different amino acids. Assuming that a minimum of two amino acids must be combined to produce a protein, have each group determine how many different proteins could be created by four amino acids. Evaluate each group on its ability to solve the problem together and find the 32 possible combinations.

TEACHING STRATEGIES, continued

● **Process Skills:** *Inferring, Applying*

Ask the students to define the term *catalyst*. (A catalyst is a substance that speeds up a chemical reaction but does not take part in it.) Remind the students that many chemical reactions require a catalyst. Point out that with respect to proteins, enzymes are special protein catalysts produced by living things. These protein catalysts allow chemical reactions to occur quickly and efficiently. Explain to the students that although vitamins are necessary for good health, vitamins do not actually enter into the chemical reactions of cells. However, they must be present for chemical reactions to occur in cells, because many vitamins form part of the structure of coenzymes. Coenzymes are chemical compounds that work together with enzymes as catalysts in metabolic processes.

● **Process Skills:** *Comparing, Expressing Ideas Effectively*

Ask the students to compare and contrast water-soluble and fat-soluble vitamins. (Both types of vitamins are necessary in small quantities for good health and proper growth and maintenance of the body. Both types of vitamins work with enzymes to

MEETING SPECIAL NEEDS

Gifted

Challenge the students to use research materials and discover what the health and medical communities think about people taking large quantities, or "megadoses," of vitamins on a regular basis. Encourage the students to share their findings with their classmates.

BACKGROUND INFORMATION

Many vitamins were first discovered when their role in preventing deficiency diseases became known. Scurvy is a disease caused by a lack of vitamin C. British sailors often suffered from scurvy because most food carried on long voyages did not contain enough vitamin C. As early as the 1500s, the British recognized that a brew made of certain pine needles reversed the symptoms of scurvy. It was then discovered that supplementing shipboard diets with limes supplied the needed vitamin. By 1795, all British navy ships carried limes, and from that time on British sailors became known as "limeys."

Protein makes up about three-quarters of the solid material in your body. Growth, repair, and replacement of cells require a continuous supply of new protein molecules.

Some protein molecules are *enzymes*. Enzymes are molecules that speed up certain chemical reactions. Without enzymes, most chemical reactions would not occur fast enough to keep the cell alive.

A, B, C, and More

Vitamins are molecules that work with certain enzymes to speed up chemical reactions. Without vitamins, many enzymes cannot work properly. Vitamins are necessary for proper growth and repair of your body. Because your body cannot make the vitamins it needs for good health, you must get the vitamins from the food you eat. Vegetables, fruits, and dairy products are good sources of vitamins.

Vitamins are divided into two groups. The vitamins in one group dissolve in water and are called water soluble. Your body does not store these vitamins, so they should be in your diet each day. Any extra water-soluble vitamins eaten in your food leave the body in your urine. Some of the vitamins in this group are B_1, B_{12}, and C.

The vitamins in the second group dissolve in fat but not in water. Fat-soluble vitamins that are not immediately used can be stored by the body. When your diet does not include needed fat-soluble vitamins, your body uses those you have stored in your fatty tissues. Because your body stores extra fat-soluble vitamins, they can build up in your tissues. Excess fat-soluble vitamins can collect in your liver and harm it. Vitamins A, D, E, and K are fat soluble. Through the following activity you will gain more information about vitamins.

Figure 17–13. Fruits and vegetables are good sources of vitamins.

speed up chemical reactions. Water-soluble vitamins dissolve in water, cannot be stored in large amounts in the body, must be ingested daily, and leave the body in urine. Fat-soluble vitamins dissolve in fat, can be stored in body tissues, and collect in the liver.)

● **Process Skills:** *Inferring, Applying*

Remind the students that although water is not used as an energy source or to build tissue in the human body, it is nonetheless essential to life processes. Ask the students to suggest various functions of water in life processes. (Water is the medium in which chemicals are transported in the body; water is an essential ingredient of many chemical reactions, allowing those reactions to occur; water helps regulate body temperature.)

GUIDED PRACTICE

 Ask each student to use posterboard and colored markers to create a detailed diagram of the digestive tract. Each diagram should identify the primary digestive organs and detail the important function(s) of each organ. Have the students place their diagrams in their science portfolios.

DISCOVER BY Researching

Choose a vitamin—A, B_1, B_2, B_6, B_{12}, pantothenic acid, biotin, C, D, E, folic acid, K, or niacin—to research. Gather information about sources of the vitamin, what the vitamin does for the body, and problems that may result from a deficiency of that vitamin. Share your findings with your classmates, and together compile a chart that includes information for all the vitamins. Illustrate your chart with photographs or drawings.

Cool, Clear Water Water makes up about two-thirds of your body. Water is not a source of energy, and it is not a building block for living tissue. However, water is a liquid that dissolves most other nutrients. Solutions of water and other nutrients move throughout your body and surround every cell.

Each day, you lose about 2 L of water through your skin and lungs and during waste removal. Unless you replace this water, you will become dehydrated. You could probably live without food for several weeks surviving on the nutrients stored in your body. However, you could live without water for only a few days. Even a 10 percent loss of water from your body would cause dehydration. A 20 percent loss could cause death.

Mineral Lode *Minerals* form parts of your body and help perform some body processes. You need most minerals in only small amounts. Calcium, phosphorus, and sodium are typical minerals your body needs to help blood clot, to allow nerves to function properly, and to build strong bones and teeth. Green vegetables and milk are good sources of minerals.

Figure 17–14. Water can be obtained from milk, fruit juices, and other liquids.

DISCOVER BY Researching

In order to avoid duplication, encourage the students to select different vitamins. A typical response from a student who chooses vitamin A, for example, should include: sources—dark green vegetables such as spinach and broccoli, deep orange vegetables and fruits such as carrots, squash, and sweet potatoes; function—essential for tissue and cell membrane maintenance, night vision, healthy bones, teeth, and skin, may help prevent certain forms of cancer; deficiency—night blindness, poor skin quality, susceptibility to infections.

 LASER DISC
2348
Spinach

 LASER DISC
2344
Milk

SECTION 1 **469** ◀

DISCOVER BY
Problem Solving

A balanced diet for one week: milk, yogurt, and cheese—10%; meat, poultry, fish, dry beans, eggs, and nuts—10%; fruit group—20%; vegetable group—25%; bread, cereal, rice, and pasta—35%.

ONGOING ASSESSMENT
▼ **ASK YOURSELF**

A well-balanced diet can decrease the frequency with which sickness occurs, increase the likelihood of achieving the maximum physical potential, and increase the likelihood of living a long life.

SECTION 1 REVIEW AND APPLICATION

Reading Critically

1. The path of food through the digestive system includes the mouth, esophagus, stomach, small intestine, and large intestine.

2. Mechanical and chemical digestion are processes that prepare food for absorption. Mechanical digestion physically breaks large pieces of food into smaller pieces in the mouth. Chemical digestion chemically breaks large molecules of food into smaller molecules in the mouth and other organs of the digestive system.

3. Carbohydrates—provide energy; fats—provide energy, vitamin storage; proteins—tissue maintenance and repair; vitamins—catalysts of chemical reactions within a cell; minerals—form parts of the body and perform some body processes; water—maintains cell hydration.

Thinking Critically

4. It would be easier to survive without a stomach because most nutrient absorption occurs in the small intestine.

5. Plant proteins can be substituted for the animal proteins obtained through eating meat.

6. Nutrition guidelines help people select a balanced diet, maintain good health, and prevent disease.

▶ **470** CHAPTER 17

INDEPENDENT PRACTICE

Have the students provide written answers to the Section Review and Application questions. Ask the students to write a paragraph that describes the importance of a well-balanced diet to their health.

EVALUATION

Ask the students to summarize the six basic nutrient groups and identify one important function of each nutrient group.

EXTENSION

Challenge the students to design an experiment that tests the effectiveness of different brands of stomach antacids.

RETEACHING

Ask a volunteer to name a food. Have the other students discuss and choose the basic food group to which the suggested food belongs. Repeat the activity until all students have had an opportunity to name a food.

Figure 17-15. Food pyramid

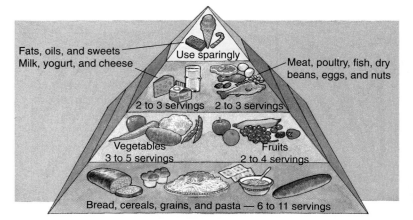

What's for Dinner? Your daily diet should include foods that provide all six types of nutrients. In the early 1990s, the Department of Agriculture of the United States revised its guidelines for nutrition. The food pyramid shows the recommended number of daily servings for each type of food. Use the pyramid and the information about nutrients to help you complete the following activity.

Using information from the food pyramid, plan a balanced diet for one week. Be sure you include foods that provide all the nutrients your body needs to be healthy.

▼ **ASK YOURSELF**

Give three reasons why a well-balanced diet is important.

SECTION 1 REVIEW AND APPLICATION

Reading Critically
1. Trace the path of food through the digestive tract.
2. Compare and contrast mechanical digestion and chemical digestion.
3. State the functions of the six basic nutrients needed for good health.

Thinking Critically
4. Would it be easier to survive without your stomach or without your small intestine? Explain.
5. Can a diet be balanced if it does not contain meat? Explain.
6. Why do you think the Department of Agriculture recommends guidelines for nutrition?

CLOSURE

Cooperative Learning Ask one member from each group to ask a question about the section material, challenging the other group members to discuss and answer the question. Questioner and respondent roles should be rotated until the content has been covered.

SKILL
Classifying Nutritional Information

Process Skills: Interpreting Data, Comparing, Analyzing

Grouping: Groups of 3 or 4

Objectives
- **Communicate** using a table.
- **Interpret data** about nutrition.

Hints
Prior to the activity, ask the students to bring in empty food packages with nutrition information labels.

Discussion
Point out to the students that the ingredients of a product are listed on the product label by quantity; the first ingredient of the list is the most abundant ingredient and subsequent ingredients are present in lesser amounts. Also point out that vitamins and minerals are often added to foods and consequently are listed among the ingredients.

▶ Application 1–4
Answers will vary according to the food products being analyzed. The students should be able to identify the products that contain the greatest variety of vitamins and minerals, and the products that have the highest percentage of the RDA for each vitamin or mineral. Using this information, have the students suggest a balanced meal that includes the recommended amounts of as many vitamins and minerals as possible.

✳ Using What You Have Learned
Check to ensure that the students follow the procedure given in the activity.

Have the students place their tables in their science portfolios.

LASER DISC
4142

Ingredient panel

SKILL Classifying Nutritional Information

▶ MATERIALS
- food products (2 or 3) • paper • pencil

▼ PROCEDURE

1. Choose two or three food products to study. Make a table like the one shown.

TABLE 1: NUTRITION INFORMATION

Food Product	Vitamin	Percentage of RDA	Mineral

2. List the name of each product on separate lines down the side of your paper. List the vitamins contained in each food product.
3. Use the "Nutrition Information" on the label of each product to study the product's vitamin content. On your table, record the percentage of the recommended daily allowance (RDA) for each vitamin listed on the label.
4. Study the food products for mineral content in the same way. Add mineral information to your table.

▶ APPLICATION
1. Which of the food products contains many different kinds of vitamins?
2. For each vitamin, which food product supplies the largest percentage of the RDA?
3. Which of the food products contains many different kinds of minerals?
4. For each mineral, which food product supplies the largest percentage of the RDA?

✳ Using What You Have Learned
Choose five additional food products to study. Read the labels carefully, and record their vitamin and mineral content as outlined in the Procedure. Expand your table to include substances such as preservatives and other additives that are not considered nutrients.

Section 2: BLOOD

FOCUS
This section presents the liquid connective tissue known as blood. Both the function and the components of blood are discussed.

MOTIVATING ACTIVITY

Cooperative Learning Have the class initiate a blood drive to be held at the school or other accessible location in the community. Ask the class to work together and develop a comprehensive and thoughtful plan that outlines the concept and delegates responsibilities such as obtaining approvals from school and local officials, contacting blood donor agencies, and developing public service advertisements for local radio stations and newspapers. Encourage the students to volunteer as helpers during the drive.

PROCESS SKILLS
- Comparing • Observing
- Inferring

POSITIVE ATTITUDES
- Curiosity • Openness to new ideas

TERMS
plasma

PRINT MEDIA
How Did We Find Out About Blood?
by Isaac Asimov
(see p. 429b)

ELECTRONIC MEDIA
The River of Life,
Britannica
(see p. 429b)

BLACKLINE MASTERS
Study and Review Guide

INTEGRATION—
Social Studies

For centuries bloodletting was a standard practice in the curing of diseases. It was believed that when blood was drained from a sick person, the disease would drain out with the blood. The practice was based on the theories of the Greek physician Hippocrates, who believed that diseases were caused by an imbalance of the four *humors* or body fluids—black bile, yellow bile, phlegm, and blood. In the Middle Ages, barbers performed bloodlettings. The practice was still being recommended by reputable physicians in the early 1800s.

▶ 472 CHAPTER 17

SECTION 2

Blood

Objectives

Summarize the functions of blood.

Compare and contrast red blood cells, white blood cells, and platelets.

Describe ways in which the human body develops immunity to disease.

As your meal concludes, you remember your occasionally bumpy journey through the digestive system, and you wonder if your upcoming trip through the blood and circulatory system will be similar or different. You are sure of one thing—if you decided to travel all of the paths through the circulatory system, you would travel a journey of more than 96 000 km! So instead of traveling the entire system, you decide to take a trip that will show you the highlights of the circulatory system.

As your journey begins, you remember shrinking your capsule before you were absorbed by blood vessels in the villi. To be able to travel through many of the tubes that carry blood, the capsule must be much smaller than it was in the digestive system.

You will be moving in a red, living river. A muscular organ that weighs about as much as a large orange keeps this river moving. This organ propels the river through the tubes about 2.5 billion times in one lifetime. Hour after hour, day after day, year after year, the river keeps flowing.

The Functions of Blood

Blood connects all the other tissues of the body. It is the river of life that helps keep all other cells alive. As you watch from the window of your capsule, you see that blood transports nutrients and oxygen throughout the body. Blood delivers this important cargo to the cells. After the delivery, blood picks up a load of waste materials. It carries these waste materials to the lungs and kidneys for removal from the body.

Blood also helps control your body temperature. When you are hot, vessels open, or dilate, allowing more blood to flow through the skin and heat to escape. When you feel cold, the reverse happens. Vessels close, or constrict, limiting the flow of blood to the skin. Your skin loses less heat and your body stays warmer.

TEACHING STRATEGIES

- **Process Skills:** *Inferring, Generating Ideas*

Introduce the students to the section by asking them to describe what they know about capillaries in the skin. (The skin is well supplied with tiny blood vessels known as capillaries.) Ask the students to explain why a person's skin often appears red when he or she is hot. (In an effort to cool the body, blood moves toward the surface of the skin.) Also ask the students to describe what they think might happen to the blood when a person is cold. (The blood moves away from the surface of the skin, helping prevent heat loss and conserve body heat.) Have the students recall the six major groups of essential nutrients they learned about earlier in this chapter (carbohydrates, lipids, proteins, vitamins, minerals, and water), then remind the students that blood is the vehicle by which these nutrients are transported throughout the human body.

- **Process Skills:** *Inferring, Expressing Ideas Effectively*

Ask the students whether blood is a solid or a liquid. (The liquid portion of blood contains solid components.) Point out that even though life depends on blood, blood makes up less than 10 percent of the total mass of the human body.

Figure 17–16. Blood samples are often tested in laboratories.

ONGOING ASSESSMENT
ASK YOURSELF

Responses might suggest that all functions of blood are important. The loss of any of its functions would likely result in death.

SCIENCE BACKGROUND

Plasma is a complex liquid that consists of many organic and inorganic substances dissolved in water. Proteins found in plasma are the most abundant solutes by weight. Plasma proteins can be classified into three general groups: albumins, globulins, and fibrinogen. Albumins help regulate osmotic pressure. Globulins contain antibodies. Fibrinogen is needed for blood clotting.

MEETING SPECIAL NEEDS

Mainstreamed

Point out the phrase "river of life" in the last paragraph on the page. Encourage the students to reread the opening pages of the section, then write a paragraph explaining why blood can be called a "river of life" and draw a picture to accompany their explanation.

Blood also helps you to fight disease. You see specialized cells and chemicals patrolling the bloodstream, attacking disease- or allergy-causing substances 24 hours a day.

 ASK YOURSELF

Blood has many functions. Which is most important? Why?

Components of Blood

Although blood flows like a river, it is part liquid and part solid. Water and other molecules make up the **plasma**, which is the liquid part of blood. You feel surprised when you notice that plasma is not red but a clear, yellowish liquid. Water makes up about 90 percent of plasma. Dissolved substances, such as proteins, form the remaining 10 percent and give plasma its color. Plasma carries dissolved nutrient molecules, such as amino acids and sugars, to all the cells of the body. It also carries carbon dioxide and other wastes away from the cells.

Much of the plasma can pass through the walls of the blood vessels. Outside of the blood vessels, this liquid is called *lymph*. You see this colorless lymph when a blister breaks open. Lymph surrounds every cell bringing the dissolved nutrient molecules closer to the cell membrane for easier transfer.

Blood Is Thicker Than Water Soon you realize that this river of life is much "thicker" than water. Plasma makes up only about 55 percent of the blood. Your capsule bounces from side to side as you travel through the blood.

> ✧ **Did You Know?**
> Blood is 4.5 to 5.5 times "thicker," or more dense, than water.

TEACHING STRATEGIES, continued

● **Process Skills:** *Comparing, Expressing Ideas Effectively*

Have the students examine the normal and sickled red blood cells shown in the illustration on page 475. Ask the students to describe the appearance of a normal red blood cell and a sickle-shaped cell. (Normal red blood cells are disk-shaped, and slightly narrower at the center than at the edges; sickled red blood cells are shaped somewhat like a crescent moon.) Have the students discuss the role of red blood cells in oxygen and carbon dioxide transport and the problems associated with sickle-cell disease.

● **Process Skills:** *Inferring, Applying*

Discuss the importance of sufficient iron intake with the students. Explain that since iron is a component of hemoglobin, it is needed for carrying oxygen. Point out to the students that iron intake is often low in the diets of many people, including teenagers, and in particular, teenage girls. Have the students recall what they learned previously in this chapter about vitamins and minerals, then ask the students to list some foods that are known to contain iron. (Liver, dark green leafy vegetables, lean meats, shellfish, whole grains, beans, and nuts)

DISCOVER BY *Calculating*

One day contains $24 \times 60 \times 60$, or 86 400 seconds. Given that the human spleen and bone marrow produce red blood cells at a rate of 2 500 000 per second, $86\,400 \times 2\,500\,000$, or 2.16×10^{11} red blood cells will be produced each day by the human body.

INTEGRATION— *Mathematics*

Explain to the students that there are approximately 5 billion red blood cells per mL of blood in the human body, and an average person has approximately 5000 mL of blood. Encourage interested students to estimate the number of red blood cells that travel through the human bloodstream. ($5 \times 10^9 \times 5 \times 10^3 = 2.5 \times 10^{13}$, or 25 000 000 000 000)

 After the students have completed the mathematical calculations, have them place their work in their science portfolios.

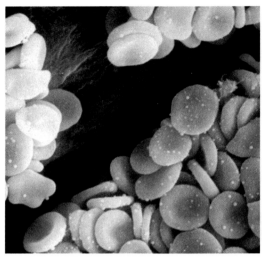

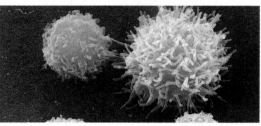

Figure 17–17. Red blood cells (left), platelets (top right), and white blood cells (bottom right) make up the solid part of blood.

In the Red Red blood cells make up about 44 percent of your blood. These cells are disk shaped, with the center of each side pushed in. They look like doughnuts with the holes not fully formed. Red blood cells act like barges loaded down with oxygen. Each cell carries millions of oxygen molecules from the lungs to other parts of the body. The unusual shape of the cell allows oxygen to move into and out of it quickly.

As you look more closely at red blood cells, you discover that they lack a nucleus. Without a nucleus, red blood cells cannot reproduce. A normal red blood cell lives only about four months. Don't worry! You won't run out of red blood cells because your bone marrow and your spleen produce new red blood cells at a rate of 2.5 million per second. A tremendous number of these cells are produced in your lifetime.

DISCOVER BY *Calculating*

Given that your body produces new red blood cells at a rate of 2.5 million per second, determine how many new red blood cells will be produced by your body today.

Red blood cells get their color from an oxygen-carrying chemical called *hemoglobin* (HEE muh gloh buhn), which contains iron. Hemoglobin molecules take the oxygen you obtain by inhaling and transport it throughout the body. Each hemo-

- **Process Skills:** *Inferring, Generating Ideas*

Explain to the students that platelets also release special proteins called *clotting factors*, which form fiberlike threads in the area of a cut. Ask the students to discuss and determine how blood clotting can harm the human body. (Blood clots that form in blood vessels will prevent oxygen and nutrients from reaching the cells of the affected area. Without oxygen and nutrients, cells will die. If a blood vessel clot prevents blood from reaching the brain, a stroke will occur.)

INDEPENDENT PRACTICE

Have the students answer the Section Review and Application questions. Have the students diagram and label a healthy red blood cell and a sickled red blood cell and write a description of the effects of sickle-cell disease.

EXTENSION

Have the students research the Rh system of typing blood and find out why knowledge of a woman's Rh blood type is medically important if she becomes pregnant. Ask the students to share their findings with the class.

---

globin molecule can carry more than one oxygen molecule. When an iron-containing hemoglobin molecule attaches itself to one or more oxygen molecules, the blood becomes bright red in color. The same thing happens when you receive a small cut—the blood you lose turns bright red when it contacts the oxygen in the atmosphere. Which cells in your body have the greatest need for oxygen at this moment? ①

People with sickle-cell disease have abnormally shaped red blood cells. As a result, these cells move through the blood vessels with much more difficulty. Often they stick to one another or to the walls of the vessels. A person with sickle-cell disease often has pain because much of the needed oxygen does not reach all the cells throughout the body quickly enough.

Time to Close Soon the red river carries you and your capsule to a small cut in the skin. You watch as tiny bits of cytoplasm from the blood burst open and begin to clot and seal the damaged area. These small, colorless fragments of cytoplasm, called *platelets,* are another solid part of the blood. Platelets break apart in a wound and cause the blood to clot. The clot plugs the injured blood vessel, and the bleeding stops.

To the Rescue Although only a small amount of blood escaped through the cut, you notice that many bacteria entered the body. While there are millions of red blood cells in a drop of blood, they seem helpless when it comes to defending the body. Then a white giant—a white blood cell—appears. Although larger than red blood cells, white blood cells are much fewer in number. Only one white blood cell exists for every 700 red blood cells. Unlike the red cells, the white cells have a nucleus. As you watch, this giant white blood cell surrounds a bacterium. Slowly, the white cell kills and consumes the invader. Then it goes looking for more.

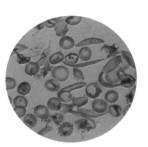

Figure 17–18. Normal red blood cells carry oxygen. Sickled red blood cells often block blood vessels.

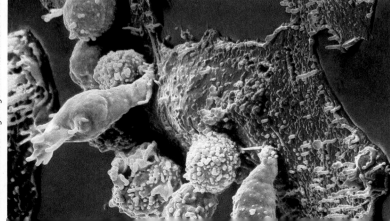

Figure 17–19. White blood cells also attack abnormal cells. This cancer cell is being attacked by white blood cells.

① The students should give reasons for their choice. For example, if they name the cells of the brain, they might say that the brain cells require more oxygen because they are busy working while the students are studying their textbook.

SCIENCE BACKGROUND

Sickle-cell disease is characterized by hemoglobin molecules that stack together when there is a low oxygen concentration in a red blood cell. This stacking gives the red blood cells a sickle shape. This shape causes the red blood cells to get stuck in the capillaries, which are only large enough to allow normal red blood cells to pass in single file. Sickle-cell disease is a very serious health problem that can result in serious injury or death.

INTEGRATION— *Language Arts*

Write the terms *erythrocyte* and *leukocyte* on the chalkboard. Ask the students to look up the meanings and derivations of the two terms in a dictionary. (*Erythrocyte*, which is another name for a red blood cell, is derived from the Greek words *erythros* meaning "red" and *kytos* meaning "hollow vessel"; *leukocyte*, which is another name for a white blood cell, is derived from the Greek words *leukos* meaning "white" and *kytos*.)

DISCOVER BY *Doing*

Use only a prepared slide from a supply house and if necessary assure the students that the blood slide presents no threat to them.

PERFORMANCE ASSESSMENT

Display a red blood cell, a white blood cell, and a platelet at three different microscope stations. Evaluate each student's ability to identify each type of cell.

ONGOING ASSESSMENT
ASK YOURSELF

Blood is part liquid and part solid. Water and dissolved substances make up plasma, the liquid portion of blood. Red blood cells, white blood cells, and platelets make up the solid portion.

SECTION 2 REVIEW AND APPLICATION

Reading Critically

1. Blood transports oxygen and nutrients throughout the body, carries waste materials to the lungs and kidneys, regulates body temperature, and fights disease.

2. Red blood cells are relatively small, numerous, disk-shaped, and slightly narrower at the center than at the edges, and do not have nuclei; white blood cells are relatively large and few and have nuclei.

3. Platelets are bits of cytoplasm; they are not cells. Red blood cells and white blood cells are cells.

Thinking Critically

4. The primary function of both red blood cells and plasma is the transport of materials. The primary function of white blood cells is to fight invading organisms.

5. After receiving a mumps vaccination, your body develops antibodies for mumps that prevent you from getting that disease.

▶ **476** CHAPTER 17

GUIDED PRACTICE

Ask the students to name and describe the three solid components of human blood.

EVALUATION

Ask the students to describe five functions of human blood.

RETEACHING

Give a lump of modeling clay to each student. Have the students use the descriptions of healthy red blood cells and sickled red blood cells in the text to make clay models of each kind of red blood cell. Ask the students how the shapes affect the blood cells' ability to move through blood vessels.

CLOSURE

Cooperative Learning Divide the class into three groups. Ask one group to make a detailed picture of a red blood cell, the second group a picture of a white blood cell, and the third group a picture of a platelet. Have the groups display their pictures in the classroom.

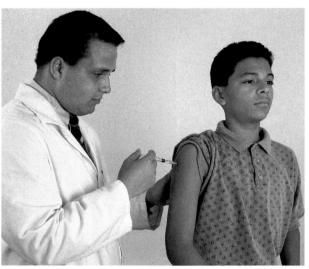

Figure 17–20. Vaccination is an important method of providing individuals with immunity to many diseases.

Some white blood cells cause the body to make proteins, known as *antibodies,* that attack and kill invading organisms. These antibodies make you immune to a disease. Have you ever had chicken pox? If you have, your body has built up antibodies that now make you immune to the disease. You can never again contract chicken pox. Vaccinations are now available for many diseases. The vaccinations help your body develop antibodies that prevent you from getting the diseases. Some vaccinations develop antibodies that last a lifetime; others develop antibodies that eventually disintegrate. You have to be vaccinated more than once to be protected from diseases for which only temporary vaccinations are available.

DISCOVER BY *Doing*

Observe a prepared slide of human blood, using the microscope's low and high power. Try to find red blood cells, white blood cells, and platelets. Describe each type of cell. In your journal, draw and label what you see. 📎

ASK YOURSELF

Describe the composition of blood.

SECTION 2 REVIEW AND APPLICATION

Reading Critically
1. What are four functions of blood?
2. How do red blood cells differ from white blood cells?
3. How are platelets different from red and white blood cells?

Thinking Critically
4. Is the function of red blood cells more like the function of white blood cells or the function of plasma? Explain your answer.
5. If you have received a vaccination for mumps, will you get the mumps? Why or why not?

INVESTIGATION

Observing Hemoglobin

Process Skills: Comparing, Observing, Inferring

Grouping: Groups of 3 or 4

Objectives
- **Observe** and **compare** how the availability of oxygen affects the amount of hemoglobin.
- With respect to hemoglobin, **infer** the response of an organism to an oxygen-deficient.

Pre-Lab
Have the students recall what they have learned about the relationship between hemoglobin and oxygen. Then ask the students to consider these questions: What is the function of hemoglobin? How might a hemoglobin deficiency affect the life processes of an organism?

Hints
When *Daphnia* are deprived of oxygen, they manufacture more hemoglobin. Since hemoglobin is red, a culture of oxygen-deprived *Daphnia* should change color from pale yellow to red as the organisms produce more hemoglobin. Pumping air into a culture with a pipette can cause *Daphnia* to reduce their hemoglobin production.

Analyses and Conclusions
1. A color change from pale yellow to red should have occurred in beaker A. The plastic wrap that covered beaker A prevented oxygen from entering the culture, and the *Daphnia* responded by increasing the amount of hemoglobin in their bodies.

2. The color of hemoglobin is red. Because of a lack of oxygen, the *Daphnia* increased their production of hemoglobin, consequently changing color from pale yellow to red.

Application
An organism will increase its hemoglobin content if there is a lack of oxygen in its environment. An increased supply of hemoglobin increases the likelihood that an organism can obtain oxygen from its environment.

✻ Discover More
The students should observe that as the supply of oxygen decreases, heartbeat rate increases, and vice versa.

Post-Lab
The students should conclude that hemoglobin transports oxygen to the body cells of an organism and infer that a hemoglobin deficiency would threaten the continued existence of the organism.

INVESTIGATION

Observing Hemoglobin

▶ MATERIALS
- beakers, 250 mL (2)
- water (dechlorinated)
- *Daphnia*
- medicine dropper
- yeast suspension
- plastic food wrap

▼ PROCEDURE

1. Obtain two beakers from your teacher. Label one beaker *A* and the other beaker *B*.
2. Fill each beaker with 5 cm of dechlorinated water or with water that has been standing for at least 24 hours.
3. Add a culture of *Daphnia* to each container. Arrange the containers so that they are not in direct light and the temperature of the water remains fairly constant.
4. Using a medicine dropper, feed the *Daphnia* a few drops of a yeast suspension. Be careful not to add so much yeast that the water becomes cloudy and foul smelling. If you add too much, replace most of the clouded water with fresh, dechlorinated water. Feed the *Daphnia* again in two or three days.
5. Cover beaker *A* with plastic food wrap to reduce the amount of oxygen entering the container. If you need to add food or water to beaker *A*, do so by lifting only a small area of the covering. Replace the covering quickly. Leave beaker *B* open to the air.
6. Care for your *Daphnia* and observe them daily. Record your observations daily. After five or six days, look for a color change. As you look at the containers, place a white sheet of paper behind them so you can see a color change more easily.

▶ ANALYSES AND CONCLUSIONS
1. Was there a color change in either one of the containers? If so, explain why.
2. Use what you know about hemoglobin's reaction to oxygen to help you explain the color change in the *Daphnia*.

▶ APPLICATION
If there is a lack of oxygen in the environment, would you expect an organism to increase or decrease its hemoglobin content? Why?

✻ Discover More
Use a hand lens to observe the *Daphnia* in both containers. Watch the heartbeat of the *Daphnia*. Count how many times the *Daphnia's* heart pulses in one minute. Are the heartbeats of the *Daphnia* in both beakers the same? If not, make a hypothesis to explain the difference. How would you test your hypothesis?

Section 3: CIRCULATION

FOCUS

This section presents the human circulatory system. The structure of the human heart, the different types of circulatory blood vessels, and the path by which blood moves through the human circulatory system are examined in detail.

MOTIVATING ACTIVITY

Cooperative Learning Working in small groups, have the students make a basic stethoscope using rubber tubing and a small funnel. Place one end of a 30- to 50-cm length of rubber tubing over the narrow end of a small funnel. The funnel can then be placed on a student's chest while the free end of the rubber tube is held to, not in, the ear. Although such a stethoscope is not as efficient as a doctor's stethoscope, a heartbeat can be detected in a quiet setting.

PROCESS SKILLS
- Comparing • Inferring
- Applying

POSITIVE ATTITUDES
- Curiosity • Enthusiasm for science and scientific endeavor

TERMS
- arteries • capillaries
- veins

PRINT MEDIA
The Circulatory System: From Harvey to the Artificial Heart by Tom McGowen (see p. 429b)

ELECTRONIC MEDIA
Circulatory System, Coronet (see p. 429b)

Science Discovery Circulatory system; human

BLACKLINE MASTERS
Study and Review Guide
Laboratory Investigation 17.2

DISCOVER BY Calculating

Estimates will vary depending on the heartbeat rate and birthday of each student. Assuming 75 beats per minute, a heart will beat 551 880 000 times in 14 years.

PERFORMANCE ASSESSMENT

Have the students determine the number of times their heart will have beaten on their 100th birthday, assuming an average rate of 75 heartbeats per minute throughout their lifetime.
(3 942 000 000)

▶ 478 CHAPTER 17

SECTION 3

Objectives

Describe the structure of the heart and **trace** the flow of blood through the heart and lungs.

Explain what causes a heartbeat and pulse.

Compare and contrast the three types of blood vessels found in the circulatory system.

Circulation

Your movement through the circulatory system is not smooth and constant—your capsule seems to move in spurts. It almost feels like a giant wave picks you up and hurls you forward. Soon you hear sounds like Lub-Dup! Lub-Dup! The sounds grow louder and louder until your capsule races out of a blood vessel into the heart.

DISCOVER BY Calculating

Use a watch or a clock that reads in seconds to determine how many times per minute your heart is beating. Determine how many times your heart will beat in a day. There are 1440 minutes in a day. Find out how many days there are until your next birthday. Then estimate how many times your heart will have beaten by your next birthday. Calculate how many times your heart has beaten in your life so far. Record your results in your journal.

A Story with Heart

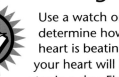

Your heart is a muscle located between your lungs, almost in the center of your chest. Protected by your breastbone, your heart produces muscular contractions that push blood throughout your body. During each minute, almost all of the blood in your body moves through your heart. During an average lifetime, a heart will beat about 2.5 billion times.

The Heart of the Matter Figure 17–21 shows how the right side of your heart receives blood from the body and pumps it to your lungs. In the lungs, blood picks up oxygen and releases its load of waste carbon dioxide. The left side of your heart receives blood from the lungs and pumps it to the rest of the body. This blood is rich with oxygen and very red. A thick, muscular wall separates these two sides of the heart.

TEACHING STRATEGIES

● **Process Skills:** *Observing, Applying*

Refer the students to Figure 17–21 and discuss the structure of the human heart. Have the students locate the atria [singular: atrium] and the ventricles of the heart, then explain that the left side of the heart is on the right side of the illustration and the right side of the heart is on the left side of the illustration. Point out to the students that if they turn their books around and place them on their chests, the left side of the illustration is on the left side of their bodies and the right side of the illustration is on the right side of their bodies. Then ask the students to explain why blood needs to be constantly moved throughout the body of a human being by the heart. (Blood supplies oxygen and distributes nutrients to, and removes wastes from, all parts of the human body.)

● **Process Skills:** *Inferring, Generating Ideas*

Remind the students that valves between the various chambers of the heart allow blood to flow in only one direction. Ask the students to describe the effects on the rest of the body if these heart valves did not exist or did not function properly.

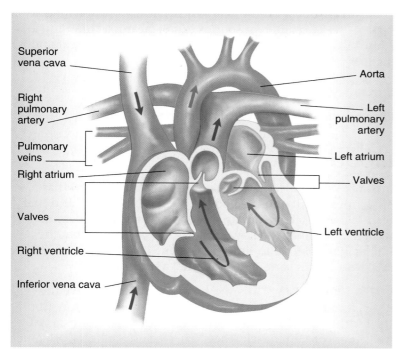

Figure 17–21. The heart is a two-sided pump. One side of the heart pumps blood to the lungs. The other side of the heart pumps blood to the rest of the body.

MEETING SPECIAL NEEDS

Second Language Support

To help the students remember the structure and function of the heart, encourage them to create their own diagram of the heart. Suggest a simple design such as a four-room house with doors. Have the students trace the path that blood would take and use arrows to show the direction of the flow. Encourage the students to place their diagrams in their science portfolios.

SCIENCE BACKGROUND

If possible, obtain an electrocardiogram and show it to the students. Explain that an electrocardiogram, also known as an ECG or EKG, is a tool for evaluating the electrical events in the heart. A normal electrocardiogram shows three waves: a first wave, called the *P wave*, indicates atrial excitation; a second wave, called the *QRS wave*, shows ventricular excitation; and a third wave, called the *T wave*, represents ventricular relaxation. Abnormal waves indicate to doctors that something is wrong.

DISCOVER BY *Researching*

A heartbeat rate will typically drop 10 to 20 beats per minute during sleep. It will increase during times when a sleeping individual is actively dreaming.

Each side of your heart contains two chambers. An *atrium* (AY tree uhm) forms the upper chamber of each side. Each lower chamber is a *ventricle* (VEHN trih kuhl). Blood flows through these chambers in only one direction. Valves between the chambers act like one-way swinging doors. The valves open to allow blood to flow in one direction, then close to prevent blood from flowing backward.

The Beat Goes On Your heart beats many times in a day without you thinking about it. While you rest, your heart usually beats about 60 to 70 times each minute. When you exercise, your heart can beat as many as 200 times each minute. The number of times your heart beats each minute is called your *heartbeat rate*.

 Researching

Have you ever wondered if your heartbeat rate changes when you sleep? Use reference materials to find out the heartbeat rate for an average person during sleep. Do you think your heartbeat rate increases or decreases when you dream?

TEACHING STRATEGIES, continued

- **Process Skills:** *Comparing, Expressing Ideas Effectively*

Discuss the meaning of *heartbeat* with the students, then ask them to describe and explain the relationship that exists between exercise and heartbeat rate. (During periods of exercise, muscles consume additional oxygen and generate additional waste products. As a result, the heart beats faster to supply the muscles with oxygenated blood and remove the waste products produced by the activity of the muscles.)

- **Process Skills:** *Inferring, Expressing Ideas Effectively*

Explain that the "lub" sound of the heart is made by the closing of the valves between the atria and the ventricles. The "dup" sound is made by the closing of the valves between the ventricles and the large arteries that carry blood away from the heart. Ask the students to think about how these sounds are made, and then consider the meaning of the term *heart murmur*. (If the valves of the heart are not closing properly, some blood may flow backward in the chambers, making a sloshing sound. This condition is known as a heart murmur.)

LASER DISC

4160

Circulatory system; human

ONGOING ASSESSMENT
ASK YOURSELF

Valves prevent blood from flowing backward in the heart.

BACKGROUND INFORMATION

William Harvey (1578–1657) published his famous work on blood circulation in mammals in 1628. His findings were based on his firsthand observations and dissections of humans and other animals. His theory displaced the theory posited by the Greek physician Galen 1500 years earlier. Harvey was the physician to two kings of England, James I and Charles I. He is also known for his work on the reproductive process, in particular, the role of the egg.

① Arteries have thicker walls than veins, veins have thicker walls than capillaries, and capillaries have a smaller diameter than either arteries or veins.

The heartbeat can be heard by using an instrument called a *stethoscope*. Two different sounds can be heard, a "lub" and a "dup." Each lub-dup sound represents a single heartbeat and is made by the closing of valves within the heart.

 ASK YOURSELF

What function do the valves in the heart serve?

Blood Vessels

In 1868, William Harvey, an English scientist, showed that blood moves through the body in a series of "tubes," and that blood flows in a continuous, closed system. Harvey demonstrated, in front of the Royal College of Physicians, that the tubes allowed blood to flow in only one direction. Today, we call these "tubes" arteries, capillaries, and veins.

From the Heart In order for your capsule to leave the heart, it must enter an artery. **Arteries** carry blood away from the heart to other parts of the body. Arteries have thick, elastic walls that contain muscle fibers. These muscle fibers work much like rubber bands. As blood spurts from the heart, an artery stretches in diameter to make room for it. When the artery returns to its normal size, the blood is pushed or squeezed farther along the vessel.

The farther you travel from the heart, the narrower you notice the artery becoming. Now the speed your capsule is traveling decreases. The red river of blood is flowing more slowly. The artery is beginning to divide into smaller and smaller vessels, or **capillaries.** In fact, capillaries are so narrow that your capsule and red blood cells must move through them in single file. As you observe capillary walls, you seem to be able to see right

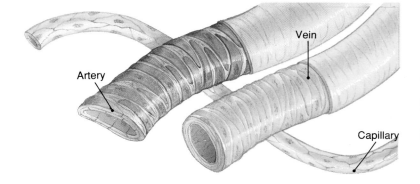

Figure 17–22. How do these three types of blood vessels differ from each other? ①

- **Process Skills:** *Inferring, Applying*

Ask the students to describe the relationship among the heart muscle, mitochondria, and oxygenated blood. (The heart muscle contains many mitochondria and a rich supply of oxygenated blood.)

- **Process Skills:** *Classifying/Ordering, Generating Ideas*

Write the term *cardiovascular* on the chalkboard and ask the students if they have ever used or heard the term, and if possible, to describe the context in which it was used or heard.

- **Process Skills:** *Comparing, Applying*

Have the students examine and discuss the blood vessels shown in Figures 17–22 and 17–23. Ask the students to compare the structures of the different vessels. Make sure that they note the thick walls of arteries and the thin walls of capillaries. Ask the students to relate the thin capillary walls to the capillaries' function. (Materials pass easily between the body cells and the blood through the thin walls of capillaries.) You may wish to point out to the students that *arteries* carry blood *away* from the heart—both words begin with the letter *a*.

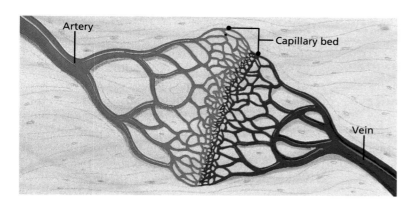

Figure 17–23. Capillaries connect arteries to veins.

REINFORCING THEMES— Systems and Structures

Stress the roles that the heart, arteries, veins, and capillaries play in providing all parts of the human body with oxygenated blood and nutrients.

> ✧ **Did You Know?**
> When the human heart stops beating, blood tends to accumulate in the lowest portions of the body due to gravity. One of the ways criminal investigators determine the time of death of a crime victim is to estimate the amount of blood that has accumulated in the lower portions of the victim's body.

through them, because the tiny capillaries have walls only one cell thick. The movement of nutrients, oxygen, and waste molecules between the blood and the tissues takes place through these thin capillary walls. You observe nutrients and oxygen moving from the blood into the fluid spaces and on into body cells. At the same time, waste products, including carbon dioxide, move from the body cells into the blood.

A Journey Made in "Vein" Your capsule flows through the capillary into a small vein. Small veins join to form larger veins. **Veins** carry the blood back to the heart. Veins have thinner walls than arteries. Unlike arteries, veins contain one-way valves. These valves prevent the blood from falling backward as it is pushed toward the heart.

Skeletal muscles surround many veins. As a person moves, these muscles squeeze the veins and push blood through the blood vessel. This squeezing helps the blood return to the heart. People who must stand in one place for a long time learn to flex the muscles of their legs. This helps to keep the blood moving. Otherwise, the blood may pool in the legs, and the flow of blood to the brain may decrease. Eventually, the person might faint from inadequate oxygen in the brain. What kinds of jobs require people to stand in one place for a long period of time? ② Astronauts in the space shuttle do not have this problem. Because of the lack of gravity, their circulatory systems do not have the tendency to pool blood in the lowest portions of their bodies. Their circulatory systems function quite well whether they are right side up or upside down!

② Responses might include cashiers and check-out clerks in stores, tollbooth operators, and traffic officers.

ONGOING ASSESSMENT
▼ **ASK YOURSELF**

Arteries have thick elastic walls that contain muscle fibers. Veins have thinner walls than arteries, contain one-way valves, and are surrounded by voluntary muscle tissue. Capillaries have walls that are only one cell thick and allow the movement of nutrients, oxygen, and waste materials between body tissues and the blood.

▼ **ASK YOURSELF**

How are the types of blood vessels different?

TEACHING STRATEGIES, continued

● **Process Skills:** *Comparing, Applying*

Explain to the students that the heart is really two separate pumps. The right side of the heart functions in pulmonary circulation, and the left side of the heart functions in systemic circulation. Remind the students that blood travels through the human body in a continuous path, then ask the students to identify which system, pulmonary or systemic, transports oxygen-rich blood throughout the body. (systemic)

● **Process Skills:** *Comparing, Applying*

Direct the students' attention to Figure 17–24 and discuss the path of pulmonary circulation. Ask the students to infer which blood vessels would contain bright red blood, and which vessels would contain darker, bluish blood. (The pulmonary arteries move oxygen-poor and carbon dioxide-rich blood from the heart to the lungs; pulmonary arteries contain darker, bluish blood. The pulmonary veins move oxygen-rich and carbon dioxide-poor blood from the lungs to the heart; pulmonary veins contain bright red blood.)

SCIENCE BACKGROUND

Cardiovascular disease is a leading cause of death in the United States. Point out to the students that there are certain risk factors associated with cardiovascular disease. Some of these risk factors can be changed by modifying a person's lifestyle; others cannot be changed. Risk factors that cannot be changed include heredity, sex, age, and race. Risk factors that can be changed include smoking, high blood cholesterol level, high blood pressure, obesity, lack of exercise, and stress.

MEETING SPECIAL NEEDS

Gifted

 Have the students research the names of at least ten arteries that supply different parts of the body with blood. For example, the renal arteries supply the kidneys and the ureters with blood. Ask the students to create a table that lists the names of the arteries and the regions of the body supplied by those arteries. Remind them to include their tables in their science portfolios.

Path of Blood

Blood travels through the body in a continuous path. This continuous path is divided into two parts—the pulmonary system and the systemic system. Blood in the pulmonary system moves from the heart to the lungs through pulmonary arteries. In the lungs, the blood obtains oxygen and releases waste carbon dioxide. Then the blood moves to the heart again through pulmonary veins. The oxygen-rich blood then is pumped throughout the body through the systemic system. The systemic system supplies oxygen to all parts of your body, including the heart.

Clogged Pipes   Proper movement of blood through the blood vessels of the heart is necessary to keep the heart tissue healthy. If arteries or veins in the heart become blocked, the supply of nutrients and oxygen that the heart receives will be reduced. Such blockage can damage the heart muscle and is called a "heart attack."

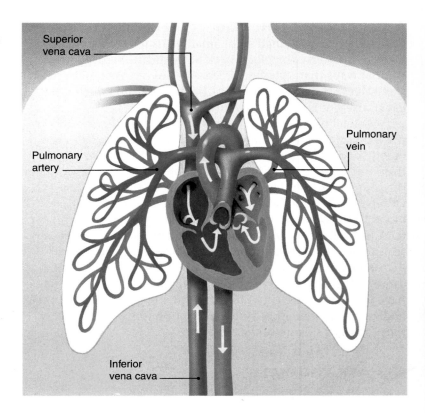

Figure 17–24. Pulmonary and systemic circulation are shown in this illustration.

- **Process Skills:** *Inferring, Expressing Ideas Effectively*

Discuss the concept of blood pressure with the students. Explain that blood moves in the body from an area of greatest pressure when it leaves the ventricles to an area of least pressure when it empties into the atria. The blood pressure of an individual person is determined by the heart rate, the volume of blood, and the resistance to the flow of blood in the blood vessels. Ask the students to predict some of the factors that may contribute to high blood pressure. (Responses should include obesity and high salt intake.)

GUIDED PRACTICE

Have the students describe the path of blood through the body as it enters the right atrium, moves into the right ventricle, is pumped to the lungs, and so on, finally returning to the right atrium.

INDEPENDENT PRACTICE

Have the students provide written answers to the Section Review and Application questions. Ask the students to write their reaction to the statement "True or false: The heart is the hardest working organ of the human body."

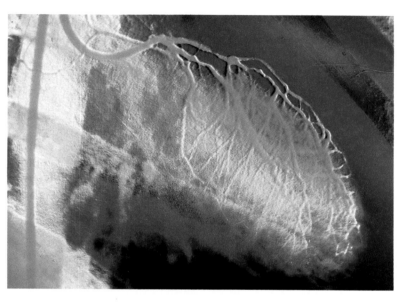

Figure 17–25. This X-ray picture shows the blood vessels of the heart.

As shown in Figure 17–26, arteries can become clogged with deposits of cholesterol and other substances. *Arteriosclerosis* is a condition characterized by abnormal thickening and clogging of artery walls. As you grow older, your arteries will gradually begin to thicken naturally—it's a part of the aging process. However, you can increase the rate at which your arteries become clogged through a diet heavy in fats and cholesterol. Cholesterol is a fatty substance that can stick to the walls of arteries. When this happens, the flow of blood through that artery is restricted to a smaller area. This restriction causes blood pressure to increase and can lead to blockages that cause heart attacks. These blockages result from deposits breaking off from the artery wall and blocking a smaller vessel elsewhere in the body.

Under Pressure A physician is sometimes interested in the force at which the blood moves through the arteries. This force is known as the *blood pressure*. Blood pressure increases as the ventricles push the blood through the arteries. Blood pressure then decreases while the ventricles refill with blood. Having a normal blood pressure is an important health concern. Blood pressure that is too high makes the kidneys work harder and can damage valves and muscle tissue of the heart. Blood pressure that is too low can cause poor circulation. In a cold climate, poor circulation is especially dangerous because a person might not realize when extremities of the body are becoming frostbitten.

Figure 17–26. Arteries can become clogged with deposits of cholesterol.

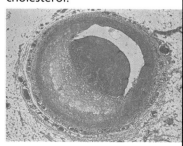

THE NATURE OF SCIENCE

Angioplasty is a procedure used by cardiologists to encourage the flow of blood through clogged arteries using an inflated balloon-like device. Diets high in cholesterol can lead to serious health complications later in life. Drugs and alcohol that are present in the circulatory system of a pregnant woman are also present in the circulatory system of the baby she is carrying. Although innovations and research findings such as these are relatively recent, they are likely to be "old news" to many students. Point out to the students that it might seem as if research related to health care is the only kind of research being performed by scientists and researchers. It may seem that way because many of these innovations and findings can directly impact the length and quality of people's lives and are therefore news. However, explain that the nature of scientific research is not to specialize in one domain, and scientific research is not limited to health-related issues. Remind the students that scientific research exists in every subject area they could possibly think of.

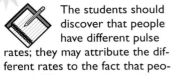

DISCOVER BY *Calculating*

The students should discover that people have different pulse rates; they may attribute the different rates to the fact that people of different ages have different metabolic rates.

ONGOING ASSESSMENT

▼ **ASK YOURSELF**

Blood must circulate to provide all parts of the body with oxygen and nutrients and to remove waste products.

SECTION 3 REVIEW AND APPLICATION

Reading Critically

1. A valve allows blood to flow in only one direction.

2. The "lub-dup" sound is caused by the closing of the heart valves.

3. Arteries have thick elastic walls that contain muscle fibers; they carry blood away from the heart. Veins have one-way valves and thinner walls than arteries; they carry blood toward the heart.

Thinking Critically

4. High blood pressure makes the heart work much harder and can damage the valves by stretching and weakening them.

5. Because all areas of the body require oxygen-rich blood, and oxygen-rich blood is delivered by the systemic system, the systemic system will contain more kilometers of blood vessels than the pulmonary system.

6. The walls of capillaries are very thin compared to the walls of veins and arteries. As a result, a bruise is more likely to be the result of damage to capillaries.

RETEACHING

Have the students recall the structure and function of arteries, veins, and capillaries, then draw pictures of an artery, a vein, and a capillary.

EVALUATION

Have the students compare and contrast the function of arteries, veins, and capillaries.

EXTENSION

One effect of serious injury is called *shock*. Encourage the students to investigate how shock affects a person's circulatory system and describe what can be done for a person who is in shock or is suspected to be in shock. Ask the students to share their findings with the class.

CLOSURE

 Cooperative Learning Working in groups, have the students play a game of "Where Does the Blood Go Next?" One member in each group names a chamber of the heart or a body region, and the other group members name the chamber or region that the blood travels to next.

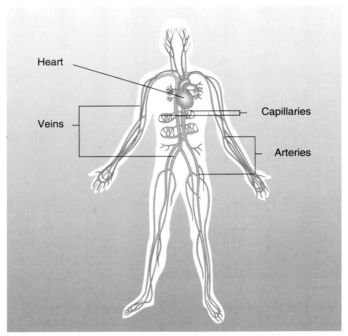

Figure 17–27. The human circulatory system is a complex, continuous pathway.

Your journey through the complex, continuous circulatory system reaches a conclusion with a note about your pulse. Your pulse can be detected at various locations throughout your body. A *pulse* is felt as an artery expands, then relaxes after blood surges through the artery. Your pulse can be found at points where an artery is near the surface of your body. Common pulse points are found on the insides of the wrists, ankles, and thighs. The pulse can also be felt on the temples and the sides of the neck. How do you think your pulse rate compares to the pulse rate of someone else?

DISCOVER BY *Calculating*

Survey the pulse rate for some of your friends or family members. In your journal, record each person's age, sex, and pulse rate. Graph your results. Compare age to pulse rates. Do all of your family members have the same pulse rate? What factors might explain some of the differences?

▼ **ASK YOURSELF**

Why must blood circulate through the body?

SECTION 3 REVIEW AND APPLICATION

Reading Critically

1. Explain the function of a valve.
2. What causes the "lub-dup" sound in the heart?
3. How are arteries different from veins?

Thinking Critically

4. What kind of damage might high blood pressure do to valves of the heart?
5. Which system—pulmonary or systemic—contains more kilometers of blood vessels? Why?
6. Is a bruise more likely to be the result of damage to an artery, a vein, or a capillary? Explain.

CHAPTER 17 HIGHLIGHTS

The Big Idea—
SYSTEMS AND STRUCTURES

Help the students understand that the cells of their body require a constant supply of nutrients and oxygen and a means by which the waste products that the cells produce are regularly removed. Supplying the cells of the body with their basic needs is no small task, for the human body is composed of an enormous number of cells. The digestive system and the circulatory system of the human body work together to create and supply nutrients and oxygen to every cell of the body, as well as to remove the waste products generated by each cell. Explain to the students that when the digestive and circulatory systems are functioning efficiently, they sometimes tend to be forgotten or ignored, although we could not exist without them.

CHAPTER 17 HIGHLIGHTS

The Big Idea

All parts of the human body require a consistent source of energy, which is provided by nutrients. Nutrients are processed by the digestive system. The circulatory system absorbs the nutrients and oxygen and supplies them to the entire body. Powered by the heart and flowing through capillaries, veins, and arteries, blood transports nutrients and oxygen to the body and carries waste products in return. The normal operation of the systems and their interactions enable the body to function properly. As a result, people are able to lead active and healthy lives.

At the beginning of the chapter, you wrote your ideas about traveling to the brain through arteries, the relationship of the digestive and circulatory systems, and the role of nutrients in maintaining health. Reread your journal entry. Revise it to reflect changes in your understanding. Be sure to include a list of things you could do to take better care of your digestive and circulatory systems.

The ideas expressed by the students should reflect not only revisions in their thinking based on what they learned in the chapter, but also an understanding that it makes good sense to take better care of themselves and, in particular, of their circulatory and digestive systems.

Connecting Ideas

Copy this unfinished concept map into your journal. Complete the concept map by writing the correct term in each blank.

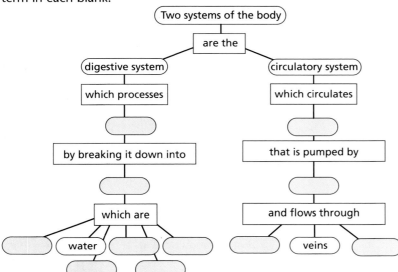

CONNECTING IDEAS

 The concept map should be completed with words similar to those shown here.

Row 5: food; blood
Row 7: nutrients; the heart
Row 9: lipids; proteins; vitamins; arteries; capillaries
Row 10: water; minerals

CHAPTER 17 **485**

CHAPTER 17 REVIEW ANSWERS

Understanding Vocabulary

1. Ingestion, or eating, must occur before digestion can begin. Digestion is the process that changes food into a form that the body can use.

2. Mechanical digestion and chemical digestion both occur in the mouth. Mechanical digestion physically breaks large pieces of food into smaller pieces. Chemical digestion breaks large food molecules into smaller ones.

3. Absorption is the movement of molecules of nutrients into the blood vessels in the walls of the small intestine.

4. Plasma is the liquid component of blood. Red blood cells are one of the solid components of blood.

5. White blood cells and platelets are two components of the solid portion of human blood.

6. Arteries, veins, and capillaries are the vessels through which blood, nutrients, oxygen, and waste products move through the human body.

7. The pulmonary and systemic systems make up the circulatory system. The pulmonary system obtains the oxygen that is delivered to the cells of the body by the systemic system.

Understanding Concepts

Multiple Choice
8. c   10. d
9. a   11. d

Short Answer

12. The small intestine contains an enormous amount of surface area for efficient nutrient absorption due to the folds and villi found inside its walls.

13. A contraction of the heart forces blood into the arteries. The force of the blood against the walls of the arteries causes the walls to expand. A pulse can be felt as the blood is pumped through the arteries. Blood pressure rises during the contraction of the ventricles and falls when the ventricles are relaxed and filling with blood.

14. Carbohydrates supply energy. Lipids store energy and protect cells. Proteins are the primary building blocks of body tissues. Vitamins combine with enzymes to increase the rate at which chemical reactions occur. Minerals build bones and teeth and are necessary for many other body processes. Water provides a liquid medium in which all nutrients are dissolved.

15. Chewing food slowly and completely increases the surface area of the food. Increased surface area provides more sites on the food for digestive enzymes to act, speeding the digestive process.

CHAPTER 17 REVIEW

Understanding Vocabulary

For each set of terms, explain how they are related.

1. digestion (458), ingestion (459)
2. chemical digestion (459), mechanical digestion (459)
3. nutrients (458), absorption (463)
4. plasma (473), red blood cells (474)
5. white blood cells (475), platelets (475)
6. arteries (480), veins (481), capillaries (480)
7. pulmonary system (482), systemic system (482)

Understanding Concepts

MULTIPLE CHOICE

8. A small bruise is most likely to show damage to which type of blood vessel?
   a) artery
   b) vein
   c) capillary
   d) Each vessel is equally likely to show damage.

9. Why do people sometimes take a drink of something while chewing food?
   a) to aid mechanical digestion
   b) to aid chemical digestion
   c) to aid the absorption of nutrients
   d) to speed the digestive process

10. A diet that is high in cholesterol should be a concern for which group of people?
    a) infants
    b) youths
    c) adults
    d) people of all ages

11. A well-balanced diet provides
    a) cell energy.
    b) cell growth.
    c) cell repair.
    d) all of these.

SHORT ANSWER

12. In what way is the structure of the small intestine well suited for the absorption of nutrients?

13. How are heart contraction, pulse, and blood pressure related?

14. What function does each of the six nutrient groups provide for the human body?

15. Why is it often suggested that food be chewed slowly and completely?

Interpreting Graphics

16. The illustration shows how a network of fibers grows over a wound to stop the flow of blood. Explain how these fibers stop blood flow.

17. Some people must take medication that thins their blood. Do you think such people have more or less difficulty than usual stopping the flow of blood from a wound? Why?

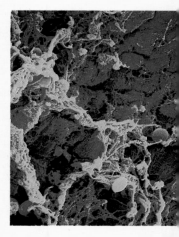

Interpreting Graphics

16. Clotting fibers form a fine seal or net over a wound through which the solid and liquid components of blood cannot pass.

17. Accept reasonable, supported conclusions. Thin blood will clot less quickly than thick blood

Reviewing Themes

18. The digestive system provides basic nutrients, and the circulatory system delivers those nutrients and oxygen to all parts of the human body.

19. Proper functioning of the digestive and circulatory systems requires varying quantities of the six basic nutrients. If these nutrients are not provided through a balanced diet, the digestive and circulatory systems will not function properly, and the health of the individual will be affected.

Thinking Critically

20. The digestive system will effectively digest an infrequent meal that is not balanced because it is using energy for digestion that was previously supplied by balanced meals.

21. The stomach and esophagus are lined with a layer of mucus. The mucus layer protects the walls of the stomach and esophagus by preventing digestive juices from damaging its tissues. The throat does not have that protection.

22. The placement of arteries far beneath the skin offers a greater degree of protection from external hazards.

23. Dietary variety is important to ensure that the body receives all of the nutrients necessary for good health. Consumption of large quantities of food that are lacking in a comprehensive variety of nutrients may result in malnutrition.

24. The buildup of fatty substances on the walls of arteries narrows the vessels and renders them less elastic. Since inelastic vessels cannot expand when blood is pumped through them, pressure against the vessel walls increases. Also less blood can flow through the narrow portions of the vessels. Narrower vessels make it easier for a blockage to occur, which can cause a heart attack or a stroke.

25. When water cannot be absorbed to replace that lost through perspiration and excretion, the body begins to dehydrate. Severe dehydration can lead to death.

26. Damage to an artery is likely to result in more rapid blood loss than damage to a vein because blood pressure in arteries is greater than blood pressure in veins.

27. Responses should reflect an understanding of the relationship between red blood cells and oxygen. If red blood cells are destroyed, the cells of the body will not receive oxygen and consequently will die.

Reviewing Themes

18. *Systems and Structures*
Explain how the digestive and circulatory systems work together to supply nutrients and oxygen to all parts of the body.

19. *Energy*
Explain why a balanced diet is necessary for the proper functioning of the digestive and circulatory systems.

Thinking Critically

20. Will your digestive system function effectively if you eat a meal that is not well balanced? Explain.

21. A person may vomit when ill. Why does the mixture of food and gastric juice produce a burning sensation in the throat but not in the stomach?

22. Arteries are usually found farther from the surface of the body than veins. Why do you think this is true?

23. How is it possible for a person to eat large amounts of food and still suffer malnutrition?

24. Arteriosclerosis causes the walls of the arteries to become less elastic. How would this disease affect the heart?

25. An intestinal virus often prevents the large intestine from absorbing water. What effect might this have on a person if the condition lasted for several days or longer?

26. Explain why a wound that cuts an artery is more dangerous than a wound that cuts a vein.

27. This photograph shows red blood cells infected with the protozoans that cause malaria. Using your knowledge of what red blood cells do, explain why their breaking open is harmful.

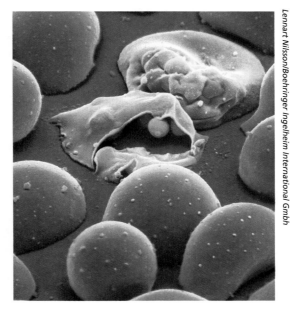

Lennart Nilsson/Boehringer Ingelheim International Gmbh

Discovery Through Reading

Lambourne, Mike. *Down the Hatch: Find Out about Your Food.* Millbrook Press, 1992. This book about food and digestion is illustrated with colored drawings and diagrams and includes investigative activities.

CHAPTER 18
RESPIRATION AND EXCRETION

PLANNING THE CHAPTER

Chapter Sections	Page	Chapter Features	Page	Program Resources	Source
Chapter Opener	488	*For Your Journal*	489		
Section 1: RESPIRATION	490	Discover By Doing (A)	492	*Science Discovery**	SD
• The Respiratory System (B)	490	Discover By Doing (A)	493	Investigation 18.1: Measuring Carbon Dioxide Released During Respiration (H)	TR, LI
• Breathing (A)	492	Section 1 Review and Application	495	Connecting Other Disciplines: Science and Mathematics, Examining the Composition of Air (A)	TR
• Exchange of Gases (H)	494	Investigation: Measuring Lung Capacity During Exercise (A)	496	The Human Respiratory System and Skin (A)	IT
				Record Sheets for Textbook Investigations (A)	TR
				Study and Review Guide, Section 1 (B)	TR, SRG
Section 2: EXCRETION	497	Discover By Researching (A)	499	Investigation 18.2: Vertebrate Kidneys (A)	TR, LI
• Waste Removal (A)	497	Discover By Researching (A)	500	Reading Skills: Predicting Outcomes (A)	TR
• The Excretory System (B)	498	Section 2 Review and Application	500	The Human Excretory System (B)	IT
		Skill: Sequencing Data (A)	501	Study and Review Guide, Section 2 (B)	TR, SRG
Section 3: THE SKIN	502	Activity: What are some characteristics of skin? (A)	504	*Science Discovery**	SD
• The Structure of the Skin (B)	502	Discover By Observing (B)	505	Thinking Critically (A)	TR
• The Functions of the Skin (A)	505	Section 3 Review and Application	506	Extending Science Concepts: Perspiration and Body Temperature (A)	TR
				The Human Respiratory System and Skin (A)	IT
				Study and Review Guide, Section 3 (B)	TR, SRG
Chapter 18 HIGHLIGHTS	507	The Big Idea	507	Study and Review Guide, Chapter 18 Review (B)	TR, SRG
Chapter 18 Review	508	For Your Journal	507	Chapter 18 Test	TR
		Connecting Ideas	507	Test Generator	

B = Basic A = Average H = Honors
The coding Basic, Average, and Honors indicates subsections, features, and resources that might be appropriate for different levels of learners. For additional suggestions regarding choice of topic and depth of coverage, see the Pacing Chart on pages T26–T29.

*Frame numbers at point of use
(TR) Teaching Resources, Unit 6
(IT) Instructional Transparencies
(LI) Laboratory Investigations
(SD) *Science Discovery* Videodisc Correlations and Barcodes
(SRG) Study and Review Guide

▶ 487A

CHAPTER MATERIALS

Title	Page	Materials
Discover By Doing	492	(per group of 3 or 4) three different-sized boxes with lids, elastic bands (3), scissors
Discover By Doing	493	(per individual) paper, pencil
Teacher Demonstration	494	balloon, plastic detergent bottle (clear)
Investigation: Measuring Lung Capacity During Exercise	496	(per group of 2) mouthpiece holder, lung volume bag, rubber band, mouthpiece (disposable), paper towel, pencil, paper
Discover By Researching	499	(per class) reference materials, journal
Discover By Researching	500	(per individual) paper, pencil
Skill: Sequencing Data	501	(per group of 2) paper, pencil
Activity: What are some characteristics of skin?	504	(per group of 2 or 3) plain white paper, pen or pencil, ruler, ink pad, hand lens
Discover By Observing	505	(per group of 3 or 4) microscope slide, microscope, journal

ADVANCE PREPARATION

For the *Discover By Doing* on page 492, the students will need three different-sized boxes with lids and three equal-sized elastic bands. Obtain a balloon and clear plastic detergent bottle for the *Demonstration* on page 494. You will need mouthpiece holders, lung volume bags, and disposable mouthpieces for the *Investigation* on page 496. For the *Activity* on page 504, the students will need ink pads.

TEACHING SUGGESTIONS

Field Trip
Arrange a tour of the local area, encouraging the students to identify places with air pollution as well as open, clean-air spaces. Ask the students to write a journal entry describing how respiration is affected by the air quality and comparing the air quality in different locations on the tour.

Outside Speaker
Have a local physician or medical technician speak to the class about the function of kidney-dialysis machines and heart-lung machines. Another speaker could be a physician or technician from a burn unit of a local hospital. Ask the person to talk about damage to the skin and treatment of burns.

CHAPTER 18 RESPIRATION AND EXCRETION

CHAPTER THEME—SYSTEMS AND STRUCTURES

This chapter introduces the students to the structures and the functions of the respiratory and excretory systems. The students explore the processes of breathing and gas exchange and learn about the process of excretion. The major theme **Systems and Structures** is also developed through concepts in Chapters 4, 5, 7, 8, 9, 10, 12, 13, 16, 17, 19, and 20. A supporting theme of this chapter is **Energy.**

MULTICULTURAL CONNECTION

The Sherpas are a people of Nepal who have gained fame as mountaineers. Living in the shadow of Mount Everest, the world's highest mountain, some Sherpas live at an altitude of 6100 m. At this altitude, many people experience mountain sickness because the amount of oxygen in the air is so low. The lack of oxygen causes disorientation, dizziness, headaches, and weakness. Acclimatization to high altitudes may take up to 2 weeks. Some people do not become acclimatized and must have additional oxygen administered to them. The Sherpas, however, have been accustomed to low levels of oxygen since birth. As a result, they are able to survive well at altitudes that are dangerous to others. Even the Sherpas, however, must become acclimatized to the very small amount of oxygen available at the top of Mount Everest. In 1953, Tenzing Norgay, a Sherpa, and Edmund Hillary, a mountain climber from New Zealand, became the first people to reach the top of Mount Everest, which rises 8840 m above sea level.

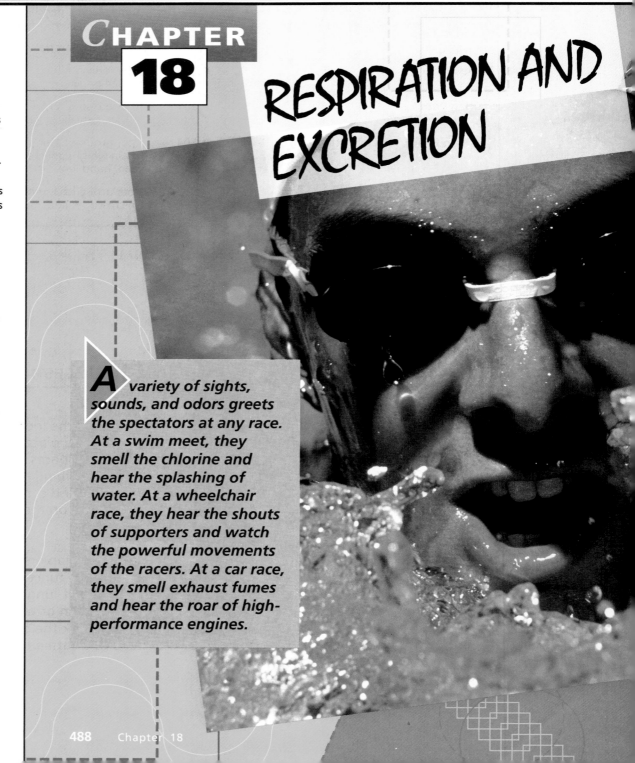

CHAPTER 18 RESPIRATION AND EXCRETION

A variety of sights, sounds, and odors greets the spectators at any race. At a swim meet, they smell the chlorine and hear the splashing of water. At a wheelchair race, they hear the shouts of supporters and watch the powerful movements of the racers. At a car race, they smell exhaust fumes and hear the roar of high-performance engines.

CHAPTER MOTIVATING ACTIVITY

Prepare a solution of 0.1 g bromothymol blue in 2000 mL of water. If the solution is green, add 1 drop of 0.4 percent sodium hydroxide to make the solution blue. Explain that this blue solution will turn yellow with the addition of carbon dioxide. Give each student three straws and a small flask containing about 25 mL of the bromothymol blue solution. Have the students use the straws to blow into their solutions. **CAUTION: Warn the students not to inhale the solutions.** The solutions should turn yellow in 2 or 3 minutes. Ask the students to explain their observations. (The carbon dioxide in exhaled breath causes the indicator to change color.)

For Your Journal

Answering the journal questions provides the students with an opportunity to display their knowledge of the structures and the functions of the respiratory and excretory systems. From the entries, you may note whether the students have some misconceptions about these systems. You may wish to address these misconceptions as the respiratory and excretory systems are studied.

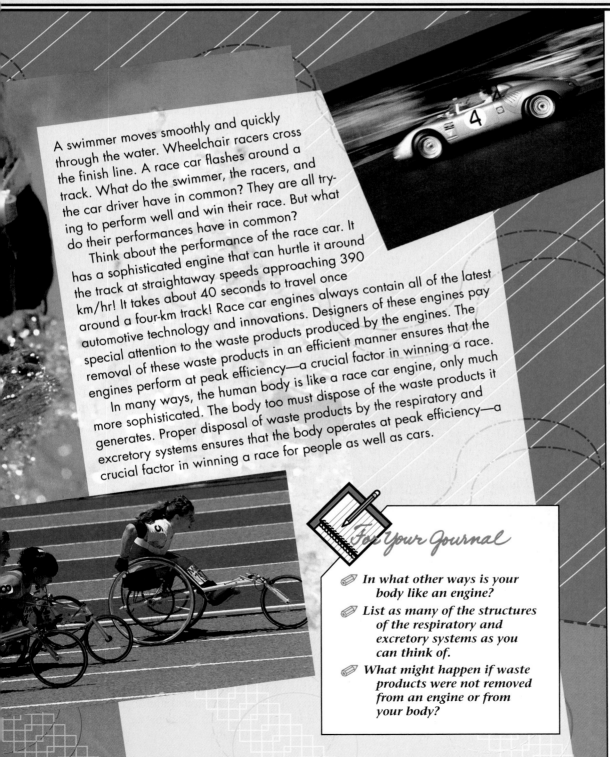

A swimmer moves smoothly and quickly through the water. Wheelchair racers cross the finish line. A race car flashes around a track. What do the swimmer, the racers, and the car driver have in common? They are all trying to perform well and win their race. But what do their performances have in common?

Think about the performance of the race car. It has a sophisticated engine that can hurtle it around the track at straightaway speeds approaching 390 km/hr! It takes about 40 seconds to travel once around a four-km track! Race car engines always contain all of the latest automotive technology and innovations. Designers of these engines pay special attention to the waste products produced by the engines. The removal of these waste products in an efficient manner ensures that the engines perform at peak efficiency—a crucial factor in winning a race.

In many ways, the human body is like a race car engine, only much more sophisticated. The body too must dispose of the waste products it generates. Proper disposal of waste products by the respiratory and excretory systems ensures that the body operates at peak efficiency—a crucial factor in winning a race for people as well as cars.

For Your Journal

- In what other ways is your body like an engine?
- List as many of the structures of the respiratory and excretory systems as you can think of.
- What might happen if waste products were not removed from an engine or from your body?

ABOUT THE PHOTOGRAPH

Racing is a contest of speed in which participants compete against one another to determine who is the fastest. Whether driving an automobile, propelling a wheelchair, or swimming, all racers train and practice for their events. Peak performances depend on the racers' physical fitness and alertness. Just as all the systems of an automobile work together to keep the racing car operating efficiently, all the systems of the human body work together to enable the racers to perform efficiently. Their respiratory and excretory systems play important roles by providing the oxygen needed to produce energy and by eliminating waste products.

MEETING SPECIAL NEEDS

Second Language Support

To reinforce the names, functions, and locations of parts of the respiratory and the excretory systems, have the students draw and label diagrams of the systems in their journals. Then ask them to write the name of each part with its English pronunciation and an identification of its function. The students may use their diagrams and notes as study guides.

CHAPTER 18

Section 1: RESPIRATION

FOCUS
This section identifies the structures that make up the respiratory system. The pathway of air from the nasal passages to the lungs is described, and the functions performed by each organ along this pathway are explained. In addition, the mechanism of breathing and the exchange of gases are explored.

MOTIVATING ACTIVITY
Have the students press their hands to their sides just above their waists. Ask them to inhale deeply and exhale slowly several times. Ask the students what they felt. (The students might suggest that they could feel their chests expand as they inhaled and decrease as they exhaled.)

PROCESS SKILLS
- Comparing • Constructing/Interpreting Models
- Solving Problems/Making Decisions

POSITIVE ATTITUDES
- Openness to new ideas
- Precision

TERMS
- trachea • epiglottis
- bronchi • bronchioles
- alveoli • diaphragm

PRINT MEDIA
The Lungs and Breathing by Steve Parker (see p. 429b)

ELECTRONIC MEDIA
The Lungs and Respiratory System, Britannica (see p. 429b)

Science Discovery
Respiratory system; human

BLACKLINE MASTERS
Study and Review Guide
Laboratory Investigation 18.1
Connecting Other Disciplines

SCIENCE BACKGROUND
Mucus and the cilia of specialized cells in the respiratory tract help protect the human body from disease. Mucus and cilia trap germ-laden dust and spores before they can enter the lungs.

① The students may suggest that just as a car engine uses air and fuel to run and gives off exhaust, the human body uses oxygen and nutrients to generate energy and produces waste products.

SECTION 1 | Respiration

Objectives

List each structure of the respiratory system and **describe** its function.

Relate the structure of lungs to their function.

Summarize the process of breathing.

Imagine that you are standing next to the engine shown in the photograph. Although you don't look like the engine, you and the engine are alike in many ways. Both of you are machines. Both of you burn fuel to release energy. Both of you take in oxygen and give off waste products. If the waste products produced by each machine are not released, neither machine will run efficiently. In fact, if the waste products from the machines are allowed to build up, the machines will stop functioning.

Figure 18-1. How is your body like this engine? ①

The Respiratory System

Your digestive system changes food into nutrient molecules that you use for fuel. Your circulatory system then transports oxygen and these nutrients to cells in all parts of your body. Inside each cell, the powerhouses of the cell—mitochondria—use the nutrients and oxygen in a process called *cellular respiration*. This process releases energy and produces carbon dioxide and water as waste.

Breathe In, Breathe Out Your respiratory system brings oxygen into your body and removes carbon dioxide. Your nose, your lungs, and a number of tubes that connect your nose and lungs make up the major components of your respiratory system. However, to bring air into your body, your respiratory system needs help from your skeletal and muscular systems. The systems work together to fill your lungs with air.

Air enters your body through either your nose or your mouth. Even though both ways bring oxygen into your body, breathing through your nose is healthier than breathing through your mouth. When air enters through your nose, your nasal passages warm, moisten, and clean the air. A layer of wet, sticky mucus coats the inner walls of your nose. Most of the dust and dirt in the air sticks to the mucus. Your nose acts like the air filter in an engine; it cleans particles and debris from the air that don't belong in the system. The tissues in your mouth cannot perform these functions.

TEACHING STRATEGIES

● **Process Skills:** *Inferring, Applying*

Remind the students that the purpose of the respiratory system is to supply the body with oxygen for aerobic cellular respiration and to remove carbon dioxide and other waste products. Point out that through the process of aerobic respiration, oxygen is used to release energy from food molecules; water, carbon dioxide, and heat are produced as waste products. Discuss the role of the respiratory system in obtaining oxygen from the air and making it available to every cell in the body. Then ask the students to describe how oxygen is used in aerobic respiration to provide body cells with energy. (Oxygen is necessary for the breakdown of glucose, which results in the release of energy.)

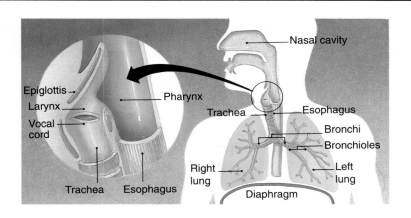

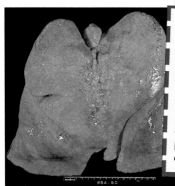

Figure 18-2. The healthy lung (right) is an essential component of the human respiratory system.

✦ **Did You Know?**
During normal breathing, only about 10 to 25 percent of the air in the lungs of humans is replaced with each breath. A person breathes in over 18 900 L of air every day. Oxygen makes up about 20 percent of that air, so a person takes in about 3780 L of oxygen daily.

SCIENCE BACKGROUND

Several organs merge in the back of the mouth—the pharynx; the two nasal cavities; the two Eustachian tubes that lead to the middle ear; the glottis, which leads to the larynx; the epiglottis, which covers the trachea when swallowing; the trachea, which leads to the bronchi and then the lungs; and the esophagus, which leads into the stomach.

 LASER DISC
4169

Respiratory system; human

Notice in the illustration that the passages of the nose and the mouth come together at the back of your throat. The place where your nasal passages and mouth meet is called the *pharynx* (FAR inks). As you swallow, food pushes up the cartilage at the back of your mouth. This cartilage covers the opening that leads to your nose and keeps food from entering your nasal passages.

Two tubes lead deeper into the body from the pharynx. One tube, the esophagus, carries food down to the stomach. Air moves to the lungs through the other tube, called the windpipe, or **trachea** (TRAY kee uh). You can feel the trachea beneath your skin by gently rubbing the front of your neck. You can feel the rings of cartilage that hold the trachea open like a stiff but flexible garden hose. These cartilage rings hold the trachea open so that air can move through it more easily. A flap of tissue called the **epiglottis** (ep uh GLAH tis) swings down over the opening to the trachea when you swallow. This helps keep food from blocking airflow.

Say "Ahhh" The *larynx* (LAR inks), or voicebox, forms the upper part of the trachea. Thin ligaments stretch across the opening of the larynx, forming your vocal cords. As air passes through the space between the vocal cords, they vibrate and make sound. This is the sound of your voice. You change the pitch of the sound by changing the length and thickness of your vocal cords. Rest your fingers on your throat while you make an "Ahhh" sound. Change the pitch of the sound, first higher and then lower.

The action of the vocal cords works in a way similar to the tightening and loosening of a rubber band or guitar string. A model of the vocal cords can be made in the next activity.

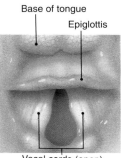

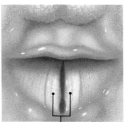

Figure 18-3. To create sounds, vocal cords open and close to allow air to pass through them.

SECTION 1 **491**

TEACHING STRATEGIES, continued

● **Process Skills:** *Inferring, Expressing Ideas Effectively*

Ask the students to describe how the vocal cords produce sound and to explain why the vocal cords are composed of ligaments rather than other materials such as bones. (Air rushing past the vocal cords causes the cords to vibrate, and sound is generated by these vibrations. The vocal cords must be composed of a material that is flexible enough to vibrate and possesses sufficient elasticity to vary pitch as the cords are lengthened.)

● **Process Skills:** *Comparing, Generating Ideas*

Have the students hum several different tones while placing their fingertips on the front of their throats. Point out that they should be able to feel the vibrations caused by air passing over the vocal cords. Remind the students that the boxes used in the *Discover By Doing* activity allowed the tones generated by the elastic bands to be more easily heard. Ask the students to identify the structures of their respiratory system that serve the same purpose as the boxes. (the sinuses in the head and the larynx)

DISCOVER BY *Doing*

When comparing the vocal cord model to the actual vocal cords of the human body, the students should note that, like the rubber bands, human vocal cords are elastic. The students should discover that as the length of an elastic band decreases, its pitch increases, and vice versa.

ONGOING ASSESSMENT
▼ **ASK YOURSELF**

Different pitches can be produced by changing the length and thickness of the vocal cords.

INTEGRATION— Language Arts

Explain to the students that several English idioms refer to the process of breathing and the respiratory system. Write this idiom on the board: save one's breath. Ask the students to explain the meaning of the idiom. (to keep silent) Then ask the students to think of other idioms that refer to breathing and to provide their meaning. (The students might suggest "under one's breath" meaning "in a low voice" and "take one's breath away" meaning "to surprise or flabbergast.")

▶ **492** CHAPTER 18

Figure 18–4. People produce sound as air is forced past the vocal cords.

DISCOVER BY *Doing*

Obtain three different-sized boxes with lids and three elastic bands that are the same size. Cut a square out of each box lid. Stretch an elastic band around the length of each box so that the elastic band passes over the square hole in the lid. Pluck each elastic band and listen to the sound. Describe its pitch. Why is there a difference in pitch? Which elastic band has the highest pitch? Which has the lowest pitch? Compare the model to your vocal cords.

The sounds made by the boxes in the activity are simple. The sounds you make are much more sophisticated. Your tongue, cheeks, and lips help shape the sounds into words. The volume of the sound depends on the force of the air as it flows over the vocal cords.

Beneath the larynx, the trachea widens and branches into two smaller tubes called **bronchi** (BRAHN kee). Each tube leads to a lung where it branches into smaller tubes called **bronchioles** (BRAHN kee ohlz). As the bronchioles go deeper into the lungs, they branch into still smaller and smaller tubes like the limbs of a mature tree. For this reason, the tubes are often called the bronchial tree. Each of these tiny tubes ends in a group of tiny air sacs called **alveoli** (al VEE uh ly). These tiny air sacs make up the tissue of the lungs. About 150 million alveoli are found in each of your lungs.

The trachea, bronchi, and bronchioles are lined with mucus that traps dust, dirt, and other foreign material brought in with the air. Cilia, which are tiny, hairlike projections, continually sweep the mucus upward and out of the body.

▼ **ASK YOURSELF**

How do your vocal cords produce sounds of different pitch?

Breathing

When you breathe, your chest first expands, then contracts. It might seem as if your lungs pull air in and then push air out when you breathe, but this is not the case. Your lungs are not made of muscle tissue, so they cannot pull air in, any more than a balloon can pull air in. Your respiratory system works together with your muscles and your bones to move air into and out of your body.

- **Process Skills:** *Comparing, Analyzing*

Help the students relate changes in pressure to the breathing process by explaining that air fills the lungs during inhalation because of pressure changes inside the chest cavity. When the diaphragm contracts, the air pressure inside the chest cavity is less than the air pressure outside the chest cavity, so air rushes into the lungs from the environment. Ask the students to describe what causes air to be expelled from the respiratory system during exhalation. (When the diaphragm relaxes, the space in the chest cavity lessens. The pressure in the chest cavity then reaches a higher level than the pressure outside, and air rushes out of the lungs.)

- **Process Skills:** *Comparing, Applying*

Refer the students to Figure 18–5 and ask them to explain how the shapes of the diaphragm and chest cavity change when air is inhaled and exhaled. (When contracted, the diaphragm is lowered and flattened, causing the chest cavity to enlarge. When relaxed, the diaphragm rises and assumes a domelike shape, which reduces the volume of the chest cavity.)

Take a Deep Breath

When the size of your chest cavity changes, it causes you to breathe. However, you need muscles to change the size of your chest cavity. One of these muscles, the **diaphragm** (DY uh fram), is a thick sheet of muscle that forms the floor of your chest cavity. When relaxed, your diaphragm is dome-shaped, but it flattens when it contracts. Contraction moves the diaphragm down, making the chest cavity larger. When your diaphragm relaxes, it rises to its original position and your chest cavity gets smaller.

As your diaphragm moves down, other muscles pull your rib cage up and away from your backbone. This action further increases the size of your chest cavity. Your chest cavity is a flexible container that is sealed from the outside environment. The lungs hang inside this container like two balloons.

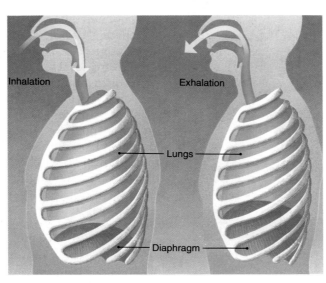

Figure 18–5. How do the positions of the rib cage and the diaphragm change during breathing? How do these changes affect the pressure in the chest cavity? ①

① During inhalation, the diaphragm contracts, increasing the volume and decreasing the pressure inside the chest cavity; during exhalation, the diaphragm relaxes, decreasing the volume and increasing the pressure inside the chest cavity.

MEETING SPECIAL NEEDS

Gifted

Have the students discuss the structure and function of capillaries. Then ask the students to use reference materials to discover the relationship between the alveoli of the lungs and the capillaries of the circulatory system. Encourage volunteers to orally present a brief summary of their findings to their classmates. (A vast network of capillaries surrounds each alveolus.)

REINFORCING THEMES— *Systems and Structures*

Remind the students that each structure of the respiratory system performs a specific function. These respiratory structures work together with the muscular and skeletal systems to exchange oxygen and carbon dioxide with the environment.

 **Doing**

Using the description given in the text, design and create a model of the human lungs. What materials did you use? How does your model demonstrate the breathing process? Share your model with your classmates.

All Bottled Up

What happens when you squeeze a container such as a plastic detergent bottle? Squeezing the container increases the pressure on its walls. This reduces the volume of the container and increases the pressure inside. With less volume inside, there is less space for the contents. The increased pressure forces some of the soap out through the open top.

What happens when the top is closed and the bottle is squeezed? Since the contents cannot escape from the container, the pressure inside increases. The bottle pushes back on your hand, but the soap cannot escape. If you could increase the size of the container by pulling the walls outward, the reverse would happen. The pressure inside the container would decrease and, if the bottle were open, air would rush in to fill the extra space.

TEACHING STRATEGIES, continued

● **Process Skills:** *Expressing Ideas Effectively, Applying*

Ask the students to describe the exchange of gases in respiration. (Gas exchange moves oxygen from the alveoli to the blood and carbon dioxide from the blood to the alveoli. Oxygen is delivered to cells, and carbon dioxide is removed from cells.)

GUIDED PRACTICE

Ask the students to describe the respiratory functions of the following structures: nasal passages, epiglottis, trachea, alveoli, diaphragm.

INDEPENDENT PRACTICE

 Have the students answer the Section Review and Application questions in their journals. Ask them to write a paragraph describing how the respiratory system filters dust and other contaminants from inhaled air before it reaches the lungs.

Demonstration

To demonstrate and observe the function of the lungs, you will need a balloon and a clear plastic detergent bottle. Place the balloon inside the bottle as a volunteer squeezes it. Extend the open end of the balloon over the rim of the bottle, and screw the cap in place, and leave the cap valve open while the volunteer is still squeezing the bottle. Then have the students take turns alternately squeezing and releasing the bottle. Ask them to observe what happens to the balloon. Squeezing and releasing the bottle should empty and fill the balloon. Tell the students that the model functions in much the same way as the diaphragm, chest cavity, and lungs of the human body do during breathing. When muscles expand the chest cavity, the lungs fill with air; when muscles decrease the size of the chest cavity, air is pushed out of the lungs.

① When the bottle is squeezed, the volume decreases and the pressure increases inside the bottle, and air is pushed out. In breathing, the volume of the chest cavity decreases and the pressure increases inside the chest cavity, and air is pushed out.

ONGOING ASSESSMENT
▼ **ASK YOURSELF**

During inhalation, the diaphragm contracts, allowing air pressure from outside the body to fill the lungs with air.

▶ **494** CHAPTER 18

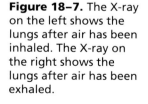

Figure 18–6. How is the process of breathing like squeezing this liquid detergent bottle? ①

Figure 18–7. The X-ray on the left shows the lungs after air has been inhaled. The X-ray on the right shows the lungs after air has been exhaled.

Your chest cavity acts in somewhat the same way as the plastic detergent bottle. Your diaphragm pulls downward and your rib cage rises. This increases the volume of your chest cavity and decreases the pressure. Since the chest cavity is sealed, air cannot rush in to fill the extra space. However, the lungs, which are open to the environment, hang inside the chest cavity. The pressure on the lungs inside your chest cavity is lower than the air pressure outside your body. Thus, air outside the body forces its way through your nose and down the trachea into the lungs. The higher air pressure outside the body fills the lungs with air. The lungs expand and fill the extra space in the chest cavity.

You breathe out by relaxing your diaphragm and lowering your rib cage. The volume of the chest cavity decreases and the pressure inside the chest cavity increases. This increased pressure squeezes the air from your lungs. In other words, as you inhale, air is pushed into your body from the outside. As you exhale, your body pushes the air back out.

 ASK YOURSELF

What causes you to breathe in?

Exchange of Gases

The pie graph on page 495 shows that air is a mixture of gases. Nitrogen gas makes up the largest component of air. Oxygen makes up only about 21 percent, and carbon dioxide about 0.03 percent of air. All of the air that you breathe in does not reach your bloodstream. Your blood can carry only oxygen and carbon dioxide in large quantities.

When you breathe, fresh air moves into your lungs. Oxygen moves from the air in the alveoli into the surrounding capillaries. Red blood cells absorb the oxygen and carry it through

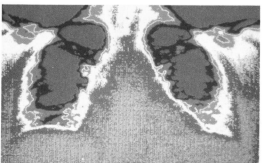
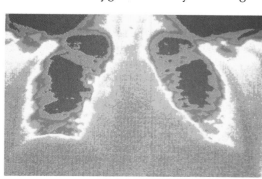

EVALUATION

Have the students describe how the skeletal and muscular systems work with the respiratory system in the breathing process.

RETEACHING

Have the students draw and label the pathways from the nasal passages to the lungs. Encourage them to place their illustrations in their science portfolios.

EXTENSION

Ask the students to use reference materials to research the reaction of the respiratory system to an increased level of carbon dioxide in the blood.

CLOSURE

Cooperative Learning Ask small groups of students to discuss and summarize the important concepts presented in this section. Have the groups share their summaries with each other.

---

your body. At the same time, carbon dioxide moves from the blood's plasma into the alveoli. The carbon dioxide moves out of your body when you breathe out. This process of taking in oxygen and giving off carbon dioxide is called *gas exchange.*

Gas exchange also takes place through the walls of capillaries in your body tissues. Oxygen in the blood diffuses into the cells. At the same time, carbon dioxide diffuses from the cells into the blood.

Gas exchange occurs because there is a difference in the concentration of the gases on each side of a membrane. In the lungs, the concentration of oxygen is greater in the air inside the alveoli than it is in the blood. The opposite is true for carbon dioxide; the concentration of carbon dioxide is greater in the blood than in the air inside the alveoli. The gases move from areas of high concentration to areas of low concentration. When blood reaches the body tissues, the difference in concentrations of these gases is reversed. As a result, oxygen moves into the cells while carbon dioxide moves into the blood.

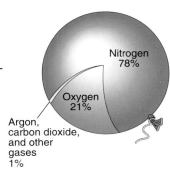

Figure 18–8. This pie graph shows the percentages of different gases in the air you breathe.

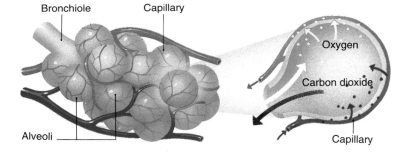

Figure 18–9. In gas exchange, oxygen moves from the alveoli to the blood, and carbon dioxide moves from the blood to the alveoli.

ASK YOURSELF

Why does gas exchange occur?

SECTION 1 REVIEW AND APPLICATION

Reading Critically
1. Why does your body need oxygen?
2. How is the air cleaned as it travels to the lungs?
3. Through what structures does air pass as it moves from the nose to the lungs?

Thinking Critically
4. Why does the health of the respiratory system depend, in part, on the health of the muscular system?
5. People sometimes take several deep breaths before swimming underwater. Why might this practice be dangerous?
6. Why is the image of human lungs as a large balloon not accurate?

ONGOING ASSESSMENT
ASK YOURSELF

Gas exchange occurs because the concentration of oxygen and carbon dioxide in the bloodstream is different from the concentration of those gases in the alveoli of the lungs.

SECTION 1 REVIEW AND APPLICATION

Reading Critically

1. Oxygen is needed by the human body to release energy from nutrient molecules through aerobic respiration.

2. Foreign material is trapped by the hairlike cilia and the mucus lining of the air passages.

3. Air moves through the nasal passages, pharynx, larynx, trachea, bronchi, and bronchioles to the alveoli of the lungs.

Thinking Critically

4. The size of the chest cavity would not increase and decrease if the diaphragm and other muscles did not function.

5. Taking several deep breaths creates artificially high levels of oxygen in the body, altering the brain's ability to sense carbon dioxide levels in the bloodstream. Consequently, the normal breathing process is interfered with.

6. The human lungs are filled with tiny air sacs called alveoli; a single balloon might better represent just one of these air sacs in the lungs.

INVESTIGATION

Measuring Lung Capacity During Exercise

Process Skills: Measuring, Observing, Applying

Grouping: Groups of 2

Objectives

- **Measure** the effect of exercise on the volume of air a person can expel after taking the deepest possible breath.
- **Interpret** data from an investigation and **relate** lung capacity to exercise.

Pre-Lab

Ask the students to predict what the lung capacities of sprinters and long-distance runners might be.

Hints

Be sure to use disposable mouthpieces during the investigation. Students who cannot run in place can still participate by doing other forms of exercise.
CAUTION: The students should not share mouthpieces. Have the students discard their mouthpieces immediately after the investigation. Also check that the students do not have medical conditions that would prohibit them from participating in the activity.

Analyses and Conclusions

1. Students who are taller and in better physical condition may have lung capacities greater than those of other students.

2. Exercise will generally maintain or increase lung capacity as long as a point of tiredness is not reached. Shortness of breath will temporarily decrease lung capacity.

3. Conditioning activities, such as aerobic exercises, will improve muscle tone of the respiratory system and increase lung capacity.

Application

The effect of exercise on the lung capacity of different students varies. The lung capacity or the increase in the ability of the respiratory system to perform should increase if an individual exercises on a regular basis.

✳ Discover More

Before they perform their experiments, remind the students of the hypotheses they made during the Application.

Post-Lab

Ask the students to explain whether their data supported their predictions about the lung capacities of sprinters and distance runners.

 PORTFOLIO ASSESSMENT Have the students place their results in their science portfolios.

INVESTIGATION

Measuring Lung Capacity During Exercise

▼ MATERIALS
- mouthpiece holder • lung volume bag • rubber band
- mouthpiece (disposable) • paper towel • pencil • paper

▼ PROCEDURE

1. **CAUTION: Advise your teacher of any medical condition that prevents your participation in this activity. Also be sure to use only your own mouthpiece and then throw it away.**
2. Working with a partner, insert the mouthpiece holder into the end of the lung volume bag. Secure it by wrapping a rubber band around it tightly. Insert the disposable mouthpiece into the mouthpiece holder.
3. Remove as much air from the bag as possible by flattening it.
4. Take a deep breath and blow all of your breath into the bag. Immediately trap the air by rolling the mouthpiece end of the bag in your hand until the bag becomes stiff with your trapped air.

5. Read the numbers on the bag to find your lung capacity in liters. On a sheet of paper, make a chart and record the number of liters.
6. Repeat steps 2-4 three times and calculate your average lung capacity.
7. Remove your mouthpiece and save it by placing it on a clean paper towel.
8. Rest while your partner is measuring his or her lung capacity.
9. Run in place for three minutes. Immediately following this exercise, again measure your lung capacity and record it on your chart. **CAUTION: Be sure to use your own mouthpiece and then throw it away.**

▶ ANALYSES AND CONCLUSIONS

1. Before exercising, how did your average lung capacity compare to your partner's? Explain any similarities or differences.
2. After exercising, was your lung capacity larger, smaller, or the same? Why?
3. Your lung capacity can change. Name several things you could do to increase your lung capacity.

▶ APPLICATION

How does exercise affect your lung capacity? If you exercised every day, how do you think your lung capacity would change?

✳ Discover More

Design an experiment that measures the lung capacities of volunteer members of one of your school's athletic teams. Record their capacities before the season begins, on several occasions during the season, and after the season ends. Graph and display your results. Do your results support the hypothesis you made for the Application question?

Section 2: EXCRETION

FOCUS

This section presents the process of waste removal by the excretory system of the human body. The structure, function, and role of the kidneys in the elimination of waste products from the human body are explored.

MOTIVATING ACTIVITY

Cooperative Learning Display a 1-L bottle filled with water. Ask small groups of students to predict how many 1-L bottles would be filled by the blood filtered by the kidneys in one day. Have the groups share their predictions; then inform them that the kidneys of the average person filter approximately 1600 L of blood every day.

PROCESS SKILLS
- Comparing • Inferring
- Evaluating

POSITIVE ATTITUDES
- Curiosity • Enthusiasm for science and scientific endeavor

TERMS
- excretion • kidneys

PRINT MEDIA
Is My Sister Dying?
by Alida E. Young
(see p. 429b)

ELECTRONIC MEDIA
Work of the Kidneys, Britannica
(see p. 429b)

BLACKLINE MASTERS
Study and Review Guide
Laboratory Investigation 18.2
Reading Skills

Excretion

SECTION 2

Race cars, like all vehicles, burn fuel and give off gases. Have you ever seen an automobile, truck, bus, or train engine giving off clouds of exhaust fumes? Have you ever traveled down a street or road that was littered with empty bottles, cans, and paper? An environment polluted with trash and exhaust fumes is not only ugly, but unhealthy. To help make the environment a cleaner place, the government requires cars to meet emission standards, and many workers and volunteers regularly pick up trash from streets, highways, and lots. Just like a city or an engine, your body also produces wastes that need to be removed.

Objectives

Name the waste products typically eliminated by the excretory system.

Describe the way in which the excretory system works.

Relate the importance of the excretory system to the overall health of the human body.

SCIENCE BACKGROUND

Several organ systems are involved in the process of excretion. The lungs excrete carbon dioxide and water, both products of respiration. The circulatory system transports wastes to organs of excretion.

① City streets can accumulate waste materials and so can the human body.

Waste Removal

The chemical activities that your body constantly performs are called your *metabolism*, and the waste products of these activities are called *metabolic wastes*. As your body performs the chemical activities that keep you alive, waste products such as carbon dioxide, nitrogen, water, and heat are produced. In order for you to remain healthy, your body must remove the metabolic wastes that you generate. **Excretion** (ihks KREE shuhn) is the process by which these wastes are removed from your body.

A number of different organs and systems work together to remove wastes from your body. Most of the waste your body produces is removed by your lungs and skin. Your lungs remove carbon dioxide and water vapor. When you breathe on a mirror or cool window glass, you can see the water vapor that an exhaled breath contains as the mirror or glass becomes cloudy. Your skin removes the waste product nitrogen, as well as water, and helps control your body temperature by removing extra heat. The following table shows the approximate amount of water gained and lost each day by an adult.

Figure 18–10. How are city streets and the human body alike? ①

TEACHING STRATEGIES

● **Process Skills:** *Formulating Hypotheses, Generating Ideas*

Review with the students some of the metabolic processes that produce waste products in cells. Then ask the students to hypothesize what might happen to body cells if metabolic wastes were not removed. (The concentration of waste products and toxic substances in body tissues would interfere with the normal functions of the cells, resulting in the death of the cells.)

● **Process Skills:** *Inferring, Applying*

Ask the students to explain why the blood in the human body must be filtered by the kidneys. (Waste products of cellular metabolism diffuse or are transported out of cells into the bloodstream. The filtering of the blood by the kidneys prevents these wastes from accumulating in the bloodstream and poisoning the human body.)

ONGOING ASSESSMENT
 ASK YOURSELF

Carbon dioxide, water, water vapor, nitrogen, and excess heat are waste products removed by the lungs and the skin.

THE NATURE OF SCIENCE

Ironically, tragedy often leads to new medical breakthroughs. For example, World War II led to medical improvements as scientists tried to improve the survival rate of war casualties. Recognizing the need to help an injured person survive kidney failure caused by war wounds, a Dutch doctor William Kolff devised a machine that imitated the filtering process of the kidneys. An American doctor John Merrill read of Kolff's work and encouraged him to emigrate to the United States. Together, they devised a filtering system that cleansed the body of waste materials just as the kidneys do. Dialysis proved an effective but costly treatment for patients with chronic kidney failure.

In the 1950s, Merrill and David Hume pioneered kidney transplant surgery in which a kidney from a live donor or a cadaver was implanted in an individual. Kidney transplants proved to be successful with the use of antirejection treatments such as radiation and drugs. The combination of artificial kidneys and kidney transplants has led to prolonged lives for chronic kidney patients. Today, better tissue matching and surgical procedures increase the chances of survival for persons with transplanted kidneys.

Table 18-1 **Daily Adult Water Gain and Loss**

Input:	mL	Output:	mL
Produced by metabolism	400	Feces	100
Eating	900	Skin and lungs	1000
Drinking	1300	Urine	1500
Total:	**2600**	**Total:**	**2600**

The table shows that the excretory system removes most of the water that is lost by the human body.

ASK YOURSELF

Name the waste products removed by your lungs and skin.

The Excretory System

The **kidneys** are the main organs of the excretory system. Your two kidneys are located above your waist on either side of your spine. To find them, place your hands on your back just below the ribs. Your kidneys are located just under your hands.

Finicky Filters When cells of your body break down certain compounds and proteins, a nitrogen waste called *urea* (yoo REE uh) is produced. Your kidneys work constantly to filter your blood and produce *urine* (YOOR ihn). Urine is a yellow liquid that contains urea and excess water, minerals, and salt filtered out of your blood. The following activity will help you discover the serious problems that might occur if a person puts an additional strain on his or her kidneys and excretory system.

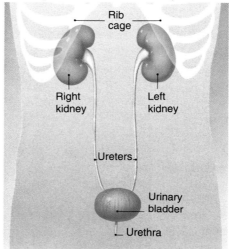

Figure 18–11. The excretory system is responsible for removing some of the waste products produced by the human body.

- **Process Skills:** *Interpreting Data, Synthesizing*

Refer the students to Table 18–1. Ask the students to determine the percentage of daily fluid loss through urine produced by the kidneys. (about 58 percent)

- **Process Skills:** *Inferring, Applying*

Explain to the students that their kidneys are located above their waists, just below the ribs, on both sides of their spine. Have the students use their hands to locate their kidneys. Based on the location of the kidneys, ask the students to infer an occupation that might cause damage to these relatively unprotected organs. (The students might suggest contact sports, such as football.)

- **Process Skills:** *Inferring, Generating Ideas*

Have several volunteers describe the structure and the function of nephrons. (Nephrons consist of long, thin, folded tubules that filter the blood.) Ask the students to describe what happens to the liquid that passes through the kidneys but does not become urine. (The liquid is reabsorbed into the bloodstream.)

DISCOVER BY *Researching*

When some people feel like they are getting a cold, they take very large doses of vitamin C in hopes of warding off the cold. Other people take massive doses of various vitamins on a regular basis. Use reference materials to discover what serious effects these megadoses of vitamins can have on the excretory system. Record your findings in your journal.

Your kidneys work in a way that is similar to the way an air filter works in a car engine or a coffee filter works in a coffeepot. Each filter traps unwanted materials. Your kidneys filter your blood, helping to keep it clean. In an average person, about 1600 L of blood pass through the kidneys each day, creating about 1.0 to 1.5 L of urine.

Each kidney contains more than a million tiny filtering units called *nephrons* (NEF rahnz). The illustration shows that a nephron consists of a long, thin, folded tubule surrounded by a group of capillaries. Liquid from the blood flows through the capillary wall into the nephron tubule. The liquid contains urea as well as water, sugars, and important minerals. As the liquid moves through the tubule, sugars and needed minerals and water move back into the capillaries, leaving the urine behind. Ninety-nine percent of the liquid that leaves the capillaries returns to the bloodstream.

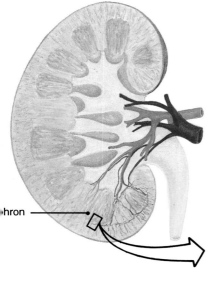
Kidney

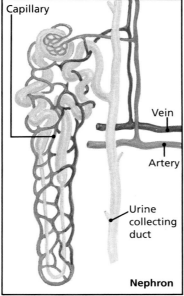
Nephron

Figure 18–12. Each kidney contains more than a million nephrons, which filter the blood.

DISCOVER BY *Researching*

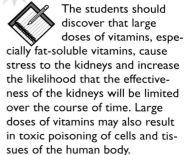

The students should discover that large doses of vitamins, especially fat-soluble vitamins, cause stress to the kidneys and increase the likelihood that the effectiveness of the kidneys will be limited over the course of time. Large doses of vitamins may also result in toxic poisoning of cells and tissues of the human body.

MEETING SPECIAL NEEDS

Mainstreamed

Several of the terms in this chapter—for example, nephrons and diaphragm—spell /f/ with *ph*. The students may find pronouncing and spelling these words difficult. Review the words with the students and reinforce the *ph* spelling of /f/.

SCIENCE BACKGROUND

In Britain, Sir William Bowman was a nationally known ophthalmologist (eye doctor) during the mid- and late-nineteenth century. Yet, some of his most important medical contributions stem from his study of the structure and function of the kidneys. Bowman established that a capsule surrounding the glomeruli, or group of capillaries, in each nephron forms into a urine-collecting duct that eventually drains urine into the bladder. The capsule is known as Bowman's capsule.

GUIDED PRACTICE

Help the students explain the role of the kidneys in removing waste products from the body.

EVALUATION

Have the students summarize the process by which metabolic wastes are removed from the blood and then from the body.

INDEPENDENT PRACTICE

Have the students write answers to the Section Review and Application questions in their journals. Then ask them to write a paragraph that describes the importance to the human body of a properly functioning excretory system.

RETEACHING

Explain to the students that a typical kidney measures about 11 cm in length, 6 cm in width, and 2.5 cm in thickness. Have the students use modeling clay and a metric ruler to create a model of a human kidney.

EXTENSION

Challenge the students to research information about kidney stones and the treatments for patients with kidney stones.

ONGOING ASSESSMENT
ASK YOURSELF

Nephrons, which function as filtering units in the kidneys, trap urea and excess water, minerals, and salts from the blood and send the waste material to a drainage area.

SECTION 2 REVIEW AND APPLICATION

Reading Critically

1. Excretion rids the body of toxic metabolic waste material.

2. The excretory system includes the kidneys, the ureters, the urinary bladder, and the urethra.

3. As liquid from the blood filters through nephrons, urea and excess sugars, minerals, and water are retained and combined to form urine.

Thinking Critically

4. Undigested material removed by the digestive tract is not a product of cellular metabolism because it has not been inside cells of the body. Waste material that is excreted by the kidneys is, however, the end product of material that was once inside the cells of the body.

5. The kidneys are more like a swimming-pool filter because dissolved materials are filtered. A noodle strainer simply filters out undissolved materials that are too large to pass through.

▶ **500** CHAPTER 18

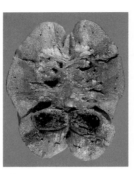 

Figure 18–13. A diseased kidney (left) cannot filter wastes as efficiently as a healthy kidney (right).

Moving On All the nephron tubules of the kidney connect to a larger tubule called the *ureter* (yoo REET uhr). The ureter from each kidney carries the urine to a large sac, the *urinary bladder*. The bladder stores urine until it can be removed from the body. Once the bladder is filled, its muscular walls contract, forcing the urine down the *urethra* (yoo REE thruh) and out of the body.

The composition of urine is often checked during a physical examination. This process, known as a urinalysis, tests the urine for signs of kidney malfunction or diseases of other body organs. It can also detect the presence of drugs.

Sometimes, due to disease or injury, a person's kidneys do not function as they should. A person with kidneys that do not function will die because waste products are not being removed from the bloodstream. Too many waste products in the bloodstream will poison the human body. Fortunately, something can be done to help these people, and the following activity will give you the chance to explore what can be done to help.

DISCOVER BY Researching

Kidney diseases can be very serious illnesses. Prepare a brief oral report on kidney dialysis. In your report, explain what a dialysis machine is, how it works, who uses dialysis, and how often dialysis must be done.

▼ ASK YOURSELF

Describe how your kidneys remove wastes from your blood.

SECTION 2 REVIEW AND APPLICATION

Reading Critically

1. Why is excretion important to the body?
2. What structures make up the excretory system?
3. How do the nephrons produce urine?

Thinking Critically

4. Undigested material from the digestive tract is "eliminated" rather than "excreted" from the body. How are these two processes different?
5. Are the kidneys more like a swimming-pool filter or a strainer used to drain noodles? Explain your answer.

CLOSURE

Cooperative Learning Ask small groups of students to create and exchange a brief true-false quiz about the excretory process. Have each group identify the correct answers to a quiz that was created by another group.

SKILL

Sequencing Data

Process Skills: Classifying/Ordering, Communicating, Evaluating

Grouping: Groups of 2

Objectives
- **Order** the steps of different body processes.
- **Create** flow charts.
- **Evaluate** different methods of organizing data.

Discussion
Point out that putting things in their proper order is called *sequencing*, and remind the students that being able to correctly sequence various events is an important science and life skill.

▶ **Application**
The flow charts designed by each group should include the major organs of the respiratory, circulatory, and excretory systems.

✶ **Using What You Have Learned**

1. A flow chart presents and organizes information in a more visual manner than a written account. It is often easier to trace a complex series of events by using a flow chart than by reading a written account.

2. A flow chart does not explain a process thoroughly but only highlights the important points of the process.

3. Specific details of a process are easier to visualize with a flow chart than with a written account.

4. The ease with which a flow chart can be visualized can act as an aid in remembering detailed information.

5. Information in flow charts may be easier to combine than information in written accounts.

SKILL   Sequencing Data

▼ PROCEDURE

A flow chart is a type of diagram that shows the steps in a process or problem. Arrows lead the reader from one step to the next. Flow charts can be used to make sequencing much easier. A flow chart of pulmonary circulation is shown here. Make flow charts that show the sequence involved in each of the following processes:

- the movement of oxygen from the environment to the lungs and the movement of carbon dioxide back to the environment
- the circulation of blood
- the excretion of urea from the blood and its removal from the body as urine

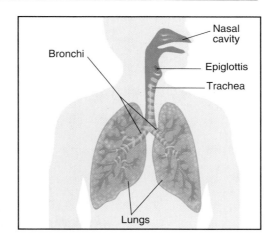

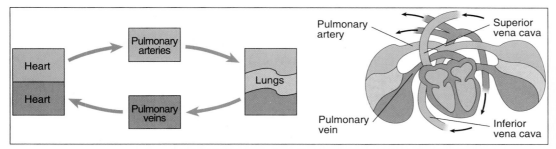

▶ **APPLICATION**
Combine the information from your three flow charts and the flow chart of pulmonary circulation into one flow chart. This flow chart should show the structures involved in:

- the movement of oxygen from the environment to the cells
- the movement of carbon dioxide from the cells to the environment
- the movement of urea from the cells to the environment

✶ **Using What You Have Learned**

1. How is a flow chart different from a written passage in which you might describe a process?
2. Is a flow chart a good substitute for a written description of a process?
3. Which method of presenting information allows the reader to visualize the details of the process more easily?
4. Can a flow chart act as an aid in remembering detailed information?
5. Which method of presenting information allows the reader to combine the information more easily?

★ PERFORMANCE ASSESSMENT

Ask the students to display their flow charts and to explain the sequence of steps in them. Evaluate the performance of the students based on their ability to sequence and display information in a flow chart correctly.

Section 3: THE SKIN

FOCUS
This section presents the structures and the functions of the skin. How the skin helps to regulate the temperature of the human body, protects against invasive organisms and radiation from the sun, senses stimuli in the external environment, and serves as an excretory organ are some of the functions explored.

MOTIVATING ACTIVITY
Ask the students to use a hand lens to observe the skin on one of their hands. Help them identify the epidermis and the pores of the sweat glands. Then have them wear a plastic surgical glove for 10 or 15 minutes.

After they remove the gloves, have the students once again observe their hands with the hand lens. Ask what the moisture is. (perspiration from the sweat glands in the skin)

PROCESS SKILLS
- Comparing • Observing
- Applying

POSITIVE ATTITUDES
- Precision • Curiosity

TERMS
- epidermis • dermis

PRINT MEDIA
Healthy Skin by Richard A. Walzer, M.D.
(see p. 429b)

ELECTRONIC MEDIA
The Skin: Its Structure and Function, Britannica
(see p. 429b)

Science Discovery Desert; survival time for humans

BLACKLINE MASTERS
Study and Review Guide
Thinking Critically
Extending Science Concepts

Two layers of tissue make up the skin.

MEETING SPECIAL NEEDS

Second Language Support
Explain that *dermis* is derived from the Greek word *derma* meaning "skin." *Epidermis* is derived from the Greek words *epi* and *derma* meaning "over" and "skin"; so *epidermis* refers to the outer layer of skin. Point out that when the students see *derm* in a word, the word probably has something to do with skin.

▶ 502 CHAPTER 18

SECTION 3

The Skin

Objectives

Describe the structure of the skin.

Relate each component of the skin to its function.

Explain the different functions of the skin.

If someone asked you to write a list of the excretory organs of your body, how long would your list be? If your list didn't include skin, you would not have listed all of the excretory organs of your body! It's easier to think of excretory organs such as the kidneys and bladder. But your skin is also an organ that performs excretory tasks. Just as the exhaust system of a race car engine helps eliminate waste products, your skin helps eliminate waste products from your body.

The Structure of the Skin

Your skin performs many important tasks that help keep you healthy and alive. Many kinds of cells and tissues are needed to perform the jobs of the skin. These different cells and tissues are arranged in two layers. The thinner outer layer of skin is called the **epidermis** (ep uh DUHR mis), and the thicker inner layer is called the **dermis** (DUHR mis).

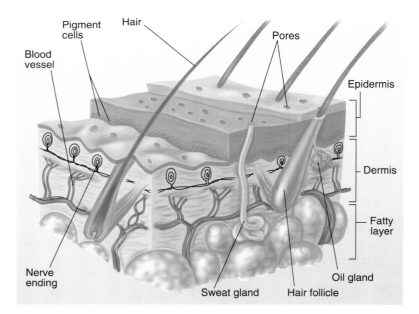

Figure 18–14. How many layers of tissue make up the skin?

TEACHING STRATEGIES

- **Process Skills:** *Comparing, Generating Ideas*

  Explain that the skin is composed of many different kinds of cells and tissues with distinct characteristics that enable them to perform specific functions. Ask the students to name cells or tissues they think might be found in the skin. (The students might name nerve cells, protein cells, and fatty tissues.)

- **Process Skills:** *Comparing, Evaluating*

  Explain that the epidermis is composed of flat cells that overlap in much the same way as shingles overlap on a roof. Ask the students to suggest why this kind of cell arrangement might be advantageous. (Flattened, overlapping cells form a more complete barrier against invasive materials.)

- **Process Skills:** *Comparing, Applying*

  Have the students compare the epidermis and dermis and then name the layer that performs the greatest number of sophisticated functions. (The dermis has structures that produce protective oils and sweat and contains a rich supply of capillaries and nerves that enable it to perform more diverse functions than the epidermis.)

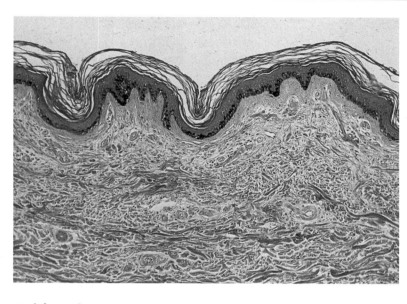

Figure 18–15. The pattern of ridges and valleys on the skin of the fingertips makes each person's fingerprints unique.

② Accept any reasonable response. The outer layer of skin is replaced every 2 to 4 weeks.

③ The students might suggest that the thicker dermis gives added protection to those areas of high contact.

MULTICULTURAL CONNECTION

For thousands of years, people have decorated and protected their skin by using skin paints, perfumes, and oils. Ancient Egyptians, Romans, and Greeks used powdered minerals as facial makeup and hair dyes. By the 1100s, Africans were using cosmetics to paint their bodies for war and special ceremonies. Native Americans used animal fats to protect their skin from insects and cold and to serve as a base for body paints made from plant dyes.

BACKGROUND INFORMATION

In the United States, the Food and Drug Administration (FDA), a federal agency, regulates cosmetics. According to the FDA, a cosmetic is any product designed solely to improve a person's appearance or to promote attractiveness. Skin cosmetics include makeup, perfumes, and deodorants. Products, such as shampoo, may be classified as cosmetics or as drugs depending on their purpose. A shampoo that simply cleans hair is a cosmetic. An antidandruff shampoo, however, is classified as a drug because it cleans hair and works to prevent dandruff.

Epidermis The epidermis has an outer layer of dead cells over an inner layer of living cells. Cells of the outer epidermis die because the supply of nutrients and oxygen is poor in that area. Capillaries do not reach into the epidermis. The epidermis gets its nutrients from the fluid that surrounds the dermis. Therefore, only the cells closest to the dermis can survive. As those cells reproduce, the new cells push the older cells farther away from the dermis, where they die. These dry, dead cells form a waterproof barrier between the living cells and the environment outside the body.

Dead cells are constantly being shed from the body. Each time you scratch your skin or rub something against it, these dead cells break free. The epidermis sheds over a million dead cells each hour! Cells from below replace these lost cells. This loss of cells keeps the body's surface fresh and clean. How long do you think it takes the skin of your body to completely replace itself? ②

Dermis The dermis is thicker, tougher, more elastic, and more complex than the epidermis. The dermis gets its food from the blood in the capillaries. Within the dermis are many glands, nerve endings, and hair follicles.

The dermis of your fingers, palms, toes, and soles of your feet is thicker than the dermis on other areas of your body. Why do you think this is so? In this thicker dermis, the upper layer ③ forms ridges and valleys. In the next activity, you can find out more about the ridges and valleys on the skin of your fingers.

ACTIVITY

What are some characteristics of skin?

Process Skills: Observing, Comparing

Grouping: Groups of 2 or 3

Hints

The students may wish to practice making several fingerprints on scrap paper before beginning the activity. Use water-soluble ink for the activity, and make sure there are plenty of paper towels available so that the students can wipe their hands.

▶ **Application**

1. The creases show where the skin of the hand folds when the hand flexes and unflexes.

2. Both fingerprint ridges and tire treads provide traction.

3. Constant friction builds up calluses. A person with calluses probably perform manual labor with his or her hands on a regular basis; a person with smooth skin does not.

 Have the students include their fingerprints and answers to the *Activity* questions in their science portfolios.

SCIENCE BACKGROUND

The ability of muscles to contract at hair follicles is useful for many animals. Mammals and birds fluff out their fur or feathers to trap air and keep warm in cold weather. They may also fluff out their fur or feathers to increase their apparent body size during courtship displays or disputes over territory.

TEACHING STRATEGIES, continued

● **Process Skills:** *Inferring, Applying*

The students may be interested to learn that hair, fingernails, feathers, hoofs, claws, and scales are all derivatives of the skin. These structures are formed from a tough, fibrous protein called *keratin*. Explain that keratin is waterproof and is found throughout the outer epidermis. Ask the students to explain why it is important that the skin is composed of a waterproof material such as keratin. (Keratin protects organisms from saturation by not allowing water to penetrate and from dehydration by preventing excessive water loss.)

● **Process Skills:** *Inferring, Generating Ideas*

Point out that the skin works with several organ systems in the body. Ask the students what functions of the skin complement the work of other systems. (It excretes water and waste products and receives and transmits nerve impulses.)

ACTIVITY

What are some characteristics of skin?

MATERIALS
plain white paper, pen or pencil, ruler, ink pad, hand lens

PROCEDURE

1. Trace around each of your hands on a plain white sheet of paper. Then examine your hands and make a list of the locations and causes of any blisters or calluses you have.

2. Using a hand lens, examine the skin creases or folds around the joints of your hands. Find the longest line on your palm and draw it on your sketch. Record its length in centimeters on the line.

3. Using the following procedure, add your fingerprints to the bottom of your drawing.
   a. Roll your thumb onto the ink pad from left to right until the area from joint to tip is covered with an even layer of ink.
   b. Placing the left side down first, lay the inked thumb on your paper and roll the thumb until it is resting on its right side.
   c. Lift the thumb straight up off the paper to prevent smudging.
   d. Repeat this procedure for your other fingers.

4. Use a hand lens to compare your fingerprints to those of your classmates. Do you see any similarities or differences?

APPLICATION

1. Why do you think your hands have so many creases on them?
2. What do fingerprints have in common with tire treads?
3. Explain how the calluses on a person's hands can be used to guess the occupation of the person.

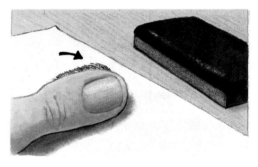

Your fingernails and toenails are produced by certain cells in your skin. Your hair is made of dead cells in the epidermis. Tiny pits in the epidermis push deep into the dermis and form tubelike hair follicles. Dead skin cells are pushed up through the follicle and form a strand of hair. Muscles fastened to the hair follicle can contract, pulling the hair upright. This causes the "gooseflesh," or "goose bumps," you get when you are frightened or cold. In the next activity, you can examine a hair more closely.

▶ 504   CHAPTER 18

- **Process Skills:** *Inferring, Applying*

Remind the students that the skin helps control and maintain the temperature of the human body by providing a surface on which the sweat, or perspiration, released by sweat glands can evaporate. Ask the students to infer what happens to capillaries in the skin on a hot day. (Capillaries dilate and large quantities of sweat are produced. Water and salts move from the blood in the capillaries into the sweat glands and then out through the openings of these glands.)

RETEACHING

 Ask the students to draw and label structures in the skin layers. Have the students place their illustrations in their science portfolios.

EXTENSION

Melanoma is the most serious type of skin cancer. The occurrence of melanoma has doubled in the past 30 years. Have the students research information about melanoma and its causes.

Discover by Observing

Mount a hair from your head or eyebrow on a clean microscope slide. Examine the hair under the low and high powers of a microscope. In your journal, make a diagram of what you see. A typical hair has a hair shaft and a root. Label these on your diagram.

The bottom of the hair shaft you looked at in the activity was pulled out of an oil gland. Oil glands are found around the bottom of each of your hair follicles. These glands produce an oily mixture that keeps your skin and your hair soft. A thin coat of oil on the skin also reduces the amount of water given off through perspiration. Acne occurs when these oils mix with dead skin cells, plug up the hair follicles, and allow bacteria to be trapped and to multiply.

Sweat glands are coiled tubes connected to the surface of your skin by pores. Capillaries surround these tubes. Water, salts, and even small amounts of urea move from the blood in your capillaries into the sweat glands and then out through your pores.

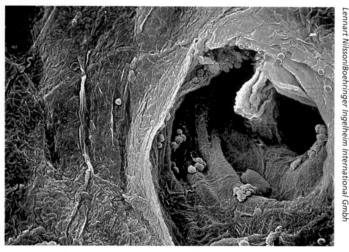

Figure 18–16. Sweat glands, such as this one, give off excess water and salts from the body.

 ASK YOURSELF

Compare the epidermis and the dermis.

The Functions of the Skin

Your skin performs several important functions for you. Skin protects other tissues in your body from harmful bacteria and other organisms. Skin keeps your body from drying out and protects it from the harmful effects of the sun. Skin helps to control your body temperature, senses your external environment, and manufactures vitamin D. In addition to all of these, your skin also functions as an excretory organ. It removes excess water, salt, urea, and heat from your body.

 Treatment of patients with extensive third-degree burns may include skin grafting. The patient's own skin provides the best grafting material because the body does not reject it. The skin is usually taken from unburned parts of the body. Patients who have burns over most of their body may not have enough skin for grafting. Scientists, however, have developed methods for taking small amounts of skin tissue from a patient, grinding it up, mixing it with substances that promote growth, and growing skin tissue in cultures. Scientists have also developed an artificial skin that is seldom rejected by the body. Ask the students to research information about the use of artificial skin for skin grafting. They should find out what artificial skin is made from and how it is manufactured.

PERFORMANCE ASSESSMENT

Use the *Discover By Observing* activity to evaluate the students' ability to prepare a slide and to use a microscope properly. Have the students describe the steps in both procedures to you as they perform them.

ONGOING ASSESSMENT
ASK YOURSELF

The epidermis is composed of living and dead cells; the dermis layer is composed of living cells.

SECTION 3   **505**

 **LASER DISC**
4322

Desert; survival time for humans

ONGOING ASSESSMENT
▼ **ASK YOURSELF**

The skin helps the body maintain a constant temperature, provides a barrier against invading organisms and foreign substances, blocks harmful radiation from the sun and other sources, produces vitamin D, senses the external environment, protects the body from drying out, and acts as an excretory organ.

SECTION 3 REVIEW AND APPLICATION

Reading Critically

1. The epidermis, or outer layer of skin, is partially composed of dead cells.

2. The cells of the epidermis do not receive food or oxygen because capillaries do not extend into this layer of the skin. As a result, the cells die and flatten as they are pushed from below by newly forming cells.

3. "Gooseflesh," or "goose bumps," form when the muscles attached to hair follicles contract, pulling the hair upright.

Thinking Critically

4. Like all organs, the skin contains many different tissues that work together to perform specific functions.

5. Hair is composed of dead cells and does not have sensory receptors.

6. Some students may suggest that a person could live without skin in very antiseptic conditions if room temperature were controlled and vitamin D supplements were provided. Other students may suggest that an individual could not survive without skin even under artificial conditions because the skin performs too many vital functions.

▶ **506** CHAPTER 18

INDEPENDENT PRACTICE

 Have the students write answers to the Section Review and Application questions in their journals. Then ask the students to describe several ways that their skin protects them.

GUIDED PRACTICE

Have the students define these terms: dermis, epidermis, and melanin.

EVALUATION

Have the students list seven functions of the skin.

CLOSURE

 Cooperative Learning Have the students work in small groups to construct a table or a chart showing the major structures and functions of the skin.

Figure 18–17. The amount of melanin manufactured in the epidermis determines the shade of your skin.

The outer layer of cells forms your body's first line of defense. Dead cells, salty sweat, and oil form a protective outer shield that keeps bacteria, other microbes, and many harmful chemicals away from your living cells. Most microscopic organisms die when they come in contact with this outer surface.

Your skin also keeps out harmful radiation. The chemical responsible for your skin color is called *melanin*. Melanin helps protect your body by blocking the ultraviolet radiation from the sun. More melanin is made when the skin is exposed to sunlight. This results in a "tan" or darkened skin color. However, scientific studies have shown that repeated exposure to ultraviolet radiation can cause skin cancer and may cause premature wrinkles.

 ASK YOURSELF

Name several important functions of the skin.

SECTION 3 REVIEW AND APPLICATION

Reading Critically
1. Which layer of the skin is partially composed of dead cells?
2. Why do the cells of the outer skin layer die?
3. What causes "gooseflesh," or "goose bumps"?

Thinking Critically
4. Why is the skin considered to be an organ?
5. Why doesn't it hurt when you get a haircut?
6. Explain under what conditions, if any, a person could live without skin.

CHAPTER 18 HIGHLIGHTS

The Big Idea— SYSTEMS AND STRUCTURES

Lead the students to understand that individual organs and tissues within a system are interdependent. When individual organs of the various systems, such as the respiratory system and excretory system, work together in an efficient manner, the entire body is able to function at peak efficiency like a well-maintained engine.

CHAPTER 18 HIGHLIGHTS

The Big Idea

The respiratory system is responsible for bringing oxygen into your body and removing carbon dioxide. But to fulfill this vital function, the respiratory system must interact with the skeletal system and the muscular system. Lungs are not muscle tissue and, therefore, cannot expand and contract on their own to bring air into the body. In the breathing process, the lungs are helped by muscles, such as the diaphragm, and by bones, such as the rib cage.

Cells eventually use the oxygen supplied by the lungs to perform chemical activities and create energy. The creation of energy, however, results in the creation of waste products, such as carbon dioxide, nitrogen, water, and heat. Both the respiratory and the excretory systems work to eliminate these waste products. Through their combined efforts, your body can function efficiently and remain healthy.

Look back at the ideas you wrote in your journal at the beginning of the chapter. Revise your journal to include more ways in which your body is like an engine. Also include things that could be done to keep the skin and respiratory system of your body in peak operating condition.

The ideas expressed in the journal of each student should reflect the understanding that the waste products produced by the human body must be eliminated from the body if the body is to function in a healthy way. You may wish to suggest that volunteers share their comparisons of the human body and the automobile engine.

CONNECTING IDEAS

The concept map should be copied and completed with words similar to those shown here.

Row 1: kidney
Row 2: urine, ureter
Row 3: urinary bladder, urethra

Connecting Ideas

Copy the unfinished flow chart into your journal. Complete the flow chart by writing the correct term in each blank. Make a similar flow chart for the respiratory system.

Parts of excretory system

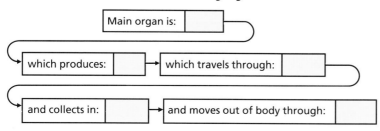

CHAPTER 18 REVIEW

ANSWERS

Understanding Vocabulary

1. The epiglottis is a flap of tissue that covers the opening to the trachea, or windpipe, during swallowing.

2. Metabolism, or the chemical reactions of the human body, produces waste products that are removed from the body through excretion.

3. Kidneys filter the blood, producing urine that is stored in the urinary bladder until it is eliminated from the body.

4. The actions of the muscular diaphragm allow oxygen to enter and carbon dioxide to be removed from the alveoli, or air sacs, of the lungs.

5. The epidermis is the simple, outermost layer of the skin of the human body; the dermis is the complex, inner layer of skin.

Understanding Concepts

Multiple Choice

6. a
7. d
8. d
9. c

Short Answer

10. Breathing through your nose is healthier because it allows the nasal passages to clean, warm, and moisten the air before it moves into the lungs.

11. The epiglottis folds or swings over the opening to the trachea while food is being swallowed.

12. The kidneys are the main excretory organs of the human body. However, the lungs and the skin also excrete metabolic wastes. Urea, excess water, minerals, and salts are removed by the kidneys. Urea, excess water, salts, and heat are removed by the skin. The lungs remove carbon dioxide and excess water.

13. Pores of the skin can become clogged when oils released from glands mix with dead skin cells, plugging hair follicles.

Interpreting Graphics

14. Skin prevents bacteria and other organisms from entering the body, helps regulate the temperature of the body, and prevents the body from drying out.

15. There is a chance that artificial skin could be rejected, thus prolonging recovery.

Reviewing Themes

16. During the process of breathing, the diaphragm and other muscles change the size of the chest cavity, and the rib cage expands or contracts to accommodate the change in size.

CHAPTER 18 REVIEW

Understanding Vocabulary

For each set of terms, explain the relationships in function.

1. trachea (491), epiglottis (491)
2. metabolism (497), excretion (497)
3. kidney (498), urinary bladder (500)
4. alveoli (492), diaphragm (493)
5. epidermis (502), dermis (502)

Understanding Concepts

MULTIPLE CHOICE

6. Clogged alveoli will result in
   a) shortness of breath.
   b) increased excretory rate.
   c) decreased heart rate.
   d) nothing; alveoli cannot clog.

7. When the water intake of a person increases,
   a) urinary output increases.
   b) urinary output remains constant.
   c) urinary output decreases.
   d) urinary output is determined by the activity level of the person.

8. The dead cell layer of the epidermis protects people from
   a) water.
   b) germs.
   c) harmful rays of the sun.
   d) all of these.

9. The epiglottis functions in a way similar to which of these?
   a) a screen door
   b) a revolving door
   c) a hinged door
   d) a sliding door

SHORT ANSWER

10. Why is breathing through your nose healthier than breathing through your mouth?

11. How is food prevented from entering your lungs?

12. What are the main excretory organs of the body? Which metabolic wastes are removed by each of these organs?

13. How can pores in the skin become clogged?

Interpreting Graphics

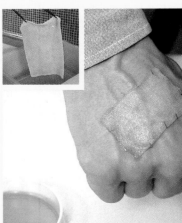

14. The illustrations show a piece of artificial skin that has been used on the hand of a burn patient. Why is not replacing burned epidermis and dermis tissue dangerous?

15. How do you think the recovery of the patient with artificial skin compares to the recovery of a patient using natural skin? Explain.

508 CHAPTER 18

17. The energy required for the life processes of the human body is produced by aerobic cellular respiration. Aerobic cellular respiration requires oxygen, which is supplied by breathing.

Thinking Critically

18. Inflamed bronchi and bronchioles may restrict the flow of oxygen into the alveoli of the lungs and the flow of carbon dioxide into the environment. As a result, efficient gas exchange is less likely to occur.

19. Increased perspiration is often associated with a fever. Drinking water or other fluids during a time of fever will help replace the fluids and balance the level of fluids in the body.

20. Responses might suggest that hiccups are caused by the diaphragm contracting involuntarily and the larynx closing as air is inhaled during the involuntary contraction.

21. Skin that does not "breathe" may not allow the area of the body covered by that skin to perspire, adversely affecting the temperature of nearby tissues.

22. A respiratory system coated with tar will result in inefficient gas exchange and an inability to properly clean and moisten air before it enters the lungs.

Reviewing Themes

16. **Systems and Structures**
    Explain how the muscular system and the skeletal system work together in the process of respiration.

17. **Energy**
    Describe how breathing provides fuel for the cells of your body.

Thinking Critically

18. Bronchitis is characterized by inflammation of one or more bronchi. Explain how bronchitis might affect the breathing process.

19. People are often told to drink large amounts of water when they have a fever. Why is this good advice?

20. Hiccups occur when there is a lack of coordination in the respiratory system. Suggest a possible reason for the occurrence of hiccups.

21. Natural skin "breathes" in the sense that it has pores. Describe the complications that might result from covering an area of a burn victim with artificial skin that does not "breathe."

22. Among other things, tobacco smoke contains tar. Describe how a respiratory system coated with tar might function when compared to a clean respiratory system.

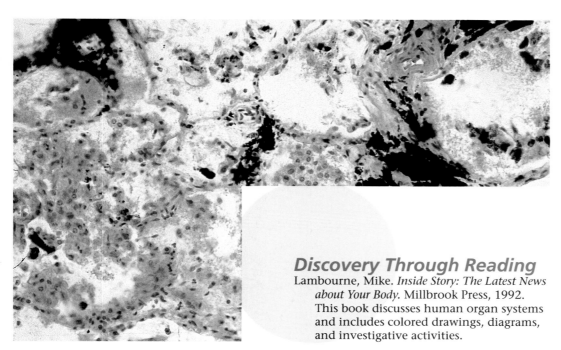

Discovery Through Reading

Lambourne, Mike. *Inside Story: The Latest News about Your Body*. Millbrook Press, 1992. This book discusses human organ systems and includes colored drawings, diagrams, and investigative activities.

CHAPTER 19

COORDINATION AND CONTROL

PLANNING THE CHAPTER

Chapter Sections	Page	Chapter Features	Page	Program Resources	Source
Chapter Opener	510	*For Your Journal*	511		
Section 1: THE NERVOUS SYSTEM	512	Discover By Calculating (B)	513	Science Discovery*	SD
• The Parts of the Nervous System (B)	512	Discover By Doing (A)	514	Connecting Other Disciplines: Science and Computer Science, Communicating with Computers (H)	TR
• Nerve Cells (A)	514	Discover By Doing (B)	517		
• Types of Neurons (A)	516	Discover By Problem Solving (A)	518	Thinking Critically (A)	TR
• Reflex Actions (A)	516	Activity: Where are the greatest number of skin receptors located? (A)	519	The Human Brain and Nerves (A)	IT
• The Brain (A)	517	Section 1 Review and Application	519	Study and Review Guide, Section 1 (B)	TR, SRG
Section 2: THE SENSES	520	Discover By Observing (B)	521	Science Discovery*	SD
• Vision (B)	521	Discover By Observing (B)	523	Investigation 19.1: Taste Perception (A)	TR, LI
• Hearing (B)	524	Discover By Doing (B)	527		
• Taste and Smell (B)	526	Discover By Researching (A)	528	The Senses (A)	IT
• Touch (B)	527	Section 2 Review and Application	528	Study and Review Guide, Section 2 (B)	TR, SRG
		Skill: Observing Sound Waves (A)	529		
Section 3: ALCOHOL, TOBACCO, AND OTHER DRUGS	530	Discover By Researching (A)	532	Science Discovery*	SD
• Drugs and Medicines (B)	530	Discover By Researching (A)	533	Investigation 19.2: Stimulants and Depressants (A)	TR, LI
• Drug Abuse (B)	531	Discover By Calculating (A)	534		
• Tobacco (B)	533	Section 3 Review and Application	537	Reading Skills: Determining Cause and Effect (B)	TR
• Alcohol (B)	536	Investigation: Mapping Your Sensory Receptors (A)	538	Extending Science Concepts: Tobacco and Disease (A)	TR
				Record Sheets for Textbook Investigations (A)	TR
				Study and Review Guide, Section 3 (B)	TR, SRG
Chapter 19 HIGHLIGHTS	539	The Big Idea	539	Study and Review Guide, Chapter 19 Review (B)	TR, SRG
Chapter 19 Review	540	For Your Journal	539	Chapter 19 Test	TR
		Connecting Ideas	539	Test Generator	

B = Basic   A = Average   H = Honors
The coding Basic, Average, and Honors indicates subsections, features, and resources that might be appropriate for different levels of learners. For additional suggestions regarding choice of topic and depth of coverage, see the Pacing Chart on pages T26–T29.

*Frame numbers at point of use
(TR) Teaching Resources, Unit 6
(IT) Instructional Transparencies
(LI) Laboratory Investigations
(SD) *Science Discovery* Videodisc Correlations and Barcodes
(SRG) Study and Review Guide

CHAPTER MATERIALS

Title	Page	Materials
Discover By Calculating	513	(per individual) paper, pencil
Discover By Doing	514	(per group of 2) ruler
Teacher Demonstration	514	stranded copper wire
Discover By Doing	517	none
Discover By Problem Solving	518	(per individual) newspaper, smooth ball, paper, pencil, journal
Activity: Where are the greatest number of skin receptors located?	519	(per group of 2) cork, wire, ruler, blindfold
Discover By Observing	521	(per group of 3 or 4) mirror, stopwatch
Teacher Demonstration	522	none
Discover By Observing	523	none
Discover By Doing	527	(per group of 2) small pieces of apple, potato, onion
Discover By Researching	528	(per individual) paper, pencil
Skill: Observing Sound Waves	529	(per group of 3 or 4) empty soup can, thin rubber membrane, rubber band, glue, piece of mirror (small), flashlight, can opener
Discover By Researching	532	(per individual) paper, pencil
Discover By Researching	533	(per individual) paper, pencil
Discover By Calculating	534	(per individual) paper, pencil
Investigation: Mapping Your Sensory Receptors	538	(per group of 3 or 4) fine-tipped felt pen, paper clips (2), straight pin, dull pencil, ice cube, hot water, small cups (2)

ADVANCE PREPARATION

For the *Demonstration* on page 514, obtain a few feet of stranded copper wire. You will need to obtain smooth balls for the *Discover By Problem Solving* on page 518. For the *Activity* on page 519, the students will need corks and blindfolds. Obtain empty soup cans and thin rubber membranes for the *Skill* on page 529. For the *Investigation* on page 538, the students will need ice cubes and water-soluble fine-tipped felt pens.

TEACHING SUGGESTIONS

Field Trip
Many local hospitals, universities, and clinics have special facilities for performing hearing tests. Arrange to have some of these tests demonstrated to your class. The staff can also discuss proper ear care and how to avoid hearing damage.

Outside Speaker
Contact a local university or medical center, and ask a neurologist to speak to the class. The specialist should concentrate on discussing, in simple terms, the way in which the nerve cells transmit information to the brain.

A visit from a local optometrist may be interesting. He or she can discuss proper eye care and common eye defects.

CHAPTER 19
COORDINATION AND CONTROL

CHAPTER THEME—SYSTEMS AND STRUCTURES

This chapter introduces the students to the structures of the central nervous system and their functions. The students will examine how receptors receive information and the central nervous system processes that information. The students will also explore how drugs, tobacco, and alcohol affect the body. This theme is also developed through concepts in Chapters 4, 5, 7, 8, 9, 10, 12, 13, 16, 17, 18, and 20.

MULTICULTURAL CONNECTION

Baseball has long been a popular sport in the United States, and since 1947 when Jackie Robinson became the first African American player in the professional major leagues, it has become a truly multicultural sport. Today, major league players come from many different ethnic and cultural backgrounds. For example, Fernando Valenzuela, a pitcher who won the National League Cy Young award in 1981, hails from Mexico. Carlos Quintana, who shares the major league record of 6 RBIs in a single inning, comes from Venezuela. Andre Dawson, who was named the National League Most Valuable Player in 1987, is an African American. Tony Pena, who won the American League Gold Glove at catcher in 1991, is from the Dominican Republic. Other players come from Jamaica, Cuba, and Canada. Although the cultural backgrounds of the players vary, the players all share a love for the game of baseball. The love of baseball has spread to many other cultures as well. Today, professional or semiprofessional baseball is played in Japan, Italy, and many other nations.

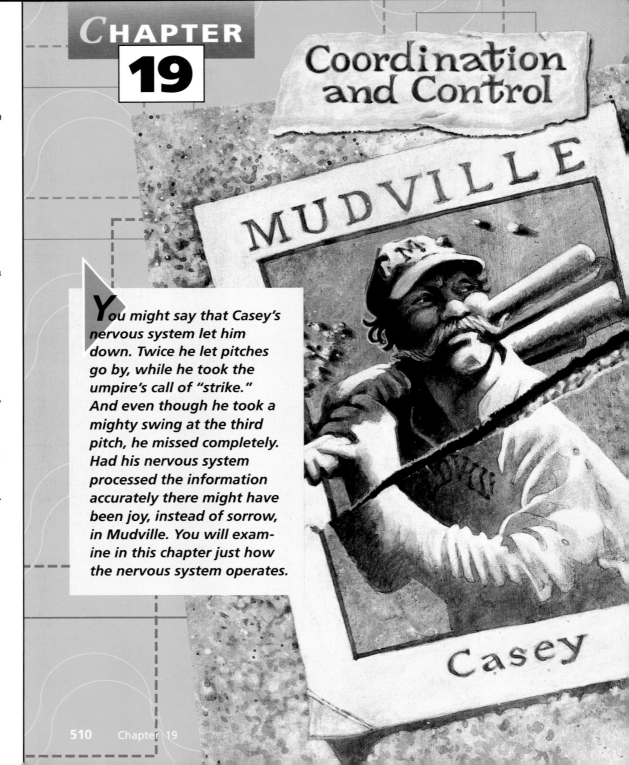

CHAPTER 19

Coordination and Control

You might say that Casey's nervous system let him down. Twice he let pitches go by, while he took the umpire's call of "strike." And even though he took a mighty swing at the third pitch, he missed completely. Had his nervous system processed the information accurately there might have been joy, instead of sorrow, in Mudville. You will examine in this chapter just how the nervous system operates.

CHAPTER MOTIVATING ACTIVITY

If possible, have the students play several innings of indoor baseball, using a soft foam-rubber bat and ball. If some students do not wish to participate in the game, ask them to watch the activities of the players carefully. After all the students have had a turn at bat, encourage the students to discuss the game. Ask such questions as: How did you know when to swing the bat? How could you tell where to run to catch the ball?

By answering the journal questions, the students can demonstrate their knowledge of the nervous system and its functions. Their answers may reveal that some students have misconceptions about the structures and functions of the nervous system. You may wish to address these misconceptions when discussing the nervous system and the sensory organs.

Ten thousand eyes were on him as he rubbed his hands with dirt;
Five thousand tongues applauded when he wiped them on his shirt.
Then while the writhing pitcher ground the ball into his hip,
Defiance gleamed in Casey's eye, a sneer curled Casey's lip.

And now the leather-covered sphere came hurtling through the air,
And Casey stood a-watching it in haughty grandeur there.
Close by the sturdy batsman the ball unheeded sped—
"That ain't my style," said Casey. "Strike one," the umpire said....

The sneer is gone from Casey's lip, his teeth are clenched in hate;
He pounds with cruel violence his bat upon the plate.
And now the pitcher holds the ball, and now he lets it go,
And now the air is shattered by the force of Casey's blow.

Oh, somewhere in this favored land the sun is shining bright;
The band is playing somewhere, and somewhere hearts are light,
And somewhere men are laughing, and somewhere children shout;
But there is no joy in Mudville—mighty Casey has struck out.

from "Casey At the Bat"
by Ernest Lawrence Thayer

- What caused Casey to swing at the ball?
- How do you think messages travel to and from the various parts of your body?
- Describe how you think the sense organs gather information from your environment.

ABOUT THE AUTHOR

Ernest Lawrence Thayer was born in 1863 in Lawrence, Massachusetts, and was educated at Harvard University. While at Harvard, Thayer wrote for and served as president of the *Lampoon*, the university's humorous magazine. Through his work on the *Lampoon*, Thayer met William Randolph Hearst. This acquaintanceship eventually led to Thayer's accepting a job writing for the San Francisco *Examiner*, which Hearst had taken over. "Casey at the Bat" was just one of the many poems Thayer wrote for the newspaper. After working on the newspaper for almost two years, Thayer returned to Massachusetts and worked for his family's business. By the time Thayer died in 1940, "Casey at the Bat" had become an established American classic.

MEETING SPECIAL NEEDS

Second Language Support

Fluent English-speaking students should be paired with students who have limited English proficiency to help them read and understand the chapter content. The student pairs might outline important information and provide definitions for scientific terms in their journals. The outline and definitions can then be used as study aids.

CHAPTER 19 **511**

Section 1:
THE NERVOUS SYSTEM

FOCUS

This section describes how the central nervous system exerts control over all body parts and coordinates all their activities. The section also discusses the physiology and anatomy of neurons and explains how the neurons carry impulses from one part of the body to another.

MOTIVATING ACTIVITY

Without alerting the students to what you are going to do, slam a door, blare loud music, or bang a book on your desk. Then have the students discuss their reactions to the sudden and unexpected noise.

Encourage the students to talk about their immediate physical reactions. Startled students may have jumped or gasped. Lead the students to understand that their reactions were triggered by the nervous system.

PROCESS SKILLS
- Comparing • Inferring
- Interpreting Data

POSITIVE ATTITUDES
- Curiosity • Precision

TERMS
- receptors • nerves
- neuron • reflex
- cerebrum • brain stem
- cerebellum

PRINT MEDIA
World of the Brain
by Alvin and Virginia Silverstein
(see p. 429b)

ELECTRONIC MEDIA
The Nervous System, Britannica
(see p. 429b)

Science Discovery
Stimulus; on a worm

BLACKLINE MASTERS
Study and Review Guide
Connecting Other Disciplines
Thinking Critically

---

① The brain must process the information supplied by the receptors and designate a reaction that will allow the player to use muscles to swing the bat and hit the ball.

▶ 512   CHAPTER 19

SECTION 1

The Nervous System

Objectives

Describe the three major parts of the nervous system and their functions.

Compare and contrast the three types of neurons.

Relate the three major areas of the brain to the functions each performs.

Whether or not you hit a ball when you swing a bat depends on how well your nervous system does its job. Your nervous system is both a control system and a communication system. Let's look at the nervous system in terms of the basic jobs it does. Your nerves are responsible for gathering information and delivering it to your brain or spinal cord. The information is then processed, and a response is sent to various body parts telling them how to react. This important job of the nervous system is what enables you to swing a bat.

Another job of your nervous system is controlling the internal organs of your body. Your breathing, heartbeat, and digestion are a few of the functions of internal organs that are controlled by your nervous system.

A small, white sphere sails toward the batter at more than 90 km/h. In an instant, her eye tracks the ball and her brain must select an action. Swing! The bat begins to move.

Figure 19–1. What must the batter's body do to hit the ball? ①

The Parts of the Nervous System

Your nervous system functions much like a video computer game. A video game receives information from the player. By using the keyboard or control stick, the player sends information through wires to the processor, which is the computer's "brain." The processor then interprets this information and selects a response. More wires carry the processor's response to an output device, such as a computer monitor or a TV screen. The screen produces a picture, which represents the processor's response.

Like the wires and processor of the computer, your nervous system has sense organs to collect information, a brain to interpret this information, and nerves to deliver responses to other body parts. Muscles and glands are the body's output devices that respond to directions from the nervous system.

Bring the tips of your thumb and middle finger together so that they just touch. A signal travels up your fingers and arm to your brain, and you feel a slight sensation on your skin. Specialized cells in your nervous system act like the computer's keyboard or control stick. These cells, called **receptors,** receive information from their surroundings. Some receptors are individual cells, such as those in your skin that produce the sense of touch. Other receptors are part of other sense organs, such as your eyes and ears.

Each kind of receptor is affected by a certain kind of stimulus. A *stimulus* is something that causes the nervous system to react. Stimuli such as light, heat, sound, chemicals, and pressure affect different kinds of receptors, causing them to send messages to the brain.

Two outs, bases loaded, bottom of the eighth, and down by four runs! The batter waits as the pitcher winds up for the pitch. For an instant, the batter sees the tiny, white sphere as it sails toward the plate. Is it in the strike zone?

The brain and spinal cord collect and interpret information for the receptors. In the case of the batter, light reflects off the ball, enters her eyes, and stimulates receptor cells in the back of them. Information about the size, speed, and position of the ball is gathered by the receptors. This information is then sent to the brain for processing. The brain along with the spinal cord make up the *central nervous system,* or the CNS. The CNS is responsible for interpreting all information gathered by receptors and selecting the proper response.

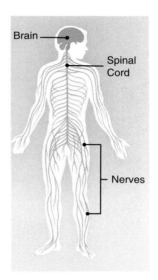

Figure 19–2. The human nervous system enables the body to respond to stimuli.

DISCOVER BY Calculating

If a pitcher's mound is 20 m from home plate and a ball is pitched at 40 m/s, how much time does the batter have to decide whether or not to swing?

Should she swing? Can she hit it?

So how does information travel to and from the brain and spinal cord? The answer lies with chains of specialized cells that form long, thin fibers called **nerves.** Nerves carry messages from receptors to the central nervous system. Nerves also carry responses from the CNS to the muscles, organs, and glands of your body. A stimulus causes a receptor to send messages along a nerve to the CNS. The messages that travel along

BACKGROUND INFORMATION

Baseball is an adaptation of an old British game called "rounders," which was played by British colonists in America as early as the 1700s. In both games, the players hit a ball with a bat and circle bases. However, in rounders, fielders threw the ball at the base runners. If the ball hit a runner, the runner was out. In baseball, the runner is out if tagged with the ball. In 1845, Alexander Cartwright founded the first organized baseball team and established rules for playing the game. The rules have changed since that time, and baseball has become a professional and amateur sport as well as a recreational activity.

LASER DISC
37215–37898

Stimulus; on a worm

SCIENCE BACKGROUND

A weak stimulus may not trigger an impulse. The amount of stimulus needed to activate an impulse is known as the *threshold*. A stimulus that is weaker than the threshold will not cause an impulse; a stimulus that is equal to or greater than the threshold will cause an impulse.

DISCOVER BY Calculating

A ball traveling at a speed of 40 m/s will reach a destination 20 m away in 0.5 s.

SECTION 1 **513**

TEACHING STRATEGIES, continued

● **Process Skills:** *Applying, Comparing*

To help the students understand just how fast impulses travel, have them determine how fast a race car would have to travel in k/h to travel 90 m/s. Remind the students that they will have to multiply 90 m/s by 60 seconds and then by 60 minutes to calculate m/h. They then must divide m/h by 1000 to determine k/h. (324 k/h)

● **Process Skills:** *Inferring, Generating Ideas*

Remind the students that a neuron is a specialized cell of the human body. Ask them to describe what it means to be specialized. (The function of a specialized cell is restricted to a very specific role.) Point out to the students that nerve cells represent only one example of cells in the human body that are specialized. Ask the students to name the specialized cells of the human body that make movement possible. (muscle cells) Challenge the students to think of other examples of specialized cells in the human body. (The students might name cells of the circulatory system—red blood cells, which carry oxygen throughout the body; and white blood cells, which protect the body against invasion by foreign cells or substances.)

DISCOVER BY *Doing*

The students' data may indicate that as the number of repetitions performed by each student increased, reaction time decreased. In other words, their data is likely to lend credence to the expression "Practice makes perfect."

ONGOING ASSESSMENT
▼ ASK YOURSELF

The central nervous system, or CNS, is formed by the spinal cord and the brain; it is responsible for interpreting information gathered by receptors and producing an appropriate response to that information.

Demonstration

Distribute a short length of stranded copper wire to each student. If the wire is insulated, remove some of the insulation from each end of the wire. **CAUTION: Remind the students to handle the wires carefully. The ends of the wire might be sharp from cutting.** Ask the students to fray each end of the wire by unwinding the large strand. When the ends of the wires have been frayed and individual strands no longer touch each other, ask the students to describe how their wires might be analogous to neurons of the human nervous system.

Figure 19-3. Your body receives, processes, and acts on information in much the same way as the components of this computer system do.

nerves are in the form of weak electrical signals called *impulses*. The stimuli may be in the form of light, sound waves, or temperature changes gathered by receptors in your eyes, ears, or skin. Whatever the stimulus, nerve impulses are sent racing along nerves to the CNS at speeds of about 90 m/s. While this is a lot slower than electricity, which travels through space at more than 300 000 km/s, it is still fast enough to deliver messages throughout your body in a very short time. In fact, it would take an impulse less than $\frac{1}{49}$ s to move from the toe to the brain of a 183-cm-tall person. Now that is fast!

Let's take another look at the batter. The receptors in her eyes gathered information about the pitch and converted the information into an impulse that was sent to the brain. The brain then determined that the ball was in the strike zone and directed her muscles to go into action and take a swing.

 **Strike One!**

DISCOVER BY *Doing*

Work with a partner in this activity. Hold out your hand with your thumb and forefinger separated. Have your partner hold a ruler vertically above your hand so that the zero mark is between your fingers. Try to catch the ruler as soon as your partner drops it. Record the distance the ruler falls before you catch it. Repeat the procedure several times. What does the distance the ruler falls tell you about your reaction time? What happened to your reaction time as you did more repetitions? Compare the reaction times of several of your classmates.

▼ ASK YOURSELF
What is the central nervous system?

Nerve Cells

The basic unit of the nervous system is a specialized cell called a nerve cell, or **neuron.** There are billions of neurons in your body. While not all neurons are alike, they do have the same basic structure and perform the same function—carrying impulses to and from different parts of the body.

▶ **514** CHAPTER 19

- **Process Skills:** *Inferring, Hypothesizing*

The chemicals involved in the transmission of nerve impulses across a synapse are called *neurotransmitters*. One neurotransmitter called *serotonin* requires iron for its manufacture. Ask the students to consider the possible effects of a dietary iron deficiency on the nervous system. (An iron-deficient diet may reduce the amount of serotonin that the nerve cells can manufacture. Too little serotonin may slow the thinking process and interfere with learning.)

- **Process Skills:** *Inferring, Applying*

Have the students observe the neuron shown in Figure 19–4. Point out that neurons have many dendrites; some neurons have hundreds of dendrites. In contrast, a neuron usually has only one axon. The axon branches out at the end so that it can contact many dendrites. The ends of the branches of the axon are called the *axon terminals*. Ask the students to explain why the axon terminals are aptly named. (*Terminal* is a term meaning "related to the end" or "at the end." The axon terminals are at the end of the axon.)

The Nerve of It All

As with all cells in your body, the structure of a neuron is well suited to its job. Look at the drawing of the neuron. You can see that there is a main body to the cell. The main body houses the nucleus and other cell parts that are needed to keep the cell alive. Extending from the cell body are two types of branches. One type of branch, called the *dendrite* (DEN dryt), brings impulses into the cell. Usually, several dendrites branch from the cell body just as limbs branch from a tree trunk.

The other kind of branch is called an *axon* (AK sahn). It is much longer than a dendrite and is responsible for carrying impulses away from the cell to other neurons. There is usually only one axon extending from a cell body. Bundles of these axons make up what we call nerves. A single nerve can be a fraction of a centimeter long or more than a meter long. The long axons are often wrapped in a protective covering that speeds the nerve impulse.

Going My Way?

A neuron acts like a one-way street. Impulses move through the neuron in only one direction. As a result, one set of nerves sends information to the CNS, while other nerves carry impulses away from the CNS back to the body parts. Dendrites receive information from receptor cells or from other neurons. The impulse moves down the dendrites to the cell body. From the cell body, the impulse travels down the axon toward the next neuron in the chain. Neurons are not connected. A small space, or *synapse* (SIN aps), separates one neuron from another. For the impulse to continue down the nerve, it must cross the synapse. When the impulse reaches the end of the axon, it causes the axon to release chemicals. These chemicals spread across the synapse and trigger a new electrical impulse in the dendrite of the next neuron. In this way, information travels from neuron to neuron through the nervous system.

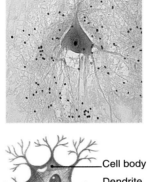

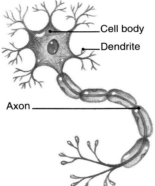

Figure 19–4. The neuron shown here is a single cell. The illustration shows the parts of a neuron.

MEETING SPECIAL NEEDS

Mainstreamed

Help the students who are finding it difficult to differentiate the functions of dendrites and axons in neurons by providing memory clues. Explain that dendrites = deliver and axons = away. Dendrites receive impulses and "deliver" them into the nerve cell, and axons send the impulses "away" to other nerve cells.

THE NATURE OF SCIENCE

At times, scientific research in one area leads to scientific advancement in other areas as well. The research of Luigi Galvani, an Italian physician and researcher of the eighteenth century, led to advancements in both physics and physiology. Galvani is credited with discovering that electrical currents could cause muscle contractions. Although Galvani incorrectly believed that muscle contractions resulted from "animal electricity," his research served as the foundation for neurophysiology. His research also contributed to the field of physics. His discoveries about electrical conduction led to the invention of the voltaic pile, which was a type of battery that was a source of constant current.

ONGOING ASSESSMENT
 ASK YOURSELF

The great surface areas of axons, dendrites, and cell bodies of neurons allow them to receive and efficiently relay information to other neurons of the body.

 ASK YOURSELF

How is the structure of a nerve cell suited to the job it does?

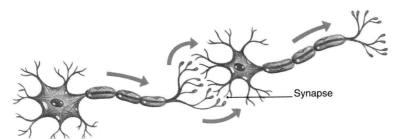

Figure 19–5. Chemicals carry nerve impulses across the gap that separates neurons.

TEACHING STRATEGIES, continued

● **Process Skills:** *Comparing, Applying*

Have the students compare and contrast the meanings of the terms *stimulus* and *response* as they apply to the nervous system. (A stimulus is an event that causes a reaction by the nervous system. A response is the reaction initiated by the central nervous system to a stimulus.) Ask the students to suggest several examples of a stimulus and a response. (Suggestions might include turning in the direction of a person who has called your name or going outside on a sunny day and squinting because of the brightness.)

● **Process Skills:** *Comparing, Communicating*

Cooperative Learning Remind the students that neurons carry messages throughout the nervous system and that these messages are known as impulses. Point out to the students that neurons are classified according to the direction in which impulses move through them. Ask small groups of students to develop a table comparing and contrasting the functions of motor neurons, sensory neurons, and interneurons. (Motor neurons carry impulses from the brain and spinal cord to muscles and

---

ONGOING ASSESSMENT
▼ **ASK YOURSELF**

Motor neurons carry impulses from the brain and spinal cord to muscles; sensory neurons carry impulses from sense organs to the spinal cord and brain; interneurons connect and carry impulses between sensory and motor neurons.

① The size and shape of each cell is different; the structure and function of each cell is similar.

② The hand moves before the body is aware of the pain because the reflex action is controlled by interneurons in the spinal cord, whereas the pain is interpreted by the brain.

SCIENCE BACKGROUND

The spinal cord links the brain with the nerves of most of the body. In the center of the cord is gray matter, which contains the cell bodies of motor neurons that send impulses to the body's muscles. Around the gray matter is a layer of white matter, which contains nerves that pass impulses to and from the brain. Layers of connective tissues, known as meninges, surround the spinal cord and help protect it from injury. Branching out from the spinal cord along its entire length are 31 pairs of nerves, which contain both sensory and motor neurons.

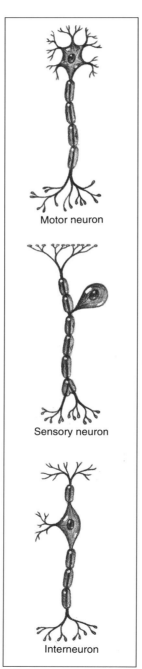

Figure 19–6. How are the three types of nerve cells different? How are they alike? ①

Types of Neurons

The nervous system contains three types of neurons: motor neurons, sensory neurons, and interneurons. Each kind of neuron has a unique function. Information sent from your brain or spinal cord to other cells in your body moves through *motor neurons*. In order for you to wiggle your big toe, an impulse must travel from your brain and down your spinal cord to motor neurons that lead to your leg. Some of these motor neurons connect to muscles that move your big toe. The impulse from your brain causes the muscles to contract, and your toe moves.

Try pressing a finger against the corner of your textbook cover. Information travels through *sensory neurons* from receptors in your sense organs to your spinal cord and brain. Your brain and spinal cord contain *interneurons* that connect sensory neurons to other neurons. Interneurons in your brain, for example, form connections that allow information to be transferred between different parts of your brain. Your memory is related, in part, to the number of connections between interneurons in the brain. About 97 percent of all the neurons in your body are interneurons.

▼ **ASK YOURSELF**

Describe the three types of neurons and explain how they differ from each other in function.

Reflex Actions

You are constantly making decisions and choices about all sorts of things. For example, if you are up to bat and the pitcher throws a fast ball high and inside, your eyes send a message to your brain. The brain processes the information. Messages are then sent to your muscles telling them to respond in a manner that will allow you to duck out of the way of the pitch. These kinds of responses are *voluntary responses*. In other words, you think about an action and decide whether or not to respond in a certain way. But not all responses to stimuli are voluntary.

Have you ever grabbed a very hot pan? In an emergency, response time is very important. A fraction of a second can mean the difference between an injury and no injury. In the case of the hot pan, you move your hand away from the pan much faster than you could move it away if you had to think about it. Actually, your hand begins to move even before your brain processes information about the heat and pain. Why? ②

Your response to the hot pan is very fast because it does not require thought. Rather than relying on your brain to process a response, the impulse is interpreted by the spinal cord. These impulses travel a shorter distance, so you are able to respond much more quickly than you would if you were relying on your brain. Because your spinal cord, not your brain, sends a signal telling your muscles to release your grip on the pan, you are able to avoid a painful burn. This type of response is called a **reflex**. Reflexes are automatic; they do not require thought.

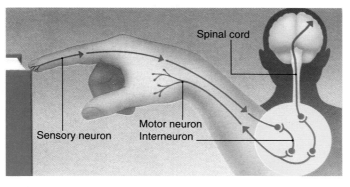

Figure 19–7. A reflex action involves three types of neurons—sensory neurons, interneurons, and motor neurons.

 **Doing**

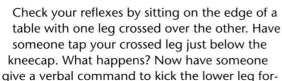

Check your reflexes by sitting on the edge of a table with one leg crossed over the other. Have someone tap your crossed leg just below the kneecap. What happens? Now have someone give a verbal command to kick the lower leg forward. How did the speed of the reaction differ from the tap on the knee? Which response was faster? Why?

 ASK YOURSELF

Why are you able to move your foot so quickly if you step on a sharp object?

The Brain

The brain is your body's main control center. Not only does it control the activities of your body, but it also controls your thoughts and feelings. In addition, your brain serves as a storehouse of information known as *memory*. Your brain uses the information it has stored to evaluate new information and select present and future actions.

The average human brain is about the size of a large grapefruit and weighs about 1.3 kg. That's about as much as a head of cabbage weighs. The brain is the most complex organ in your body and contains over 10 billion neurons. Since the brain is so active, it requires a rich supply of blood to bring it fuel and oxygen. In fact, the brain uses one-fifth of the body's energy and

SECTION 1 **517**

ACTIVITY

Where are the greatest number of skin receptors located?

Process Skills: Comparing, Applying

Grouping: Groups of 2

Hints

Tell the students to gently press the wires against their partner's skin.

▶ **Application**

1. The concentration of receptors in some parts of the body is greater than the concentration of receptors in other parts of the body.

2. The results from similar locations should be relatively consistent across groups.

3. The students might suggest that an increased awareness of the locations of receptors creates a more complete understanding of the human body.

4. The students might suggest that different parts of the body perform different functions, and the number of receptors necessary to perform those functions may vary.

DISCOVER BY *Problem Solving*

The students can calculate the surface area and volume of the smooth ball by using the formulas $SA = 4\pi r^2$ and $V = 4\pi r^3 \div 3$ and estimate the surface area and volume of the crumpled-up newspaper. Help them conclude that although the ball and the newspaper have similar volumes, the surface area of the newspaper is greater than that of the ball. The students should infer that the wrinkled surface of the brain increases its surface area, which is important since the brain is encased within the skull.

▶ **518** CHAPTER 19

TEACHING STRATEGIES, continued

● **Process Skills:** *Comparing, Expressing Ideas Effectively*

Have the students discuss and describe the differences between the functions of the cerebellum and those of the cerebrum. Then ask the students to identify the portion of the brain that controls speech and problem solving (cerebrum) and the portion of the brain that controls muscle coordination (cerebellum).

INDEPENDENT PRACTICE

 Have the students provide written answers to the Section Review and Application questions in their journals. Then ask the students to describe how the nervous system acts as a controller and a communicator.

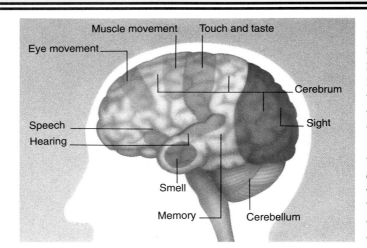

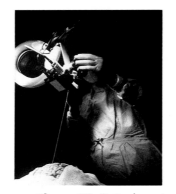

Figure 19–8. Each part of the brain (top) has its own job. Sometimes, delicate brain surgery (bottom) is needed to correct a problem. Here, a laser cuts the tissue and stops the bleeding at the same time.

nearly one-fourth of its oxygen supply. The brain has three major parts: the cerebrum (sehr REE bruhm), the brain stem, and the cerebellum (sehr uh BEHL uhm). You can see these parts in the illustration.

The **cerebrum** fills most of the skull, making up about 80 percent of the brain tissue. It is divided into halves, or hemispheres. Your thoughts, memories, emotions, and decisions begin in the thin outer portion of your cerebrum. This outer layer is often called gray matter because of its color. The gray matter also directs the voluntary movements of the skeletal muscles. In addition, it contains the centers for evaluating information from the sense organs and for speech. A large mass of white tissue lies below the gray matter. This white matter of the cerebrum acts much like a switchboard, controlling the movement of information between the gray matter and other parts of the brain.

DISCOVER BY *Problem Solving*

Get a sheet of newspaper and crumple the newspaper into a tight ball. Estimate the size of the ball. Then find a smooth ball of approximately the same size. In your journal, compare the surface area and volume of the crumpled-up newspaper ball to the surface area and volume of the smooth ball. Which has the larger surface area? Why? Then compare the crumpled-up newspaper to the wrinkled outer surface of the brain. In relation to surface area and volume, why is the wrinkled surface of the cerebrum an advantage?

Below the cerebrum lies the **brain stem.** The brain stem is the place where the brain and spinal cord meet. The brain stem relays information received from the spinal cord to the rest of the brain. Many of your body's functions are controlled by this part of the brain. Heartbeat rate, breathing rate, coughing, and swallowing are all controlled by the brain stem. The **cerebellum** controls muscle coordination, balance, and muscle tone. It is located below the cerebrum and behind the brain stem.

GUIDED PRACTICE

Have the students describe the function of each structure of the central nervous system.

EVALUATION

Ask the students to explain why human life depends on the nervous system and the action of involuntary muscles.

RETEACHING

 Cooperative Learning Have the students work in small groups to create and label models of the three types of neurons. Each model should include a cell body, an axon, and several dendrites. Allow the students to choose the art materials to use in creating their models.

EXTENSION

Have the students research information about a disease of the nervous system, such as Parkinson's disease or multiple sclerosis. Ask the students to identify symptoms of and treatments for the disease.

CLOSURE

 Cooperative Learning Ask each group to create a diagram of the human brain that identifies its three major parts and describes the important functions of those parts.

ACTIVITY

Where are the greatest number of skin receptors located?

MATERIALS
cork, wire, centimeter ruler, blindfold

PROCEDURE
1. Cut two pieces of wire about 4 cm long.
2. Stick one piece of wire into one of the ends of the cork. Measure a distance about 5 mm away from the first wire and stick a second wire into the cork. Check your measurement to make sure the ends of the wire are about 5 mm apart.
3. Blindfold your partner. Tell your partner that you will press one or both tips of the wire against his or her skin. Then press either one or both tips of the wire gently against the skin of the forearm. Ask your partner whether he or she feels one or two points. Do several tests in different regions of the forearm and record your results.
4. Next, test other areas of the body. Try the palm of the hand, fingertips, back of the hand, and the elbow. Record the results of each test.
5. Next, have your partner do the testing while you are blindfolded.

APPLICATION
1. What conclusions were you able to draw about the number of receptors in the different parts of the body?
2. Compare the recorded results of the tests of several classmates and report your findings.
3. How might the results of your testing be useful to you?
4. Why do you think some body parts have more receptors than others?

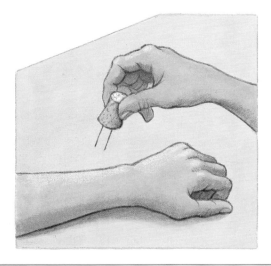

▼ ASK YOURSELF

List and describe the functions of each part of the brain.

SECTION 1 REVIEW AND APPLICATION

Reading Critically
1. What are sensory neurons?
2. Describe how an impulse moves from one neuron to another.
3. What structures of the nervous system are involved in a reflex?

Thinking Critically
4. Describe the nervous system as both a communication system and a control system.
5. Why might a doctor test reflexes such as the knee-jerk reflex during a general examination of your health?

ONGOING ASSESSMENT
▼ ASK YOURSELF

The cerebellum controls muscle coordination, tone, and balance; the cerebrum controls thoughts, emotions, memories, and decisions; and the brain stem controls heartbeat rate, breathing, coughing, and swallowing.

SECTION 1 REVIEW AND APPLICATION

Reading Critically

1. Sensory neurons are nerve cells that respond to stimuli by carrying impulses to the central nervous system.

2. When an electrical impulse reaches the end of a neuron axon, the axon releases chemicals that spread across the synapse and trigger a new electrical impulse in the dendrites of the next neuron.

3. Receptors, sensory neurons, interneurons in the spinal cord, and motor neurons are involved in a reflex.

Thinking Critically

4. The nervous system communicates information from one part of the body to another. The nervous system controls the activities of the human body.

5. Reflexes allow a physician to observe a function of the nervous system. Reflexes are impaired or nonexistent when the body has certain diseases or injuries.

Section 2:
THE SENSES

FOCUS
In this section, the specialized senses that allow humans to monitor their environment—vision, hearing, smell, taste, and touch—are explored. The sense organs and their functions are identified.

MOTIVATING ACTIVITY
 Cooperative Learning Ask each student to make an eye patch, using string and paper. Have the students wear their paper eye patch as they and a partner play a short game of one-handed catch with a soft object. Have them remove the patch and play another game of catch. Ask the students to describe their experiences. (The students should note that the ability to recognize depth is significantly diminished when one eye is covered.)

PROCESS SKILLS
- Inferring • Measuring
- Observing • Applying

POSITIVE ATTITUDES
- Enthusiasm for science and scientific endeavor
- Initiative and persistence
- Precision

TERMS
- cornea • lens • retina

PRINT MEDIA
The Senses
by Mary Kittredge
(see p. 429b)

ELECTRONIC MEDIA
Sense Organs, Coronet
(see p. 429b)

Science Discovery
Senses; hearing
Ear; human

BLACKLINE MASTERS
Study and Review Guide
Laboratory Investigation 19.1

SCIENCE BACKGROUND
Eyesight and hearing are called *distance-receiving senses* because the eyes and ears need very little stimulus to respond and can detect a stimulus that may occur far away from the sense organs. Touch, taste, and smell are called *contact senses* because they need a relatively large stimulus to respond and must come in direct contact with the stimulus.

SECTION 2

The Senses

Objectives

List and **describe** the five senses.

Identify the major sense organs of the body and **relate** each to the appropriate stimulus.

Explain how each of the major sense organs functions.

Your brain is an amazing organ that performs many functions to keep you alive. Yet, your brain would be useless without receptor cells. Receptor cells provide your brain with information about your internal and external environment.

The buzz of a bee near her ear causes the pitcher to move off the mound for a moment. The batter squeezes the bat tightly and feels every ripple in the wood's grain. The umpire cries, "Let's play ball!" All eyes focus on the pitcher.

Some receptors are scattered throughout your body, while others are concentrated in your sense organs. These receptors provide humans with five senses: touch, taste, smell, sight, and hearing. Let's take a closer look at the sense organs and their functions.

Figure 19–9. When one of the senses is impaired, people devise ways to function without it. The person shown here is visually impaired. The guide dog serves as her eyes.

TEACHING STRATEGIES

● **Process Skills:** *Inferring, Applying*

Explain to the students that the loss of one or more senses tends to increase the sensitivity of the remaining senses. For example, people who are visually impaired often compensate by becoming more aware of sounds and by using their sense of touch. People who are hearing impaired communicate with one another through sight and sign language. Ask the students to describe how a person who cannot see or hear could communicate. (The person might communicate through touch.)

● **Process Skills:** *Inferring, Expressing Ideas Effectively*

Explain to the students that vision is closely related to brain function. Ask the students to describe the relationship between the brain and the eyes of a human being. (Images detected by the eyes must be made sense of, or interpreted. One function of the brain is to interpret the images that are detected by the eyes and relayed as electrochemical impulses by the nervous system. Once these images have been received, the brain can act upon those images and even store them for later recall.)

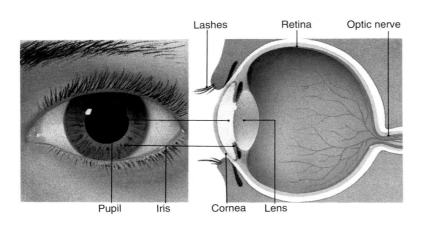

Figure 19–10. The eye is a complex sense organ that is stimulated by light.

DISCOVER BY Observing

CAUTION: Advise the students to avoid extremely bright light while performing this activity. The students should discover that gradual changes in light intensity cause gradual, or slow, changes in pupil size, and significant changes in light intensity cause significant, or fast, changes in pupil size.

MULTICULTURAL CONNECTION

In about A.D. 1000, Alhazen, an Arabian mathematician and physicist, became the first person to give an accurate description of vision. He stated that the eye sees because of light that comes from the object seen. Alhazen compared the eye to a primitive camera obscura. The camera obscura allowed light to shine through a hole to make a picture on a screen. Alhazen correctly stated that vision is caused by light reflected from an object that then passes through the eye and forms an image.

Vision

A white blur leaves the pitcher's hand and streaks toward the plate. Can the batter's eyes follow this blazing pitch?

Your ability to see things begins with your eyes. Each eye is a slightly flattened ball about the size of a Ping-Pong ball. Three layers of tissue form the hollow eyeball. The center is filled with a clear, jellylike fluid. This fluid gives the eyeball its shape and helps keep the inner surface moist. A tough covering of tissue forms the outer layer of the eye. Only the front portion of this layer is clear; the rest is white. The clear portion, or **cornea**, lets light pass into your eye.

The second layer of tissue absorbs stray light. This layer has many capillaries that bring blood to the eye tissue. The front part of this layer is a colored ring called the *iris*. The *pupil* is the opening in the center of the iris.

DISCOVER BY Observing

Enter a dimly lit room and look at your pupils in a mirror. Now turn on a bright lamp or flashlight. Look at your pupils again. What did you observe? Repeat the test, using a stopwatch to time how long it takes for your pupils to get larger and smaller as you change the lighting.

The opening and closing of the iris changes the size of the pupil. When the light is dim, the pupil gets larger. In bright light, the pupil gets smaller. The changes in the size of the pupil allow for optimal vision.

TEACHING STRATEGIES, continued

● **Process Skills:** *Constructing/Interpreting Models, Applying*

 Cooperative Learning Have small groups of students work together to construct models of the eyeball. Encourage the students to use the illustration in the text as well as illustrations in reference books to help them identify and label different parts of the eye on their models.

● **Process Skills:** *Comparing, Applying*

Explain to the students that two types of specialized cells are arranged on the surface of the retina. These specialized cells, known as rods and cones, contain a light-sensitive pigment. Point out that cones distinguish between different colors of light and rods do not. Ask the students to describe the time of day when color is not important or the environment lacks color. (night) Ask the students to infer whether rods or cones would be most functional at night, when there is a distinct lack of color. (rods) Explain to the students that as light intensity

---

① Nearsightedness and farsightedness occur when light waves traveling into the eye do not focus on the retina.

INTEGRATION—
Physical Science

Explain to the students that the cornea and the lens of the human eye change the speed and the direction of light so that it focuses on the retina. The process of bending waves due to a change in speed is called *refraction*. You can demonstrate the principle of refraction by placing a pencil in a clear glass partly filled with water. The pencil looks broken at the water line because light travels slower in water than it does in air.

Demonstration
Before the students arrive in class, draw two circles of equal size about 1 m apart on the chalkboard. Around one circle draw six smaller circles; around the other circle draw six larger circles. Ask the students if one of the circles in the middle is larger than the other and if so, which one. Ask a volunteer to measure the diameter of the circles to show that they are the same size. Point out that the sizes of the surrounding circles help create an optical illusion.

Focus on This A lens is located behind the pupil. The **lens** bends light rays entering the eye. The lens works in much the same way as the lens of a camera, enabling you to focus on objects any distance from your eye.

The **retina** (RET ih nuh) is the layer of light-sensitive cells that lines the inside of the eyeball. Millions of these cells make up the surface of the retina. An image is made on the retina in much the same way as an image is made on film. A camera lens moves back and forth to focus the image on the film. The lens in your eye focuses an image onto the retina by changing shape. The lens becomes rounder and thicker for closer vision. As an object moves closer to your eye, the lens thickens to keep the image clearly in focus. Eventually, the lens cannot thicken enough and the image blurs. Extend your right arm and look at your index finger. Slowly move your finger toward one eye until the image is blurred. This is the point at which the lens can no longer thicken.

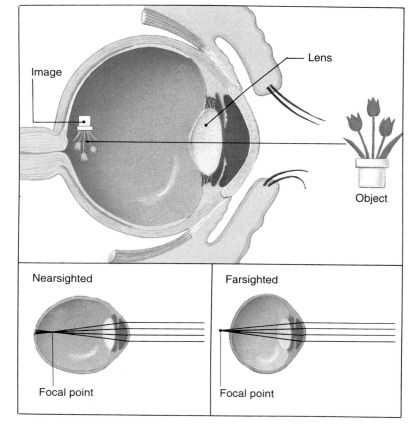

Figure 19–11. Light rays from an object cross after they pass through the lens. The image of the object that forms on the retina is upside-down. What conditions result when the light rays do not form on the retina? ①

522 CHAPTER 19

Sometimes the lens is not able to focus an image on your retina. In such cases, the focused image falls in front of or behind the retina. If the image is in front of the retina, the condition is called *nearsightedness*. If it falls behind the retina, the condition is known as *farsightedness*. A nearsighted person sees close objects much better than far objects. It is just the opposite for farsighted people. Fortunately, both conditions can be corrected with eyeglasses or contact lenses.

Seeing Red and Blue How are you able to see the color of a ball? Two types of receptors make up the outer surface of the retina. Receptors called *rods* are most sensitive to light. Receptors called *cones* are most sensitive to color.

Have you ever noticed that you cannot see colors very well by moonlight? Your inability to see color in dim light is due to the fact that cones work only in bright light. Rods, on the other hand, work well in dim light but produce only black-and-white vision. Night vision depends on rods.

In Sight Once it is formed on the retina, an image is sent to the brain as a nerve impulse. Sensory neurons attached to each rod and cone come together to form the *optic nerve*. Impulses move along the optic nerve to your brain for processing. Vision is a combination of the image seen by the eye and your brain's interpretation of the information it receives about that image. The brain's interpretation does not always agree with reality. This is why illusions are possible.

Figure 19–12. Even if you have normal vision, your perception may be different from someone else's perception. In this optical illusion, some people see an old woman while others see a young woman. If you look first for one and then for the other, you will probably see both.

You can do a simple test to find out which eye is your dominant eye. Hold both of your arms out in front of your face. Now use your thumbs and index fingers to form a triangle as shown. With both eyes open, look at a distant object through the triangle. Next, shut one eye and then the other, looking at the same distant object each time. What did you observe? The eye that allowed you to see the same as you did with both eyes is your dominant eye.

ASK YOURSELF

Name the parts of the eye and describe the function of each part.

TEACHING STRATEGIES, continued

● **Process Skills: Observing, Expressing Ideas Effectively**

Have the students examine the diagram of the ear depicted in Figure 19–14. Ask the students to trace the path of a sound wave through the ear to the brain, naming all the structures through which the waves pass on the way to the brain. (ear canal, eardrum, hammer, anvil, stirrup, cochlea, auditory nerve, brain) Ask the students to describe how sound waves change when they reach the auditory nerve. (The sound waves are changed to nerve impulses.)

● **Process Skills: Comparing, Applying**

To help the students better understand the role of the brain in processing sensory information from sense organs, such as the ear, perform the following activity. Ask a student to volunteer to be blindfolded. Have that student then stand and face the class. Ask another student to stand about 1 m behind the blindfolded student and to snap his or her fingers in front of, behind, to the right of, to the left of, and directly above the blindfolded student's head. After each snap, have the blindfolded student describe where he or she thinks the sound originated

---

LASER DISC
4143
Senses; hearing

Sensorineural disorders result from defects in the inner ear or in the auditory nerve leading to the brain. In recent years, two implanted devices have been developed that can restore some hearing to people with hearing loss resulting from conductive disorders. The XOMED Auditory Bone Conductor is an implanted magnet that holds a sound processor to the skull. It allows sound waves to bypass damaged eardrums and bones and go directly to the inner ear. A newer device, the cochlear implant, is a microphone transmitter that is connected to a receiver transmitter implanted inside the ear. The receiver changes sound waves to electrical signals that are then sent through wires to the nerve endings in the inner ear.

> ✧ **Did You Know?**
> Vibrations from your own voice travel through the bones of your skull as well as in the air. For this reason, your voice sounds different to you when you listen to a recording of it.

Figure 19–13. The sense of hearing can be damaged or destroyed by repeated exposure to sounds that are too loud.

Figure 19–14. Sound causes the air to vibrate. The ear translates these vibrations into nerve impulses, which are sent to the brain.

Hearing

⚾ The murmur of the crowd dies away. The crowd can almost hear the whoosh of the ball as it cuts through the air. It's a fly ball. The fielder goes deep to right field and . . . she catches it!

Your ears receive two types of information. You know that your ears enable you to hear. But did you know your ears also help control your sense of balance?

Through the Canal Your ear is divided into three parts: the *outer ear,* the *middle ear,* and the *inner ear.* The outer ear is what you most often think of as your ear because it is the part you see. The shape of the outer ear helps it collect sound waves and direct them into the ear canal, which leads deeper into the skull. A thin piece of tissue called the *eardrum* stretches across the canal.

Drumbeats Sound waves travel down the ear canal and strike the surface of the eardrum, causing it to vibrate. The middle ear is found on the other side of the eardrum. Three tiny bones—the hammer, the anvil, and the stirrup—form a bridge across the middle ear. The vibrating eardrum vibrates the hammer. The hammer transfers the vibrations to the anvil, and the anvil, in turn, relays the vibrations to the stirrup. The vibrations then move into the inner ear.

Under Pressure The middle ear is connected to the throat by the Eustachian (yoo STAY shuhn) tube. The purpose of this tube is to equalize pressure in the middle ear with the pressure of the atmosphere. If you hold your nose and swallow, the tube closes and you can feel the pressure change. You can also feel pressure changes when you dive deep into water or fly in an airplane. If you did not have this tube to equalize pressure on both sides of your eardrum, the eardrum would rupture, causing severe pain.

524 CHAPTER 19

from. After several volunteers have had an opportunity to perform the activity, ask the students to describe their observations and relate them to the function of the brain with respect to sensory input. (The brain can determine the direction of sound when sound waves strike the ear at the same time—snaps that are directly above, behind, or in front of the head—or at different times—snaps that are to the left or right.)

- **Process Skills:** *Inferring, Applying*

Ask the students to name qualities of sound that their ears enable them to detect. (pitch, loudness, and tone quality) Remind the students that loud noises can cause permanent damage to the ears, and as a result, permanent hearing loss. Ask the students to suggest careers that expose people to continual or loud noises. (Careers suggested by the students might include musician, pneumatic drill operator, lawn maintenance worker, and airline repair crew member.)

Message Received A series of curved, fluid-filled tubes makes up the inner ear. The spiral-shaped *cochlea* (KAHK lee uh), which looks somewhat like a snail, forms one part of the inner ear. Receptor cells with tiny hairs on them line the inner walls of the cochlea. Movement of the bones of the middle ear vibrates the membrane of the inner ear, which causes the liquid in the cochlea to vibrate. The vibrating liquid moves the tiny hairs of the receptor cells. The receptors change the vibrations to nerve impulses that travel along the auditory nerve to the brain. The brain processes these signals and interprets them.

Your ears are amazingly sensitive organs. They can distinguish over 300 000 different sounds. Your ears are made of delicate parts that can be easily damaged. Loud noises, especially repeated and prolonged exposure to them, can lead to a loss of hearing. Noise is measured in units called *decibels*. Noise at 120 decibels can be painful. Frequent exposure to noise levels as low as 85 decibels can damage delicate structures in your ears.

TABLE 19-1: MEASUREMENT OF SOUND

Jet takeoff at close range	140	
	120	Live rock band
Power mower	100	
	85	Subway train
Vacuum cleaner	80	
	70	Telephone bell
Normal talk	60	
	20	Whisper
Normal breathing	10	

Balancing Act The sensory organs that control balance are also located in your inner ear. Three fluid-filled tubes, called the *semicircular canals,* are lined with receptors. When you bend over, lie down, spin around, or do a flip, the fluid in these canals moves. The receptors detect this motion and send nerve impulses to your brain. Your brain sends signals telling your muscles to move in ways that help control your balance.

Tip your head to the left. The liquid within the semicircular canals is now pressing on a different group of receptors. Your brain interprets this information and tells you that you have changed position. Sometimes the information from the inner ear does not agree with the information gathered by your eyes.

Have you ever been dizzy or had an upset stomach while you were in a car or an airplane? Information from your eyes indicates that you are not moving. At the same time, information from your inner ear tells your brain that you are moving. This conflict sometimes causes motion sickness.

Figure 19-15. Your sense of balance is located in the inner ear. This acrobat must have an excellent sense of balance.

▼ **ASK YOURSELF**

What happens to sound waves that enter the ear?

LASER DISC
4180
Ear; human

INTEGRATION—
Physical Science

Sound is produced by the vibrations of an object. When the object vibrates, it makes its surrounding *medium*, or substance such as air or water, vibrate. A pebble dropped into a pond creates waves. Sound travels in waves, too. As the vibrating object moves outward, it creates a region of compression called a *condensation*. As the object moves inward, it creates a region of expansion called a *rarefaction*. Sound waves are a series of condensations and rarefactions produced by the vibrating object. The waves in the pond travel away from the point where the pebble hit the water. In the same way, sound waves travel away from the vibrating object.

ONGOING ASSESSMENT
▼ **ASK YOURSELF**

Sound waves that enter the ear travel through the ear canal and vibrate the surface of the eardrum, which vibrates the bones of the middle ear. The bones vibrate the membrane of the inner ear, which vibrates the liquid in the cochlea. The liquid vibrates the hairs of receptor cells, which change the vibrations into nerve impulses that travel along the auditory nerve to the brain.

TEACHING STRATEGIES, continued

● **Process Skills:** *Inferring, Applying*

Ask the students to identify on which part of the tongue they would expect to taste each of the following substances: pickle juice (sides of the tongue), soda cracker (toward the front and sides of the tongue), lemon juice (sides of the tongue), orange juice (near the tip of the tongue), honey (near the tip of the tongue), salt water (toward the front and sides of the tongue)

● **Process Skills:** *Experimenting, Comparing*

Collect a number of items with distinct odors, for example, hand soap, cinnamon, flowers, garlic, and so on. Ask a volunteer to come to the front of the room and sit on a chair facing the class. Blindfold the volunteer and then ask him or her to identify the various items by using the sense of smell. Give a number of students an opportunity to volunteer. Explain that if the students are unfamiliar with the odor of a particular substance, they will not be able to identify it.

SCIENCE BACKGROUND

The receptors in the nasal cavity that are responsible for the detection of different odors are called *chemoreceptors*. The chemoreceptors are covered by a thin layer of mucus. When a molecule of a substance dissolves in the mucus and touches cilia growing on the chemoreceptor, the chemoreceptor reacts.

> ✦ **Did You Know?**
> About 2 million Americans cannot smell anything. They suffer from a condition known as anosmia, which is the inability to smell.

REINFORCING THEMES— *Systems and Structures*

Emphasize as often as possible the interrelatedness of the human body systems. Point out that although for convenience the systems are described individually, actually each system could accomplish little all on its own. All the body systems must work together. For example, the sense organs, brain, and nerves of the nervous system collect and interpret information and send messages to other body parts, specifically, to muscles (muscular system) and glands (endocrine system).

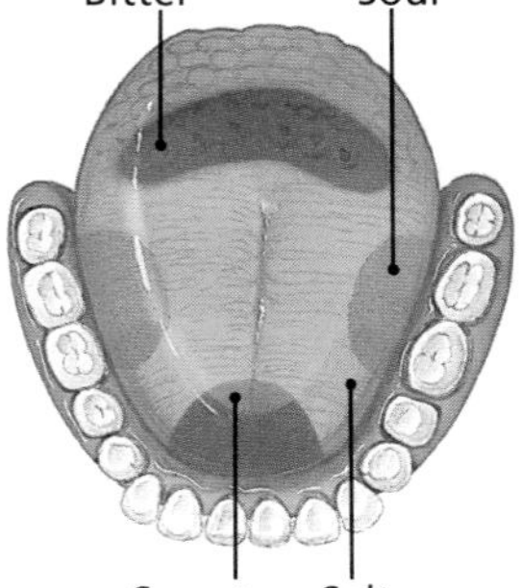

Figure 19–16. Note the location and distribution of taste buds on the tongue.

Taste and Smell

⚾ **As you wait for the next batter to come to the plate, you notice her sipping a cool drink. She then chooses a bat, swings it, and stands up at the plate. She's ready for her turn at bat.**

The batter can taste the drink because of her senses of smell and taste. These senses depend on chemicals in the food you eat and in the air you breathe. These chemicals affect taste and smell receptors that are found mainly on your tongue and in your nose. The receptors of the tongue are called *taste buds*. There are four kinds of taste buds, and each responds to one of four basic tastes: sweet, sour, salty, and bitter. These four kinds of receptors are grouped in different areas on the tongue. Sweet receptors are found near the tip of the tongue, while bitter receptors are near the back of the tongue and sour receptors are at the sides. Receptors for salty taste are located between the sweet and sour receptors. You can see the location of these receptors in the drawing.

Your sense of smell works closely with your sense of taste. You might wonder how you can taste so many different flavors with only four kinds of taste receptors. The answer lies in your nose, which senses chemicals. A small area, about the size of a postage stamp, at the top of each nostril has many more receptor cells than the tongue. These receptors can sense more than 10 000 different chemicals, giving you your sense of smell. Your sense of smell greatly improves your sense of taste. With the help of your nose, you taste the flavor of strawberries as well as their sweetness. The next activity can show you how much your sense of taste depends on smell.

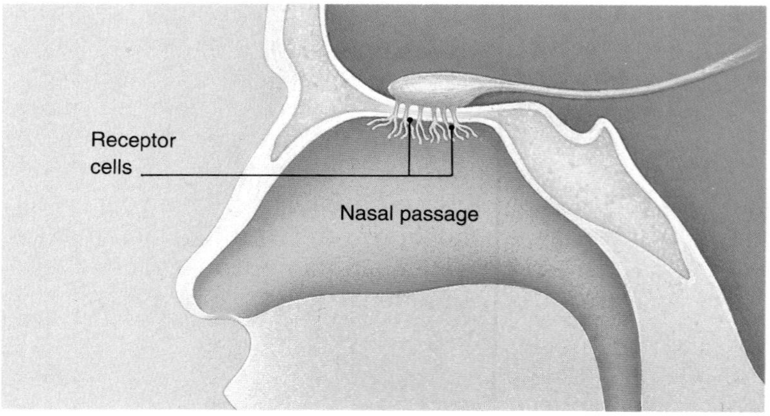

Figure 19–17. Receptor cells high in your nose allow you to detect odors.

- **Process Skills:** *Expressing Ideas Effectively, Applying*

Point out to the students that certain smells can trigger memories or emotions in some people. For example, the smell of baking bread may remind someone of a specific person who baked bread or of a place where bread is baked. Ask the students to recall smells that remind them of a person, place, or event. Encourage the students to share their memories.

- **Process Skills:** *Observing, Comparing*

Explain to the students that they can use their sense of touch to help them identify objects that they cannot see. Once again ask for volunteers to come to the front of the classroom and be blindfolded. Give the volunteers different objects to identify by using the sense of touch. Remind the volunteers to feel each item for its shape, size, and texture. You might have the students try to identify such objects as a classroom globe, a wool sweater, a ruler, a pencil, different types of fruit, and so on.

 **Discover by Doing**

Close your eyes and tightly hold your nose. Have your partner feed you small pieces of apple, potato, and onion. Try to identify each piece without looking. Describe the taste of each piece. Why was it so difficult to identify each food?

As molecules are released from the food you eat, some move into the nasal passages. These molecules dissolve in a sticky substance called *mucus,* which lines the inside of your nose. The dissolved chemicals trigger chemical receptors in the nose. The receptors produce nerve impulses that move along sensory neurons to the brain for processing. Taste results from the brain's evaluation of information received from the nose and tongue. How does food taste when you have a stuffed nose due to a cold? Without a sense of smell, your sense of taste is limited.

 ASK YOURSELF

Where are the receptors that detect chemicals located?

Touch

The bat meets the ball with a crack, and a tingle moves up the batter's arms from the bat. The batter doesn't even move as the ball is going, going, gone over the center field fence.

If you were to shut your eyes and close off your ears and nose, you could still learn a lot about your surroundings by using your sense of touch. The sense of touch comes from individual receptors scattered over the entire surface of your body. Along with the sense of touch, the body's surface receptors also detect heat, pressure, and pain. You can feel even the lightest contact on your skin. Touch your finger to an ice cube, and the receptors tell you that it lacks heat. Very hot water can stimulate heat receptors and sometimes pain receptors as well.

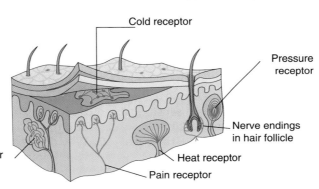

Figure 19–18. There are many different types of receptors.

BACKGROUND INFORMATION

Helen Keller is an outstanding example of a person who overcame physical disabilities. She lost her sight and hearing as a result of a serious illness when she was 1½ years old. With the help of her devoted teacher Anne Sullivan, Keller learned to communicate through touch. She listened to people by placing her hand on their nose, lips, and throat. At first, she communicated to others just through sign language, but later she also learned to speak. She became an international lecturer and spent much of her time working on behalf of visually impaired people throughout the world.

DISCOVER BY *Doing*

The students should discover that the ability to differentiate tastes is diminished when receptors in the nasal cavity are not used. Explain to the students that their senses of taste and smell work together, enabling them to detect the flavors of foods.

ONGOING ASSESSMENT
ASK YOURSELF

Taste and smell chemical receptors are located on the tongue and in the nasal cavity.

GUIDED PRACTICE

Ask the students to define each of the following terms: cornea, retina, lens, Eustachian tube, eardrum, taste buds, and chemical receptors.

INDEPENDENT PRACTICE

Have the students provide written answers to the Section Review and Application questions in their journals. Then ask the students to describe the relationship between the brain and the sensory receptors of the human body.

EVALUATION

Have the students draw and label the structures of the eye and ear.

RETEACHING

Cooperative Learning Ask small groups of students to make a chart identifying the structures of the eye and their functions.

---

① The pain receptors enable a person to feel pain and to seek help in identifying the source of the pain and relieving it.

ONGOING ASSESSMENT
ASK YOURSELF

The skin has touch, temperature, pressure, and pain receptors.

SECTION 2 REVIEW AND APPLICATION

Reading Critically

1. The human ear is involved in hearing, the eye in seeing, the mouth and the nose in tasting, the nose in smelling, and the skin in touching.

2. Areas of the human body that are most prone to injury are more sensitive, or contain more sensory receptors, than other areas of the body.

3. Responses should detail how the ear hears, the eye sees, the mouth and the nose taste, the nose smells, and the skin feels.

Thinking Critically

4. A lack of structures known as rods or poorly functioning rods in the eyes of humans are related to night blindness.

5. Thin gloves allow surgeons to reduce the likelihood of contamination while retaining their sense of touch.

6. The workers need to protect their eyes from debris and their ears from prolonged loud noise to avoid damaging these sense organs.

▶ **528** CHAPTER 19

Figure 19–19. Why is it important that you have pain receptors in most tissues of your body? ①

Why are your fingertips and lips more sensitive than the backs of your hands? What could cause this difference? Receptors for touch, pressure, and temperature are spread unevenly over the body. A small square area on the front side of your fingertip may have 100 touch receptors, while an equal area on the back side has only ten. Areas of your body that have closely spaced receptors are more sensitive than areas with fewer receptors. Touch receptors lie near the surface of the skin, but pressure receptors lie much deeper. Pain receptors can be found in almost every part of the body. These receptors help protect the body from injury by sending a warning of possible danger to the brain. In the next activity, you can discover an ancient art that is used to control pain.

DISCOVER BY Researching

For hundreds of years, the Chinese have practiced a method of pain relief called *acupuncture*. This method involves placing needles into various parts of the body. Go to the library and look up acupuncture. Write a report on this procedure to relieve pain.

ASK YOURSELF

What types of sense receptors are in your skin?

SECTION 2 REVIEW AND APPLICATION

Reading Critically
1. Name the sense organs involved with each sense.
2. Why are some areas of the body more sensitive to touch than others?
3. Explain how each sense organ functions.

Thinking Critically
4. What might cause a condition, called night blindness, in which a person has difficulty seeing at night?
5. Why do surgeons wear such thin rubber gloves when doing surgery?
6. Why should workers operating power grinders wear eye and ear protection?

EXTENSION

Ask the students to research information about color blindness. Encourage them to find out about the different types of color blindness and the tests that are used for identifying people who are color blind.

CLOSURE

 Cooperative Learning Assign each of five groups a different sense. Ask each group to describe what a day would be like for a person who lacked that sense, and the difficulties he or she might encounter. Then have each group share its ideas with the other groups.

SKILL

Observing Sound Waves

Process Skills: Observing, Comparing

Grouping: Groups of 3 or 4

Objectives
- **Construct** models that change sound waves into light waves.
- **Observe** and **compare** waves created by different characteristics of sound.
- **Evaluate** the nature of sound waves with respect to loudness and pitch.

Discussion
This activity will work best in a dimly lit environment. **CAUTION: The ends of the can or mirror may be sharp.** Remind the students to handle any sharp edges of the can or the mirror with care.

Application
1. The vertical images produced by the models should vary with loudness—the louder the noise, the greater the height of the vertical images, and vice versa. Also the horizontal images produced by the models should vary with pitch—the greater the pitch, the faster the speed with which the horizontal images move, and vice versa.

2. The vibrations will vary with the loudness and pitch of each speaker's voice.

Using What You Have Learned
The "sound visualizer" and an eardrum both function by vibration.

SKILL Observing Sound Waves

▶ **MATERIALS**
- empty soup can • piece of thin rubber membrane • rubber band • glue
- small piece of mirror • flashlight • can opener

▼ **PROCEDURE**

1. Using the can opener, remove both ends of the soup can. **CAUTION: Be careful of the sharp edges of the can.**
2. Stretch the rubber membrane over one end of the can. Hold the piece of rubber in place with a rubber band.
3. Glue the piece of mirror, slightly off center, to the rubber membrane.
4. Hold the open end of the can near your mouth and shine the light toward the mirror. Position the can so that the light reflecting from the mirror strikes a shaded wall. You should be able to see the mirror's reflection easily.
5. Talk into the open end of the can. Vary the volume and the pitch of your voice.

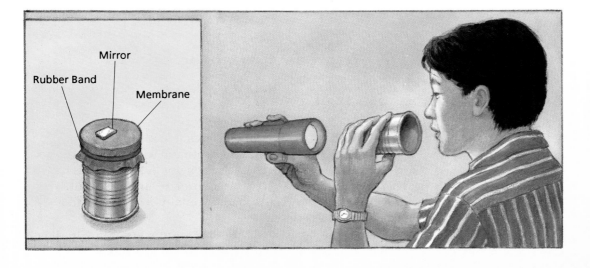

▶ **APPLICATION**
1. How do the patterns made by the sound change with volume and pitch?
2. Have two people talk normally into their "sound visualizers" at the same time and with their reflections striking the wall at about the same place. Can you detect any difference in the vibrations?

✳ **Using What You Have Learned**
Describe any changes in the vibrations. How is your "sound visualizer" like an eardrum?

SECTION 2 **529** ◀

Section 3:

ALCOHOL, TOBACCO, AND OTHER DRUGS

FOCUS

This section introduces the subject of drugs and explains their potential for misuse. These drugs include over-the-counter drugs, prescription drugs, and illegal drugs—drugs that cannot be bought or sold legally. Tobacco and alcohol, their effects, and their addictive properties are also examined.

MOTIVATING ACTIVITY

Bring in a filter from an unsmoked cigarette and a filter from a smoked cigarette. Have the students examine and compare the two filters. Ask the students to describe the filters. (The one from the unsmoked cigarette should be clean and white in color; the one from the smoked cigarette should be yellow and/or brown in appearance.) Point out to the students that the filter traps only a small amount of the substances given off by cigarettes; most of the material goes into the smoker's lungs.

PROCESS SKILLS
- Comparing • Inferring
- Measuring • Applying

POSITIVE ATTITUDES
- Caring for the environment
- Initiative and persistence

TERMS
- drug • drug abuse
- nicotine

PRINT MEDIA
Using Medicines Wisely by the editors of **Prevention** magazine (see p. 429b)

ELECTRONIC MEDIA
Drug Abuse: Sorting It Out, Britannica (see p. 429b)

Science Discovery Fleming; discovery of penicillin

BLACKLINE MASTERS
Study and Review Guide
Laboratory Investigation 19.2
Reading Skills
Extending Science Concepts

SECTION 3 | Alcohol, Tobacco, and Other Drugs

Objectives

Distinguish between drugs and medicines.

Contrast drug use and drug abuse.

Describe the effects of tobacco and alcohol on the body.

Cold medicines can make a person feel better, but they can also affect the person's reaction time. Most people use such medicines only as needed. But some people abuse medicines and other substances. Substance abuse is a problem that touches every part of society. It is a problem of the young, the old, the rich, and the poor. It costs the United States over 50 billion dollars in lost productivity, medical expenses, and crime.

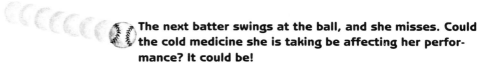

The next batter swings at the ball, and she misses. Could the cold medicine she is taking be affecting her performance? It could be!

Drugs and Medicines

Generally speaking, a **drug** is any chemical substance, other than water, oxygen, and most foods, that alters the way your body works. Drug use is not new. People have been using drugs for thousands of years. Many drugs are made in laboratories, prescribed by doctors, and sold by pharmacies. When prescribed by doctors, these drugs can be very helpful. Many foods naturally contain drugs. Coffee, tea, chocolate, and cola contain a drug called *caffeine*. Tobacco also contains a substance that affects the nervous system.

Drugs can be divided into three groups according to their availability and use. The first group of drugs includes those that cannot legally be bought or sold by anyone in this country for any reason. They are illegal for any use, including medical use, and are extremely dangerous. Examples of drugs in this group are heroin and LSD.

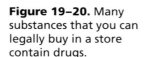

Figure 19–20. Many substances that you can legally buy in a store contain drugs.

TEACHING STRATEGIES

- **Process Skills:** *Synthesizing, Expressing Ideas Effectively*

Have the students discuss what they know about drugs, then lead them in a discussion of where drugs come from and why people use them.

- **Process Skills:** *Observing, Applying*

Ask the students to examine the pharmaceutical picture shown in Figure 19–21. Explain to the students that many different kinds of drugs may be purchased at a pharmacy similar to the one shown in the picture. Ask the students to name several over-the-counter drugs they have seen displayed in drug stores or pharmacies. (The students are likely to name cold medicines, headache remedies, cough drops and syrups, aspirin, and inhalants for clogged sinuses.) After reminding the students that many over-the-counter drugs can be abused, ask the students to speculate about how these drugs can be abused. (People can abuse the drugs by taking a larger dose than recommended, using them more frequently than recommended, or taking drugs that are not compatible at the same time.)

Drugs in the second group have medical uses, but they can be just as dangerous as those in the first group if used improperly. These drugs can be used when prescribed by a doctor. Examples of these drugs are tranquilizers, steroids, and morphine. Although these drugs have some legal uses, they are often bought illegally and used without a doctor's supervision.

Drugs in the third group can be bought by anyone and used by following the directions on the package. These drugs are often called over-the-counter drugs. Most headache and cold medicines are examples of over-the-counter drugs. However, even these drugs can be harmful if the directions for their use are not properly followed.

Medicines are drugs that can cure illness, heal the body, and relieve symptoms of disease or injury. Some medicines may poison and kill disease-causing organisms. For example, your doctor may give you *antibiotics* to help your body recover from an infection. They kill harmful organisms that invade your body.

Other medicines contain *psychoactive drugs*. These drugs affect the nervous system and can change a person's behavior. Some psychoactive drugs may cause a person to feel relaxed or sleepy. Others may make a person feel more energetic. If used improperly, psychoactive drugs are extremely dangerous.

Figure 19–21. A pharmacist can give advice on the action and effects of over-the-counter drugs. The label on a prescription drug gives instructions for its use. These instructions should be followed carefully.

LASER DISC
2811, 2812

Fleming; discovery of penicillin

BACKGROUND INFORMATION

In the United States, the Food and Drug Administration (FDA) has the responsibility of approving the drugs that can be legally sold. According to law, drug companies must prove that new drugs are safe and effective before the FDA will approve them for distribution in the United States. The FDA also regulates the advertising of prescription drugs. The Federal Trade Commission regulates the advertising of nonprescription drugs.

ONGOING ASSESSMENT
ASK YOURSELF

Medicines are drugs that cure an illness of the body or relieve the symptoms of sickness or disease. These drugs include prescription drugs and over-the-counter drugs. Other drugs are illegal for any use, medical or nonmedical, and cannot be bought or sold for any reason.

ASK YOURSELF

How are medicines different from other drugs?

Drug Abuse

The incorrect and unsafe use of a drug is called **drug abuse**. Drug abuse includes the taking of drugs for which you have no prescription or no medical reason for using. Taking more of a prescribed drug or taking it more often than recommended also is a form of drug abuse. Almost any drug can be abused. The drug does not have to be illegal or dangerous to be abused. Even over-the-counter drugs that people take for a cold can be abused. Drug abuse can damage body tissues, have a negative effect on behavior, and even lead to death.

TEACHING STRATEGIES, continued

● **Process Skills:** *Inferring, Applying*

Explain to the students that aspirin may seem like a harmless drug, but in certain situations, it can be very dangerous. Obtain a bottle of aspirin and read the comprehensive set of warnings that is included on the label. Have the students note all of the warnings, especially the warning that children and teenagers should not take aspirin to alleviate chicken pox or flu symptoms without first consulting a doctor about Reye's syndrome, a rare but serious illness. Explain that recent research seems to indicate a link between Reye's syndrome and the taking of aspirin by young people suffering from a viral infection. Ask the students to describe why it is extremely important that people take time to read and understand the labels of any prescription and nonprescription drugs they intend to take. (To be effective, prescription drugs should be administered according to the frequency and dosage given on their labels. Labels on nonprescription drugs provide dosage information and indicate circumstances in which the drug should not be taken or a doctor should be consulted.)

DISCOVER BY *Researching*

Encourage the students to use more than one reference source during their research. The students might present a short oral report summarizing their research findings.

 Have the students place their research notes and their written reports in their science portfolios.

ONGOING ASSESSMENT
ASK YOURSELF

Proper use of a drug involves using the drug in recommended dosages for the medical reason it was manufactured. Drug abuse involves using drugs in improper dosages or using illegal or unneeded drugs.

SCIENCE BACKGROUND

During the nineteenth century, cocaine was an additive in several patent medicines. In an article published in 1885, a major drug company explained the benefits of cocaine: "Cocaine, through its stimulant properties, can take the place of food, make the coward brave and the silent eloquent, free victims of alcohol and opium from their bondage, and render the sufferer insensitive to pain." The article failed to recognize the addictive nature of cocaine and the problems that its use could cause.

▶ **532** CHAPTER 19

Figure 19–22. People who abuse drugs need help, such as counseling, to overcome their habit.

Many drugs that are safe in small quantities are *toxic*, or poisonous, in larger amounts. High doses or prolonged use of many drugs can damage the liver, kidneys, lungs, heart, and brain. This damage can increase the risk of disease and death.

If a drug is taken over a long period, the person will gradually get used to the drug, or build up a *tolerance*. This means that a higher dose of the drug is needed to produce the same effect. As a result, the person takes a larger, more dangerous quantity, or *overdose*, of the drug. An overdose may result in serious physical or mental harm or even death.

Some drugs are dangerous because they are habit-forming. A person may use a drug so frequently that using the drug becomes a habit. A drug habit can lead to drug *addiction*, or dependence on a drug. A person with a drug addiction must continue taking the drug to feel good. Such a condition is known as a physical addiction. Psychological addiction exists when a person is emotionally dependent on the drug. A person with this type of addiction depends on a drug to produce a false sense of well-being.

When a person is physically addicted to a drug, his or her body needs the drug. A condition, called *withdrawal*, occurs when an addict does not use the drug. Withdrawal symptoms may include extreme nervousness, uncontrollable shaking, excessive sweating, nausea, and muscle cramps. Extreme feelings of sadness or depression may also occur with drug withdrawal.

DISCOVER BY *Researching*

There are many types of illegal drugs abused by people throughout the world. These include hallucinogens, narcotics, stimulants, and depressants. Select one of these groups of drugs, go to the library, and research the group of drugs. Write about what you learn.

ASK YOURSELF

How is drug abuse different from using drugs properly?

- **Process Skills:** *Inferring, Analyzing*

Have the students find pictures of cigarette advertisements in magazines. Ask the students to infer what kind of message about smoking they think the copy and the picture convey in each ad. Ask them to locate the health warning in the ad and to estimate the amount of page space the warning occupies. Ask the students to discuss how the message of the ad and the health warning may be in conflict with one another.

- **Process Skills:** *Expressing Ideas Effectively, Applying*

Cooperative Learning Divide the class into small groups and explain that each group is to design an antismoking poster. Encourage the students to discuss the focus of the message they want to convey on the poster. Explain that they might want to focus the message on health, social, or personal issues. Ask the students to include an antismoking slogan as well as graphics on their posters.

Tobacco

More than 80 percent of all Americans are aware that smoking can cause cancer. All tobacco products and advertisements carry warnings about the harmful effects of tobacco, yet billions of cigarettes are smoked each year. In the United States alone, between 40 and 50 million people smoke cigarettes. Why do people start or continue smoking when they know it can damage their health?

Many people smoke for social and emotional reasons. Some young people begin because they observe adults smoking or they are influenced by advertisements. Some people think that smoking is a sign of independence or rebellion. Other people smoke because they want to be like their friends. Smoking is a behavior that people try for many different reasons. The next activity can give you some idea as to why some people start to smoke.

Figure 19–23. By law in the United States, warnings about the hazards of smoking cigarettes must be printed on the packages.

Researching

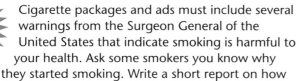

Cigarette packages and ads must include several warnings from the Surgeon General of the United States that indicate smoking is harmful to your health. Ask some smokers you know why they started smoking. Write a short report on how to deal differently with these reasons or pressures. Be sure to include how a person could avoid starting to smoke.

People often smoke while doing something else. Soon a person associates one behavior with the other. Once this happens, the person feels the need to smoke whenever the other behavior occurs. Such association makes a smoker psychologically dependent on smoking. Eventually, the smoking habit leads to a physical dependence. Smoking is a costly habit not only in terms of health but also in terms of money. The following activity will help you calculate the monetary costs of smoking.

BACKGROUND INFORMATION

The tobacco plant is an annual plant that can grow to about 1.6 m high. Its leaves are used mainly to make cigarettes and cigars. China is the leading producer of tobacco, harvesting about 2 000 000 metric tons annually. The United States produces about 700 000 metric tons every year.

DISCOVER BY *Researching*

Ask the students to explain that they are working on a science project before they ask the smokers questions. Remind the students that they are simply to gather information from the smokers.

Did You Know? Although people of all ages abuse drugs, most drug abusers are between 18 and 25 years old.

SECTION 3   **533**

TEACHING STRATEGIES, continued

● **Process Skills:** *Inferring, Applying*

Explain to the students that one of the gases in cigarette smoke is carbon monoxide. Carbon monoxide in the blood prevents oxygen from reaching the brain and the heart. Ask the students what effect continual exposure to high levels of carbon monoxide from cigarette smoking might have on the heart. (Continual exposure to carbon monoxide might lead to heart disease or heart attack.)

● **Process Skills:** *Comparing, Generating Ideas*

Have the students describe how tar and nicotine damage the health of the human body and then compare the healthy lung tissue to the cancerous lung tissue shown in Figure 19-24. Ask the students to describe why a person would continue to smoke after seeing a picture of the cancerous lung and being told of all of the other harmful effects of smoking. (Responses might indicate that smokers simply enjoy smoking, despite the consequences; smokers might feel that the terrible health problems associated with smoking will not happen to them; or smokers are addicted to the nicotine in the cigarettes.)

DISCOVER BY Calculating

With an average of $2.50 per pack, savings for one week (7 days) would total $17.50, for one month (30 days) would total $75.00, and for one year (365 days) would total $912.50.

SCIENCE BACKGROUND

Researchers have found that smokers who quit smoking often significantly lower the risk of cancer. After 15 or 20 years of not smoking, an ex-smoker will have the same risk of contracting lung cancer as a person who has never smoked.

MEETING SPECIAL NEEDS
Second Language Support

Some terms in the section, such as *caffeine*, *tranquilizers*, *antibiotics*, and *nicotine*, may be unfamiliar to students who are acquiring English. Suggest that the students list these terms in their journals and add the equivalent term from their first language. Remind the students that they can refer to their journal list when they encounter these terms to help them remember the meanings.

Figure 19-24. One common effect of smoking tobacco is lung cancer. Shown here are a healthy lung (top) and a cancerous lung (bottom).

DISCOVER BY Calculating

Smoking is an expensive habit. Investigate the cost of a package of cigarettes in three types of stores. Then use an average of these costs to determine how much money you would save in a week, a month, and a year if you never started a one-pack-a-day habit.

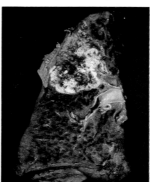

Tobacco smoke contains over 300 different chemical substances. At least 15 of these are cancer causing. And one of these chemicals, **nicotine,** is a psychoactive drug. In small doses, it acts as a stimulant to the nervous system. It speeds the heartbeat rate and raises blood pressure. In large doses, nicotine acts as a depressant. It depresses the nervous system and slows the heartbeat. In doses as small as 70 mg, pure nicotine is poisonous and can cause death. An addicted smoker develops both a psychological and a physical addiction to nicotine. Withdrawal symptoms include nervousness, anxiety, shakiness, and loss of sleep.

Nicotine addiction is only one of the harmful effects of smoking. Tobacco smoke contains gases and tiny particles that damage the body of a smoker as well as those of nonsmokers who breathe the smoke. The respiratory, circulatory, and digestive systems are all harmed by smoke from tobacco products. But smoking does not harm just the body. Careless smoking can result in fires in homes, hotels, and forests. Cigarette butts and wrappers are a major source of litter.

The effects of smoking on the respiratory system are well known. Smoking makes breathing more difficult and increases the chances of respiratory infection and lung cancer. The mucus that lines the air passages traps smoke particles. As these particles cool, they form a sticky substance called *tar*. Tiny, hairlike cilia usually sweep the mucus up the air passages and out of the body so that the air passages are kept clean. However, the cilia have a very difficult time moving mucus that is heavy with tar.

● **Process Skills:** *Applying, Expressing Ideas Effectively*

Explain that research has shown that breathing secondhand smoke is harmful to the nonsmoker. Point out that many states and communities have banned smoking in many places. Ask the students to find out whether their local or state governments have laws prohibiting or restricting smoking in public places.

● **Process Skills:** *Applying, Analyzing*

Point out to the students that smoking causes 30 percent of all cancer deaths. Provisional data indicated that about 202 000 people died of cancer in the United States in 1990. Ask the students to calculate how many of those deaths were related to smoking. (60 600)

● **Process Skills:** *Applying, Analyzing*

Explain to the students that, in addition to causing cancer and heart disease, smoking may lead to a serious condition called *emphysema*. Ask the students to use reference materials to discover more about emphysema. (Emphysema results in the destruction of the alveoli in the lungs. The body does not get enough oxygen, and death may eventually occur.)

The buildup of tar and other dirt particles damages the cells and cilia that line the air passages. New cells, more resistant to damage but less able to keep the air passages clean, are formed to replace damaged cells.

The dirt and tar that pass through the air passages also collect in the lungs. These sticky substances damage the delicate tissues of the air sacs, or alveoli. Tougher, less flexible cells form scar tissue in the lungs. This new tissue cannot expand as normal tissue does.

Tobacco smoke also affects the digestive tract. Smoke that is swallowed irritates the lining of the stomach and intestines. Swallowed smoke can cause indigestion and constipation and increases the chance of ulcers forming. Stomach and liver cancer are also more common in smokers than in nonsmokers.

Recently, more attention has been paid to the harmful effect of tobacco smoke on nonsmokers. Breathing smoke from other people's tobacco products is known as *passive smoking*. Studies show that more respiratory infections occur in children of smokers than in children of nonsmokers. Passive smoking may also cause other diseases of the lungs, including cancer, in nonsmokers. Husbands and wives of smokers have 3.4 times the risk of having a heart attack that husbands and wives of nonsmokers have. Concerns about passive smoking have resulted in a ban on smoking in airplanes and in many other public places.

Smoking is not the only dangerous use of tobacco. Thousands of people use smokeless tobacco, such as snuff and chewing tobacco. The National Cancer Society and other agencies have found that the use of such tobacco increases the risk of cancer of the lips and jaws and of other diseases of the mouth. Smokeless tobacco can be as habit-forming and addictive as tobacco that is smoked.

Figure 19–25. Many harmful effects are hidden in cigarette smoke. Don't start!

BACKGROUND INFORMATION

Two smokeless forms of tobacco are chewing tobacco and snuff. Chewing tobacco is made of low-grade tobacco and is usually treated with honey or licorice. Snuff is a powder of finely ground tobacco leaves and stems, which is flavored with spices and oil.

ONGOING ASSESSMENT
▼ **ASK YOURSELF**

Nicotine is the addictive ingredient in tobacco.

▼ **ASK YOURSELF**

What is the addictive drug in tobacco?

TEACHING STRATEGIES, continued

● **Process Skills:** *Inferring, Applying*

Explain to the students that alcohol interferes with all the processes carried out by the central nervous system. Some effects of alcohol in the human body include blurred vision, poor eye-hand coordination, double-vision, and reduced reaction time. Ask the students to describe how these effects could contribute to motor vehicle accidents. (Each of the effects would contribute to the erosion of the skills needed to operate a motor vehicle safely.)

GUIDED PRACTICE

Have the students review the effects of tobacco and alcohol on the human body.

EVALUATION

Ask the students to differentiate medicines from drugs and explain the behaviors that constitute drug abuse.

INDEPENDENT PRACTICE

 Have the students provide written answers to the Section Review and Application questions in their journals. Then ask the students to discuss how drug education in schools could be made more effective.

SCIENCE BACKGROUND

Many studies have been performed with respect to alcoholism. Some studies indicate that alcoholism may be an inherited disease. In one study, a Harvard scientist found an unusual chemical in the blood of alcoholics after they consumed hard liquor. This chemical was not found in the blood of nonalcoholics after they consumed hard liquor. The presence of the chemical indicates that some people may not be able to metabolize alcohol normally and, as a result, may be potential alcoholics.

✧ **Did You Know?**
Alcohol is the oldest known sedative. Beer was brewed in ancient Egypt as long ago as 4000 B.C.

Figure 19–26. It is estimated that well over one half of all accidents involve drivers who have been drinking alcoholic beverages.

Alcohol

Like smoking, drinking alcoholic beverages is a learned behavior. People drink for many reasons. Alcoholic drinks are commonly served at social gatherings and business affairs. Some people drink because they feel pressured by others or because they want to feel like one of the crowd. Other people drink because they like the taste or the feeling alcohol provides. As a result, alcohol is the most commonly abused drug in our society.

Alcoholic beverages contain water, flavorings, and ethyl alcohol. Ethyl alcohol is a psychoactive drug that acts as a depressant, slowing impulses in the nervous system.

When a person drinks, about 20 percent of the alcohol passes directly into the bloodstream through the stomach wall. The rest enters the bloodstream through the wall of the small intestine. Once in the circulatory system, it is carried throughout the body. Alcohol in the body is slowly broken down by the liver. It takes the liver a little less than one hour to break down the alcohol in two bottles of beer.

Pure alcohol leaves a strong taste and a burning sensation in the mouth and throat. People who drink alcoholic beverages often say that they experience a freer and happier feeling. For these reasons, even though alcohol is a depressant, it is mistakenly thought to be a stimulant.

Alcohol interferes with the function of the brain. Its effects begin with a feeling of relaxation. As more alcohol reaches the brain, drinkers become less aware of their environment and themselves. Most activities that depend on the senses and muscular movement are impaired by the use of alcohol. Alcohol also causes people to lose their self-control and impairs their judgment. Reaction time becomes much slower. Increased consumption usually leads to blurred vision and loss of coordination. It is easy to understand why people who drink alcohol and drive motor vehicles are responsible for thousands of deaths every year.

RETEACHING

Cooperative Learning Ask the students to obtain empty packaging from several different over-the-counter drugs. Have small groups of students read the warning label on each product and explain the implication(s) of those warnings.

EXTENSION

Have the students write a letter requesting literature from the group Students Against Drunk Driving (SADD). The address for SADD is P.O. Box 800, Marlboro MA 01752. If the school has a SADD chapter, ask a member to talk to the class or encourage the students to attend a SADD meeting.

CLOSURE

Ask the students to compare the effects of alcohol and tobacco on the human body and to explain why people may become addicted to these substances.

ONGOING ASSESSMENT
ASK YOURSELF

Alcohol interferes with brain function by impairing the senses and muscular coordination.

SECTION 3 REVIEW AND APPLICATION

Reading Critically

1. Drugs can be divided into three groups based on their availability and use: drugs that cannot be legally bought, sold, or used for any reason; drugs that are prescribed by trained medical professionals; and over-the-counter drugs that can be purchased by anyone. Medicines are drugs that can cure illness, heal the body, and relieve symptoms of disease or injury.

2. Drug abuse includes ingesting drugs for no medical reason, increasing the frequency with which a drug is ingested, exceeding the recommended dosage of a drug, and prolonging the use of a drug.

3. In small doses, the nicotine in tobacco acts as a stimulant; in large doses, it acts as a depressant. Alcohol is a depressant and causes impaired brain, sensory, and muscle function.

Thinking Critically

4. A tolerance requires an increased dosage of a drug to obtain the desired effect and increases the likelihood of an overdose.

5. Both caffeine and nicotine can act as stimulants and cause addiction.

6. The impaired brain, sensory, and muscular functions of a person under the influence of alcohol can affect the safety and well-being of other people.

The effects of alcohol on a person's behavior depend on how much alcohol reaches the brain through the bloodstream. The alcohol level in the blood depends on several factors, including (1) how much and how fast alcohol is consumed, (2) how much the person drinking weighs, and (3) whether or not the stomach is empty. When the percentage of alcohol in the blood reaches 0.10 percent, a person is usually considered to be drunk and should not drive or operate any tool or machine that could cause injury.

As with other drugs, alcohol can create a psychological dependence. Occasionally, it can even lead to physical dependence. People who become dependent on alcohol are called *alcoholics*. They use alcohol as an escape from stress and other situations that they have difficulty facing. Heavy use of alcohol can lead to serious damage of the liver. Alcoholism is the most serious drug problem in the United States, which has over 5 million alcoholics.

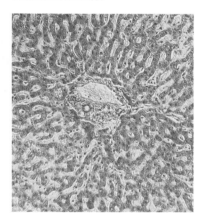 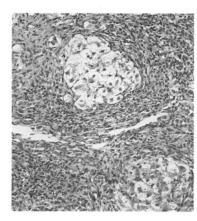

Figure 19–27. Cirrhosis of the liver can result from drinking too much alcohol. The liver on the right was removed from the body of an alcoholic.

 ASK YOURSELF

How does alcohol affect the functions of the brain?

SECTION 3 REVIEW AND APPLICATION

Reading Critically

1. Distinguish between drugs and medicines.
2. What are some types of drug abuse?
3. How do tobacco and alcohol affect a person?

Thinking Critically

4. Why is the ability to develop a tolerance to a drug dangerous?
5. What effects do nicotine and caffeine have in common?
6. Why could it be dangerous to work at a construction site with an alcoholic?

INVESTIGATION

Mapping Your Sensory Receptors

Process Skills: Comparing, Measuring, Interpreting Data

Grouping: Groups of 3 or 4

Objectives
- **Compare** the different sensory receptors of the human hand.
- **Interpret** data generated by a sensory investigation.

Pre-Lab
Ask the students to predict whether the backs of their hand will contain more sensory receptors for pain, pressure, cold, or heat.

Hints
Encourage the students to perform several repetitions per square if they have difficulty distinguishing between pain, pressure, cold, or heat.
CAUTION: Remind the students to avoid breaking the surface of their skin when performing tests with pins.

Analyses and Conclusions
1. Each map should display varying concentrations of different receptors. Receptors for pain, pressure, and heat typically outnumber cold receptors in the human body.
2. The concentration of receptors in body parts is related to the function of those parts.
3. Although receptor maps will vary from person to person, the maps should reveal tendencies of general concentrations.

Application
A sensor map might allow a patient to encourage a physician to dispense a shot in an area that contains a low concentration of pain receptors. A receptor map would enable a person to provide extra protection to those places that have been shown to be more sensitive to cold, although all areas of the body should be protected in cold weather.

✷ Discover More
The students should discover that different sensory receptors are much more concentrated in certain areas of the body than in others. For example, the toes of the foot are more sensitive to touch than the heel, and the inside of the forearm is more sensitive to touch than the palm of the hand.

Post-Lab
Have the students recall their earlier predictions and explain their findings.

 PORTFOLIO ASSESSMENT After the students have completed the Investigation, ask them to place their recorded observations and answers in their science portfolios.

INVESTIGATION

Mapping Your Sensory Receptors

▼ MATERIALS
- water-soluble, fine-tipped felt pen • two paper clips • straight pin
- dull pencil • ice cube • hot water • two small cups

▼ PROCEDURE

1. Use a fine-tipped pen to draw a 2-cm square on the back of your hand. Divide the square into a grid of 16 equal-sized squares as shown.
2. Make a similar but larger grid on a sheet of paper. This grid will serve as a recording device.
3. Place an ice cube in one cup and hot water in the other. Unbend one section of each paper clip. Put the tip of the straight end of one clip on the ice cube and put the straight end of the other clip in the hot water.
4. You are now ready to locate receptors for pain, pressure, and heat. You will test each part of the grid to determine whether pain, pressure, or hot and cold sensory receptors are located there. As you test each part, you should invent a symbol for each receptor. You will then use these symbols to record the results on your paper grid.
5. **CAUTION: Be careful when using the pin so that you do not break the surface of the skin.** Use a pin for pain, a dull pencil for pressure, the paper clip touched to an ice cube for cold, and the paper clip dipped in hot water for hot. Use each tool in each part of the grid and record your findings.

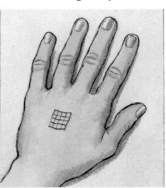 

▶ ANALYSES AND CONCLUSIONS
1. What does your skin receptor map look like? Which kind of receptor seems to be most common? Least common?
2. Why do you think you have more of one receptor than another?
3. Compare your receptor map with those of other students. How are they different?

▶ APPLICATION
Suppose you were to go to the doctor for a shot. How might you use a sensor map to make your visit more pleasant? How could a receptor map help you decide which parts of your body would be most sensitive to cold weather?

✷ Discover More
Mark grids on other areas of your body, such as the heel of your foot, the palm of your hand, and your forearm. Make predictions about the types of receptors most common in these areas. Test to find out how accurate your predictions are.

Chapter 19 HIGHLIGHTS

The Big Idea—SYSTEMS AND STRUCTURES

Lead the students to understand that the microscopic and macroscopic structures of the nervous system work together to control the functions and regulate the activities of the human body. Also remind the students that substances such as tobacco, alcohol, and certain drugs contain ingredients that do not allow the nervous system to function effectively. These substances deteriorate the general health of the human body, and their use for an extended period of time can result in addiction and death.

Chapter 19 HIGHLIGHTS

The Big Idea

The nervous system is a control and message system that operates throughout the body. The structure of the system enables the senses to send and receive messages to and from the CNS. The system interprets information in order to control body functions and to assure well-being. The interaction of microscopic receptors and larger sense organs make it possible to see, hear, smell, taste, and touch.

Alcohol, tobacco, and drugs interfere with the proper functioning of the nervous system as well as other systems in the body. This interference can have long-term and short-term effects on individuals who use or abuse these substances. People who use and abuse such substances can cause great harm to themselves and to other people.

You have probably changed your thinking about how the nervous system works. Review the journal entry you wrote about the nervous system before you began your study of this chapter. How would you explain why Casey struck out? What role might the nervous system have played in his failure to hit the ball?

The ideas expressed by the students should describe how their understanding of the nervous system has changed. The students should also use their knowledge of the structures and the functions of the nervous system to infer a reason why the mighty Casey might have struck out while at bat.

CONNECTING IDEAS

The concept map should be completed with words similar to those shown here.

Row 3: central nervous system; sense organs
Row 5: brain; interneurons; sensory neurons; skin; ears
Row 6: motor neurons; nose; mouth

Connecting Ideas

Copy the unfinished concept map into your journal. Complete the map by filling in the blank boxes.

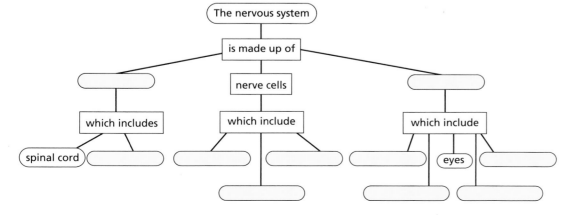

CHAPTER 19 REVIEW ANSWERS

Understanding Vocabulary

1. A receptor is a specialized cell in the nervous system that receives information in the form of a stimulus from the environment.

2. The basic component of the nervous system is a specialized cell called a neuron.

3. The cerebrum is the largest portion of the brain and is responsible for all of the voluntary activities of the human body.

4. The cerebellum is the portion of the brain that coordinates the activities of the muscles.

5. The retina is the layer of light-sensitive cells that line the inside of the eyeball.

6. The lens of the eye functions to bend light waves entering the eye so that the light waves focus on the retina.

7. The semicircular canals are fluid-filled tubes of the ear that control balance.

8. Drug abuse occurs when a drug or medicine is used incorrectly or in an unsafe way.

9. Nicotine is a psychoactive, addictive ingredient of tobacco.

Understanding Concepts

Multiple Choice
10. a
11. b
12. d
13. b
14. a
15. b
16. d

Short Answer
17. Voluntary actions require thought; reflex actions do not.

18. An addiction requires the presence of a drug for normal function.

19. Sound waves that enter the ear travel through the ear canal and vibrate the surface of the eardrum, which in turn vibrates the bones of the middle ear. The bones of the middle ear vibrate the membrane of the inner ear, which in turn vibrates the liquid in the cochlea. The moving liquid of the cochlea then vibrates the hairs of receptor cells, which change the vibrations into nerve impulses that travel along the auditory nerve to the brain.

Interpreting Graphics
20. A: the lens of the eye bends and slows the light waves that pass through it; B: the light waves

---

CHAPTER 19 REVIEW

Understanding Vocabulary

Demonstrate your understanding of these terms by using each in a sentence. You may use more than one term in a sentence.

1. receptor (513)
2. neuron (514)
3. cerebrum (518)
4. cerebellum (518)
5. retina (522)
6. lens (522)
7. semicircular canals (525)
8. drug abuse (531)
9. nicotine (534)

Understanding Concepts

MULTIPLE CHOICE

10. Which of the following is *not* a type of neuron?
    a) dendrite
    b) sensory
    c) motor
    d) interneuron

11. Imagine being a police officer trying to determine if a driver has had too much to drink. Which of the following symptoms would *not* be helpful in determining if a person has had too much alcohol?
    a) lack of coordination
    b) age
    c) a slower reaction rate
    d) slurred speech

12. The organ most likely to be damaged by excessive use of alcohol is the
    a) kidneys. b) lungs. c) brain. d) liver.

13. A distance sense is one that could be used to identify something that is 10 m away. Which of the following would *not* be considered a distance sense?
    a) hearing b) touch
    c) vision d) all of the above

14. How are dendrites and axons similar?
    a) They both are parts of nerve cells.
    b) They both carry messages to another structure.
    c) They both carry messages away from another structure.
    d) All of the above are true.

15. Which part of your brain is responsible for thought processes?
    a) brain stem   b) cerebrum
    c) cerebellum   d) receptors

16. When you see a car coming at you and you have to get out of the way, which neurons carry the message to your muscles, telling them to act?
    a) sensory neurons   b) spinal neurons
    c) interneurons   d) motor neurons

SHORT ANSWER

17. Describe the difference between a voluntary action and a reflex action.

18. Explain why it is more difficult to give up drugs after you have become addicted.

19. Trace what happens to a sound wave that enters your outer ear.

Interpreting Graphics

20. Look at the diagram of the eyeball. Describe what happens at each lettered point as the light from the apple enters the eye.

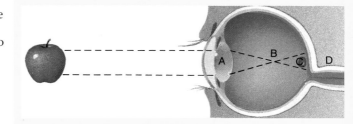

540 CHAPTER 19

focus before the retina—a condition known as nearsightedness; C: the retina image is not focused, and the rods, cones, and sensory neurons of the retina change the light waves into nerve impulses; D: the nerve impulses travel along the optic nerve to the brain.

Reviewing Themes

21. Receptors, sense organs, and the brain are components of the nervous system that demonstrate wide differences in scale. Dendrites, axons, and the cell bodies of neurons are structures that are well suited to their functions; the branches of dendrites concentrate impulses from a great area into a cell body; a cell body contains the nucleus of the neuron and structures to transfer the impulse to the axon; the axon efficiently carries an impulse away from a cell body.

22. Touching a hot object with a finger will cause a sensory neuron in the finger to transmit impulses to an interneuron in the central nervous system. The impulse then moves through the interneuron to a motor neuron that causes a muscle or muscles to pull the finger away from the hot object.

Thinking Critically

23. The brain stem coordinates all information destined for the brain. The brain stem also controls the involuntary muscles of the human body that perform essential life processes, such as heartbeat rate and breathing.

24. Dilation of blood vessels near the surface of the skin will allow blood to flow near the surface of the skin, causing a loss of heat and a decrease in body temperature, despite the sensation of warmth.

25. A long-time smoker is likely to be physically and psychologically addicted to smoking.

26. Your eyes and ears can detect stimuli that are far away from your sense organs. Many human activities, from driving a car to walking down a city street, rely on information received by your ears and eyes from a distance.

27. Large doses of an antibiotic may not be able to be metabolized by the human body and may pass unused from the body by the filtering action of the kidneys.

28. Nocturnal animals need to be able to see in dim light; they have little need to distinguish colors in their nighttime surroundings. Therefore, their eyes have an abundance of light-sensitive rods and an absence of color-sensitive cones.

29. When a cat falls, the semicircular canals in its ears quickly provide the cat's brain with information about gravity and body orientation. This information enables the cat to right itself and land on its feet.

Reviewing Themes

21. **Systems and Structures**
A close look at the nervous system helps you understand that it consists of structures that range in size from microscopic to large enough to be seen by the naked eye. You also learn that the structure of each part is adapted to its function. Give three examples of structures in the nervous system that demonstrate wide differences in scale. Also give three examples in which structure is particularly well suited to function. Explain how the parts are suited to the job they do.

22. **Systems and Structures**
Select two examples within the nervous system of how parts interact to perform a bodily function.

Thinking Critically

23. Neurons damaged by injury cannot be repaired or replaced. Why would serious damage to your brain stem almost certainly result in death?

24. Tiny blood vessels near the skin dilate (widen) when a person consumes a lot of alcoholic beverages. This causes heat to escape from the body through the skin. The escaping heat gives a warm sensation. Some people, therefore, believe that drinking alcohol will keep them warm on a cold day. What is wrong with this reasoning?

25. Why is it so difficult for a long-time smoker to quit?

26. Why are your eyes and ears referred to as the distance receptors? Why is it important to have distance receptors?

27. Why might taking a prescribed amount of antibiotic in half the amount of time as prescribed be less effective than taking the antibiotic in the prescribed amount of time?

28. Most animals that are more active at night are colorblind. Explain why.

29. Cats have the ability to land on their feet. Based on what you have learned in this chapter, explain why this is so.

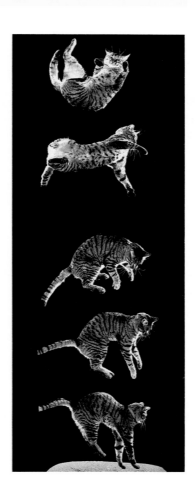

Discovery Through Reading

Parker, Steve. *The Brain and Nervous System*. Watts, 1990. Describes the control system of the body: its structure and function, sleep mechanism, reflexes and autonomic nervous system, the two brain hemispheres, and health care.

CHAPTER 20
REPRODUCTION AND DEVELOPMENT

PLANNING THE CHAPTER

Chapter Sections	Page	Chapter Features	Page	Program Resources	Source
Chapter Opener	542	For Your Journal	543		
Section 1: THE ENDOCRINE SYSTEM	544	Discover By Researching (A)	547	Investigation 20.1: Determining the Best Insulator for Your Body (H)	TR, LI
• Endocrine Glands (A)	544	Section 1 Review and Application	547	The Human Endocrine and Reproductive Systems (A)	IT
• Feedback Control (H)	546	Investigation: Predicting Changes in Heartbeat and Breathing Rates (A)	548	Record Sheets for Textbook Investigations (A)	TR
				Study and Review Guide, Section 1 (B)	TR, SRG
Section 2: THE REPRODUCTIVE SYSTEMS	549	Activity: What do sperm and ova look like? (B)	552	Science Discovery*	SD
• Male Reproductive System (B)	549	Discover By Calculating (A)	553	Reading Skills: Taking Study Notes (B)	TR
• Female Reproductive System (B)	551	Section 2 Review and Application	553	The Human Endocrine and Reproductive Systems (A)	IT
• The Menstrual Cycle (B)	552			Study and Review Guide, Section 2 (B)	TR, SRG
Section 3: PRENATAL DEVELOPMENT AND BIRTH	554	Discover By Researching (A)	555	Science Discovery*	SD
• Fertilization (B)	554	Section 3 Review and Application	559	Investigation 20.2: Embryo Development in Amphibians (H)	TR, LI
• The Developing Embryo (A)	555	Skill: Graphing Human Fetal Growth (B)	560	Connecting Other Disciplines: Science and Social Studies, Studying the Development of Human Babies (A)	TR
• Birth and Development (A)	558			Thinking Critically (H)	TR
				Extending Science Concepts Development of a Chicken Egg (H)	TR
				Study and Review Guide, Section 3 (B)	TR, SRG
Chapter 20 HIGHLIGHTS	561	The Big Idea	561	Study and Review Guide, Chapter 20 Review (B)	TR, SRG
Chapter 20 Review	562	For Your Journal	561	Chapter 20 Test	TR
		Connecting Ideas	561	Test Generator	
				Unit 6 Test	TR

B = Basic A = Average H = Honors
The coding Basic, Average, and Honors indicates subsections, features, and resources that might be appropriate for different levels of learners. For additional suggestions regarding choice of topic and depth of coverage, see the Pacing Chart on pages T26–T29.

*Frame numbers at point of use
(TR) Teaching Resources, Unit 6
(IT) Instructional Transparencies
(LI) Laboratory Investigations
(SD) *Science Discovery* Videodisc Correlations and Barcodes
(SRG) Study and Review Guide

▶ 541A

CHAPTER MATERIALS

Title	Page	Materials
Discover By Researching	547	(per class) reference books, journal
Investigation: Predicting Changes in Heartbeat and Breathing Rates	548	(per group of 2) watch with a second hand
Activity: What do sperm and ova look like?	552	(per group of 3 or 4) compound microscope, prepared slides of human sperm and ova
Discover By Calculating	553	(per individual) paper, pencil
Discover By Researching	555	(per class) reference books, paper, pencil
Teacher Demonstration	555	several chicken eggs, glass bowls
Skill: Graphing Human Fetal Growth	560	(per group of 2) graph paper, colored pencils

ADVANCE PREPARATION

For the *Activity* on page 552, you will need to obtain prepared slides of human sperm and ova. They can be purchased from a biological supply house. Obtain several chicken eggs and glass bowls for the *Demonstration* on page 555.

TEACHING SUGGESTIONS

Field Trip
Arrange a field trip to a gym or health club where exercise equipment is available. Ask a staff trainer to demonstrate and teach several exercises. Have the students monitor their heartbeat rate, temperature, and breathing rate as well as noting perspiration.

Outside Speaker
You may wish to invite a gynecologist or a urologist to speak to the class about the development of the male and female reproductive systems during puberty. However, check with your supervisor before scheduling this presentation. Some school districts require that such presentations be given separately to male students and female students.

CHAPTER 20 REPRODUCTION AND DEVELOPMENT

CHAPTER THEME—SYSTEMS AND STRUCTURES

This chapter introduces the students to the endocrine and reproductive systems of the human body and the processes of prenatal development and birth. In their studies, the students will explore how the endocrine system helps control the body's internal activities through the release of chemicals. They also examine the biology of human reproduction and development. This theme is also developed through concepts in Chapters 4, 5, 7, 8, 9, 10, 12, 13, 16, 17, 18, and 19.

MULTICULTURAL CONNECTION

Twins play an important role in the mythology and legends of several cultures. For example, in the migration mythology of some Southwest Native Americans, the gods known as the Twins, or the Little War Gods, led the people south. The Twins taught the people about their culture and held the earth's surface steady. Greek mythology includes stories of the twins Castor and Pollux. These twins helped shipwrecked sailors and were known for their athletic ability. When one was killed, the other refused immortality. Zeus allowed them to live together by spending time alternately in the heavens and in the world of the dead. Later, Zeus changed them into the constellation Gemini. In Roman mythology, the twins Romulus and Remus were abandoned at birth by order of their uncle, nurtured by a wolf, raised by a farming family, and later defeated their uncle and founded Rome.

MEETING SPECIAL NEEDS

Second Language Support

Recommend that the students prepare index cards for scientific terms used in the chapter. The cards should include the scientific term, its English pronunciation, the term in the students' native language, and a definition. The students can refer to the cards throughout their study of the chapter.

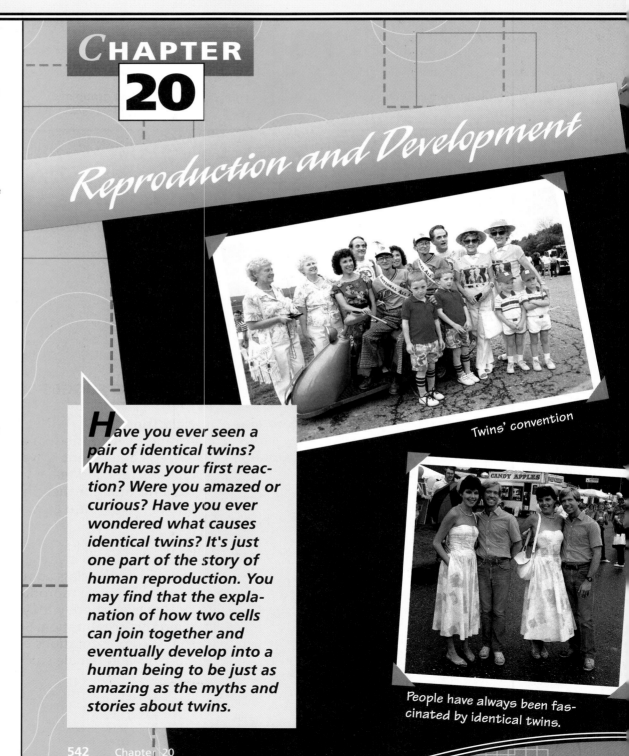

Chapter 20
Reproduction and Development

*H*ave you ever seen a pair of identical twins? What was your first reaction? Were you amazed or curious? Have you ever wondered what causes identical twins? It's just one part of the story of human reproduction. You may find that the explanation of how two cells can join together and eventually develop into a human being to be just as amazing as the myths and stories about twins.

Twins' convention

People have always been fascinated by identical twins.

CHAPTER MOTIVATING ACTIVITY

Ask the students to share any experiences they may have had with the births of pets or farm animals. Ask each student to describe the events that led up to the birth, as well as the actual birth and the delivery of the afterbirth. This activity is a good icebreaker and sets a mature tone for the discussions in the chapter. However, when discussing human reproduction and development, use other organisms only for comparison, not as substitutes for humans.

For Your Journal

The journal questions provide the students with an opportunity to demonstrate their knowledge of the development and growth process of human beings and to separate myths from the realities of human reproduction. Their answers may also provide you with an opportunity to note any misconceptions that they may have about human reproduction and development. You can address these misconceptions at appropriate times during the study of the chapter material.

What is it like to have an identical twin?

Twins appear as characters in many myths, legends, and stories. From Shakespeare to Walt Disney, story plots have revolved around identical twins being mistaken for one another. Dozens of movies have focused on the relationship between twins, often characterizing one as "good" and the other as "evil."

In some cultures, twins are regarded as a sign of good fortune. Often twins appear in the legends of these cultures, sometimes as the sun and the moon, or as darkness and light.

Other cultures feared twins, believing that their unlooked-for arrival was a curse and even going so far as to kill one or both twins to protect the community. For some nomadic tribes or societies with limited resources, two babies were a burden for the mother, and one child was killed to ensure the survival of the other. Some cultures associated twins with animals' multiple births. Depending on the culture's attitude toward animals, this association could help or harm twins.

As science has revealed more about the process of human reproduction, superstitions and fears about twins have largely disappeared. In fact, identical twins have been important in scientific research. Because identical twins are born with identical genes, research into the effects of heredity versus environment has often involved pairs of identical twins. Also interesting to scientists are pairs of identical twins who have been raised apart. How similar will two people be if they have the same genetic material but different environmental influences? These studies have provided some intriguing data, proving that twins have not completely lost their power to amaze us!

ABOUT THE PHOTOGRAPH

Twin birth is the birth of two babies from the same pregnancy. Two common types of twins are identical twins and fraternal twins. Identical twins may appear to be mirror images of one another. They develop from a single fertilized egg that split early in its development. Identical twins have exactly the same genetic makeup. Fraternal twins develop from two eggs that were released from the ovaries about the same time and were fertilized by two different sperm. The genetic makeup of fraternal twins differs just as the genetic makeup of other siblings differs.

For Your Journal

- What myths have you heard about how babies are born? Why do such myths exist?
- What do you think causes identical twins?
- What do you think it would be like to have an identical twin?

Section 1:
THE ENDOCRINE SYSTEM

FOCUS

This section explores how endocrine, or ductless, glands control the activities of body systems. It also examines how endocrine glands, in concert with the nervous system, regulate the human body's internal environment.

MOTIVATING ACTIVITY

Read a passage from a book, such as *What Do You Do When Your Mouth Won't Open?* by Susan Beth Pfeffer, that describes physical reactions to strong emotions. Use the passage as a springboard to a discussion of the students' own responses to strong emotions, such as excitement, fear, anticipation, and nervousness. Have the students discuss whether they can control the reactions and why or why not. Point out that in this section, the students will learn about the endocrine system, which produces chemicals, or hormones.

PROCESS SKILLS
- Comparing • Inferring
- Applying

POSITIVE ATTITUDES
- Cooperativeness
- Curiosity

TERMS
- endocrine system • gland
- hormones

PRINT MEDIA
The Endocrine System by Marjorie Little (see p. 429b)

ELECTRONIC MEDIA
The Endocrine System, Britannica (see p. 429b)

BLACKLINE MASTERS
Study and Review Guide
Laboratory Investigation 20.1

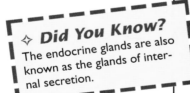

✧ **Did You Know?**
The endocrine glands are also known as the glands of internal secretion.

SCIENCE BACKGROUND
Generally, as cells and organs use hormones that have been secreted by the endocrine system, the hormones become partially or completely disabled. For this reason, hormones are not reusable substances and must be constantly secreted by organs of the endocrine system.

SECTION 1 — The Endocrine System

Objectives

Distinguish between endocrine glands and other glands in the body.

Identify the glands of the endocrine system and **state** their functions.

Explain what is meant by a feedback control system.

Your body systems do not act alone; they function together to ensure the health and smooth functioning of your body. For example, the skeletal and muscular systems are both responsible for your body's ability to move. The respiratory and excretory systems both work to remove wastes from your body. Two systems—the nervous system and the endocrine system—help regulate your body's functions. Recall that the nervous system works through electrical impulses. The **endocrine** (EHN duh krihn) **system** controls body functions with chemicals. Your growth rate, the amount of sugar in your blood, and the development of your reproductive organs are three functions controlled by your endocrine system. The nervous system responds to a stimulus in an instant. The endocrine system can take minutes, hours, or even months to respond. The nervous system and the endocrine system work together to control your body systems.

Endocrine Glands

A **gland** is a group of cells that makes chemicals for your body. Salivary glands produce saliva. Gastric glands that line the walls of the stomach produce digestive juices. Sweat glands produce perspiration. Chemicals leave these glands through small tubes, or ducts, and flow directly to where they are needed. The chemicals of endocrine glands, however, go directly into the bloodstream because endocrine glands do not have ducts.

Figure 20–1. The endocrine system works together with the other systems of the body to help you react quickly to fear or excitement.

544 CHAPTER 20

TEACHING STRATEGIES

● **Process Skills:** *Comparing, Evaluating*

Direct the students' attention to Figure 20–2 and have them identify the name and the location of the endocrine glands of the human body. Explain that endocrine glands secrete chemicals called *hormones* directly into the bloodstream. Have the students note that hormones help control the cells and organs of the human body. Contrast the secretions of endocrine glands with the secretions of other glands such as sweat glands, salivary glands, and gastric glands, in which secreted chemicals leave the glands through small tubes, or ducts, and flow directly to the part of the body where they are needed. Ask the students to describe the advantages of bloodstream transport versus duct transport. (Bloodstream transport allows a hormone to affect different parts of the body simultaneously.)

● **Process Skills:** *Comparing, Applying*

Explain to the students that the action of some hormones is faster than the action of others. For example, the human growth hormone is much slower than adrenalin, which will cause the heart to beat faster in only a few seconds.

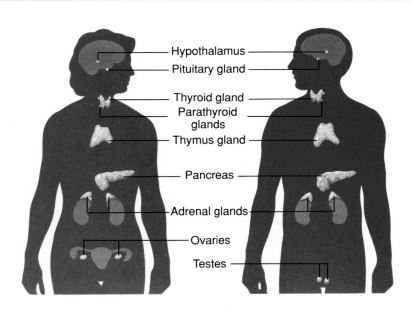

Figure 20–2. The endocrine glands monitor and control many of your body functions.

MEETING SPECIAL NEEDS

Gifted

Challenge the students to use reference materials to identify disorders of the glands of the endocrine system. Have the students write a short report identifying a disorder, its symptoms, and treatments for the disorder.

 Encourage the students who have written reports on glandular disorders to place their reports in their science portfolios.

SCIENCE BACKGROUND

The adrenal gland consists of two distinct portions, each with different functions. The inside of the gland (the medulla) secretes adrenalin, or epinephrine, and another hormone, noradrenalin, or norepinephrine. These hormones help the body adjust to stress. The outside of the gland (the adrenal cortex) secretes hormones that help regulate the balance of sodium and potassium and the use of digested food in the body. The adrenal cortex also produces small amounts of sex hormones.

ONGOING ASSESSMENT
ASK YOURSELF

Produced in the body, hormones are chemicals that regulate the activity of certain cells or organs of the body.

Chemical Messengers Endocrine glands produce chemicals called *hormones*. **Hormones** are chemical messengers that control other cells and organs. Hormones enter the bloodstream and travel throughout the body. This method of transport allows an endocrine gland to signal cells and organs some distance away. An endocrine gland in the head can control the actions of an organ located elsewhere in your body. By traveling through the bloodstream, a hormone can affect many organs at the same time.

Stressed Out The endocrine system includes the *adrenal* (uh DREE nuhl) *glands,* which are located on the top of each kidney. They prepare your body to deal with stress. The hormone *adrenalin,* which is released by the adrenal glands, speeds up your heartbeat and breathing rates. Adrenalin also opens your air passages and decreases your rate of digestion. Each of these actions prepares your body either to respond to danger or to run from it. This response is sometimes call the "fight or flight" response. You may have noticed these effects when you were angry or frightened.

There are several other endocrine glands. The following table lists the names and functions of the endocrine glands.

What are hormones?

SECTION 1   **545**

TEACHING STRATEGIES, continued

● **Process Skills:** *Comparing, Evaluating*

Have the students examine the information presented in Table 20–1 and ask the students to name the gland that causes a young man to begin growing a beard (testes); the glands that cause the greatest and most immediate overall body response (adrenals); and the gland that controls the release of hormones by other glands (pituitary).

GUIDED PRACTICE

Help the students compare and contrast the following terms: hormones and chemical messengers, endocrine system and feedback control system, and thyroid and parathyroid.

INDEPENDENT PRACTICE

Have the students write answers to the Section Review and Application questions. Ask the students to describe in their journals how their life might change if a particular gland did not function properly.

REINFORCING THEMES—
Systems and Structures

When discussing the specific structures and functions of the endocrine system, remind the students that the endocrine system works with all the body systems to regulate the activities of the human body.

MULTICULTURAL CONNECTION

Scientists from many different backgrounds have contributed significantly to our knowledge of hormones through their research efforts. Choh Hao Li studied the endocrine system, and his research of a pituitary hormone ACTH led to a major medical breakthrough in the treatment of arthritis. Ask interested students to find out more about Choh Hao Li or other endocrine researchers including Sir Frederick Banting, Sir Edward Sharpey-Schafer, and Ernest Starling.

INTEGRATION—*Health*

Cooperative Learning Encourage small groups of students to explore the role of insulin in regulating the body's blood sugar level and in controlling diabetes.

Table 20-1   The Endocrine System

Gland	Hormone	Function
Thyroid	Thyroxine	Regulates the release of energy in body; iodine is needed to make thyroxine
Parathyroid	Parathyroid hormone	Controls the use of calcium
Pituitary	Growth hormone and other pituitary hormones	Regulates bone growth; controls the release of hormones from other glands; controls kidney function; regulates blood pressure
Adrenals	Adrenalin	Increases heartbeat rate, blood flow, and the amount of sugar in the blood; activates the nervous system; regulates water and mineral balance in body tissues
Pancreas	Insulin	Allows liver to store sugar; regulates use of sugar
Ovaries (female)	Estrogen (female sex hormone)	Controls the development of reproductive organs and female characteristics
Testes (male)	Testosterone (male sex hormone)	Controls the development of reproductive organs and male characteristics

Feedback Control

The *pituitary gland,* located just below the brain, is often called the body's "master gland." This gland is about the size of a kidney bean, yet it is very important. Hormones given off by the pituitary gland control the actions of the other endocrine glands. The part of the brain called the *hypothalamus* (hy puh THAL uh muhs) controls the pituitary gland. This area of the brain is, in turn, affected by hormones produced by other glands in the endocrine system. For the endocrine glands to work properly, they should produce hormones only when needed. A process called a *feedback control system* controls endocrine glands. Let's look at an example of a feedback system.

Hot Enough for You?   Think about how the temperature inside your home is regulated. Often a home has a furnace with a thermostat. The thermostat can be set at a certain temperature so that it will automatically turn the furnace on and off. A low temperature causes the thermostat to send a message that turns the furnace on. Heat from the furnace then warms the house. Once the house is warm, the thermostat stops sending its message. The furnace turns off and the cycle begins again. The temperature of the house regulates the furnace in this feedback system.

EVALUATION

Ask the students to identify one function of each gland in the endocrine system.

RETEACHING

Use a human anatomy text to supply the students with a picture and the typical dimensions of each gland in the endocrine system. Then ask the students to select a gland and use modeling clay and the provided information to create an accurate model of their chosen endocrine gland.

EXTENSION

Store-bought salt typically contains iodine. Challenge the students to discover the relationship between lack of iodine and thyroid function.

CLOSURE

Cooperative Learning Ask small groups of students to create a colorful diagram of a feedback control system other than a thermostat and furnace. Ask the students to relate the feedback system they drew to the feedback system of the thyroid gland.

A Personal Furnace The functioning of the thyroid gland shows how a feedback control system works in your body. When you go outside on a cold day, your cells must produce more heat to keep you warm. A hormone called *thyroxine* (thy RAHK sihn) controls the amount of heat produced by your body's cells.

Large quantities of thyroxine cause the cells to work harder and produce more heat. The amount of thyroxine decreases as the cells work harder. The hypothalamus senses this decrease in thyroxine and sends a message to the pituitary gland. The pituitary, in turn, releases a hormone that increases the activity of the thyroid gland. As a result, the thyroid increases production of thyroxine. Increased thyroxine causes the body's cells to produce more heat.

When the hypothalamus senses the higher level of thyroxine, it stops sending its message to the pituitary gland. If your body temperature falls again, the process begins again. In this feedback control system, the level of thyroxine turns the system on and off. Use the following activity to learn more about the thyroid gland.

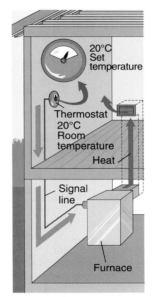

Figure 20–3. A thermostat is an important part of the feedback system that controls the furnace in most homes.

DISCOVER BY *Researching*

Use reference books to find out more about the function of the thyroid gland. How is the body affected by too much thyroxine? By too little thyroxine? How can these conditions be treated? Write a short summary of your research in your journal. Illustrate your report with diagrams.

ASK YOURSELF

Diagram a feedback control system.

SECTION 1 REVIEW AND APPLICATION

Reading Critically

1. How are endocrine glands different from glands in the digestive system?
2. Why is the pituitary gland called the master gland?

Thinking Critically

3. Why would endocrine glands be unable to function as they do if they had ducts?
4. Give another example of a feedback control system.

 DISCOVER BY *Researching*

A thyroxine deficiency can retard the development of infants and cause a loss of mental and physical vigor, hair loss, and weight gain in adults. Thyroxine can be given orally to correct a deficiency. Too much thyroxine causes a high metabolic rate, weight loss, nervousness, and protruding eyeballs and is corrected by the surgical removal of a portion of the thyroid gland or other treatments.

ONGOING ASSESSMENT
ASK YOURSELF

The nature of each change should be labeled wherever a change occurs in each diagram.

SECTION 1 REVIEW AND APPLICATION

Reading Critically

1. Endocrine glands are ductless; digestive glands have ducts.

2. The pituitary gland controls the release of hormones by the other endocrine glands.

Thinking Critically

3. Hormones produced by the endocrine glands could not reach distant target cells and organs as efficiently or effectively if they moved through ducts.

4. The students might describe a light-sensitive device that turns a light on or off or a motion-sensitive device that opens and closes a door.

INVESTIGATION

Predicting Changes in Heartbeat and Breathing Rates

Process Skills: Measuring, Comparing, Predicting

Grouping: Groups of 2

Objectives
- **Measure** the effect of aerobic exercise on heartbeat and breathing rates.
- **Interpret** data from an investigation.
- **Compare** changes caused by aerobic exercise with adrenal changes.

Pre-Lab
Ask the students to predict how exercise might affect the heartbeat and breathing rates of a person.

Hints
Students who cannot run in place or skip rope can still participate in the investigation by doing other forms of exercise. **CAUTION: Ensure that none of the students participating in this activity have known medical histories that would prohibit them from running in place or skipping rope.**

Analyses and Conclusions
1. Heartbeat and breathing rates will increase with aerobic exercise.
2. The rate at which the students perspire is likely to increase as aerobic exercise is performed. The students may also note tired or aching muscles.

Application
Changes caused by aerobic exercise occur gradually; changes caused by adrenalin occur very quickly.

✳ Discover More
Changes caused by stress might include increased heartbeat and breathing rates, disrupted thinking processes, or increased awareness of surroundings.

Post-Lab
Have the students recall their predictions about the effect of exercises on breathing and heartbeat rates. Ask them to describe how the investigation supports or does not support their predictions.

INVESTIGATION

Predicting Changes in Heartbeat and Breathing Rates

▼ MATERIALS
- watch with second hand

▼ PROCEDURE

1. To find your heartbeat rate, press your fingers to your neck just below your ear and jaw. Using a watch with a second hand, count the number of beats per minute. You can also count for 10 seconds and then multiply by 6. Repeat the count several times and then find and record your average heartbeat rate while at rest.
2. Now find and record your breathing rate. Count the number of breaths you take in a minute. Again, you can count for 10 seconds and then multiply by 6. Repeat the count several times to find and record your average breathing rate while at rest.
3. **CAUTION: Notify your teacher of any medical condition that prevents you from doing this activity.** Perform five minutes of aerobic exercise, such as running in place or skipping rope.
4. Now find your heartbeat and breathing rates again as you did in steps 1 and 2. Record your heartbeat and breathing rates after exercise.

TABLE 1: HEARTBEAT AND BREATHING RATES

	At Rest	After Exercise
Heartbeat Rate		
Breathing Rate		

▶ ANALYSES AND CONCLUSIONS
1. How did the aerobic exercise affect your heartbeat and breathing rates?
2. What other changes, if any, occurred?

▶ APPLICATION
Compare the changes caused by the aerobic exercise with the changes caused by adrenalin when you are excited or frightened.

✳ Discover More
Working in a group, list as many stressful situations as you can. Examples might be giving a speech in class, preparing for an important game, or watching a scary movie. Describe the physical changes that occur in the body before, during, and after stress-causing events.

Section 2:
THE REPRODUCTIVE SYSTEMS

FOCUS
This section presents the anatomy of the male and female reproductive systems and compares the structures and functions of these systems. The relationship of hormones to the reproductive systems is explained, and the details of the menstrual cycle are described and related to reproduction.

MOTIVATING ACTIVITY
Focus the students' attention on the importance of the reproductive systems to the survival of a species rather than to the survival of an individual member of the species. Ask a series of questions similar to the following: Could an organism survive if its reproductive system were destroyed? (probably) If organisms can survive without reproductive systems, why are reproductive systems important? (Reproductive systems provide the means for the continuation of a species.)

PROCESS SKILLS
- Comparing • Inferring
- Applying

POSITIVE ATTITUDES
- Openness to new ideas
- Precision

TERMS
- puberty • menstruation

PRINT MEDIA
Menstruation by Alan E. Nourse, M.D.
(see p. 429b)

ELECTRONIC MEDIA
Reproductive System, Coronet
(see p. 429b)

Science Discovery
Reproductive system; male
Reproductive system; female

BLACKLINE MASTERS
Study and Review Guide

The Reproductive Systems
SECTION 2

A single sex cell from each parent fuses together to form a single cell. This single cell divides again and again. Although the cells start the same, they gradually become different. First, tissues form, then organs, then systems. About 266 days after the formation of the first cell, a new individual made of trillions of cells emerges. A new human being enters the world. How can such an event take place? Although all the body's systems help produce the new individual, the endocrine and reproductive systems are among the most important.

Objectives
Compare and contrast the male and female reproductive systems.

Describe the changes that occur during puberty.

Explain the process of menstruation.

Male Reproductive System

The male reproductive system has two functions in the process of reproduction. It produces sex cells, or *sperm,* and it delivers the sperm to the female. Both the nervous system and endocrine glands control the male reproductive system.

The *testes* (TEHS teez) are endocrine glands that produce hormones and sperm cells. The testes are located outside the body cavity in a sac called the *scrotum* (SKROHT uhm). Sperm are sensitive to high temperatures and survive better at the slightly cooler temperatures found outside the body cavity.

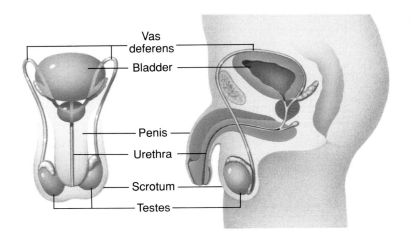

Figure 20-4. Hormones produced by the endocrine glands affect or control the functions of the organs in the male reproductive system.

SCIENCE BACKGROUND
The nuclei of sex cells contain only half the number of chromosomes as the nuclei of body cells. A specialized cell division, meiosis, reduces the number of chromosomes in the nucleus by half. Meiosis is involved in the formation of sex cells, and each sex cell contains only half of the genetic information.

SECTION 2   **549**

TEACHING STRATEGIES

● **Process Skills:** *Inferring, Synthesizing*

Have the students observe the structures of the sperm cell shown in Figure 20–6. Ask the students to explain how the large number of mitochondria relates to the presence of a large amount of sugar in semen. (Responses should identify two relationships. First, energy is released from sugar in the mitochondria. Second, the large number of mitochondria indicates that sperm require a significant amount of energy. Therefore, the sugar present in the semen provides the sperm with a source of energy.)

● **Process Skills:** *Comparing, Applying*

Remind the students that the process of puberty causes changes in various structures of the human body and causes the physical appearance of the body to change. Have the students observe the young men shown in Figure 20–5 and describe how the effects of the hormone testosterone are apparent in the young man on the right when compared to the young man on the left. (Testosterone has caused the young man on the right to develop larger and stronger muscles.) Explain to the students that these and other characteristics, such as

---

LASER DISC

4192

Reproductive system; male

MEETING SPECIAL NEEDS

Second Language Support

Explain to the students that the word *sperm* is derived from a word meaning "seed." Help the students compare and contrast "sperm" and "seed."

REINFORCING THEMES— *Systems and Structures*

Reinforce the understanding that the male reproductive system functions to produce sperm and to deliver those cells to the female reproductive tract. The female reproductive system functions to produce ova, receive sperm, and nourish and protect a developing embryo.

ONGOING ASSESSMENT ASK YOURSELF

The testes produce the hormone testosterone and manufacture sperm.

Figure 20–5. Muscle development is one change that occurs during puberty in males. The changes that occur during puberty are called *secondary sex characteristics*.

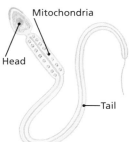

Figure 20–6. The male reproductive system produces sex cells called *sperm*.

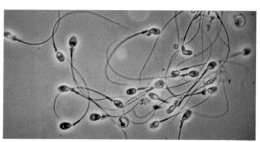

Hormones produced by the pituitary gland and the testes control the male reproductive system. The pituitary hormones control the production of sperm in the testes. A hormone called *testosterone* (tehs TAHS tuh rohn) is produced by the testes. Testosterone causes the changes that occur in the male body during **puberty** (PYOO buhr tee), the stage of life during which a person matures sexually. In males, these changes include the growth of a beard and body hair, as well as the deepening of the voice. During this period, a male's shoulders broaden in proportion to the hips, and the muscles increase in size and strength.

As you can see in the illustration, a sperm cell consists of three regions: the head, middle piece, and tail. The head of the sperm contains DNA, the hereditary material. The middle piece contains many mitochondria, which release the energy needed for movement. The tail section propels the cell forward.

Within the testes are tiny, tightly coiled tubes that produce the sperm. Together, the tubes of the two testes would stretch almost 7 m in length, if they were uncoiled. These sperm-producing tubes join to other sperm ducts, or *vas deferens* (vas DEHF uh rehnz), which lead back into the body cavity. Cilia, which line the inner surfaces of the sperm ducts, sweep the sperm forward through the ducts to the urethra.

As the sperm move through the vas deferens, they mix with a thick fluid produced by glands in the reproductive system. This fluid contains large amounts of sugar and other chemicals. The sugar provides nourishment to the sperm. The other chemicals moisten the sperm ducts and help the sperm move. Together, the sperm and the thick fluid form *semen* (SEE muhn). The urethra carries both semen and urine out of the body through the *penis* but never at the same time. Typically, about 250 million sperm cells are released from the body during each emission of sperm, which is called an *ejaculation*.

 ASK YOURSELF

What are two functions of the testes?

The female reproductive system produces hormones and female sex cells, called *ova* [singular, *ovum*], or eggs. It also receives the sperm and provides a place for the development of the offspring. Ova develop near the surface of two *ovaries,* which are located in the abdomen. The ovum is the largest cell in the human body and can be seen without a microscope. Generally, one ovary releases an ovum every 28 days. Once released from the ovary, the ovum enters the *Fallopian* (fah LOH pee uhn) *tube,* which leads from the ovary to the *uterus* (YOO tuh ruhs). Since there are two ovaries, there are also two Fallopian tubes.

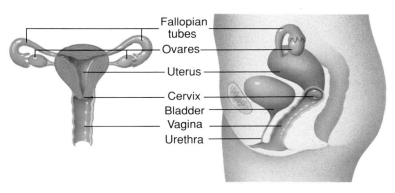

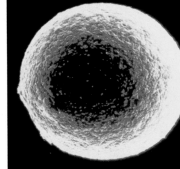

Figure 20–7. The female reproductive system produces ova and provides a place for the offspring to develop.

Thick walls of smooth muscle form the pear-shaped uterus. The lower end of the uterus narrows into the *cervix.* The cervix acts as a muscular gate that guards the uterus. The lower end of the cervix opens into the *vagina* (vuh JY nuh).

Hormones play an important role in controlling the female reproductive system. Hormones produced by the pituitary gland, the ovaries, and the uterus control the formation and release of ova. Hormones are also involved in the changes that take place in the female during puberty. These changes include the development of breasts, an increase in body hair, and the widening of the hips.

Sex cells, either male or female, are called *gametes.* In the following activity you can compare male and female gametes.

Figure 20–8. Secondary sex characteristics of females include the widening of the hips and the development of breasts.

ACTIVITY

What do sperm and ova look like?

Process Skills: Observing, Comparing, Inferring

Grouping: Groups of 3 or 4

Hints
Prepared slides are available from science supply houses.

▶ **Application**

1. Sperm cells and egg cells both contain nuclei and hereditary material. Sperm cells contain very little cytoplasm, are much smaller than egg cells, and have a "tail" for locomotion; egg cells have a significant amount of cytoplasm, are much larger than sperm cells, and do not have a means of locomotion.

2. The ovum must carry large amounts of nutrient material to support a developing zygote. This material must last until the fertilized egg implants itself in the uterine wall.

3. Sperm must swim up the Fallopian tube, where fertilization usually takes place. The ovum must possess sufficient cytoplasm to provide energy for the developing zygote.

★ **PERFORMANCE ASSESSMENT**

Ask the students how an egg is able to move through a Fallopian tube. Evaluate the students' answers. Explain that an egg is moved by waving cilia, muscle contractions, and fluid secretions.

ONGOING ASSESSMENT
▼ **ASK YOURSELF**

The ovaries store eggs and release hormones that control the formation and release of ova and the development of female secondary sex characteristics.

▶ **552** CHAPTER 20

TEACHING STRATEGIES, continued

● **Process Skills:** *Inferring, Applying*

Lead the students to understand that hormones in the female body coordinate the timing of ovulation with the thickening of the uterine lining. Ask the students to describe what would happen if the timing of ovulation were not coordinated with the thickening of the uterine lining. (If a fertilized ovum reached the uterus and could not attach to the thickened lining of the uterus, the fertilized ovum could not develop.)

INDEPENDENT PRACTICE

 Have the students write answers to the Section Review and Application questions in their journals. Then ask the students to speculate about whether other animals experience puberty and explain their reasoning in their journals.

ACTIVITY

What do sperm and ova look like?

MATERIALS
compound microscope, prepared slides of human sperm and ova

PROCEDURE

1. Place the slide of the sperm on the microscope stage and focus on low power. Draw what you see.

2. Switch to high power and examine the sperm more closely. Draw what you see.

3. Examine the slide of the ova as you did the slide of the sperm.

APPLICATION

1. Describe the structure of the sperm and the ova. How are they alike? How are they different?

2. Why do you think an ovum is so much larger than a sperm?

3. How does the design of these cells help them carry out their functions?

 ASK YOURSELF

What are the functions of the ovaries?

The Menstrual Cycle

After puberty, several events prepare the body of a female for the possible development of a baby. These events occur during a process called the *menstrual* (MEHN struhl) *cycle,* which begins about every 28 days.

Three events take place within each menstrual cycle. The first two events occur at the same time. First, during the first 14 days of each cycle, an ovum matures within an ovary. Second, the inner lining of the uterus thickens. This lining is made of a spongy tissue that has a rich supply of blood.

At about day 14 of the cycle, the ovum bursts free from the surface of the ovary. The release of a mature ovum is called *ovulation.* The ovum then enters the funnel-shaped opening of the Fallopian tube. As the mature ovum travels down the Fallopian tube, the lining of the uterus continues to thicken. If fertilization occurs, it will take place in the Fallopian tube. However, if fertilization does not take place, the third event of the menstrual cycle occurs. If a sperm does not fertilize the ovum, the thickened lining of the uterus begins to break down. The unfertilized ovum and the tissues of the lining of the uterus leave the body through the vagina. This process is called **menstruation** (mehn STRAY shuhn). Menstruation usually lasts from three to five days. As menstruation begins, hormones cause another ovum to begin maturing in the ovary. The cycle begins again.

GUIDED PRACTICE

Have the students identify and locate the structures of the male and the female reproductive systems.

EVALUATION

Ask the students to highlight the important functions of the male and female reproductive structures in a list.

RETEACHING

 Cooperative Learning Have the students work in groups to outline or diagram the stages of the menstrual cycle.

CLOSURE

 Cooperative Learning Ask each small group of students to create, then exchange and answer, a brief true-false or multiple-choice quiz highlighting important structures and functions of human reproductive systems.

DISCOVER BY *Calculating*

Assuming that a woman ovulates once every 28 days, ovulation would occur 13 times during each year (365/28); 500 ovulations in a lifetime divided by 13 ovulations each year yields 38.5 productive years. If ovulation begins at age 13, a woman will be approximately 51 years old when her child-bearing years are over.

ONGOING ASSESSMENT
ASK YOURSELF

The menstrual cycle typically extends for 28 days and consists of ovulation and preparation of the uterus for the growth and development of a fertilized egg. In the event an egg does not implant in the uterine wall, the lining of the uterus and the ovum are discharged from the female body, and the cycle begins again.

SECTION 2 REVIEW AND APPLICATION

Reading Critically

1. Sperm are heat-sensitive and would be damaged by body heat if the testes were located within the body cavity.

2. The female reproductive system is designed to produce ova, receive sperm, and nourish and protect a developing embryo; the male reproductive system is designed to produce sperm cells and deliver them to the female reproductive tract.

3. Hormones control the menstrual cycle.

Thinking Critically

4. The onset of puberty is controlled by hormones produced by the pituitary and other glands. Many factors influence the activities of these glands and cause variations in the time that puberty begins.

5. The primary function of the reproductive system is the production of sex cells; the structures and the physical characteristics that are not relevant to the production of sex cells are secondary.

DISCOVER BY *Calculating*

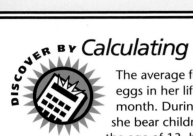

The average female produces about 500 mature eggs in her lifetime, at the rate of about one per month. During how many years of her life can she bear children? If a female begins to ovulate at the age of 13, how old will she be when her possible child-bearing years are over?

The average length of the menstrual cycle is 28 days. However, when menstruation first begins after puberty, it might not be regular. Even after the cycle becomes regular, it varies for each person. Cycles can be as short as 21 days or as long as 35 days.

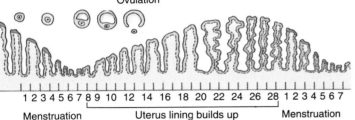

Figure 20-9. The events of the average menstrual cycle occur over a period of 28 days.

If a sperm fertilizes the ovum before it reaches the uterus, menstruation does not occur. Instead, the fertilized egg attaches to the lining of the uterus, where it begins to grow. This attachment of the fertilized ovum is the beginning of pregnancy. Hormones produced by the uterus tell the reproductive system of this change. During pregnancy, the ovaries do not release new ova, and menstruation does not occur.

ASK YOURSELF
Describe the menstrual cycle.

SECTION 2 REVIEW AND APPLICATION

Reading Critically
1. Why are the testes found outside the body cavity?
2. How are the functions of the male and female reproductive systems different?
3. What controls the menstrual cycle?

Thinking Critically
4. Why do you think the time of puberty is different for different people?
5. The changes that occur during puberty are called *secondary sex characteristics.* Explain why this is a good name.

Section 3:
PRENATAL DEVELOPMENT AND BIRTH

FOCUS
This section explains the process of fertilization and focuses on the development of the resulting zygote. Implantation of the embryo in the uterine wall is described. The structure and function of the placenta and the three stages of the birthing process are explained. Prenatal development and postnatal growth are also discussed.

MOTIVATING ACTIVITY
Display a number of photographs of infants. If you wish, display your own baby picture and those of your students as well. Ask the students to compare and contrast the babies in the pictures. Ask them to consider the babies' physical needs and how they are met as well as physical characteristics. Encourage discussion about what a baby will learn in his or her first year of life. (hold up the head, roll over, sit up, crawl) Point out that before and after birth, humans go through different stages of development.

PROCESS SKILLS
- Classifying/Ordering
- Comparing • Applying

POSITIVE ATTITUDES
- Curiosity • Enthusiasm for science and scientific endeavor

TERMS
- fertilization • zygote
- embryo • fetus

PRINT MEDIA
Growth and Development: The Span of Life by Torstar Books (see p. 429b)

ELECTRONIC MEDIA
Human Reproduction, Britannica (see p. 429b)

Science Discovery Embryo, human; 40 days Fetus, human; with dyed skeleton

BLACKLINE MASTERS
Study and Review Guide
Laboratory Investigation 20.2
Connecting Other Disciplines
Thinking Critically
Extending Science Concepts

BACKGROUND INFORMATION
Conjoined twins are physically connected and often share some organs. At one time, conjoined twins were commonly called *Siamese twins*. The name *Siamese twins* was first used in the nineteenth century to refer to Chang and Eng, conjoined twins from Siam, now Thailand.

▶ 554   CHAPTER 20

SECTION 3

Prenatal Development and Birth

Objectives

Explain how fertilization takes place.

Describe human development from conception to old age.

Summarize the stages of the birth process.

Lennart Nilsson/Boehringer Ingelheim International Gmbh

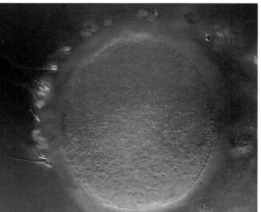

Figure 20–10. Although many sperm surround the ovum, the ovum can be fertilized by only one sperm.

Day 14 of the menstrual cycle: An ovum, which has been maturing in an ovary for the last 14 days, breaks away from the surface of the ovary and begins to travel down the Fallopian tube. What happens to the ovum now may be the first in a series of events that will result in the development and birth of a new human being.

Fertilization

The male releases millions of sperm cells from his penis into the female's vagina. The sperm swim through the cervix, into the uterus, and up the Fallopian tube in search of an ovum. Contractions of the uterine muscles help move the sperm along. Tiny cilia, which line the Fallopian tube, sweep the ovum toward the uterus. When a sperm cell and an ovum join, **fertilization** occurs. The new cell that results from fertilization is called a **zygote** (ZY goht).

One at a Time Of the millions of sperm that begin the journey, only a few hundred ever reach the ovum. Of those, only one sperm can fertilize the ovum. Immediately after fertilization, a membrane forms around the zygote. This membrane keeps other sperm cells from entering the zygote.

Or Maybe Two! Occasionally, a woman gives birth to twins. There are two kinds of twins: identical and fraternal.
Identical twins come from a single fertilized ovum. As in the development of a single baby, the zygote goes through a series of cell divisions to form a ball of cells. During the formation of identical twins, two different sections of this ball begin to act as if each were a separate entity. Each section continues to develop as a normal individual. These two sections eventually separate into two individuals. Since all the cells came from the same sperm and ovum, the two infants have identical genes. Identical twins are always the same sex.

Fraternal twins are formed differently. Occasionally, two eggs are released during ovulation. Each egg is then fertilized by a different sperm cell. The zygotes continue to develop separately. Except for their age, fraternal twins are no more similar than any other two children in a family.

▼ **ASK YOURSELF**

How and where does fertilization occur?

Figure 20–11. Twins may be either identical (left) or fraternal (right).

The Developing Embryo

The process of development that takes place between fertilization and birth lasts about 266 days. After fertilization, the zygote passes down the Fallopian tube and into the uterus. During this five- to seven-day trip, cell division occurs several times. By the time it reaches the uterus, the zygote has become a ball of tightly packed cells called an **embryo**. After one week of development, the embryo is about the size of the period at the end of this sentence. The new organism is called an embryo from the second to the eighth week. From the ninth week until birth, it is called a **fetus.**

 **DISCOVER BY Researching**

Using reference books, prepare a chart or table that describes human development between fertilization and birth. Illustrate your descriptions.

For development to continue, the embryo must attach itself to the wall of the uterus. Enzymes, given off by the embryo, break down a tiny spot in the thick wall of the uterus. The embryo attaches itself into this tiny spot. In a few days, the wall of the uterus covers the embryo.

Figure 20–12. Shown here are four stages of human embryonic development—fertilized ovum, two cells, four cells, and eight cells.

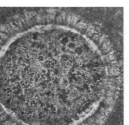 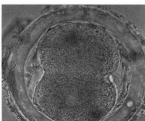  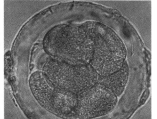

TEACHING STRATEGIES, continued

● **Process Skills:** *Inferring, Applying*

Remind the students of their observations of chicken eggs earlier in this section. Point out that a chicken completes its prenatal development in about three weeks, whereas a human being completes its prenatal development in about nine months. Ask the students to explain why fertilized chicken eggs have much more stored food than fertilized human eggs, which take longer to develop. (The developing chicken relies completely on the yolk and egg white for nutrients during the gestation period. The developing human being receives a constant supply of nutrients through the placenta.) Help the students realize that the placenta is vital to the development of the embryo and the fetus.

● **Process Skills:** *Comparing, Applying*

Explain to the students that by the end of the sixth month, a fetus is usually 30 to 36 cm long and weighs about 0.5 kg. Also point out that a full-term baby is approximately 50 cm long and weighs between 3.0 and 3.6 kg

---

 Medical science has developed alternatives for parents who try to have children but find they cannot. *In vitro* fertilization, or IVF, is one such alternative. In IVF, a mature egg is fertilized by sperm in a culture dish, and the developing zygote is later injected into the uterus. Newer procedures have been developed in the past few years. In one procedure, sperm and mature eggs are injected directly into a Fallopian tube, where an egg is fertilized and moves to the uterus for implantation. Another procedure involves injecting an already fertilized egg into the Fallopian tube and allowing it to move naturally through the tube to the uterus for implantation. The success rates of the two newer procedures have proven to be higher than the rate for IVF.

✦ **Did You Know?**
Three hundred years ago, some scientists believed that a tiny, completely formed human being, called an *ovicule*, was carried within an ovum. Other scientists believed that a tiny, completely formed human being, called a *homunculus*, was carried within a sperm cell.

 **LASER DISC**
4220
Embryo, human; 40 days

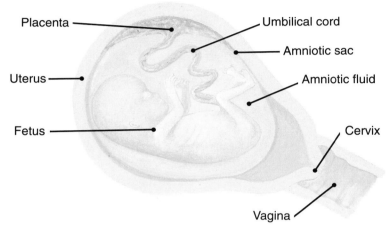

Figure 20–13. While in the uterus, the fetus is attached to the mother by the umbilical cord. However, the blood supplies of the fetus and the mother do not mix.

The Placenta Cells from the embryo combine with cells from the uterus to form the *placenta* (pluh SEHN tuh), which nourishes the developing embryo. An exchange of nutrients and wastes between the mother and the embryo takes place through the placenta. The developing circulatory system of the embryo forms a network of capillaries in the placenta. The blood-rich, spongy tissue of the uterus surrounds these capillaries, but the blood of the embryo and the blood of the mother do not mix.

The placenta nourishes the embryo throughout its development. Food and oxygen molecules pass through the placenta from the mother to the embryo by diffusion. Large blood vessels collect the blood and carry it to the embryo's tissues. These blood vessels are located in the *umbilical cord*, which connects the embryo with the placenta. Wastes move through another blood vessel in the umbilical cord from the embryo back to the placenta.

Organ Development Cell division occurs rapidly during embryonic development. At the same time, groups of cells gradually begin to differ from each other. Each cell group eventually develops into a different organ or body part.

The embryo's heart develops early. By the end of the fourth week, it begins to pump blood. Blood circulates through tiny arteries carrying nutrients and oxygen to the embryo's tissues. Carbon dioxide and other waste molecules diffuse from the body tissues into the blood. This blood moves through veins back to the placenta for removal of wastes.

The nervous system also begins to develop during this period. Since the brain requires more time to develop than other structures, at first the brain grows faster than the rest of the embryo. This uneven growth makes the head look large compared to the body. By the end of the second month, all the body's systems have begun development. The embryo is now called a fetus. By the end of the third month, the eyes, ears, and nose begin to form, and the arms and legs become visible as tiny bumps. The fetus is about 7 cm long with its head making up about half of its length.

During the next three months, the body catches up with the growth of the head. Development of the body's systems continues. During this time, the fetus begins its first movements in the uterus, and it responds to stimuli. The mother usually feels movement for the first time during the fifth month of pregnancy. By the end of the sixth month, the fetus is usually 30 to 36 cm long and weighs about as much as a head of lettuce (0.5 kg). The fetus can hear sounds and may even respond to music. The fetus floats in a sac of clear fluid. This fluid protects and cushions the fetus from sudden shocks.

By the seventh month, all the body's parts are formed. A baby born at this stage can survive with help. The fetus is active and has already begun sucking its thumb as practice for nursing after birth. The last two months are largely a period of growth. At birth, the newborn weighs about 36 000 times as much as an embryo at two months of development.

Figure 20–14. Early development is rapid. Shown here are an embryo at five weeks (left), a fetus at nine weeks (center), and a fetus at four months (right).

Care of Mother and Fetus During pregnancy, a woman must be especially careful about what she takes into her body. Some chemical substances can easily damage the developing organs of the fetus. Drugs can pass through the placenta and harm the developing fetus, possibly resulting in mental and physical defects. Even commonly used medicines can be harmful. Aspirin, for example, can cause bleeding from embryonic tissues and can lead to severe birth defects.

The use of alcohol and tobacco can also harm the developing fetus. Alcohol crosses the placenta and collects in the fetal brain and liver where it can severely damage these organs. The use of tobacco reduces the flow of blood to the placenta.

GUIDED PRACTICE

Have the students summarize the process of fertilization and the stages of embryonic development.

INDEPENDENT PRACTICE

Have the students provide written answers to the Section Review and Application questions in their journals. Also ask the students to write a paragraph summarizing the importance of prenatal care.

EVALUATION

Ask the students to provide a detailed description of the birth process.

RETEACHING

Cooperative Learning Organize the students into small groups. Ask them to make a set of index cards with important terms from this section written on one side and the definitions of the terms on the other. Have the students use the cards to study and quiz one another.

ONGOING ASSESSMENT
ASK YOURSELF

During the second to the eighth week of development, a new organism is called an *embryo*. From week nine until birth, the organism is called a *fetus*.

THE NATURE OF SCIENCE

The heredity work pioneered by Gregor Mendel continued as other scientists and researchers used it as the foundation for further study. Our current knowledge of human heredity is the result of a continuous scientific pursuit of information that began with Mendel's experiments. Today, scientists are identifying specific genes on human chromosomes. They hope identification will help them better understand and perhaps cure genetic diseases. Scientists hope to eventually locate and identify all the genes on the chromosomes within a cell. In other words, they want to map the entire human *genome*. So far, out of about 100 000 genes, about 2000 have been isolated and identified. Research begun by Mendel continues as scientists search for answers to questions about genetics.

▶ **558** CHAPTER 20

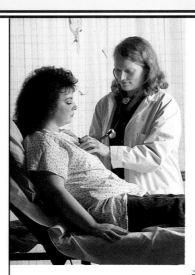

Figure 20–15. It is important for a woman to receive medical advice as soon as she knows she is pregnant.

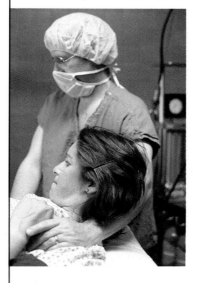

Figure 20–16. Labor is the first stage of the birth process.

Reduced blood flow means less oxygen and fewer nutrients reach the fetus. The use of either alcohol or tobacco can also result in premature birth.

Diet is also important during pregnancy. Both the mother and the fetus need a balanced diet, rich in nutrients such as calcium and iron. Calcium is needed for bone growth, and iron for blood cells. The fetus produces large amounts of protein during this period. Muscle and brain cells require large amounts of amino acids to make the proteins required for growth. If the mother's diet has been poor, nutrients are removed from her tissues to supply the fetus. This can weaken the mother and make her less able to carry the fetus. To protect her health and the health of the fetus, a pregnant woman should start prenatal care as soon as she suspects pregnancy.

ASK YOURSELF

What is the difference between an embryo and a fetus?

Birth and Development

Birth usually takes place between 256 and 276 days after fertilization. During birth, the baby is forced from the mother's body by strong contractions of the uterine muscles. Although birth is a continuous process, it is often described in three stages.

The Birth Process The first stage of birth, called *labor*, begins as the muscles of the uterus begin to contract and relax. This process causes the opening of the cervix to enlarge. At first, these contractions last for about 30 seconds and occur about 15 minutes apart. As labor continues, the contractions become stronger. This part of the birth process may require from 2 to 20 hours. Once the cervix is open enough, the contractions of the uterus can push the baby's head through the opening and into the vagina.

Delivery, the second stage of birth, may take from a few minutes to several hours. This stage ends when the contractions have pushed the baby completely out of the mother's body. The third stage of the birth process begins after delivery. The umbilical cord, which still connects the baby to the placenta, is tied and cut. Ten to fifteen minutes later, another series of contractions begins in the uterus. At this time, the *afterbirth,* which consists of the placenta and other membranes, is pushed out of the uterus. After the birth is complete, the smooth muscles of the uterus begin to contract, returning the uterus to its normal size.

EXTENSION

Ask the students to research information about postnatal development from newborn baby through adulthood. Have them use the information to construct a developmental time line.

CLOSURE

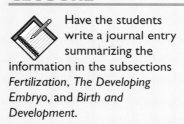

Have the students write a journal entry summarizing the information in the subsections *Fertilization*, *The Developing Embryo*, and *Birth and Development*.

Development After Birth The baby's development does not end at birth. In comparison with many other newborn animals, newborn humans are helpless. Much care is necessary in order for the baby to survive. During pregnancy, the mother's body prepares itself to feed the baby during the early months of life. The mammary glands are prepared by certain hormones to produce milk to feed the newborn child.

During the nine months of development in the uterus, a human being grows from a single cell weighing less than a gram to a individual weighing an average of 3.18 kg. At no time during the rest of his or her life will growth be as rapid. However, the changes that occur during the first two years of life are very dramatic. The baby soon develops into a toddler. No longer satisfied to just sleep, eat, and be held, the toddler explores his or her world.

The toddler quickly develops into a young child. As development continues, physical and mental abilities increase. Within a space of a few years, the child becomes a teenager and then an adult. In a period of 14 or 15 years, an individual develops from a baby into a physically mature person. Within just a few more years, the individual becomes emotionally and socially mature enough for the life cycle to begin again.

Figure 20–17. Humans continue to develop as they become older, from (left to right) infant, to toddler, to child, to adolescent. What stages come after adolescence? ①

ASK YOURSELF

Describe the three stages of birth.

SECTION 3 REVIEW AND APPLICATION

Reading Critically

1. Where does fertilization of the egg usually take place?

2. When does the heart of an embryo begin to function?

3. How does a pregnant woman know that labor has begun?

Thinking Critically

4. Why are such large numbers of sperm cells released, if only one sperm is needed for fertilization?

5. When astronauts walk in space, they are sometimes attached to the spacecraft by a line called an *umbilical*. Why do you think this line has been given this name?

① Adulthood and old age occur after adolescence.

ONGOING ASSESSMENT
ASK YOURSELF

During labor, the muscles of the uterus begin to contract and relax, causing the opening of the cervix to enlarge, or dilate. During delivery, birth occurs. During the delivery of the afterbirth, the placenta and other material is expelled from the uterus, and the uterus begins to return to its normal size.

SECTION 3 REVIEW AND APPLICATION

Reading Critically

1. Fertilization usually occurs in a Fallopian tube.

2. The heart of an embryo begins to function after the fourth week of development.

3. Labor begins when the muscles of the uterus begin to contract and relax.

Thinking Critically

4. Many sperm die as they move through the vagina, uterus, and Fallopian tubes of the female reproductive system.

5. The umbilical cord carries the oxygen needed to sustain life for both the embryo and the astronauts.

SKILL

Graphing Human Fetal Growth

Process Skills: Communicating, Comparing, Interpreting Data

Grouping: Groups of 2

Objectives
- **Communicate** using a line graph.
- **Interpret** data from a line graph.
- **Interpolate** data from a line graph.

Discussion
Remind the students that the growth that occurs between fertilization and birth constitutes the most rapid period of growth in the life of a human being. In a period of 40 weeks, a new human being grows from a single cell into an organism composed of trillions of cells.

Application
1. The rate at which fetal mass increases is not constant—it is slow initially, but as the number of cells increases, fetal mass begins to increase significantly. Early in the development of a fetus, there are few cells dividing to double the number of cells. Later in its development, there are many cells dividing to double the number of cells.

2. The rate at which fetal length increases is not constant. The length of a fetus is a reflection, to some degree, of its mass. As additional cells divide, the increase in length becomes progressively more significant.

Using What You Have Learned
A fetus has a mass of 2500 g near week 36 of its development.

PERFORMANCE ASSESSMENT
Have the students compare their graphs for similarities and differences. Evaluate the students' performances based on their ability to note that mass and length increase significantly in the final weeks of pregnancy.

SKILL — Graphing Human Fetal Growth

▶ MATERIALS
- graph paper
- colored pencils

▼ PROCEDURE

1. Make two graphs like those shown. Use intervals of 25 mm on the length graph. Extend the graph to 500 mm. Use intervals of 100 g on the mass graph. Extend this graph to 3300 g. Use intervals of 2 weeks for time on both graphs. Both graphs should extend to 40 weeks.

2. Study the table of data. Plot the data in the table onto your graphs. Use a colored pencil to draw the curved line that joins the points.

TABLE I: INCREASE OF MASS AND LENGTH OF AVERAGE HUMAN FETUS

Time (weeks)	Mass (g)	Length (mm)
2	0.1	1.5
3	0.3	2.3
4	0.5	5.0
5	0.6	10.0
6	0.8	15.0
8	1.0	30.0
13	15.0	90.0
17	115.0	140.0
21	300.0	250.0
26	950.0	320.0
30	1500.0	400.0
35	2300.0	450.0
40	3300.0	500.0

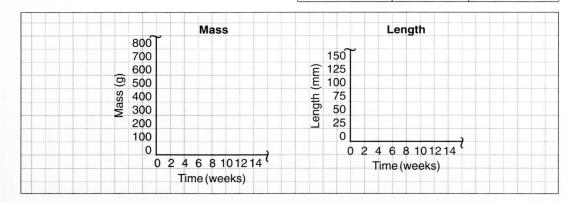

▶ APPLICATION
1. Describe the change in mass of a developing fetus. How can you explain this change?
2. Describe the change in length of a developing fetus. How can you explain this pattern?

✳ Using What You Have Learned
The average infant with a mass less than 2500 g at birth is 40 times more likely to die than an infant who has a greater mass. Look at your graph and determine in which week a fetus would have a mass of 2500 g.

CHAPTER 20 HIGHLIGHTS

The Big Idea—SYSTEMS AND STRUCTURES

Help the students understand that the systems of the human body interact and work together, allowing it to grow, develop, and reproduce. Remind the students that the endocrine system and the nervous system regulate the activities of the human body. Also remind them that the male reproductive system functions to produce sperm and to deliver those cells to the female reproductive system; the female reproductive system functions to produce ova, receive sperm, and nourish and protect a developing embryo.

CHAPTER 20 HIGHLIGHTS

The Big Idea

All organisms are born, grow, reproduce, and die. Because the process of reproduction ensures the continuation of all species, including humans, it is one of the most important life processes. All the systems of the human body work together during the production of a new individual, but the endocrine and reproductive systems perform the most important roles.

Hormones produced by glands in the endocrine system control the functions of male and female reproductive systems. Each reproductive system has its own separate but related functions. The structures of the male reproductive system produce and deliver sperm. The structures of the female reproductive system produce ova, receive sperm, and shelter the developing offspring from fertilization to birth.

Review the answers you wrote in your journal before you read the chapter. How accurate was your explanation of the process that causes identical twins? Summarize your understanding of the chapter by creating a diagram that shows how the endocrine and reproductive systems are related.

For Your Journal

The ideas expressed by the students should reflect the understanding that identical twins are formed from the fertilization of one egg cell by one sperm cell. The diagrams drawn by the students should reflect an understanding of the relationship between the endocrine and reproductive systems of the human body. The students should note that the release of hormones triggers the reproductive processes within the body. They should also note that some of the organs of the reproductive system are also part of the endocrine system.

CONNECTING IDEAS

The words that complete the diagram should be selected in this order: gamete, fertilization, zygote, embryo, fetus, newborn baby, puberty, adolescent, adult.

Connecting Ideas

Copy this diagram of the human life cycle into your journal. Complete the diagram by writing the terms from the box in the correct order.

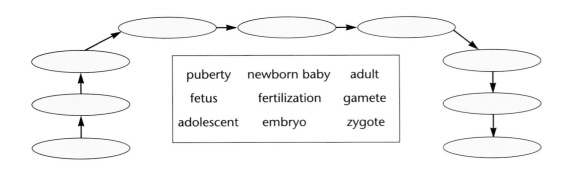

puberty   newborn baby   adult
fetus   fertilization   gamete
adolescent embryo   zygote

CHAPTER 20 REVIEW

ANSWERS

Understanding Vocabulary

1. Hormones secreted by glands of the endocrine system are the chemical messengers of the human body.

2. Hormones secreted by glands of the endocrine system initiate and regulate the process of puberty.

3. After puberty, menstruation prepares a woman's body for possible fertilization and pregnancy.

4. Fertilization occurs when male and female gametes, or sex cells, combine.

5. A fertilized ovum is known as an embryo during the first eight weeks of its development. From the ninth week until birth, it is known as a fetus.

6. The new cell formed from the fertilization of an ovum by a sperm cell is known as a zygote. A zygote, after it has undergone several cell divisions and implants in the uterine wall, is known as an embryo.

Understanding Concepts

Multiple Choice

7. b
8. c
9. a
10. b
11. d
12. d

Short Answer

13. A fetus obtains nourishment and eliminates wastes through the placenta.

14. Effects of adrenalin include increased heartbeat rate, increased blood flow, increased level of blood sugar, dilation of air passages, activation of the nervous system, and decreased rate of digestion.

15. The menstrual cycle typically extends for 28 days and consists of ovulation and preparation of the uterus for the growth and development of a fertilized egg. In the event an egg is not fertilized and does not implant in the uterine wall, the lining of the uterus and the ovum are discharged from the female body, and the cycle begins again.

Interpreting Graphics

16. Identical twins are developing in the illustration on the left, and fraternal twins are developing in the illustration on the right.

CHAPTER 20 REVIEW

Understanding Vocabulary

Explain the relationship between each pair of terms.

1. hormones (545), endocrine system (544)
2. hormones (545), puberty (550)
3. puberty (550), menstruation (552)
4. fertilization (554), gametes (551)
5. embryo (555), fetus (555)
6. zygote (554), embryo (555)

Understanding Concepts

MULTIPLE CHOICE

7. The endocrine system is an example of a feedback control system because
   a) hormones are needed for the digestive system.
   b) hormones are produced only when needed.
   c) hormones help keep emotions under control.
   d) hormones are released during puberty.

8. Hormones travel through the human bloodstream so that they
   a) create the changes that occur during puberty.
   b) enable the reproductive system to produce gametes.
   c) can affect many organs at the same time.
   d) can easily be eliminated as wastes when they are no longer needed.

9. One ovum is released in the female reproductive system during
   a) ovulation. b) menstruation.
   c) labor.   d) puberty.

10. Before the zygote attaches itself to the wall of the uterus,
    a) menstruation must occur.
    b) it goes through several cell divisions.
    c) the placenta develops.
    d) it develops into a fetus.

11. Functions controlled by the endocrine glands include
    a) the amount of sugar in the blood.
    b) growth rate.
    c) development of reproductive organs.
    d) all of these.

12. The function of the placenta is to
    a) prepare the uterus to receive a fertilized ovum.
    b) release the ova.
    c) prevent more than one sperm from fertilizing the ovum.
    d) nourish the developing embryo.

SHORT ANSWER

13. How does a fetus get nourishment and remove wastes?

14. Describe the effects of adrenalin on the body.

15. Describe the menstrual cycle.

Interpreting Graphics

16. Explain what is happening in each of these illustrations.

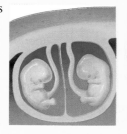 

562 CHAPTER 20

Reviewing Themes

17. The endocrine system and the nervous system regulate functions of the human body. The nervous system works through electrochemical impulses; the endocrine system works through the secretion of hormones.

18. The structures of the male reproductive system include the testes, vas deferens, urethra, and penis. The structures of the female reproductive system include the ovaries, Fallopian tubes, uterus, cervix, and vagina.

Thinking Critically

19. Menstruation is an example of a feedback control system because the menstrual cycle is controlled or regulated by hormones produced by the endocrine and the reproductive systems.

20. Because sperm cells are heat-sensitive, a fever might alter the development and maturation of the sperm cells. The cells of a developing embryo or fetus are also heat-sensitive and might be damaged by a fever.

21. The development and release of a new ovum that was not fertilized would begin the menstrual cycle again, causing the uterine lining and the developing embryo to be lost.

22. Constricted blood vessels carry smaller quantities of blood. As a result, less nourishment and oxygen are likely to reach the developing fetus. Also carbon dioxide levels are more likely to increase. These conditions can negatively affect fetal development.

23. In a normal pregnancy, the blood of the mother and the blood of the fetus do not mix.

24. A reaction or reflex caused by a secretion of adrenalin into the bloodstream may have allowed human beings to successfully escape from life-threatening situations.

25. The signs are designed to remind pregnant women of their responsibility to the developing embryo and fetus.

Reviewing Themes

17. *Systems and Structures*
How are the endocrine and nervous systems similar? How are they different?

18. *Systems and Structures*
Compare the structures of the male and female reproductive systems.

Thinking Critically

19. Explain why menstruation is an example of a feedback control system.

20. How might a high fever affect the male and female reproductive systems?

21. During pregnancy, the placenta produces hormones that keep the ovaries from ovulating. Why is this necessary?

22. Smoking of tobacco can cause blood vessels to narrow. How might this affect a developing fetus?

23. A person with type A blood cannot safely receive a transfusion of type B blood. Why is it safe for a woman with type A blood to bear a fetus that has type B blood?

24. What important role might adrenalin have played in the survival of human beings over thousands of years?

25. In some areas of the country, new laws require signs, such as the one in the photograph, to be displayed in some businesses. Why do you think laws of this type have been passed? Do you think the signs are necessary? Do you think they are a good idea?

Discovery Through Reading

Avraham, Regina. *The Reproductive System.* Chelsea House Publishers, 1991. Discusses aspects of reproduction, including heredity and pregnancy, and explains the stages of growth from fertilization to birth.

Science PARADE

SCIENCE APPLICATIONS

Discussion

● **Process Skills:** *Inferring, Applying*

Have the students infer the meaning of the word *breakfast*. (*Breakfast* means "break the fast"; that is, eat after not eating all night.) Ask the students whether they have heard the assertion that breakfast is the most important meal of the day, and ask them whether they agree or disagree with that assertion. (Help the students realize that every meal of the day is important; however, breakfast is the most important because it is the first meal of the day.) Explain that nutritionists and dietitians recommend that one meal of the day should be larger than the others, then ask the students which meal they think should be the largest. (Since most people are more active during the day than at night, ideally breakfast should be the largest meal of the day.)

SCIENCE APPLICATIONS

DAILY DECISIONS ABOUT YOUR BODY

Young people have to make dozens of choices every day about their bodies. They must decide what to eat, how to stay fit, and how to feel about their bodies. Nutrition science and exercise science provide information for good choices for your body.

STARTING OUT RIGHT

Take one of your first decisions every day: eating breakfast. Overnight your body's metabolism has fallen, and your body has used up its temporary store of fluid. Nutrition experts advise everyone to drink something soon after waking. A glass of fruit juice or water restores your digestive and excretory systems.

Even more important, scientists say, eat something for breakfast. Even a small bowl of cereal, some toast, or fruit

Drinking something soon after waking is advised.

▶ 564 UNIT 6

- **Process Skills:** *Inferring, Applying*

Given that breakfast should be the largest meal of the day, ask the students why lunch is also an important meal. (Digestion of food takes approximately four hours. By midday, the energy stores of the human body are depleted. The function of lunch is to refuel the body for the remaining activities of the day.) Explain to the students that given the daytime versus nighttime energy needs of a typical person, lunch should provide more energy (Calories) than the evening meal.

- **Process Skills:** *Applying, Expressing Ideas Effectively*

Remind the students that the benefits of exercise are numerous, then ask the students to name several benefits of regular exercise. (Among other benefits, exercise helps reduce and maintain weight; strengthens muscles, including the heart; lowers cholesterol, blood sugar, and blood pressure; reduces stress; and helps the digestive system work more efficiently.)

Regular exercise is beneficial.

stimulates the body's functions. Eating breakfast can raise metabolism by four to five percent for the entire day. A good decision about breakfast helps your body burn food more efficiently all day.

LUNCH CHOICE

And lunch? A young person trying to lose weight might decide to skip lunch—not a good decision, according to nutrition science. Skipping meals can trigger a "starvation response" in your body. When deprived of food, the body lowers its metabolism and burns calories more slowly. Eating regular meals will do more to control weight than skipping meals. The scientific choice is a lunch high in complex carbohydrates and low in fat. An apple and a whole-wheat roll, or a salad with low-fat dressing will keep the body's metabolic fires burning.

When school's out, what are you ready to do? Scientists who study the mind and the body recommend that you make time for moderate exercise. Scientific research has documented the benefits of physical fitness. Physical activity raises metabolism and stimulates the body's circulatory and respiratory systems. Regular exercise increases your body's immunity to illness.

Fitness also increases your mental capacity. A Canadian research study found a link between physical fitness and ability to learn math and writing skills. Perhaps best of all, studies show that young people who exercise regularly feel better about themselves. Exercise enhances self-esteem at a time in life when young people are deciding how to feel about themselves.

NEW ROUTINES

Exercise science advises that to get benefits from exercise, you don't have to sprint around a track. Benefits will come from any daily activity that raises your heart rate for 20 to 40 minutes. You might decide to bicycle to the library or store instead of going by car or bus. You can have that talk with your best friend during a brisk walk. Instead of sunning by the pool, walk through the water at chest-high depth. Or take up roller-skating, an exercise that tones hips and thighs and reduces body fat.

It helps to find easy and pleasant ways to get physical exercise. You don't need to subscribe to the philosophy of "No pain, no gain!" Exercise physiologists advise that pain during exercise may be a warning of impending injury. Besides, if exercise is dull or painful, you won't do it regularly. And regular exercise is a key to a healthy body and a positive attitude.

Sooner or later, most young people come face-to-face with the menu at a fast-food restaurant. Does anyone think of science at a

Background Information

Fiber is a nondigestive component of plants that is not absorbed by the human body. Fiber increases the mobility of food as it passes through the digestive system. In addition, fiber helps to prevent certain diseases such as colon cancer and heart disease.

Journal Activity

Ask the students to keep a record of the meals they eat every day for a week in their journals. Have them check the size and the fat and fiber content of the meals. Interested students can use Calorie counters to add up the Caloric content of each meal. Encourage the students to use the information they compile to adjust the size and content of their meals. They might wish to continue their journal record and to note the results of any changes they made. Emphasize that their journal record is for their private use only and need not be shared with others.

---

time like that? Nutritionists would give you simple advice. Whenever possible, stay away from foods high in fat and choose low-fat, high-fiber foods.

To begin with, notice how a fast-food item is cooked. A flame-grilled hamburger is lower in fat than a burger fried on a griddle. A roast turkey sandwich may contain only 9 grams of fat and 300 calories. A fried fish sandwich may pack a whopping 30 grams of fat and 540 calories.

Another fast-food tip is to balance high-fat foods against low-fat foods. A broiled chicken sandwich and low-fat milk help balance that fat-ladened order of French fries. Finally, be sparing with toppings like mayonnaise and ketchup. You don't gain much by ordering a salad and then drenching it in high-fat dressing.

DIET KNOW-NO'S

Finally, a scientific word about dieting. In their early teenage years, many people start to worry about body size and shape. They often find appeal in specialized diet programs and sometimes suffer from eating disorders. The teenage years can mark the beginning of "yo-yo" dieting, a pattern of rapid weight loss followed by regained weight.

Scientific research has good advice for choices about dieting. First, low-calorie "starvation" diets may simply lower the body's metabolism. In a month, metabolism can fall by 25 percent. When calorie con-

Starvation diets do more harm than good.

sumption returns to normal, the body prepares for the next "famine" by storing extra fat. So starvation diets may actually increase body fat, not reduce it. Also, repeated weight loss and gain increase the chances of heart disease in both males and females.

Few diet programs achieve consistent, long-term success. Diets that encourage eating large amounts of a single food may actually harm your health. There are scientific alternatives. Nutritionists suggest you eat the same amount of food daily but eat less fat. Have a snack of raw vegetables instead of chips or cheese and crackers. Eat more mashed potatoes at dinner and less fried chicken. Also, try dividing your normal diet into five meals instead of three. Start an exercise routine of three to four moderate workouts per week.

The decisions that young people make about fitness and diet can last a lifetime. Today's science recommends a few basic principles for deciding how to care for your body. ◆

Is this a balanced meal?

Read About It!

Discussion

● **Process Skills:** *Inferring, Generating Ideas*

Ask the students to define *sleep* and describe what they think happens during sleep. (The students are likely to suggest that sleep is a resting period for the human body and that functions of the human body tend to slow during sleep.) Explain to the students that current research suggests that sleep is actually a very dynamic state of being, almost, but not quite, as active as ordinary consciousness when a person is awake and alert. Many different processes occur during sleep, of which dreams represent only one process.

● **Process Skills:** *Inferring, Predicting*

Ask the students to estimate, in minutes, how long it usually takes them to fall asleep, then ask them to predict how long it takes a typical person to fall asleep each night. (Estimates will vary depending on the student; a typical person takes between 30 seconds and seven minutes to fall asleep.)

Read About It!

I SLEPT FOR SCIENCE:
A Study of Sleep Problems

"How are you doing in there, Andy? Think you're about done sleeping?" The voice from the ceiling drifted down softly, gently, like the soothing sounds of a cool summer rain.

It's a good thing too. I was in no mood for screaming. I had just spent the night with 13 wires attached to various parts of my body, a metal chain wrapped around my rib cage, a clamp stuck on my fingertip, and a plastic sensor clipped to my nostril.

"Are you sure I'll be able to sleep with all these wires?" I asked Joe Brown, sleep technologist, as he "wired" my body the night before.

"Absolutely," Joe replied. "You'd be surprised at how well most people sleep when they're here."

I was more than surprised; I was shocked. Falling asleep under such strange conditions proved to be as easy for me as falling asleep in my own bed. Not everyone is so lucky. Some people—including many kids—have trouble sleeping. To find out more about kids with sleeping problems and what can be done to treat them, I decided to "sleep for science."

So I went to the New Haven (Conn.) Sleep Disorders Center. There, Dr. Robert Watson, sleep expert and director of the center, tested me for sleep disorders.

I learned two important facts. First, I learned that I don't like wearing clips in my nose. Second, I learned that about 50 million people in the U.S. suffer from one or more sleep disorders, many of which can be easily treated. If you have a sleeping problem—perhaps you snore or feel unusually tired during the day—read on. This article is for you.

I Get "Hooked Up"

When I arrived at the sleep center, Joe "hooked" me to a *polysomnograph*, a machine that records brain, heart, lung, and muscle activity during sleep. Joe first rubbed certain

Discussion

● **Process Skills:** *Applying, Generating Ideas*

Lead the class in a discussion of whether sleep is or is not a necessary function of the human body. (The discussion is likely to suggest that because sleep refreshes and invigorates people, it is a necessary function of the human body.) Ask the students whether they have ever experienced a shortage of sleep, such as staying up all night, and ask those who have to describe what it felt like to function when the amount of sleep they received was insufficient. (Lack of sleep typically makes a person experience sluggishness, and depending on the amount of sleep loss, the inability to think clearly.)

Explain to the students that scientists and researchers have studied the effects of long-term sleep deprivation. In a well-publicized event, a New York disc jockey stayed awake and broadcasted continuously for 200 hours, or more than 8 days, while submitting to medical research. Ask the students to predict what the behavior of the disc jockey was like after 200

---

spots on my skin with a special solution. This solution made the wires Joe glued to my skin more sensitive to electrical signals coming from my body.

All body tissues "give off" weak electrical signals during activity. For instance, whenever a muscle moves, muscle cells produce electrical energy. The wires, or electrodes, that must be worn sense this energy and send it to the polysomnograph.

Joe also attached a small clip to my nostril to measure air flow, and a larger clip to my index finger to measure oxygen in the blood. Then he wrapped a metal chain and spring around my chest. Each time I breathed, the chain would stretch and send a signal to the polysomnograph.

Electrical signals from all the wires and electrodes produce a line of ink on a continuous length of lined paper. This is the *polysomnogram*, or sleep recording. Doctors use polysomnograms to interpret sleep patterns.

Before letting me fall asleep, Dr. Watson explained that there are two stages of sleep: Rapid-eye-movement, or REM; and Non-rapid-eye-movement, or Non-REM.

Non-REM sleep begins with stage one, that strange period between being awake and being asleep. If you've ever told your parents that you were "just about to fall asleep" when they woke you, you may have been floating through stage one sleep.

I vaguely recall going through this stage when I slept at the center. My thoughts sort of drifted away from me, and I became less and less aware of the wires attached to my skin. Stage one usually lasts from 30 seconds to about seven minutes in persons who don't have a sleep disorder. Dr. Watson says that my stage one lasted about two minutes.

Stage two is the first real sleep stage. It lasts about 15 to 45 minutes in adolescents and young adults. In stage two, brain activity begins to slow down. Thoughts come in short bursts and tend not to make much sense.

Next comes deep, or *delta*, sleep. Experts believe this stage affords the brain an opportunity to "rest." The body, on the other hand, often moves about during this stage. In fact, despite common belief to the contrary, we shift positions about every 15 minutes dur-

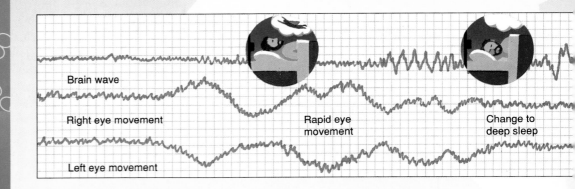

hours with no sleep. (The students are likely to suggest atypical behavior of some type.) Explain that the disc jockey experienced severe hallucinations and other bizarre delusions and lengthy episodes of depression.

● **Process Skills:** *Inferring, Applying*

Explain to the students that stress can occur in different ways in people's lives. Ask the students to describe stressful situations that people might experience. (The students might suggest loss of employment for adults and test-taking situations for students.) Have the students describe the link, if any, between stressful situations and quality of sleep. (The students should infer that stress, regardless of its cause, can negatively affect the sleep mechanism of people, resulting in sleep loss.)

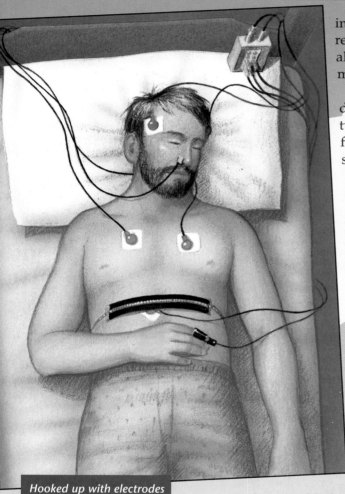

Hooked up with electrodes

ing sleep. Although I don't remember shifting position at all, Joe convinced me that I had moved several times.

After about 30 minutes of delta sleep, I returned to stage two sleep. Then I entered the first of several periods of REM sleep.

Dreaming in REM Sleep

REM sleep is a period of intense activity within the brain that experts say relates to the time we spend dreaming. Sleep experts know exactly when a person is dreaming because the person's eyes move in a typically rolling pattern, called rapid eye movement.

Unlike Non-REM, only the eyes move during REM. Most of the rest of our muscles become paralyzed, or unable to move, during REM sleep.

Under normal conditions, we drift in and out of REM sleep several times a

This sample from my polysomnogram shows the switch from dream sleep to deep sleep. Notice how the tracings suddenly jerk up and down. Noise from an ambulance siren started a burst of electrical activity in my brain and muscles. I awakened with a start and lurched forward in bed.

UNIT 6 **569**

Discussion

● **Process Skills:** *Predicting, Applying*

Ask the students to discuss why people dream. (Some students might suggest that dreaming provides exercise and stimulation for the brain.) Point out that sleep experts disagree about the reason for dreams. Some speculate that dreams allow the brain to consolidate and assimilate the experiences of the day. Others speculate that the brain might begin to atrophy if it were not used during sleep periods. Point out to the students that one thing is certain—the real reason for dreams is not known.

Journal Activity Ask the students to explain the meaning of the saying "Eat right to sleep tight" in a journal entry. Encourage the students to investigate any other advice about sleep that they can recall. For example, is there any truth to the idea that drinking warm milk will induce sleep?

EXTENSION

Explain to the students that the different stages of sleep have varying degrees of importance to the human body. Challenge interested students to use reference materials to discover more about the different stages of sleep and the connection between those stages and a person's health and well-being.

---

night. The first REM cycle lasts about five minutes. Experts say dreams in this first cycle tend to be rather boring.

As the night progresses, however, REM cycles lengthen and dreams become more and more vivid. The final REM cycle may last as long as one hour. Dreams at this time are the most active of the night (or day, depending on when you sleep).

After I woke up, Joe showed me the REM activity on my polysomnogram. He told me if he had awakened me during my last REM cycle, I would

Joe Brown monitors a polysomnogram. Each foot (30 centimeters) of polysomnograph paper equals one minute of time. My polysomnogram was 773 feet (236 meters) long and weighed 7 pounds (3 kilograms)!

have rambled on and on about the dream I was having. "You probably wouldn't have made much sense," Joe said, "but you would have remembered the dream. It was a whopper."

Whopper dream or not, I learned that my sleep patterns were normal. "You're lucky," Dr. Watson told me later. "Almost everyone who comes here has some kind of sleep disorder. Many have more than one." ◆

Then and Now

Discussion

● **Process Skills:** *Inferring, Applying*

After the students have had an opportunity to read the selections, ask them to describe how the research of Charles Drew, although some of it was performed more than 50 years ago, is still useful and will affect human beings for many years to come. (Because of people's need for blood, Drew's work not only has saved lives in the past but will continue to save countless lives in the future.)

Journal Activity

Have the students recall the work of Irene Duhart Long, then ask them to imagine a scenario in which they are asked to work as a doctor aboard a space station. Have the students describe the trip to the space station and life aboard the station. What might a day in the life of a space-station doctor be like? Encourage the students to write their descriptions in the form of a daily log.

Then and Now

Charles Drew (1904–1950)

Millions of people are alive today thanks to the blood plasma work of Charles Drew. His blood preservation discoveries led to the establishment of today's blood banks.

Charles Richard Drew was born in 1904 in Washington, D.C. Drew was a star athlete in both high school and college. After graduating from Amherst College, he coached, and taught biology and chemistry at Morgan State College.

In 1933, Drew received his medical degree from McGill Medical College and began his blood research at Canada's Montreal General Hospital. Drew discovered that blood plasma could be given to any person and could be stored for long periods. This knowledge helped save millions of lives in World War II.

Drew became the first director of the blood plasma collection program of the American Red Cross. He then was professor of surgery and director of Freedman's Hospital at Howard University Medical School. For his achievements, Drew received the National Association for the Advancement of Colored People's (NAACP) Spingarn Medal. ◆

Irene Duhart Long (1951–)

Dr. Irene Duhart Long plans to be there when the space station of the National Aeronautics and Space Administration's (NASA) goes into operation. As an aerospace physician, Long's goal is to be the space station's medical officer.

Irene Duhart Long was born in Cleveland, Ohio, in 1951. As a child, she was fascinated with air travel. At the age of nine, Long decided that she wanted to become a NASA physician.

Long attended Northwestern University and in 1977 received her medical degree from the St. Louis University School of Medicine. Long then earned her masters of science in aerospace medicine from the Wright University School of Medicine. Long began work with NASA in 1982 and is now Chief of the Medical and Environmental Health Office in the Biomedical Operations and Research Office at the John F. Kennedy Space Center.

Long studies the effects of gravity on the health of astronauts and also provides emergency medical care. She is doing experiments on cardiovascular and endocrine reactions in space.

Long's discoveries might make it possible for you to live in space one day. ◆

SCIENCE AT WORK

Discussion

● **Process Skills:** *Applying, Generating Ideas*

Before reading the selection, ask the students to describe the different tasks they think a pharmacist might perform during the course of a typical day.

● **Process Skills:** *Inferring, Expressing Ideas Effectively*

After reading the selection, ask the students to describe how Mamie Lou uses a knowledge of mathematics and a knowledge of chemistry in her work. (Descriptions might detail how dosages of medicines must be determined on a patient-by-patient basis using mathematical proportions and how a knowledge of chemistry enables Lou to understand the ways in which medications interact with other medications or with the human body.) Explain to the students that one of the many responsibilities of a pharmacist is to check that a prescription ordered by a doctor is correct; the pharmacist helps guarantee that a patient is receiving a correct medicine in the correct dosage.

SCIENCE AT WORK

Mamie Lou, Pharmacist

Many people think that all a pharmacist does is prepare the medicine prescribed by a doctor for people who are ill. However, a pharmacist's job is more that just counting out pills and filling medicine bottles. Today's pharmacist is an important part of a health care team.

Mamie Lou is a pharmacist at a medical teaching hospital in Houston, Texas. As a pharmacist, she performs many jobs. "I counsel patients about the medicines prescribed by their doctors as well as the medicines that they buy without a prescription," says Lou.

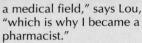

Communicating with doctors and nurses about a patient's medical treatment is also a very important part of a pharmacist's job. Lou reviews a patients' charts and computer profiles to verify that they receive the correct medications that the doctor ordered. Computers are very important in helping pharmacists keep accurate records on patients and their medications.

Lou instructs nurses and technicians on the skills and measuring tools needed for the preparation of medicines. "I use a balance to accurately measure dry medicines and a graduated cylinder to measure liquids whenever I make medicines," says Lou. Pharmacists also use a mortar and pestle to grind medicines to make sure the ingredients are evenly mixed.

A pharmacist is a licensed health care professional. To become a pharmacist, a person must complete college courses in biology, chemistry, math, and physics. He or she must also pass a state examination before receiving a license. "I have always wanted to help people and work in a medical field," says Lou, "which is why I became a pharmacist."

Pharmacists can work in other areas besides hospitals. They can work in nursing homes, sales, government, research, education, and home health care. Pharmacists can also choose specialty fields such as oncology, emergency medicine, nutrition, and pain management. Being a pharmacist is a great profession for students who are interested in people and science. ◆

DISCOVER MORE

For more information about a career as a pharmacist, write to the

American Association of Colleges of Pharmacy
1426 Prince Street
Alexandria, VA 22314

SCIENCE/TECHNOLOGY/SOCIETY

Discussion

● **Process Skills:** *Applying, Expressing Ideas Effectively*

Before reading the selection, ask the students to suggest reasons why it is extremely unlikely that people will live for more than 100 years. (The students will probably suggest that the human body and its systems tend to deteriorate and become more susceptible to sickness and disease over time.) After reading the selection, ask the students to debate whether a "cure" for the aging process will ever be developed.

Journal Activity
Have the students describe what they can do to help increase the probability that they will lead lengthy, robust lives. Encourage them to write their description in the form of a poem or song.

EXTENSION

Challenge interested students to research the Spanish conquistador and explorer, Juan Ponce de León, and his search for the legendary "Fountain of Youth." Why has this idea had such appeal for people throughout history? How do people today still search for a "Fountain of Youth"?

SCIENCE/TECHNOLOGY/SOCIETY

Aging

A person born in 1900 could expect to live an average of about 45 years. Infants born in 1986, however, could expect to live an average of about 74 years. Few people can expect to live beyond 100 years.

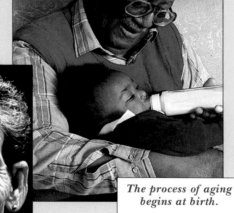

The process of aging begins at birth.

Modern medicine and technology now enable people to live longer lives.

The Study of Aging

Scientists called *gerontologists* study aging. They are trying to answer the question, "What causes people to age?" Gerontologists are not sure exactly how aging takes place, but most agree that genes are somehow involved. They also agree that only a few people reach a very old age.

Studies show that proteins produced by old cells are like those produced by young cells. Research also shows that DNA replicates more slowly in old cells than it does in young cells. This may be due to a molecule blocking reproduction of cells at one of the phases of mitosis. In addition, repair of DNA takes longer as we get older. The combination of slower replication and slower repair of DNA could result in aging. The studies imply that there is no difference in the DNA itself, only in the amount and use of it in young and old cells.

Victims of Nature

Systems become more disorganized as time goes by. This law of nature is called *entropy*. Some scientist think that cells are victims of entropy. The disorder in cells appears in the form of "mistakes" by enzymes that are not doing their jobs. These mistakes might cripple cells and could bring on changes we see as people age.

Scientists today are searching for the causes of these mistakes. Their research includes the study of certain molecules that are byproducts of cell processes. These molecules, called *free radicals*, react with nearly every other molecule in a cell, including DNA, and they can change a harmless form of cholesterol into the form that clogs arteries. But free radicals are not the only threat to cells. Studies also show that the cell system that breaks down harmful proteins works more slowly in old cells than in young cells.

Scientists do not yet fully understand what causes aging, but progress is being made as research and discoveries continue. In the future, people can expect to live longer and healthier lives. ◆

Reference Section

Safety Guidelines	**576**
Laboratory Procedures	**578**
Reading a Metric Ruler	578
Converting SI Units	579
SI Conversion Table	580
Reading a Graduate	581
Using a Laboratory Balance	581
Using Dissecting Tools	582
Making a Wet Mount	582
Using a Compound Light Microscope	582
Five-Kingdom System of Classification	**584**
Glossary	**586**
Index	**596**

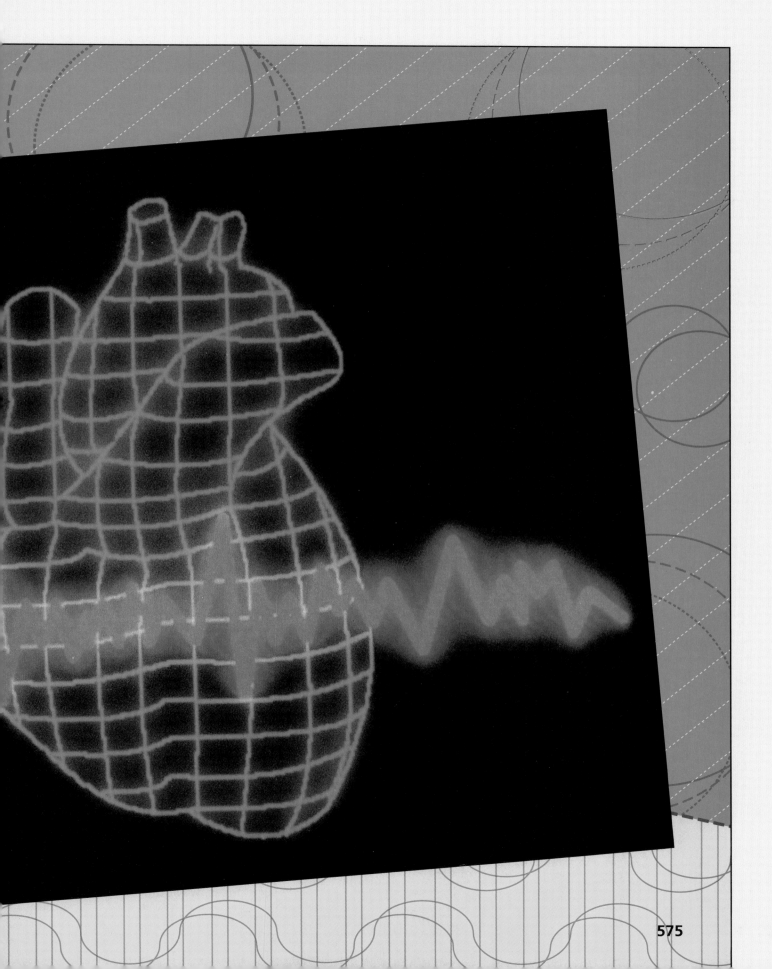

SAFETY GUIDELINES

Participating in laboratory investigations should be an enjoyable learning experience. You can ensure both learning and enjoyment from the experience by making the laboratory a safe place in which to work. Carelessness, lack of attention, and showing off are the major causes of laboratory accidents. It is, therefore, important that you follow safety guidelines at all times. If an accident should occur, you should know exactly where to locate emergency equipment. Practicing good safety procedures means being responsible for your classmates' safety as well as your own.

You will be expected to practice the following safety guidelines whenever you are in the laboratory.

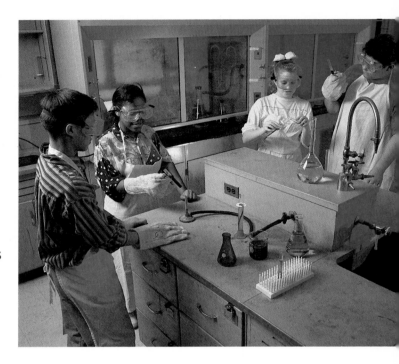

1. **Preparation** Study your laboratory assignment in advance. Before beginning your investigation, ask your teacher to explain any procedures you do not understand.
2. **Neatness** Keep work areas clean. Tie back long, loose hair and button or roll up long sleeves when working with chemicals or near an open flame.
3. **Eye Safety** Wear goggles when handling liquid chemicals, using an open flame, or performing any activity that could harm the eyes. If a solution is splashed into the eyes, wash the eyes with plenty of water and notify your teacher at once. Never use reflected sunlight to illuminate a microscope. This practice is dangerous to the eyes.
4. **Chemicals and Other Dangerous Substances** Some chemicals can be dangerous if they are handled carelessly. If any solution is spilled on a work surface, wash the solution off at once with plenty of water.
   - Never taste chemicals or place them near your eyes. Never eat in the laboratory. Counters and glassware may contain substances that can contaminate food. Handle toxic substances in a well-ventilated area or under a ventilation hood.
   - Never pour water into a strong acid or base. The mixture produces heat. Sometimes the heat causes splattering. To keep the mixture cool, pour the acid or base slowly into the water.
   - When noting the odor of chemical substances, wave the fumes

toward your nose with your hand rather than putting your nose close to the source of the odor.
  • Do not use flammable substances near a flame.

5. **Safety Equipment** Know the location of all safety equipment, including fire extinguishers, fire blankets, first-aid kits, eyewash fountains, and emergency showers. Report all accidents and emergencies to your teacher immediately.

6. **Heat** Whenever possible, use an electric hot plate instead of an open flame. If you must use an open flame, shield the flame with a wire screen that has a ceramic center. When heating chemicals in a test tube, do not point the test tube toward anyone.

7. **Electricity** Be cautious around electrical wiring. Do not let cords hang loose over a table edge in a way that permits equipment to fall if the cord is tugged. Do not use equipment with frayed cords.

8. **Knives** Use knives, razor blades, and other sharp instruments with extreme care. Do not use double-edged razor blades in the laboratory.

9. **Glassware** Examine all glassware before heating. Glass containers for heating should be made of borosilicate glass or some other heat-resistant material. Never use cracked or chipped glassware.
  • Never force glass tubing into rubber stoppers.
  • Broken glassware should be swept up immediately, never picked up with the fingers. Broken glassware should be discarded in a special container, never into a sink.

10. **Unauthorized Experiments** Do not perform any experiment that has not been assigned or approved by your teacher. Never work alone in the laboratory.

11. **Cleanup** Wash your hands immediately after any laboratory activity. Before leaving the laboratory, clean up all work areas. Put away all equipment and supplies. Make sure water, gas, burners, and electric hot plates are turned off.

Remember at all times that a laboratory is a safe place only if you regard laboratory work as serious work.

The instructions for your laboratory investigations will include cautionary statements when necessary. In addition, you will find that the following safety symbols appear whenever a procedure requires extra caution:

 Wear safety goggles

 Biohazard/disease-causing organisms

 Electrical hazard

 Wear laboratory apron

 Flame/heat

 Rubber gloves

 Sharp/pointed object

 Dangerous chemical/poison

 Radioactive material

Safety Guidelines

LABORATORY PROCEDURES

READING A METRIC RULER

1. Examine your metric ruler. The numbers on it represent lengths in centimeters. The usual metric ruler is about 30 cm long. There are 10 marked spaces within each centimeter, which represent tenths of centimeters (0.1 cm).

2. To measure the width of a piece of paper, place the ruler on the paper. The zero end of the ruler must line up exactly with one edge of the paper. Look at the other edge of the paper to see which of the marks on the ruler is closest to that edge. In Figure A, for example, the edge of the paper is nearest to the second line beyond the 7. Therefore, the width of the paper is 7.2 cm.

3. The edge of the paper might fall exactly on one of the centimeter marks. In Figure B, the edge is just on the 5-cm mark. The width of this paper is 5.0 cm. You must write in the .0 to indicate that the measurement is accurate to the nearest tenth of a centimeter; that is, it is more than 4.9 cm and less than 5.1 cm.

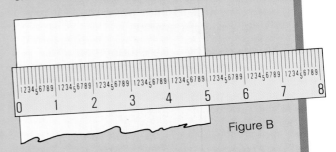

Figure B

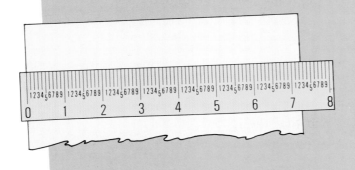

4. Sometimes you may want to make a reading with more accuracy. It is possible to estimate readings to the nearest hundredth of a centimeter, but you must be very careful. Look at Figure A again. You can guess the number of tenths in the distance between the marks. The edge of the paper is about 3 tenths of the space between 7.2 and 7.3. The best estimate, then, is that the width of the paper is 7.23 cm.

5. In Figure C, the edge of the paper falls exactly on the 8.6 mark. If you are taking careful readings, accurate to the nearest hundredth of a centimeter, you must record the width as 8.60 cm.

6. Note the general rule: You can estimate scale readings to the nearest tenth of a scale division. If the scale is marked in tenths, you can estimate the hundredths place but never more than that.

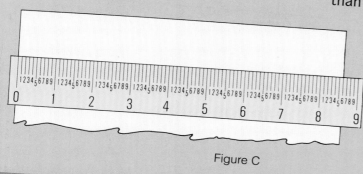

Figure C

CONVERTING SI UNITS

In SI, it is easy to convert from unit to unit. To convert from a larger unit to a smaller unit, move the decimal to the left. To convert from a smaller unit to a larger unit, move the decimal to the right. Figure D shows you how to move the decimals to convert in SI.

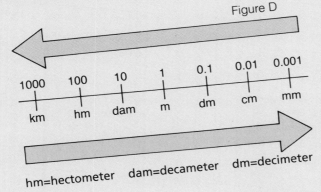

Figure D

hm=hectometer   dam=decameter   dm=decimeter

Laboratory Procedures

LABORATORY PROCEDURES

SI Conversion Table

SI Units		Converting SI to Customary		Converting Customary to SI	
Length					
kilometer (km)	= 1000 m	1 km	= 0.62 mile	1 mile	= 1.609 km
meter (m)	= 100 cm	1 m	= 1.09 yards	1 yard	= 0.914 m
			= 3.28 feet	1 foot	= 0.305 m
centimeter (cm)	= 0.01 m	1 cm	= 0.394 inch		= 30.5 cm
millimeter (mm)	= 0.001 m	1 mm	= 0.039 inch	1 inch	= 2.54 cm
micrometer (μm)	= 0.000 001 m				
nanometer (nm)	= 0.000 000 001 m				
Area		$1\ km^2$	= 0.3861 square mile	1 square mile	= $2.590\ km^2$
square kilometer (km^2)	= 100 hectares	1 ha	= 2.471 acres	1 acre	= 0.4047 ha
hectare (ha)	= 10 000 m^2	$1\ m^2$	= 1.1960 square yards	1 square yard	= $0.8361\ m^2$
square meter (m^2)	= 10 000 cm^2			1 square foot	= $0.0929\ m^2$
square centimeter (cm^2)	= 100 mm^2	$1\ cm^2$	= 0.155 square inch	1 square inch	= $6.4516\ cm^2$
Mass		1 kg	= 2.205 pounds	1 pound	= 0.4536 kg
kilogram (kg)	= 1000 g	1 g	= 0.0353 ounce	1 ounce	= 28.35 g
gram (g)	= 1000 mg				
milligram (mg)	= 0.001 g				
microgram (μg)	= 0.000 001 g				
Volume of Solids		$1\ m^3$	= 1.3080 cubic yards	1 cubic yard	= $0.7646\ m^3$
1 cubic meter (m^3)	= 1 000 000 cm^3		= 35.315 cubic feet	1 cubic foot	= $0.0283\ m^3$
1 cubic centimeter (cm^3)	= 1000 mm^3	$1\ cm^3$	= 0.0610 cubic inch	1 cubic inch	= $16.387\ cm^3$
Volume of Liquids		1 kL	= 264.17 gallons	1 gallon	= 3.785 L
kiloliter (kL)	= 1000 L	1 L	= 1.06 quarts	1 quart	= 0.94 L
liter (L)	= 1000 mL	1 mL	= 0.034 fluid ounce	1 pint	= 0.47 L
milliliter (mL)	= 0.001 L			1 fluid ounce	= 29.57 mL
microliter (μL)	= 0.000 001 L				

READING A GRADUATE

1. Examine the graduate and note how the scale is marked. The units are milliliters (mL). A milliliter is a thousandth of a liter and is equal to a cubic centimeter. Note carefully how many milliliters are represented by each scale division on the graduate.

2. Pour some liquid into the cylinder and set the cylinder on a level surface. Notice that the upper surface of the liquid is flat in the center and curved at the edges. This curve is called the *meniscus* and may be either upward or downward. In reading the volume, you must ignore the curvature and read the scale at the flat part of the surface.

3. Bring your eye to the level of the surface and read the scale at the level of the flat surface of the liquid.

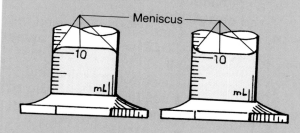

USING A LABORATORY BALANCE

1. Make sure the balance is on a level surface. Use the leveling screws at the bottom of the balance to make any necessary adjustments.
2. Place all the countermasses at zero. The pointer should be at zero. If it is not, adjust the balancing knob until the pointer rests at zero.
3. Place the object you wish to mass on the pan. **CAUTION: Do not place hot objects or chemicals directly on the balance pan, because they can damage its surface.**
4. Move the largest countermass along the beam to the right until it is at the last notch that does not tip the balance. Follow the same procedure with the next largest countermass. Then move the smallest countermass until the pointer rests at zero.
5. Determine the readings on all beams and add them together to determine the mass of the object.
6. When massing crystals or powders, use a piece of filter paper. First, mass the paper; then add the crystals or powders and remass. The actual mass is the total minus the mass of the paper. When massing liquids, first mass the empty container, then mass the liquid and container. Finally, subtract the mass of the container from the mass of the liquid and the container to get the mass of the liquid.

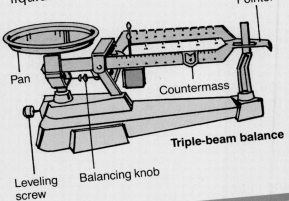

Laboratory Procedures

LABORATORY PROCEDURES

USING DISSECTING TOOLS

1. Dissecting tools are used to examine the internal and external features of an organism.
2. **CAUTION: Some dissecting tools are sharp and can cause injuries when not used properly.**
3. The most commonly used dissecting tools are the dissecting pan, scissors, scalpel, forceps, needle probe, and blunt probe.
4. The dissecting pan holds the specimen in place during examination. **CAUTION: Never attempt to dissect a specimen that is not secured to the dissecting pan.**
5. The scissors and scalpel are used for cutting. **CAUTION: Always cut away from yourself.**
6. Forceps are used for grasping and holding. The needle probe is used to move delicate parts of the specimen. The blunt probe is used to move larger, less delicate parts of the specimen.
7. Always clean and dry each of your dissecting tools after using them.

MAKING A WET MOUNT

1. Use lens paper to clean a glass slide and a coverslip.
2. Place the specimen you wish to observe in the center of the slide.
3. Using a medicine dropper, place one drop of water on the specimen.
4. Hold the coverslip at the edge of the water and at a 45° angle to the slide. Position the coverslip so that it is at the edge of the drop of water. Make sure that the water runs along the edge of the coverslip.
5. Lower the coverslip slowly to avoid trapping air bubbles.
6. Water might evaporate from the slide as you work. Add more water to keep the specimen fresh. Place the tip of the medicine dropper next to the edge of the coverslip. Add a drop of water. (You also can use this method to add stain or solutions to a wet mount.) Remove excess water from the slide by using the corner of a paper towel as a blotter. Do not lift the coverslip to add or remove water.

USING A COMPOUND LIGHT MICROSCOPE

Parts of the Compound Light Microscope

- The *eyepiece* magnifies the image 10X.
- The *low-power objective* magnifies the image 10X.
- The *high-power objective* magnifies the image either 40X or 43X.
- The *revolving nosepiece* holds the objectives and can be turned to change from one magnification to the other.
- The *body tube* maintains the correct distance between eyepiece and objectives.
- The *coarse adjustment* moves the body tube up and down to allow focusing of the image.
- The *fine adjustment* moves the body tube slightly to bring the image into sharper focus.
- The *stage* supports a slide.

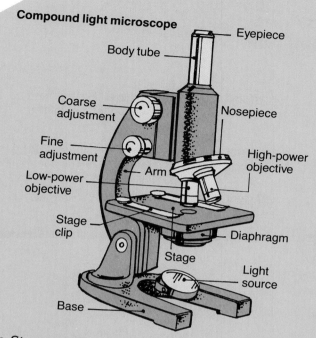

Compound light microscope

- *Stage clips* hold the slide in place for viewing.
- The *diaphragm* controls the amount of light coming through the stage.
- The *light source* provides light for viewing the slide.
- The *arm* supports the body tube.
- The *base* supports the microscope.

Proper Use of the Compound Light Microscope

1. Carry the microscope to your lab table, using both hands. Place one hand beneath the base and use the other hand to hold the arm of the microscope. Hold the microscope close to your body while moving it to your lab table.
2. Place the microscope on the lab table, at least 5 cm from the edge of the table.
3. Check to see what type of light source is used by your microscope. If the microscope has a lamp, plug it in, making sure that the cord is out of the way. If the microscope has a mirror, adjust it to reflect light through the hole in the stage. **CAUTION: If your microscope has a mirror, do not use direct sunlight as a light source. Direct sunlight can damage your eyes.**
4. Always begin work with the low-power objective in line with the body tube. Adjust the revolving nosepiece.
5. Place a prepared slide over the hole in the stage. Secure the slide with the stage clips.
6. Look through the eyepiece. Move the diaphragm to adjust the amount of light coming through the stage.
7. Now, look at the stage from eye level. Slowly turn the coarse adjustment to lower the objective until it almost touches the slide. Do not allow the objective to touch the slide.
8. Look through the eyepiece. Turn the coarse adjustment to raise the low-power objective until the image is in focus. Always focus by raising the objective away from the slide. *Never focus the objective downward.* Use the fine adjustment to sharpen the focus. Keep both eyes open while viewing a slide.
9. Make sure that the image is exactly in the center of your field of vision. Then switch to the high-power objective. Focus the image, using only the fine adjustment. *Never use the coarse adjustment at high power.*
10. When you are finished using the microscope, remove the slide. Clean the eyepiece and objectives with lens paper. Return the microscope to its storage area. Remember, you should use both hands to carry the microscope correctly.

Laboratory Procedures

Five-Kingdom Classification of Organisms*

Kingdom Monera

Organisms in this kingdom have cells that lack a true nucleus and membrane-bound organelles; mostly unicellular.

Phylum Schizophyta: Bacteria; about 2500 species, including eubacteria (true bacteria), rickettsias, mycoplasmas, and spirochetes

Phylum Cyanophyta: Blue-green algae, or cyanobacteria; about 200 species

Kingdom Protista

Organisms in this kingdom include a diverse group of unicellular and simple multicellular organisms whose cells have a true nucleus and membrane-bound organelles.

Phylum Euglenophyta: Euglenoids; about 800 species

Phylum Zoomastigina: Flagellates, about 2500 species

Phylum Sarcodina: Sarcodines; about 11 500 species; includes amoebas

Phylum Ciliophora: Ciliates; about 7200 species; includes paramecia

Phylum Sporozoa: Sporozoans; about 6000 species; includes *Plasmodia*, the cause of malaria

Phylum Chrysophyta: Golden algae; about 12 000 species

Phylum Pyrrophyta: Fire algae; 1100 species; major component of marine phytoplankton

Phylum Myxomycota: Slime molds; about 600 species

Phylum Chlorophyta: Green algae; about 7000 species; probable ancestor of modern land plants

Phylum Phaeophyta: Brown algae; about 1500 species; includes kelps

Phylum Rhodophyta: Red algae; about 4000 species; includes multicellular seaweeds

Kingdom Fungi

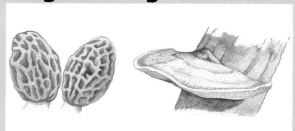

Organisms in this kingdom obtain food by absorption; most are multicellular, composed of intertwined filaments.

Division Basidiomycota: Mushrooms, bracket fungi, rusts, smuts; about 13 000 species

Division Deuteromycota: Fungi Imperfect; about 16 000 species; includes some *Penicillium*, athlete's foot fungus

* This is not a complete list of all known phyla or divisions. For each kingdom, only representative examples are given.

Kingdom Plantae

Organisms in this kingdom are multicellular and carry out photosynthesis in chloroplasts; mostly land dwellers; cell walls contain cellulose; body has distinct tissues; life cycle of alternating sporophyte and gametophyte generations.

Division Bryophyta: Bryophytes; about 15 600 species that lack vascular tissues and true roots, stems, and leaves; includes liverworts and mosses

Division Psilophyta: Whisk ferns; a few species of seedless plants lacking roots and leaves

Division Sphenophyta: Horsetails; about 15 species of seedless plants with hollow stems

Division Lycophyta: Club mosses; about 1000 diverse species of seedless plants with leafy sporophytes

Division Pterophyta: Ferns; about 12 000 diverse species of seedless plants

Division Cycadophyta: Cycads; about 100 species of palmlike plants; gymnosperms

Division Ginkgophyta: Ginkgo; 1 species only; fan-shaped leaves

Division Gnetophyta: Seed plants similar to angiosperms; about 70 species

Division Coniferophyta: Conifers; about 550 species of gymnosperms; most species are evergreens

Division Anthophyta: Angiosperms; about 235 000 species of plants that produce enclosed seeds; reproductive structures are flowers; mature seeds are enclosed in fruits

Kingdom Animalia

Organisms in this kingdom are multicellular and obtain food by ingestion; most are motile; reproduction is predominantly sexual.

Phylum Porifera: Sponges; about 5000 aquatic, mostly marine

Phylum Cnidaria (also called Coelenterata): Coelenterates; about 9000 aquatic species; tentacles armed with stinging cells; includes jellyfish and coral

Phylum Ctenophora: Sea walnut and comb jellies; about 90 species; gelatinous marine animals

Phylum Platyhelminthes: Flatworms; about 13 000 species; includes tapeworms

Phylum Nematoda: Roundworms; about 12 000 parasitic species

Phylum Acanthocephala: Spiny-headed worms; about 500 species

Phylum Rotifera: Rotifers or "wheel" animals; wormlike or spherical

Phylum Bryozoa: "Moss" animals

Phylum Brachiopoda: Lamp shells; about 250 species, 30 000 extinct

Phylum Mollusca: Mollusks; about 47 000 species of soft-bodied animals; includes snails and clams

Phylum Annelida: Segmented worms; about 9000 species; includes earthworms

Phylum Arthropoda: Arthropods; at least 1 million species; includes insects, spiders, crustaceans

Phylum Echinodermata: Echinoderms; about 6000 marine species; includes starfish, sand dollars, and sea urchins

Phylum Hemichordata: Acorn worms; about 80 species

Phylum Chordata: Chordates; about 43 000 species that at some stage have gill slits and tail; includes fish, amphibians, reptiles, birds, and mammals

Glossary

Pronunciation Key

Symbol	As In	Phonetic Respelling	Symbol	As In	Phonetic Respelling
a	bat	a (bat)	ô	dog	aw (dawg)
ā	face	ay (fays)	oi	foil	oy (foyl)
â	careful	ai (CAIR fuhl)	ou	mountain	ow (MOWN tuhn)
ä	argue	ah (AHR gyoo)	s	sit	s (siht)
ch	chapel	ch (CHAP uhl)	sh	sheep	sh (sheep)
e	test	eh (tehst)	u	love	uh (luhv)
ē	eat	ee (eet)	ủ	pull	u (pul)
	ski	ee (skee)	ü	mule	oo (myool)
ėr	fern	ur (furn)	zh	treasure	zh (TREH zhuhr)
i	bit	ih (biht)	ə	medal	uh (MEHD uhl)
ī	ripe	y (ryp)		effect	uh (uh FEHKT)
	idea	eye (eye DEE uh)		serious	uh (SIHR ee uhs)
k	card	k (kahrd)		onion	uh (UHN yuhn)
o	lock	ah (lahk)		talent	uh (TAL uhnt)
ō	over	oh (OH vuhr)			

A

absorption (uhb ZAWRP shuhn) the movement of nutrient molecules into blood vessels **(463)**

active transport the movement of molecules across a membrane from an area of low concentration to an area of high concentration **(122)**

algae plantlike protists that contain chlorophyll and carry out photosynthesis **(236)**

alveoli (al VEE uh ly) air sacs in the lungs **(492)**

antibiotic chemical substance used to kill or slow the growth of bacteria **(221)**

arteries blood vessels that carry blood away from the heart to other parts of the body **(480)**

asymmetry absence of symmetry **(330)**

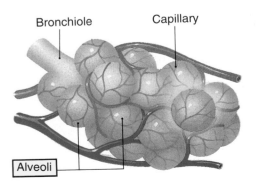

B

bilateral symmetry the arrangement of body parts the same way on both sides of an animal's body; two sides that are mirror images **(328)**

binomial nomenclature system of naming organisms using two Latin names **(177)**

biome large-scale land ecosystem that has a similar climate and vegetation throughout **(62)**

biosphere narrow layer near Earth's surface in which life can exist **(60)**

brain stem the place where the brain and the spinal cord meet **(518)**

bronchi (BRAHN kee) two small tubes that branch off from the trachea and carry air into the lungs; lined with mucus to trap dirt, dust, and any unwanted material brought in with the air **(492)**

bronchioles (BRAHN kee ohlz) small tubes in the lungs that branch from bronchi **(492)**

Bilateral symmetry

C

camouflage (KAM uh flahj) any marking or coloring that helps an animal hide from other animals **(400)**

capillaries small blood vessels with walls that are only one cell thick and that carry blood to all body cells **(480)**

cartilage (KAHRT uh lihj) the tough, flexible tissue from which most bones are formed **(438)**

cell the smallest unit of life; the basic unit of structure and function of all living things **(106)**

cell membrane the covering that surrounds the cell **(108)**

cell theory theory that states that all organisms are made of cells, that cells are basic units of structure and function for all living things, and that all cells come from other cells **(107)**

cellular respiration process in which energy is released when oxygen combines with sugar molecules, forming carbon dioxide and water as waste products **(134)**

cerebellum (sehr uh BEHL uhm) part of the brain that controls muscle coordination, balance, and muscle tone **(518)**

Camouflage

Glossary 587

Climate

cerebrum (seh REE bruhm) the largest part of the mammal brain; involved with intelligence and the organizing of information **(518)**

chlorophyll (KLAWR uh fihl) the green material in plants that traps energy from sunlight and uses it to break down water molecules into atoms of hydrogen and oxygen **(270)**

chromosomes (KROH muh sohmz) threadlike structures made of DNA that are in the nuclei of all cells **(127)**

chromosome theory theory that states that genes are located on chromosomes, that traits are passed to offspring by the chromosomes, and that each gamete contains chromosomes in the nucleus **(165)**

cilia short, hairlike structures used for locomotion in certain protozoans **(232)**

climate the general weather pattern that occurs in an area **(62)**

coldblooded having a body temperature that changes with the temperature of the environment **(372)**

community different populations living together in an area **(34)**

complete metamorphosis the changes in insect body form through four stages of development—egg, larva, pupa, adult **(360)**

compound light microscope microscope in which two lenses are used to magnify an image **(102)**

conservation the careful use of the earth's resources **(19)**

consumers organisms that eat other organisms for food **(44)**

cornea the clear part of the eye; lets light pass into the eye **(521)**

cotyledons the first leaves, or seed leaves, to develop on a seedling **(289)**

D

dermis (DUR mihs) the thick inner layer of skin **(502)**

diaphragm (DY uh fram) a thick muscle found at the bottom of the chest cavity that helps in breathing **(493)**

Glossary

diffusion (dih FYOO zhuhn) the movement of molecules from an area of high concentration to an area of low concentration **(121)**

digestion (dih JEHS chuhn) the process that changes food into a form that the body can use **(458)**

DNA hereditary material found within the nucleus **(127)**

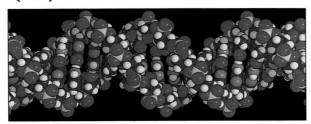

dominant (DAHM uh nuhnt) a strong factor in determining a trait; prevents the recessive trait from showing up in offspring **(162)**

drug any substance other than food, air, or water that can affect the way the body functions **(530)**

drug abuse the incorrect and unsafe use of a drug **(531)**

E

ecology the study of the relationships between organisms and their environment **(35)**

ecosystem the combination of a community and its nonliving environment **(35)**

embryo (EHM bree oh) in animals, a ball of tightly packed cells that develops from a zygote **(555)**

endocrine (EHN duh krihn) **system** group of glands that produce hormones **(544)**

epidemic the rapid spread of a disease through a large area **(213)**

epidermis (ehp uh DUR mihs) in animals, the outer layer of skin **(502)**

epiglottis (ehp uh GLAH tihs) a flap of tissue that moves over the opening to the trachea during swallowing **(491)**

erosion the carrying away of topsoil by water, wind, or glaciers **(20)**

evolution slow changes in living organisms **(150)**

excretion (ihks KREE shuhn) the process by which wastes are removed from the body **(497)**

exoskeleton (ehk soh SKEHL uh tuhn) an outer skeleton covering an arthropod's body **(358)**

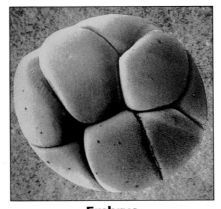

Embryo

Glossary **589**

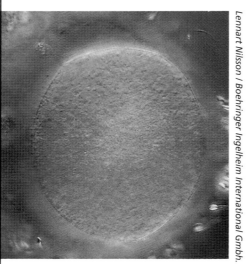
Fertilization

F

fermentation (fur muhn TAY shuhn) process that gives off energy without using oxygen **(135)**

fertilization the joining of a sperm cell and an ovum **(554)**

fetus in humans, the embryo after two months of development **(555)**

flagella (fluh JEHL uh) long, whiplike structures used for locomotion in certain protozoans **(233)**

fossils traces of once-living organisms **(152)**

fungi (FUHN jy) one of the five kingdoms of organisms; organisms that cannot move from place to place, have no chlorophyll, and absorb food from their surroundings **(240)**

G

gamete a reproductive cell; an egg or a sperm cell **(165)**

genes factors that determine hereditary characteristics **(163)**

gestation the period during which a young mammal is developing within its mother's body **(393)**

gland a group of cells that make special chemicals for the body **(544)**

gram the basic unit of mass in SI **(100)**

H

habitat (HAB uh tat) the place in which a population lives **(25)**

hormones (HOHR mohnz) chemicals produced by endocrine glands and certain body functions **(545)**

host an organism invaded by a virus or by another organism **(212)**

hypothesis (hy PAHTH uh sihs) a possible answer to a question **(5)**

Glossary

I

immune system system by which the body fights infection **(214)**

incomplete metamorphosis changes in insect body form through three stages of development—egg, nymph, adult **(360)**

incubation keeping eggs warm until they hatch **(388)**

invertebrate any animal without a backbone **(332)**

involuntary muscles muscles that work without signals from the brain **(448)**

J

joint place where two or more bones come together **(442)**

Invertebrate

K

kidney main organ of the excretory system **(498)**

kingdom the largest category in the classification system of living things **(185)**

L

lens the part of the eye that focuses light entering the eye **(522)**

lichen organism that is part fungus and part alga **(244)**

ligaments (LIHG uh muhnts) tough strips of connective tissue that act like strong rubber bands to hold together the bones in movable joints **(442)**

liter the basic unit of volume in SI **(100)**

M

marrow (MAR oh) the soft, living part of bone **(435)**

meiosis (my OH sihs) the type of cell division in which gametes are formed **(165)**

menstruation (mehn STRAY shuhn) the breakdown of the thickened lining of the uterus and the movement of that material out of the body through the vagina **(552)**

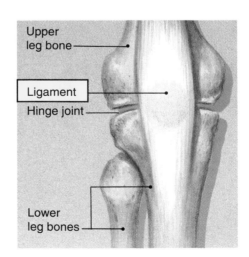

meter the basic unit of length or distance in SI **(99)**

mimicry (MIHM ihk ree) an animal's ability to copy the appearance or behavior of another unrelated organism **(403)**

mitosis (my TOH sihs) the type of cell division by which two identical daughter cells are formed **(127)**

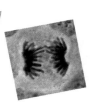

monerans one of the five kingdoms of organisms; organisms that do not have nuclei in their cells **(218)**

muscular system muscles of the body working together to coordinate the movements of the body **(445)**

N

natural selection the process by which those organisms best suited to their environment survive and reproduce **(155)**

nerves groups of neurons that carry messages between the different parts of the body and the central nervous system **(513)**

neuron individual nerve cell **(514)**

niche (NIHCH) the function an organism plays in its habitat **(43)**

nicotine (NIHK uh teen) a psychoactive drug found in tobacco **(534)**

nucleus that part of a cell that controls the cell's activities **(109)**

nutrient (NOO tree uhnt) a chemical substance that the body needs to build new cells and to keep those cells alive **(458)**

O

organ group of tissues that work together **(113)**

organelle (AWR guh nehl) structure within a cell that has a certain job to do in the cell **(108)**

osmosis (ahs MOH sihs) a special type of diffusion that occurs when water moves across a selectively permeable membrane **(123)**

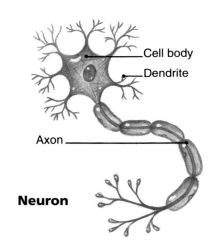

Neuron

P

passive transport process that requires no energy for the movement of molecules across a membrane from an area of high concentration to an area of low concentration **(122)**

pheromones (FEHR uh mohnz) chemicals serving as signals that are picked up by animals of the same species **(412)**

phloem (FLOH ehm) in vascular plants, tissue that transports nutrients that are made in the leaves to all parts of the plants **(260)**

photosynthesis (foht uh SIHN thuh sihs) the process by which green plants use chemicals from the environment and energy from the sun to make their own food **(263)**

pigment chemical that gives color to the tissue of living organisms **(223)**

pistil the female part of a flower **(278)**

plasma (PLAZ muh) the liquid part of blood **(473)**

pollution making the environment unclean with waste products **(10)**

population organisms of the same species living together in a particular place and at a particular time **(34)**

producers green plants that produce their own food **(44)**

protozoan (proht uh ZOH uhn) microscopic organism that is a member of the protist kingdom **(230)**

pseudopods (SOO duh pahdz) false feet; projections made of cytoplasm used for locomotion and food getting in certain protozoans **(231)**

puberty the stage of life during which a human matures sexually **(550)**

Pollution

R

radial symmetry the arrangement of body parts around a center point **(327)**

receptors specialized nerve cells that receive information from their surroundings **(513)**

recessive (rih SEHS ihv) a weak factor in determining a trait; prevented from showing up in offspring by the dominant trait **(162)**

reflex an automatic response **(517)**

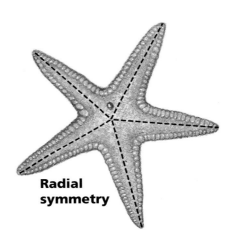

Radial symmetry

Glossary 593

regeneration in plants, the ability to grow or replace missing parts **(294)**; in animals, the growth of a part of an organism into a complete organism **(344)**

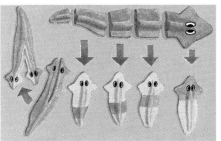

replication (rehp luh KAY shuhn) the process by which DNA makes copies of itself **(132)**

retina (REHT uhn uh) the inside layer of tissue on the back of the eyeball **(522)**

S

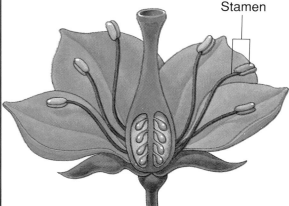

Stamen

skeletal system the framework of the body **(434)**

slime mold organism that looks and reproduces like a fungus, moves and eats like an amoeba **(244)**

species a group of organisms that naturally mate with one another and produce fertile offspring **(184)**

stamen the male part of a flower **(278)**

succession the slow changes that take place when one community replaces another **(48)**

symmetry (SIHM uh tree) balanced arrangement of body parts around a center point or line **(327)**

system group of organs that work together **(113)**

T

Torpor

taxonomy the science of classification **(176)**

tendon (TEHN duhn) type of connective tissue that holds muscles to bones and keeps different muscles together **(447)**

theory (THIR ee) a hypothesis that is supported by the work of many scientists **(8)**

tissue a group of similar cells working together to perform a specific job **(112)**

torpor an inactive state in which an animal's body functions slow down **(413)**

Glossary

trachea (TRAY kee uh) in vertebrates, tube that transports air to the lungs **(491)**

tropism (TROH pihz uhm) the response of a plant to a stimulus **(291)**

vaccine a substance made of dead or weakened viruses used to prevent a specific disease **(215)**

vascular cambium in a plant, the growth tissue that produces the xylem and the phloem **(263)**

vegetative propagation asexual reproduction in which cuttings of roots, stems, or leaves grow into new plants **(292)**

veins blood vessels that carry blood to the heart **(481)**

vertebrate (VUHR tuh briht) any animal that has a backbone **(332)**

virus a very small particle that can reproduce only inside a living cell **(211)**

Vaccine

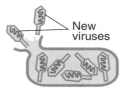

voluntary muscles muscles that can be controlled by the brain; skeletal muscles **(447)**

warmblooded having a nearly constant body temperature, regardless of the temperature of the environment **(384)**

xylem (ZY luhm) in vascular plants, tissue that transports water and dissolved minerals from the roots to the leaves **(260)**

zygote (ZY goht) the new cell resulting from fertilization **(554)**

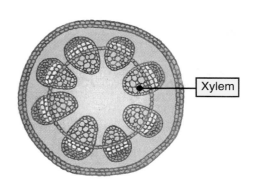

Index

A

Absorption, 463
Abyss, 66
Acetic acid, 135
Acid rain, 13–14, **13**
Acne, 505
Acquired immune deficiency syndrome. *See* AIDS
Active transport, 122, **122**
Adaptation, 155
Addiction, 532
Adenine, 131
Adolescent, **559**
Adrenal glands, 545, **545**
Adrenalin, 545
African elephant, **184**
African sleeping sickness, 233
African violets, **277**
Afterbirth, 558
Aggregate fruits, 281, **281**
Aging, 573, **573**
AIDS
 research, 248–249, **248, 249**
 transmission of, 214
 virus, 214–215, **214**
Air filter, 490
Air, gases in, 494–495, **495**
Albumen, 387
Alcohol, 536–537
 during pregnancy, 557
 effects of, 537
Alcoholism, 537
Alder trees, 50
Algae, 222, 236–238, **236**
 characteristics of, 236
 types of, 237–238
Algal blooms, 254–255, **254, 255**
Alligator, 381, **381**
Alveoli, 492, 494–495, **495**
 effect of smoking on, 535
Amber, 152, **152**
American robin, **325**
Amino acids, 467
 in blood, 473
 source of, 467
Ammonite, **152**
Amoeba, **126**, **187**, 231–232, **231, 232**
 pseudopods of, 231
Amoebic dysentery, 232
Amphibians, 376–380
 brain of, 391
 characteristics of, 376
Anaphase, 127, 128, **128**
Angiosperms, 265, **265**, 277–281
 life cycle of, **279**
 species of, 265
Animal cell, **110, 326**
 discovery of, 106
 need of, for food, 121, 134
Animals, 189
 awareness of, 321–322
 characteristics of, 325–330
 classification of, 332–335
 communication among, 410–412
 contributions of, 323
 defense, means of, 400–405
 and food-getting, 325
 and growth controls, 326
 mimicry of, 403
 multicellular organisms, 326
 as pests, 324
 protective coloration of, and defense, 401–404
 role in succession, 51
 social behavior of, 407–409
Ankle, 441
 tendons of, **447**
Annuals, 262
Anopheles mosquito, **233**
Antennae, 364
Anterior, 328
Anther, 278
Antibiotics, 221, 531
Antibodies, 476
Ants, **412**
 use of pheromones, 412
Anus
 of earthworm, 352
 of roundworm, 350
Anvil, 524
Apical bud, 263
Apical meristem, 263, **263**
Appendages, 363, 441
Apples, **185, 280, 281**
Apple tree, grafted, **294**
Arachnids, 362–364
Archaeological discoveries, 194–197, **195, 196, 197**
Arctic fox, **401**
Aristotle, 175, **175**
Armadillos, 391, **405**
Arms, 441
Arteries, 480, **480, 483**
 cholesterol buildup in, 483, **483**
Arteriosclerosis, 483
Arthropods, 358–365
 characteristics of, 358–359
 classification of, 359
 meaning of name, 358
Asexual reproduction, 292
Asian elephant, **184**
Aspirin, during pregnancy, 557
Asymmetry, 330
 of sponges, 330
Athlete's foot, 40, 243
Atoms, 120
Atrium, 479
Attenborough, David, 320, 325, 332
Auditory nerve, 525
Axon, 515, **515**

B

Backbone, **333**
Bacteria, 44, **181**, 187, **187**, 218, 219–222, **222**
 benefits of, 221–222
 and decay, 222
 of large intestine, 463
 and recycling, 222
 reproduction of, 219
Balance, 525
Bald eagle, **182, 384**
Ball-and-socket joints, **443**
Bark, 262
Bats, **189, 322**
Beaded lizard, 381
Beech-maple forest, **53**
Bees, **37, 324, 411, 464**
 communication among, 411
Betta fish, **372**
Bilateral symmetry, 328, **329,** 334
Bile, 462–463
Binomial nomenclature, 177
Biodegradable wastes, 11
Biomes
 defined, 62
 kinds of, 62, 63, 71–79
 of the world, **63**
Biosphere, 60–63
Biotechnology, 203–205, **205**
Biotic potential, 38–39
Birds, 384–389
 bones of, 386
 brain of, **391**
 characteristics of, 384–385
 development of, 387–389
 digestive system of, 386, **386**
 eggs of, 387–393, **387, 388**
 feathers of, 384–385, **385**
 heart of, 387

respiratory system of, 387, **387**
structure of, 386–387
wing of, 386, **386**
Birth, 558–559
Birth defects, 557
Birthrate, 36–37
Black walnuts, 299, **299**
Blake, Michael, 77
Blood, 214, 472–476
components of, 473–476
of embryo, 556
functions of, 472–473
movement of, 482
testing, **473**
vessels, 480–481, **480**
Blood cells, **112**
Blood clot, 475
Blood pressure, 483
and cholesterol, 483
Blue-footed booby, **147**
Boa constrictor, **400**
Body temperature, 101
of birds, 384
and blood, 472
of coldblooded animals, 372
controlled by skin, 505
of humans, 384
of mammals, 384
Body tube, 103
Boils, **221**
Bone marrow, 435, **435**, 437, **437**, 474
Bones, 126
cells of, **438**
composition of, 436–437
flat, 436, **436**
formation of, 438
function of, 434–436
growth of, 438
long, 436, **436, 437**
making new, 438
membrane of, 436, **437**
mineral layer of, 437, **437**
short, 436, **436**
structure of, 436–437
types of, 436
Book lungs, 359
Boreal forests, 75, **75**
Borton, Helen, 72
Botanic gardens, **260, 270**
Brain, 517–518
of amphibian, **391**
of bird, **391**
human, **518**
of mammals, 391, **391**
and nervous system, 513
of reptiles, **391**
Brain stem, 518
Brain surgery, **518**
Bread mold, **240**, 241, **241**
Breathing, 492–494, **493**
Breeding animals in captivity, 26–27, 420–421, 428–429

Bronchi, 492
Bronchioles, 492, **495**
Brown algae, 238, **238**
Buds, 263
Bufo toad, **378**
Bulbs, 293, **293**
Butter, 466
Butterfly, **329**
color of, 412
eggs of, **360**
metamorphosis of, 360, **360**

C

Cactus
spines, 300, **300**
Caecilians, 380, **380**
Caffeine, 530
Calcium, 469
for healthy bones, 436
need of, for bones, 435
and pregnancy, 558
Camouflage
defined, 400
types of, 401–402
Canada goose, **388**
Cancer-causing agents, 533
Cancer cell, **475**
Capillaries, 480–481, **480, 481**, 494–495, **495**
Carbohydrates, 464–466
and digestion, 466
and energy, 464–465
Carbon dioxide, 271, 273
in blood, 473, 494–495, **495**
from fermentation, 135
greenhouse effect and, 15–17
Cardiac muscle, 445, 449, **449**
Careers
as conservationist, 92
as geneticist, 202
as marine biologist, 320, **320**
as pharmacist, 572
as plant scientist, 313–314
as public health physician, 253
as science teacher, 427
Carnivores, 44, 386
top, 46
Carrying capacity, 39
Carson, Rachel, 362
biography of, 91, **91**
Cartilage, 438
as disks, 439, **439**
of jawless fishes, 373
in joints, 442
in mouth, 491
in trachea, 491
Carver, George Washington, biography of, 312, **312**
Castings, 352
Cat, **333**
Caterpillar, 360, **360**
Catfish, 375

Caudal fins, 373, **373**
Cell body, 515, **515**
Cell membrane, 108–109, **109**, 121, 326
and diffusion, 121
layers of, **109**
Cell plate, 129, **129**
Cell reproduction, 126–129
methods of, 126–127
Cells, **121**
of animals, **110**, 326, **326**
of cartilage, **438**
discovery of, 105–107
division of, 127, 555
energy in, 111, 134–137
levels of organization, 111–113, **113**
of plants, **107, 112**
production of nails and hair by, 504
and salt, 122
of skin, 503
structure of, 108–113
transportation, 122
Cell theory, 106–107
Cell walls, 111, 326
Cellular respiration, 134, **134**, 490
Cellulose, 111, 466
Celsius scale, 101, **101**
Centipedes, 365, **365**
Central nervous system, 513
Central skeleton, 439–440
Centrioles, 128
Centromere, 127, 128
Cephalothorax, 362
Cerebellum, 518, **518**
Cerebrum, 391, 518, **518**
Cervix, 551, **551**
Chameleons, 401
protective coloring of, 401
Cheese making, **222**
Chemical communication among animals, 412
Chemical digestion, 459
Chemical energy, **137**
Chemicals
and endocrine glands, 544–545
and endocrine system, 545
recycling of, 46
and taste and smell, 526
Chest cavity, 493–494
Chestnut blight, **40**, 242
Chewing tobacco, 535
Chicken pox, 213, **213**, 216, 476
Chimpanzee, **329**
Chitin, **334**
Chlorofluorocarbons (CFCs), 15
Chlorophyll, 111, 188, 223
and photosynthesis, 270, 271, 272, 273
Chloroplasts, 111, **111**, 236, 237

Index

and photosynthesis, 271
Cholera, **221**
Cholesterol, 483
Chromosomes, 127
   during mitosis, 127
   fruit fly, **165**
   theory, 165
Chrysalis, 360
Cigarette warning, **533**
Cilia, 232, 492
   of vas deferens, 550
Ciliates, 232–233
Circulation, 478–484
Circulatory system, **482**
Cirrhosis, **537**
Clams, 354, **354**, 355, **355**
   anatomy of, **355**
Class, 185, 333
   of vertebrates, table 333
Classification, 174–189, **185, 187**
   of animals, 332–335
   categories of living things, 185–187
   chart, **185**
   defined, 174
   early systems of, 174–175
   kingdoms of organisms, **187**
   modern, 181–189
Climate, 62, **62**
Climax community, 51–53
Clitellum, 351
Cloaca, of birds, 386
Clotting, 475
Clown fish, 41
Club fungi, 242, **242**
Club mosses, **267**
   as seedless plants, 267
Cochlea, 525
Codfish, 375
Coelenterates
   characteristics of, 344–345
   life cycle of, 346–348
Coffee, 530
Coldblooded animals, 372
Collecting data, 6–7
Colony
   of coelenterates, 345
   of coral, 345
Colorado Blue Spruce, **275**
Commensalism, 41
Common cold, 211
Communities, 34–35
Complete metamorphosis, 360, **360**
Complex carbohydrates, 466
Compound eyes, 359, **359**
Compound microscope, 102–103, **102**
   invention of, 105–106
Condor, 26
   breeding in captivity, 26–27
Cones, of eye cells, 523

Conifers, 75, 276, **276**
Conservation, defined, 19
Consumers (food), 44
Contamination, 232
Contour plowing, 21, **21**
Contractile vacuoles, 232
Control group, 6
Controlled experiment, 7
Copperhead, 382
Coral, 345
Coral snake, 382, **382, 402**
Cork, under microscope, **106**
Cork
   bark, **300**
Corn oil, 466
Corn smut, **242**
Cornea, 521
Cotyledons, 289
Countershading, 402
Cousteau, Jacques, 66
Cowing, Sheila, 78
Cowpox, 216
Cowrie, **335**
Cows, **176**
Coyotes, 5
   experiments with, 5–8
   sheep killed by, table 7
Crab, **358**
Crayfish, 364, **364**
Crick, Francis, 130, **130**, 131
Crocodiles, 381, **381**
Crop
   of birds, 386
   rotation, 21
Crossbreeding, 184
Cross-pollination, 279
Crustaceans, 364–365, **364**
Ctenosaur lizard, **410**
Cyanobacteria, 44, 222–223, **223**
   reproduction of, 223
Cycads, 276, **277**
Cyst (amoeba), **232**
Cytoplasm, 110, 124, 128, 129, 231
Cytosine, 131

D

Dachshund, **183**
Darwin, Charles, 144–150, 152, 155–158
Data, 6
Daughter cell, 126
Day-neutral plants, 292
Death rate, 36–37
Decay, 222
Deci-, 99
Decibels, 525, **525**
Deciduous forests, 74, **74**
Decimeter, 99
Decomposers, 46, **46**, 67

Degrees Celsius, 101
Dehydration, 469
De la Mare, Walter, 289
Delivery (birth), 558
Dendrites, 515, **515**
Depressants, 536
Depression, 536
Dermis, 502, 502, 503–504
Desert plants, **296**, 300, **300**
Desert plants and animals, **79**
Deserts, **71,** 78–79, **79**
Diabetes, 133
Diaphragm, 493–494, **493**
Diatomaceous earth, 238
Diatoms, 238, **238**
Dicots, 265, **265**
Diet, 564–566
   during pregnancy, 558
Diffusion, 120–122, **120**
   and cell membrane, 121
   and osmosis, 123
   as passive transport, 122
Digestion, 458–463
   defined, 458
Digestive process, 459
Digestive system, 458–463
   of birds, 386, **386**
   of fish, 375
   human, **458**
Digestive tract, 459
   effect of smoking on, 535
Dinoflagellates, 237, **237**
Dinosaurs, 380, **380**
Diphtheria, 221
Diseases
   and blood, 473
   caused by bacteria, table 221
Disks, 439, **439**
DNA, 127, 130–133, 166, 550
   description of, 127
   replication of, 132–133
   research in, 133
DNA model, **133**
   developers of, 130–131
DNA molecule, **130, 131**
Dogs, **329, 393**
Dolphins, 391, 408, **409**
Dominant factor, 162
Dorsal fins, 373, **373**
Dorsal side, 329
Double helix, 130–131
Double sugar, 465
Drew, Charles, biography of, 571, **571**
Drugs, 530–531
   abuse of, 531–532
   and medicine, 530–531
   and pregnancy, 557–558
Drunk driving, **536**
Duck-billed platypus, 392, **392**
Ducts, 544
Dust mites, **103**
Dutch elm disease, **40**, 242

E

Ear, anatomy of, 524, **524**
Ear canal, 524
Eardrum, 524
Earphones, **524**
Earth, **210**
Earthworms, 351–352, **351, 352**
  anatomy of, **352**
Echinoderms, 356–357
Echolocation, **322**
Ecology, 35
Ecosystems, 35, 42
  changing, 48–53
  freshwater, 66–69
  land, 62–63
  marine, 63–69
Egg-eating snake, **382**
Eggs
  of amphibians, 378
  of birds, 387–389, **387, 388**
  of chickens, **388,** 389
  fertilization of, 554
  of fishes, 376
  of flamingos, 389
  of frogs and toads, 378, 379, **379**
  of green sea turtle, **39**
  incubating, 388–389
  of insects, 360
  of larks, 389
  of mallard ducks, 389
  of reptiles, 381
  of royal albatross, 389
  of salamanders, 380
Ejaculation, 550
Electrical impulses
  and nervous system, 544
Electron microscope, 103, **103, 181**
Elephants, **25,** 391
  behavior of, 405
Embryo, 393
  care of mother and, 557–558
  described, 555
  development of, 555–558, 555
  placenta of, 556
  plant development, 288–289, **288**
Emigration, 37
Endangered species, 24
Endocrine glands, 544–547, **545,** table 546
  of female reproductive system, 551
  of male reproductive system, 549
Endocrine system, 544–547
Endoplasmic reticulum (ER), 110, **110**
Energy
  from carbohydrates, 464–466
  and cellular work, 134
  and fermentation, 135
  flow of, 43–46
  from lipids, 466
  from respiration, 134
  from simple sugars, 464–465
Energy cycle, 137
Energy pyramid, **46**
Energy transfer
  of plants, 274
Environment
  community dependence on, 35
  pollution of, 10–17
  and population size, 38–39
  protecting the, 18–27
Enzymes, 468
  of saliva, 459
  of stomach, 461
Epidemic, 213
Epidermal tissue, 112
Epidermis, 264, 502, **502,** 503, 504
Epiglottis, 491, **491**
Erosion, 20
  defined, 20
Esophagus
  of birds, 386
  of earthworms, 352
  of fish, 375
  human, 460, **460,** 491, **491**
Estivation, 413
Estuary, 69, **69**
Ethyl alcohol, 536
Euglena, **187,** 237, **237**
Eustachian tube, 524
Evening primrose, **178**
Evergreens, 75, 264
Evolution, 150
  defined, 150
  process of, 150
  theories of, 155, 158
Excretion, 497–500
  defined, 497
Excretory system, 498–500, **498**
Exercise, 136
  and muscle tone, 451
Exoskeleton
  of arthropods, 358
Experimental group, 6
Experiments
  conducting, 6
Extinction, 23
Eyes
  anatomy of humans, 521, **521**
  of crayfish, **364**
Eyespots, 348

F

Fallopian tubes, 551, **551**
  function of, 554
Family, 185
Farsightedness, 523
Fat-soluble vitamins, 468
Fats, 466
  and digestion, 463
Feathers, 384–385, **385**
Feces, 463
Feedback control, 546–547, **547**
Felis concolor. See Mountain lion
Female reproductive system, 551
Fermentation, 135, 242
  comparison with respiration, table 136
Ferns, 266–267, **266**
  life cycle of, **266**
Fertilization, 279, 554–555
Fertilizer, 21, 124, 223
Fetus, 555, 556–557, **556**
Fever blister, 213
Fiber, 466
Fibrous roots, 260, **261**
Filament, 278
Finches, 149, **149**
Fingerprints, **503**
Fins
  of sharks, **373**
Fish, 372–376
  anatomy of, **375**
  bony, 375–376
  cartilaginous, 373–374
  jawless, 373
Five-kingdom classification, 582–583
Fixed joints, 442, **443**
Flagella, 233, 237, 343
Flagellates, 233, **233**
Flamingo, **155**
Flat bones, 436, **436**
Flatworms, 333, 348–349
Fleming, Alexander, 243
Floating ribs, 441
Flounder, 375
Flowers
  parts of, 278, **278**
Flukes, 348
  life cycle of, **349**
Fly, **359**
Food chain, 45, **45**
Foods
  and health, 470
  high fiber, 466
  protein-rich, **467**
  pyramid, **470**
Food supply, 38–39
Food vacuole, 231, **232**
Food web, 45–46, **45**
Forest ecosystem, **35**
Forests, 71–75
Forsyth, Adrian, 400, 402, 403, 407–408, 410, 414, 415
Fossil fuels
  greenhouse effect and, 17
Fossils, 152, **152**
Fragmentation, 344

Index **599**

Fraternal twins, 555, **555**
Fresh water, Earth's supply of, 22
Freshwater ecosystems, 63, 66–69, **68**
Frogs, 376–379
   development of, 378–379
   heart of, **378**
   internal anatomy, **378**
   life cycle of, **379**
   and toads grouped, 378–379
Fructose, 464
Fruits, 280–281
   development of, 280–281
   waxy, **300**
Fungi, 188, **188**, 240–244
   characteristics of, 240
   types of, 241–243
Fur seal, 327

G

Galápagos Islands, 147
Galápagos tortoises, **148, 150**
Gall bladder, **462**
Gametes, 266, 551
   and chromosomes, 165
   formation of, 165
Gametophyte, 267
Gases, exchange of, 494–495, **495**
Gastric juice, 461
Gecko, 321, **321**
Gemmules, 344
Genes, 163, 165
Genetics
   discovery of, 165
   principles of, 165
Genetic traits, 162, 165
Genus, 185
Geotropism, 291
German measles, 216
Germination, 289
Gestation
   human, 393
   of placental mammals, 393
Giant pandas
   breeding in captivity, 420–421, **420, 421**
Gila monster, 381
Gills, 376
Ginkgoes, 277, **277**
Giraffe, **155, 389**
Gizzard
   of birds, 386
   of earthworms, 352
Glacier Bay, **49, 50**
Glacier Bay National Park, 49–50, 51
Glands
   of endocrine system, *table* 546
   of female reproductive system, 551

of male reproductive system, 549–550
   of mouth, 459
   of small intestine, 462
Gliding joints, **443**
Glucose, 273
Golgi bodies, 110
Gonzalez, Elma, biography of, 312, **312**
Goodall, Jane, biography of, 426, **426**
"Gooseflesh," 504
Grafting (plant), 294, **294**
Gram (g), 100
Grand Canyon, 153, **153**
Grasshopper, **359**, 360
Grasslands, 77, **77**
Gray matter, 518
Green algae, 238, **238**
Greenhouse effect, 15–16
Green tree frog, **376**
Green turtles, **38, 39**
   biotic potential of, 38–39
Grizzly bears, **34**
Ground squirrel, 413, **413**
Growth rings, 262
Grunion, **376**
Guanine, 131
Guard cells, 264
Gymnosperms, 264, 275–277
   life cycle of, **276**
   species of, 264

H

Habitats, 25, 42, 43
   destruction of, 25–26
Hagfish, 373, **373**
Hair follicle, 503, 504
Hammer (ear), 524
Hammerhead shark, **372**
Harvey, William, 480
Hatchet-footed mollusks, 355
Head (sperm), **550**
Head-foot
   of mollusks, 354
Hearing, 524–525
Heart
   of bird, 387
   four-chambered, 387
   of frog, 378
   of mollusks, 354
   three-chambered, 378, **378**
   two-chambered, 376
Heart (human), 478–479, **479**
   blood vessels of, **483**
   of embryo, 556
   location of, 478
   structure of, 479
Heart attack, 482
Heartbeat, 479
Heartbeat rate, 479
"Hearts," of earthworms, 352

Heat
   and thyroxine, 547
Helmont, J.B. van, 270
Hemoglobin, 474
Hepatitis, 216
Herbaceous stems, 262, **262**
Herbivores, 44, 386
Herd, animals in, **404**
Hereditary code, **132**, 166, **166**
Hereditary material. *See* DNA
Heroin, 530
Hibernation, 413
High concentration (area), 120
High-power objective, 102, 103
Hinge joints, **442, 443**
Hips, 441
Hoberman, Mary Ann, 362
Honey, **462**
Hooke, Robert, 106
Hookworm, 350
Hormones, 545
Hornworts, 268, **268**
Horse, **175, 176**
Horsetails, 266, **266**, 267, **267**
Host, 40, 212
   of flatworms, **349**
   of tapeworms, **349**
Howler monkeys, 407, **407, 408, 415**
Human development, 559
Human population, 36
Humpback whale, **389**
Humus, 20
Hybrid, 162
Hydra
   life cycle of, 347
Hydroponics, 306–307, **307**
Hyoid bone, **440**
Hyphae, 240, **241**
Hypothalamus, 546, 547
Hypothesis, defined, 5

I

Identical twins, 554, **555**
Iguana, **147**
Immigration, 37
Immune system, 214
   and AIDS, 214
Immunity, 476, **476**
Imperfect fungi, 243
Impulses, 514
Incomplete metamorphosis, 360
Incubation, 388–389
Infant, **559**
Influenza, 211, 216
Information
   processing, 512–513
   reception of, 514–516
Ingestion, 459
Inner ear, 524
Insects, 358–361
   adult, 360

benefits of, 361
characteristics of, 359
classifying, 359
control of, by pheromones, 412
development of, 360
Insulin, 133
International System of Units (SI). *See* SI
Interneurons, 516, **516**
Interphase, 128, **128**
Intertidal zone, 64
organisms of, **64**
Inuit, 301
Invertebrates, 332, 334–335
phylums of, *table* 334
Involuntary muscle, 448
Iris, 521
Iron
in hemoglobin, 475
and pregnancy, 558

J

Janssen, Zacharias, 105
Jaspersohn, William, 74
Jellyfish, 327, **328,** 344, 345, 347–348
life cycle of, **346**
Jenner, Edward, 216, **216**
Jointed appendages, 358
Joints, 442–443
ankle, 441
ball-and-socket, **443**
fixed, 442, **443**
gliding, **443**
hinge, **443**
pivot, **443**
Journal of Range Management, 8
Jungle, 72
Just, Ernest, biography of, 426, **426**

K

Kalanchoe, **293**
Kangaroo, 392, **392**
Kangaroo mice, 43–44, **43**
Kelp, 238
Kelvin, 101
Kidneys, 483, 498–500, **498, 499, 500**
Killer whale, camouflage of, 402, **402**
Kilo-, 99
Kilometer, 99
Kingdom, 185–189
of organisms (chart), 187
King snake, **382**
Knee, **442**
Koch, Robert, biography of, 252, **252**
Komodo dragons, **381**

L

Labor (birth), 558, **558**
Laboratory procedures
using a balance, 580
using dissecting tools, 580
using graduates, 579
and metric conversion, 578–579
using metric ruler, 578
using a microscope, 581
making a wet mount, 580
Lactic acid, 136
Lactose, 391
Lamarck, Jean Baptiste, 155
Lamprey, 373, **373**
Land biomes, 71–79
Landfill, 12, **12**
Lard, 466
Large intestine, function of, 463
Larva
of sponges, 343, **343**
of insects, 360
Larynx, 491, **491**
Laser, used for surgery, **518**
Lateral line, 375
Latin, 177
Leaf, 263–264, **300**
cross section, **264**
palisade layer of, **264**
spongy layer of, **264**
waxy, **300**
Leapfrog. *See* Frogs
Leeuwenhoek, Anton van, 105
in *Through the Microscope*, 250–251, **250, 251**
Legs, 441
Lend-a-Pet, **323**
Length (measurement), 99
Lens
of eye, 522
of microscope, 102, 103
Leprosy, 221
Lettuce plants, **211**
Levi-Montalcini, Rita, biography of, 201, **201**
Lichens, 49, 244, **244**
Life, creation of, 154
Life scientist
measurements used by, 98–101
tools of, 98, 102–103
Life zones of oceans, 64
Ligaments, 442, **442**
Liger, **184**
Light
and photosynthesis, 270–271, 273, 274
stimulus of, 291
Limiting factors (population), 39
Linnaeus, Carolus, 177–178, **177**
biography of, 201, **201**
Lipids, 466, **466**

Liter (L), 100
Liver, **537**
and digestion, 462, **462**
Liver flukes
life cycle of, **349**
Liverworts, 268, **268**
Living fossils, 277
Living things
categories of (chart), 185
interaction of, 34–35
Lizards, 381
communication among, 410
Lobsters, 364
Locomotion, 327
London, Jack, 76
Long, Irene Duhart, biography of, 571, **571**
Long bones, 436, **436,** 437
Long-day plants, 292
Low concentration (area), 122
Lower jaw, 441
Low-power objective, 102, 103
LSD, 530
Lung cancer, from smoking, 534
Lungs, **221**
cancerous, **534**
and excretion, 497
healthy, **491**
X-ray of, **494**
Lyell, Charles, 144–145
Lymph, 473

M

Malaria, 234
Male reproductive system, 549–550, **549**
Mallard duck, **384**
Malthus, Thomas, 150
Mammals, 389–393
brain of, 391, **391**
characteristics of, 391
parenting among, 414–415
types of, 392–393
Mammary glands, 391, 559
Mandibles, 364
Manta ray, 374, **374**
Mantle, of mollusks, 354
Marbled orb-weaver spider, **335**
Margarine, 466
Margulis, Lynn, biography of, 252, **252**
Marsupials, 392
Mass, 100, 101
Matter, 120
May beetle, **325**
Meader, Stephen W., 75
Measles, 213, 216
Measurement, system of, 98–101
Mechanical digestion, 459
Medicines, 531
during pregnancy, 557
Medusa, 347

Meiosis, 165, **165**
   defined, 165
Melanin, 506
Memory, 517
Mendel, Gregor, 158, **158,** 160–163, **161**
Menstrual cycle, 552–553, **553**
Menstruation, 552
Meristem, 261, **261**
Metabolic wastes, 497
Metabolism, 497
Metamorphosis, 379
   of insects, 360
Metaphase, 127, 128, **128**
Meter (m), 99
Microscope, 102, **102,** 103, 181, **326**
   carrying, 102
   focusing, 103
Middle ear, 524
Migration, 414
Milkweeds, 298
Miller, Stanley, 154, **154**
Millipedes, 365, **365**
Milt, 376
Mimicry, 403
Minerals
   and bones, 437
   as nutrient, 469
Mississippi River, 35
Mistletoe, 297
Mites, 364
Mitochondria, 111, **111,** 134, **134,** 490, 550, **550**
Mitosis, 127–129, **127, 128**
   and meiosis compared, table 166
   preparing for, 127
Molecules, 120
Mollusks
   types of, 354–356, **354, 355, 357**
Molting, 358
Monarch butterfly, **403, 414**
Monerans, 187, 218–223
   characteristics of, 218–219
Monocots, 265, **265**
Monotremes, 392
Moray eel, **41**
Morels, **242**
Morphine, 531
Mosquito, **189, 359**
Mosses, 50, 268, **268**
Motile, **327**
Motion sickness, 525
Motor neurons, 516, **516**
Mountain lion, 177, **177**
Mouse
   gestation of, 393
Mouth
   of coelenterates, 344
   human, 459, 490–491
   of lamprey, **373**
   of planarian, 348
   of roundworm, 350
Mucus
   affected by smoking, 534
   in nasal passages, 527
   in stomach, 461
Mule, **184**
Multicellular organisms, 112, **112, 113**
Multiple fruit, 281, **281**
Mumps, 213, 216
Muscle cells, **112**
Muscle fiber, 448, 449, 451
Muscles
   and ankle, **447**
   cardiac, **445,** 449, **449**
   development, 550
   diaphragm, 493, **493,** 494
   and exercise, 136, 451
   functions of, 445
   heart, **478**
   skeletal, **445, 446,** 447–448, **447**
   smooth, **445,** 448, **448**
   of stomach, 460–461, **461**
   types of, 445–449
   of upper arm, 448, **448**
   of uterus, 558
   well-exercised, 451
   working of, 449–451
Muscle tone, 451
Muscular contraction, 447, 449, 451
Muscular system, 445–451, **446**
   and respiratory system, 490
Mushrooms, **43, 46, 188,** 242, **242**
Mutations, 133
Mutualism, 41
Mycelium, 240
Myriapods, 365

N

Nash, Ogden, 356, 361, 381
Natural resources, 18–19
Natural selection, 155–157
   theory of, 157
Nearsightedness, 523
Nephrons, 499, **499**
Neritic zone, 65
Nerve cells, 514–515
Nerves, 513–514
Nervous system, 512–518, **513**
   of embryo, 556
   and endocrine system, 544
   parts of, 512–514
Nests, 388
Neurons, 514–515, **515**
   types of, 516, **516**
Niches, 43–44
Nicotine, 534
Night vision, 523
Nitrogen, 223
Nitrogen base (of DNA), 131
Nitrogen oxides, 13
Nonrenewable resources, 19
Nonvascular plants, 268
Nose, 490–491
Nostril, 526
Nuclear envelope, 109, **109**
Nucleic acid, 212, **212**
Nucleotides, 131, 166
Nucleus (cell), 109, **109**
Nutrient molecules, **134**
Nutrients
   carbohydrates as, 464–466
   defined, 458
   lipids as, 466
   minerals as, 469
   proteins as, 467
   vitamins as, 468
   water as, 469
Nutrition, 464–470, 564–568
Nymphs, 360

O

Objective lens, 102
Ocean life, zones of, **64**
Octopus, 356, **356**
Ocular lens, 103
Oil glands, 505
Oils, 466
Okubo, Akira, biography of, 91, **91**
Omnivores, 44, 386
Oparin, Aleksandr, 154
Open sea zone, 65
Operculum, 375
Opossum, **389,** 392
Optical illusion, 523, **523**
Optic nerve, 523
Oral groove, **233**
Order, 185
Organ development, 556–557
Organelles, 108, 110–111
Organisms, 113
   levels of classification (chart), 187
   levels of organization, 34, 111–113, **113**
   multicellular, 112
   relatedness of, 107
   single-celled, 112
Organs, 113
Osmosis, 123–124
Ostrich, **182**
Outer ear, 524
Ova, 551
Ovaries, 551, **551,** 552
Ovary
   of angiosperm, 278
Overdose, 532
Over-the-counter drugs, 531
Oviduct, 387

Ovulation, 552
Ovum
　during menstrual cycle, 552–553
　fertilized, 553, 554, **554**
Oxen, **175**
Oxygen
　and breathing, 494–495
　and cells, 134
　and exercise, 136
　and heart, 478
　from lungs to cells, 482
　and photosynthesis, 271, 273
　and red blood cells, 474
　and respiration, 134
　and respiratory system, 490
Oyster, **189**, 355
Ozone, defined, 14
Ozone layer, hole in, 14–15, **15,** 93, **93**

P

Pain, 528
Paint pots, **219**
Palamedes swallowtail butterfly, **360**
Pancreas, 462, 463
Paramecium, **187**, 232, **233**
Parasites, 40, **40**, 233
　flatworms as, 348
　tapeworms as, 349, **349**
　ticks and mites, 364
Parasitism, 40
Parent cell, 126
Passenger pigeon, **23**
Passive smoking, 535
Passive transport, 122, **122**
Pauraque, **400**
Pea plants, 160–163, **160, 162**
Pecans, **280**
Pectoral fins, 373, **373**
Pelvic fins, 373, **373**
Penicillin, 243
Penicillium, 243, **243**
Penis, **549,** 550
Perch, 375
Perennials, 262
Peristalsis, 460, **460**
Permafrost, 76
Permeable membrane, 121
Perspiration, 505
Pests
　animals as, 324
Petals, 278
Petrified forests, 153
Petunias, **275**
Pharmacist, **531**
Pharynx, 491
　of planarian, 348
Pheasants, **183**
Pheromones, 412
Phloem, defined, 260

Phosphorus, 469
Photosynthesis, 136–137, **137,** 236, 263, 270–274, **273**
　defined, 263
　and food chain, 274
　how it occurs, 273
　process of, 270–271
Phototropism, 291
Phylum, 185, 333
　of invertebrates, *table* 334
Physical addiction, 532, 533, 534, 537
Pigeons, 410, **410**
Pigments, 223, 237, 272, 385
Pill bug, 364
Pineapples, **281**
Pink flower mantis, **403**
Pioneer plants, 49
Pioneer trees, 53
Pistils, 278
Pituitary gland, 546, 547
Pivot joints, **443**
Placenta, 392, 556, **556**
Placental mammals, 392–393
Planarian, 348, **348**
Plankton, **65**
Plant adaptations, 296–301
Plant cells, **107, 112, 129, 326**
　discovery of, 106
　production of food and, 136–137
Plants, 188, **188,** 260–281
　energy transfer of, 274
　growth of, 288–294
　growth of, in space, 315
　land, **271**
　nonvascular, 268
　reaction to light, **273**
　responses of, 290–292, **291**
　seedless, 266–268
　seeds of, 275–281
　as source of energy, 274
　types of, 260–268
　vascular, 260–267
　water, **271**
Plasma, 473, 495
Platelets, **474,** 475
Platyhelminthes, 333
Pneumonia, **221**
Pods, of whales, 409
Poinsettias, **292**
Poison. *See also* Venom
　of arachnids, 363
Poison-arrow frog, **376**
Poison ivy, **298**
Polar bear, **189**
　camouflage of, 401
Polio, 213, 216
Pollen grains, 279
Pollen tube, 279
Pollination, 279, **279,** 324, **324**
Pollution, 10, **10**
　causes of, 10–11

Polyp, 347
Pond communities, 68, **68**
Poodle, **183**
Population density, 35, **36**
Populations
　characteristics of, 35–37
　human, 36
　increase of, 36–37
　relationship between, 40–41
　size of, 35–36
Porcupine, **405**
Pores, 342, **342,** 505
Portuguese man-of-war, 344, **344,** 345
Posterior, 328
Potatoes, **292**
　and starch production, 465
Pouches, 392
Prairies, 77
Predator-prey relationship, 40, **40**
Predators, 40
Pregnancy, 393, 553
Prenatal care, 557–558, **558**
Prenatal development, 554–558
Prescription drugs, **531**
Pressure, 528
Prey, 40
Primary succession, 49
Problem solving, 4–6
Producers (food), 44, 67
Prophase, 127, 128, **128**
Protein, 273, 467–468
　body's need for, 468
　and pregnancy, 558
　source of, 467, **467**
　storing, 467
Protein molecule, **467**
Protists, 187, **187,** 237, 326
Protozoans, 41, 230–234, **231**
　characteristics of, 230
　types of, 231–234, *table* 231
Pseudopods, 231, **232**
Psychoactive drugs, 531, 536
Psychological addiction, 532, 534, 537
Puberty, 550, 551
Puffball, **242**
Pulmonary artery, 482
Pulmonary circulation, **482**
Pulmonary system, 482
Pulmonary vein, 482
Pulse, 484
Pupa, 360
Pupil, 521

Q

Quill, 384

R

Rabies, 216

Rachis, 384, **385**
Radial symmetry, 327, **328,** 334
Radiation, 506
Raspberries, **281**
Rattlesnake, 382
Ray, John, 176
Ray-finned fish, 375
Rays, 373, 374
Reaction time, 536
Receptacle, 278
Receptor cells, 513
　in nasal passages, 526, **526**
　of skin, 527, **527**
Recessive factor, 162
Recombinant DNA, 133
　uses of, 133
Red algae, 238, **238**
Red blood cells, **109,** 126, 472, 474–475, **474,** 494–495
Red Hawaiian lobster, **358**
Red meats, 466
Red tide, 237
Red-winged blackbirds, 36, **36**
Reflex, 516–517, **517**
Reforestation
　in *Andy Lipkis and the Tree People*, 308–311, **308, 309, 310, 311**
Regeneration, 344
　of planarian, 348
　of sponges, 344
　of starfish, 357
Region of elongation, 261
Remora, **41**
Renewable resources, 18–19
Replication, 132
Reproduction
　of angiosperms, 279–280
　asexual, 266, 292
　of birds, 387
　of bony fishes, 376
　cell, 126–129, 165
　of coelenterates, 346–348
　of earthworms, 352
　of ferns, 266
　of flatworms, 348
　of frogs, 379
　of gymnosperms, 275–277
　of insects, 360
　of mammals, 392–393
　of mollusks, 355
　of reptiles, 381
　of roundworms, 350
　of spiders, 363
　of sponges, 343–344
　of starfish, 357
　of toads, 379
Reproductive system (human), 549–553
　female, 551
　male, 549–550
Reptiles, 380–382
　ancient, 380
　brain of, **391**
　modern, 381–382
Respiration, 134, 136, 490–495
　comparison with fermentation, *table* 136
Respiratory system, **491**
　of birds, 387, **387**
　structure of, 490–492
Response time, 516
Retina, 522
Rhinoceros, **176**
Rib cage, 493, 494
Ribosomes, 110, **110**
Ribs, 441
Rieu, E.V., 374
Ringworm, 243
Robin, **182**
Rods, of eye cells, 523
Roosevelt, Franklin D., 213, **213**
Root cap, 261
Root hairs, 261
Roots, 260–261
Root tip, **261**
Roses
　thorns, **299**
Roundworms, 350
　anatomy of, **350**
Rubella (German measles), 216
Rusts, 242

S

Sac fungi, 242, **242**
Sadness, **532**
Safety guidelines, 576–577
Salamanders, 380, **380**
Saliva, 214, 459, 544
Salivary glands, 544
Salmon, 375
Salt
　and water transport, 122
Saltwater ecosystems, 63, 64–66
Sanitary landfills, 12
Saplings, 290, **290**
Saprophytes, 240
Sarcodines, 231–232
Scales
　of fish, 373
Scavengers, 44, **44**
Schleiden, Matthias, 106
Schools, of fish, 409, **409**
Schwann, Theodor, 106
Scientific method, 4–8
　and collection of data, 6–7
　defined, 4
　and drawing conclusions, 7
　experiments and, 6–7
　problem solving in, 4–6
　reporting results and, 8
Scientists
　tools of, 98, 102–103
Scorpions, 364, **415**
Scrotum, 549, **549**
Sea anemones, 41, 345, **345**
Sea level, 61
Secondary sex characteristics
　female, 551, **551**
　male, 550, **550**
Secondary succession, 52, **52**
Seed development, 280–281, 288–289, **288, 289**
Seed germination, 281
Seeds, 264, **288, 289, 296, 297**
　dispersal of, 296–297
Segmented worms, 350–352, **351**
Selectively permeable
　membrane, 121
Self-pollination, 279
Semen, 214, 550
Semicircular canals, 525
Sense organs, 513
Senses, 520–528
Sensory neurons, 516, **516**
Sepals, 278
Sessile, 342
Setae, 351
Sharks, 373–374, **374**
　in *Shark Lady*, 198–200, **198, 200**
Sheep killed by coyotes, 5–8
Shell gland
　of birds, 387
Short bones, 436, **436**
Short-day plants, 292
Shoulders, 441
Shrub community, 52, **52**
SI, 98–101
　prefixes of, 99, *table* 99
Sickle-cell disease, 475, **475**
Sight, 521–523
Silverfish, 360
Simple eyes, 362–363
Simple fruit, 281, **281**
Simple sugar, 464
Skates, 373
Skeletal muscle, **445, 446,** 447–448, **447**
　reaction of, 450
Skeletal system, 434–443, **440**
　and muscular system, **446**
　and respiratory system, 490
Skeleton, 439–441
　appendages, 441
　of cat, **333**
　central, 439–440
Skin, 502–506, **502**
　cancer, 506
　cells, 126, 503
　and excretion, 505
　of fingertips, **503**
　functions of, 505–506
　of mammals, 391
　pores, 505
　receptors, 527
　shades of, **506**
　structure of, 502–505

Skull
    of human adult, 441, **441**
    of human infant, 441, **441**
Skunks, 405
Sleep disorders
    in *I Slept for Science*, 567–570, **569, 570**
Sleeping sickness, 233
Slime molds, 244, **244**
Small intestine, human, 462
Smallpox, 216
Smell, 526–527
Smoking, 533–535
    dangers of, 533–535
    during pregnancy, 557–558
    effects of smoking, 534–535
Smooth muscles, **445,** 448, **448**
Smuts, 242
Snails, 355, **355**
Snakes, 382. *See also* Reptiles
Snapping turtle, **325**
Snuff, 535
Sodium, 469
Soil
    conservation, 19–22
    defined, 19
    erosion of, 20–21
    layers of, **20**
Solar energy, 137
Solid waste disposal, 11–12
Spadefoot toad, 413, **413**
Sparrow, 384
Species, 182–184, 185
Sperm cells, 549, 550, **550,** 554, **554**
Spicules, 342
Spider, 362, **362,** 363–364
    anatomy of, **363**
    feeding, 363
    female with young, 363
    web of, 363
Spinal column, **439**
Spinal cord, 333
    and nervous system, 513, **513**
Spindle fibers, 128
Spine, 439
Spinnerets, 363
Spiny anteater, 392
Spleen, 474
Sponge, 330, **330**
    characteristics of, 342–343
    freshwater, 342
    life cycle of, 343–344, **343**
    natural, 342, **343**
    synthetic, **343**
Spongin, 342
Sporangia, 241, **241**
Sporangium, 241
Sporangium fungi, 241, **241**
Spores, 240, **241, 242,** 266
Sporophyte generation, 266
Sporozoans, 234, **234**
Spruce trees, 50, 51, **51**

Squid, **355,** 356
Stage (microscope), **102**
Stamens, 278
Starches, 273, 465, **465**
Starfish, **189,** 327, **328,** 356–357, **357**
    reproduction of, 357
Statues, and acid rain, **13**
Stems, 262–263, **262, 263**
Steroids, 531
Stethoscope, 480
Stigma, 278
Stimulus, 291, 513
Stingray, 374
Stirrup, 524
Stomach
    of fish, 375
    human, 460–461, **461**
    of starfish, 357
Stomata, 264
Strep throat, 218, **218,** 221
Stress
    responses to, 545
Style, 278
Subsoil, 20
Succession, 48–50
    final stage, 50
    grassy stage, 50
    mossy stage, 50
    pioneer stage, 49
    primary, 48
    secondary, 52
Sugar, 273, 464
    in blood, 473
Sugar cane, **464**
Sulfur oxides, 13
Sunlight
    source of energy, 137, 274
Sutton, Walter, 165a
Sweat, 505
Sweat glands, 505, **505,** 544
Swensen, Peter, 66
Swim bladder, 375
Swimmerets, 364
Symmetry, 327–330
Synapse, 515
Syrinx
    of birds, 387
System, defined, 113
Systemic circulation, **482**
Systemic system, 482

T

Tadpole, 379, **379**
Taiga. *See* Boreal forests
Tail (sperm), 550, **550**
Tapeworms, 349, **349**
Taproots, 260, **261**
Tar, 534
Taste, 526–527
Taste buds, 526, **526**
Taxonomy, 176

Tears, 214
Teenager, **559**
Teeth, of humans, 459, **459**
Telophase, 127, 128, **128**
Temperature. *See also* Body temperature
    measurement of, 101
Tendons, 447
Termites, 41, 233, 361
Testes, 549, **549,** 550
    of birds, 387
Testosterone, 550
Tetanus, 221
Theophrastus, 175, **175**
Theory, defined, 8
Threatened species, 25
Thrush, 384
Thymine, 131
Thyroid gland, 547
Thyroxine, 549
Ticks, 364
Tiger, **389,** 401
    camouflage of, 401
Tiger swallowtail butterfly, **335**
Tissues, 112
Toads
    development of, 379
    and frogs grouped, 378–379
    tadpoles of, 379
Tobacco, 530, 533–535
    during pregnancy, 557–558
Toddler, **559**
Tolerance, 532
Tomatoes, **280**
Topsoil, 20
Torpor, defined, 413
Tortoises, **148, 150,** 380
Toucan, **182**
Touch, 527–528
Toxic, 532
Trachea, 491, **491**
Traits, 160, **160,** 162–163
Tranquilizers, 531
Transport (in cells), 120–124
Trees
    bark of, 262
Trilobite, **152**
Troop, of baboons, 409
Tropical rain forests, 71–73, **72,** 84–86, **85,** 145
    animals of, **86, 423, 424, 425**
    destruction of, 17, 73, **73,** 84–86, **86**
    in *Watch Your Step*, 422–425, **422**
Tropisms, 291
Trout, 375
Tsetse fly, 233
Tuberculosis, 221, **221**
"Tube-within-a-tube" body plan, 352
Tundra, 76, **76,** 301, **301**
Twins, 554–555, **555**

Index   **605**

U

Ulcer, 461
Umbilical cord, 392, 556, **556,** 558
Unicorn beetle, **334**
United States Census Bureau, 37
Urea, 498, 505
Ureter, 500
Urethra, 500, 550, **551**
Urinalysis, 500
Urinary bladder, 500
Urine, 498, 550
Uropod, 364
Uterus, 393, 551, **551, 556**

V

Vaccination, **215,** 476, **476**
Vaccine, 215
  for viral diseases, *table* 216
Vacuoles, 111
  food, 231–232
Vagina, 551, **551**
Valves, of heart, 479, 480
Vane, of feather, 384, **385**
Variable, 6
Vascular cambium, 263
Vascular plants, 260–267
  complex, 264–265
  simple, 266–267
Vas deferens, 550
Vegetative propagation, 292–293
Veins, **480,** 481
Venom
  of beaded lizard, 381
  of gila monster, 381
  of snakes, 382
  of stingray, 374
Ventral side, 329
Ventricles, 479
Venus' flytrap, 290–291, **290**
Vertebrae, 332, 333, 439, **439**

Vertebrates, 332–333
  classes of, *table* 333
  coldblooded, 372–382
  species of, 333
  warmblooded, 384–393
Viceroy butterfly, **403**
Villi, of small intestine, 463, **463**
Violets, **292**
Viral diseases, *table* 216
Virchow, Rudolf, 106
Viruses, 210–216, **210, 212, 213**
  AIDS, 214
  characteristics of, 211–212
  and diseases, 213–216
  reproduction of, 212, **212**
  shapes of, 211–212
Visceral mass, of mollusks, 354
Vision, 521–523
Vitamins, 468
  overdosing of, 468
  source of, **468**
Vocal cords, 491–492, **491**
Volume, 99–100, **101**
Voluntary muscle, 447
Voluntary responses, 516
Vultures, **44**

W

Walking sticks, 403
Wallace, Alfred, 158
Walnuts, **280**
Warmblooded animals, 384
Warts, 213
Waste removal, 497–498
Water
  consumption and excretion, *table* 498
  as life essential, 469
  recycling of, 23, 67
  from respiration, 134
  sources of, **469**
Water conservation, 22–23

Water cycle, **22, 67**
Water moccasin, 382
Watershed, defined, 23, **23**
Water-soluble vitamins, 468
Watson, James, 130, **130,** 131
Wax (plants), 300
Webs, 363
Weeds, 52
Whales, 391
Wheat rust, **242**
White blood cells, 474, **474,** 475, **475**
White fly virus, **211**
White pelican, **408**
Whooping cough, 221
Wildlife conservation, 23–27, 428–429, **428, 429**
  in *An Eagle to the Wind*, 87–90
Wilting, 124
Windpipe, 460, 491
Wings, 386
  of insects, 359
Witch-hazels, seeds, **297**
Withdrawal, 532
Wood frog, **376**
Woody stems, 262, **262**
Worms, 348–352
Worth, Valerie, 351

X

Xylem, defined, 260

Y

Yeast, 135, 242, **242**
Yolk, 387

Z

Zebras, 401
  camouflage of, 401
Zygote, 279, 393, 554

Credits

PHOTOS

Abbreviations used: (t) top, (c) center, (b) bottom, (l) left, (r) right, (bkgrd) background.

Page: ii, Laurence Parent; vi (t), NASA/Science Source/Photo Researchers; vi (bl), Ed Reschke/Peter Arnold; vi (br), Roger Wilmshurst/Bruce Coleman, Inc.; vii (tl), David Scharf/Peter Arnold; vii (tr), HRW photo by James Newberry; vii (b), Dwight Kuhn; viii (tl), Art Wolfe/Allstock; viii (tr), Montagnier/Instit Pasteur/SPL/Science Source/Photo Researchers; viii (b), David Doubilet; ix (t), Alfred Pasieka/Bruce Coleman, Ltd.; ix (bl), Kerry T. Givens/Tom Stack & Associates; ix (br), Runk/Schoenberger/Grant Heilman; x (tl), Michael P. Gadomski/Bruce Coleman, Inc.; x (tr), Art Wolfe/Allstock; x (b), Hans Reinhard/Okapia/Photo Researchers; xi, Breck Kent/Animals Animals; xii (tl), Frans Lanting/Allstock; xii (tr), Shelby Thorner/David Madison; xii (b), Mary Gow; xiii, Park Street; xiv (tl), HRW photo by Richard Haynes; xiv (tc), Don Fawcett/Science Source/Photo Researchers; xiv (tr), M. Wurtz/Biozentrum, University of Basel/Photo Researchers; xiv (b), John Walsh/Photo Researchers; xv (t), Eric Grave/Photo Researchers; xv (b), HRW photo by Richard Haynes; xvi (tl), M. I. Walker/Photo Researchers; xvi (tr), Terry Domico/Earth Images; xvi (b), Marty Snyderman; xviii (t), HRW photo by James Newberry; xviii (b), HRW photo by Henry Friedman; xix, HRW photo by Dennis Carlyle Darling; xxii, David Parker/Science Photo Library/Photo Researchers; 1 (inset), Will & Deni McIntyre/Allstock; 2, Photo Library International/Nawrocki Stock Photo; 3 (t), Susan McCartney/Photo Researchers; 3 (b), Simon Fraser/Science Photo Library/Photo Researchers; 4(l), Frederica Georgia/Photo Researchers; 4 (r), Dave Brown/Nawrocki Stock Photo; 5 (t), E.R. Degginger; 5 (b), Joe Branney/Tom Stack & Associates; 6, both HRW photos by Richard Haynes; 8, HRW photo by Griff Smith; 10 (l), Gerhard Gscheidle/Peter Arnold, Inc.; 10 (c), Craig Aurness/Woodfin Camp & Associates; 10 (r), Chuck O'Rear/Woodfin Camp & Associates; 11, Paul Dix/Tony Stone Worldwide; 12 (l), Tompix/Peter Arnold; 12 (r), Ray Pfortner/Peter Arnold; 13 (l), Ray Pfortner/Peter Arnold, Inc.; 13(r), Runk/Schoenberger/Grant Heilman Photography; 15, NASA/Science Source/Photo Researchers; 17, Dr. Nigel Smith/Earth Scenes; 18 (l), Grant Heilman/Grant Heilman Photography; 18 (r), Laurence Parent; 19, Peter Miller/Photo Researchers; 20, Ray Ellis/Photo Researchers; 21, Grant Heilman/Grant Heilman Photography; 23 (t), Randall Hyman/Stock Boston; 23 (b), Culver Pictures; 24 (t), Rick Sullivan/Bruce Coleman, Inc.; 24 (b), Tom & Pat Leeson/Photo Researchers; 25 (l), Bruce Davidson/Animals Animals; 25 (r), Steve Turner/Oxford Scientific Films/Animals Animals; 26 (t), Tom McHugh/Photo Researchers; 26 (b), 27, Ron Garrison/Zoological Society of San Diego; 29, Photo Library International/Nawrocki Stock Photo; 31, Park Street; 32, William Townsend/Photo Researchers; 33, Tim Childs/Leo de Wys; 34, Johnny Johnson/DRK Photo; 35 (l), David Cavagnaro/DRK Photo; 35 (r), John Trott/Animals Animals; 36, D. & M. Zimmerman/VIREO; 37 (l), John Walsh/Photo Researchers; 37 (r), David Cavagnaro/DRK Photo; 38, Kevin Schafer/Peter Arnold; 39, J.A.L. Cooke/Oxford Scientific Films/Animals Animals; 40 (tl), Arthur Panzer/Photo Researchers; 40 (tr), Norman Myers/Bruce Coleman, Inc.; 40 (bl), Peter Ward/Bruce Coleman, Inc.; 40 (br), Adrian Davies/Bruce Coleman, Inc.; 41, Bill Wood/Robert Harding Picture Library; 43 (tl), E. R. Degginger; 43 (tr), Jeff Lepore/Photo Researchers; 43 (bl), Hans & Judy Beste/Animals Animals; 43 (br), Keith Gillett/Animals Animals; 44, Stephen J. Krasemann/DRK Photo; 46, E. R. Degginger; 48 (l), Jed Wilcox/Tony Stone Worldwide; 48 (r), William E. Ferguson; 49, Larry Ulrich/DRK Photo; 50 (l), Tom Bean; 50 (c), (r), Tom & Susan Bean/DRK Photo; 51, Tom Bean; 52, William E. Ferguson; 53, R. Carr/Bruce Coleman Ltd.; 55, Wm. Townsend/Photo Researchers; 57 (l), Tom & Susan Bean/DRK Photo; 57 (r), John Gerlach/Tom Stack & Associates; 58, both photos by NASA; 60, Park Street; 61, NASA; 62 (l), Laurence Parent; 62 (tr), Steve Vidler/Nawrocki Stock Photo; 62 (br), David C. Fritts/Earth Scenes; 64, Frans Lanting/Photo Researchers; 65, both photos by D.P. Wilson/Science Source/Photo Researchers; 66, Peter David/Photo Researchers; 67 (l), Jeff Apoian/Nawrocki Stock Photo; 67 (r), Ronald Toms/Oxford Scientific Films/Earth Scenes; 69, Doug Wechsler/Earth Scenes; 71 (l), Laurence Parent; 71 (r), Brian Lovell/Nawrocki Stock Photo; 72, Peter Veit/DRK Photo; 73, David Hiser/Photographers Aspen; 74, Roger Tully/Tony Stone Worldwide; 75, Jim Brompton/Valan Photos; 76, Roger Wilmshurst/Bruce Coleman, Inc.; 77, Linda Dufurrena/Grant Heilman; 79 (l), John Cancalosi/Peter Arnold; 79 (r), Ed Reschke/Peter Arnold; 81, NASA; 82, Charlie Ott/Photo Researchers; 83 (l) Bill Bachman/Earth Images; 83 (r), Grant Heilman/Grant Heilman Photography; 84, Gregory G. Dimijian, M.D./Photo Researchers; 85, David Barnes/Allstock; 85 (inset), Joy Spurr/Bruce Coleman, Inc.; 86 (l), Gregory G. Dimijian, M.D./Photo Researchers; 86 (tr), Michael Fogden/Animals Animals; 86 (cr), E.R. Degginger, Bruce Coleman, Inc.; 86 (br), Tom McHugh/Photo Researchers; 91 (t), Erich Hartman/Magnum; 91 (b), Marine Science Research Center; 92, both HRW photos by James Newberry; 93, NASA; 94, 95 (inset), Figaro Magazine/Gamma–Liaison; 96, Dr. Jeremy Burgess/Science Photo Library/Photo Researchers; 98, HRW photo by Eric Beggs; 99, 102, HRW photos by Richard Haynes; 103, Science Photo Library/Photo Researchers; 104 (t), (b), Jim Zuckerman/Westlight; 105, HRW photo by Eric Beggs; 106 (l), Bettmann Archives; 106 (r), Runk/Schoenberger/Grant Heilman Photography; 107, Arthur M. Siegelman; 109 (tl), Don Fawcett/Photo Researchers; 109 (tr), Bill Longcore/Science Source/Photo Researchers; 109 (b), Don Fawcett/Photo Researchers; 110, Don Fawcett/Photo Researchers; 112 (t), Michael Abbey/Photo Researchers; 112 (c), Kevin Morris/Allstock; 112 (bl), (bc), G.W. Willis, M.D./Biological Photo Service; 112 (br), Jim Solliday/Biological Photo Service; 114 (l), Alfred Owczarzak/Biological Photo Service; 114 (r), Runk/Schoenberger/Grant Heilman; 115, Dr. Jeremy Burgess/Science Library/Photo Researchers; 116, P. Dayanandan/Photo Researchers; 117, David Scharf/Peter Arnold, Inc.; 120, all HRW photos by Richard Haynes; 121, HRW photo by Eric Beggs; 122, both HRW photos by Richard Haynes; 124, both HRW photos by Eric Beggs; 126, Dwight Kuhn; 127, Dr. Lloyd M. Beidler/Science Photo Library/Photo Researchers; 128, 129, all photos by Arthur M. Siegelman; 130, Barr-Brown/Camera Press; 131, Dan McCoy/Robert

Langridge/Rainbow; 133, HRW photo by Richard Haynes; 134, Leonard Harris/Leo de Wys, Inc.; 135 (l), Uselmann/H. Armstrong Roberts, Inc.; 135 (r), HRW photo by Richard Haynes; 136, Paul J. Sutton/ Duomo; 141, Tom Stack & Associates; 147, both photos by Tom Brakefield; 150 (l), Breck P. Kent; 150 (r), Eric Hosking/Bruce Coleman, Inc.; 152 (t), Runk-Schoenberger/Grant Heilman Photography; 152 (bl), Stephen J. Krasemann/DRK Photo; 152 (br), J. Koivula/Photo Researchers; 153 (l), Ed Cooper; 153 (r), William E. Ferguson; 154, UPI/Bettmann Newsphotos; 155 (t), Stephen J. Krasemann/Photo Researchers, Inc.; 155 (b), M. P. Kaul/VIREO; 156, Zig Leszczynski/Animals Animals; 158, Culver Pictures; 160, Ken Brate/Photo Researchers; 161, Culver Pictures; 163, HRW photo by Mary Gow; 165, Manfred Kage/Peter Arnold; 167, Culver Pictures; 169 (l), Tom McHugh/Photo Researchers; 169 (cl), Art Wolfe/Allstock; 169 (cr), (r), Park Street; 171, Tom McHugh/Photo Researchers; 172, HRW photo by James Newberry; 174, HRW photo by Mary Gow; 175 (tl), Ralph A. Reinhold/Animals Animals; 175 (tr), M. Wendler/Okapia/Photo Researchers; 175 (c), Alinari/Art Resource; 175 (b), Culver Pictures; 176 (l), Larry Lefever/Grant Heilman Photography; 176 (l, inset), Leonard Lee Rue III/Animals Animals; 176 (c), Park Street; 176 (c, inset), Kristin Finnegan/Allstock; 176 (r), Vince Streano/Allstock; 176 (r, inset), Charles Krebs/Allstock; 177 (t), The Mansell Collection; 177 (b), Robert Winslow/Tom Stack & Associates; 178, E. R. Degginger; 179, all HRW photos by Eric Beggs; 181 (t), A. B. Dowsett/Science Photo Library/Photo Researchers; 181 (b), HRW photo by Eric Beggs; 182 (tl), Rod Planck/Photo Researchers, Inc.; 182 (tr), Renee Lynn/Photo Researchers, Inc.; 182 (bl), Jeff Foott/Bruce Coleman, Inc.; 182 (br), Laura Riley/Bruce Coleman, Inc.; 183 (tl), Laurence Parent; 183 (tr), George E. Jones III/Photo Researchers; 183 (b), Art Wolfe/Allstock; 184 (tl), L. L. T. Rhodes/Animals Animals; 184 (tr), Rod Allin/Tom Stack & Associates; 184 (c), Porterfield/Chickering/Photo Researchers; 184 (b), Grant Heilman Photography, Inc.; 185, HRW photo by Eric Beggs; 187 (t), Dr. Tony Brain/Photo Researchers; 187 (bl), Eric V. Grave/Photo Researchers; 187 (bc), (br), E.R. Degginger/Bruce Coleman, Inc.; 188 (t), E.R. Degginger; 188 (b), Laurence Parent; 189 (l), (c), E.R. Degginger; 189 (tr), Jeff Rotman; 189 (cr), John Shaw/Tom Stack & Associates; 189 (br), Breck P. Kent/Animals Animals; 191, HRW photo by James Newberry; 194, Figaro Magazine/Gamma–Liaison; 194 (bkgr), Ron Slenzak/Westlight; 195, Gamma-Liaison; 196 (tl), (tc), (tr), David Brill; 196 (b), Yoichi R. Okamoto/Photo Researchers; 197, all photos by David Brill; 198, (l), (r), David Doubilet; 198 (bkgrd), Ralph A. Clevenger/Westlight; 200, David Doubilet; 201 (t), The Mansell Collection; 201 (b), Erich Hartmann/Magnum; 202, HRW photo by Henry Friedman; 203 (l), Runk/Schoenberger/Grant Heilman Photography; 203 (r), Louis Bencze Photo/Allstock; 204 (t), Grant Heilman/Grant Heilman Photography; 204 (b), Thomas C. Boyden; 205 (t), Chip Porter/Allstock; 205 (b), Ken Graham/Allstock; 206, CDC/Science Source/Photo Researchers; 208 (tl), CNRI/SPL/Photo Researchers; 208 (tr), Dwight Kuhn; 208 (b), Jean–Loup Charmet/Science Photo Library/Photo Researchers; 209, Jane Burton/Bruce Coleman, Inc.; 210 (t), Alfred Pasieka/Bruce Coleman, Ltd.; 210 (bl), (bc), HRW photos by Eric Beggs; 210 (br), NASA; 211, Dr. James E. Duffus/U.S. Department of Agriculture; 211 (inset), U.S. Department of Agriculture; 212 (l), CDC/RG/Peter Arnold; 212 (c), M. Wurtz/Biozentrum, University of Basel/Photo Researchers; 212 (r), Omikron/Photo Researchers; 213 (t), Susan Gibler/Tom Stack & Associates; 213 (b), Bettmann Archives; 214, Montagnier/Instit Pasteur/SPL/Science Source/Photo Researchers; 215 (t), Nathan Benn/Woodfin Camp & Associates; 215 (b), Zeva Oelbaum/Peter Arnold; 216, Bettmann Archives; 218, David M. Phillips/Visuals Unlimited; 219 (l), J. Robert Stottlemyer/Biological Photo Service; 219 (r), R. Knauft/Biology Media/Photo Researchers; 221 (tl), CNRI/Science Photo Library/Photo Researchers; 221 (tr), Laurence Parent; 221 (bl), Centers for Disease Control; 221 (bcl), (bcr), Arthur M. Siegelman; 221 (br), Martin Rotker/Phototake; 222, Hank Morgan/Rainbow; 223 (l), T. E. Adams/Visuals Unlimited; 223 (r), Sinclair Stammers/Science Photo Library/Photo Researchers; 225, CNRI/SPL/Photo Researchers; 226 (l), J. Robert Stottlemeyer/Biological Photo Service; 226 (r), Park Street; 228 (l), C. James Webb/Bruce Coleman, Inc.; 228 (r), Stephen Dalton/Photo Researchers; 229, Brown Brothers; 230, Laurence Parent; 231, Manfred Kage/Peter Arnold, Inc.; 232 (tl), (tr), (cl), (cr), Michael Abbey/Science Source/Photo Researchers; 232 (b), Manfred Kage/Peter Arnold, Inc.; 233, Arthur M. Siegelman; 235 (l), M. I. Walker/Photo Researchers; 235 (c), E. R. Degginger/Bruce Coleman, Inc.; 235 (r), E. R. Degginger; 236, Ed Degginger/Bruce Coleman, Inc.; 237, Eric Grave/Photo Researchers; 238 (t), Manfred Kage/Peter Arnold, Inc.; 238 (bl), Brian Parker/Tom Stack & Associates; 238 (bc), Tom Stack/Tom Stack & Associates; 238 (br), E.R. Degginger/Bruce Coleman, Inc.; 240, Runk/Schoenberger/Grant Heilman Photography; 241, Barry L. Runk/Grant Heilman Photography; 242 (tl), (tr), (cl), E.R. Degginger; 242 (cr), Grant Heilman/Grant Heilman Photography; 242 (bl), Manfred Kage/Peter Arnold, Inc.; 242 (br), Kerry T. Givens/Tom Stack & Associates; 243 (l), Arthur M. Siegelman; 243 (r), A. & F. Michler/Peter Arnold, Inc.; 244 (t), Grant Heilman/Grant Heilman Photography; 244 (b), Kerry T. Givens/Tom Stack & Associates; 245, Stephen Dalton/Photo Researchers; 246, Grant Heilman/Grant Heilman Photography; 247, D.C. Lowe/Allstock; 248 (t), Will & Deni McIntyre/Allstock; 248 (bl), Will & Deni McIntyre/Photo Researchers; 248 (br), NIBSC/Science Photo Library/Photo Researchers; 249 (l), Wil Phinney; 249 (c), Reuters/Bettmann; 249 (r), Will & Deni McIntyre/Photo Researchers; 250, Science Photo Library/Photo Researchers; 250 (bkgrd), Ed Reschke/Peter Arnold; 251 (t), (c), Biophoto Associates/Photo Researchers; 251 (b), Mary Evans Picture Library/Photo Researchers; 251 (bkgrd), Manfred Kage/Peter Arnold; 252 (t), Dan McCoy/Rainbow; 252 (b), Bettmann Archives; 253, all HRW photos by James Newberry; 254 (t), William E. Ferguson; 254 (c), (b), David J. Sams/Texas Imprint; 255 (t), (b) , D.P. Wilson/Science Source/Photo Researchers; 255 (c), M. I. Walker/Photo Researchers; 256, M. Thonig/Allstock; 258, Terry Madison/The Image Bank; 259, HRW photo by Eric Beggs; 260, Laurence Parent; 261 (l), Runk/Schoenberger/Grant Heilman; 261 (r), Walter Chandoha; 265, David Ball/Allstock; 266, Phil Degginger; 266 (inset), Wayne Lankinen/Bruce Coleman, Inc.; 267 (l), Bob & Clara Calhoun/Bruce Coleman, Inc.; 267 (r), Professor R. C. Simpson/Valan Photos; 268 (l), Larry West/Bruce Coleman, Inc.; 268 (c), E. R. Degginger/Bruce Coleman, Inc.; 268 (r), E. R. Degginger; 270, Will McIntyre/Photo Researchers; 271 (l), Marty Snyderman; 271 (r), E.R. Degginger; 272 (t), Runk/Schoenberger/Grant Heilman Photography; 272 (bl), Ed Cooper; 272 (bc), Norman Owen Tomalin/Bruce Coleman, Inc.; 272 (br), Michael P. Gadomski/Bruce Coleman, Inc.; 274 (l), Dan McCoy/Rainbow; 274 (r), Paul Conklin/TexaStock; 275 (l), Hans Reinhard/Okapia/Photo Researchers;

275 (r), Laurence Parent; 276 (l), Dwight Kuhn; 276 (r), Grant Heilman/Grant Heilman Photography; 277 (tl), Julia Brooke-White/Photo Researchers; 277 (tc), JC Carton/Bruce Coleman, Inc.; 277 (tr), Biophoto Associates/Photo Researchers; 277 (b), E.R. Degginger; 279, John Shaw/Tom Stack & Associates; 280, HRW photo by Mary Gow; 281 (l), (c), E.R. Degginger; 281 (r), Phil Degginger; 283, Terry Madison/The Image Bank; 285, all photos by E.R. Degginger; 286 Dr. Charles Steinmetz/Photo/Nats; 286 (bkgrd), Cameron Davidson/Wingstock; 287, Valerie Hodgson/Photo/Nats; 288, all photos by Terry Domico/Earth Images; 289, Dr. G.J. Chafaris; 290 (tl), (bl), (br), E.R. Degginger; 290 (tr), Laurence Parent; 291, Runk/Schoenberger/Grant Heilman; 292 (tl), (tr), E.R. Degginger; 292 (b), Dwight Kuhn; 293 (t), Darrell Gulin/Allstock; 293 (bl), Laurence Parent; 293 (br), Dwight Kuhn; 294 (tl), (tr), Muriel Orans; 294 (b), Roxie Olsen/Gurney Seed & Nursery; 296 (t), Laurence Parent; 296 (b), E.R. Degginger; 297 (tl), (tc), Dwight Kuhn; 297 (tr), E.R. Degginger; 297 (b), Dave Spier/Natural Selection; 298, E.R. Degginger; 299 (l), John Shaw/Bruce Coleman, Ltd.; 299 (r), Laurence Parent; 300 (tl), Gary Holscher/Allstock; 300 (tc), Gary Braasch/Allstock; 300 (tr), Dr. E.R. Degginger; 300 (c), Carr Clifton/Allstock; 300 (b), Tom Bean; 301, Charles Mauzy/Allstock; 303, Valerie Hodgson/Photo/Nats; 304 (l), Dwight R. Kuhn; 304 (c), (r), Charles Krebs/Allstock; 305, Michael Tweedie/NHPA; 306, Grant Heilman/Grant Heilman Photography; 307 (t), G. I. Bernard/Earth Scenes; 307 (b), Barry L. Runk/Grant Heilman Photography; 308, Mark Wexler; 309, Treepeople; 310, 311, Mark Wexler; 312 (t), Courtesy of Tuskegee Institute; 312 (b), Scott Hutchinson; 313, 314, HRW photos by Dennis Carlyle Darling; 315 (t), NASA; 315 (b), Plantek/Photo Researchers; 315 (bkgd), NASA: 316 (1), Robert Young Pelton/Westlight; 316 (r), George B. Schaller; 318 (l), Thomas C. Boyden; 318 (r), Frans Lanting/Minden Pictures; 319, Thomas C. Boyden; 320, FourByFive/Superstock; 321, David Hughes/Bruce Coleman, Ltd.; 322 (t), Stephen Dalton/Oxford Scientific Films/Animals Animals; 322 (b), Laurence Parent; 323, Mary Gow; 324, A.J. Deane/Bruce Coleman, Ltd.; 325 (tl), Tom Brakefield; 325 (bl), Hans Pfletschinger/Peter Arnold; 325 (r), Dwight R. Kuhn; 326, Harry Holloway/Carolina Biological Supply Co.; 327, Stephen J. Krasemann/Photo Researchers; 328 (l), Mike Neuman/Photo Researchers; 328 (r), William Curtsinger/Photo Researchers; 330, HRW photo by Mary Gow; 332 (t), Nadia Mackenzie/Nawrocki Stock Photo; 332 (b), Marty Snyderman; 333, Radiographic image courtesy of the College of Veterinary Medicine, Texas A & M University; 334, J.H. Robinson; 335 (l), John Gerlach/Tony Stone Worldwide; 335 (tr), Robert Harding Picture Library; 335 (br), J.H. Robinson; 336, James Newberry; 337, Thomas C. Boyden; 339 (l), Breck P. Kent; 339 (r), Walter Chandoha; 340, Jane Burton/Bruce Coleman, Ltd.; 341, both photos by Marty Snyderman; 342, Jeff Rotman; 343 (l), Russ Kinne/Comstock; 343 (r), HRW photo by Richard Haynes; 344, Runk/ Schoenberger/Grant Heilman; 345 (l), Robert Lee/Photo Researchers; 345 (r), Robert Harding Picture Library; 346, NHAP; 347, Kim Taylor/Bruce Coleman, Ltd.; 350, Mary Gow; 351, E.R. Degginger; 354, James Carmichael/ NHPA; 356, Tom McHugh/Photo Researchers; 357, Russ Kinne/ Comstock; 358 (l), Rod Borland/Bruce Coleman, Inc.; 358 (r), E.R. Degginger/ Bruce Coleman, Inc.; 360, all photos by E.R. Degginger, 362, John Shaw; 364, Tom McHugh/Photo Researchers; 365 (t), A. Kerstitch/Bruce Coleman, Inc.; 365 (b), John MacGregor/Peter Arnold, Inc.; 367, Jane Burton/Bruce Coleman, Ltd.; 369 (l), Al Grotell; 369 (r), Dwight R. Kuhn; 370 (l), Craig Blacklock; 370 (r), Nicholas deVore III/Photographers Aspen; 371 (t), Tom & Pat Leeson; 371 (c), Tom Brakefield; 371 (b), Stephen J. Krasemann/Allstock; 372 (l), Marty Snyderman/The Waterhouse; 372 (r), Dwight Kuhn; 373 (l), Russ Kinne/Comstock; 373 (r), Heather Angel; 374 (tl), David Doubilet; 374 (tr), Tom McHugh/Photo Researchers, Courtesy of Steinhart Aquarium; 374 (b), Mike Neumann/Photo Researchers; 376 (tl), Tom McHugh/Photo Researchers; 376 (cl), Jeff March; 376 (bl), Gwen Fidler/Comstock; 376 (r), Breck P. Kent; 378, Breck Kent/Animals Animals; 380 (t), Raymond A. Mendez/Animals Animals; 380 (b), David M. Dennis; 381, George Holton/Photo Researchers; 382 (tl), (bl), Breck P. Kent; 382 (r), John Visser/Bruce Coleman, Inc.; 384, both photos by Art Wolfe/Allstock; 385, Walter Hodges/Westlight; 388 (tl), (cl), J. H. Robinson; 388 (bl), Breck P. Kent; 388 (r), Jim Brandenburg/Minden Pictures; 389 (tl), George Holton/Photo Researchers; 389 (tr), Jeff Lepore/Photo Researchers; 389 (bl), C. Allan Morgan/Peter Arnold, Inc.; 389 (br), William & Marcia Levy/Photo Researchers; 390, Laurence Parent; 392 (t), NHPA; 392 (b), Adrienne T. Gibson/ Animals Animals; 393, Kent & Donna Dannen; 394, Hans Reinhard/Bruce Coleman, Inc.; 395, Tom Brakefield; 397, Geopress/Allstock; 398 (t), Turid Forsyth; 398 (b), Adrian Forsyth; 400 (l), Doug Wechsler/VIREO; 400 (r), Cosmos Blank/Photo Researchers; 401 (l), Breck P. Kent; 401 (r), E.R. Degginger/Bruce Coleman, Inc.; 402 (t), Stuart Westmoreland/The Stock Market; 402 (b), M.P.L. Fogden/Bruce Coleman, Inc.; 403 (t), P. Ward/Bruce Coleman, Inc.; 403 (b), Breck P. Kent; 404 (l), Norman Myers/Bruce Coleman, Inc.; 404 (r), Mitch Reardon/Photo Researchers, Inc.; 405 (l), Tom Brakefield; 405 (r), Tom Brakefield/ Bruce Coleman, Inc.; 406, M. Reardon/ Photo Researchers; 407, D & R Sullivan/Bruce Coleman Ltd.; 408 (tl), Rod Williams/Bruce Coleman, Inc.; 408 (tr), Tom McHugh/Photo Researchers; 408 (b), Calvin Larsen/Photo Researchers; 409 (l), Francois Gohier/Photo Researchers; 409 (r), Burt Jones & Maurine Shimlock/Photographers Aspen; 410 (t),Tom McHugh/Photo Researchers; 410 (b), Thomas C. Boyden; 411, Terry Domico/Earth Images; 412, Sven–Olof Lindblad/Photo Researchers; 413 (t), Wolfgang Bayer/Bruce Coleman, Inc.; 413 (b), Jeff Lepore/Photo Researchers; 414 (l), Laurence Parent; 414 (r), Francois Gohier/Photo Researchers; 415 (t), Thomas C. Boyden; 415 (b), Carol Hughes/Bruce Coleman, Inc.; 417, Turid Forsyth; 418 (l), Erwin & Peggy Bauer/Bruce Coleman, Inc.; 418 (r), Stephen Dalton/Photo Researchers; 419 (l), Frans Lanting/Allstock; 419 (tr), Gregory D. Dimijian, M.D./Photo Researchers; 419 (br), M.P.L. Fogden/Bruce Coleman, Inc.; 420, Superstock; 420 (bkgrd), Robert Young Pelton/Westlight; 421, both photos by Ralph Reinhold/Animals Animals; 426 (t), Marine Biological Laboratory Archives, Woods Hole, MA; 426 (b), P. Breese/Gamma-Liaison; 427, HRW photo by James Newberry; 428, Dr. Betsy Dresser/Cincinnati Zoo; 428 (bkgrd), Kjell Sandved/Allstock; 429, Susan Jones/Animals Animals; 430, HRW photo by Eric Beggs; 432, Jerry Wachter/Photo Researchers; 435, Prof. P. Motta/Dept. of Anatomy/University "La Sapienza", Rome/Science Photo Library/Photo Researchers; 438 (t), Michael Abbey/Photo Researchers, Inc.; 438 (b), Biophoto Associates/Photo Researchers; 439, David York/Medichrome; 443, all photos by Park Street; 447, Biophoto Associates/Science Source/Photo Researchers; 448, Biophoto Associates/Science Source/Photo Researchers; 449 (t), Michael Abbey/Photo Researchers, Inc.; 449 (b), Warren Morgan/Focus on Sports; 453, Jerry Wachter/Photo Researchers; 461, Manfred Kage/Peter Arnold; 463, both photos by Dwight Kuhn; 464 (tl), HRW Photo by Richard Haynes; 464 (bl), Grant Heilman/Grant Heilman Photography; 464 (r), K. Gunnar/Bruce

Credits 609

Coleman, Inc.; 465, 466, Barry L. Runk/Grant Heilman Photography; 467 (t), HRW photo by James Newberry; 467 (b), Will McIntyre/Photo Researchers; 468, HRW photo by Park Street; 469, James Newberry; 471, Laurence Parent; 472, Manfred Kage/Peter Arnold, Inc.; 473, Steve Dunwell/The Image Bank; 474 (l), Sklar & Peiper/Photo Researchers; 474 (tr), James White/University of Minnesota; 474 (br), Don Fawcett/Science Source/Photo Researchers; 475, Phillip A. Harrington/Peter Arnold; 476, Park Street; 478, Jean Claude Revy/Phototake; 483 (t), CNRI/Science Photo Library/Photo Researchers; 483 (b), Biophoto Associates/Photo Researchers; 486, Manfred Kage/Peter Arnold, Inc.; 488, David Lissy/Nawrocki Stock Photo; 489 (t), Cindy Lewis; 489 (b), David Madison; 490, Guido Alberto Rossi/The Image Bank; 491, Clark Overton/Phototake; 492, Carol Rosegg/Martha Swope Associates; 494 (t), Mary Gow; 494 (bl), (br), Herbert Wagner/Phototake; 496, H. Armstrong Roberts; 497, HRW photo by Richard Haynes; 500 (l), (r), Dianora Niccolini/Medichrome; 503 (l), Runk/Schoenberger/Grant Heilman Photography; 503 (r), Martin Dohrn/Science Photo Library/Photo Researchers; 506, Peter Menzel/Stock Boston; 507, David Lissy/Nawrocki Stock Photo; 508, both photos by Ira Wyman/Sygma; 509, Astrid & Hanns-Frieder Michler/Science Photo Library/Photo Researchers; 512, Shelby Thorner/David Madison; 514, Everett C. Johnson/Leo de Wys Inc.; 515, Biophoto Association/Photo Researchers; 518, Alexander Tsiaras/Science Source/Photo Researchers; 520, P.R. Dunn; 524, Mike Maple/Woodfin Camp & Associates; 525, Paolo Koch/Photo Researchers; 528, Gabe Palmer/Tony Stone Worldwide; 530, 531, all HRW photos by Richard Haynes; 532, 533, James Newberry; 534 (t), A. Glauberman/Photo Researchers; 534 (b), James Stevenson/Science Photo Library/Photo Researchers; 536, L. O'Shaughnessy/Allstock; 537 (l), Frederick C. Skvara/Peter Arnold; 537 (r), Alfred Pasieka/Bruce Coleman, Ltd.; 541, NHPA; 542 (both), 543, John Ficara/Woodfin Camp & Associates; 544, Renee Lynn/David Madison; 548, Mary Gow; 550 (t), HRW photo by Richard Haynes; 550 (b), John Walsh/Photo Researchers; 551 (t), John Giannicchi/Science Source/Photo Researchers; 551 (b), HRW photo by Richard Haynes; 555 (tl), Tony Freeman/PhotoEdit; 555 (tr), Mary Kate Denny/PhotoEdit; 555 (bl), (bcl), (bcr), (br), 557 (all), Petit Format/Nestle/Science Source/Photo Researchers; 558 (t), Kindra Clineff/Allstock; 558 (b), Jeff Reed/Medichrome; 559 (l), Tony Freeman/PhotoEdit; 559 (cr), Myrleen Ferguson/PhotoEdit; 559 (cr), Mary Kate Denny/PhotoEdit; 559 (r), Gabe Palmer/Tony Stone Worldwide; 561, John Ficara/Woodfin Camp & Associates; 563, 564, James Newberry; 565, 566 (t), (b), Mary Gow; 571 (t), NIH Visitors Information Center; 571 (b), NASA; 572, all HRW photos by P.R. Dunn; 573 (l), HRW photo by James Newberry; 573 (r), Judy Allen-Newberry; 574, David Wagner/Phototake; 576, HRW by James Newberry; 585, Cosmos Blank/Photo Researchers; 586 (t), David C. Fritts/Earth Scenes; 586 (b), Steve Vidler/Nawrocki Stock Photo; 587 (t), Tom Stack & Associates; 587 (b), Dr. Lloyd M. Beidler/Science Photo Library/Photo Researchers; 588, Mitch Reardon/Photo Researchers; 589, NHPA; 590, all photos by Arthur Siegelman; 591, Craig Aurness/Woodfin Camp & Associates; 592, Wolfgang Bayer/Bruce Coleman, Inc.

ILLUSTRATIONS

Bego, Dolores 333, 335(t)
Benner, Beverly 241, 242, 343, 349(b), 386, 387, 391
Bordelon, Melinda 118, 119
Botzis, Ka 349(t), 351, 352, 355(b), 373, 441, 452
Warren Budd & Associates 154, 295, 377, 383
Byers, Scott 434
Collins, Don 42, 74, 76, 77, 78, 82, 113, 149, 162, 163, 166, 170, 180, 233, 266, 273, 276, 278, 279, 284, 346, 348, 350, 353, 355, 366, 396, 409, 416, 470
Colrus, Bill 385
Cooper, Holly 7, 142, 143, 144, 145, 146, 148, 149, 150, 316
David, Susan 22, 68
Erickson, Barry 14
Evans, Tom 456, 457
Fischer, David 47, 87, 88, 89, 90, 108, 157, 193, 432, 433, 436, 437, 438, 440, 442, 443, 445, 446, 447, 448, 454, 455, 462, 482, 491, 493, 495, 498, 515, 516, 517, 518, 521, 524, 540, 545, 549, 551, 568, 569
Frank, Robert 359, 363, 364, 380, 381
Gardner, Sharon Carter 101, 168, 173, 308, 309, 310, 312, 313, 314, 318, 319, 363, 390, 398, 399, 437, 451, 469, 481, 504, 512, 513, 514, 521, 527, 530, 535
JAK Graphics 16, 30, 49, 63, 64, 67, 81, 101, 165, 107, 110, 111, 128, 186, 187, 212, 264, 282, 291, 326, 404, 406, 458, 459, 460, 461, 463, 479, 484, 495, 501, 502, 525, 527, 538, 547
Katz, Joel 132, 231, 233, 237
Kelvin, George 20, 52, 121, 126, 134, 212, 239
Lebo, Narda 513, 522, 526(b)
LeMonnier, Joe 100, 102, 103, 224
Longacre, Jimmy 510, 511
Merrilees, Rebecca 261, 262, 263, 265, 269
Morgan-Cain & Associates 9, 28, 38, 54, 70, 78, 123, 125, 159, 164, 190, 217, 220, 243, 298, 302, 331, 336, 411, 416, 450, 465, 477, 504, 519, 529, 538
Network Graphics 36, 137, 444, 553
Phillips, Harriet 480, 481, 491(t), 526(t)
Reid, Fiona 45, 46, 329
Skorpil, Judy 375, 378, 379, 499, 550, 556
Smith-Griswold, Wendy 160, 161

For permission to reprint copyrighted material, grateful acknowledgment is made to the following sources:

Ballantine Books, a division of Random House, Inc.: From *Dances with Wolves* by Michael Blake. Copyright © 1988 by Michael Blake.

Bantam Books, a division of Bantam Doubleday Del Publishing Group, Inc: From *Fantastic Voyage* by Isaac Asimov. Copyright © 1966 by Bantam Books. A novel by Isaac Asimov based on the screenplay be Harry Kleiner from original story by Otto Klement and Jay Lewis Bixby.

Bonnier Fakta Bokförlag AB: From *Close to Nature: An Exploration of Nature's Microcosm* by Lennart Nilsson. Copyright © 1984 by Lennart Nilsson and Bonnier Fakta Bokförlag AB, Stockholm.

Children's Better Health Institute, Benjamin Franklin Literary & Medical Society, Inc., Indianapolis, IN: Adaptation of "An Eagle to the Wind" by Nancy Ferrell from *Young World*, June/July 1979. Copyright © 1979 by The Saturday Evening Post Company.

Current Science ®: Adapted from "I Slept for Science: A Study of Sleep Problems" by Andy T. McPhee from *Current Science ®*, October 7, 1988. Copyright © 1988 by Weekly Reader Corporation. Published by Weekly Reader Corporation.

Danbury Press, a division of Grolier Inc.: From *The Ocean World of Jacques Cousteau: The Quest for Food*. Volume 3. Copyright © 1973 by Jacques-Yves Cousteau.

Farrar, Straus & Giroux, Inc.: "earthworms" from *More Small Poems* by Valerie Worth. Copyright © 1976 by Valerie Worth.

Greenwillow Books, a division of William Morrow & Company, Inc.: Illustrations and adaptation of text from *How the Forest Grew* by William Jaspersohn, illustrated by Chuck Eckart. Text copyright © 1980 by William Jaspersohn; illustration copyright © 1980 by Chuck Eckart.

Harcourt Brace & Company: Illustration and text from *The Jungle* by Helen Borton. Copyright & 1968 by Helen Borton. Illustration and text from *Lumberjack* by Stephen W. Meader, illustrated by Henry C. Pitz. Copyright 1934 by Harcourt Brace & Company; copyright renewed & 1961 by Stephen Meader.

Holt, Rinehart and Winston, Inc.: Adapted text from "Controlling Algal Blooms," adapted text from "Growing Plants in Space," and adapted text from "Saving Endangered Wildlife" from *Biology Today*, by Thomas C. Emmel, Harvey Goodman, Linda E. Graham, and Yaakov Shechter. Copyright & 1991 by Holt, Rinehart and Winston, Inc.

Little, Brown and Company: From *Life on Earth: A Natural History* by David Attenborough. Copyright & 1979 by David Attenborough Productions Ltd. "The Octopus," "The Purist," and "The Termite" from *Verses from 1929 On* by Ogden Nash. Copyright 1935, 1942 by Ogden Nash.

Lodestar Books, an affiliate of Dutton Children's Books, a division of Penguin Books USA Inc.: From *If You Lived On Mars* by Melvin Berger. Copyright & 1989 by Melvin Berger.

Gina Maccoby Literary Agency: "Spider" from *Bugs: Poems* by Mary Ann Hoberman. Copyright & 1976 by Mary Ann Hoberman.

Macmillan Publishing Company: From "Through the Microscope" from *A Short History of Science: Man's Conquest of Nature from Ancient Times to the Atomic Age* by Arthur C. Gregor. Copyright & 1963 by Arthur C. Gregor.

Margaret K. McElderry Books, an imprint of Macmillan Publishing Company: From *Searches in the American Desert* by Sheila Cowing, with photographs and maps by Walter C. Cowing. Copyright & 1989 by Sheila Cowing.

National Wildlife Federation: "Shark Lady" by Bet Hennefrund from *Ranger Rick*, March 1989. Copyright & 1989 by the National Wildlife Federation. From "China's Precious Pandas" by Claire Miller from *Ranger Rick*, July 1989. Copyright & 1989 by the National Wildlife Federation. "Andy Lipkis and the Tree People" by Mark Wexler from *Ranger Rick*, May 1984. Copyright & 1984 by the National Wildlife Federation.

W. W. Norton & Company, Inc.: From *The Winter of the Fisher* by Cameron Langford. Copyright © 1971 by Cameron Langford.

Oxford University Press, Inc.: From *The Sea Around Us, Revised Edition*, by Rachel L. Carson. Copyright © 1950, 1951, 1961 by Rachel L. Carson; copyright renewed © 1979, 1989 by Roger Christie; Golden Press edition copyright © 1958 by Western Publishing Company, Inc.

Prentice-Hall, Inc.: From *order: in life* by Edmund Samuel. Copyright © 1972 by Prentice-Hall, Inc.

Richard Rieu: "The Flattered Flying Fish" by E. V. Rieu. Copyright © 1983.

Simon and Schuster Books For Young Readers, New York: From *Journey Through a Tropical Jungle* by Adrian Forsyth. Copyright © 1988 by Adrian Forsyth.

The Society of Authors as the representative of The Literary Trustees of Walter de la Mare: "Seeds" from *Rhymes and Verses: Collected Poems for Children* by Walter de la Mare. Copyright © 1947 by Henry Holt and Company, Inc.

Doug Stewart: From "These Germs Work Wonders" by Doug Stewart from *Reader's Digest*, January 1991. Copyright © 1991 by Doug Stewart.

Time-Life Books Inc.: From *The Everglades: The American Wilderness* by Archie Carr and The Editors of Time-Life Books. Copyright © 1973 by Time-Life Books Inc.

U.S. News & World Report: Quote by Magic Johnson from "Stunned by Magic" by Tom Callahan from *U.S. News & World Report*, November 18, 1991. Copyright © by U.S. News & World Report Inc.

Franklin Watts, Inc., New York: From *The Black Plague* by Walter Oleksy. Copyright © 1982 by Walter Oleksy.

Wayland Publishers Limited and Franklin Watts, New York: From *The Voyage of the Beagle* by Kate Hyndley. Copyright © 1989 by Wayland Publishers Limited.

Credits

Annotated Teacher's Edition

PHOTO

Abbreviations used:(t) top,(c) center,(b) bottom,(l) left,(r) right.

Page T6(l), Erich Hartmann/Magnum; T6(r), J.H. Robinson; T20-T25(border), Frans Lanting/Photo Researchers; T21, David Hughes/Bruce Coleman, Ltd.; T22(t), HRW photo by Richard Haynes; T22(b), Stuart Westmoreland/ The Stock Market; T-23(t), Mark Wexler; T23(b), Dan McCoy/Rainbow; T24(t), Breck P. Kent; T24(c), Burt Jones & Maurine Shimlock/Photographers Aspen; T24(b), Breck P. Kent; T25(t), Everett C. Johnson / Leo de Wys; T-25(b), Breck P. Kent; T26-T29(border), Robert Young Pelton/ Westlight; T27, NASA; T29, Keith Gillett/Animals Animals; T30(border), Tony Stone Worldwide; T30, all photos by Terry Domico/Earth Images; T31(border), Laurence Parent; T31, Peter Menzel/Stock Boston; T32-T33(border),Peter Veit/DRK Photo; T32, HRW photo by Richard Haynes; T33, Mike Maple/Woodfin Camp & Associates; T34-T35(border), Tony Stone Worldwide; T-35, HRW photo by Richard Haynes; T-36(border), Kjell Sandved/Allstock; T-36, both photos by Park Street; T-37(border), Dr. Tony Brain/Photo Researchers; T38(border), Gregory G. Dimijian/Photo Researchers; T-38, HRW photo by James Newberry; T-39 (border), Frans Lanting/Allstock; T-40(border),Tom Bean; T-40, Marine Biological Laboratory Archives, Woods Hole, MA; T-41(border), R.C.Simpson/Valan Photos; T-41,Laurence Parent; T42-T43,Walter Hodges/Westlight; T-42, Will & Deni McIntyre/Photo Researchers;T44-T45(border), HRW photo by James Newberry; T44(l), HRW photo by Eric Beggs; T44(r), Harry Holloway/Carolina Biological Supply Co.; T-45(tl), HRW photo by Eric Beggs; T-45(tr), Breck P. Kent; T45(b), John Gerlach/Tom Stack & Associates; T46-T50, Laurence Parent; T46(t), Leonard Harris/Leo de Wys; T46(b), John Shaw/Tom Stack & Associates; T47(t), Norman Myers/Bruce Coleman,Inc.; T47(c), Tom Stack & Associates; T47(b), Mike Neumann/Photo Researchers; T48(tl), Ed Cooper; T48(tr), Renee Lynn/Photo Researchers; T48(b), Gwen Fidler/Comstock; T49(t), Louis Bencze Photo/Allstock; T49(b), Ed Degginger; T50, William E. Ferguson.

ART

Using the Program Design: Robin Bouvette
Icons: Rosario Cosgrove, Joan Rivers
Page T25, Holly Cooper; **T34**, Morgan-Cain & Assoc.; **T49**, Graphic Concern.